Classical Mechanics

MIT 8.01 Course Notes

Peter Dourmashkin

Massachusetts Institute of Technology

Wiley Custom Learning Solutions

Cover Image © Pathetique no. 1, Camille Dourmashkin-Cagol

To order books or for customer service, please call 1(800)-CALL-WILEY (225-5945).

Printed in the United States of America.

ISBN 978-1-119-91802-8
Printed and bound by Lightning Source.

10 9 8 7 6 5 4 3 2 1

Classical Mechanics

MIT 8.01 Course Notes

Peter Dourmashkin

MIT

Classical Mechanics: MIT 8.01 Course Notes

For Dorothea

Acknowledgements

I would like to acknowledge the many contributions of my friends, colleagues, and family, without which the completion of these course notes would never have been possible.

Craig Watkins, whose efforts have contributed to all aspect of these course notes, especially to the appendices of various chapters, and to the celestial mechanics chapter in particular.

John Belcher, for his tireless efforts in improving undergraduate education at MIT, especially by creating the TEAL (Technology Enabled Active Learning) program for first-year physics students at MIT.

Thomas Greytak, for his dedication both in the classroom and to improving physics education at MIT. His presentations of mechanics were an inspiration for many parts of these course notes.

Deepto Chakrabarty, for our many discussions prompted by his insightful understanding of classical mechanics.

Andy Neely, and his years of devotion towards making TEAL work.

Peter Katsumi Montag for his insightful editorial suggestions.

Camille Dourmashkin-Cagol, for her painting, "Beethoven's Pathétique No. 1," on the front and back cover of this book.

I am eternally grateful to the Physics Department at MIT, especially Edwin Bertschinger, for his devotion to undergraduate education.

I would also like to thank the following people for their contributions to TEAL and to the development of these course notes: Sen-Ben Liao, Sahana Murthy, Saif Rayyan, Eric Mazur, Daniel Kleppner, Marin Soljačić, David Pritchard, Eric Hudson, Enectali Figueroa-Feliciano, Joseph Formaggio, Jeff Gore, Christoph Paus, Marcus Klute, David Litster, Alan Guth, Walter Lewin, Matthew Strafuss, Daniel Kelleher, Nancy Savioli, Nancy Boyce, and Catherine Modica.

And finally, I would like to thank my wife, Dorothea, and my two daughters, Clarissa and Camille, for their patience and understanding.

Chapter 1 The History and Limitations of Classical Mechanics

Chapter 1 The History and Limitations of Classical Mechanics

Classical mechanics is the mathematical science that studies the displacement of bodies under the action of forces. Gailieo Galilee initiated the modern era of mechanics by using mathematics to describe the motion of bodies. His *Mechanics,* published in 1623, introduced the concepts of force and described the constant accelerated motion of objects near the surface of the Earth. Sixty years later Newton formulated his Laws of Motion, which he published in 1687 under the title, *Philosophiae Naturalis Principia Mathematica* (*Mathematical Principles of Natural Philosophy*). In the third book, subtitled *De mundi systemate* (*On the system of the world*), Newton solved the greatest scientific problem of his time by applying his Universal Law of Gravitation to determine the motion of planets. Newton established a mathematical approach to the analysis of physical phenomena in which he stated that it was unnecessary to introduce final causes (hypothesis) that have no experimental basis, *"Hypotheses non fingo* (I frame no hypotheses), but that physical models are built from experimental observations and then made general by induction. This led to a great century of applications of the principles of *Newtonian mechanics* to many new problems culminating in the work of Leonhard Euler. Euler began a systematic study of the three dimensional motion of rigid bodies, leading to a set of dynamical equations now known as *Euler's Equations of Motion*.

Alongside this development and refinement of the concept of force and its application to the description of motion, the concept of energy slowly emerged, culminating in the middle of the nineteenth century in the discovery of the *principle of conservation of energy* and its immediate applications to the laws of *thermodynamics*. Conservation principles are now central to our study of mechanics; the conservation of momentum, energy, and angular momentum enabled a new reformulation of classical mechanics.

During this period, the experimental methodology and mathematical tools of Newtonian mechanics were applied to other non-rigid systems of particles leading to the development of *continuum mechanics.* The theories of *fluid mechanics*, *wave mechanics*, and *electromagnetism* emerged leading to the development of the wave theory of light. However there were many perplexing aspects of the wave theory of light, for example does light propagate through a medium, the "ether". A series of optics experiments, culminating in the Michelson-Morley experiment in 1887 ruled out the hypothesis of a stationary medium. Many attempts were made to reconcile the experimental evidence with classical mechanics but the challenges were more fundamental. The basics concepts of absolute time and absolute space, which Newton had defined in the *Principia,* were themselves inadequate to explain a host of experimental observations. Einstein, by insisting on a fundamental rethinking of the concepts of space and time, and the relativity of motion, in his *special theory of relativity* (1905) was able to resolve the apparent conflicts between optics and Newtonian mechanics. In particular, special relativity provides the necessary framework for describing the motion of rapidly moving objects (speed greater than $v > 0.1\,c$).

A second limitation on the validity of Newtonian mechanics appeared at the microscopic length scale. A new theory, *statistical mechanics*, was developed relating the microscopic properties of individual atoms and molecules to the macroscopic or bulk thermodynamic properties of materials. Started in the middle of the nineteenth century, new observations at very small scales revealed anomalies in the predicted behavior of gases. It became increasingly clear that classical mechanics did not adequately explain a wide range of newly discovered phenomena at the atomic and sub-atomic length scales. An essential realization was that the language of classical mechanics was not even adequate to qualitatively describe certain microscopic phenomena. By the early part of the twentieth century, *quantum mechanics* provided a mathematical description of microscopic phenomena in complete agreement with our empirical knowledge of all non-relativistic phenomena.

In the twentieth century, as experimental observations led to a more detailed knowledge of the large-scale properties of the universe, Newton's Universal Law of Gravitation no longer accurately modeled the observed universe and needed to be replaced by *general relativity.* By the end of the twentieth century and beginning of the twenty-first century, many new observations, for example the accelerated expansion of the Universe, have required introduction of new concepts like *dark energy* that may lead once again to a fundamental rethinking of the basic concepts of physics in order to explain observed phenomena.

Chapter 2 Units, Dimensional Analysis, Problem Solving, and Estimation

Chapter 2 Units, Dimensional Analysis, Problem Solving, and Estimation

But we must not forget that all things in the world are connected with one another and depend on one another, and that we ourselves and all our thoughts are also a part of nature. It is utterly beyond our power to measure the changes of things by time. Quite the contrary, time is an abstraction, at which we arrive by means of the change of things; made because we are not restricted to any one definite measure, all being interconnected. A motion is termed uniform in which equal increments of space described correspond to equal increments of space described by some motion with which we form a comparison, as the rotation of the earth. A motion may, with respect to another motion, be uniform. But the question whether a motion is in itself uniform, is senseless. With just as little justice, also, may we speak of an "absolute time" --- of a time independent of change. This absolute time can be measured by comparison with no motion; it has therefore neither a practical nor a scientific value; and no one is justified in saying that he knows aught about it. It is an idle metaphysical conception.[1]

Ernst Mach

2.1 The Speed of light

When we observe and measure phenomena in the world, we try to assign numbers to the physical quantities with as much accuracy as we can possibly obtain from our measuring equipment. For example, we may want to determine the speed of light, which we can calculate by dividing the distance a known ray of light propagates over its travel time,

$$\text{speed of light} = \frac{\text{distance}}{\text{time}}. \tag{2.1.1}$$

In 1983 the General Conference on Weights and Measures defined the *speed of light* to be

$$c = 299{,}792{,}458 \text{ meters/second}. \tag{2.1.2}$$

This number was chosen to correspond to the most accurately measured value of the speed of light and is well within the experimental uncertainty.

2.2 International System of Units

The system of units most commonly used throughout science and technology today is the *Système International* (SI). It consists of seven *base quantities* and their corresponding *base units*:

---

[1] E. Mach, *The Science of Mechanics*, translated by Thomas J. McCormack, Open Court Publishing Company, La Salle, Illinois, 1960, p. 273.

Base Quantity	Base Unit
Length	meter (m)
Mass	kilogram (kg)
Time	second (s)
Electric Current	ampere (A)
Temperature	kelvin (K)
Amount of Substance	mole (mol)
Luminous Intensity	candela (cd)

We shall refer to the *dimension* of the base quantity by the quantity itself, for example

$$\text{dim length} \equiv \text{length} \equiv \text{L}, \text{dim mass} \equiv \text{mass} \equiv \text{M}, \text{dim time} \equiv \text{time} \equiv \text{T}. \qquad (2.2.1)$$

Mechanics is based on just the first three of these quantities, the MKS or meter-kilogram-second system. An alternative metric system to this, still widely used, is the so-called CGS system (centimeter-gram-second).

2.2.1 Standard Mass

The unit of mass, the kilogram (kg), remains the only base unit in the International System of Units (SI) that is still defined in terms of a physical artifact, known as the "International Prototype of the Standard Kilogram." George Matthey (of Johnson Matthey) made the prototype in 1879 in the form of a cylinder, 39 mm high and 39 mm in diameter, consisting of an alloy of 90 % platinum and 10 % iridium. The international prototype is kept at the Bureau International des Poids et Mesures (BIPM) at Sevres, France under conditions specified by the 1st Conférence Générale des Poids et Mèsures (CGPM) in 1889 when it sanctioned the prototype and declared "This prototype shall henceforth be considered to be the unit of mass." It is stored at atmospheric pressure in a specially designed triple bell-jar. The prototype is kept in a vault with six official copies.

The 3rd Conférence Générale des Poids et Mesures CGPM (1901), in a declaration intended to end the ambiguity in popular usage concerning the word "weight" confirmed that:

> *The kilogram is the unit of mass; it is equal to the mass of the international prototype of the kilogram.*

There is a stainless steel one-kilogram standard that can travel for comparisons with standard masses in other laboratories. In practice it is more common to quote a conventional mass value (or weight-in-air, as measured with the effect of buoyancy), than the standard mass. Standard mass is normally only used in specialized measurements wherever suitable copies of the prototype are stored.

Example 2.1 The International Prototype Kilogram

Determine the type of shape and dimensions of the platinum-iridium prototype kilogram such that it has the smallest surface area for a given volume. The standard kilogram is an alloy of 90 % platinum and 10 % iridium. The density of the alloy is $\rho = 21.56\ \text{g} \cdot \text{cm}^{-3}$. You may want to consider the following questions. (a) Is there any reason that the surface area of the standard could be important? (b) What is the appropriate density to use? (c) What shape (that is, sphere, cube, right cylinder, parallelepiped, etc.) has the smallest surface area for a given volume? (d) Why was a right-circular cylinder chosen?

Solution: The standard kilogram is an alloy of 90% platinum and 10% iridium. The density of platinum is $21.45\ \text{g} \cdot \text{cm}^{-3}$ and the density of iridium is $22.55\ \text{g} \cdot \text{cm}^{-3}$. Thus the density of the standard kilogram, $\rho = 21.56\ \text{g} \cdot \text{cm}^{-3}$, and its volume is

$$V = m / \rho \cong 1000\ \text{g} / 22\ \text{g} \cdot \text{cm}^{-3} \cong 46.38\ \text{cm}^3 . \tag{2.2}$$

Corrosion would affect the mass through chemical reaction; platinum and iridium were chosen for the standard's composition as they resist corrosion. To further minimize corrosion, the shape should be chosen to have the least surface area. The volume for a cylinder or radius r and height h is a constant and given by

$$V = \pi r^2 h . \tag{2.3}$$

The surface area can be expressed in terms of the radius r as

$$A = 2\pi r^2 + 2\pi r h = 2\pi r^2 + \frac{2V}{r} . \tag{2.4}$$

To find the smallest surface area, minimize the area with respect to the radius

$$\frac{dA}{dr} = 4\pi r - \frac{2V}{r^2} = 0 . \tag{2.5}$$

Solve for the radius

$$r^3 = \frac{V}{2\pi} = \frac{\pi r^2 h}{2\pi} . \tag{2.6}$$

Thus the radius is one half the height,

$$r = \frac{h}{2} . \tag{2.7}$$

For the standard mass, the radius is

$$r = \left(\frac{V}{2\pi}\right)^{1/3} = \left(\frac{46.38\ \text{cm}^3}{2\pi}\right)^{1/3} \cong 1.95\ \text{cm}\ . \qquad (2.8)$$

Twice this radius is the diameter of the standard kilogram.

Because the prototype kilogram is an artifact, there are some intrinsic problems associated with its use as a standard. It may be damaged, or destroyed. The prototype gains atoms due to environment wear and cleaning, at a rate of change of mass corresponding to approximately $1\ \mu\text{g}\ /\ \text{year}$, ($1\ \mu\text{g} \equiv 1\text{microgram} \equiv 1\times 10^{-6}\ \text{g}$).

Several new approaches to defining the SI unit of mass (kg) are currently being explored. One possibility is to define the kilogram as a fixed number of atoms of a particular substance, thus relating the kilogram to an atomic mass. Silicon is a good candidate for this approach because it can be grown as a large single crystal, in a very pure form.

Example 2.2 Mass of a Silicon Crystal

A given standard unit cell of silicon has a volume V_0 and contains N_0 atoms. The number of molecules in a given mole of substance is given by Avogadro's constant $N_A = 6.02214129(27)\times 10^{23}\ \text{mol}^{-1}$. The molar mass of silicon is given by M_{mol}. Find the mass m of a volume V in terms of V_0, N_0, V, M_{mol}, and N_A.

Solution: The mass m_0 of the unit cell is the density ρ of silicon cell multiplied by the volume of the cell V_0,

$$m_0 = \rho V_0\ . \qquad (2.9)$$

The number of moles in the unit cell is the total mass, m_0, of the cell, divided by the molar mass M_{mol},

$$n_0 = m_0\ /\ M_{\text{mol}} = \rho V_0\ /\ M_{\text{mol}}\ . \qquad (2.10)$$

The number of atoms in the unit cell is the number of moles n_0 times the Avogadro constant, N_A,

$$N_0 = n_0 N_A = \frac{\rho V_0 N_A}{M_{\text{mol}}}\ . \qquad (2.11)$$

The density of the crystal is related to the mass m of the crystal divided by the volume V of the crystal,

$$\rho = m\ /\ V\ . \qquad (2.12)$$

The number of atoms in the unit cell can be expressed as

$$N_0 = \frac{mV_0N_A}{VM_{\text{mol}}}. \tag{2.13}$$

The mass of the crystal is

$$m = \frac{M_{\text{mol}}}{N_A}\frac{V}{V_0}N_0 \tag{2.14}$$

The molar mass, unit cell volume and volume of the crystal can all be measured directly. Notice that M_{mol} / N_A is the mass of a single atom, and $(V / V_0)N_0$ is the number of atoms in the volume. This approach is therefore reduced to the problem of measuring the Avogadro constant, N_A, with a relative uncertainty of 1 part in 10^8, which is equivalent to the uncertainty in the present definition of the kilogram.

2.2.2 The Atomic Clock and the Definition of the Second

Isaac Newton, in the *Philosophiae Naturalis Principia Mathematica* ("Mathematical Principles of Natural Philosophy"), distinguished between time as duration and an absolute concept of time,

> "*Absolute true and mathematical time, of itself and from its own nature, flows equably without relation to anything external, and by another name is called duration: relative, apparent, and common time, is some sensible and external (whether accurate or unequable) measure of duration by means of motion, which is commonly used instead of true time; such as an hour, a day, a month, a year.* "[2].

The development of clocks based on atomic oscillations allowed measures of timing with accuracy on the order of 1 part in 10^{14}, corresponding to errors of less than one microsecond (one millionth of a second) per year. Given the incredible accuracy of this measurement, and clear evidence that the best available timekeepers were atomic in nature, the *second* (s) was redefined in 1967 by the International Committee on Weights and Measures as a certain number of cycles of electromagnetic radiation emitted by cesium atoms as they make transitions between two designated quantum states:

[2] Isaac Newton. *Mathematical Principles of Natural Philosophy*. Translated by Andrew Motte (1729). Revised by Florian Cajori. Berkeley: University of California Press, 1934. p. 6.

The second is the duration of 9,192,631,770 periods of the radiation corresponding to the transition between the two hyperfine levels of the ground state of the cesium 133 atom.

2.2.3 The Meter

The meter was originally defined as 1/10,000,000 of the arc from the Equator to the North Pole along the meridian passing through Paris. To aid in calibration and ease of comparison, the meter was redefined in terms of a length scale etched into a platinum bar preserved near Paris. Once laser light was engineered, the meter was redefined by the 17th Conférence Générale des Poids et Mèsures (CGPM) in 1983 to be a certain number of wavelengths of a particular monochromatic laser beam.

The meter is the length of the path traveled by light in vacuum during a time interval of 1/299 792 458 of a second.

Example 2.3 Light-Year

Astronomical distances are sometimes described in terms of *light-years* (ly). A light-year is the distance that light will travel in one year (yr). How far in meters does light travel in one year?

Solution: Using the relationship $\text{distance} = (\text{speed of light}) \cdot (\text{time})$, one light year corresponds to a distance. Because the speed of light is given in terms of meters per second, we need to know how many seconds are in a year. We can accomplish this by converting units. We know that

1 year = 365.25 days, 1 day = 24 hours, 1 hour = 60 minutes, 1 minute = 60 seconds

Putting this together we find that the number of seconds in a year is

$$1\text{ year} = (365.25\text{ day})\left(\frac{24\text{ hours}}{1\text{ day}}\right)\left(\frac{60\text{ min}}{1\text{ hour}}\right)\left(\frac{60\text{ s}}{1\text{ min}}\right) = 31{,}557{,}600\text{ s}. \qquad (2.2.15)$$

The distance that light travels in a one year is

$$1\text{ ly} = \left(\frac{299{,}792{,}458\text{ m}}{1\text{ s}}\right)\left(\frac{31{,}557{,}600\text{ s}}{1\text{ yr}}\right)(1\text{ yr}) = 9.461\times 10^{15}\text{ m}. \qquad (2.2.16)$$

The distance to the nearest star, a faint red dwarf star, Proxima Centauri, is 4.24 light years. A standard astronomical unit is the parsec. One parsec is the distance at which there is one arcsecond = 1/3600 degree angular separation between two objects that are separated by the distance of one astronomical unit, $1\text{AU} = 1.50\times 10^{11}\text{ m}$, which is the mean distance between the earth and sun. One astronomical unit is roughly equivalent to eight light minutes, $1\text{AU} = 8.31\text{-min}$ One parsec is equal to 3.26 light years, where one light year is the distance that light travels in one earth year, $1\,\text{pc} = 3.26\,\text{ly} = 2.06\times10^5\,\text{AU}$ where $1\,\text{ly} = 9.46\times10^{15}\,\text{m}$.

2.2.4 Radians and Steradians

Consider the triangle drawn in Figure 2.1. The basic trigonometric functions of an angle θ in a right-angled triangle ONB are $\sin(\theta) = y/r$, $\cos(\theta) = x/r$, and $\tan(\theta) = y/x$.

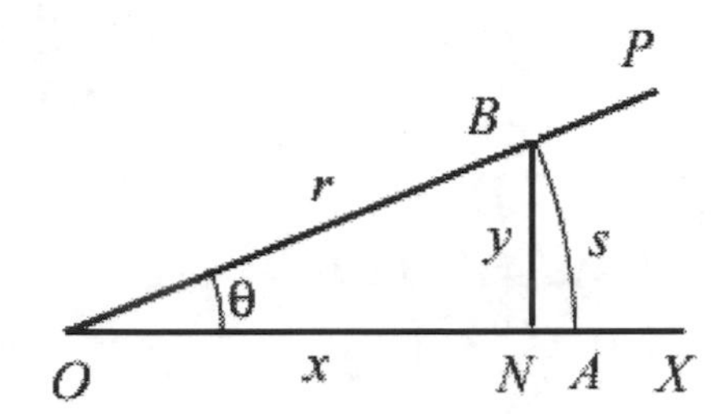

Figure 2.1 Trigonometric relations

It is very important to become familiar with using the measure of the angle θ itself as expressed in *radians* [rad]. Let θ be the angle between two straight lines OX and OP. Draw a circle of radius r centered at O. The lines OP and OX cut the circle at the points A and B where $OA = OB = r$. Denote the length of the arc AB by s, then the radian measure of θ is given by

$$\theta = s/r, \tag{2.2.17}$$

and the ratio is the same for circles of any radii centered at O -- just as the ratios y/r and y/x are the same for all right triangles with the angle θ at O. As θ approaches 360°, s approaches the complete circumference $2\pi r$ of the circle, so that $360^\circ = 2\pi\ \text{rad}$.

Let's compare the behavior of $\sin(\theta)$, $\tan(\theta)$ and θ itself for small angles. One can see from Figure 1.1 that $s/r > y/r$. It is less obvious that $y/x > \theta$. It is very instructive to plot $\sin(\theta)$, $\tan(\theta)$, and θ as functions of θ [rad] between 0 and $\pi/2$ on the same graph (see Figure 2.2). For small θ, the values of all three functions are almost equal. But how small is "small"? An acceptable condition is for $\theta << 1$ in radians. We can show this with a few examples.

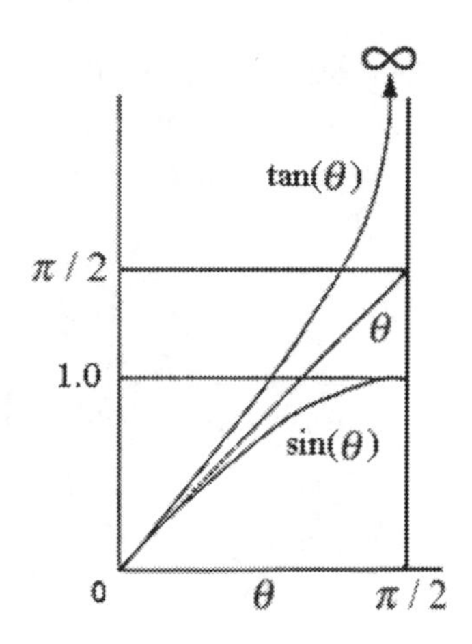

Figure 2.2 Radians compared to trigonometric functions.

Recall that $360^{\circ} = 2\pi \text{ rad}$, $57.3^{\circ} = 1 \text{ rad}$, so an angle $6^{\circ} \cong (6^{\circ})(2\pi \text{ rad} / 360^{o}) \cong 0.1 \text{ rad}$ when expressed in radians. In Table 2.1 we compare the value of θ (measured in radians) with $\sin(\theta)$, $\tan(\theta)$, $(\theta - \sin\theta)/\theta$, and $(\theta - \tan\theta)/\theta$, for $\theta = 0.1 \text{rad}$, 0.2rad, 0.5rad, and 1.0rad.

Table 2.1 Small Angle Approximation

θ [rad]	θ [deg]	$\sin(\theta)$	$\tan(\theta)$	$(\theta - \sin\theta)/\theta$	$(\theta - \tan\theta)/\theta$
0.1	5.72958	0.09983	0.10033	0.00167	-0.00335
0.2	11.45916	0.19867	0.20271	0.00665	-0.01355
0.5	28.64789	0.47943	0.54630	0.04115	-0.09260
1.0	57.29578	0.84147	1.55741	0.15853	-0.55741

The values for $(\theta - \sin\theta)/\theta$, and $(\theta - \tan\theta)/\theta$, for $\theta = 0.2 \text{ rad}$ are less than $\pm 1.4\%$. Provided that θ is not too large, the approximation that

$$\sin(\theta) \simeq \tan(\theta) \simeq \theta, \qquad (2.2.18)$$

called the *small angle approximation*, can be used almost interchangeably, within some small percentage error. This is the basis of many useful approximations in physics calculations.

The *steradian* [sr] is the unit of solid angle that, having its vertex in the center of a sphere, cuts off an area of the surface of the sphere equal to that of a square with sides of length equal to the radius of the sphere. The conventional symbol for steradian measure is Ω the uppercase Greek "Omega." The total solid angle Ω_{sph} of a sphere is then found by dividing the surface area of the sphere by the square of the radius,

$$\Omega_{\text{sph}} = 4\pi r^2 / r^2 = 4\pi \qquad (2.2.19)$$

Note that this result is independent of the radius of the sphere. Note also that it was implied that the solid angle was measured from the center of the sphere (the radius r is constant). It turns out that the above result does not depend on the position of the vertex as long as the vertex is inside the sphere.

2.2.5 Radiant Intensity

> *"The SI unit, candela, is the luminous intensity of a source that emits monochromatic radiation of frequency* $540\times10^{12}\ \mathrm{s}^{-1}$, *in a given direction, and that has a radiant intensity in that direction of 1/683 watts per steradian."*

Note that "in a given direction" cannot be taken too literally. The intensity is measured per steradian of spread, so if the radiation has no spread of directions, the luminous intensity would be infinite.

2.3 Dimensions of Commonly Encountered Quantities

Many physical quantities are derived from the base quantities by set of algebraic relations defining the physical relation between these quantities. The dimension of the derived quantity is written as a power of the dimensions of the base quantities. For example velocity is a derived quantity and the dimension is given by the relationship

$$\text{dim velocity} = (\text{length})/(\text{time}) = \mathrm{L}\cdot\mathrm{T}^{-1}. \tag{2.2.20}$$

where $\mathrm{L}\equiv\text{length}$, $\mathrm{T}\equiv\text{time}$. Force is also a derived quantity and has dimension

$$\text{dim force} = \frac{(\text{mass})(\text{dim velocity})}{(\text{time})}. \tag{2.2.21}$$

where $M\equiv\text{mass}$. We can also express force in terms of mass, length, and time by the relationship

$$\text{dim force} = \frac{(\text{mass})(\text{length})}{(\text{time})^2} = \mathrm{M}\cdot\mathrm{L}\cdot\mathrm{T}^{-2}. \tag{2.2.22}$$

The derived dimension of kinetic energy is

$$\text{dim kinetic energy} = (\text{mass})(\text{dim velocity})^2, \tag{2.2.23}$$

which in terms of mass, length, and time is

$$\text{dim kinetic energy} = \frac{(\text{mass})(\text{length})^2}{(\text{time})^2} = \mathrm{M}\cdot\mathrm{L}^2\cdot\mathrm{T}^{-2}. \tag{2.2.24}$$

The derived dimension of work is

$$\text{dim work} = (\text{dim force})(\text{length}), \tag{2.2.25}$$

which in terms of our fundamental dimensions is

$$\text{dim work} = \frac{(\text{mass})(\text{length})^2}{(\text{time})^2} = \text{M} \cdot \text{L}^2 \cdot \text{T}^{-2}. \tag{2.2.26}$$

So work and kinetic energy have the same dimensions. Power is defined to be the rate of change in time of work so the dimensions are

$$\text{dim power} = \frac{\text{dim work}}{\text{time}} = \frac{(\text{dim force})(\text{length})}{\text{time}} = \frac{(\text{mass})(\text{length})^2}{(\text{time})^3} = \text{M} \cdot \text{L}^2 \cdot \text{T}^{-3}. \tag{2.2.27}$$

Table 2.2 Dimensions of Some Common Mechanical Quantities

$\text{M} \equiv \text{mass}, \ \text{L} \equiv \text{length}, \ \text{T} \equiv \text{time}$

Quantity	Dimension	MKS unit
Angle	dimensionless	Dimensionless = radian
Solid Angle	dimensionless	Dimensionless = sterradian
Area	L^2	m^2
Volume	L^3	m^3
Frequency	T^{-1}	s^{-1} = hertz = Hz
Velocity	$\text{L} \cdot \text{T}^{-1}$	$\text{m} \cdot \text{s}^{-1}$
Acceleration	$\text{L} \cdot \text{T}^{-2}$	$\text{m} \cdot \text{s}^{-2}$
Angular Velocity	T^{-1}	$\text{rad} \cdot \text{s}^{-1}$
Angular Acceleration	T^{-2}	$\text{rad} \cdot \text{s}^{-2}$
Density	$\text{M} \cdot \text{L}^{-3}$	$\text{kg} \cdot \text{m}^{-3}$
Momentum	$\text{M} \cdot \text{L} \cdot \text{T}^{-1}$	$\text{kg} \cdot \text{m} \cdot \text{s}^{-1}$
Angular Momentum	$\text{M} \cdot \text{L}^2 \cdot \text{T}^{-1}$	$\text{kg} \cdot \text{m}^2 \cdot \text{s}^{-1}$
Force	$\text{M} \cdot \text{L} \cdot \text{T}^{-2}$	$\text{kg} \cdot \text{m} \cdot \text{s}^{-2}$ = newton = N
Work, Energy	$\text{M} \cdot \text{L}^2 \cdot \text{T}^{-2}$	$\text{kg} \cdot \text{m}^2 \cdot \text{s}^{-2}$ = joule = J
Torque	$\text{M} \cdot \text{L}^2 \cdot \text{T}^{-2}$	$\text{kg} \cdot \text{m}^2 \cdot \text{s}^{-2}$
Power	$\text{M} \cdot \text{L}^2 \cdot \text{T}^{-3}$	$\text{kg} \cdot \text{m}^2 \cdot \text{s}^{-3}$ = watt = W
Pressure	$\text{M} \cdot \text{L}^{-1} \cdot \text{T}^{-2}$	$\text{kg} \cdot \text{m}^{-1} \cdot \text{s}^{-2}$ = pascal = Pa

In Table 2.2 we include the derived dimensions of some common mechanical quantities in terms of mass, length, and time.

2.3.1 Dimensional Analysis

There are many phenomena in nature that can be explained by simple relationships between the observed phenomena.

Example 2.4 Period of a Pendulum

Consider a simple pendulum consisting of a massive bob suspended from a fixed point by a string. Let T denote the time (period of the pendulum) that it takes the bob to complete one cycle of oscillation. How does the period of the simple pendulum depend on the quantities that define the pendulum and the quantities that determine the motion?

Solution: What possible quantities are involved? The length of the pendulum l, the mass of the pendulum bob m, the gravitational acceleration g, and the angular amplitude of the bob θ_0 are all possible quantities that may enter into a relationship for the period of the swing. Have we included every possible quantity? We can never be sure but let's first work with this set and if we need more than we will have to think harder! Our problem is then to find a function f such that

$$T = f(l,m,g,\theta_0) \tag{2.2.28}$$

We first make a list of the dimensions of our quantities as shown in Table 2.3. Choose the set: mass, length, and time, to use as the base dimensions.

Table 2.3 Dimensions of quantities that may describe the period of pendulum

Name of Quantity	Symbol	Dimensional Formula
Time of swing	t	T
Length of pendulum	l	L
Mass of pendulum	m	M
Gravitational acceleration	g	$\mathrm{L \cdot T^{-2}}$
Angular amplitude of swing	θ_0	No dimension

Our first observation is that the mass of the bob cannot enter into our relationship, as our final quantity has no dimensions of mass and no other quantity can remove the dimension of the pendulum mass. Let's focus on the length of the string and the gravitational acceleration. In order to eliminate length, these quantities must divide each other in the above expression for T must divide each other. If we choose the combination l/g, the dimensions are

$$\dim[l/g] = \frac{\text{length}}{\text{length}/(\text{time})^2} = (\text{time})^2 \quad (2.2.29)$$

It appears that the time of swing is proportional to the square root of this ratio. We have an argument that works for our choice of constants, which depend on the units we choose for our fundamental quantities. Thus we have a candidate formula

$$T \sim \left(\frac{l}{g}\right)^{1/2}. \quad (2.2.30)$$

(in the above expression, the symbol " $\sim$ " represents a proportionality, not an approximation).

Because the angular amplitude θ_0 is dimensionless, it may or may not appear. We can account for this by introducing some function $y(\theta_0)$ into our relationship, which is beyond the limits of this type of analysis. Then the time of swing is

$$T = y(\theta_0)\left(\frac{l}{g}\right)^{1/2}. \quad (2.2.31)$$

We shall discover later on that $y(\theta_0)$ is nearly independent of the angular amplitude θ_0 for very small amplitudes and is equal to $y(\theta_0) = 2\pi$,

$$T = 2\pi\left(\frac{l}{g}\right)^{1/2} \quad (2.2.32)$$

2.4 Significant Figures, Scientific Notation, and Rounding

2.4.1 Significant Figures

We shall define significant figures by the following rules.[3]

1. The leftmost nonzero digit is the most significant digit.
2. If there is no decimal place, the rightmost nonzero digit is the least significant digit.
3. If there is a decimal point then the right most digit is the least significant digit even if it is a zero.

[3] Philip R Bevington and D. Keith Robinson, Data Reduction and Error Analysis for the Physical Sciences, 2nd Edition, McGraw-Hill, Inc., New York, 1992.

4. All digits between the least and most significant digits are counted as significant digits.

When reporting the results of an experiment, the number of significant digits used in reporting the result is the number of digits needed to state the result of that measurement (or a calculation based on that measurement) without any loss of precision.

There are exceptions to these rules, so you may want to carry around one extra significant digit until you report your result. For example if you multiply $2 \times 0.56 = 1.12$, not 1.1. There is some ambiguity about the number of significant figure when the rightmost digit is 0, for example 1050, with no terminal decimal point. This has only three significant digits. If all the digits are significant the number should be written as 1050., with a terminal decimal point. To avoid this ambiguity it is wiser to use scientific notation.

2.4.2 Scientific Notation

Careless use of significant digits can be easily avoided by the use of decimal notation times the appropriate power of ten for the number. Then all the significant digits are manifestly evident in the decimal number. Therefore the number $1050 = 1.05 \times 10^3$ while the number $1050. = 1.050 \times 10^3$.

2.4.3 Rounding

To round off a number by eliminating insignificant digits we have three rules. For practical purposes, rounding will be done automatically by a calculator or computer, and all we need do is set the desired number of significant figures for whichever tool is used.

1. If the fraction is greater than 1/2, increment the new least significant digit.
2. If the fraction is less than 1/2, do not increment.
3. If the fraction equals 1/2, increment the least significant digit only if it is odd.

The reason for Rule 3 is that a fractional value of 1/2 may result from a previous rounding up of a fraction that was slightly less than 1/2 or a rounding down of a fraction that was slightly greater than 1/2. For example, 1.249 and 1.251 both round to three significant digits 1.25. If we were to round again to two significant digits, both would yield the same value, either 1.2 or 1.3 depending on our convention in Rule 3. Choosing to round up if the resulting last digit is odd and to round down if the resulting last digit is even reduces the systematic errors that would otherwise be introduced into the average of a group of such numbers.

2.5 Problem Solving

Solving problems is the most common task used to measure understanding in technical and scientific courses, and in many aspects of life as well. In general, problem solving requires factual and procedural knowledge in the area of the problem, plus knowledge of numerous schema, plus skill in overall problem solving. Schema is loosely defined as a "specific type of problem" such as principal, rate, and interest problems, one-dimensional kinematic problems with constant acceleration, etc. In most introductory university courses, improving problem solving relies on three things:

1. increasing domain knowledge, particularly definitions and procedures
2. learning schema for various types of problems and how to recognize that a particular problem belongs to a known schema
3. becoming more conscious of and insightful about the process of problem solving.

To improve your problem solving ability in a course, the most essential change of attitude is to focus more on the *process of solution* rather than on *obtaining the answer.* For homework problems there is frequently a simple way to obtain the answer, often involving some specific insight. This will quickly get you the answer, but you will not build schema that will help solve related problems further down the road. Moreover, if you rely on insight, when you get stuck on a problem, you're stuck with no plan or fallback position.

2.5.1 General Approach to Problem Solving

A great many physics textbook authors recommend overall problem solving strategies. These are typically four-step procedures that descend from George Polya's influential book, *How to Solve It*, on problem solving[4]. Here are his four steps:

1. Understand – get a conceptual grasp of the problem

What is the problem asking? What are the given conditions and assumptions? What domain of knowledge is involved? What is to be found and how is this determined or constrained by the given conditions?

What knowledge is relevant? E.g. in physics, does this problem involve kinematics, forces, energy, momentum, angular momentum, equilibrium? If the problem involves two different areas of knowledge, try to separate the problem into parts. Is there motion or is it static? If the problem involves vector quantities such as velocity or momentum, think of these geometrically (as arrows that add vectorially). Get conceptual understanding: is some physical quantity (energy, momentum, angular momentum, etc.) constant? Have you done problems that involve the same concepts in roughly the same way?

Model: Real life contains great complexity, so in physics (chemistry, economics…) you actually solve a model problem that contains the essential elements of the real problem. The bike and rider become a point mass (unless

[4] G. Polya, *How to Solve It, 2nd ed.*, Princeton University Press, 1957.

angular momentum is involved), the ladder's mass is regarded as being uniformly distributed along its length, the car is assumed to have constant acceleration or constant power (obviously not true when it shifts gears), etc. Become sensitive to information that is implicitly assumed (Presence of gravity? No friction? That the collision is of short duration relative to the timescale of the subsequent motion? …).

Advice: Write *your own* representation of the problem's stated data; *redraw* the picture with your labeling and comments. Get the problem into your brain! Go systematically down the list of topics in the course or for that week if you are stuck.

2. **Devise a Plan - set up a procedure to obtain the desired solution**

General - Have you seen a problem like this – i.e., does the problem fit in a schema you already know? Is a part of this problem a known schema? Could you simplify this problem so that it is? Can you find *any* useful results for the given problem and data even if it is not the solution (e.g. in the special case of motion on an incline when the plane is at $\theta = 0$)? Can you imagine a route to the solution if only you knew some apparently not given information? If your solution plan involves equations, count the unknowns and check that you have that many independent equations.

In Physics, exploit the freedoms you have: use a particular type of coordinate system (e.g. polar) to simplify the problem, pick the orientation of a coordinate system to get the unknowns in one equation only (e.g. only the x-direction), pick the position of the origin to eliminate torques from forces you don't know, pick a system so that an unknown force acts entirely within it and hence does not change the system's momentum… Given that the problem involves some particular thing (constant acceleration, momentum) think over *all* the equations that involve this concept.

3. **Carry our your plan – solve the problem!**

This generally involves mathematical manipulations. Try to keep them as simple as possible by not substituting in lengthy algebraic expressions until the end is in sight, make your work as neat as you can to ease checking and reduce careless mistakes. Keep a clear idea of where you are going and have been (label the equations and what you have now found), if possible, check each step as you proceed. Always check dimensions if analytic, and units if numerical.

4. **Look Back – check your solution and method of solution**

Can you see that the answer is correct now that you have it – often simply by retrospective inspection? Can you solve it a different way? Is the problem equivalent to one you've solved before if the variables have some specific values?

Check special cases (for instance, for a problem involving two massive objects moving on an inclined plane, if $m_1 = m_2$ or $\theta = 0$ does the solution reduce to a simple expression that you can easily derive by inspection or a simple argument?) Is the scaling what you'd expect (an energy should vary as the velocity squared, or linearly with the height). Does it depend sensibly on the various quantities (e.g. is the acceleration less if the masses are larger, more if the spring has a larger k)? Is the answer physically reasonable (especially if numbers are given or reasonable ones substituted).

Review the schema of your solution: Review and try to remember the outline of the solution – what is the model, the physical approximations, the concepts needed, and any tricky math manipulation.

2.6 Order of Magnitude Estimates - Fermi Problems

Counting is the first mathematical skill we learn. We came to use this skill by distinguishing elements into groups of similar objects, but we run into problems when our desired objects are not easily identified, or there are too many to count. Rather than spending a huge amount of effort to attempt an exact count, we can try to estimate the number of objects in a collection. For example, we can try to estimate the total number of grains of sand contained in a bucket of sand. Because we can see individual grains of sand, we expect the number to be very large but finite. Sometimes we can try to estimate a number, which we are fairly sure but not certain is finite, such as the number of particles in the universe.

We can also assign numbers to quantities that carry dimensions, such as mass, length, time, or charge, which may be difficult to measure exactly. We may be interested in estimating the mass of the air inside a room, or the length of telephone wire in the United States, or the amount of time that we have slept in our lives. We choose some set of units, such as kilograms, miles, hours, and coulombs, and then we can attempt to estimate the number with respect to our standard quantity.

Often we are interested in estimating quantities such as speed, force, energy, or power. We may want to estimate our natural walking speed, or the force of wind acting against a bicycle rider, or the total energy consumption of a country, or the electrical power necessary to operate a university. All of these quantities have no exact, well-defined value; they instead lie within some range of values.

When we make these types of estimates, we should be satisfied if our estimate is reasonably close to the middle of the range of possible values. But what does "reasonably close" mean? Once again, this depends on what quantities we are estimating. If we are describing a quantity that has a very large number associated with it, then an estimate within an order of magnitude should be satisfactory. The number of molecules in a breath of air is close to 10^{22}; an estimate anywhere between 10^{21} and 10^{23} molecules is close enough. If we are trying to win a contest by estimating the number of marbles in a glass container, we cannot be so imprecise; we must hope that our estimate is within 1% of the real quantity.

These types of estimations are called ***Fermi Problems***. The technique is named after the physicist Enrico Fermi, who was famous for making these sorts of "back of the envelope" calculations.

2.6.1 Methodology for Estimation Problems

Estimating is a skill that improves with practice. Here are two guiding principles that may help you get started.

(1) You must identify a set of quantities that can be estimated or calculated.

(2) You must establish an approximate or exact relationship between these quantities and the quantity to be estimated in the problem.

Estimations may be characterized by a precise relationship between an estimated quantity and the quantity of interest in the problem. When we estimate, we are drawing upon what we know. But different people are more familiar with certain things than others. If you are basing your estimate on a fact that you already know, the accuracy of your estimate will depend on the accuracy of your previous knowledge. When there is no precise relationship between estimated quantities and the quantity to be estimated in the problem, then the accuracy of the result will depend on the type of relationships you decide upon. There are often many approaches to an estimation problem leading to a reasonably accurate estimate. So use your creativity and imagination!

Example 2.5 Lining Up Pennies

Suppose you want to line pennies up, diameter to diameter, until the total length is 1 kilometer . How many pennies will you need? How accurate is this estimation?

Solution: The first step is to consider what type of quantity is being estimated. In this example we are estimating a dimensionless scalar quantity, the number of pennies. We can now give a precise relationship for the number of pennies needed to mark off 1 kilometer

$$\#\text{ of pennies} = \frac{\text{total distance}}{\text{diameter of penny}}. \qquad (2.2.33)$$

We can estimate a penny to be approximately 2 centimeters wide. Therefore the number of pennies is

$$\#\text{ of pennies} = \frac{\text{total distance}}{\text{length of a penny}} = \frac{(1\text{ km})}{(2\text{ cm})(1\text{ km}/10^5\text{ cm})} = 5\times10^4\text{ pennies}. \qquad (2.2.34)$$

When applying numbers to relationships we must be careful to convert units whenever necessary. How accurate is this estimation? If you measure the size of a penny, you will find out that the width is $1.9\,\mathrm{cm}$, so our estimate was accurate to within 5%. This accuracy was fortuitous. Suppose we estimated the length of a penny to be 1 cm. Then our estimate for the total number of pennies would be within a factor of 2, a margin of error we can live with for this type of problem.

Example 2.6 Estimation of Mass of Water on Earth

Estimate the mass of the water on the Earth?

Solution: In this example we are estimating mass, a quantity that is a fundamental in SI units, and is measured in kg. We start by approximating that the amount of water on Earth is approximately equal to the amount of water in all the oceans. Initially we will try to estimate two quantities: the density of water and the volume of water contained in the oceans. Then the relationship we want is

$$(\mathrm{mass})_{\mathrm{ocean}} = (\mathrm{density})_{\mathrm{water}} (\mathrm{volume})_{\mathrm{ocean}} . \tag{2.2.35}$$

One of the hardest aspects of estimation problems is to decide which relationship applies. One way to check your work is to check dimensions. Density has dimensions of mass/volume, so our relationship is

$$(\mathrm{mass})_{\mathrm{ocean}} = \left(\frac{\mathrm{mass}}{\mathrm{volume}} \right) (\mathrm{volume})_{\mathrm{ocean}} . \tag{2.2.36}$$

The density of fresh water is $\rho_{\mathrm{water}} = 1.0\ \mathrm{g \cdot cm^{-3}}$; the density of seawater is slightly higher, but the difference won't matter for this estimate. You could estimate this density by estimating how much mass is contained in a one-liter bottle of water. (The density of water is a point of reference for all density problems. Suppose we need to estimate the density of iron. If we compare iron to water, we estimate that iron is 5 to 10 times denser than water. The actual density of iron is $\rho_{\mathrm{iron}} = 7.8\ \mathrm{g \cdot cm^{-3}}$).

Because there is no precise relationship, estimating the volume of water in the oceans is much harder. Let's model the volume occupied by the oceans as if they completely cover the earth, forming a spherical shell (Figure 2.3, which is decidedly not to scale). The volume of a spherical shell of radius R_{earth} and thickness d is

$$(\mathrm{volume})_{\mathrm{shell}} \cong 4\pi R_{\mathrm{earth}}^2 d , \tag{2.2.37}$$

where R_{earth} is the radius of the earth and d is the average depth of the ocean.

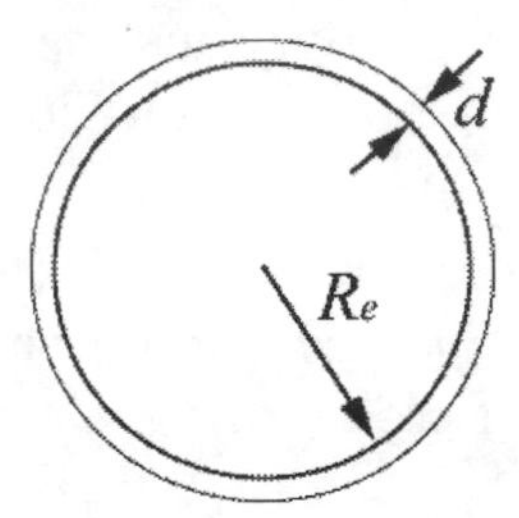

Figure 2.3 A model for estimating the mass of the oceans.

We first estimate that the oceans cover about 75% of the surface of the earth. So the volume of the oceans is

$$\text{volume}_{\text{ocean}} \cong (0.75)(4\pi R_{\text{earth}}^2 d). \tag{2.2.38}$$

We therefore have two more quantities to estimate, the average depth of the ocean, which we can estimate the order of magnitude as $d \cong 1\,\text{km}$, and the radius of the earth, which is approximately $R_{\text{earth}} \cong 6 \times 10^3\,\text{km}$. (The quantity that you may remember is the circumference of the earth, about $25{,}000$ miles. Historically the circumference of the earth was defined to be $4 \times 10^7\,\text{m}$). The radius R_{earth} and the circumference s are exactly related by

$$s = 2\pi R_{\text{earth}}. \tag{2.2.39}$$

Thus

$$R_{\text{earth}} = \frac{s}{2\pi} = \frac{\left(2.5 \times 10^4\ \text{mi}\right)\left(1.6\ \text{km} \cdot \text{mi}^{-1}\right)}{2\pi} = 6.4 \times 10^3\ \text{km} \tag{2.2.40}$$

We will use $R_{\text{earth}} \cong 6 \times 10^3\,\text{km}$; additional accuracy is not necessary for this problem, since the ocean depth estimate is clearly less accurate. In fact, the factor of 75% is not needed, but included more or less from habit. Altogether, our estimate for the mass of the oceans is

$$(\text{mass})_{\text{ocean}} = (\text{density})_{\text{water}} (\text{volume})_{\text{ocean}} \cong \rho_{\text{water}} (0.75)(4\pi R_{\text{earth}}^2 d)$$

$$(\text{mass})_{\text{ocean}} \cong \left(\frac{1\text{g}}{\text{cm}^3}\right)\left(\frac{1\,\text{kg}}{10^3\ \text{g}}\right)\left(\frac{(10^5\ \text{cm})^3}{(1\,\text{km})^3}\right)(0.75)(4\square)(6 \times 10^3\ \text{km})^2(1\text{km}) \tag{2.2.41}$$

$$(\text{mass})_{\text{ocean}} \cong 3 \times 10^{20}\ \text{kg} \cong 10^{20}\ \text{kg}.$$

Chapter 3 Vectors

Chapter 3 Vectors

Philosophy is written in this grand book, the universe which stands continually open to our gaze. But the book cannot be understood unless one first learns to comprehend the language and read the letters in which it is composed. It is written in the language of mathematics, and its characters are triangles, circles and other geometric figures without which it is humanly impossible to understand a single word of it; without these, one wanders about in a dark labyrinth.[1]

Galileo Galilee

3.1 Vector Analysis

3.1.1 Introduction to Vectors

Certain physical quantities such as mass or the absolute temperature at some point only have magnitude. Numbers alone can represent these quantities, with the appropriate units, and they are called scalars. There are, however, other physical quantities that have both magnitude and direction: the magnitude can stretch or shrink, and the direction can reverse. These quantities can be added in such a way that takes into account both direction and magnitude. Force is an example of a quantity that acts in a certain direction with some magnitude that we measure in newtons. When two forces act on an object, the sum of the forces depends on both the direction and magnitude of the two forces. Position, displacement, velocity, acceleration, force, momentum and torque are all physical quantities that can be represented mathematically by vectors. We begin by defining precisely what we mean by a vector.

3.1.2 Properties of Vectors

A vector is a quantity that has both direction and magnitude. Let a vector be denoted by the symbol $\vec{\mathbf{A}}$. The magnitude of $\vec{\mathbf{A}}$ is $|\vec{\mathbf{A}}| \equiv A$. We can represent vectors as geometric objects using arrows. The length of the arrow corresponds to the magnitude of the vector. The arrow points in the direction of the vector (Figure 3.1).

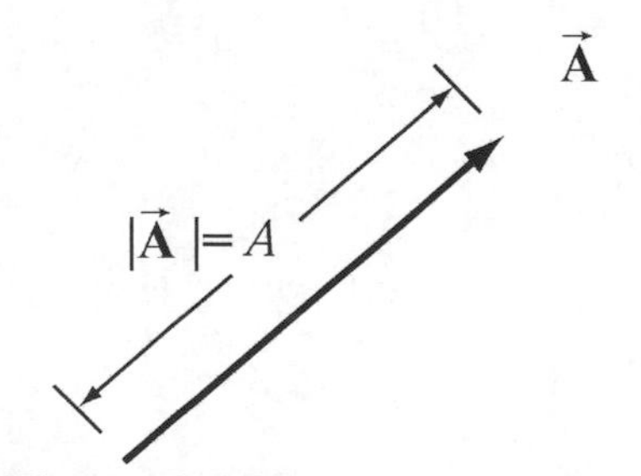

Figure 3.1 Vectors as arrows.

[1] Galileo Galilei, *The Assayer*, tr. Stillman Drake (1957), *Discoveries and Opinions of Galileo* pp. 237-8.

There are two defining operations for vectors:

(1) Vector Addition:

Vectors can be added. Let $\vec{\mathbf{A}}$ and $\vec{\mathbf{B}}$ be two vectors. We define a new vector, $\vec{\mathbf{C}} = \vec{\mathbf{A}} + \vec{\mathbf{B}}$, the "vector addition" of $\vec{\mathbf{A}}$ and $\vec{\mathbf{B}}$, by a geometric construction. Draw the arrow that represents $\vec{\mathbf{A}}$. Place the tail of the arrow that represents $\vec{\mathbf{B}}$ at the tip of the arrow for $\vec{\mathbf{A}}$ as shown in Figure 3.2a. The arrow that starts at the tail of $\vec{\mathbf{A}}$ and goes to the tip of $\vec{\mathbf{B}}$ is defined to be the "vector addition" $\vec{\mathbf{C}} = \vec{\mathbf{A}} + \vec{\mathbf{B}}$. There is an equivalent construction for the law of vector addition. The vectors $\vec{\mathbf{A}}$ and $\vec{\mathbf{B}}$ can be drawn with their tails at the same point. The two vectors form the sides of a parallelogram. The diagonal of the parallelogram corresponds to the vector $\vec{\mathbf{C}} = \vec{\mathbf{A}} + \vec{\mathbf{B}}$, as shown in Figure 3.2b.

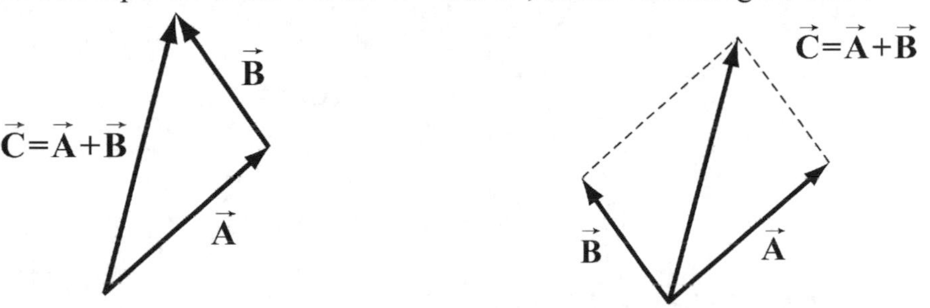

Figure 3.2 (a) Geometric sum of vectors. **Figure 3.2 (b)** Geometric sum of vectors.

Vector addition satisfies the following four properties:

(i) Commutativity:

The order of adding vectors does not matter;

$$\vec{\mathbf{A}} + \vec{\mathbf{B}} = \vec{\mathbf{B}} + \vec{\mathbf{A}}. \qquad (3.1.1)$$

Our geometric definition for vector addition satisfies the commutative property (i) since in the parallelogram representation for the addition of vectors, it doesn't matter which side you start with, as seen in Figure 3.3.

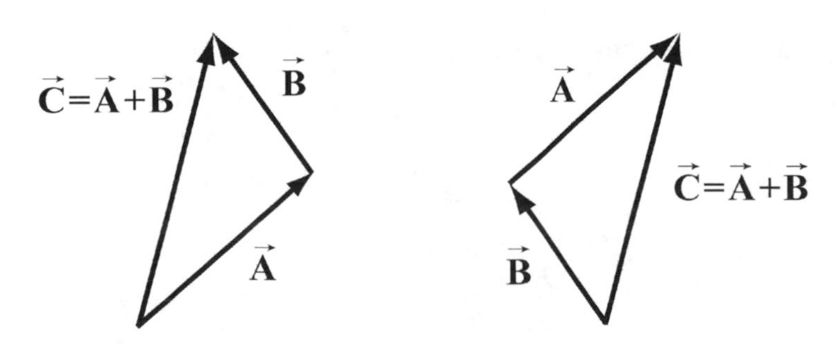

Figure 3.3 Commutative property of vector addition.

(ii) Associativity:

When adding three vectors, it doesn't matter which two you start with

$$(\vec{\mathbf{A}}+\vec{\mathbf{B}})+\vec{\mathbf{C}}=\vec{\mathbf{A}}+(\vec{\mathbf{B}}+\vec{\mathbf{C}}). \tag{3.1.2}$$

In Figure 3.4, we add $(\vec{\mathbf{A}}+\vec{\mathbf{B}})+\vec{\mathbf{C}}$, and $\vec{\mathbf{A}}+(\vec{\mathbf{B}}+\vec{\mathbf{C}})$ to arrive at the same vector sum in either case.

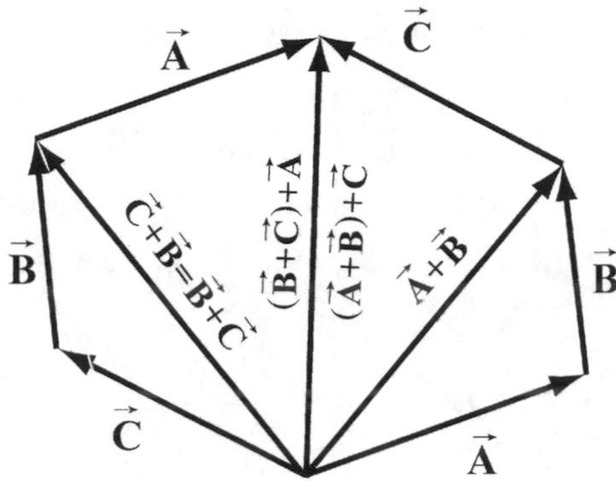

Figure 3.4 Associative law.

(iii) Identity Element for Vector Addition:

There is a unique vector, $\vec{\mathbf{0}}$, that acts as an identity element for vector addition. For all vectors $\vec{\mathbf{A}}$,

$$\vec{\mathbf{A}}+\vec{\mathbf{0}}=\vec{\mathbf{0}}+\vec{\mathbf{A}}=\vec{\mathbf{A}}. \tag{3.1.3}$$

(iv) Inverse Element for Vector Addition:

For every vector $\vec{\mathbf{A}}$, there is a unique inverse vector

$$(-1)\vec{\mathbf{A}}\equiv-\vec{\mathbf{A}}, \tag{3.1.4}$$

such that

$$\vec{\mathbf{A}}+(-\vec{\mathbf{A}})=\vec{\mathbf{0}}.$$

The vector $-\vec{\mathbf{A}}$ has the same magnitude as $\vec{\mathbf{A}}$, $|\vec{\mathbf{A}}|=|-\vec{\mathbf{A}}|=A$, but they point in opposite directions (Figure 3.5).

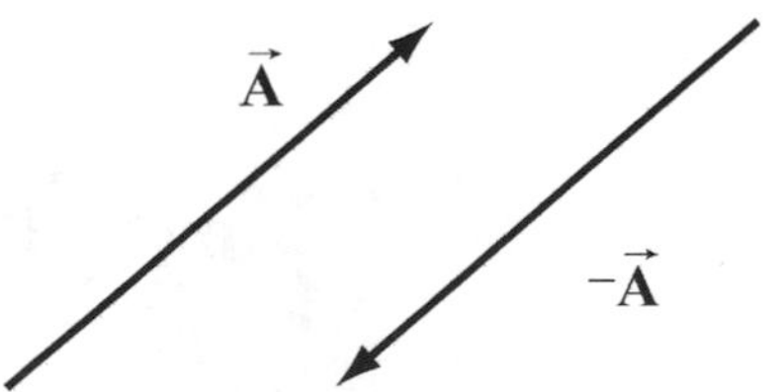

Figure 3.5 Additive inverse

(2) Scalar Multiplication of Vectors:

Vectors can be multiplied by real numbers. Let $\vec{\mathbf{A}}$ be a vector. Let c be a real positive number. Then the multiplication of $\vec{\mathbf{A}}$ by c is a new vector, which we denote by the symbol $c\vec{\mathbf{A}}$. The magnitude of $c\vec{\mathbf{A}}$ is c times the magnitude of $\vec{\mathbf{A}}$ (Figure 3.6a),

$$\left|c\vec{\mathbf{A}}\right| = c\left|\vec{\mathbf{A}}\right|. \tag{3.1.5}$$

Let $c > 0$, then the direction of $c\vec{\mathbf{A}}$ is the same as the direction of $\vec{\mathbf{A}}$. However, the direction of $-c\vec{\mathbf{A}}$ is opposite of $\vec{\mathbf{A}}$ (Figure 3.6b).

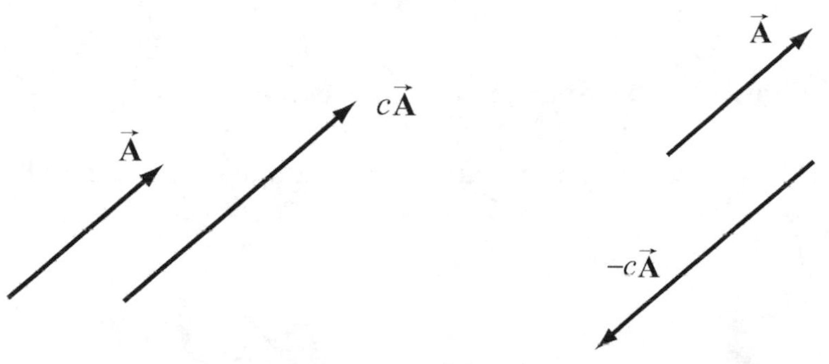

Figure 3.6 Multiplication of vector $\vec{\mathbf{A}}$ by (a) $c > 0$, and (b) $-c < 0$.

Scalar multiplication of vectors satisfies the following properties:

(i) Associative Law for Scalar Multiplication:

The order of multiplying numbers is doesn't matter. Let b and c be real numbers. Then

$$b(c\vec{\mathbf{A}}) = (bc)\vec{\mathbf{A}} = (cb\vec{\mathbf{A}}) = c(b\vec{\mathbf{A}}). \tag{3.1.6}$$

(ii) Distributive Law for Vector Addition:

Vector addition satisfies a distributive law for multiplication by a number. Let c be a real number. Then

$$c(\vec{\mathbf{A}} + \vec{\mathbf{B}}) = c\vec{\mathbf{A}} + c\vec{\mathbf{B}}. \tag{3.1.7}$$

Figure 3.7 illustrates this property.

(iii) Distributive Law for Scalar Addition:

The multiplication operation also satisfies a distributive law for the addition of numbers. Let b and c be real numbers. Then

$$(b+c)\vec{\mathbf{A}} = b\vec{\mathbf{A}} + c\vec{\mathbf{A}} \qquad (3.1.8)$$

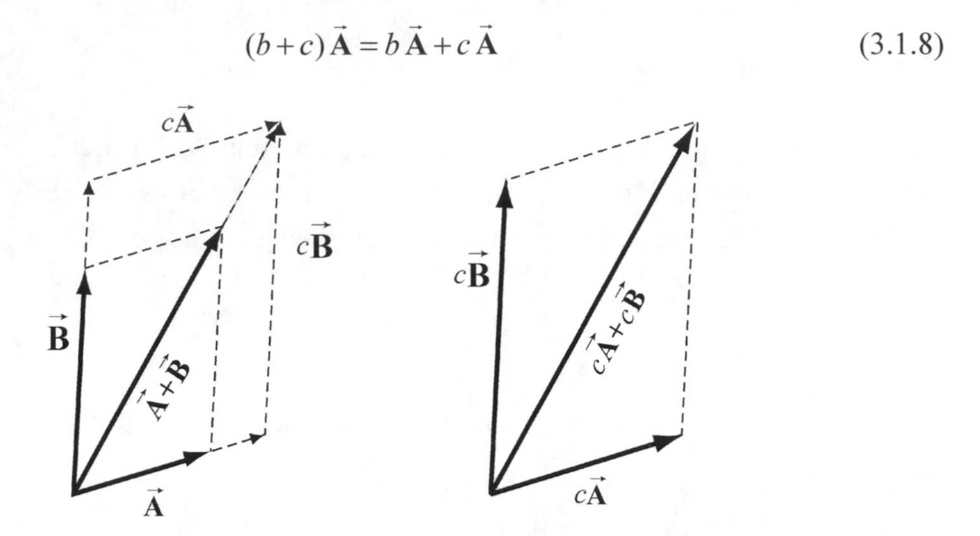

Figure 3.7 Distributive Law for vector addition.

Our geometric definition of vector addition satisfies this condition as seen in Figure 3.8.

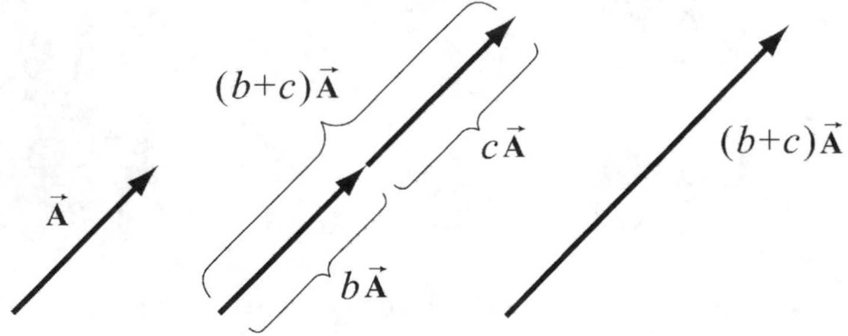

Figure 3.8 Distributive law for scalar multiplication.

(iv) Identity Element for Scalar Multiplication:

The number 1 acts as an identity element for multiplication,

$$1\,\vec{\mathbf{A}} = \vec{\mathbf{A}}\,. \qquad (3.1.9)$$

3.2 Cartesian Coordinate System

Physics involve the study of phenomena that we observe in the world. In order to connect the phenomena to mathematics we begin by introducing the concept of a coordinate system. A coordinate system consists of four basic elements:

(1) Choice of origin

(2) Choice of axes

(3) Choice of positive direction for each axis

(4) Choice of unit vectors for each axis

There are three commonly used coordinate systems: Cartesian, cylindrical and spherical. What makes these systems extremely useful is the associated set of infinitesimal line, area, and volume elements that are key to making many integration calculations in classical mechanics, such as finding the center of mass and moment of inertia.

3.2.1 Cartesian Coordinates

Cartesian coordinates consist of a set of mutually perpendicular axes, which intersect at a common point, the origin O. We live in a three-dimensional spatial world; for that reason, the most common system we will use has three axes, for which we choose the directions of the axes and position of the origin are.

(1) Choice of Origin

Choose an origin O. If you are given an object, then your choice of origin may coincide with a special point in the body. For example, you may choose the mid-point of a straight piece of wire.

(2) Choice of Axis

Now we shall choose a set of axes. The simplest set of axes is known as the Cartesian axes, x-axis, y-axis, and the z-axis. Once again, we adapt our choices to the physical object. For example, we select the x-axis so that the wire lies on the x-axis, as shown in Figure 3.9

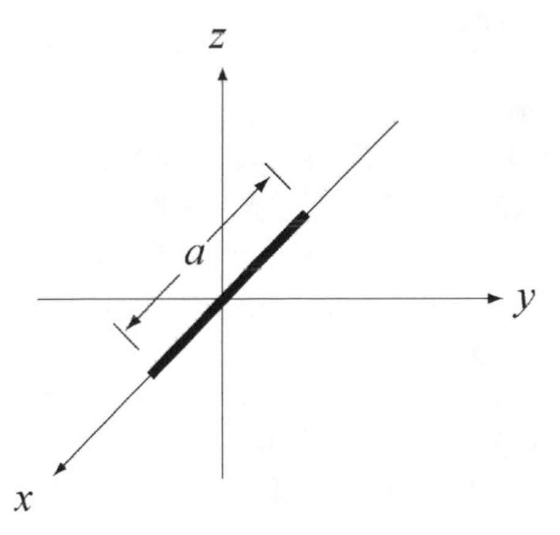

Figure 3.9 A segment of wire of length a lying along the x-axis of a Cartesian coordinate system.

Then each point P in space our S can be assigned a triplet of values (x_P, y_P, z_P), the Cartesian coordinates of the point P. The ranges of these values are: $-\infty < x_p < +\infty$, $-\infty < y_P < +\infty$, $-\infty < z_P < +\infty$.

The collection of points that have the same the coordinate y_P is called a level surface. For example, the set of points with the same value of $y = y_P$ is

$$S_{y_P} = \{(x, y, z) \in S \text{ such that } y = y_P\}. \tag{3.2.1}$$

This set S_{y_P} is a plane, the xz-plane (Figure 3.10), called a ***level set*** for constant y_P. Thus, the y-coordinate of any point actually describes a plane of points perpendicular to the y-axis.

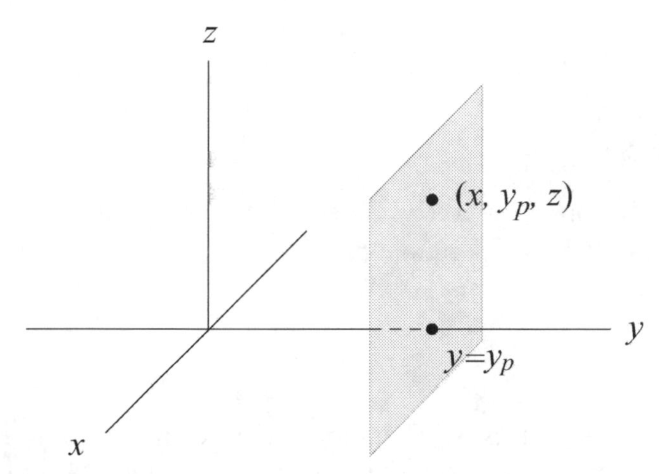

Figure 3.10 Level surface set for constant value y_P.

(3) Choice of Positive Direction

Our third choice is an assignment of positive direction for each coordinate axis. We shall denote this choice by the symbol + along the positive axis. Conventionally, Cartesian coordinates are drawn with the yz-plane corresponding to the plane of the paper. The horizontal direction from left to right is taken as the positive y-axis, and the vertical direction from bottom to top is taken as the positive z-axis. In physics problems we are free to choose our axes and positive directions any way that we decide best fits a given problem. Problems that are very difficult using the conventional choices may turn out to be much easier to solve by making a thoughtful choice of axes. The endpoints of the wire now have coordinates $(a/2, 0, 0)$ and $(-a/2, 0, 0)$.

(4) Choice of Unit Vectors

We now associate to each point P in space, a set of three unit directions vectors $(\hat{\mathbf{i}}_P, \hat{\mathbf{j}}_P, \hat{\mathbf{k}}_P)$. A unit vector has magnitude one: $|\hat{\mathbf{i}}_P| = 1$, $|\hat{\mathbf{j}}_P| = 1$, and $|\hat{\mathbf{k}}_P| = 1$. We assign the direction of $\hat{\mathbf{i}}_P$ to point in the direction of the increasing x-coordinate at the point P. We define the directions for $\hat{\mathbf{j}}_P$ and $\hat{\mathbf{k}}_P$ in the direction of the increasing y-coordinate and z-coordinate respectively, (Figure 3.11). If we choose a different point S, the units vectors $(\hat{\mathbf{i}}_S, \hat{\mathbf{j}}_S, \hat{\mathbf{k}}_S)$ at S, are equal to the unit vectors $(\hat{\mathbf{i}}_P, \hat{\mathbf{j}}_P, \hat{\mathbf{k}}_P)$ at P. This fact only holds true for a Cartesian coordinate system and does not hold for cylindrical coordinates, as we shall soon see. We therefore can drop the reference to the point and use $(\hat{\mathbf{i}}, \hat{\mathbf{j}}, \hat{\mathbf{k}})$ to represent the unit vectors in a Cartesian coordinate system.

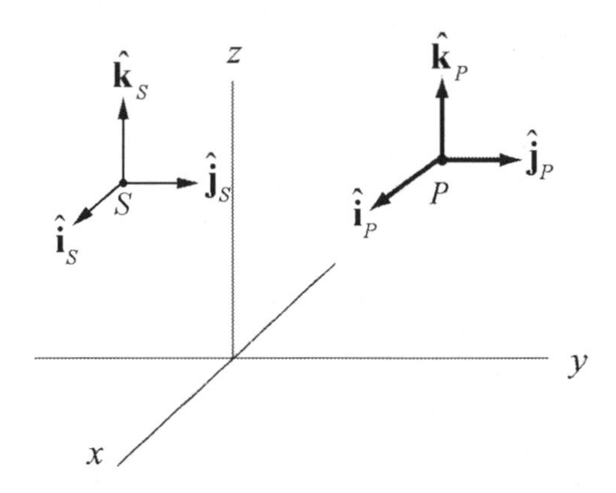

Figure 3.11 Choice of unit vectors at points P and S.

3.3 Application of Vectors

When we apply vectors to physical quantities it's nice to keep in the back of our minds all these formal properties. However from the physicist's point of view, we are interested in representing physical quantities such as displacement, velocity, acceleration, force, impulse, momentum, torque, and angular momentum as vectors. We can't add force to velocity or subtract momentum from torque. We must always understand the physical context for the vector quantity. Thus, instead of approaching vectors as formal mathematical objects we shall instead consider the following essential properties that enable us to represent physical quantities as vectors.

(1) Vectors can exist at any point P in space.

(2) Vectors have direction and magnitude.

(3) Vector Equality: Any two vectors that have the same direction and magnitude are equal no matter where in space they are located.

(4) Vector Decomposition: Choose a coordinate system with an origin and axes. We can decompose a vector into component vectors along each coordinate axis. In Figure 3.12 we choose Cartesian coordinates for the x-y plane (we ignore the z-direction for simplicity but we can extend our results when we need to). A vector $\vec{\mathbf{A}}$ at P can be decomposed into the vector sum,

$$\vec{\mathbf{A}} = \vec{\mathbf{A}}_x + \vec{\mathbf{A}}_y, \tag{3.3.1}$$

where $\vec{\mathbf{A}}_x$ is the x-component vector pointing in the positive or negative x-direction, and $\vec{\mathbf{A}}_y$ is the y-component vector pointing in the positive or negative y-direction (Figure 3.12).

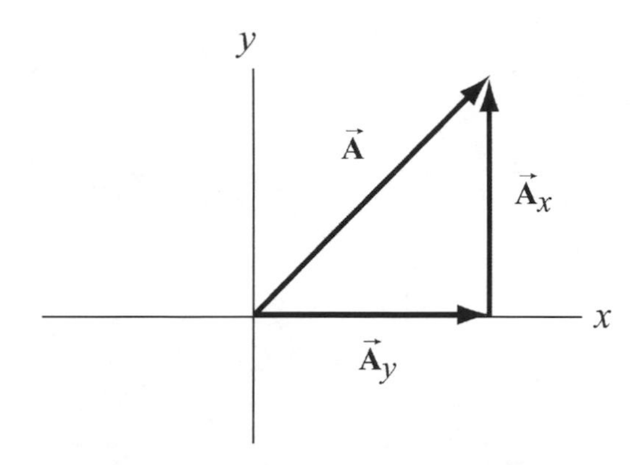

Figure 3.12 Vector decomposition.

(5) Vector Components: Once we have defined unit vectors $(\hat{\mathbf{i}},\hat{\mathbf{j}},\hat{\mathbf{k}})$, we then define the x-component and y-component of a vector. Recall our vector decomposition, $\vec{\mathbf{A}} = \vec{\mathbf{A}}_x + \vec{\mathbf{A}}_y$. We define the x-component vector, $\vec{\mathbf{A}}_x$, as

$$\vec{\mathbf{A}}_x = A_x\,\hat{\mathbf{i}}. \qquad (3.3.2)$$

In this expression the term A_x, (without the arrow above) is called the x-component of the vector $\vec{\mathbf{A}}$. The x-component A_x can be positive, zero, or negative. It is not the magnitude of $\vec{\mathbf{A}}_x$ which is given by $(A_x^{\,2})^{1/2}$. Note the difference between the x-component, A_x, and the x-component vector, $\vec{\mathbf{A}}_x$.

In a similar fashion we define the y-component, A_y, and the z-component, A_z, of the vector $\vec{\mathbf{A}}$

$$\vec{\mathbf{A}}_y = A_y\,\hat{\mathbf{j}}, \quad \vec{\mathbf{A}}_z = A_z\,\hat{\mathbf{k}}. \qquad (3.3.3)$$

A vector $\vec{\mathbf{A}}$ can be represented by its three components $\vec{\mathbf{A}} = (A_x, A_y, A_z)$. We can also write the vector as

$$\vec{\mathbf{A}} = A_x\,\hat{\mathbf{i}} + A_y\,\hat{\mathbf{j}} + A_z\,\hat{\mathbf{k}}. \qquad (3.3.4)$$

In Figure 3.13, we show the vector components $\vec{\mathbf{A}} = (A_x, A_y, A_z)$.

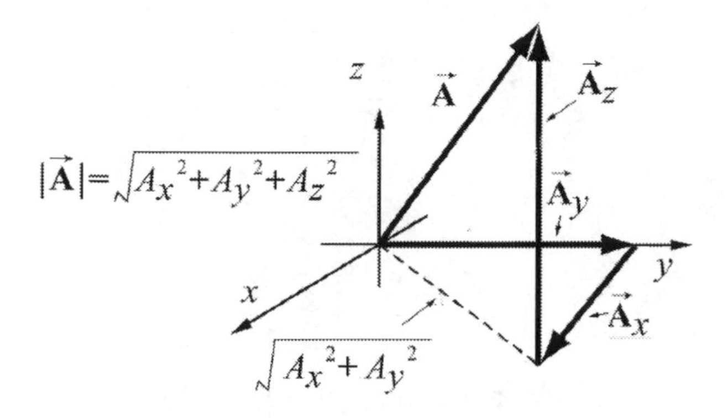

Figure 3.13 Component vectors in Cartesian coordinates.

(7) Magnitude: Using the Pythagorean theorem, the magnitude of $\vec{\mathbf{A}}$ is,

$$A = \sqrt{A_x^2 + A_y^2 + A_z^2}\,. \tag{3.3.5}$$

(8) Direction: Let's consider a vector $\vec{\mathbf{A}} = (A_x, A_y, 0)$. Since the z-component is zero, the vector $\vec{\mathbf{A}}$ lies in the x-y plane. Let θ denote the angle that the vector $\vec{\mathbf{A}}$ makes in the counterclockwise direction with the positive x-axis (Figure 3.14). Then the x-component and y-component are

$$A_x = A\cos(\theta), \quad A_y = A\sin(\theta)\,. \tag{3.3.6}$$

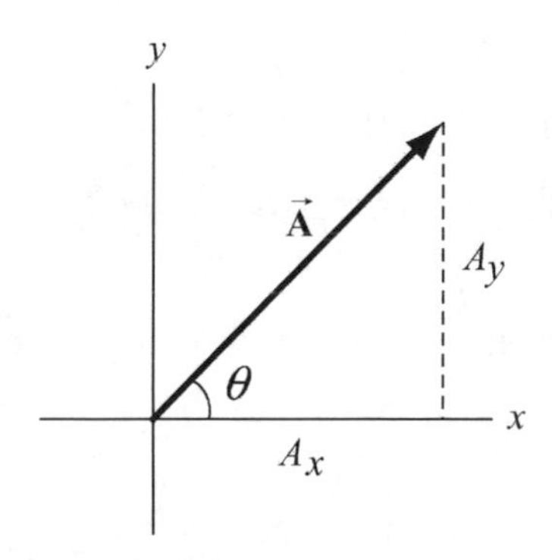

Figure 3.14 Components of a vector in the xy-plane.

We now write a vector in the xy-plane as

$$\vec{\mathbf{A}} = A\cos(\theta)\ \hat{\mathbf{i}} + A\sin(\theta)\ \hat{\mathbf{j}} \tag{3.3.7}$$

Once the components of a vector are known, the tangent of the angle θ can be determined by

$$\frac{A_y}{A_x} = \frac{A\sin(\theta)}{A\cos(\theta)} = \tan(\theta), \tag{3.3.8}$$

and hence the angle θ is given by

$$\theta = \tan^{-1}\left(\frac{A_y}{A_x}\right). \tag{3.3.9}$$

Clearly, the direction of the vector depends on the sign of A_x and A_y. For example, if both $A_x > 0$ and $A_y > 0$, then $0 < \theta < \pi/2$, and the vector lies in the first quadrant. If, however, $A_x > 0$ and $A_y < 0$, then $-\pi/2 < \theta < 0$, and the vector lies in the fourth quadrant.

(9) Unit vector in the direction of $\vec{\mathbf{A}}$: Let $\vec{\mathbf{A}} = A_x\,\hat{\mathbf{i}} + A_y\,\hat{\mathbf{j}} + A_z\,\hat{\mathbf{k}}$. Let $\hat{\mathbf{A}}$ denote a unit vector in the direction of $\vec{\mathbf{A}}$. Then

$$\hat{\mathbf{A}} = \frac{\vec{\mathbf{A}}}{\left|\vec{\mathbf{A}}\right|} = \frac{A_x\,\hat{\mathbf{i}} + A_y\,\hat{\mathbf{j}} + A_z\,\hat{\mathbf{k}}}{(A_x^{\,2} + A_y^{\,2} + A_z^{\,2})^{1/2}}. \tag{3.3.10}$$

(10) Vector Addition: Let $\vec{\mathbf{A}}$ and $\vec{\mathbf{B}}$ be two vectors in the *x*-*y* plane. Let θ_A and θ_B denote the angles that the vectors $\vec{\mathbf{A}}$ and $\vec{\mathbf{B}}$ make (in the counterclockwise direction) with the positive x-axis. Then

$$\vec{A} = A\cos(\theta_A)\,\hat{\mathbf{i}} + A\sin(\theta_A)\,\hat{\mathbf{j}}, \tag{3.3.11}$$

$$\vec{\mathbf{B}} = B\cos(\theta_B)\,\hat{\mathbf{i}} + B\sin(\theta_B)\,\hat{\mathbf{j}} \tag{3.3.12}$$

In Figure 3.15, the vector addition $\vec{\mathbf{C}} = \vec{\mathbf{A}} + \vec{\mathbf{B}}$ is shown. Let θ_C denote the angle that the vector $\vec{\mathbf{C}}$ makes with the positive *x*-axis.

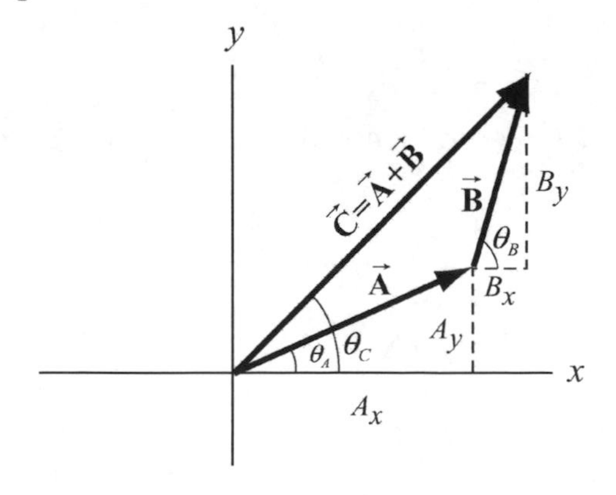

Figure 3.15 Vector addition with using components.

Then the components of $\vec{\mathbf{C}}$ are

$$C_x = A_x + B_x, \quad C_y = A_y + B_y. \tag{3.3.13}$$

In terms of magnitudes and angles, we have

$$\begin{aligned} C_x &= C\cos(\theta_C) = A\cos(\theta_A) + B\cos(\theta_B) \\ C_y &= C\sin(\theta_C) = A\sin(\theta_A) + B\sin(\theta_B). \end{aligned} \tag{3.3.14}$$

We can write the vector $\vec{\mathbf{C}}$ as

$$\vec{\mathbf{C}} = (A_x + B_x)\,\hat{\mathbf{i}} + (A_y + B_y)\,\hat{\mathbf{j}} = C\cos(\theta_C)\,\hat{\mathbf{i}} + C\sin(\theta_C)\,\hat{\mathbf{j}}, \tag{3.3.15}$$

Example 3.1 Vector Addition

Given two vectors, $\vec{\mathbf{A}} = 2\,\hat{\mathbf{i}} + -3\,\hat{\mathbf{j}} + 7\,\hat{\mathbf{k}}$ and $\vec{\mathbf{B}} = 5\hat{\mathbf{i}} + \hat{\mathbf{j}} + 2\hat{\mathbf{k}}$, find: (a) $|\vec{\mathbf{A}}|$; (b) $|\vec{\mathbf{B}}|$; (c) $\vec{\mathbf{A}} + \vec{\mathbf{B}}$; (d) $\vec{\mathbf{A}} - \vec{\mathbf{B}}$; (e) a unit vector $\hat{\mathbf{A}}$ pointing in the direction of $\vec{\mathbf{A}}$; (f) a unit vector $\hat{\mathbf{B}}$ pointing in the direction of $\vec{\mathbf{B}}$.

Solution:

(a) $|\vec{\mathbf{A}}| = \left(2^2 + (-3)^2 + 7^2\right)^{1/2} = \sqrt{62} = 7.87$. (b) $|\vec{\mathbf{B}}| = \left(5^2 + 1^2 + 2^2\right)^{1/2} = \sqrt{30} = 5.48$.

(c)
$$\begin{aligned} \vec{\mathbf{A}} + \vec{\mathbf{B}} &= (A_x + B_x)\,\hat{\mathbf{i}} + (A_y + B_y)\,\hat{\mathbf{j}} + (A_z + B_z)\,\hat{\mathbf{k}} \\ &= (2+5)\,\hat{\mathbf{i}} + (-3+1)\,\hat{\mathbf{j}} + (7+2)\,\hat{\mathbf{k}} \\ &= 7\,\hat{\mathbf{i}} - 2\,\hat{\mathbf{j}} + 9\,\hat{\mathbf{k}}. \end{aligned}$$

(d)
$$\begin{aligned} \vec{\mathbf{A}} - \vec{\mathbf{B}} &= (A_x - B_x)\,\hat{\mathbf{i}} + (A_y - B_y)\,\hat{\mathbf{j}} + (A_z - B_z)\,\hat{\mathbf{k}} \\ &= (2-5)\,\hat{\mathbf{i}} + (-3-1)\,\hat{\mathbf{j}} + (7-2)\,\hat{\mathbf{k}} \\ &= -3\,\hat{\mathbf{i}} - 4\,\hat{\mathbf{j}} + 5\,\hat{\mathbf{k}}. \end{aligned}$$

(e) A unit vector $\hat{\mathbf{A}}$ in the direction of $\vec{\mathbf{A}}$ can be found by dividing the vector $\vec{\mathbf{A}}$ by the magnitude of $\vec{\mathbf{A}}$. Therefore

$$\hat{\mathbf{A}} = \vec{\mathbf{A}} / |\vec{\mathbf{A}}| = \left(2\,\hat{\mathbf{i}} + -3\,\hat{\mathbf{j}} + 7\,\hat{\mathbf{k}}\right) / \sqrt{62}.$$

(f) In a similar fashion, $\hat{\mathbf{B}} = \vec{\mathbf{B}} / |\vec{\mathbf{B}}| = \left(5\hat{\mathbf{i}} + \hat{\mathbf{j}} + 2\hat{\mathbf{k}}\right) / \sqrt{30}$.

Example 3.2 Sinking Sailboat

A Coast Guard ship is located 35 km away from a checkpoint in a direction 42° north of west. A distressed sailboat located in still water 20 km from the same checkpoint in a direction 36° south of east is about to sink. Draw a diagram indicating the position of both ships. In what direction and how far must the Coast Guard ship travel to reach the sailboat?

Solution: The diagram of the set-up is Figure 3.16.

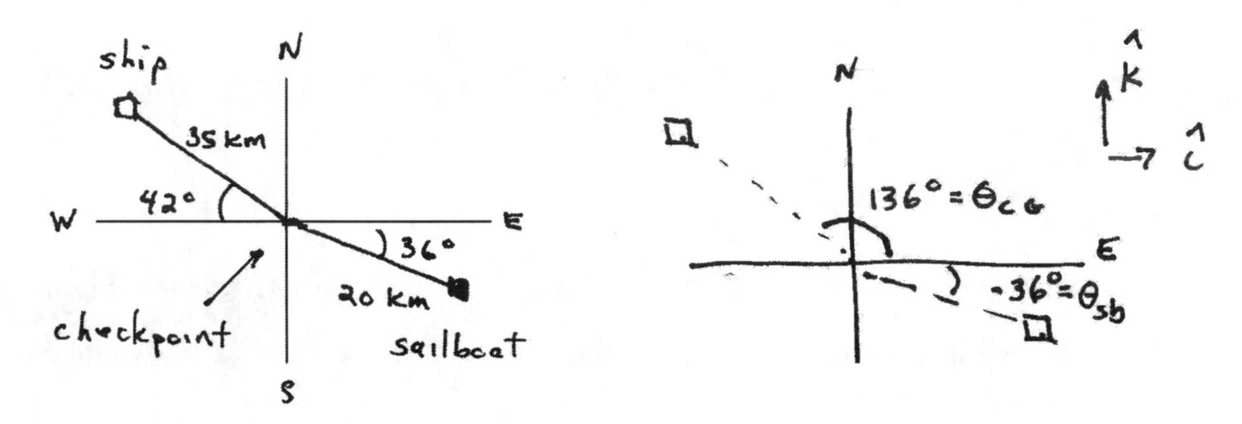

Figure 3.16 Example 3.2

Figure 3.17 Coordinate system for sailboat and ship

Choose the checkpoint as the origin, with North as the positive $\hat{\mathbf{k}}$-direction and East as the positive $\hat{\mathbf{i}}$-direction (see Figure 3.17). The Coast Guard ship is then at an angle $\theta_{\text{CG}} = 180^{\circ} - 42^{\circ} = 138^{\circ}$ from the checkpoint, and the sailboat is at an angle $\theta_{\text{sb}} = -36^{\circ}$ from the checkpoint. The position of the Coast Guard ship is then

$$
\begin{aligned}
\vec{\mathbf{r}}_{\text{CG}} &= r_{\text{CG}}(\cos\theta_{\text{CG}}\,\hat{\mathbf{i}} + \sin\theta_{\text{CG}}\hat{\mathbf{k}}) \\
&= -26.0\,\text{km}\,\hat{\mathbf{i}} + 23.4\,\text{km}\,\hat{\mathbf{k}},
\end{aligned}
$$

and the position of the sailboat is

$$
\begin{aligned}
\vec{\mathbf{r}}_{\text{sb}} &= r_{\text{sb}}(\cos\theta_{\text{sb}}\,\hat{\mathbf{i}} + \sin\theta_{\text{sb}}\hat{\mathbf{k}}) \\
&= 16.2\,\text{km}\,\hat{\mathbf{i}} - 11.8\,\text{km}\,\hat{\mathbf{k}}.
\end{aligned}
$$

Note that an extra significant figure has been kept for the intermediate calculations. The position vector from the Coast Guard ship to the sailboat is (Figure 3.18)

$$\vec{\mathbf{r}}_{sb} - \vec{\mathbf{r}}_{CG} = (16.2\,\text{km}\,\hat{\mathbf{i}} - 11.8\,\text{km}\,\hat{\mathbf{k}}) - (-26.0\,\text{km}\,\hat{\mathbf{i}} + 23.4\,\text{km}\,\hat{\mathbf{k}})$$
$$= 42.2\,\text{km}\,\hat{\mathbf{i}} - 35.2\,\text{km}\,\hat{\mathbf{k}}.$$

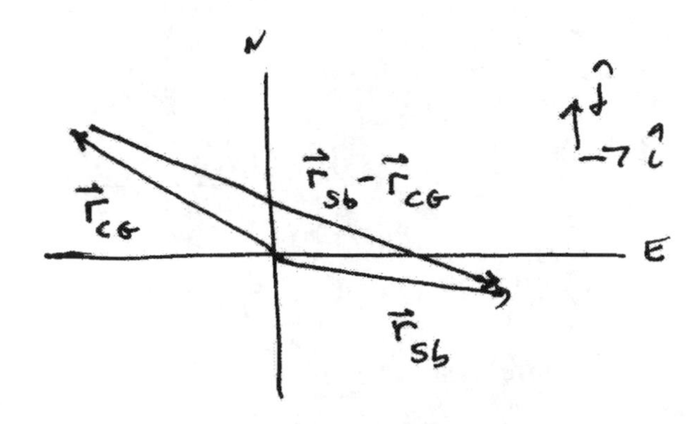

Figure 3.18 Relative position vector from ship to sailboat

The rescue ship's heading would be the inverse tangent of the ratio of the North and East components of the relative position,

$$\theta_{\text{rescue}} = \tan^{-1}(-35.2/42.2) = -39.8^\circ,$$

roughly 40° South of East.

Example 3.3 Vector Description of a Point on a Line

Consider two points located at $\vec{\mathbf{r}}_1$ and $\vec{\mathbf{r}}_2$, separated by distance $r_{12} = |\vec{\mathbf{r}}_1 - \vec{\mathbf{r}}_2|$. Find a vector $\vec{\mathbf{A}}$ from the origin to the point on the line between $\vec{\mathbf{r}}_1$ and $\vec{\mathbf{r}}_2$ at a distance xr_{12} from the point at $\vec{\mathbf{r}}_1$, where x is some number.

Solution: Consider the unit vector pointing from $\vec{\mathbf{r}}_1$ and $\vec{\mathbf{r}}_2$ given by $\hat{\mathbf{r}}_{12} = \vec{\mathbf{r}}_1 - \vec{\mathbf{r}}_2 / |\vec{\mathbf{r}}_1 - \vec{\mathbf{r}}_2| = \vec{\mathbf{r}}_1 - \vec{\mathbf{r}}_2 / r_{12}$. The vector $\vec{\boldsymbol{\alpha}}$ in Figure 3.19 connects $\vec{\mathbf{A}}$ to the point at $\vec{\mathbf{r}}_1$, therefore we can write $\vec{\boldsymbol{\alpha}} = xr_{12}\hat{\mathbf{r}}_{12} = xr_{12}(\vec{\mathbf{r}}_1 - \vec{\mathbf{r}}_2 / r_{12}) = x(\vec{\mathbf{r}}_1 - \vec{\mathbf{r}}_2)$. The vector $\vec{\mathbf{r}}_1 = \vec{\mathbf{A}} + \vec{\boldsymbol{\alpha}}$. Therefore $\vec{\mathbf{A}} = \vec{\mathbf{r}}_1 - \vec{\boldsymbol{\alpha}} = \vec{\mathbf{r}}_1 - x(\vec{\mathbf{r}}_1 - \vec{\mathbf{r}}_2) = \vec{\mathbf{r}}_1(1-x) + x\vec{\mathbf{r}}_2$.

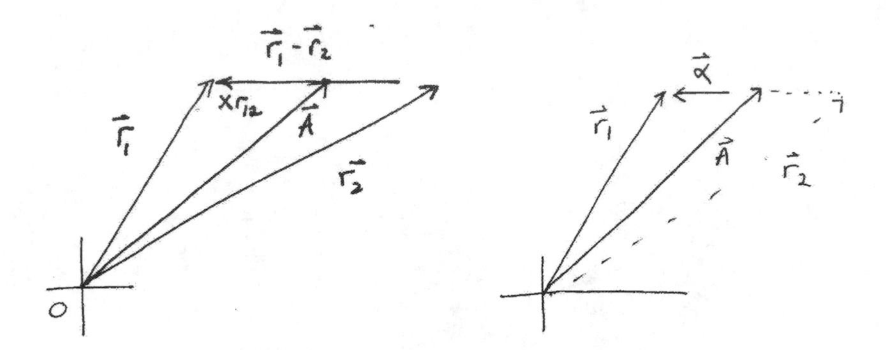

Figure 3.19 Vector geometry for Example 3.3

Example 3.4 Rotated Coordinate Systems

Consider two Cartesian coordinate systems S and S' with the same origin (Figure 3.20). Show that if the axes x', y' axes are rotated by an angle θ relative to the axes x, y, then the corresponding unit vectors are related according to $\hat{\mathbf{i}}' = \hat{\mathbf{i}}\cos\theta + \hat{\mathbf{j}}\sin\theta$, and $\hat{\mathbf{j}}' = \hat{\mathbf{j}}\cos\theta - \hat{\mathbf{i}}\sin\theta$.

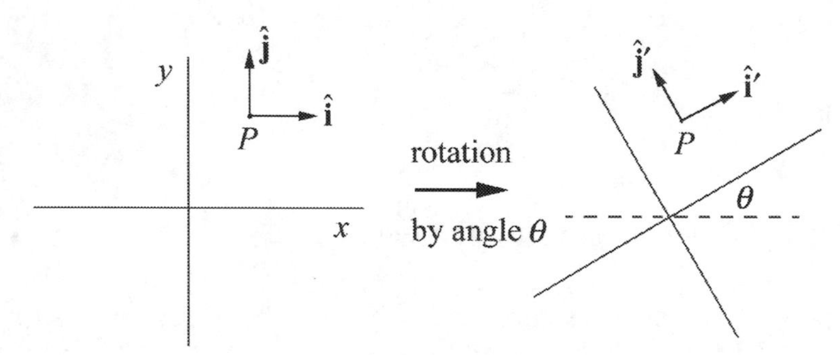

Figure 3.20 Example 3.4

Solution: This is simple vector decomposition. The components of $\hat{\mathbf{i}}'$ in the $\hat{\mathbf{i}}$ and $\hat{\mathbf{j}}$ direction are given by $\hat{i}'_x = |\hat{\mathbf{i}}'|\cos\theta = \cos\theta$ and $\hat{i}'_y = |\hat{\mathbf{i}}'|\sin\theta = \sin\theta$. Therefore

$$\hat{\mathbf{i}}' = \hat{i}'_x\hat{\mathbf{i}} + \hat{i}'_y\hat{\mathbf{j}} = \cos\theta\hat{\mathbf{i}} + \sin\theta\hat{\mathbf{j}}. \tag{3.3.16}$$

A similar argument holds for the components of $\hat{\mathbf{j}}'$. The components of $\hat{\mathbf{j}}'$ in the $\hat{\mathbf{i}}$ and $\hat{\mathbf{j}}$ direction are given by $\hat{j}'_x = -|\hat{\mathbf{j}}'|\sin\theta = -\sin\theta$ and $\hat{j}'_y = |\hat{\mathbf{j}}'|\cos\theta = \cos\theta$. Therefore

$$\hat{\mathbf{j}}' = \hat{j}'_x\hat{\mathbf{i}} + \hat{j}'_y\hat{\mathbf{j}} = -\sin\theta\hat{\mathbf{i}} + \cos\theta\hat{\mathbf{j}}. \tag{3.3.17}$$

Example 3.5 Vector Description in Rotated Coordinate Systems

With respect to a given Cartesian coordinate system S, a vector has components $A_x = 5$, $A_y = -3$, $A_z = 0$. (a) What are the components $A_{x'}$ and $A_{y'}$ of this vector in a second coordinate system S' whose x' and y' axes make angles of 60° with the x and y axes respectively of the first coordinate system S? (b) Calculate the magnitude of the vector from its A_x and A_y components and from its $A_{x'}$ and $A_{y'}$ components. Does the result agree with what you expect?

Solution: We begin with a sketch of the rotated coordinate systems and the vector $\vec{\mathbf{A}}$ (Figure 3.21).

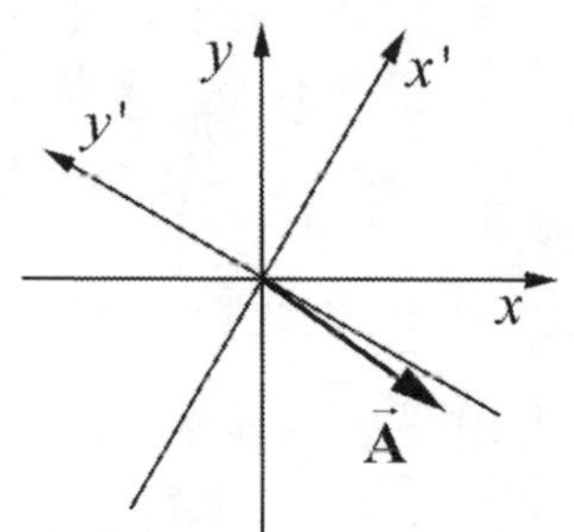

Figure 3.21 Example 3.5

We then take the vector decomposition of $\vec{\mathbf{A}}$ with respect to the xy-coordinate system,

$$\vec{\mathbf{A}} = A_x\hat{\mathbf{i}} + A_y\hat{\mathbf{j}}. \quad (3.3.18)$$

Now we can use our results from Example 3.5 to solve for the unit vectors $\hat{\mathbf{i}}$ and $\hat{\mathbf{j}}$ in terms of $\hat{\mathbf{i}}'$ and $\hat{\mathbf{j}}'$. Multiply Eq. (3.3.16) for $\hat{\mathbf{i}}'$ by $\sin\theta$ and Eq.(3.3.17) for $\hat{\mathbf{j}}'$ by $\cos\theta$ yielding

$$\sin\theta\hat{\mathbf{i}}' = \sin\theta\cos\theta\hat{\mathbf{i}} + \sin^2\theta\hat{\mathbf{j}}$$
$$\cos\theta\hat{\mathbf{j}}' = -\sin\theta\cos\theta\hat{\mathbf{i}} + \cos^2\theta\hat{\mathbf{j}}.$$

Now add these equations, using the identity $\sin^2\theta + \cos^2\theta = 1$ yielding

$$\hat{\mathbf{j}} = \sin\theta\hat{\mathbf{i}}' + \cos\theta\hat{\mathbf{j}}'. \quad (3.3.19)$$

Similarly multiply the equation for $\hat{\mathbf{i}}'$ by $\cos\theta$ and the equation for $\hat{\mathbf{j}}'$ by $-\sin\theta$ yielding

$$\cos\theta\hat{\mathbf{i}}' = \cos^2\theta\hat{\mathbf{i}} + \cos\theta\sin\theta\hat{\mathbf{j}}$$

$$-\sin\theta\hat{\mathbf{j}}' = \sin^2\theta\hat{\mathbf{i}} - \sin\theta\cos\theta\hat{\mathbf{j}}.$$

Now add these equations, using the identity $\sin^2\theta + \cos^2\theta = 1$

$$\hat{\mathbf{i}} = \cos\theta\hat{\mathbf{i}}' - \sin\theta\hat{\mathbf{j}}' \tag{3.3.20}$$

Now we can rewrite the vector $\vec{\mathbf{A}}$ as

$$\begin{aligned}\vec{\mathbf{A}} &= A_x\hat{\mathbf{i}} + A_y\hat{\mathbf{j}} = A_x(\cos\theta\hat{\mathbf{i}}' - \sin\theta\hat{\mathbf{j}}') + A_y(\sin\theta\hat{\mathbf{i}}' + \cos\theta\hat{\mathbf{j}}') \\ &= (A_x\cos\theta + A_y\sin\theta)\hat{\mathbf{i}}' + (-A_x\sin\theta + A_y\cos\theta)\hat{\mathbf{j}}' \\ &= A_{x'}\hat{\mathbf{i}} + A_{y'}\hat{\mathbf{j}},\end{aligned}$$

where

$$A_{x'} = A_x\cos\theta + A_y\sin\theta \tag{3.3.21}$$

$$A_{y'} = -A_x\sin\theta + A_y\cos\theta. \tag{3.3.22}$$

We can use the given information that $A_x = 5$, $A_y = -3$, and $\theta = 60^\circ$ to solve for the components of $\vec{\mathbf{A}}$ in the prime coordinate system

$$A_{x'} = A_x\cos\theta + A_y\sin\theta = \frac{1}{2}(5 - 3\sqrt{3}),$$

$$A_{y'} = -A_x\sin\theta + A_y\cos\theta = \frac{1}{2}(-5\sqrt{3} - 3).$$

c) The magnitude of can be calculated in either coordinate system

$$|\vec{\mathbf{A}}| = \sqrt{(A_x)^2 + (A_y)^2} = \sqrt{(5)^2 + (-3)^2} = \sqrt{34}$$

$$|\vec{\mathbf{A}}| = \sqrt{(A_{x'})^2 + (A_{y'})^2} = \sqrt{\left(\frac{1}{2}(5 - 3\sqrt{3})\right)^2 + \left(\frac{1}{2}(-5\sqrt{3} - 3)\right)^2} = \sqrt{34}.$$

This result agrees with what I expect because the length of vector is independent of the choice of coordinate system.

Example 3.6 Vector Addition

Two force vectors $\vec{\mathbf{A}}$ and $\vec{\mathbf{B}}$, such that $|\vec{\mathbf{B}}| = 2|\vec{\mathbf{A}}|$ have a resultant $\vec{\mathbf{C}} = \vec{\mathbf{A}} + \vec{\mathbf{B}}$ of magnitude $26.5\,\text{N}$, which makes an angle of 41° with respect to the smaller vector $\vec{\mathbf{A}}$. Find the magnitude of each vector and the angle between them.

Solution: We begin by making a sketch of the three vectors, choosing $\vec{\mathbf{A}}$ to point in the positive x-direction (Figure 3.22).

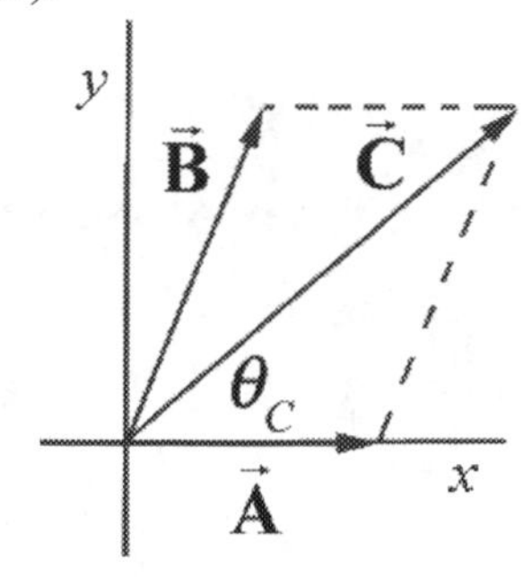

Figure 3.22 Choice of coordinates system for Example 3.6

Let's denote the magnitude of $\vec{\mathbf{C}}$ by $C \equiv |\vec{\mathbf{C}}| = \sqrt{(C_x)^2 + (C_y)^2} = 26.5\ \text{N}$. We first note that the components of $\vec{\mathbf{C}} = \vec{\mathbf{A}} + \vec{\mathbf{B}}$ are given by

$$C_x = A_x + B_x = C\cos\theta_C = (26.5\ \text{N})\cos(41^\circ) = 20\ \text{N} \tag{3.3.23}$$

$$C_y = B_y = C\sin\theta_C = (26.5\ \text{N})\sin(41^\circ) = 17.39\ \text{N}\,. \tag{3.3.24}$$

Let's denote the magnitude of $\vec{\mathbf{C}}$ by $C \equiv |\vec{\mathbf{C}}|$. From the condition that $|\vec{\mathbf{B}}| = 2|\vec{\mathbf{A}}|$, we know that

$$(B_x)^2 + (B_y)^2 = 4(A_x)^2\,. \tag{3.3.25}$$

Using Eqs. (3.3.23) and (3.3.24), Eq. (3.3.25) becomes

$$(C_x - A_x)^2 + (C_y)^2 = 4(A_x)^2$$
$$(C_x)^2 - 2C_xA_x + (A_x)^2 + (C_y)^2 = 4(A_x)^2\,.$$

This is a quadratic equation

$$0 = 3(A_x)^2 + 2C_xA_x - C^2$$

with solution

$$A_x = \frac{-2C_x \pm \sqrt{(2C_x)^2 + (4)(3)(C^2)}}{6} = \frac{-2(20\ \text{N}) \pm \sqrt{(40\ \text{N}))^2 + (4)(3)(26.5\ \text{N})^2}}{6}$$
$$= 10.0\ \text{N},$$

where we choose the positive square root because we originally choose $A_x > 0$. Then

$$B_x = C_x - A_x = 20.0\ \text{N} - 10.0\ \text{N} = 10.0\ \text{N}$$

$$B_y = 17.39\ \text{N}.$$

Note that the magnitude of $|\vec{\mathbf{B}}| = \sqrt{(B_x)^2 + (B_y)^2} = 20.0\ \text{N}$ is equal to two times the magnitude of $|\vec{\mathbf{A}}| = 10.0\ \text{N}$. The angle between $\vec{\mathbf{A}}$ and $\vec{\mathbf{B}}$ is given by

$$\theta = \sin^{-1}(B_y / |\vec{\mathbf{B}}|) = \sin^{-1}(17.39\ \text{N} / 20.0\ \text{N}) = 60^\circ.$$

Chapter 4 One Dimensional Kinematics

Chapter 4 One Dimensional Kinematics

In the first place, what do we mean by time and space? It turns out that these deep philosophical questions have to be analyzed very carefully in physics, and this is not easy to do. The theory of relativity shows that our ideas of space and time are not as simple as one might imagine at first sight. However, for our present purposes, for the accuracy that we need at first, we need not be very careful about defining things precisely. Perhaps you say, "That's a terrible thing—I learned that in science we have to define everything precisely." We cannot define anything precisely! If we attempt to, we get into that paralysis of thought that comes to philosophers, who sit opposite each other, one saying to the other, "You don't know what you are talking about!" The second one says. "What do you mean by know? What do you mean by talking? What do you mean by you?", and so on. In order to be able to talk constructively, we just have to agree that we are talking roughly about the same thing. You know as much about time as you need for the present, but remember that there are some subtleties that have to be discussed; we shall discuss them later.[1]

Richard Feynman

4.1 Introduction

Kinematics is the mathematical description of motion. The term is derived from the Greek word *kinema,* meaning movement. In order to quantify motion, a mathematical coordinate system, called a *reference frame*, is used to describe space and time. Once a reference frame has been chosen, we can introduce the physical concepts of position, velocity and acceleration in a mathematically precise manner. Figure 4.1 shows a Cartesian coordinate system in one dimension with unit vector $\hat{\mathbf{i}}$ pointing in the direction of increasing x-coordinate.

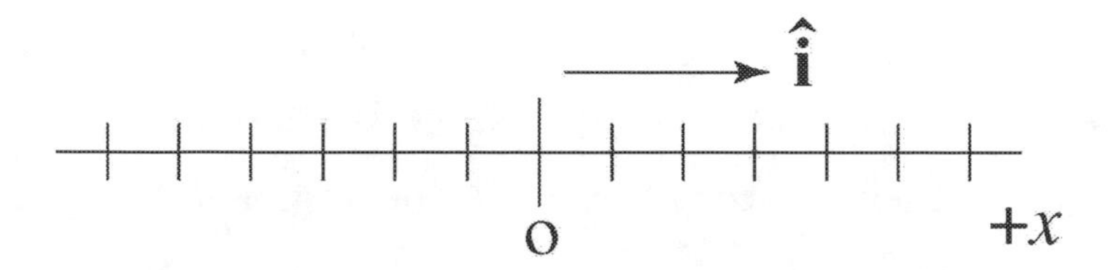

Figure 4.1 A one-dimensional Cartesian coordinate system.

[1] Richard P. Feynman, Robert B. Leighton, Matthew Sands, *The Feynman Lectures on Physics*, Addison-Wesley, Reading, Massachusetts, (1963), p. 12-2.

4.2 Position, Time Interval, Displacement

4.2.1 Position

Consider an object moving in one dimension. We denote the ***position coordinate*** of the center of mass of the object *with respect to the choice of origin* by $x(t)$. The position coordinate is a function of time and can be positive, zero, or negative, depending on the location of the object. The position has both direction and magnitude, and hence is a vector (Figure 4.2),

$$\vec{\mathbf{x}}(t) = x(t)\,\hat{\mathbf{i}}. \qquad (4.2.1)$$

We denote the position coordinate of the center of the mass at $t = 0$ by the symbol $x_0 \equiv x(t = 0)$. The SI unit for position is the meter [m].

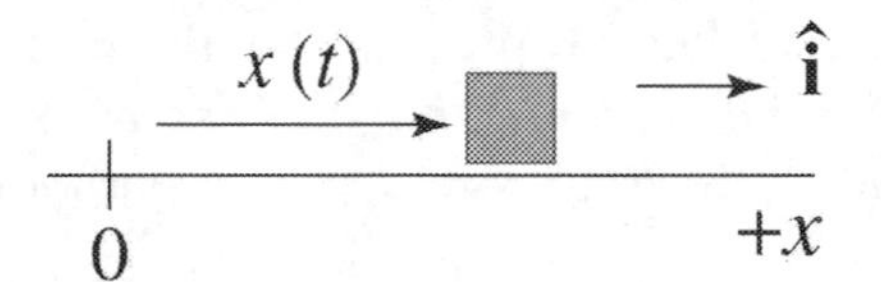

Figure 4.2 The position vector, with reference to a chosen origin.

4.2.2 Time Interval

Consider a closed interval of time $[t_1, t_2]$. We characterize this time interval by the difference in endpoints of the interval such that

$$\Delta t = t_2 - t_1. \qquad (4.2.2)$$

The SI units for time intervals are seconds [s].

4.2.3 Displacement

The ***displacement***, of a body between times t_1 and t_2 (Figure 4.3) is defined to be the change in position coordinate of the body

$$\Delta\vec{\mathbf{x}} \equiv (x(t_2) - x(t_1))\,\hat{\mathbf{i}} \equiv \Delta x(t)\,\hat{\mathbf{i}}. \qquad (4.2.3)$$

Displacement is a vector quantity.

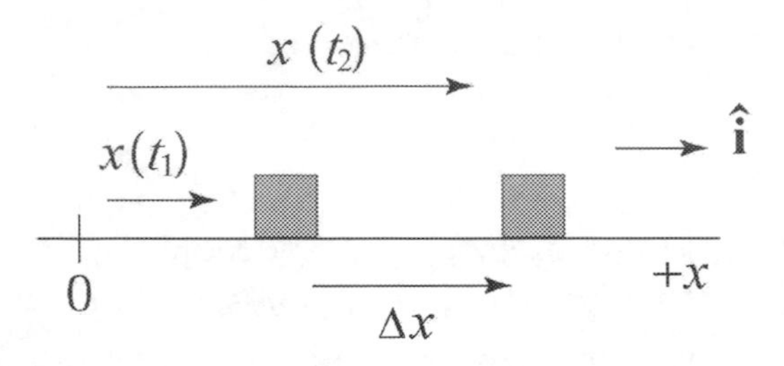

Figure 4.3 The displacement vector of an object over a time interval is the vector difference between the two position vectors

4.3 Velocity

When describing the motion of objects, words like "speed" and "velocity" are used in common language; however when introducing a mathematical description of motion, we need to define these terms precisely. Our procedure will be to define average quantities for finite intervals of time and then examine what happens in the limit as the time interval becomes infinitesimally small. This will lead us to the mathematical concept that velocity at an instant in time is the derivative of the position with respect to time.

4.3.1 Average Velocity

The component of the ***average velocity***, $\overline{v_x}$, for a time interval Δt is defined to be the displacement Δx divided by the time interval Δt,

$$\overline{v_x} \equiv \frac{\Delta x}{\Delta t}. \tag{4.3.1}$$

The average velocity vector is then

$$\overline{\vec{\mathbf{v}}}(t) \equiv \frac{\Delta x}{\Delta t}\,\hat{\mathbf{i}} = \overline{v_x}(t)\,\hat{\mathbf{i}}. \tag{4.3.2}$$

The SI units for average velocity are meters per second $\left[\mathrm{m \cdot s^{-1}}\right]$.

4.3.3 Instantaneous Velocity

Consider a body moving in one direction. We denote the position coordinate of the body by $x(t)$, with initial position x_0 at time $t = 0$. Consider the time interval $[t, t+\Delta t]$. The average velocity for the interval Δt is the slope of the line connecting the points $(t, x(t))$ and $(t, x(t+\Delta t))$. The slope, the rise over the run, is the change in position over the change in time, and is given by

$$\overline{v}_x \equiv \frac{\text{rise}}{\text{run}} = \frac{\Delta x}{\Delta t} = \frac{x(t+\Delta t) - x(t)}{\Delta t}. \tag{4.3.3}$$

Let's see what happens to the average velocity as we shrink the size of the time interval. The slope of the line connecting the points $(t, x(t))$ and $(t, x(t+\Delta t))$ approaches the slope of the tangent line to the curve $x(t)$ at the time t (Figure 4.4).

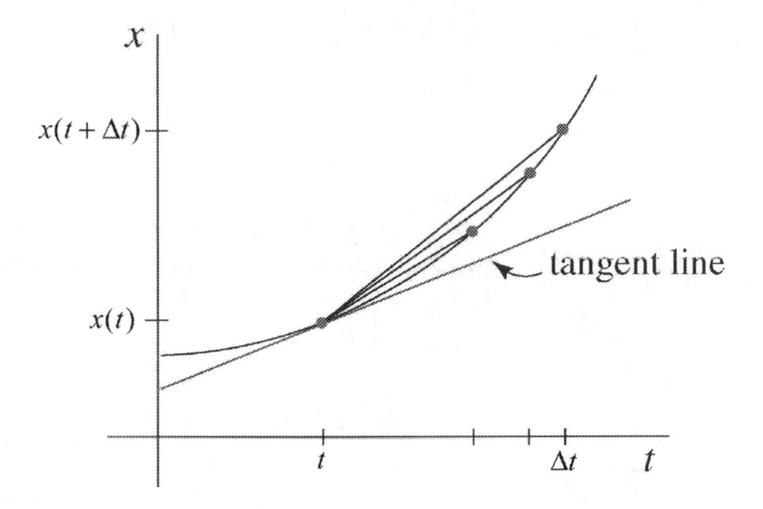

Figure 4.4 Graph of position *vs*. time showing the tangent line at time t.

In order to define the limiting value for the slope at any time, we choose a time interval $[t, t+\Delta t]$. For each value of Δt, we calculate the average velocity. As $\Delta t \to 0$, we generate a sequence of average velocities. The limiting value of this sequence is defined to be the x-component of the instantaneous velocity at the time t.

The x-component of ***instantaneous velocity*** at time t is given by the slope of the tangent line to the curve of position vs. time curve at time t:

$$v_x(t) \equiv \lim_{\Delta t \to 0} \overline{v}_x = \lim_{\Delta t \to 0} \frac{\Delta x}{\Delta t} = \lim_{\Delta t \to 0} \frac{x(t+\Delta t) - x(t)}{\Delta t} \equiv \frac{dx}{dt}. \tag{4.3.4}$$

The instantaneous velocity vector is then

$$\vec{\mathbf{v}}(t) = v_x(t)\,\hat{\mathbf{i}}. \tag{4.3.5}$$

Example 4.1 Determining Velocity from Position

Consider an object that is moving along the x-coordinate axis represented by the equation

$$x(t) = x_0 + \frac{1}{2}bt^2 \tag{4.3.6}$$

where x_0 is the initial position of the object at $t = 0$.

We can explicitly calculate the x-component of instantaneous velocity from Equation (4.3.4) by first calculating the displacement in the x-direction, $\Delta x = x(t + \Delta t) - x(t)$. We need to calculate the position at time $t + \Delta t$,

$$x(t + \Delta t) = x_0 + \frac{1}{2}b(t + \Delta t)^2 = x_0 + \frac{1}{2}b\left(t^2 + 2t\Delta t + \Delta t^2\right). \tag{4.3.7}$$

Then the instantaneous velocity is

$$v_x(t) = \lim_{\Delta t \to 0} \frac{x(t + \Delta t) - x(t)}{\Delta t} = \lim_{\Delta t \to 0} \frac{\left(x_0 + \frac{1}{2}b(t^2 + 2t\,\Delta t + \Delta t^2)\right) - \left(x_0 + \frac{1}{2}bt^2\right)}{\Delta t}. \tag{4.3.8}$$

This expression reduces to

$$v_x(t) = \lim_{\Delta t \to 0}\left(bt + \frac{1}{2}b\,\Delta t\right). \tag{4.3.9}$$

The first term is independent of the interval Δt and the second term vanishes because the limit as $\Delta t \to 0$ of Δt is zero. Thus the instantaneous velocity at time t is

$$v_x(t) = bt\,. \tag{4.3.10}$$

In Figure 4.5 we graph the instantaneous velocity, $v_x(t)$, as a function of time t.

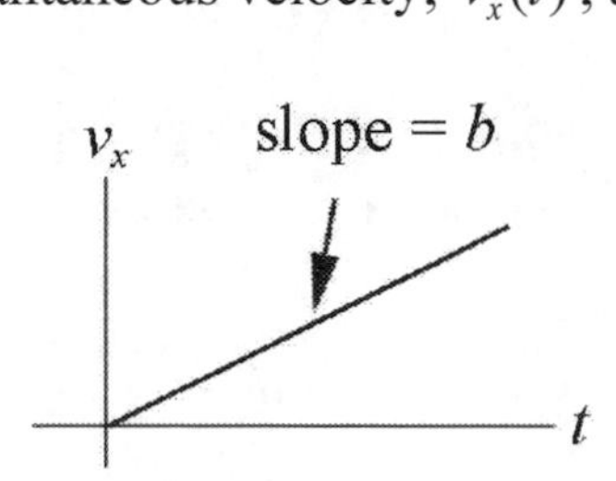

Figure 4.5 A graph of instantaneous velocity as a function of time.

4.4 Acceleration

We shall apply the same physical and mathematical procedure for defining acceleration, the rate of change of velocity. We first consider how the instantaneous velocity changes over an interval of time and then take the limit as the time interval approaches zero.

4.4.1 Average Acceleration

Acceleration is the quantity that measures a change in velocity over a particular time interval. Suppose during a time interval Δt a body undergoes a change in velocity

$$\Delta\vec{\mathbf{v}} = \vec{\mathbf{v}}(t+\Delta t) - \vec{\mathbf{v}}(t). \tag{4.4.1}$$

The change in the x-component of the velocity, Δv_x, for the time interval $[t, t+\Delta t]$ is then

$$\Delta v_x = v_x(t+\Delta t) - v_x(t). \tag{4.4.2}$$

The x*-component of the average acceleration* for the time interval Δt is defined to be

$$\bar{\vec{\mathbf{a}}} = \overline{a}_x\,\hat{\mathbf{i}} \equiv \frac{\Delta v_x}{\Delta t}\hat{\mathbf{i}} = \frac{(v_x(t+\Delta t) - v_x(t))}{\Delta t}\hat{\mathbf{i}} = \frac{\Delta v_x}{\Delta t}\hat{\mathbf{i}}. \tag{4.4.3}$$

The SI units for average acceleration are meters per second squared, $[\mathrm{m \cdot s^{-2}}]$.

4.4.2 Instantaneous Acceleration

On a graph of the x-component of velocity *vs.* time, the average acceleration for a time interval Δt is the slope of the straight line connecting the two points $(t, v_x(t))$ and $(t+\Delta t, v_x(t+\Delta t))$. In order to define the x-component of the instantaneous acceleration at time t, we employ the same limiting argument as we did when we defined the instantaneous velocity in terms of the slope of the tangent line.

The x*-component of the instantaneous acceleration* *at time* t is the limit of the slope of the tangent line at time t of the graph of the x-component of the velocity as a function of time,

$$a_x(t) \equiv \lim_{\Delta t \to 0} \overline{a}_x = \lim_{\Delta t \to 0} \frac{(v_x(t+\Delta t) - v_x(t))}{\Delta t} = \lim_{\Delta t \to 0} \frac{\Delta v_x}{\Delta t} \equiv \frac{dv_x}{dt}. \tag{4.4.4}$$

The instantaneous acceleration vector is then

$$\vec{\mathbf{a}}(t) = a_x(t)\,\hat{\mathbf{i}}. \tag{4.4.5}$$

In Figure 4.6 we illustrate this geometrical construction. Because the velocity is the derivative of position with respect to time, the x-component of the acceleration is the second derivative of the position function,

$$a_x = \frac{dv_x}{dt} = \frac{d^2x}{dt^2}. \tag{4.4.6}$$

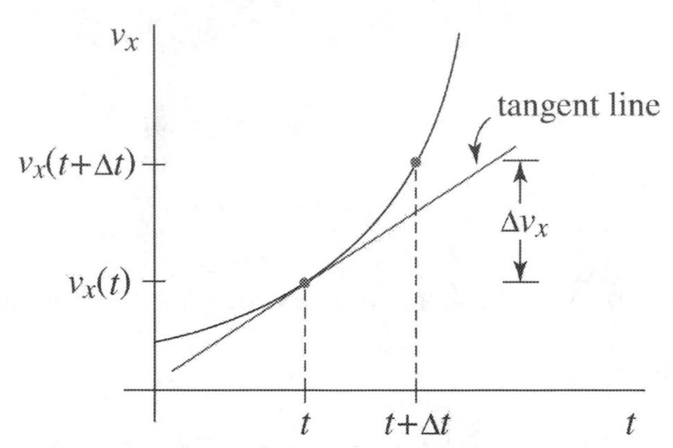

Figure 4.6 Graph of velocity *vs*. time showing the tangent line at time t.

Example 4.2 Determining Acceleration from Velocity

Let's continue Example 4.1, in which the position function for the body is given by $x = x_0 + (1/2)bt^2$, and the x-component of the velocity is $v_x = bt$. The x-component of the instantaneous acceleration at time t is the limit of the slope of the tangent line at time t of the graph of the x-component of the velocity as a function of time (Figure 4.5)

$$a_x = \frac{dv_x}{dt} = \lim_{\Delta t \to 0} \frac{v_x(t+\Delta t) - v_x(t)}{\Delta t} = \lim_{\Delta t \to 0} \frac{bt + b\Delta t - bt}{\Delta t} = b. \tag{4.4.7}$$

Note that in Equation (4.4.7), the ratio $\Delta v / \Delta t$ is independent of Δt, consistent with the constant slope of the graph in Figure 4.5.

4.5 Constant Acceleration

Let's consider a body undergoing constant acceleration for a time interval $\Delta t = [0, t]$. When the acceleration a_x is a constant, the average acceleration is equal to the instantaneous acceleration. Denote the x-component of the velocity at time $t = 0$ by $v_{x,0} \equiv v_x(t = 0)$. Therefore the x-component of the acceleration is given by

$$a_x = \overline{a_x} = \frac{\Delta v_x}{\Delta t} = \frac{v_x(t) - v_{x,0}}{t}. \tag{4.5.1}$$

Thus the velocity as a function of time is given by

$$v_x(t) = v_{x,0} + a_x t. \tag{4.5.2}$$

When the acceleration is constant, the velocity is a linear function of time.

4.5.1 Velocity: Area Under the Acceleration *vs.* Time Graph

In Figure 4.7, the x-component of the acceleration is graphed as a function of time.

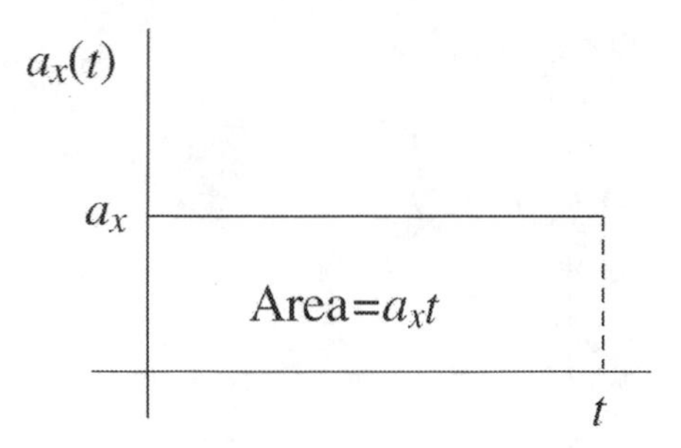

Figure 4.7 Graph of the x-component of the acceleration for a_x constant as a function of time.

The area under the acceleration *vs.* time graph, for the time interval $\Delta t = t - 0 = t$, is

$$\text{Area}(a_x, t) \equiv a_x t. \tag{4.5.3}$$

Using the definition of average acceleration given above,

$$\text{Area}(a_x, t) \equiv a_x t = \Delta v_x = v_x(t) - v_{x,0}. \tag{4.5.4}$$

4.5.2 Displacement: Area Under the Velocity *vs.* Time Graph

In Figure 4.8, we graph the x-component of the velocity *vs.* time curve.

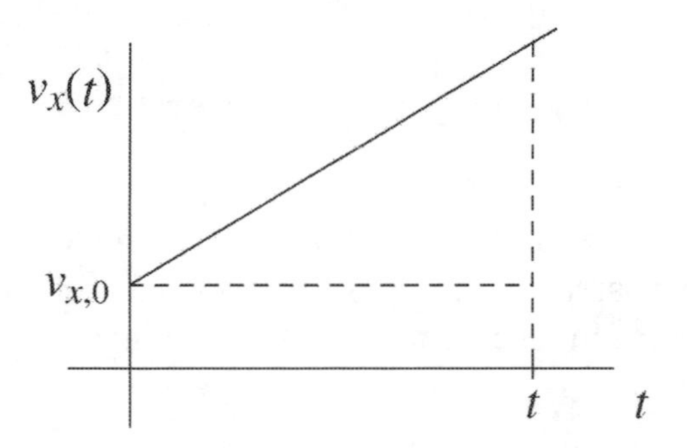

Figure 4.8 Graph of velocity as a function of time for a_x constant.

The region under the velocity *vs.* time curve is a trapezoid, formed from a rectangle and a triangle and the area of the trapezoid is given by

$$\text{Area}(v_x, t) = v_{x,0}\, t + \frac{1}{2}(v_x(t) - v_{x,0})t. \tag{4.5.5}$$

Substituting for the velocity (Equation (4.5.2)) yields

$$\text{Area}(v_x,t) = v_{x,0}\,t + \frac{1}{2}(v_{x,0} + a_x\,t - v_{x,0})t = v_{x,0}\,t + \frac{1}{2}a_x\,t^2 . \tag{4.5.6}$$

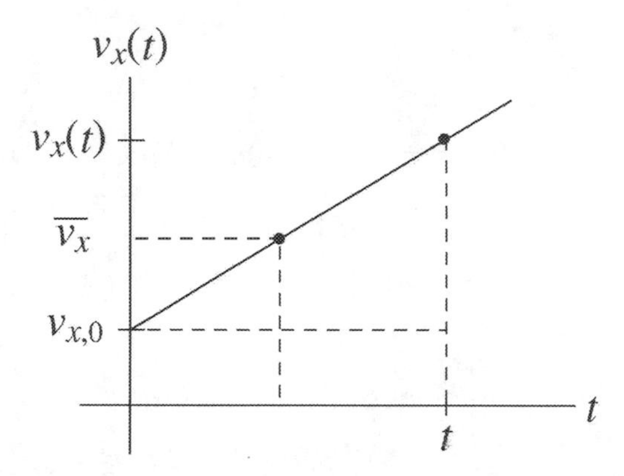

Figure 4.9 The average velocity over a time interval.

We can then determine the average velocity by adding the initial and final velocities and dividing by a factor of two (Figure 4.9).

$$\overline{v_x} = \frac{1}{2}(v_x(t) + v_{x,0}) . \tag{4.5.7}$$

The above method for determining the average velocity differs from the definition of average velocity in Equation (4.3.1). When the acceleration is constant over a time interval, the two methods will give identical results. Substitute into Equation (4.5.7) the x-component of the velocity, Equation (4.5.2), to yield

$$\overline{v_x} = \frac{1}{2}(v_x(t) + v_{x,0}) = \frac{1}{2}((v_{x,0} + a_x\,t) + v_{x,0}) = v_{x,0} + \frac{1}{2}a_x\,t . \tag{4.5.8}$$

Recall Equation (4.3.1); the average velocity is the displacement divided by the time interval (note we are now using the definition of average velocity that always holds, for non-constant as well as constant acceleration). The displacement is equal to

$$\Delta x \equiv x(t) - x_0 = \overline{v_x}\,t . \tag{4.5.9}$$

Substituting Equation (4.5.8) into Equation (4.5.9) shows that displacement is given by

$$\Delta x \equiv x(t) - x_0 = \overline{v_x}\,t = v_{x,0}\,t + \frac{1}{2}a_x\,t^2 . \tag{4.5.10}$$

Now compare Equation (4.5.10) to Equation (4.5.6) to conclude that the displacement is equal to the area under the graph of the x-component of the velocity *vs.* time,

$$\Delta x \equiv x(t) - x_0 = v_{x,0}\,t + \frac{1}{2}a_x t^2 = \text{Area}(v_x, t), \tag{4.5.11}$$

and so we can solve Equation (4.5.11) for the position as a function of time,

$$x(t) = x_0 + v_{x,0}\,t + \frac{1}{2}a_x t^2. \tag{4.5.12}$$

Figure 4.10 shows a graph of this equation. Notice that at $t = 0$ the slope may be in general non-zero, corresponding to the initial velocity component $v_{x,0}$.

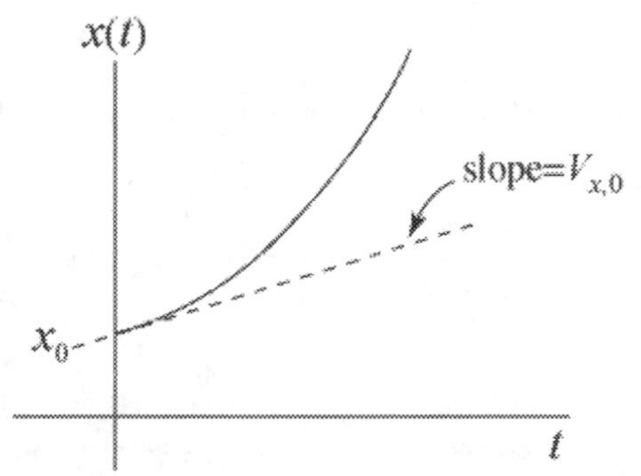

Figure 4.10 Graph of position vs. time for constant acceleration.

Example 4.3 Accelerating Car

A car, starting at rest at $t = 0$, accelerates in a straight line for $100\ \text{m}$ with an unknown constant acceleration. It reaches a speed of $20\ \text{m}\cdot\text{s}^{-1}$ and then continues at this speed for another $10\ \text{s}$. (a) Write down the equations for position and velocity of the car as a function of time. (b) How long was the car accelerating? (c) What was the magnitude of the acceleration? (d) Plot speed *vs.* time, acceleration *vs.* time, and position *vs.* time for the entire motion. (e) What was the average velocity for the entire trip?

Solutions: (a) For the acceleration a, the position $x(t)$ and velocity $v(t)$ as a function of time t for a car starting from rest are

$$\begin{aligned} x(t) &= (1/2)at^2 \\ v_x(t) &= at. \end{aligned} \tag{4.5.13}$$

b) Denote the time interval during which the car accelerated by t_1. We know that the position $x(t_1) = 100\text{m}$ and $v(t_1) = 20\ \text{m}\cdot\text{s}^{-1}$. Note that we can eliminate the acceleration a between the Equations (4.5.13) to obtain

$$x(t) = (1/2)\,v(t)\,t. \tag{4.5.14}$$

We can solve this equation for time as a function of the distance and the final speed giving

$$t = 2\frac{x(t)}{v(t)}. \tag{4.5.15}$$

We can now substitute our known values for the position $x(t_1) = 100\,\text{m}$ and $v(t_1) = 20\ \text{m}\cdot\text{s}^{-1}$ and solve for the time interval that the car has accelerated

$$t_1 = 2\frac{x(t_1)}{v(t_1)} = 2\frac{100\ \text{m}}{20\ \text{m}\cdot\text{s}^{-1}} = 10\,\text{s}. \tag{4.5.16}$$

c) We can substitute into either of the expressions in Equation (4.5.13); the second is slightly easier to use,

$$a = \frac{v(t_1)}{t_1} = \frac{20\ \text{m}\cdot\text{s}^{-1}}{10\,\text{s}} = 2.0\,\text{m}\cdot\text{s}^{-2}. \tag{4.5.17}$$

d) The x-component of acceleration vs. time, x-component of the velocity vs. time, and the position vs. time are piece-wise functions given by

$$a_x(t) = \begin{cases} 2\ \text{m}\cdot\text{s}^{-2}; & 0 < t < 10\ \text{s} \\ 0; & 10\ \text{s} < t < 20\ \text{s} \end{cases},$$

$$v_x(t) = \begin{cases} (2\ \text{m}\cdot\text{s}^{-2})t; & 0 < t < 10\ \text{s} \\ 20\ \text{m}\cdot\text{s}^{-1}; & 10\ \text{s} < t < 20\ \text{s} \end{cases},$$

$$x(t) = \begin{cases} (1/2)(2\ \text{m}\cdot\text{s}^{-2})t^2; & 0 < t < 10\ \text{s} \\ 100\ \text{m} + (20\ \text{m}\cdot\text{s}^{-2})(t - 10\ \text{s}); & 10\ \text{s} < t < 20\ \text{s} \end{cases}.$$

The graphs of the x-component of acceleration vs. time, x-component of the velocity vs. time, and the position vs. time are shown in Figure 4.11

(e) After accelerating, the car travels for an additional ten seconds at constant speed and during this interval the car travels an additional distance $\Delta x = v(t_1) \times 10\text{s} = 200\,\text{m}$ (note that this is twice the distance traveled during the $10\,\text{s}$ of acceleration), so the total distance traveled is $300\,\text{m}$ and the total time is $20\,\text{s}$, for an average velocity of

$$v_{\text{ave}} = \frac{300\,\text{m}}{20\,\text{s}} = 15\,\text{m}\cdot\text{s}^{-1}. \tag{4.5.18}$$

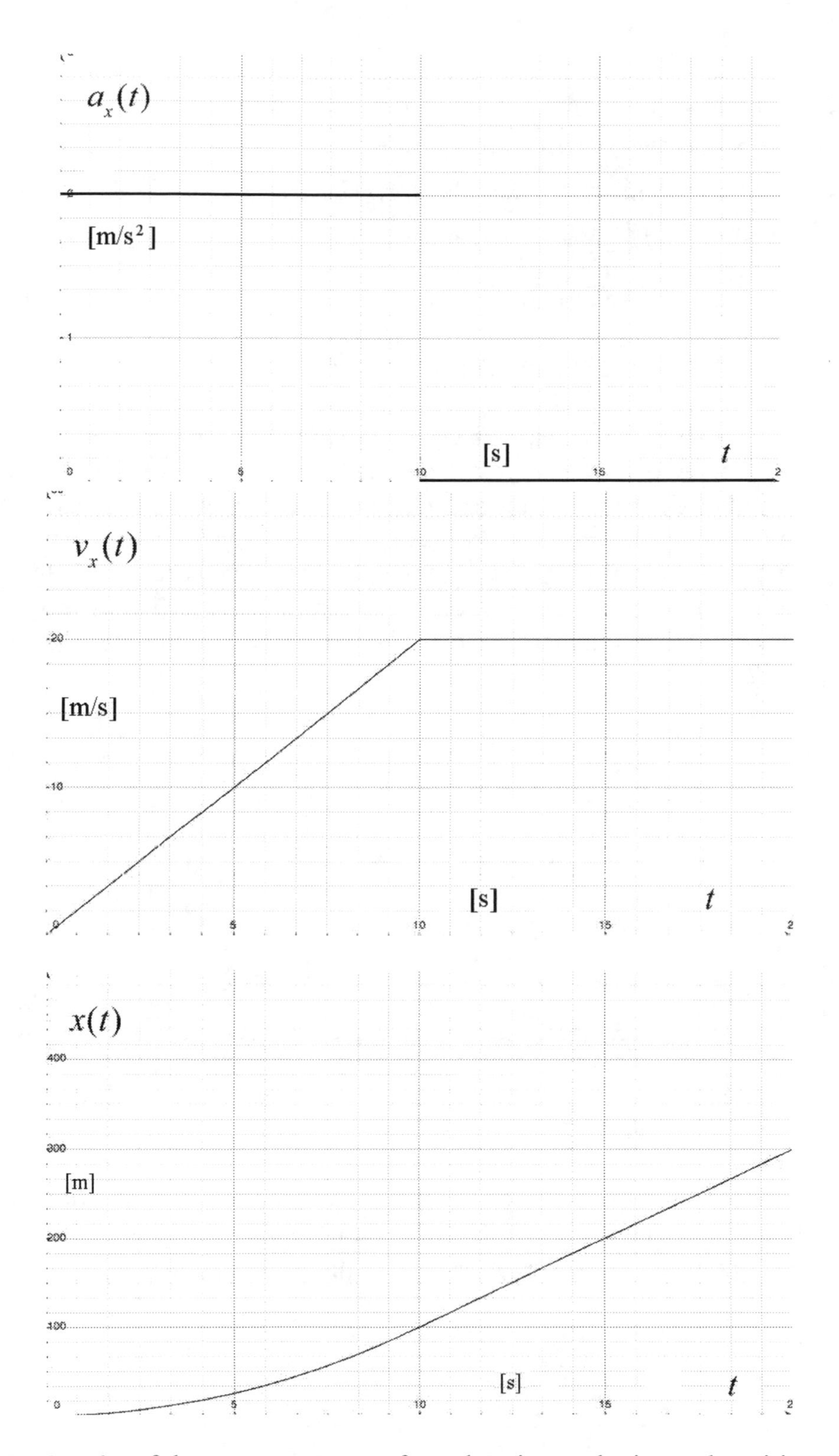

Figure 4.11 Graphs of the x-components of acceleration, velocity and position as piece-wise functions of time

Example 4.4 Catching a Bus

At the instant a traffic light turns green, a car starts from rest with a given constant acceleration, $5.0 \times 10^{-1}\,\text{m} \cdot \text{s}^{-2}$. Just as the light turns green, a bus, traveling with a given constant speed, $1.6 \times 10^{1}\,\text{m} \cdot \text{s}^{-1}$, passes the car. The car speeds up and passes the bus some time later. How far down the road has the car traveled, when the car passes the bus?

Solution: In this example we will illustrate the Polya approach to problem solving.

1. Understand – get a conceptual grasp of the problem

Think about the problem. How many objects are involved in this problem? Two, the bus and the car. How many different stages of motion are there for each object? Each object has one stage of motion. For each object, how many independent directions are needed to describe the motion of that object? We need only one independent direction for each object. What information can you infer from the problem? The acceleration of the car, the velocity of the bus, and that the position of the car and the bus are identical when the bus just passes the car. Sketch qualitatively the position of the car and bus as a function of time (Figure 4.12).

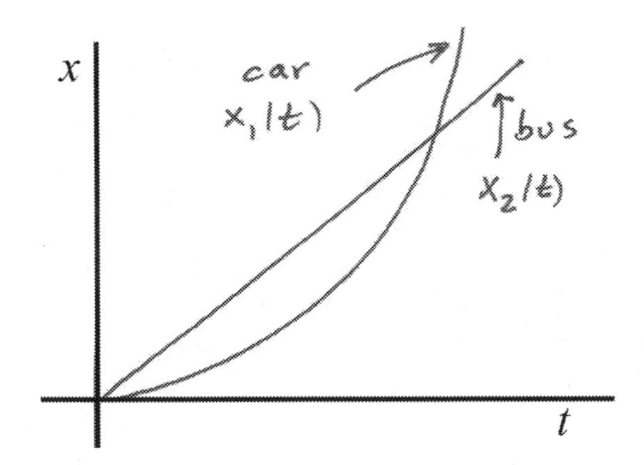

Figure 4.12 Position vs. time of the car and bus.

What choice of coordinate system best suits the problem? Cartesian coordinates with a choice of coordinate system in which the car and bus begin at the origin and travel along the positive x-axis (Figure 4.13). Draw arrows for the position coordinate function for the car and bus.

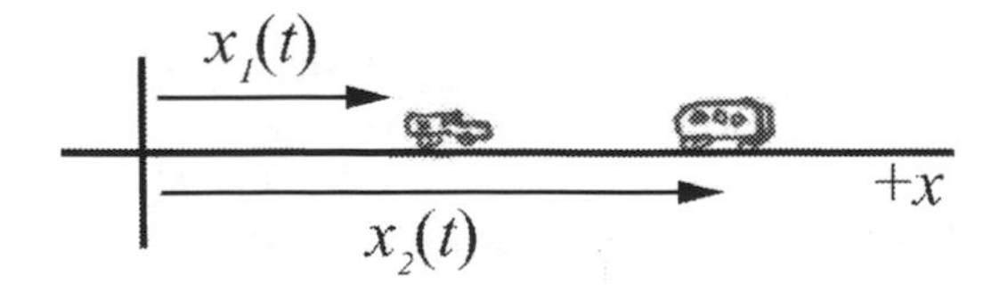

Figure 4.13 A coordinate system for car and bus.

2. Devise a Plan - set up a procedure to obtain the desired solution

Write down the complete set of equations for the position and velocity functions. There are two objects, the car and the bus. Choose a coordinate system with the origin at the traffic light with the car and bus traveling in the positive x-direction. Call the position function of the car, $x_1(t)$, and the position function for the bus, $x_2(t)$. In general the position and velocity functions of the car are given by

$$x_1(t) = x_{1,0} + v_{x10}t + \frac{1}{2}a_{x,1}t^2,$$
$$v_{x,1}(t) = v_{x10} + a_{x,1}t.$$

In this example, using both the information from the problem and our choice of coordinate system, the initial position and initial velocity of the car are both zero, $x_{1,0} = 0$ and $v_{x10} = 0$, and the acceleration of the car is non-zero $a_{x,1} \neq 0$. So the position and velocity of the car is given by

$$x_1(t) = \frac{1}{2}a_{x,1}t^2,$$
$$v_{x,1}(t) = a_{x,1}t.$$

The initial position of the bus is zero, $x_{2,0} = 0$, the initial velocity of the bus is non-zero, $x_2(t_a) = v_{x,20}t_a$ $v_{x,20} \neq 0$, and the acceleration of the bus is zero, $a_{x,2} = 0$. So the position function for the bus is $x_2(t) = v_{x,20}t$. The velocity is constant, $v_{x,2}(t) = v_{x,20}$.

Identify any specified quantities. The problem states: "The car speeds up and passes the bus some time later." What analytic condition best expresses this condition? Let $t = t_a$ correspond to the time that the car passes the bus. Then at that time, the position functions of the bus and car are equal, $x_1(t_a) = x_2(t_a)$.

How many quantities need to be specified in order to find a solution? There are three independent equations at time $t = t_a$: the equations for position and velocity of the car $x_1(t_a) = \frac{1}{2}a_{x,1}t_a^2$, $v_{x,1}(t_a) = a_{x,1}t_a$, and the equation for the position of the bus, $x_2(t) = v_{x,20}t$. There is one 'constraint condition' $x_1(t_a) = x_2(t_a)$.

The six quantities that are as yet unspecified are $x_1(t_a)$, $x_2(t_a)$, $v_{x,1}(t_a)$, $v_{x,20}$, a_x, t_a. So you need to be given at least two numerical values in order to completely specify all the quantities; for example the distance where the car and bus meet. The problem specifying the initial velocity of the bus, $v_{x,20}$, and the acceleration, a_x, of the car with given values.

3. Carry our your plan – solve the problem!

The number of independent equations is equal to the number of unknowns so you can design a strategy for solving the system of equations for the distance the car has traveled in terms of the velocity of the bus $v_{x,20}$ and the acceleration of the car $a_{x,1}$, when the car passes the bus.

Let's use the constraint condition to solve for the time $t = t_a$ where the car and bus meet. Then we can use either of the position functions to find out where this occurs. Thus the constraint condition, $x_1(t_a) = x_2(t_a)$ becomes $(1/2)a_{x,1}t_a^{\,2} = v_{x,20}t_a$. We can solve for this time, $t_a = 2v_{x,20} / a_{x,1}$. Therefore the position of the car at the meeting point is

$$x_1(t_a) = \frac{1}{2}a_{x,1}t_a^2 = \frac{1}{2}a_{x,1}\left(2\frac{v_{x,20}}{a_{x,1}}\right) = 2\frac{v_{x,20}^2}{a_{x,1}}.$$

4. Look Back – check your solution and method of solution

Check your algebra. Do your units agree? The units look good since in the answer the two sides agree in units, $\left[\mathrm{m}\right] = \left[\mathrm{m}^2 \cdot \mathrm{s}^{-2} / \mathrm{m} \cdot \mathrm{s}^{-2}\right]$ and the algebra checks. Substitute in numbers. Suppose $a_x = 5.0 \times 10^{-1}\,\mathrm{m \cdot s^{-2}}$ and $v_{x,20} = 1.6 \times 10^1\,\mathrm{m \cdot s^{-1}}$, Introduce your numerical values for $v_{x,20}$ and a_x, and solve numerically for the distance the car has traveled when the bus just passes the car. Then

$$t_a = 2\frac{v_{x,20}}{a_{x,1}} = 2\frac{\left(1.6 \times 10^1\,\mathrm{m \cdot s^{-1}}\right)}{\left(5.0 \times 10^{-1}\,\mathrm{m \cdot s^{-2}}\right)} = 6.4 \times 10^1\mathrm{s},$$

$$x_1(t_a) = 2\frac{v_{x,20}^2}{a_{x,1}} = (2)\frac{\left(1.6 \times 10^1\,\mathrm{m \cdot s^{-1}}\right)^2}{\left(5.0 \times 10^{-1}\,\mathrm{m \cdot s^{-2}}\right)} = 1.0 \times 10^3\mathrm{m}.$$

Check your results. Once you have an answer, think about whether it agrees with your estimate of what it should be. Some very careless errors can be caught at this point. Is it possible that when the car just passes the bus, the car and bus have the same velocity? Then there would be an additional constraint condition at time $t = t_a$, that the velocities are equal, $v_{x,1}(t_a) = v_{x,20}$. Thus $v_{x,1}(t_a) = a_{x,1}t_a = v_{x,20}$ implies that $t_a = v_{x,20} / a_{x,1}$. From our other result for the time of intersection $t_a = 2v_{x,20} / a_{x,1}$. But these two results contradict each other, so it is not possible.

4.6 One Dimensional Kinematics Non-Constant Acceleration

4.6.1 Change of Velocity as the Integral of Non-constant Acceleration

When the acceleration is a non-constant function, we would like to know how the x-component of the velocity changes for a time interval $\Delta t = [0, t]$. Since the acceleration is non-constant we cannot simply multiply the acceleration by the time interval. We shall calculate the change in the x-component of the velocity for a small time interval $\Delta t_i \equiv [t_i, t_{i+1}]$ and sum over these results. We then take the limit as the time intervals become very small and the summation becomes an integral of the x-component of the acceleration.

For a time interval $\Delta t = [0, t]$, we divide the interval up into N small intervals $\Delta t_i \equiv [t_i, t_{i+1}]$, where the index $i = 1, 2, \ldots, N$, and $t_1 \equiv 0$, $t_N \equiv t$. Over the interval Δt_i, we can approximate the acceleration as a constant, $\overline{a_x(t_i)}$. Then the change in the x-component of the velocity is the area under the acceleration *vs.* time curve,

$$\Delta v_{x,i} \equiv v_x(t_{i+1}) - v_x(t_i) = \overline{a_x(t_i)}\, \Delta t_i + E_i, \qquad (4.6.1)$$

where E_i is the error term (see Figure 4.14a). Then the sum of the changes in the x-component of the velocity is

$$\sum_{i=1}^{i=N} \Delta v_{x,i} = (v_x(t_2) - v_x(t_1 = 0)) + (v_x(t_3) - v_x(t_2)) + \cdots + (v_x(t_N = t) - v_x(t_{N-1})). \qquad (4.6.2)$$

In this summation pairs of terms of the form $\left(v_x(t_2) - v_x(t_2)\right) = 0$ sum to zero, and the overall sum becomes

$$v_x(t) - v_x(0) = \sum_{i=1}^{i=N} \Delta v_{x,i}. \qquad (4.6.3)$$

Substituting Equation (4.6.1) into Equation (4.6.3),

$$v_x(t) - v_x(0) = \sum_{i=1}^{i=N} \Delta v_{x,i} = \sum_{i=1}^{i=N} \overline{a_x(t_i)}\ \Delta t_i + \sum_{i=1}^{i=N} E_i. \qquad (4.6.4)$$

We now approximate the area under the graph in Figure 4.14a by summing up all the rectangular area terms,

$$\text{Area}_N(a_x, t) = \sum_{i=1}^{i=N} \overline{a_x(t_i)}\, \Delta t_i. \qquad (4.6.5)$$

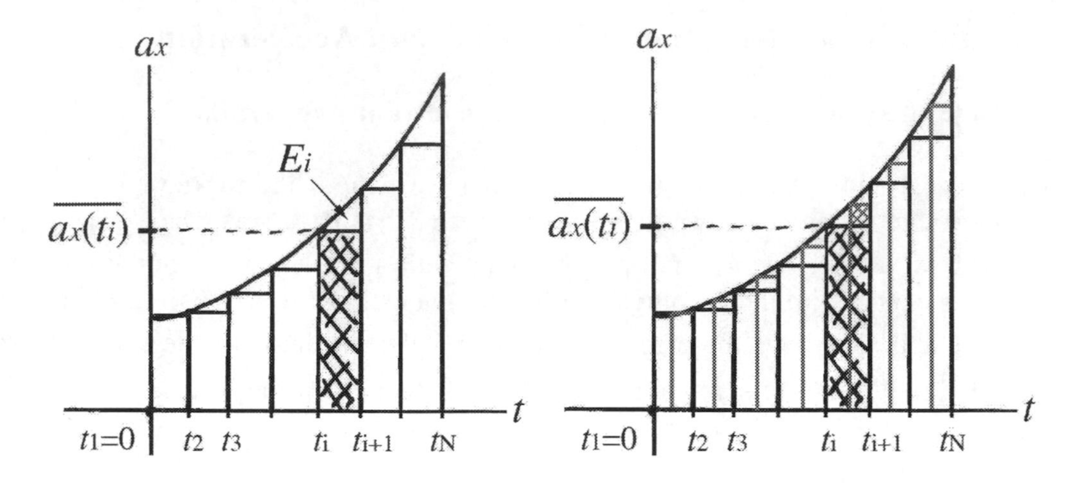

Figures 4.14a and 4.14b Approximating the area under the graph of the x-component of the acceleration *vs.* time

Suppose we make a finer subdivision of the time interval $\Delta t = [0, t]$ by increasing N, as shown in Figure 4.14b. The error in the approximation of the area decreases. We now take the limit as N approaches infinity and the size of each interval Δt_i approaches zero. For each value of N, the summation in Equation (4.6.5) gives a value for $\text{Area}_N(a_x,t)$, and we generate a sequence of values

$$\{\text{Area}_1(a_x,t), \text{Area}_2(a_x,t), ..., \text{Area}_N(a_x,t)\}. \tag{4.6.6}$$

The limit of this sequence is the area, $\text{Area}(a_x,t)$, under the graph of the x-component of the acceleration *vs.* time. When taking the limit, the error term vanishes in Equation (4.6.4),

$$\lim_{N\to\infty}\sum_{i=1}^{i=N} E_i = 0. \tag{4.6.7}$$

Therefore in the limit as N approaches infinity, Equation (4.6.4) becomes

$$v_x(t) - v_x(0) = \lim_{N\to\infty}\sum_{i=1}^{i=N}\overline{a_x(t_i)}\ \Delta t_i + \lim_{N\to\infty}\sum_{i=1}^{i=N} E_i = \lim_{N\to\infty}\sum_{i=1}^{i=N}\overline{a_x(t_i)}\ \Delta t_i = \text{Area}(a_x,t), \tag{4.6.8}$$

and thus the change in the x-component of the velocity is equal to the area under the graph of x-component of the acceleration *vs.* time.

The ***integral of the*** x***-component of the acceleration*** for the interval $[0, t]$ is defined to be the limit of the sequence of areas, $\text{Area}_N(a_x,t)$, and is denoted by

$$\int_{t'=0}^{t'=t} a_x(t')\,dt' \equiv \lim_{\Delta t_i \to 0} \sum_{i=1}^{i=N} a_x(t_i)\,\Delta t_i = \text{Area}(a_x,t)\,. \tag{4.6.9}$$

Equation (4.6.8) shows that the change in the x–component of the velocity is the integral of the x-component of the acceleration with respect to time.

$$v_x(t) - v_x(0) = \int_{t'=0}^{t'=t} a_x(t')\,dt'\,. \tag{4.6.10}$$

Using integration techniques, we can in principle find the expressions for the velocity as a function of time for any acceleration.

4.6.2 Integral of Velocity

We can repeat the same argument for approximating the area $\text{Area}(v_x, t)$ under the graph of the x-component of the velocity *vs.* time by subdividing the time interval into N intervals and approximating the area by

$$\text{Area}_N(a_x, t) = \sum_{i=1}^{i=N} \overline{v_x(t_i)}\,\Delta t_i\,. \tag{4.6.11}$$

The displacement for a time interval $\Delta t = [0, t]$ is limit of the sequence of sums $\text{Area}_N(a_x, t)$,

$$\Delta x = x(t) - x(0) = \lim_{N\to\infty} \sum_{i=1}^{i=N} \overline{v_x(t_i)}\,\Delta t_i\,. \tag{4.6.12}$$

This approximation is shown in Figure 4.15.

> The ***integral of the*** x ***-component of the velocity*** for the interval $[0, t]$ is the limit of the sequence of areas, $\text{Area}_N(a_x, t)$, and is denoted by

$$\int_{t'=0}^{t'=t} v_x(t')\,dt' \equiv \lim_{\Delta t_i \to 0} \sum_{i=1}^{i=N} v_x(t_i)\,\Delta t_i = \text{Area}(v_x,t)\,. \tag{4.6.13}$$

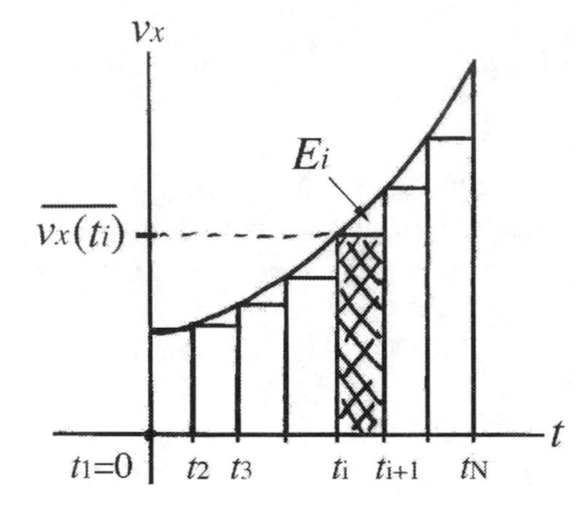

Figure 4.15 Approximating the area under the graph of the x-component of the velocity *vs.* time.

The displacement is then the integral of the x-component of the velocity with respect to time,

$$\Delta x = x(t) - x(0) = \int_{t'=0}^{t'=t} v_x(t')\,dt' . \tag{4.6.14}$$

Using integration techniques, we can in principle find the expressions for the position as a function of time for any acceleration.

Example 4.5 Non-constant Acceleration

Let's consider a case in which the acceleration, $a_x(t)$, is not constant in time,

$$a_x(t) = b_0 + b_1 t + b_2 t^2 . \tag{4.6.15}$$

The graph of the x-component of the acceleration *vs.* time is shown in Figure 4.16

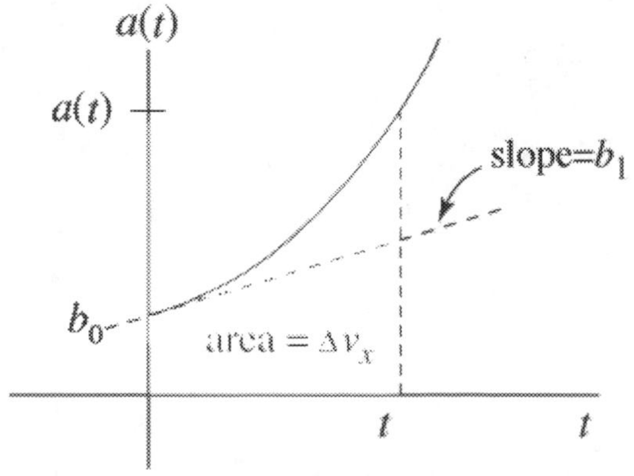

Figure 4.16 Non-constant acceleration *vs.* time graph.

Let's find the change in the x-component of the velocity as a function of time. Denote the initial velocity at $t = 0$ by $v_{x,0} \equiv v_x(t = 0)$. Then,

$$v_x(t) - v_{x,0} = \int_{t'=0}^{t'=t} a_x(t')\,dt' = \int_{t'=0}^{t'=t} (b_o + b_1 t' + b_2 t'^2)\,dt' = b_0 t + \frac{b_1 t^2}{2} + \frac{b_2 t^3}{3}. \quad (4.6.16)$$

The x-component of the velocity as a function in time is then

$$v_x(t) = v_{x,0} + b_0 t + \frac{b_1 t^2}{2} + \frac{b_2 t^3}{3}. \quad (4.6.17)$$

Denote the initial position by $x_0 \equiv x(t = 0)$. The displacement as a function of time is the integral

$$x(t) - x_0 = \int_{t'=0}^{t'=t} v_x(t')\,dt'. \quad (4.6.18)$$

Use Equation (4.6.17) for the x-component of the velocity in Equation (4.6.18) to find

$$x(t) - x_0 = \int_{t'=0}^{t'=t} \left(v_{x,0} + b_0 t' + \frac{b_1 t'^2}{2} + \frac{b_2 t'^3}{3} \right) dt' = v_{x,0}\, t + \frac{b_0 t^2}{2} + \frac{b_1 t^3}{6} + \frac{b_2 t^4}{12}. \quad (4.6.19)$$

Finally the position is then

$$x(t) = x_0 + v_{x,0}\, t + \frac{b_0 t^2}{2} + \frac{b_1 t^3}{6} + \frac{b_2 t^4}{12}. \quad (4.6.20)$$

Example 4.6 Bicycle and Car

A car is driving through a green light at $t = 0$ located at $x = 0$ with an initial speed $v_{c,0} = 12\ \text{m} \cdot \text{s}^{-1}$. The acceleration of the car as a function of time is given by

$$a_c = \begin{cases} 0; & 0 < t < t_1 = 1\ \text{s} \\ -(6\ \text{m} \cdot \text{s}^{-3})(t - t_1); & 1\ \text{s} < t < t_2 \end{cases}.$$

(a) Find the speed and position of the car as a function of time. (b) A bicycle rider is riding at a constant speed of $v_{b,0}$ and at $t = 0$ is 17 m behind the car. The bicyclist reaches the car when the car just comes to rest. Find the speed of the bicycle.

Solution: a) We need to integrate the acceleration for both intervals. The first interval is easy, the speed is constant. For the second integral we need to be careful about the endpoints of the integral and the fact that the integral is the change in speed so we must subtract $v_c(t_1) = v_{c0}$

$$v_c(t) = \begin{cases} v_{c0}; & 0 < t < t_1 = 1\,\text{s} \\ v_c(t_1) + \int_{t_1}^{t} -(6\,\text{m}\cdot\text{s}^{-3})(t - t_1); & 1\,\text{s} < t < t_2 \end{cases}.$$

After integrating we get

$$v_c(t) = \begin{cases} v_{c0}; & 0 < t < t_1 = 1\,\text{s} \\ v_{c0} - (3\,\text{m}\cdot\text{s}^{-3})(t - t_1)^2\Big|_{t_1}^{t}; & 1\,\text{s} < t < t_2 \end{cases}.$$

Now substitute the endpoint so the integral to finally yield

$$v_c(t) = \begin{cases} v_{c0} = 12\,\text{m}\cdot\text{s}^{-1}; & 0 < t < t_1 = 1\,\text{s} \\ 12\,\text{m}\cdot\text{s}^{-1} - (3\,\text{m}\cdot\text{s}^{-3})(t - t_1)^2; & 1\,\text{s} < t < t_2 \end{cases}.$$

For this one-dimensional motion the change in position is the integral of the speed so

$$x_c(t) = \begin{cases} x_c(0) + \int_{0}^{t_1} (12\,\text{m}\cdot\text{s}^{-1})dt; & 0 < t < t_1 = 1\,\text{s} \\ x_c(t_1) + \int_{t_1}^{t} \left(12\,\text{m}\cdot\text{s}^{-1} - (3\text{m}\cdot\text{s}^{-3})(t - t_1)^2\right)dt; & 1\,\text{s} < t < t_2 \end{cases}.$$

Upon integration we have

$$x_c(t) = \begin{cases} x_c(0) + (12\,\text{m}\cdot\text{s}^{-1})t; & 0 < t < t_1 = 1\,\text{s} \\ x_c(t_1) + \left((12\,\text{m}\cdot\text{s}^{-1})(t - t_1) - (1\,\text{m}\cdot\text{s}^{-3})(t - t_1)^3\right)\Big|_{t_1}^{t}; & 1\,\text{s} < t < t_2 \end{cases}.$$

We choose our coordinate system such that $x_c(0) = 0$, therefore $x_c(t_1) = (12\,\text{m}\cdot\text{s}^{-1})(1\,\text{s}) = 12\,\text{m}$. So after substituting in the endpoints of the integration interval we have that

$$x_c(t) = \begin{cases} (12\,\text{m}\cdot\text{s}^{-1})t; & 0 < t < t_1 = 1\,\text{s} \\ 12\,\text{m} + (12\,\text{m}\cdot\text{s}^{-1})(t - t_1) - (1\,\text{m}\cdot\text{s}^{-3})(t - t_1)^3; & 1\,\text{s} < t < t_2 \end{cases}.$$

(b) We are looking for the instant that t_2 the car has come to rest. So we use our expression for the speed for the interval $1\text{ s} < t < t_2$, $0 = v_c(t_2) = 12\text{ m}\cdot\text{s}^{-1} - (3\text{ m}\cdot\text{s}^{-3})(t_2 - t_1)^2$. We can solve this for t_2: $(t_2 - t_1)^2 = 4\text{ s}^2$. We have two solutions: $(t_2 - t_1) = 2\text{ s}$ or $(t_2 - t_1) = -2\text{ s}$. The second solution $t_2 = t_1 - 2\text{ s} = 1\text{ s} - 2\text{ s} = -1\text{ s}$ does not apply to our time interval and so $t_2 = t_1 + 2\text{ s} = 1\text{ s} + 2\text{ s} = 3\text{ s}$. During the position of the car at t_2 is then given by

$$\begin{aligned} x_c(t_2) &= 12\text{ m} + (12\text{ m}\cdot\text{s}^{-1})(t_2 - t_1) - (1\text{ m}\cdot\text{s}^{-3})(t_2 - t_1)^3 \\ &= 12\text{ m} + (12\text{ m}\cdot\text{s}^{-1})(2\text{ s}) - (1\text{ m}\cdot\text{s}^{-3})(2\text{ s})^3 = 28\text{ m}. \end{aligned}$$

Because the bicycle is traveling at a constant speed with an initial position $x_{b0} = -17\text{ m}$, the position of the bicycle is given by $x_b(t) = -17\text{ m} + v_b t$. The bicycle and car intersect at instant $t_2 = 3\text{ s}$: $x_b(t_2) = x_c(t_2)$. Therefore $-17\text{ m} + v_b(3\text{ s}) = 28\text{ m}$. So the speed of the bicycle is

$$v_b = \frac{(28\text{ m} + 17\text{ m})}{(3\text{ s})} = 15\text{ m}\cdot\text{s}^{-1}.$$

Chapter 5 Two Dimensional Kinematics

Chapter 5 Two Dimensional Kinematics

Where was the chap I saw in the picture somewhere? Ah yes, in the dead sea floating on his back, reading a book with a parasol open. Couldn't sink if you tried: so thick with salt. Because the weight of the water, no, the weight of the body in the water is equal to the weight of the what? Or is it the volume equal to the weight? It's a law something like that. Vance in High school cracking his fingerjoints, teaching. The college curriculum. Cracking curriculum. What is weight really when you say weight? Thirtytwo feet per second per second. Law of falling bodies: per second per second. They all fall to the ground. The earth. It's the force of gravity of the earth is the weight. [1]

James Joyce

5.1 Introduction to the Vector Description of Motion in Two Dimensions

So far we have introduced the concepts of kinematics to describe motion in one dimension; however we live in a multidimensional universe. In order to explore and describe motion in this universe, we begin by looking at examples of two-dimensional motion, of which there are many; planets orbiting a star in elliptical orbits or a projectile moving under the action of uniform gravitation are two common examples.

We will now extend our definitions of position, velocity, and acceleration for an object that moves in two dimensions (in a plane) by treating each direction independently, which we can do with vector quantities by resolving each of these quantities into components. For example, our definition of velocity as the derivative of position holds for each component separately. In Cartesian coordinates, in which the directions of the unit vectors do not change from place to place, the position vector $\vec{\mathbf{r}}(t)$ with respect to some choice of origin for the object at time t is given by

$$\vec{\mathbf{r}}(t) = x(t)\,\hat{\mathbf{i}} + y(t)\,\hat{\mathbf{j}}. \tag{5.1.1}$$

The velocity vector $\vec{\mathbf{v}}(t)$ at time t is the derivative of the position vector,

$$\vec{\mathbf{v}}(t) = \frac{dx(t)}{dt}\,\hat{\mathbf{i}} + \frac{dy(t)}{dt}\,\hat{\mathbf{j}} \equiv v_x(t)\,\hat{\mathbf{i}} + v_y(t)\,\hat{\mathbf{j}}, \tag{5.1.2}$$

where $v_x(t) \equiv dx(t)/dt$ and $v_y(t) \equiv dy(t)/dt$ denote the x- and y-components of the velocity respectively.

---

[1] James Joyce, *Ulysses*, The Corrected Text edited by Hans Walter Gabler with Wolfhard Steppe and Claus Melchior, Random House, New York.

The acceleration vector $\vec{\mathbf{a}}(t)$ is defined in a similar fashion as the derivative of the velocity vector,

$$\vec{\mathbf{a}}(t) = \frac{dv_x(t)}{dt}\,\hat{\mathbf{i}} + \frac{dv_y(t)}{dt}\,\hat{\mathbf{j}} \equiv a_x(t)\,\hat{\mathbf{i}} + a_y(t)\,\hat{\mathbf{j}}, \tag{5.1.3}$$

where $a_x(t) \equiv dv_x(t)/dt$ and $a_y(t) \equiv dv_y(t)/dt$ denote the x- and y-components of the acceleration.

5.2 Projectile Motion

Consider the motion of a body that is released at time $t = 0$ with an initial velocity $\vec{v}_0$ at a height h above the ground. Two paths are shown in Figure 5.1.

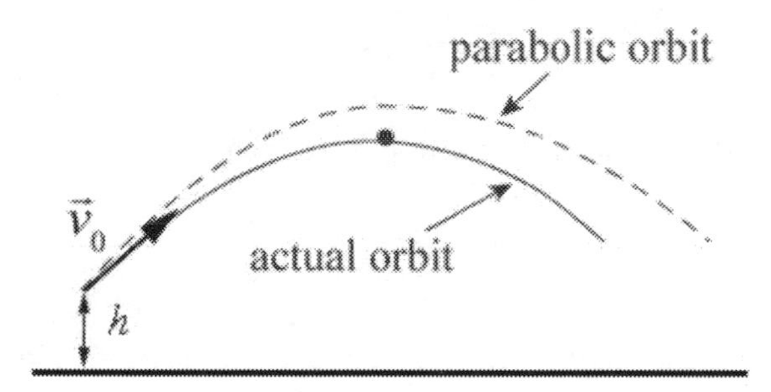

Figure 5.1 Actual orbit and parabolic orbit of a projectile

The dotted path represents a *parabolic* trajectory and the solid path represents the actual orbit. The difference between the paths is due to air resistance. There are other factors that can influence the path of motion; a rotating body or a special shape can alter the flow of air around the body, which may induce a curved motion or lift like the flight of a baseball or golf ball. We shall begin our analysis by neglecting all influences on the body except for the influence of gravity.

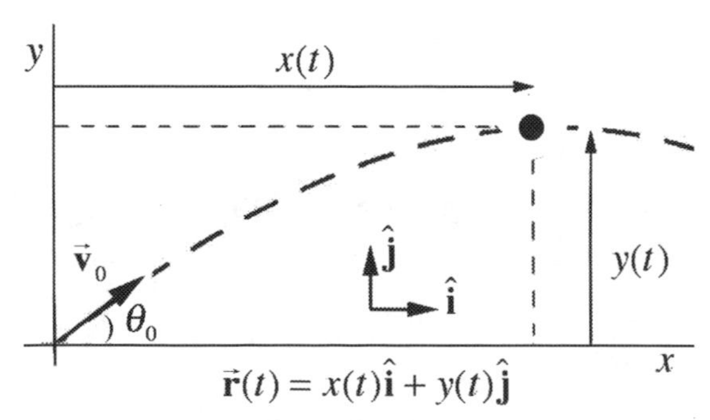

Figure 5.2 A coordinate sketch for parabolic motion.

Choose coordinates with the y-axis in the vertical direction with $\hat{\mathbf{j}}$ pointing upwards and the x-axis in the horizontal direction with $\hat{\mathbf{i}}$ pointing in the direction that the object is moving horizontally. Choose the origin to be at the point where the object is released. Figure 5.2 shows our coordinate system with the position of the object at time t and the coordinate functions $x(t)$ and $y(t)$. The coordinate function $y(t)$ represents the distance from the body to the origin along the y-axis at time t, and the coordinate function $x(t)$ represents the distance from the body to the origin along the x-axis at time t.

Initial Conditions:

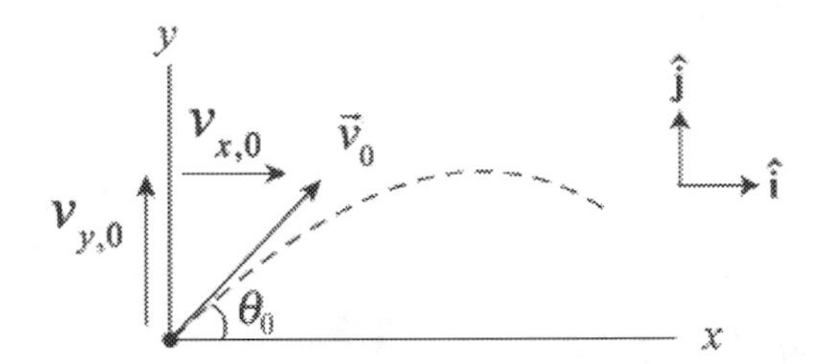

Figure 5.3 A vector decomposition of the initial velocity

We begin by making a vector decomposition of the initial velocity vector

$$\vec{\mathbf{v}}_0(t) = v_{x,0}\,\hat{\mathbf{i}} + v_{y,0}\,\hat{\mathbf{j}}. \tag{5.2.1}$$

Often the description of the flight of a projectile includes the statement, "a body is projected with an initial speed v_0 at an angle θ_0 with respect to the horizontal." The vector decomposition diagram for the initial velocity is shown in Figure 5.3. The components of the initial velocity are given by

$$v_{x,0} = v_0 \cos\theta_0 , \tag{5.2.2}$$

$$v_{y,0} = v_0 \sin\theta_0 . \tag{5.2.3}$$

Since the initial speed is the magnitude of the initial velocity, we have

$$v_0 = (v_{x,0}^2 + v_{y,0}^2)^{1/2} . \tag{5.2.4}$$

The angle θ_0 is related to the components of the initial velocity by

$$\theta_0 = \tan^{-1}(v_{y,0} / v_{x,0}). \tag{5.2.5}$$

The initial position vector appears with components

$$\vec{\mathbf{r}}_0 = x_0\,\hat{\mathbf{i}} + y_0\,\hat{\mathbf{j}}. \tag{5.2.6}$$

Note that the trajectory in Figure 5.3 has $x_0 = y_0 = 0$, but this will not always be the case.

Force Diagram:

The only force acting on the object is the gravitational interaction between the object and the earth. This force acts downward with magnitude mg, where m is the mass of the object and $g = 9.8\ \text{m}\cdot\text{s}^{-2}$. Figure 5.4 shows the force diagram on the object. The vector decomposition of the force is

$$\vec{\mathbf{F}}_{\text{grav}} = -mg\,\hat{\mathbf{j}}. \tag{5.2.7}$$

Should you include the force that gave the object its initial velocity on the force diagram? No! We are only interested in the forces acting on the object once the object has been released. It is a separate problem (and a very hard one at that) to determine exactly what forces acted on the object before the body was released. (This issue obscured the understanding of projectile motion for centuries.)

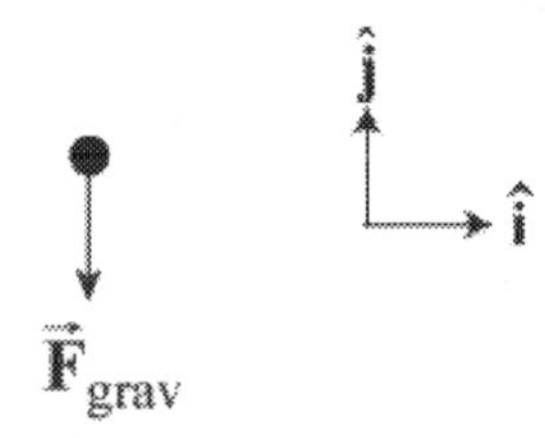

Figure 5.4 Free-body force diagram on the object with the action of gravity

Equations of Motions:

The force diagram reminds us that the only force is acting in the y-direction. Newton's Second Law states that the total vector force $\vec{\mathbf{F}}^{\text{total}}$ acting on the object is equal to the product of the mass m and the acceleration vector $\vec{\mathbf{a}}$,

$$\vec{\mathbf{F}}^{\text{total}} = m\vec{\mathbf{a}}. \tag{5.2.8}$$

This is a vector equation; the components are equated separately:

$$F_y^{\text{total}} = ma_y, \tag{5.2.9}$$

$$F_x^{\text{total}} = ma_x. \tag{5.2.10}$$

Thus the vector equation in the y-direction becomes

$$-mg = ma_y. \tag{5.2.11}$$

Therefore the y-component of the acceleration is

$$a_y = -g. \tag{5.2.12}$$

We see that the acceleration is a constant and is independent of the mass of the object. Notice that $a_y < 0$. This is because we chose our positive y-direction to point upwards. The sign of the y-component of acceleration is determined by how we chose our coordinate system. Because there are no horizontal forces acting on the object,

$$F_x^{\text{total}} = 0, \tag{5.2.13}$$

and we conclude that the acceleration in the horizontal direction is also zero,

$$a_x = 0. \tag{5.2.14}$$

This tells us that the x-component of the velocity remains unchanged throughout the flight of the object.

Newton's Second Law provides an analysis that determines that the acceleration in the vertical direction is constant for all bodies independent of the mass of the object, thus confirming Galileo's Law of Freeing Falling Bodies. Notice that the equation of motion (Equation (5.2.12)) generalizes the experimental observation that objects fall with constant acceleration. Our prediction is predicated on this force law and if subsequent observations show the acceleration is not constant then we either must include additional forces (for example, air resistance) or modify the force law (for objects that are no longer near the surface of the earth).

We can now integrate the equation of motion separately for the x- and y-directions to find expressions for the x- and y-components of position and velocity:

$$v_x(t) - v_{x,0} = \int_{t'=0}^{t'=t} a_x(t')\,dt' = 0 \Rightarrow v_x(t) = v_{x,0}$$

$$x(t) - x_0 = \int_{t'=0}^{t'=t} v_x(t')\,dt' = \int_{t'=0}^{t'=t} v_{x,0}\,dt' = v_{x,0}t \Rightarrow x(t) = x_0 + v_{x,0}t$$

$$v_y(t) - v_{y,0} = \int_{t'=0}^{t'=t} a_y(t')\,dt' = -\int_{t'=0}^{t'=t} g\,dt' = -gt \Rightarrow v_y(t) = v_{y,0} - gt$$

$$y(t) - y_0 = \int_{t'=0}^{t'=t} v_y(t')\,dt' = \int_{t'=0}^{t'=t} (v_{y,0} - gt)\,dt' = v_{y,0}t - (1/2)gt^2 \Rightarrow y(t) = y_0 + v_{y,0}t - (1/2)gt^2.$$

Then the complete vector set of equations for position and velocity for each independent direction of motion are given by

$$\vec{\mathbf{r}}(t) = x(t)\,\hat{\mathbf{i}} + y(t)\,\hat{\mathbf{j}} = (x_0 + v_{x,0}t)\,\hat{\mathbf{i}} + (y_0 + v_{y,0}t + (1/2)a_y t^2)\,\hat{\mathbf{j}} \tag{5.2.15}$$

$(x(t), y(t))$

$$\vec{\mathbf{v}}(t) = v_x(t)\,\hat{\mathbf{i}} + v_y(t)\,\hat{\mathbf{j}} = v_{x,0}\,\hat{\mathbf{i}} + (v_{y,0} + a_y t)\,\hat{\mathbf{j}}, \tag{5.2.16}$$

$$\vec{\mathbf{a}}(t) = a_x(t)\,\hat{\mathbf{i}} + a_y(t)\,\hat{\mathbf{j}} = a_y\,\hat{\mathbf{j}}. \tag{5.2.17}$$

Example 5.1 Time of Flight and Maximum Height of a Projectile

A person throws a stone at an initial angle $\theta_0 = 45^\circ$ from the horizontal with an initial speed of $v_0 = 20\ \mathrm{m\cdot s^{-1}}$. The point of release of the stone is at a height $d = 2\ \mathrm{m}$ above the ground. You may neglect air resistance. a) How long does it take the stone to reach the highest point of its trajectory? b) What was the maximum vertical displacement of the stone? Ignore air resistance.

Solution: Choose the origin on the ground directly underneath the point where the stone is released. We choose upwards for the positive y-direction and along the projection of the path of the stone along the ground for the positive x-direction. Set $t = 0$ the instant the stone is released. At $t = 0$ the initial conditions are then $x_0 = 0$ and $y_0 = d$. The initial x- and y-components of the velocity are given by Eqs. (5.2.2) and (5.2.3).

At time t the stone has coordinates $(x(t), y(t))$. These coordinate functions are shown in Figure 5.5.

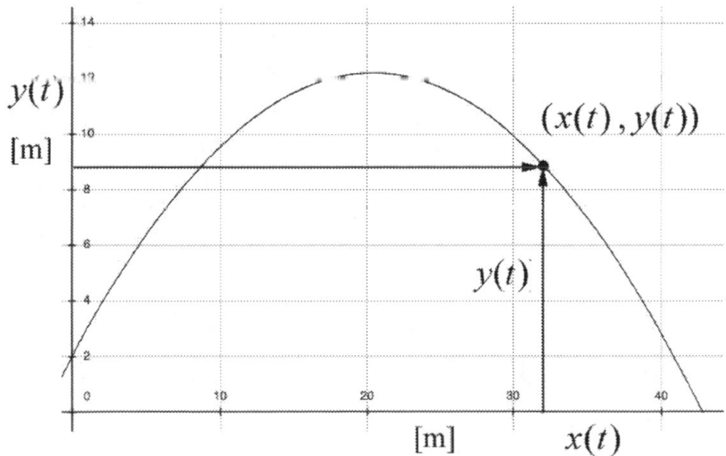

Figure 5.5: Coordinate functions for stone

The slope of this graph at any time t yields the instantaneous y-component of the velocity $v_y(t)$ at that time t.

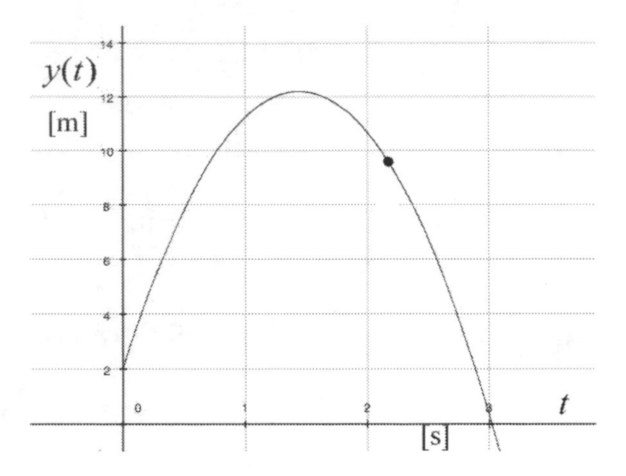

Figure 5.6 Plot of the y-component of the position as a function of time

There are several important things to notice about Figures 5.5 and 5.6. The first point is that the abscissa axes are different in both figures, Figure 5.5 is a plot of y vs. x and Figure 5.6 is a plot of y vs. t. The second thing to notice is that at $t = 0$, the slope of the graph in Figure 5.5 is equal to $(dy/dx)(t=0) = v_{y,0} / v_{x,0} = \tan\theta_0$, while at $t = 0$ the slope of the graph in Figure 5.6 is equal to $v_{y,0}$. Let $t = t_{top}$ correspond to the instant the stone is at its maximal vertical position, the highest point in the flight. The final thing to notice about Figure 5.6 is that at $t = t_{top}$ the slope is zero or $v_y(t = t_{top}) = 0$. Therefore

$$v_y(t_{top}) = v_0 \sin\theta_0 - g t_{top} = 0 .$$

We can solve this equation for t_{top},

$$t_{top} = \frac{v_0 \sin\theta_0}{g} = \frac{(20\ \text{m}\cdot\text{s}^{-1})\sin(45^\circ)}{9.8\ \ \text{m}\cdot\text{s}^{-2}} = 1.44\ \text{s} . \qquad (5.2.18)$$

The y-component of the velocity as a function of time is graphed in Figure 5.7.

Notice that at $t = 0$ the intercept is positive indicting the initial y-component of the velocity is positive which means that the stone was thrown upwards. The y-component of the velocity changes sign at $t = t_{top}$ indicating that it is reversing its direction and starting to move downwards.

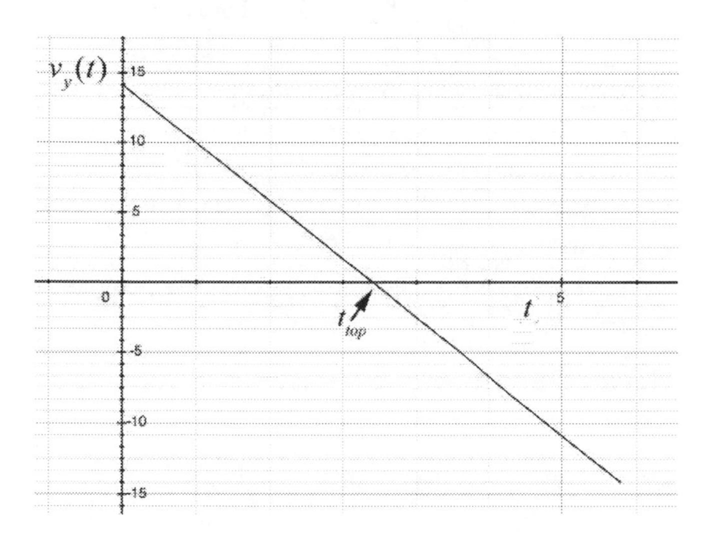

Figure 5.7 y-component of the velocity as a function of time

We can now substitute the expression for $t = t_{top}$ (Eq. (5.2.18)) into the y-component of the position in Eq. (5.2.15) to find the maximal height of the stone above the ground

$$\begin{aligned} y(t = t_{top}) &= d + v_0 \sin\theta_0 \frac{v_0 \sin\theta_0}{g} - \frac{1}{2} g \left(\frac{v_0 \sin\theta_0}{g} \right)^2 \\ &= d + \frac{v_0^{\ 2} \sin^2\theta_0}{2g} = 2\text{ m} + \frac{(20\text{ m}\cdot\text{s}^{-1})^2 \sin^2(45^\circ)}{2(9.8\text{ m}\cdot\text{s}^{-2})} = 12.2\text{ m} \end{aligned}, \qquad (5.2.19)$$

5.2.1 Orbit equation

So far our description of the motion has emphasized the independence of the spatial dimensions, treating all of the kinematic quantities as functions of time. We shall now eliminate time from our equation and find the *orbit equation* of the body. We begin with the x-component of the position in Eq. (5.2.15),

$$x(t) = x_0 + v_{x,0} t \qquad (5.2.20)$$

and solve Equation (5.2.20) for time t as a function of $x(t)$,

$$t = \frac{x(t) - x_0}{v_{x,0}}. \qquad (5.2.21)$$

The y-component of the position in Eq. (5.2.15) is given by

$$y(t) = y_0 + v_{y,0} t - \frac{1}{2} g t^2. \qquad (5.2.22)$$

We then substitute the above expression, Equation (5.2.21) for time t into our equation for the y-component of the position yielding

$$y(t) = y_0 + v_{y,0}\left(\frac{x(t)-x_0}{v_{x,0}}\right) - \frac{1}{2}g\left(\frac{x(t)-x_0}{v_{x,0}}\right)^2. \tag{5.2.23}$$

This expression can be simplified to give

$$y(t) = y_0 + \frac{v_{y,0}}{v_{x,0}}(x(t)-x_0) - \frac{1}{2}\frac{g}{v_{x,0}^2}(x(t)^2 - 2x(t)x_0 + x_0^2). \tag{5.2.24}$$

This is seen to be an equation for a parabola by rearranging terms to find

$$y(t) = -\frac{1}{2}\frac{g}{v_{x,0}^2}x(t)^2 + \left(\frac{g\,x_0}{v_{x,0}^2} + \frac{v_{y,0}}{v_{x,0}}\right)x(t) - \frac{v_{y,0}}{v_{x,0}}x_0 - \frac{1}{2}\frac{g}{v_{x,0}^2}x_0^2 + y_0. \tag{5.2.25}$$

The graph of $y(t)$ as a function of $x(t)$ is shown in Figure 5.8.

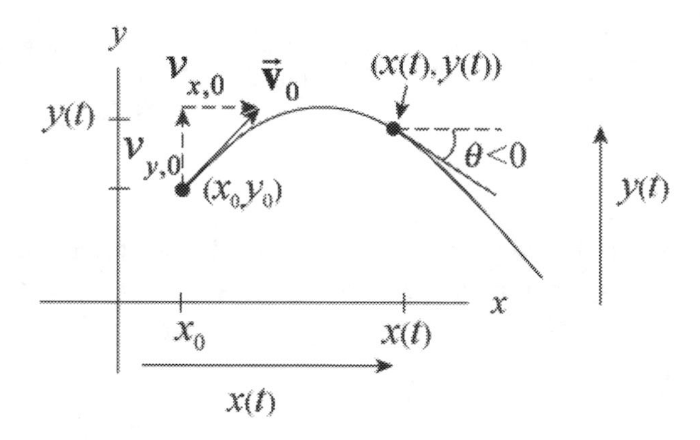

Figure 5.8 The parabolic orbit

Note that at any point $(x(t), y(t))$ along the parabolic trajectory, the direction of the tangent line at that point makes an angle θ with the positive x-axis as shown in Figure 5.8. This angle is given by

$$\theta = \tan^{-1}\left(\frac{dy}{dx}\right), \tag{5.2.26}$$

where dy/dx is the derivative of the function $y(x) = y(x(t))$ at the point $(x(t), y(t))$.

The velocity vector is given by

$$\vec{\mathbf{v}}(t) = \frac{dx(t)}{dt}\,\hat{\mathbf{i}} + \frac{dy(t)}{dt}\,\hat{\mathbf{j}} \equiv v_x(t)\,\hat{\mathbf{i}} + v_y(t)\,\hat{\mathbf{j}} \qquad (5.2.27)$$

The direction of the velocity vector at a point $(x(t), y(t))$ can be determined from the components. Let ϕ be the angle that the velocity vector forms with respect to the positive x-axis. Then

$$\phi = \tan^{-1}\left(\frac{v_y(t)}{v_x(t)}\right) = \tan^{-1}\left(\frac{dy/dt}{dx/dt}\right) = \tan^{-1}\left(\frac{dy}{dx}\right). \qquad (5.2.28)$$

Comparing our two expressions we see that $\phi = \theta$; the slope of the graph of $y(t)$ *vs.* $x(t)$ at any point determines the direction of the velocity at that point. We cannot tell from our graph of $y(x)$ how fast the body moves along the curve; the magnitude of the velocity cannot be determined from information about the tangent line.

If we choose our origin at the initial position of the body at $t = 0$, then $x_0 = 0$ and $y_0 = 0$. Our orbit equation, Equation (5.2.25) can now be simplified to

$$y(t) = -\frac{1}{2}\frac{g}{v_{x,0}^2}x(t)^2 + \frac{v_{y,0}}{v_{x,0}}x(t). \qquad (5.2.29)$$

Example 5.2 Hitting the Bucket

A person is standing on a ladder holding a pail. The person releases the pail from rest at a height h_1 above the ground. A second person standing a horizontal distance s_2 from the pail aims and throws a ball the instant the pail is released in order to hit the pail. The person releases the ball at a height h_2 above the ground, with an initial speed v_0, and at an angle θ_0 with respect to the horizontal. You may ignore air resistance.

a) Find an expression for the angle θ_0 that the person aims the ball in order to hit the pail.

b) Find an expression for the height above the ground where the collision occurred as a function of the initial speed of the ball v_0, and the quantities h_1, h_2, and s_2.

Solution: 1. Understand – get a conceptual grasp of the problem

There are two objects involved in this problem. Each object is undergoing free fall, so there is only one stage each. The pail is undergoing one-dimensional motion. The ball is undergoing two-dimensional motion. The parameters h_1, h_2, and s_2 are unspecified, so our answers will be functions of those symbolic expressions for the quantities. Figure 5.9 shows a sketch of the motion of all the bodies in this problem.

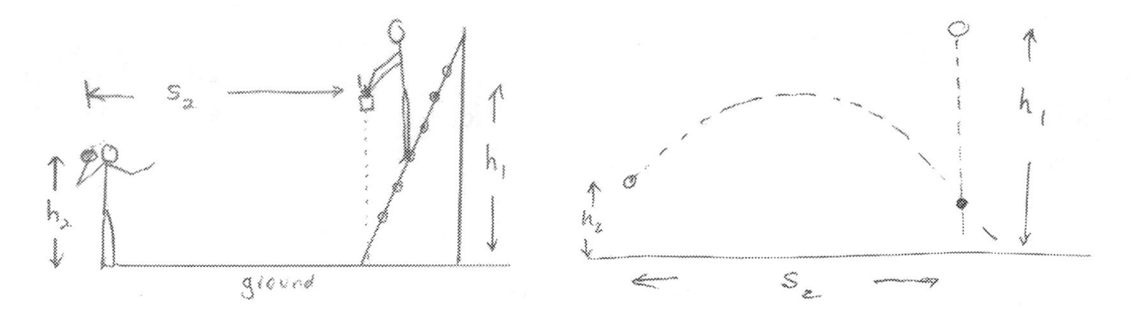

Figure 5.9: Sketch of motion of ball and bucket.

Since the acceleration is unidirectional and constant, we will choose Cartesian coordinates, with one axis along the direction of acceleration. Choose the origin on the ground directly underneath the point where the ball is released. We choose upwards for the positive y-direction and towards the pail for the positive x-direction.

We choose position coordinates for the pail as follows. The horizontal coordinate is constant and given by $x_1 = s_2$. The vertical coordinate represents the height above the ground and is denoted by $y_1(t)$. The ball has coordinates $(x_2(t), y_2(t))$. We show these coordinates in the Figure 5.10.

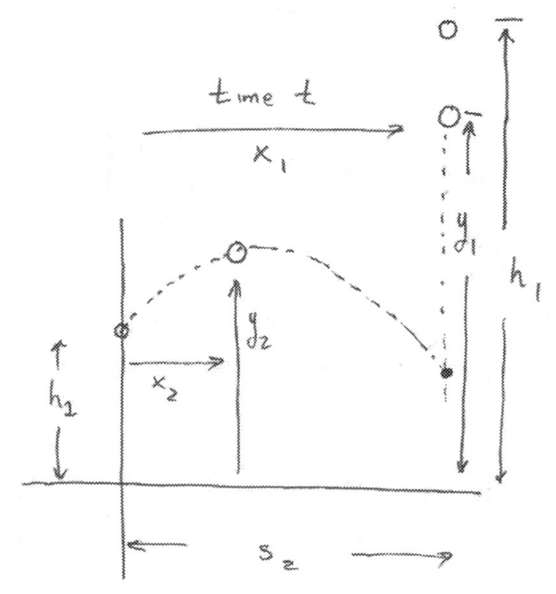

Figure 5.10: Coordinate System

2. Devise a Plan - set up a procedure to obtain the desired solution

(a) Find an expression for the angle θ_0 that the person throws the ball as a function of h_1, h_2, and s_2.

(b) Find an expression for the time of collision as a function of the initial speed of the ball v_0, and the quantities h_1, h_2, and s_2.

(c) Find an expression for the height above the ground where the collision occurred as a function of the initial speed of the ball v_0, and the quantities h_1, h_2, and s_2.

Model: The pail undergoes constant acceleration $(a_y)_1 = -g$ in the vertical direction downwards and the ball undergoes uniform motion in the horizontal direction and constant acceleration downwards in the vertical direction, with $(a_x)_2 = 0$ and $(a_y)_2 = -g$.

Equations of Motion for Pail: The initial conditions for the pail are $(v_{y,0})_1 = 0$, $x_1 = s_2$, $(y_0)_1 = h_1$. Since the pail moves vertically, the pail always satisfies the constraint condition $x_1 = s_2$ and $v_{x,1} = 0$. The equations for position and velocity of the pail simplify to

$$y_1(t) = h_1 - \frac{1}{2}gt^2 \tag{5.2.30}$$

$$v_{y,1}(t) = -gt\,. \tag{5.2.31}$$

Equations of Motion for Ball: The initial position is given by $(x_0)_2 = 0$, $(y_0)_2 = h_2$. The components of the initial velocity are given by $(v_{y,0})_2 = v_0\sin(\theta_0)$ and $(v_{x,0})_2 = v_0\cos(\theta_0)$, where v_0 is the magnitude of the initial velocity and θ_0 is the initial angle with respect to the horizontal. So the equations for position and velocity of the ball simplify to

$$x_2(t) = v_0\cos(\theta_0)t \tag{5.2.32}$$

$$v_{x,2}(t) = v_0\cos(\theta_0) \tag{5.2.33}$$

$$y_2(t) = h_2 + v_0\sin(\theta_0)t - \frac{1}{2}gt^2 \tag{5.2.34}$$

$$v_{y,2}(t) = v_0\sin(\theta_0) - gt\,. \tag{5.2.35}$$

Note that the quantities h_1, h_2, and s_2 should be treated as known quantities although no numerical values were given, only symbolic expressions. There are six independent equations with 9 as yet unspecified quantities $y_1(t)$, t, $y_2(t)$, $x_2(t)$, $v_{y,1}(t)$, $v_{y,2}(t)$, $v_{x,2}(t)$, v_0, and θ_0.

So we need two more conditions, in order to find expressions for the initial angle, θ_0, the time of collision, t_a, and the spatial location of the collision point specified by $y_1(t_a)$ or $y_2(t_a)$ in terms of the one unspecified parameter v_0. At the collision time $t = t_a$, the collision occurs when the two balls are located at the same position. Therefore

$$y_1(t_a) = y_2(t_a) \tag{5.2.36}$$

$$x_2(t_a) = x_1 = s_2 . \tag{5.2.37}$$

We shall now apply these conditions that must be satisfied in order for the ball to hit the pail.

$$h_1 - \frac{1}{2} g t_a^{\,2} = h_2 + v_0 \sin(\theta_0) t_a - \frac{1}{2} g t_a^{\,2} \tag{5.2.38}$$

$$s_2 = v_0 \cos(\theta_0) t_a . \tag{5.2.39}$$

From the first equation, the term $(1/2)gt_a^{\,2}$ cancels from both sides. Therefore we have that

$$h_1 = h_2 + v_0 \sin(\theta_0) t_a$$

$$s_2 = v_0 \cos(\theta_0) t_a .$$

We can now solve these equations for $\tan(\theta_0) = \sin(\theta_0)/\cos(\theta_0)$, and thus the angle the person throws the ball in order to hit the pail.

3. Carry our your plan – solve the problem!

We rewrite these equations as

$$v_0 \sin(\theta_0) t_a = h_1 - h_2 \tag{5.2.40}$$

$$v_0 \cos(\theta_0) t_a = s_2 . \tag{5.2.41}$$

Dividing these equations yields

$$\frac{v_0 \sin(\theta_0) t_a}{v_0 \cos(\theta_0) t_a} = \tan(\theta_0) = \frac{h_1 - h_2}{s_2} . \tag{5.2.42}$$

So the initial angle is independent of v_0, and is given by

$$\theta_0 = \tan^{-1}((h_1 - h_2)/s_2) . \tag{5.2.43}$$

From the Figure 5.11 we can see that $\tan(\theta_0) = (h_1 - h_2)/s_2$ implies that the second person aims the ball at the initial position of the pail.

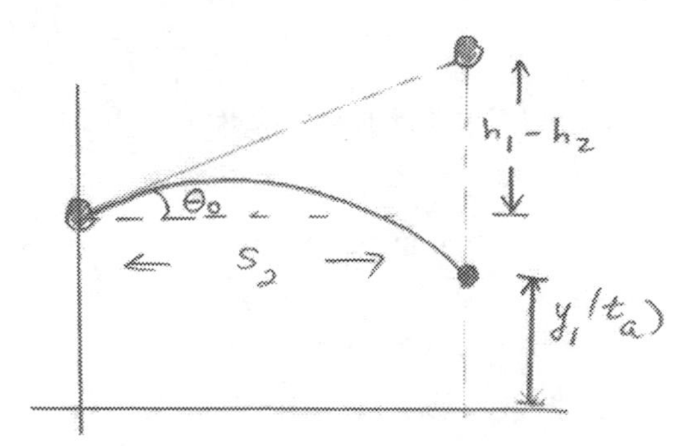

Figure 5.11: Geometry of collision

In order to find the time that the ball collides with the pail, we begin by squaring both Eqs. (5.2.40) and (5.2.41), then utilize the trigonometric identity $\sin^2(\theta_0) + \cos^2(\theta_0) = 1$. Our squared equations become

$$v_0^{\,2} \sin^2(\theta_0) t_a^{\,2} = (h_1 - h_2)^2 \tag{5.2.44}$$

$$v_0^{\,2} \cos^2(\theta_0) t_a^{\,2} = s_2^{\,2}. \tag{5.2.45}$$

Adding these equations together yields

$$v_0^{\,2}(\sin^2(\theta_0) + \cos^2(\theta_0)) t_a^{\,2} = s_2^{\,2} + (h_1 - h_2)^2, \tag{5.2.46}$$

$$v_0^{\,2} t_a^{\,2} = s_2^{\,2} + (h_1 - h_2)^2. \tag{5.2.47}$$

We can solve Eq. (5.2.47) for the time of collision

$$t_a = \left(\frac{s_2^{\,2} + (h_1 - h_2)^2}{v_0^{\,2}} \right)^{1/2}. \tag{5.2.48}$$

We can now use the y-coordinate function of either the ball or the pail at $t = t_a$ to find the height that the ball collides with the pail. Since it had no initial y-component of the velocity, it's easier to use the pail,

$$y_1(t_a) = h_1 - \frac{g(s_2^{\,2} + (h_1 - h_2)^2)}{2v_0^{\,2}}. \tag{5.2.49}$$

4. Look Back – check your solution and method of solution

The person aims at the pail at the point where the pail was released. Both undergo free fall so the key result was that the vertical position obeys

$$h_1 - \frac{1}{2} g t_a^{\ 2} = h_2 + v_0 \sin(\theta_0) t_a - \frac{1}{2} g t_a^{\ 2} .$$

The distance traveled due to gravitational acceleration are the same for both so all that matters is the contribution form the initial positions and the vertical component of velocity $h_1 = h_2 + v_0 \sin(\theta_0) t_a$. Because the time is related to the horizontal distance by $s_2 = v_0 \cos(\theta_0) t_a$, it's as if both objects were moving at constant velocity.

Chapter 6 Circular Motion

Chapter 6 Circular Motion

And the seasons they go round and round
And the painted ponies go up and down
We're captive on the carousel of time
We can't return we can only look
Behind from where we came
And go round and round and round
In the circle game [1]

Joni Mitchell

Equation Chapter 6 Section 1

6.1 Introduction

Special cases often dominate our study of physics, and circular motion is certainly no exception. We see circular motion in many instances in the world; a bicycle rider on a circular track, a ball spun around by a string, and the rotation of a spinning wheel are just a few examples. Various planetary models described the motion of planets in circles before any understanding of gravitation. The motion of the moon around the earth is nearly circular. The motions of the planets around the sun are nearly circular. Our sun moves in nearly a circular orbit about the center of our galaxy, 50,000 light years from a massive black hole at the center of the galaxy.

We shall describe the kinematics of circular motion, the position, velocity, and acceleration, as a special case of two-dimensional motion. We will see that unlike linear motion, where velocity and acceleration are directed along the line of motion, in circular motion the direction of velocity is always tangent to the circle. This means that as the object moves in a circle, the direction of the velocity is always changing. When we examine this motion, we shall see that the direction of change of the velocity is towards the center of the circle. This means that there is a non-zero component of the acceleration directed radially inward, which is called the *centripetal acceleration.* If our object is increasing its speed or slowing down, there is also a non-zero *tangential acceleration* in the direction of motion. But when the object is moving at a constant speed in a circle then only the centripetal acceleration is non-zero.

In all of these instances, when an object is constrained to move in a circle, there must exist a force $\vec{\mathbf{F}}$ acting on the object directed towards the center.

In 1666, twenty years before Newton published his *Principia*, he realized that the moon is always "falling" towards the center of the earth; otherwise, by the First Law, it would continue in some linear trajectory rather than follow a circular orbit. Therefore there must be a *centripetal force*, a radial force pointing inward, producing this centripetal acceleration.

---

[1] Joni Mitchell, *The Circle Game,* Siquomb Publishing Company.

Because Newton's Second Law $\vec{\mathbf{F}} = m\vec{\mathbf{a}}$ is a vector equality, it can be applied to the radial direction to yield

$$F_r = m a_r . \quad (6.1.1)$$

6.2 Cylindrical Coordinate System

Equation Section (Next)

We first choose an origin and an axis we call the z-axis with unit vector $\hat{\mathbf{k}}$ pointing in the increasing z-direction. The level surface of points such that $z = z_P$ define a plane. We shall choose coordinates for a point P in the plane $z = z_P$ as follows.

One coordinate, r, measures the distance from the z-axis to the point P. The coordinate r ranges in value from $0 \leq r \leq \infty$. In Figure 6.1 we draw a few surfaces that have constant values of r. These `level surfaces' are circles.

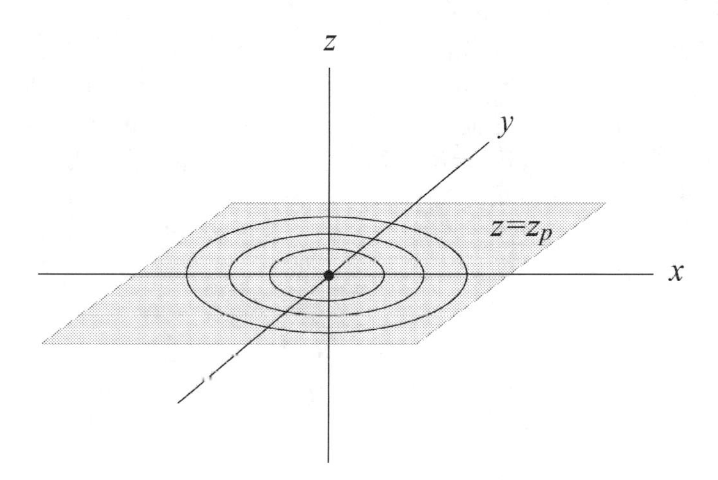

Figure 6.1 level surfaces for the coordinate r

Our second coordinate measures an angular distance along the circle. We need to choose some reference point to define the angle coordinate. We choose a 'reference ray', a horizontal ray starting from the origin and extending to $+\infty$ along the horizontal direction to the right. (In a typical Cartesian coordinate system, our 'reference ray' is the positive x-direction). We define the angle coordinate for the point P as follows. We draw a ray from the origin to the point P. We define the angle θ as the angle in the counterclockwise direction between our horizontal reference ray and the ray from the origin to the point P, (Figure 6.2). All the other points that lie on a ray from the origin to infinity passing through P have the same value as θ. For any arbitrary point, our angle coordinate θ can take on values from $0 \leq \theta < 2\pi$. In Figure 6.3 we depict other `level surfaces', which are lines in the plane for the angle coordinate. The coordinates (r, θ) in the plane $z = z_P$ are called ***polar coordinates***.

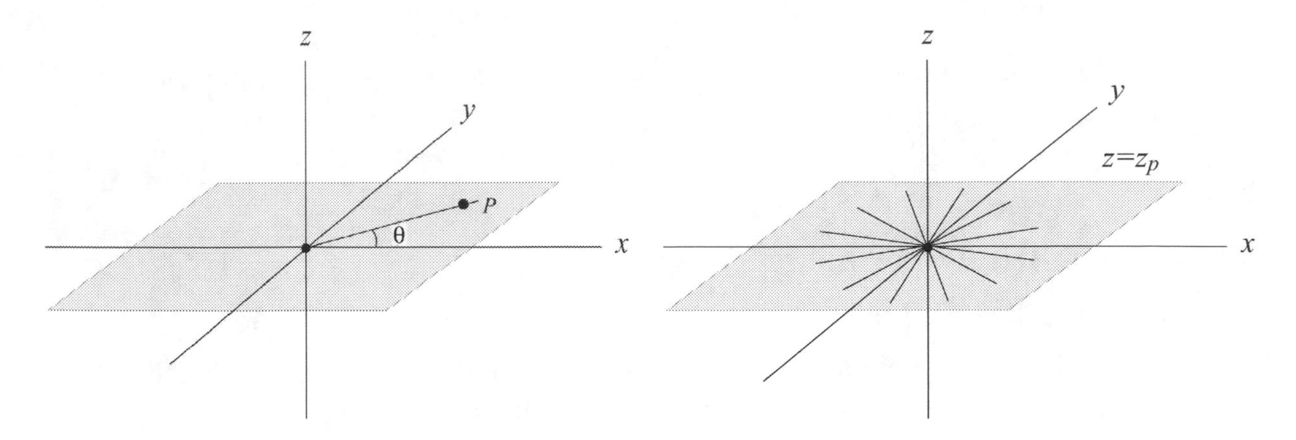

Figure 6.2 the angle coordinate

Figure 6.3 Level surfaces for the angle coordinate

6.2.1 Unit Vectors

We choose two unit vectors in the plane at the point P as follows. We choose $\hat{\mathbf{r}}$ to point in the direction of increasing r, radially away from the z-axis. We choose $\hat{\boldsymbol{\theta}}$ to point in the direction of increasing θ. This unit vector points in the counterclockwise direction, tangent to the circle. Our complete coordinate system is shown in Figure 6.4. This coordinate system is called a 'cylindrical coordinate system'. Essentially we have chosen two directions, radial and tangential in the plane and a perpendicular direction to the plane. If we are given polar coordinates (r,θ) of a point in the plane, the Cartesian coordinates (x, y) can be determined from the coordinate transformations

$$x = r\cos\theta\,, \tag{6.2.1}$$

$$y = r\sin\theta\,. \tag{6.2.2}$$

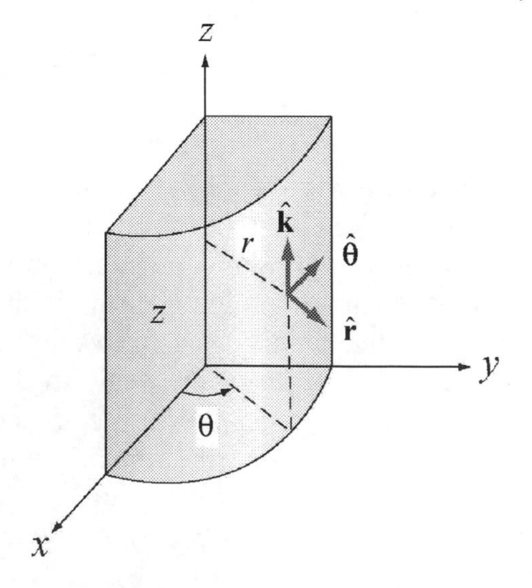

Figure 6.4 Cylindrical coordinates

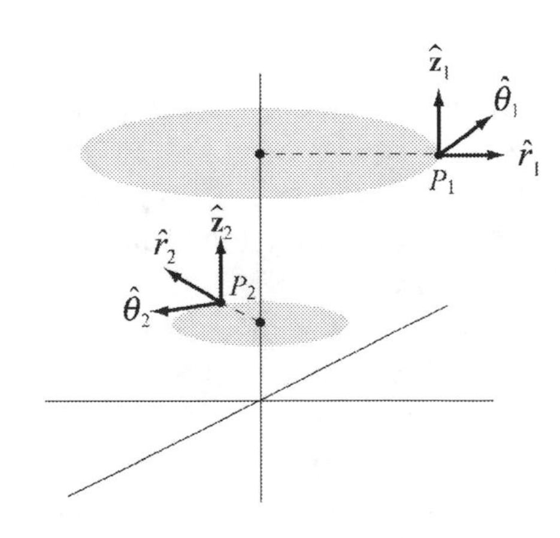

Figure 6.5 Unit vectors at two different points in polar coordinates.

Conversely, if we are given the Cartesian coordinates (x, y), the polar coordinates (r, θ) can be determined from the coordinate transformations

$$r = +(x^2 + y^2)^{1/2}, \tag{6.2.3}$$

$$\theta = \tan^{-1}(y/x). \tag{6.2.4}$$

Note that $r \geq 0$ so we always need to take the positive square root. Note also that $\tan\theta = \tan(\theta + \pi)$. Suppose that $0 \leq \theta \leq \pi/2$, then $x \geq 0$ and $y \geq 0$. Then the point $(-x, -y)$ will correspond to the angle $\theta + \pi$.

The unit vectors also are related by the coordinate transformations

$$\hat{\mathbf{r}} = \cos\theta\,\hat{\mathbf{i}} + \sin\theta\,\hat{\mathbf{j}}, \tag{6.2.5}$$

$$\hat{\boldsymbol{\theta}} = -\sin\theta\,\hat{\mathbf{i}} + \cos\theta\,\hat{\mathbf{j}}. \tag{6.2.6}$$

Similarly

$$\hat{\mathbf{i}} = \cos\theta\,\hat{\mathbf{r}} - \sin\theta\,\hat{\boldsymbol{\theta}}, \tag{6.2.7}$$

$$\hat{\mathbf{j}} = \sin\theta\,\hat{\mathbf{r}} + \cos\theta\,\hat{\boldsymbol{\theta}}. \tag{6.2.8}$$

One crucial difference between polar coordinates and Cartesian coordinates involves the choice of unit vectors. Suppose we consider a different point S in the plane. The unit vectors in Cartesian coordinates $(\hat{\mathbf{i}}_S, \hat{\mathbf{j}}_S)$ at the point S have the same magnitude and point in the same direction as the unit vectors $(\hat{\mathbf{i}}_P, \hat{\mathbf{j}}_P)$ at P. Any two vectors that are equal in magnitude and point in the same direction are equal; therefore

$$\hat{\mathbf{i}}_S = \hat{\mathbf{i}}_P, \quad \hat{\mathbf{j}}_S = \hat{\mathbf{j}}_P. \tag{6.2.9}$$

A Cartesian coordinate system is the unique coordinate system in which the set of unit vectors at different points in space are equal. In polar coordinates, the unit vectors at two different points are not equal because they point in different directions. We show this in Figure 6.5.

6.2.2 Infinitesimal Line, Area, and Volume Elements in Cylindrical Coordinates

Consider a small infinitesimal displacement $d\vec{\mathbf{s}}$ between two points P_1 and P2 (Figure 6.6). This vector can be decomposed into

$$d\vec{\mathbf{s}} = dr\,\hat{\mathbf{r}} + r d\theta\,\hat{\boldsymbol{\theta}} + dz\,\hat{\mathbf{k}}. \tag{6.2.10}$$

Consider an infinitesimal area element on the surface of a cylinder of radius r (Figure 6.7). The area of this element has magnitude

$$dA = rd\theta dz\,. \qquad (6.2.11)$$

Area elements are actually vectors where the direction of the vector $d\vec{\mathbf{A}}$ points perpendicular to the plane defined by the area. Since there is a choice of direction, we shall choose the area vector to always point outwards from a closed surface. So for the surface of the cylinder, the infinitesimal area vector is

$$d\vec{\mathbf{A}} = rd\theta dz\,\hat{\mathbf{r}}\,. \qquad (6.2.12)$$

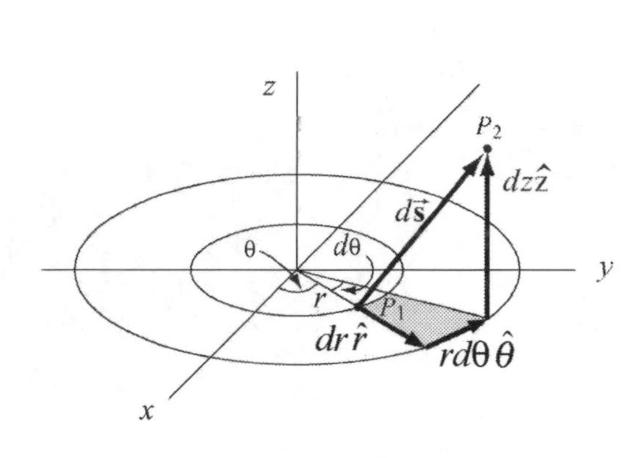

Figure 6.6 Displacement vector $d\vec{\mathbf{s}}$ between two points

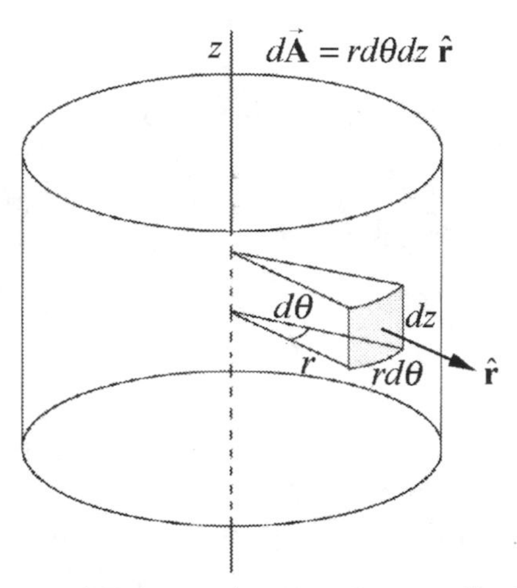

Figure 6.7 Area element for a cylinder: normal vector $\hat{\mathbf{r}}$

Example 6.1 Area Element of Disk

Consider an infinitesimal area element on the surface of a disc (Figure 6.8) in the xy-plane.

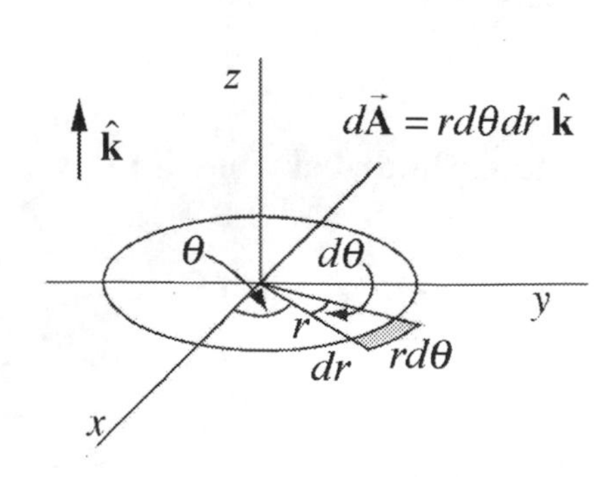

Figure 6.8 Area element for a disc: normal $\hat{\mathbf{k}}$

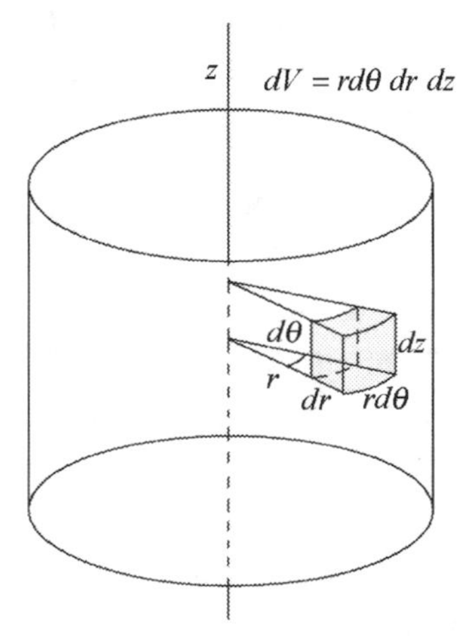

Figure 6.9 Volume element

Solution: The area element is given by the vector

$$d\vec{\mathbf{A}} = r d\theta dr\,\hat{\mathbf{k}}\,. \tag{6.2.13}$$

An infinitesimal volume element (Figure 6.9) is given by

$$dV = r d\theta\; dr\; dz\,. \tag{6.2.14}$$

The motion of objects moving in circles motivates the use of the cylindrical coordinate system. This is ideal, as the mathematical description of this motion makes use of the radial symmetry of the motion. Consider the central radial point and a vertical axis passing perpendicular to the plane of motion passing through that central point. Then any rotation about this vertical axis leaves circles invariant (unchanged), making this system ideal for use for analysis of circular motion exploiting of the radial symmetry of the motion.

6.3 Circular Motion: Velocity and Angular Velocity

Equation Section (Next)

We can now begin our description of circular motion. In Figure 6.10 we sketch the position vector $\vec{\mathbf{r}}(t)$ of the object moving in a circular orbit of radius r. At time t, the particle is located at the point P with coordinates $(r, \theta(t))$ and position vector given by

$$\vec{\mathbf{r}}(t) = r\,\hat{\mathbf{r}}(t)\,. \tag{6.3.1}$$

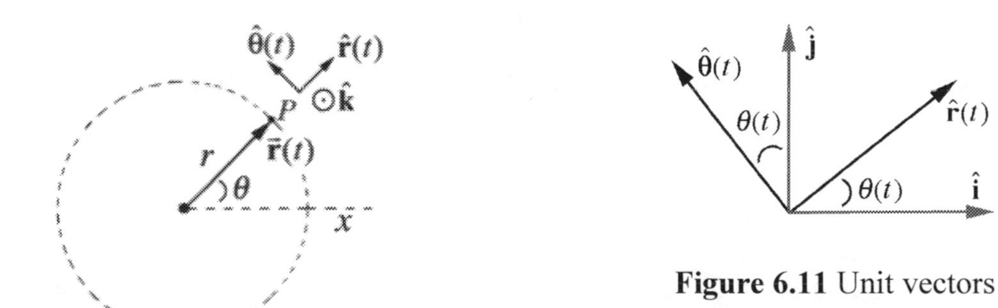

Figure 6.11 Unit vectors

Figure 6.10 A circular orbit.

At the point P, consider two sets of unit vectors ($\hat{\mathbf{r}}(t)$, $\hat{\boldsymbol{\theta}}(t)$) and ($\hat{\mathbf{i}}$, $\hat{\mathbf{j}}$). In Figure 6.11 we see that a vector decomposition expression for $\hat{\mathbf{r}}(t)$ and $\hat{\boldsymbol{\theta}}(t)$ in terms of $\hat{\mathbf{i}}$ and $\hat{\mathbf{j}}$ is given by

$$\hat{\mathbf{r}}(t) = \cos\theta(t)\,\hat{\mathbf{i}} + \sin\theta(t)\,\hat{\mathbf{j}}, \tag{6.3.2}$$

$$\hat{\boldsymbol{\theta}}(t) = -\sin\theta(t)\,\hat{\mathbf{i}} + \cos\theta(t)\,\hat{\mathbf{j}}. \tag{6.3.3}$$

We can write the position vector as

$$\vec{\mathbf{r}}(t) = r\,\hat{\mathbf{r}}(t) = r(\cos\theta(t)\,\hat{\mathbf{i}} + \sin\theta(t)\,\hat{\mathbf{j}}). \tag{6.3.4}$$

The velocity is then

$$\vec{\mathbf{v}}(t) = \frac{d\vec{\mathbf{r}}(t)}{dt} = r\frac{d}{dt}(\cos\theta(t)\,\hat{\mathbf{i}} + \sin\theta(t)\,\hat{\mathbf{j}}) = r(-\sin\theta(t)\frac{d\theta(t)}{dt}\,\hat{\mathbf{i}} + \cos\theta(t)\frac{d\theta(t)}{dt}\,\hat{\mathbf{j}}), \tag{6.3.5}$$

where we used the chain rule to calculate that

$$\frac{d}{dt}(\cos\theta(t) = -\sin\theta(t)\frac{d\theta(t)}{dt}\ , \tag{6.3.6}$$

$$\frac{d}{dt}(\sin\theta(t) = \cos\theta(t)\frac{d\theta(t)}{dt}\ . \tag{6.3.7}$$

We now rewrite Eq. (6.3.5) as

$$\vec{\mathbf{v}}(t) = r\frac{d\theta(t)}{dt}(-\sin\theta(t)\hat{\mathbf{i}} + \cos\theta(t)\,\hat{\mathbf{j}}). \tag{6.3.8}$$

Finally we substitute Eq. (6.3.3) into Eq. (6.3.8) and obtain an expression for the velocity of a particle in a circular orbit

$$\vec{\mathbf{v}}(t) = r\frac{d\theta(t)}{dt}\hat{\boldsymbol{\theta}}(t). \tag{6.3.9}$$

We denote the rate of change of angle with respect to time by the Greek letter ω,

$$\omega \equiv \frac{d\theta}{dt}, \tag{6.3.10}$$

which can be positive (counterclockwise rotation in Figure 6.10), zero (no rotation), or negative (clockwise rotation in Figure 6.10). This is often called the ***angular speed*** but it is actually the z-component of a vector called the ***angular velocity vector***.

$$\vec{\boldsymbol{\omega}} = \frac{d\theta}{dt}\hat{\mathbf{k}} = \omega\hat{\mathbf{k}}. \tag{6.3.11}$$

The SI units of angular velocity are $[\text{rad}\cdot\text{s}^{-1}]$. Thus the velocity vector for circular motion is given by

$$\vec{\mathbf{v}}(t) = r\omega\,\hat{\boldsymbol{\theta}}(t) \equiv v_\theta\,\hat{\boldsymbol{\theta}}(t), \tag{6.3.12}$$

where the $\hat{\boldsymbol{\theta}}$-component of the velocity is given by

$$v_\theta = r\frac{d\theta}{dt}. \tag{6.3.13}$$

We shall call v_θ the ***tangential component of the velocity***.

6.3.1 Geometric Derivation of the Velocity for Circular Motion

Consider a particle undergoing circular motion. At time t, the position of the particle is $\vec{\mathbf{r}}(t)$. During the time interval Δt, the particle moves to the position $\vec{\mathbf{r}}(t+\Delta t)$ with a displacement $\Delta\vec{\mathbf{r}}$.

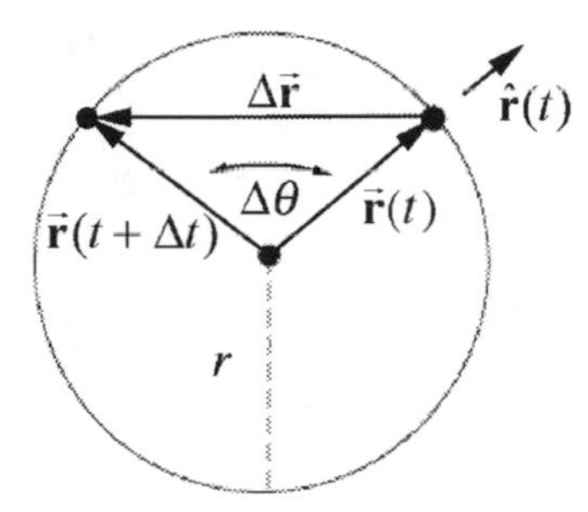

Figure 6.12 Displacement vector for circular motion

The magnitude of the displacement, $|\Delta\vec{\mathbf{r}}|$, is the represented by the length of the horizontal vector $\Delta\vec{\mathbf{r}}$ joining the heads of the displacement vectors in Figure 6.12 and is given by

$$|\Delta\vec{\mathbf{r}}| = 2r\sin(\Delta\theta/2). \tag{6.3.14}$$

When the angle $\Delta\theta$ is small, we can approximate

$$\sin(\Delta\theta/2) \cong \Delta\theta/2. \tag{6.3.15}$$

This is called the *small angle approximation*, where the angle $\Delta\theta$ (and hence $\Delta\theta/2$) is measured in radians. This fact follows from an infinite power series expansion for the sine function given by

$$\sin\left(\frac{\Delta\theta}{2}\right) = \frac{\Delta\theta}{2} - \frac{1}{3!}\left(\frac{\Delta\theta}{2}\right)^3 + \frac{1}{5!}\left(\frac{\Delta\theta}{2}\right)^5 - \cdots. \tag{6.3.16}$$

When the angle $\Delta\theta/2$ is small, only the first term in the infinite series contributes, as successive terms in the expansion become much smaller. For example, when $\Delta\theta/2 = \pi/30 \cong 0.1$, corresponding to 6°, $(\Delta\theta/2)^3/3! \cong 1.9\times10^{-4}$; this term in the power series is three orders of magnitude smaller than the first and can be safely ignored for small angles.

Using the small angle approximation, the magnitude of the displacement is

$$\left|\Delta\vec{\mathbf{r}}\right| \cong r\ \Delta\theta\ . \qquad (6.3.17)$$

This result should not be too surprising since in the limit as $\Delta\theta$ approaches zero, the length of the chord approaches the arc length $r\,\Delta\theta$.

The magnitude of the velocity, $\left|\vec{\mathbf{v}}\right| \equiv v$, is then seen to be proportional to the rate of change of the magnitude of the angle with respect to time,

$$v \equiv \left|\vec{\mathbf{v}}\right| = \lim_{\Delta t\to 0}\frac{\left|\Delta\vec{\mathbf{r}}\right|}{\Delta t} = \lim_{\Delta t\to 0}\frac{r\left|\Delta\theta\right|}{\Delta t} = r\lim_{\Delta t\to 0}\frac{\left|\Delta\theta\right|}{\Delta t} = r\left|\frac{d\theta}{dt}\right| = r\left|\omega\right|\ . \qquad (6.3.18)$$

The direction of the velocity can be determined by considering that in the limit as $\Delta t \to 0$ (note that $\Delta\theta \to 0$), the direction of the displacement $\Delta\vec{\mathbf{r}}$ approaches the direction of the tangent to the circle at the position of the particle at time t (Figure 6.13).

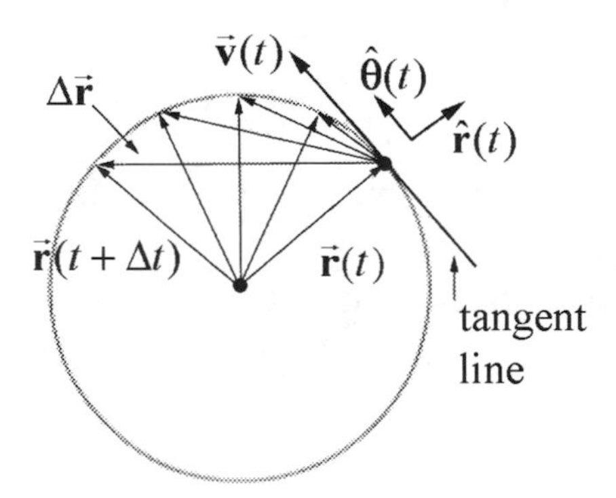

Figure 6.13 Direction of the displacement approaches the direction of the tangent line

Thus, in the limit $\Delta t \to 0$, $\Delta\vec{\mathbf{r}} \perp \vec{\mathbf{r}}$, and so the direction of the velocity $\vec{\mathbf{v}}(t)$ at time t is perpendicular to the position vector $\vec{\mathbf{r}}(t)$ and tangent to the circular orbit in the $+\hat{\boldsymbol{\theta}}$-direction for the case shown in Figure 6.13.

6.4 Circular Motion: Tangential and Radial Acceleration

Equation Section (Next)

When the motion of an object is described in polar coordinates, the acceleration has two components, the tangential component a_θ, and the radial component, a_r. We can write the acceleration vector as

$$\vec{\mathbf{a}} = a_r\,\hat{\mathbf{r}}(t) + a_\theta\,\hat{\boldsymbol{\theta}}(t). \tag{6.4.1}$$

Keep in mind that as the object moves in a circular, the unit vectors $\hat{\mathbf{r}}(t)$ and $\hat{\boldsymbol{\theta}}(t)$ change direction and hence are not constant in time.

We will begin by calculating the tangential component of the acceleration for circular motion. Suppose that the tangential velocity $v_\theta = r\omega$ is changing in magnitude due to the presence of some tangential force, where ω is the z-component of the angular velocity; we shall now consider that $\omega(t)$ is changing in time, (the magnitude of the velocity is changing in time). Recall that in polar coordinates the velocity vector Eq. (6.3.12) can be written as

$$\vec{\mathbf{v}}(t) = r\omega\,\hat{\boldsymbol{\theta}}(t). \tag{6.4.2}$$

We now use the product rule to determine the acceleration.

$$\vec{\mathbf{a}}(t) = \frac{d\vec{\mathbf{v}}(t)}{dt} = r\frac{d\omega(t)}{dt}\hat{\boldsymbol{\theta}}(t) + r\omega(t)\frac{d\hat{\boldsymbol{\theta}}(t)}{dt}. \tag{6.4.3}$$

Recall from Eq. (6.3.3) that $\hat{\boldsymbol{\theta}}(t) = -\sin\theta(t)\hat{\mathbf{i}} + \cos\theta(t)\,\hat{\mathbf{j}}$. So we can rewrite Eq. (6.4.3) as

$$\vec{\mathbf{a}}(t) = r\frac{d\omega(t)}{dt}\hat{\boldsymbol{\theta}}(t) + r\omega(t)\frac{d}{dt}(-\sin\theta(t)\hat{\mathbf{i}} + \cos\theta(t)\,\hat{\mathbf{j}}). \tag{6.4.4}$$

We again use the chain rule (Eqs. (6.3.6) and (6.3.7)) and find that

$$\vec{\mathbf{a}}(t) = r\frac{d\omega(t)}{dt}\hat{\boldsymbol{\theta}}(t) + r\omega(t)\left(-\cos\theta(t)\frac{d\theta(t)}{dt}\hat{\mathbf{i}} - \sin\theta(t)\frac{d\theta(t)}{dt}\hat{\mathbf{j}}\right). \tag{6.4.5}$$

Recall that $\omega \equiv d\theta / dt$, and from Eq. (6.3.2), $\hat{\mathbf{r}}(t) = \cos\theta(t)\,\hat{\mathbf{i}} + \sin\theta(t)\,\hat{\mathbf{j}}$, therefore the acceleration becomes

$$\vec{\mathbf{a}}(t) = r\frac{d\omega(t)}{dt}\,\hat{\boldsymbol{\theta}}(t) - r\omega^2(t)\,\hat{\mathbf{r}}(t). \tag{6.4.6}$$

We denote the rate of change of ω with respect to time by the Greek letter α,

$$\alpha \equiv \frac{d\omega}{dt}, \tag{6.4.7}$$

which can be positive, zero, or negative. This is often called the ***angular acceleration*** but it is actually the z-component of a vector called the ***angular acceleration vector***.

$$\vec{\boldsymbol{\alpha}} = \frac{d\omega}{dt}\hat{\mathbf{k}} = \frac{d^2\theta}{dt^2}\hat{\mathbf{k}} \equiv \alpha\hat{\mathbf{k}}. \tag{6.4.8}$$

The SI units of angular acceleration are $[\mathrm{rad}\cdot\mathrm{s}^{-2}]$. The ***tangential component of the acceleration*** is then

$$a_\theta = r\alpha. \tag{6.4.9}$$

The ***radial component of the acceleration*** is given by

$$a_r = -r\omega^2 < 0 . \tag{6.4.10}$$

Because $a_r < 0$, that radial vector component $\vec{\mathbf{a}}_r(t) = -r\omega^2\,\hat{\mathbf{r}}(t)$ is always directed towards the center of the circular orbit.

6.5 Period and Frequency for Uniform Circular Motion

Equation Section (Next)

If the object is constrained to move in a circle and the total tangential force acting on the object is zero, $F_\theta^{\text{total}} = 0$. By Newton's Second Law, the tangential acceleration is zero,

$$a_\theta = 0. \tag{6.5.1}$$

This means that the magnitude of the velocity (the speed) remains constant. This motion is known as ***uniform circular motion***. The acceleration is then given by only the acceleration radial component vector

$$\vec{\mathbf{a}}_r(t) = -r\omega^2(t)\,\hat{\mathbf{r}}(t) \quad \text{uniform circular motion}. \tag{6.5.2}$$

Since the speed $v = r|\omega|$ is constant, the amount of time that the object takes to complete one circular orbit of radius r is also constant. This time interval, T, is called the ***period.*** In one period the object travels a distance $s = vT$ equal to the circumference, $s = 2\pi r$; thus

$$s = 2\pi r = vT. \tag{6.5.3}$$

The period T is then given by

$$T = \frac{2\pi r}{v} = \frac{2\pi r}{r|\omega|} = \frac{2\pi}{|\omega|}. \tag{6.5.4}$$

The ***frequency*** f is defined to be the reciprocal of the period,

$$f = \frac{1}{T} = \frac{|\omega|}{2\pi}. \qquad (6.5.5)$$

The SI unit of frequency is the inverse second, which is defined as the hertz, $[\mathrm{s}^{-1}] \equiv [\mathrm{Hz}]$.

The magnitude of the radial component of the acceleration can be expressed in several equivalent forms since both the magnitudes of the velocity and angular velocity are related by $v = r|\omega|$. Thus we have several alternative forms for the magnitude of the centripetal acceleration. The first is that in Equation (6.6.3). The second is in terms of the radius and the angular velocity,

$$|a_r| = r\,\omega^2. \qquad (6.5.6)$$

The third form expresses the magnitude of the centripetal acceleration in terms of the speed and radius,

$$|a_r| = \frac{v^2}{r}. \qquad (6.5.7)$$

Recall that the magnitude of the angular velocity is related to the frequency by $|\omega| = 2\pi\, f$, so we have a fourth alternate expression for the magnitude of the centripetal acceleration in terms of the radius and frequency,

$$|a_r| = 4\pi^2 r\, f^2\;. \qquad (6.5.8)$$

A fifth form commonly encountered uses the fact that the frequency and period are related by $f = 1/T = |\omega|/2\pi$. Thus we have the fourth expression for the centripetal acceleration in terms of radius and period,

$$|a_r| = \frac{4\pi^2 r}{T^2}. \qquad (6.5.9)$$

Other forms, such as $4\pi^2 r^2 f / T$ or $2\pi r \omega f$, while valid, are uncommon.

Often we decide which expression to use based on information that describes the orbit. A convenient measure might be the orbit's radius. We may also independently know the period, or the frequency, or the angular velocity, or the speed. If we know one, we can calculate the other three but it is important to understand the meaning of each quantity.

6.5.1 Geometric Interpretation for Radial Acceleration for Uniform Circular Motion

Equation Section (Next)

An object traveling in a circular orbit is always accelerating towards the center. Any radial inward acceleration is called *centripetal acceleration*. Recall that the direction of the velocity is always tangent to the circle. Therefore the direction of the velocity is constantly changing because the object is moving in a circle, as can be seen in Figure 6.14. Because the velocity changes direction, the object has a nonzero acceleration.

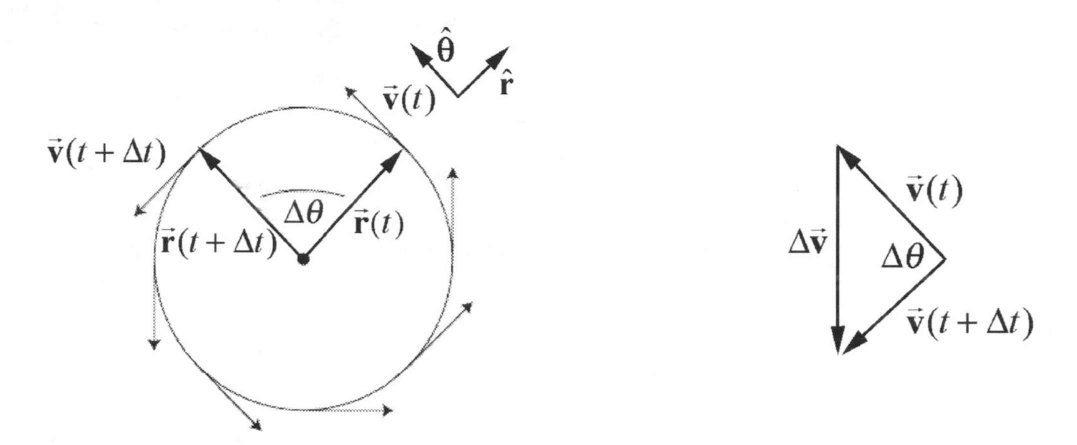

Figure 6.14 Direction of the velocity for circular motion.

Figure 6.15 Change in velocity vector.

The calculation of the magnitude and direction of the acceleration is very similar to the calculation for the magnitude and direction of the velocity for circular motion, but the change in velocity vector, $\Delta\vec{\mathbf{v}}$, is more complicated to visualize. The change in velocity $\Delta\vec{\mathbf{v}} = \vec{\mathbf{v}}(t+\Delta t) - \vec{\mathbf{v}}(t)$ is depicted in Figure 6.15. The velocity vectors have been given a common point for the tails, so that the change in velocity, $\Delta\vec{\mathbf{v}}$, can be visualized. The length $|\Delta\vec{\mathbf{v}}|$ of the vertical vector can be calculated in exactly the same way as the displacement $|\Delta\vec{\mathbf{r}}|$. The magnitude of the change in velocity is

$$|\Delta\vec{\mathbf{v}}| = 2v\sin(\Delta\theta/2). \tag{6.6.1}$$

We can use the small angle approximation $\sin(\Delta\theta/2) \cong \Delta\theta/2$ to approximate the magnitude of the change of velocity,

$$|\Delta\vec{\mathbf{v}}| \cong v|\Delta\theta|. \tag{6.6.2}$$

The magnitude of the radial acceleration is given by

$$\left|a_r\right| = \lim_{\Delta t \to 0} \frac{\left|\Delta \vec{\mathbf{v}}\right|}{\Delta t} = \lim_{\Delta t \to 0} \frac{v\left|\Delta \theta\right|}{\Delta t} = v \lim_{\Delta t \to 0} \frac{\left|\Delta \theta\right|}{\Delta t} = v\left|\frac{d\theta}{dt}\right| = v\left|\omega\right|. \qquad (6.6.3)$$

The direction of the radial acceleration is determined by the same method as the direction of the velocity; in the limit $\Delta\theta \to 0$, $\Delta\vec{\mathbf{v}} \perp \vec{\mathbf{v}}$, and so the direction of the acceleration radial component vector $\vec{\mathbf{a}}_r(t)$ at time t is perpendicular to position vector $\vec{\mathbf{v}}(t)$ and directed inward, in the $-\hat{\mathbf{r}}$-direction.

Chapter 7 Newton's Laws of Motion

Chapter 7 Newton's Laws of Motion

I have not as yet been able to discover the reason for these properties of gravity from phenomena, and I do not feign hypotheses. For whatever is not deduced from the phenomena must be called a hypothesis; and hypotheses, whether metaphysical or physical, or based on occult qualities, or mechanical, have no place in experimental philosophy. In this philosophy particular propositions are inferred from the phenomena, and afterwards rendered general by induction. [1]

Isaac Newton

7.1 Force and Quantity of Matter

In our daily experience, we can cause a body to move by either pushing or pulling that body. Ordinary language use describes this action as the effect of a person's strength or *force*. However, bodies placed on inclined planes, or when released at rest and undergo free fall, will move without any push or pull. Galileo referred to a force acting on these bodies, a description of which he published in 1623 in his *Mechanics*. In 1687, Isaac Newton published his three laws of motion in the *Philosophiae Naturalis Principia Mathematica* ("Mathematical Principles of Natural Philosophy"), which extended Galileo's observations. The First Law expresses the idea that when no force acts on a body, it will remain at rest or maintain uniform motion; when a force is applied to a body, it will change its state of motion.

Many scientists, especially Galileo, recognized the idea that force produces motion before Newton but Newton extended the concept of force to any circumstance that produces acceleration. When a body is initially at rest, the direction of our push or pull corresponds to the direction of motion of the body. If the body is moving, the direction of the applied force may change both the direction of motion of the body and how fast it is moving. Newton defined the force acting on an object as proportional to the acceleration of the object.

An impressed force is an action exerted upon a body, in order to change its state, either of rest, or of uniform motion in a right line. [2]

In order to define the magnitude of the force, he introduced a constant of proportionality, the *inertial mass*, which Newton called "quantity of matter".

---

[1] Isaac Newton (1726). *Philosophiae Naturalis Principia Mathematica*, General Scholium. Third edition, page 943 of I. Bernard Cohen and Anne Whitman's 1999 translation, University of California Press.

[2] Isaac Newton. *Mathematical Principles of Natural Philosophy*. Translated by Andrew Motte (1729). Revised by Florian Cajori. Berkeley: University of California Press, 1934. p. 2.

The quantity of matter is the measure of the same, arising from its density and bulk conjointly.

Thus air of double density, in a double space, is quadruple in quantity; in a triple space, sextuple in quantity. The same thing is to be understood of snow, and fine dust or powders, that are condensed by compression or liquefaction, and of all bodies that are by any causes whatever differently condensed. I have no regard in this place to a medium, if any such there is, that freely pervades the interstices between the parts of bodies. It is this quantity that I mean hereafter everywhere under the name of body or mass. And the same is known by the weight of each body, for it is proportional to the weight, as I have found by experiment on pendulums, very accurately made, which shall be shown hereafter.[3]

Suppose we apply an action to a body (which we refer to as the standard body) that will induce the body to accelerate with a magnitude $|\vec{\mathbf{a}}|$ that can be measured by an accelerometer (any device that measures acceleration). The magnitude of the force $|\vec{\mathbf{F}}|$ acting on the object is the product of the mass m_s with the magnitude of the acceleration $|\vec{\mathbf{a}}|$. Force is a vector quantity. The direction of the force on the standard body is defined to be the direction of the acceleration of the body. Thus

$$\vec{\mathbf{F}} \equiv m_s \vec{\mathbf{a}} \tag{7.1.1}$$

In order to justify the statement that force is a vector quantity, we need to apply two forces $\vec{\mathbf{F}}_1$ and $\vec{\mathbf{F}}_2$ simultaneously to our body and show that the resultant force $\vec{\mathbf{F}}^T$ is the vector sum of the two forces when they are applied one at a time.

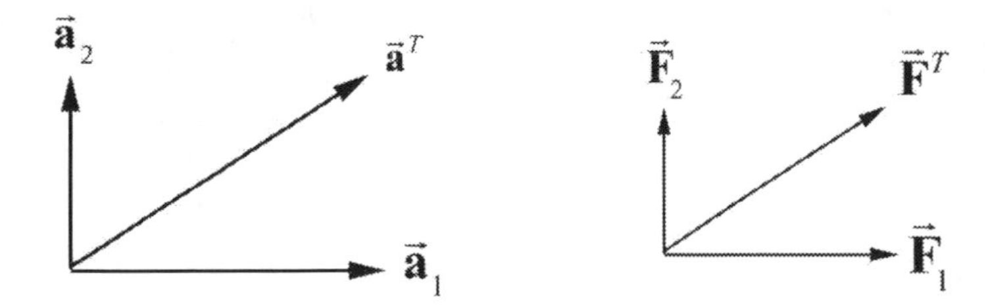

Figure 7.1 Acceleration add as vectors **Figure 7.2** Force adds as vectors.

We apply each force separately and measure the accelerations $\vec{\mathbf{a}}_1$ and $\vec{\mathbf{a}}_2$, noting that

$$\vec{\mathbf{F}}_1 = m_s \vec{\mathbf{a}}_1 \tag{7.1.2}$$

[3] Ibid. p. 1.

$$\vec{\mathbf{F}}_2 = m_s \vec{\mathbf{a}}_2 . \qquad (7.1.3)$$

When we apply the two forces simultaneously, we measure the acceleration $\vec{\mathbf{a}}$. The force by definition is now

$$\vec{\mathbf{F}}^T \equiv m_s \vec{\mathbf{a}} . \qquad (7.1.4)$$

We then compare the accelerations. The results of these three measurements, and for that matter any similar experiment, confirms that the accelerations add as vectors (Figure 7.1)

$$\vec{\mathbf{a}} = \vec{\mathbf{a}}_1 + \vec{\mathbf{a}}_2 . \qquad (7.1.5)$$

Therefore the forces add as vectors as well (Figure 7.2),

$$\vec{\mathbf{F}}^T = \vec{\mathbf{F}}_1 + \vec{\mathbf{F}}_2 . \qquad (7.1.6)$$

This last statement is not a definition but a consequence of the experimental result described by Equation (7.1.5) and our definition of force.

Example 7.1 Vector Decomposition Solution

Two horizontal ropes are attached to a post that is stuck in the ground. The ropes pull the post producing the vector forces $\vec{\mathbf{F}}_1 = 70\ \text{N}\ \hat{\mathbf{i}} + 20\ \text{N}\ \hat{\mathbf{j}}$ and $\vec{\mathbf{F}}_2 = -30\ \text{N}\ \hat{\mathbf{i}} + 40\ \text{N}\ \hat{\mathbf{j}}$ as shown in Figure 7.1. Find the direction and magnitude of the horizontal component of a third force on the post that will make the vector sum of forces on the post equal to zero.

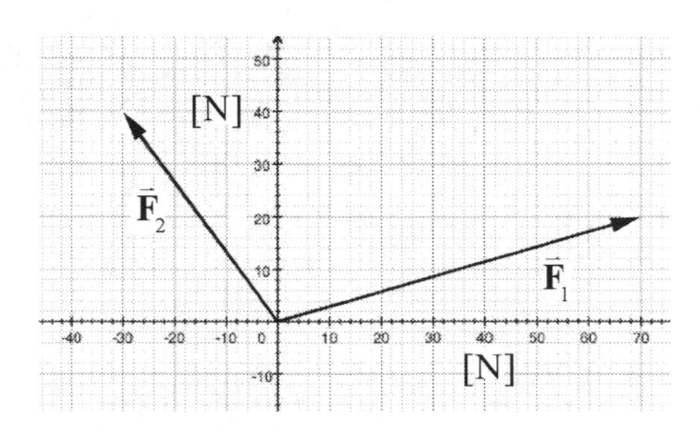

Figure 7.3 Example 7.1

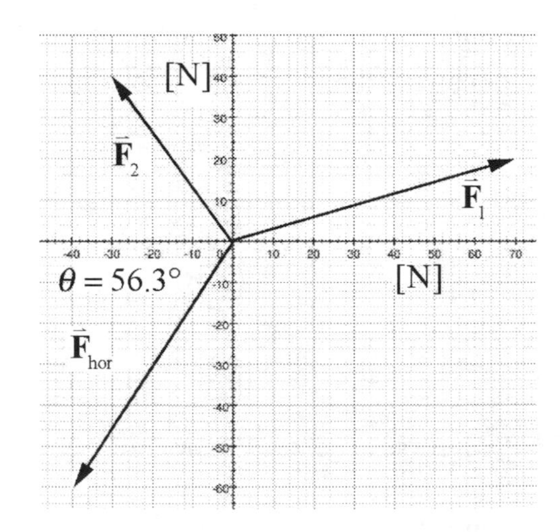

Figure 7.4 Vector sum of forces

Solution: Since the ropes are pulling the post horizontally, the third force must also have a horizontal component that is equal to the negative of the sum of the two horizontal forces exerted by the rope on the post Figure 7.4. Since there are additional vertical forces acting on the post due to its contact with the ground and the gravitational force exerted on the post by the earth, we will restrict our attention to the horizontal component of the third force. Let $\vec{\mathbf{F}}_3$ denote the sum of the forces due to the ropes. Then we can write the vector $\vec{\mathbf{F}}_3$ as

$$\vec{\mathbf{F}}_3 = (F_{1x} + F_{2x})\,\hat{\mathbf{i}} + (F_{1y} + F_{2y})\,\hat{\mathbf{j}} = (70\text{ N} + -30\text{ N})\,\hat{\mathbf{i}} + (20\text{ N} + 40\text{ N})\,\hat{\mathbf{j}}$$
$$= (40\text{ N})\,\hat{\mathbf{i}} + (60\text{ N})\,\hat{\mathbf{j}}$$

Therefore the horizontal component of the third force of the post must be equal to

$$\vec{\mathbf{F}}_{hor} = -\vec{\mathbf{F}}_3 = -(\vec{\mathbf{F}}_1 + \vec{\mathbf{F}}_2) = (-40\text{ N})\,\hat{\mathbf{i}} + (-60\text{ N})\,\hat{\mathbf{j}}.$$

The magnitude is $\left|\vec{\mathbf{F}}_{hor}\right| = \sqrt{(-40\text{ N})^2 + (-60\text{ N})^2} = 72\text{ N}$. The horizontal component of the force makes an angle

$$\theta = \tan^{-1}\left[\frac{60\text{ N}}{40\text{ N}}\right] = 56.3^\circ$$

as shown in the figure above.

7.1.1 Mass Calibration

So far, we have only used the standard body to measure force. Instead of performing experiments on the standard body, we can calibrate the masses of all other bodies in terms of the standard mass by the following experimental procedure. We shall refer to the mass measured in this way as the **inertial mass** and denote it by m_{in}.

We apply a force of magnitude F to the standard body and measure the magnitude of the acceleration a_s. Then we apply the same force to a second body of unknown mass m_{in} and measure the magnitude of the acceleration a_{in}. Since the same force is applied to both bodies,

$$F = m_{in}\,a_{in} = m_s\,a_s, \tag{1.7}$$

Therefore the ratio of the inertial mass to the standard mass is equal to the inverse ratio of the magnitudes of the accelerations,

$$\frac{m_{in}}{m_s} = \frac{a_s}{a_{in}}. \tag{1.8}$$

Therefore the second body has inertial mass equal to

$$m_{in} \equiv m_s \frac{a_s}{a_{in}}. \quad (1.9)$$

This method is justified by the fact that we can repeat the experiment using a different force and still find that the ratios of the acceleration are the same. For simplicity we shall denote the inertial mass by m.

7.2 Newton's First Law

The First Law of Motion, commonly called the "Principle of Inertia," was first realized by Galileo. (Newton did not acknowledge Galileo's contribution.) Newton was particularly concerned with how to phrase the First Law in Latin, but after many rewrites Newton perfected the following expression for the First Law (in English translation):

> *Law 1: Every body continues in its state of rest, or of uniform motion in a right line, unless it is compelled to change that state by forces impressed upon it.*
>
> *Projectiles continue in their motions, so far as they are not retarded by the resistance of air, or impelled downwards by the force of gravity. A top, whose parts by their cohesion are continually drawn aside from rectilinear motions, does not cease its rotation, otherwise than as it is retarded by air. The greater bodies of planets and comets, meeting with less resistance in freer spaces, preserve their motions both progressive and circular for a much longer time.*[4]

The first law is an experimental statement about the motions of bodies. When a body moves with constant velocity, there are either no forces present or there are forces acting in opposite directions that cancel out. If the body changes its velocity, then there must be an acceleration, and hence a total non-zero force must be present. We note that velocity can change in two ways. The first way is to change the magnitude of the velocity; the second way is to change its direction.

After a bus or train starts, the acceleration is often so small we can barely perceive it. We are often startled because it seems as if the station is moving in the opposite direction while we seem to be still. Newton's First Law states that there is no physical way to distinguish between whether we are moving or the station is, because there is essentially no total force present to change the state of motion. Once we reach a constant velocity, our minds dismiss the idea that the ground is moving backwards because we think it is impossible, but there is no actual way for us to distinguish whether the train is moving or the ground is moving.

---

[4] Ibid. p. 13.

7.3 Momentum, Newton's Second Law and Third Law

Newton began his analysis of the cause of motion by introducing the quantity of motion:

> ***Definition: Quantity of Motion:***
>
> *The quantity of motion is the measure of the same, arising from the velocity and quantity of matter conjointly.*
>
> *The motion of the whole is the sum of the motion of all its parts; and therefore in a body double in quantity, with equal velocity, the motion is double, with twice the velocity, it is quadruple.*[5]

Our modern term for quantity of motion is *momentum* and it is a vector quantity

$$\vec{\mathbf{p}} = m\vec{\mathbf{v}}\,. \tag{7.3.1}$$

where m is the inertial mass and $\vec{\mathbf{v}}$ is the velocity of the body (velocity is a vector quantity). Newton's Second Law is the most important experimental statement about motion in physics.

> *Law II: The change of motion is proportional to the motive force impressed, and is made in the direction of the right line in which that force is impressed.*
>
> *If any force generates a motion, a double force will generate double the motion, a triple force triple the motion, whether that force is impressed altogether and at once or gradually and successively. And this motion (being always directed the same way with the generating force), if the body moved before, is added or subtracted from the former motion, according as they directly conspire with or are directly contrary to each other; or obliquely joined, when they are oblique, so as to produce a new motion compounded from the determination of both.*[6]

Suppose that a force is applied to a body for a time interval Δt. The impressed force or *impulse* (a vector quantity $\vec{\mathbf{I}}$) that we denote by produces a change in the momentum of the body,

$$\vec{\mathbf{I}} = \vec{\mathbf{F}}\Delta t = \Delta\vec{\mathbf{p}}\,. \tag{7.3.2}$$

---

[5] Ibid. p. 1.
[6] Ibid. p. 13.

From the commentary to the second law, Newton also considered forces that were applied continually to a body instead of impulsively. The instantaneous action of the total force acting on a body at a time t is defining by taking the mathematical limit as the time interval Δt becomes smaller and smaller,

$$\vec{\mathbf{F}} = \lim_{\Delta t \to 0} \frac{\Delta \vec{\mathbf{p}}}{\Delta t} \equiv \frac{d\vec{\mathbf{p}}}{dt}. \tag{7.3.3}$$

When the mass remains constant in time, the Second Law can be recast in its more familiar form,

$$\vec{\mathbf{F}} = m\frac{d\vec{\mathbf{v}}}{dt}. \tag{7.3.4}$$

Because the derivative of velocity is the acceleration, the force is the product of mass and acceleration,

$$\vec{\mathbf{F}} = m\vec{\mathbf{a}}. \tag{7.3.5}$$

Because we defined force in terms of change in motion, the Second Law appears to be a restatement of this definition, and devoid of predictive power since force is only determined by measuring acceleration. What transforms the Second Law from just merely a definition is the additional input that comes from *force laws* that are based on experimental observations on the interactions between bodies. Throughout this book, we shall investigate these force laws and learn to use them in order to determine the left-hand side of Newton's Second Law.

The right-hand-side of Newton's Second Law is the product of mass with acceleration. Acceleration is a mathematical description of how the velocity of a body changes. If we know the acceleration of a body we can in principle, predict the velocity and position of that body at all future times by integration techniques.

7.4 Newton's Third Law: Action-Reaction Pairs

Newton realized that when two bodies interact via a force, then the force on one body is equal in magnitude and opposite in direction to the force acting on the other body.

> *Law III: To every action there is always opposed an equal reaction: or, the mutual action of two bodies upon each other are always equal, and directed to contrary parts.*
>
> *Whatever draws or presses another is as much drawn or pressed by that other. If you press on a stone with your finger, the finger is also pressed by the stone.*[7]

---

[7] Ibid p. 13.

The Third Law, commonly known as the "action-reaction" law, is the most surprising of the three laws. Newton's great discovery was that when two objects interact, they each exert the same magnitude of force on each other. We shall refer to objects that interact as an *interaction pair*.

Consider two bodies engaged in a mutual interaction. Label the bodies 1 and 2 respectively. Let $\vec{\mathbf{F}}_{1,2}$ be the force on body 2 due to the interaction with body 1, and $\vec{\mathbf{F}}_{2,1}$ be the force on body 1 due to the interaction with body 2. These forces are depicted in Figure 7.5.

Figure 7.5 Action-reaction pair of forces

These two vector forces are equal in magnitude and opposite in direction,

$$\vec{\mathbf{F}}_{1,2} = -\vec{\mathbf{F}}_{2,1}. \tag{7.4.1}$$

With these definitions, the three laws and force laws constitute ***Newtonian Mechanics,*** which will be able to explain a vast range of phenomena. Newtonian mechanics has important limits. It does not satisfactorily explain systems of objects moving at speeds comparable to the speed of light ($v > 0.1\,c$) where we need the theory of special relativity, nor does it adequately explain the motion of electrons in atoms, where we need *quantum mechanics*. We also need *general relativity* and *cosmology* to explain the large-scale structure of the universe.

Chapter 8 Applications of Newton's Second Law

Chapter 8 Applications of Newton's Second Law

Those who are in love with practice without knowledge are like the sailor who gets into a ship without rudder or compass and who never can be certain whether he is going. Practice must always be founded on sound theory...[1]

Leonardo da Vinci

8.1 Force Laws

There are forces that don't change appreciably from one instant to another, which we refer to as constant in time, and forces that don't change appreciably from one point to another, which we refer to as constant in space. The gravitational force on an object near the surface of the earth is an example of a force that is constant in space.

There are forces that increase as you move away. When a mass is attached to one end of a spring and the spring is stretched a distance $|x|$, the spring force increases in strength proportional to the stretch.

There are forces that stay constant in magnitude but always point towards the center of a circle; for example when a ball is attached to a rope and spun in a horizontal circle with constant speed, the tension force acting on the ball is directed towards the center of the circle. This type of attractive central force is called a ***centripetal force***.

There are forces that spread out in space such that their influence becomes less with distance. Common examples are the gravitational and electrical forces. The gravitational force between two objects falls off as the inverse square of the distance separating the objects provided the objects are of a small dimension compared to the distance between them. More complicated arrangements of attracting and repelling things give rise to forces that fall off with other powers of r: constant, $1/r$, $1/r^2$, $1/r^3$, ...,.

How do we determine if there is any mathematical relationship, a ***force law*** that describes the relationship between the force and some measurable property of the objects involved?

8.1.1 Hooke's Law

We shall illustrate this procedure by considering the forces that compressed or stretched springs exert on objects. In order to stretch or compress a spring from its equilibrium length, a force must be exerted on the spring. Attach an object of mass m to one end of a spring and fix the other end of the spring to a wall (Figure 8.1). Let x_{eq} denote the equilibrium length of the spring (neither stretched or compressed). Assume that the

[1] Notebooks of Leonardo da Vinci Complete, tr. Jean Paul Richter, 1888, Vol.1.

contact surface is smooth and hence frictionless in order to consider only the effect of the spring force.

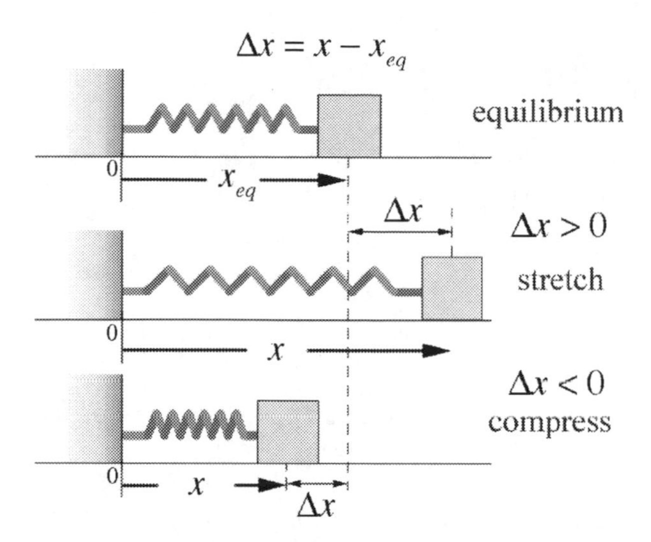

Figure 8.1 Spring attached to a wall and an object

Initially stretch the spring to a position x, corresponding to a displacement $\Delta x = x - x_{eq}$ (or compress the spring by $\Delta x < 0$), release the object, measure the acceleration, and then calculate the magnitude of the force of the spring acting on the object using the definition of force $|\vec{\mathbf{F}}| = m|\vec{\mathbf{a}}|$. Now repeat the experiment for a range of stretches (or compressions). Experiments will shown that for some range of lengths, $\Delta x_0 < \Delta x < \Delta x_1$, the magnitude of the measured force is proportional to the stretched length and is given by the formula

$$|\vec{\mathbf{F}}| \propto |\Delta x|. \tag{8.1.1}$$

In addition, the direction of the acceleration is always towards the equilibrium position when the spring is neither stretched nor compressed. This type of force is called a ***restoring force***. Let F_x denote the x-component of the spring force. Then

$$F_x = -k\Delta x = -k(x - x_{eq}). \tag{8.1.2}$$

From the experimental data, the constant of proportionality, the ***spring constant*** k, can be determined. The spring constant has units $\mathrm{N \cdot m^{-1}}$. The spring constant for each spring is determined experimentally by measuring the slope of the graph of the force *vs.* compression and extension stretch (Figure 8.2). Therefore for this one spring, the magnitude of the force is given by

$$|\vec{\mathbf{F}}| = k|\Delta x|. \tag{8.1.3}$$

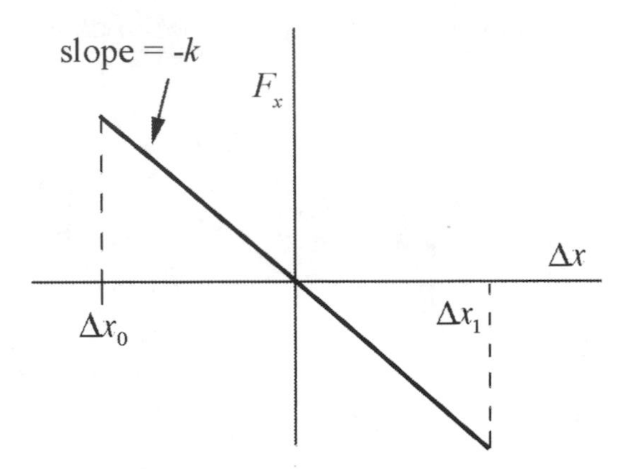

Figure 8.2 Plot of x-component of the spring force vs. compression and extension of spring

Now perform similar experiments on other springs. For a range of stretched lengths, each spring exhibits the same proportionality between force and stretched length, although the spring constant may differ for each spring.

It would be extremely impractical to experimentally determine whether this proportionality holds for all springs, and because a modest sampling of springs has confirmed the relation, we shall *infer* that all *ideal springs* will produce a restoring force, which is linearly proportional to the stretched (or compressed) length. This experimental relation regarding force and stretched (or compressed) lengths for a finite set of springs has now been inductively generalized into the above mathematical model for ideal springs, a model we refer to as a *force law*.

This inductive step, referred to as ***Newtonian induction,*** is the critical step that makes physics a predictive science. Suppose a spring, attached to an object of mass m, is stretched by an amount Δl. A prediction can be made, using the force law, that the magnitude of the force between the rubber band and the object is $|\vec{\mathbf{F}}| = k|\Delta x|$ without having to experimentally measure the acceleration. Now we can use Newton's second Law to predict the magnitude of the acceleration of the object

$$|\vec{\mathbf{a}}| = \frac{|\vec{\mathbf{F}}|}{m} = \frac{k|\Delta x|}{m}. \tag{8.1.4}$$

Now perform the experiment, and measure the acceleration within some error bounds. If the magnitude of the predicted acceleration disagrees with the measured result, then the model for the force law needs modification. The ability to adjust, correct or even reject models based on new experimental results enables a description of forces between objects to cover larger and larger experimental domains.

8.2 Fundamental Laws of Nature

Force laws are mathematical models of physical processes. They arise from observation and experimentation, and they have limited ranges of applicability. Does the linear force law for the spring hold for all springs? Each spring will most likely have a different range of linear behavior. So the model for stretching springs still lacks a universal character. As such, there should be some hesitation to generalize this observation to all springs unless some property of the spring, universal to all springs, is responsible for the force law.

Perhaps springs are made up of very small components, which when pulled apart tend to contract back together. This would suggest that there is some type of force that contracts spring molecules when they are pulled apart. What holds molecules together? Can we find some fundamental property of the interaction between atoms that will suffice to explain the macroscopic force law? This search for ***fundamental forces*** is a central task of physics.

In the case of springs, this could lead into an investigation of the composition and structural properties of the atoms that compose the steel in the spring. We would investigate the geometric properties of the lattice of atoms and determine whether there is some fundamental property of the atoms that create this lattice. Then we ask how stable is this lattice under deformations. This may lead to an investigation into the electron configurations associated with each atom and how they overlap to form bonds between atoms. These particles carry charges, which obey Coulomb's Law, but also the Laws of Quantum Mechanics. So in order to arrive at a satisfactory explanation of the elastic restoring properties of the spring, we need models that describe the fundamental physics that underline Hooke's Law.

8.2.1 Universal Law of Gravitation

At points significantly far away from the surface of the earth, the gravitational force is no longer constant with respect to the distance to the surface. ***Newton's Universal Law of Gravitation*** describes the gravitational force between two objects with masses, m_1 and m_2. This force points along the line connecting the objects, is attractive, and its magnitude is proportional to the inverse square of the distance, $r_{1,2}$, between the objects (Figure 8.3a). The force on object 2 due to the gravitational interaction between the two objects is given by

$$\vec{\mathbf{F}}_{1,2}^{G} = -G\frac{m_1 m_2}{r_{1,2}^2}\hat{\mathbf{r}}_{1,2}, \tag{8.2.1}$$

where $\vec{\mathbf{r}}_{1,2} = \vec{\mathbf{r}}_2 - \vec{\mathbf{r}}_1$ is a vector directed from object 1 to object 2, $r_{1,2} = \left|\vec{\mathbf{r}}_{1,2}\right|$, and $\hat{\mathbf{r}}_{1,2} = \vec{\mathbf{r}}_{1,2} / \left|\vec{\mathbf{r}}_{1,2}\right|$ is a unit vector directed from object 1 to object 2 (Figure 8.3b). The constant of proportionality in SI units is $G = 6.67 \times 10^{-11}\mathrm{N \cdot m^2 \cdot kg^{-2}}$.

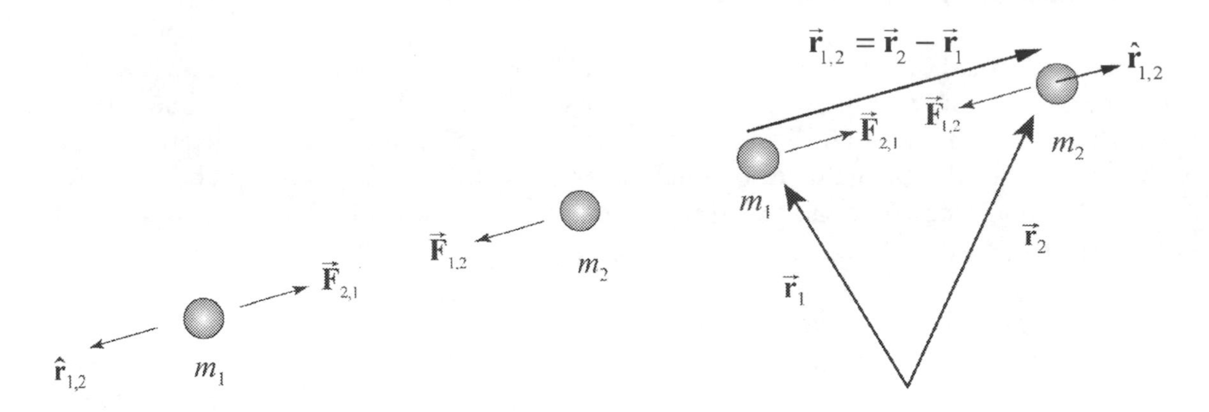

Figure 8.3 (a) Gravitational force between two objects. **Figure 8.3 (b)** Coordinate system for the two-body problem.

8.2.2 Principle of Equivalence:

The Principle of Equivalence states that the mass that appears in the Universal Law of Gravity is identical to the inertial mass that is determined with respect to the standard kilogram. From this point on, the equivalence of inertial and gravitational mass will be assumed and the mass will be denoted by the symbol m.

8.2.3 Gravitational Force near the Surface of the Earth

Near the surface of the earth, the gravitational interaction between an object and the earth is mutually attractive and has a magnitude of

$$\left|\vec{\mathbf{F}}^{G}_{earth,object}\right| = mg \tag{8.2.2}$$

where g is a positive constant.

The International Committee on Weights and Measures has adopted as a standard value for the acceleration of an object freely falling in a vacuum $g = 9.80665\ \mathrm{m\cdot s^{-2}}$. The actual value of g varies as a function of elevation and latitude. If ϕ is the latitude and h h the elevation in meters then the acceleration of gravity in SI units is

$$g = (9.80616 - 0.025928\cos(2\phi) + 0.000069\cos^2(2\phi) - 3.086\times10^{-4}h)\ \mathrm{m\cdot s^{-2}}. \tag{8.2.3}$$

This is known as Helmert's equation. The strength of the gravitational force on the standard kilogram at 42° latitude is $9.80345\ \mathrm{N\cdot kg^{-1}}$, and the acceleration due to gravity at sea level is therefore $g = 9.80345\ \mathrm{m\cdot s^{-2}}$ for all objects. At the equator, $g = 9.78\ \mathrm{m\cdot s^{-2}}$ (to three significant figures), and at the poles $g = 9.83\ \mathrm{m\cdot s^{-2}}$. This difference is primarily due to the earth's rotation, which introduces an apparent (fictitious) repulsive force that affects the determination of g as given in Equation (8.2.2) and also flattens the spherical shape earth (the distance from the center of the earth is larger at the equator than it is at the poles by about $26.5\ \mathrm{km}$). Both the magnitude and the direction of the gravitational force also show variations that depend on local features to an extent that's useful in prospecting for oil and navigating submerged nuclear submarines. Such variations in g can be measured with a sensitive spring balance. Local variations have been much studied over the past two decades in attempts to discover a proposed "fifth force" which would fall off faster than the gravitational force that falls off as the inverse square of the distance between the masses.

8.2.4 Electric Charge and Coulomb's Law

Matter has properties other than mass. As we have shown in the previous section, matter can also carry one of two types of observed ***electric charge***, positive and negative. Like charges repel, and opposite charges attract each other. The unit of charge in the SI system of units is called the *coulomb* [C].

The smallest unit of "free" charge known in nature is the charge of an electron or proton, which has a magnitude of

$$e = 1.602 \times 10^{-19}\ \mathrm{C}. \tag{8.2.4}$$

It has been shown experimentally that charge carried by ordinary objects is quantized in integral multiples of the magnitude of this free charge. The electron carries one unit of negative charge ($q_e = -e$) and the proton carries one unit of positive charge ($q_p = +e$).

In an isolated system, the charge stays constant; in a closed system, an amount of unbalanced charge can neither be created nor destroyed. Charge can only be transferred from one object to another.

Consider two objects with charges q_1 and q_2, separated by a distance $r_{1,2}$ in vacuum. By experimental observation, the two objects repel each other if they are both positively or negatively charged (Figure 8.4a). They attract each other if they are oppositely charged (Figure 8.4b). The force exerted on object 2 due to the interaction between 1 and 2 is given by Coulomb's Law,

$$\vec{\mathbf{F}}^{E}_{1,2} = k_e \frac{q_1 q_2}{r_{1,2}^2} \hat{\mathbf{r}}_{1,2} \tag{8.2.5}$$

where $\hat{\mathbf{r}}_{1,2} = \vec{\mathbf{r}}_{1,2} / |\vec{\mathbf{r}}_{1,2}|$ is a unit vector directed from object 1 to object 2, and in SI units, $k_e = 8.9875 \times 10^{9}\ \mathrm{N \cdot m^2 \cdot C^{-2}}$, as illustrated in the Figure 8.4a. This law was derived empirically by Charles Augustin de Coulomb in the late 18th century by the same methods as described in previous sections.

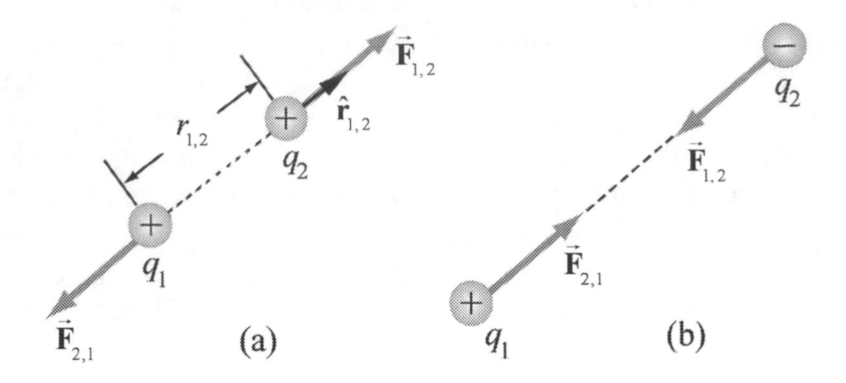

Figure 8.4 (a) and 8.4 (b) Coulomb interaction between two charges

Example 8.1 Coulomb's Law and the Universal Law of Gravitation

Show that Both Coulomb's Law and the Universal Law of Gravitation satisfy Newton's Third Law.

Solution: To see this, interchange 1 and 2 in the Universal Law of Gravitation to find the force on object 1 due to the interaction between the objects. The only quantity to change sign is the unit vector

$$\hat{\mathbf{r}}_{2,1} = -\hat{\mathbf{r}}_{1,2}. \tag{8.2.6}$$

Then

$$\vec{\mathbf{F}}^{G}_{2,1} = -G\frac{m_2 m_1}{r^2_{2,1}}\hat{\mathbf{r}}_{2,1} = G\frac{m_1 m_2}{r^2_{1,2}}\hat{\mathbf{r}}_{1,2} = -\vec{\mathbf{F}}^{G}_{1,2}. \tag{8.2.7}$$

Coulomb's Law also satisfies Newton's Third Law since the only quantity to change sign is the unit vector, just as in the case of the Universal Law of Gravitation.

8.3 Contact Forces

Pushing, lifting and pulling are ***contact forces*** that we experience in the everyday world. Rest your hand on a table; the atoms that form the molecules that make up the table and your hand are in contact with each other. If you press harder, the atoms are also pressed closer together. The electrons in the atoms begin to repel each other and your hand is pushed in the opposite direction by the table.

According to Newton's Third Law, the force of your hand on the table is equal in magnitude and opposite in direction to the force of the table on your hand. Clearly, if you push harder the force increases. Try it! If you push your hand straight down on the table, the table pushes back in a direction perpendicular (normal) to the surface. Slide your hand gently forward along the surface of the table. You barely feel the table pushing upward, but you do feel the friction acting as a resistive force to the motion of your hand. This force acts tangential to the surface and opposite to the motion of your hand. Push downward and forward. Try to estimate the magnitude of the force acting on your hand.

The force of the table acting on your hand, $\vec{\mathbf{F}}^{C} \equiv \vec{\mathbf{C}}$, is called the ***contact force.*** This force has both a normal component to the surface, $\vec{\mathbf{N}}$, called the ***normal force,*** and a tangential component to the surface, $\vec{\mathbf{f}}$, called the ***frictional force*** (Figure 8.5).

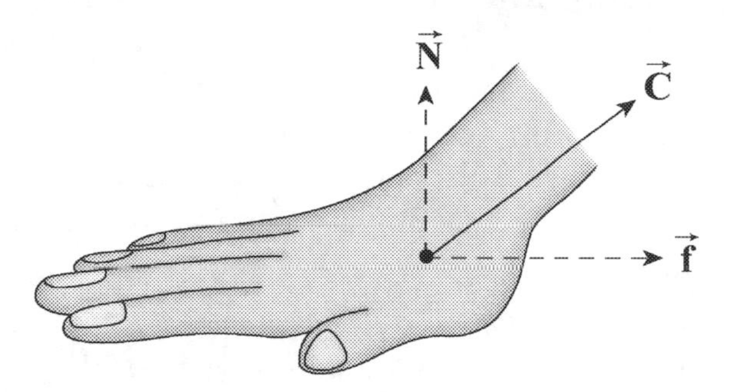

Figure 8.5 Normal and tangential components of the contact force

By the law of vector decomposition for forces,

$$\vec{\mathbf{C}} \equiv \vec{\mathbf{N}} + \vec{\mathbf{f}}. \tag{8.3.1}$$

Any force can be decomposed into component vectors so the normal component, $\vec{\mathbf{N}}$, and the tangential component, $\vec{\mathbf{f}}$, are not independent forces but the vector components of the contact force perpendicular and parallel to the surface of contact.

In Figure 8.6, the forces acting on your hand are shown. These forces include the contact force, $\vec{\mathbf{C}}$, of the table acting on your hand, the force of your forearm, $\vec{\mathbf{F}}_{\mathrm{F,H}} \equiv \vec{\mathbf{F}}_{forearm}$, acting on your hand (which is drawn at an angle indicating that you are pushing down on your hand as well as forward), and the gravitational interaction, $\vec{\mathbf{F}}^{G}_{\mathrm{E,H}}$, between the earth and your hand.

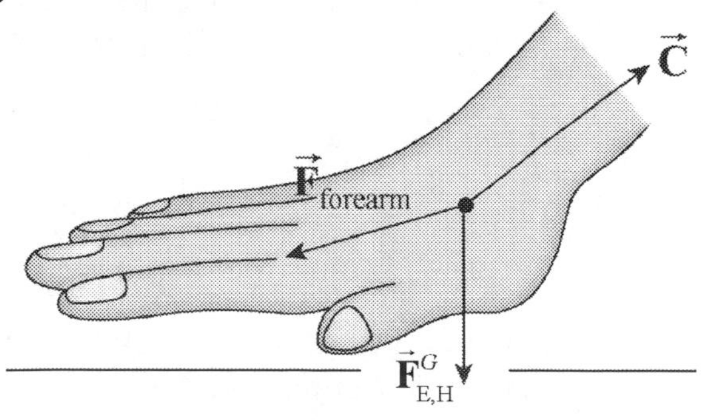

Figure 8.6 Forces on hand when moving towards the left

Is there a force law that mathematically describes this contact force? Since there are so many individual electrons interacting between the two surfaces, it is unlikely that we can add up all the individual forces. So we must content ourselves with a macroscopic model for the force law describing the contact force. One point to keep in mind is that the magnitudes of the two components of the contact force depend on how hard you push or pull your hand.

8.3.1 Free-body Force Diagram

When we try to describe forces acting on a collection of objects we must first take care to specifically define the collection of objects that we are interested in, which define our ***system.*** Often the system is a single isolated object but it can consist of multiple objects. Because force is a vector, the force acting on the system is a vector sum of the individual forces acting on the system

$$\vec{\mathbf{F}} = \vec{\mathbf{F}}_1 + \vec{\mathbf{F}}_2 + \cdots \tag{8.3.2}$$

A ***free-body force diagram*** is a representation of the sum of all the forces that act on a single system. We denote the system by a large circular dot, a "point". (Later on in the course we shall see that the "point" represents the center of mass of the system.) We represent each force that acts on the system by an arrow (indicating the direction of that force). We draw the arrow at the "point" representing the system. For example, the forces that regularly appear in free-body diagram are contact forces, tension, gravitation, friction, pressure forces, spring forces, electric and magnetic forces, which we shall introduce below. Sometimes we will draw the arrow representing the actual point in the system where the force is acting. When we do that, we will not represent the system by a "point" in the free-body diagram.

Suppose we choose a Cartesian coordinate system, then we can resolve the force into its component vectors

$$\vec{\mathbf{F}} = F_x\,\hat{\mathbf{i}} + F_y\,\hat{\mathbf{j}} + F_z\,\hat{\mathbf{k}} \tag{8.3.3}$$

Each one of the component vectors is itself a vector sum of the individual component vectors from each contributing force. We can use the free-body force diagram to make these vector decompositions of the individual forces. For example, the x-component of the force is

$$F_x = F_{1,x} + F_{2,x} + \cdots . \tag{8.3.4}$$

Example 8.2 Tug-of-War

Two people, A and B, are competing in a tug-of-war (Figure 8.7). Person A is stronger, but neither person is moving because of friction. Draw separate free-body diagrams for each person (A and B), and for the rope. For each force on your free-body diagrams, describe the action-reaction force associated with it.

Figure 8.7 Tug-of-war

Solution: The forces acting on A are (Figure 8.8):

1. The force $\vec{\mathbf{F}}_{\mathrm{R,A}}$ between A and the rope.
2. The gravitational force $\vec{\mathbf{F}}^{G}_{\mathrm{E,A}} \equiv m_{\mathrm{A}}\,\vec{\mathbf{g}}$ between A and the earth.
3. The contact force $\vec{\mathbf{C}}_{\mathrm{G,A}}$ between A and the ground.

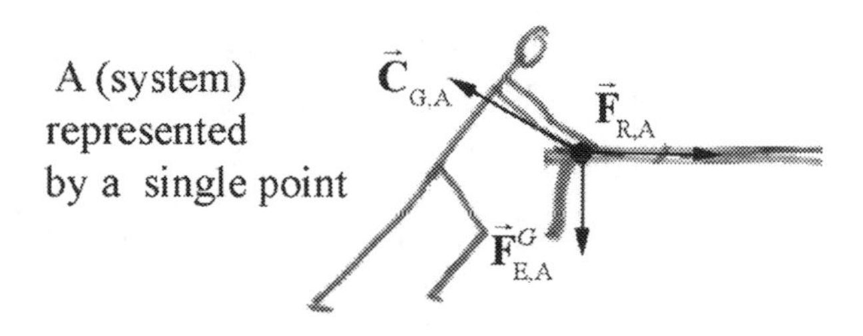

Figure 8.8 Free-body force diagram on A

The forces acting on B are (Figure 8.9):

1. The force $\vec{\mathbf{F}}_{\mathrm{R,B}}$ between B and the rope.
2. The gravitational force $\vec{\mathbf{F}}^{G}_{\mathrm{E,B}} \equiv m_{\mathrm{B}}\,\vec{\mathbf{g}}$ between B and the earth.
3. The contact force $\vec{\mathbf{C}}_{\mathrm{G,B}}$ between B and the ground.

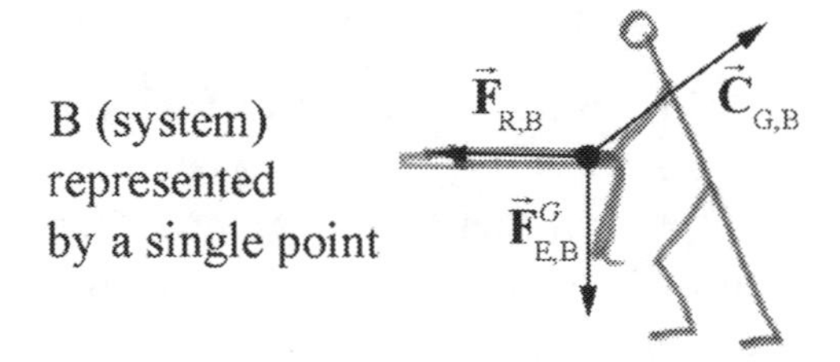

Figure 8.9 Free-body force diagram on B

The forces on the rope are (Figure 8.10)

1. The force $\vec{\mathbf{F}}_{\text{A,R}}$ between the rope and A.

2. The force $\vec{\mathbf{F}}_{\text{B,R}}$ between the rope and B.

We neglect the gravitational force on the rope (massless rope assumption).

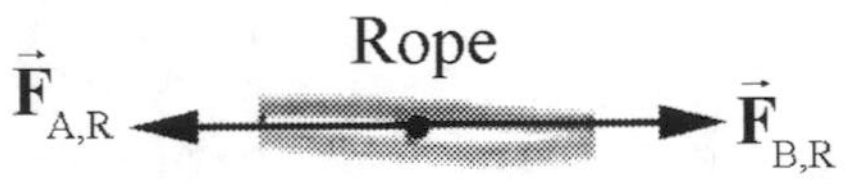

Rope

$\vec{\mathbf{F}}_{\text{A,R}}$ $\vec{\mathbf{F}}_{\text{B,R}}$

Figure 8.10 Free-body force diagram on rope

Because the rope is not moving, these forces sum to zero by Newton's Second Law but they are not Third Law pairs;

$$\vec{\mathbf{F}}_{\text{A,R}} + \vec{\mathbf{F}}_{\text{B,R}} = \vec{\mathbf{0}}.$$

The Newton's Third Law pairs are

1. The forces between A and the rope with $\vec{\mathbf{F}}_{\text{A,R}} = -\vec{\mathbf{F}}_{\text{R,A}}$.
2. The forces between B and the rope with $\vec{\mathbf{F}}_{\text{B, R}} = -\vec{\mathbf{F}}_{\text{R, B}}$.
3. The contact forces between A and the ground, $\vec{\mathbf{C}}_{\text{A, G}} = -\vec{\mathbf{C}}_{\text{G, A}}$. The contact force $\vec{\mathbf{C}}_{\text{A, G}}$ of A on the ground is not shown on the above force diagrams.
4. The contact forces between B and the ground, $\vec{\mathbf{C}}_{\text{B, G}} = -\vec{\mathbf{C}}_{\text{G, B}}$. The contact force $\vec{\mathbf{C}}_{\text{B, G}}$ of Team B on the ground is not shown on the above force diagrams.
5. The gravitational forces between the earth and A, $\vec{\mathbf{F}}^{G}_{\text{E,A}} = -\vec{\mathbf{F}}^{G}_{\text{A,E}}$. The gravitational force $\vec{\mathbf{F}}^{G}_{\text{A,E}}$ on the earth is not shown in the figures above.
6. The gravitational forces between the earth and B, $\vec{\mathbf{F}}^{G}_{\text{E,B}} = -\vec{\mathbf{F}}^{G}_{\text{B,E}}$. The gravitational force $\vec{\mathbf{F}}^{G}_{\text{B,E}}$ on the earth is not shown in the figures above.

8.3.3 Normal Component of the Contact Force and Weight

Hold an object in your hand. You can feel the "weight" of the object against your palm. But what exactly do we mean by "weight"? Consider the force diagram on the object in Figure 8.11. Let's choose the +-direction to point upward.

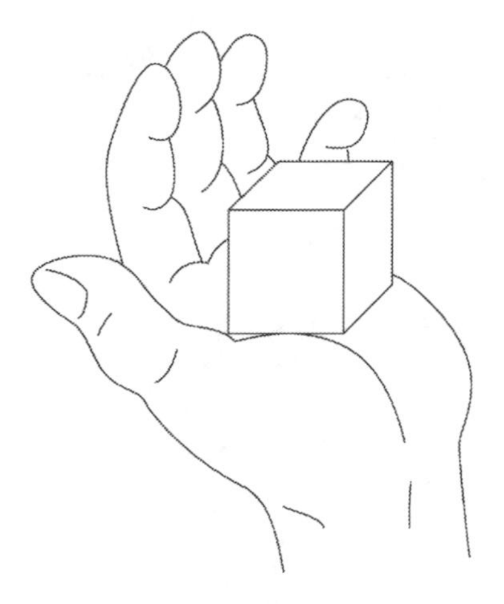

Figure 8.11 Object resting in hand

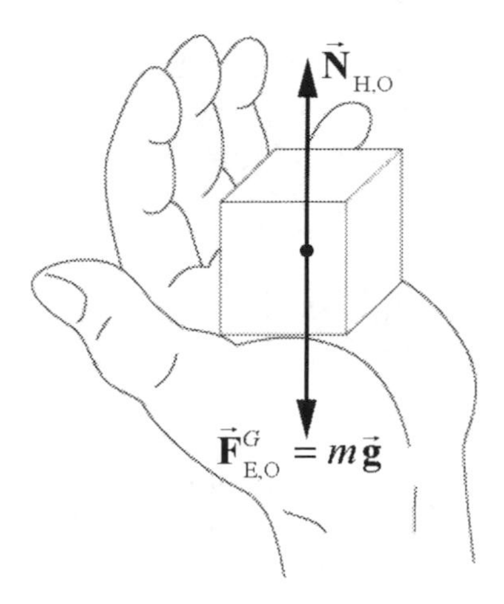

Figure 8.12 Force diagram on object

There are two forces acting on the object. One force is the gravitational force between the earth and the object, and is denoted by $\vec{\mathbf{F}}^{G}_{\mathrm{E,O}} = m\vec{\mathbf{g}}$ where $\vec{\mathbf{g}}$, also known as the gravitational acceleration, is a vector that points downward and has magnitude $g = 9.8\ \mathrm{m \cdot s^{-2}}$. The other force on the object is the contact force between your hand and the object. Because we are not pushing the block horizontally, this contact force on your hand points perpendicular to the surface, and hence has only a normal component, $\vec{\mathbf{N}}_{\mathrm{H,O}}$. Let N denote the magnitude of the normal force. The force diagram on the object is shown in Figure 8.12. Because the object is at rest in your hand, the vertical acceleration is zero. Therefore Newton's Second Law states that

$$\vec{\mathbf{F}}^{G}_{\mathrm{E,O}} + \vec{\mathbf{N}}_{\mathrm{H,O}} = \vec{\mathbf{0}}. \qquad (8.3.5)$$

Choose the positive direction to be upwards, then in terms of vertical components we have that

$$N - mg = 0, \qquad (8.3.6)$$

which can be solved for the magnitude of the normal force

$$N = mg. \qquad (8.3.7)$$

This result may give rise to a misconception that the normal force is always equal to the mass of the object times the magnitude of the gravitational acceleration at the surface of the earth. The normal force and the gravitational force are two completely different forces. In this particular example, the normal force is equal in magnitude to the gravitational force and directed in the opposite direction, which sounds like an example of the Third Law. But is it? No!

In order to see all the action-reaction pairs we must consider all the objects involved in the interaction. The extra object is your hand. The force diagram on your hand is shown in Figure 8.13.

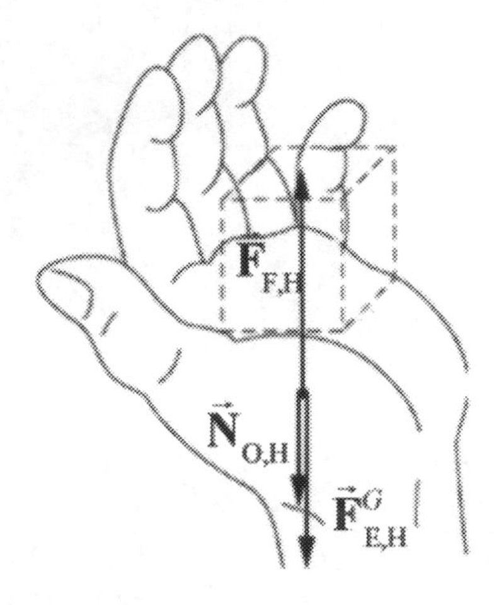

Figure 8.13 Free-body force diagram on hand

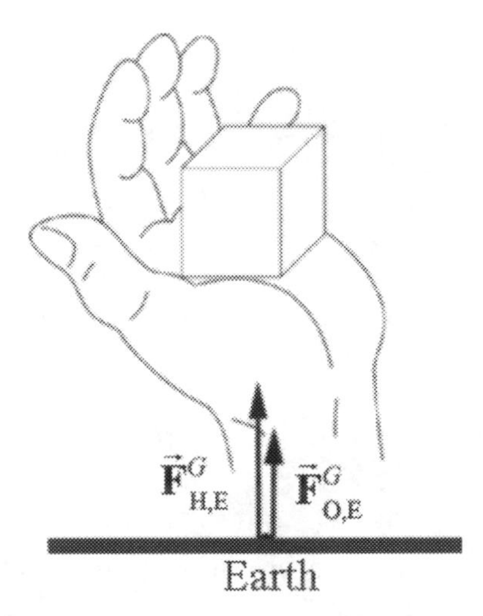

Figure 8.14 Gravitational forces on earth due to object and hand

The forces shown include the gravitational force between your hand and the earth, $\vec{\mathbf{F}}^{G}_{E,H}$ that points down, the normal force between the object and your hand, $\vec{\mathbf{N}}_{O,H}$, which also points down, and there is a force $\vec{\mathbf{F}}_{F,H}$ applied by your forearm to your hand that holds your hand up. There are also forces on the earth due to the gravitational interaction between the hand and object and earth. We show these forces in Figure 8.14: the gravitational force between the earth and your hand, $\vec{\mathbf{F}}^{G}_{H,E}$, and the gravitational force between the earth and the object, $\vec{\mathbf{F}}^{G}_{O,E}$.

There are three Third Law pairs. The first is associated with the normal force,

$$\vec{\mathbf{N}}_{O,H} = -\vec{\mathbf{N}}_{H,O}. \tag{8.3.8}$$

The second is the gravitational force between the mass and the earth,

$$\vec{\mathbf{F}}^{G}_{E,O} = -\vec{\mathbf{F}}^{G}_{O,E}. \tag{8.3.9}$$

The third is the gravitational force between your hand and the earth,

$$\vec{\mathbf{F}}^{G}_{E,H} = -\vec{\mathbf{F}}^{G}_{H,E}. \tag{8.3.10}$$

As we see, none of these three law pairs associates the "weight" of the block on the hand with the force of gravity between the block and the earth.

When we talk about the "weight" of an object, we often are referring to the effect the object has on a scale or on the feeling we have when we hold the object. These effects are actually effects of the normal force. We say that an object "feels lighter" if there is an additional force holding the object up. For example, you can rest an object in your hand, but use your other hand to apply a force upwards on the object to make it feel lighter in your supporting hand.

This leads us to the use of the word "weight," which is often used in place of the gravitational force that the earth, exerts on an object, and we will always refer to this force as the *gravitational force* instead of "weight." When you jump in the air, you feel "weightless" because there is no normal force acting on you, even though the earth is still exerting a gravitational force on you; clearly, when you jump, you do not turn gravity off! When astronauts are in orbit around the earth, televised images show the astronauts floating in the spacecraft cabin; the condition is described, rightly, as being "weightless." The gravitational force, while still present, has diminished slightly since their distance from the center of the earth has increased.

8.3.4 Static and Kinetic Friction

There are two distinguishing types of friction when surfaces are in contact with each other. The first type is when the two objects in contact are moving relative to each other; the friction in that case is called ***kinetic friction*** or ***sliding friction,*** and denoted by $\vec{\mathbf{f}}^k$.

Based on experimental measurements, the force of kinetic friction, $\vec{\mathbf{f}}^k$, between two surfaces, is independent of the relative speed of the surfaces, the area of contact, and only depends on the magnitude of the normal component of the contact force. The force law for the magnitude of the kinetic frictional force between the two surfaces can be modeled by

$$f_{\mathrm{k}} = \mu_{\mathrm{k}} N, \tag{8.3.11}$$

where μ_{k} is called the ***coefficient of kinetic friction***. The direction of kinetic friction on surface A due to the contact with a second surface B, $\vec{\mathbf{f}}^k_{\mathrm{B,A}}$, is always opposed to the relative direction of motion of surface A with respect to the surface B.

The second type of friction is when the two surfaces are static relative to each other; the friction in that case is called ***static friction***, and denoted by $\vec{\mathbf{f}}^s$. Push your hand forward along a surface; as you increase your pushing force, the frictional force feels stronger and stronger. Try this! Your hand will at first stick until you push hard enough, then your hand slides forward. The magnitude of the static frictional force, f_{s}, depends on how hard you push.

If you rest your hand on a table without pushing horizontally, the static friction is zero. As you increase your push, the static friction increases until you push hard enough that your hand slips and starts to slide along the surface. Thus the magnitude of static friction can vary from zero to some maximum value, $(f_s)_{max}$, when the pushed object begins to slip,

$$0 \le f_s \le (f_s)_{max} . \qquad (8.3.12)$$

Is there a mathematical model for the magnitude of the maximum value of static friction between two surfaces? Through experimentation, we find that this magnitude is, like kinetic friction, proportional to the magnitude of the normal force

$$(f_s)_{max} = \mu_s N . \qquad (8.3.13)$$

Here the constant of proportionality is μ_s, the ***coefficient of static friction.*** This constant is slightly greater than the constant μ_k associated with kinetic friction, $\mu_s > \mu_k$. This small difference accounts for the slipping and catching of chalk on a blackboard, fingernails on glass, or a violin bow on a string.

The direction of static friction on an object is always opposed to the direction of the applied force (as long as the two surfaces are not accelerating). In Figure 8.15a, the static friction, $\vec{\mathbf{f}}^s$, is shown opposing a pushing force, $\vec{\mathbf{F}}_{B,O}^{push}$, acting on an object. In Figure 8.15b, when a pulling force, $\vec{\mathbf{F}}_{B,O}^{pull}$, is acting on an object, static friction, $\vec{\mathbf{f}}^s$, is now pointing the opposite direction from the pulling force.

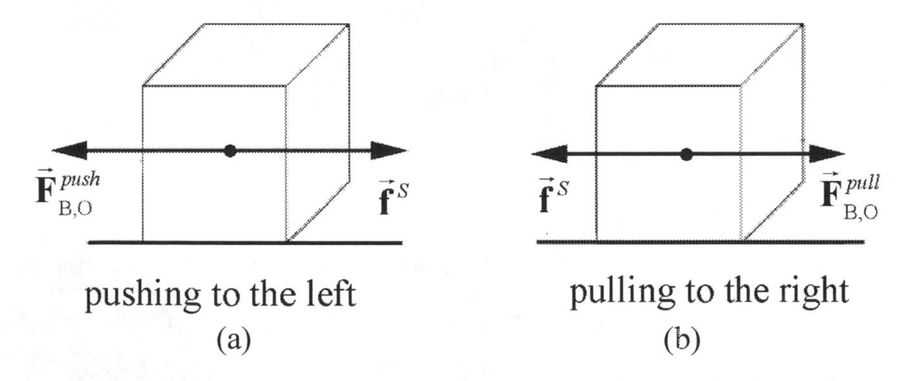

Figure 8.15 (a) and (b): Pushing and pulling forces and the direction of static friction.

Although the force law for the maximum magnitude of static friction resembles the force law for sliding friction, there are important differences:

1. The direction and magnitude of static friction on an object always depends on the direction and magnitude of the applied forces acting on the object, where the magnitude of kinetic friction for a sliding object is fixed.

2. The magnitude of static friction has a maximum possible value. If the magnitude of the applied force along the direction of the contact surface exceeds the magnitude of the maximum value of static friction, then the object will start to slip (and be subject to kinetic friction.) We call this the ***just slipping*** condition.

8.3.5 Modeling

One of the most central and yet most difficult tasks in analyzing a physical interaction is developing a physical model. A physical model for the interaction consists of a description of the forces acting on all the objects. The difficulty arises in deciding which forces to include. For example in describing almost all planetary motions, the Universal Law of Gravitation was the only force law that was needed. There were anomalies, for example the small shift in Mercury's orbit. These anomalies are interesting because they may lead to new physics. Einstein corrected Newton's Law of Gravitation by introducing General Relativity and one of the first successful predictions of the new theory was the perihelion precession of Mercury's orbit. On the other hand, the anomalies may simply be due to the complications introduced by forces that are well understood but complicated to model. When objects are in motion there is always some type of friction present. Air friction is often neglected because the mathematical models for air resistance are fairly complicated even though the force of air resistance substantially changes the motion. Static or kinetic friction between surfaces is sometimes ignored but not always. The mathematical description of the friction between surfaces has a simple expression so it can be included without making the description mathematically intractable. A good way to start thinking about the problem is to make a simple model, excluding complications that are small order effects. Then we can check the predictions of the model. Once we are satisfied that we are on the right track, we can include more complicated effects.

8.4 Tension in a Rope

How do we define "tension" in a rope? Let's return to our example of the very light rope ($m_{\text{R}} \simeq 0$) that is attached to a block B of mass m_{B} on one end, and pulled by an applied force by another object A, $\vec{\mathbf{F}}_{\text{A,R}}$, from the other end (Figure 8.16). Let's assume that the block and rope are both at rest.

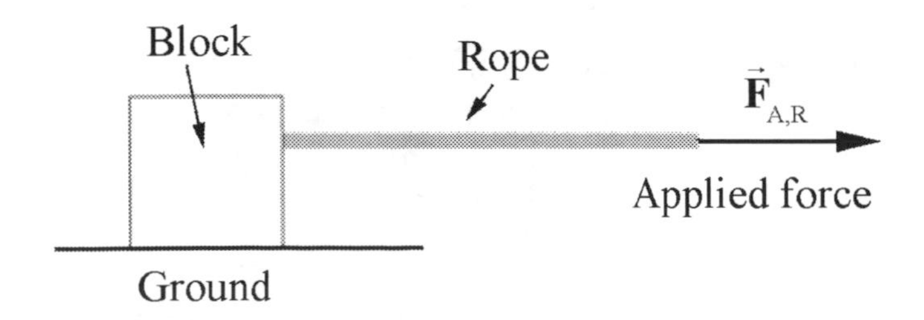

Figure 8.16 Forces acting on block and rope

Let's choose an xy-coordinate system with the $\hat{\mathbf{j}}$-unit vector pointing upward in the normal direction to the surface, and the $\hat{\mathbf{i}}$-unit vector pointing in the direction of the motion of the block. The force diagrams for the rope and block are shown in Figure 8.17.

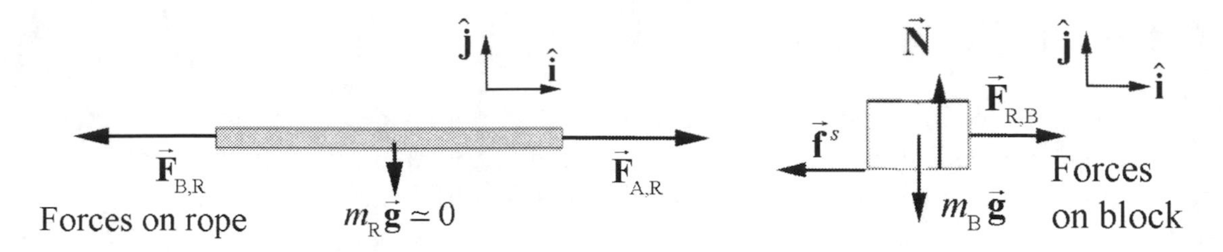

Figure 8.17 Force diagram for rope and block

The forces on the rope and the block must each sum to zero. Because the rope is not accelerating, Newton's Second Law becomes:

$$F_{\mathrm{A,R}} - F_{\mathrm{B,R}} = 0\,. \tag{8.4.1}$$

The static equilibrium conditions for the block are:

in the $+\hat{\mathbf{i}}$-direction:

$$F_{\mathrm{R,B}} - f_{\mathrm{s}} = 0\,, \tag{8.4.2}$$

in the $+\hat{\mathbf{j}}$ -direction:

$$N - m_{\mathrm{B}}g = 0\,. \tag{8.4.3}$$

We now apply Newton's Third Law, the action-reaction law,

$$\vec{\mathbf{F}}_{\mathrm{B,R}} = -\vec{\mathbf{F}}_{\mathrm{R,B}}\,, \tag{8.4.4}$$

which becomes, in terms of our magnitudes,

$$F_{\mathrm{B,R}} = F_{\mathrm{R,B}}\,. \tag{8.4.5}$$

Note there is no minus sign in Equation (8.4.5) because these are magnitudes. The sign is built into Eqs. (8.4.1) and (8.4.2). Our static equilibrium conditions now become

$$F_{\mathrm{A,R}} = F_{\mathrm{B,R}} = F_{\mathrm{R,B}} = f_{\mathrm{s}}\,. \tag{8.4.6}$$

Thus the applied pulling force is transmitted through the rope to the object since it has the same magnitude as the force of the rope on the object.

$$F_{\mathrm{A,R}} = F_{\mathrm{R,B}}\,. \tag{8.4.7}$$

In addition we see that the applied force is equal to the static friction,

$$F_{\text{A,R}} = f_s. \tag{8.4.8}$$

8.4.1 Static Tension in a Rope

We have seen that in static equilibrium the pulling force by another object A, transmits through the rope. Suppose we make an imaginary slice of the rope at a distance x from the end that the object is attached to the object (Figure 8.18).

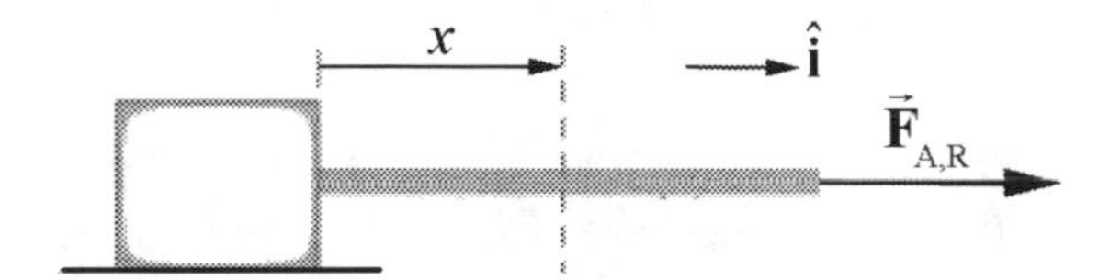

Figure 8.18 Imaginary slice through the rope

The rope is now divided into two sections, labeled left and right. Aside from the Third Law pair of forces between the object and the rope, there is now a Third Law pair of forces between the left section of the rope and the right section of the rope. We denote this force acting on the left section by $\vec{\mathbf{F}}_{\text{R,L}}(x)$. The force on the right section due to the left section is denoted by $\vec{\mathbf{F}}_{\text{L,R}}(x)$. Newton's Third Law requires that each force in this action-reaction pair is equal in magnitude and opposite in direction.

$$\vec{\mathbf{F}}_{\text{R,L}}(x) = -\vec{\mathbf{F}}_{\text{L,R}}(x) \tag{8.4.9}$$

The force diagram for the left and right sections are shown in Figure 8.19 where $\vec{\mathbf{F}}_{\text{B,L}}$ is the force on the left-segment the rope due to the block-rope interaction.

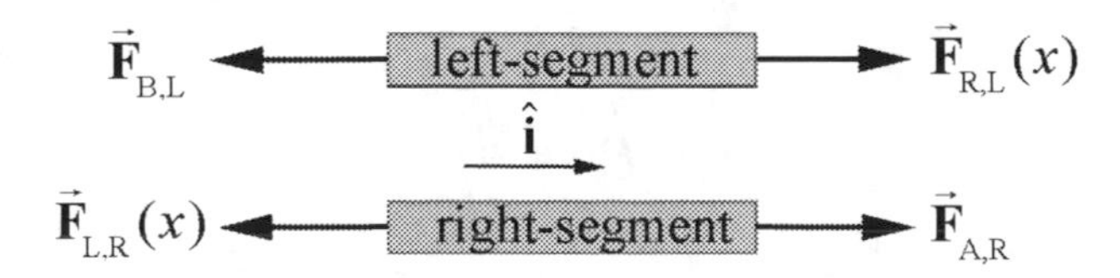

Figure 8.19 Force diagram for the left and right sections of rope

The ***tension*** $T(x)$ *in a rope at a distance* x *from one end of the rope is the magnitude of the action -reaction pair of forces acting at the point* x,

$$T(x) = \left|\vec{\mathbf{F}}_{\text{R,L}}(x)\right| = \left|\vec{\mathbf{F}}_{\text{L,R}}(x)\right|. \tag{8.4.10}$$

Special case: For a rope of negligible mass in static equilibrium, the tension is uniform and is equal to the applied force,

$$T = F_{\text{A,R}} . \tag{8.4.11}$$

Example 8.3 Tension in a Massive Rope

Suppose a uniform rope of mass m and length d is attached to an object of mass m_{B} lying on a table. The rope is pulled from the side opposite the block with an applied force of magnitude $\left|\vec{\mathbf{F}}_{\text{A,R}}\right| = F_{\text{A,R}}$. The coefficient of kinetic friction between the block and the surface is μ_k. Find the tension in the rope as a function of distance from the block.

Solution: The key point to realize is that the rope is now massive. Suppose we make an imaginary slice of the rope at a distance x from the end that the object as we did in Figure 8.18. The mass of the right slice is given by

$$m_{right} = \frac{m}{d}(d - x) . \tag{8.4.12}$$

The mass of the left slice is

$$m_{left} = \frac{m}{d}x . \tag{8.4.13}$$

We now apply Newton's Second Law to the right slice of the rope

$$F_{\text{A,R}} - T(x) = \frac{m}{d}(d - x)a_{\text{R},x} , \tag{8.4.14}$$

where $T(x)$ is the tension in the rope at a distance x from the object and $a_{\text{R},x}$ is the x-component of the acceleration of the right piece of the rope. We also apply Newton's Second Law to the left slice of the rope

$$T(x) - F_{\text{B,L}} = \frac{m}{d}xa_{\text{L},x} , \tag{8.4.15}$$

where $T(x)$ is the tension in the rope at a distance x from the object, $F_{\text{B,L}}$ is the magnitude of the force on the left-segment of the rope due to the rope-block interaction, and $a_{\text{L},x}$ is the x-component of the acceleration of the left piece of the rope. The force diagram is shown in Figure 8.20.

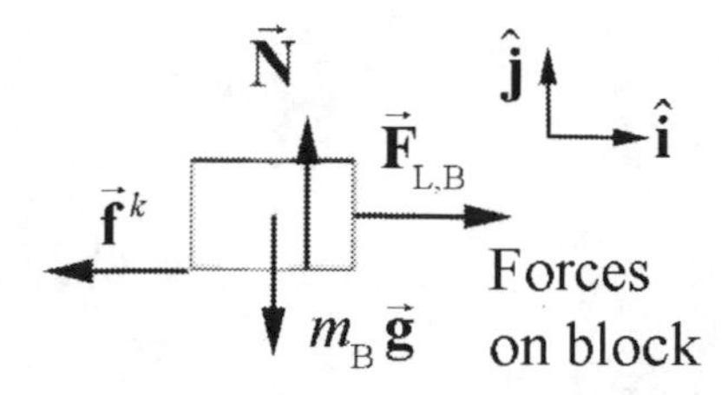

Figure 8.20 Force diagram on sliding block

Newton's Second Law on the block is now:

in the $+\hat{\mathbf{i}}$-direction:

$$F_{\mathrm{L,B}} - f_{\mathrm{k}} = m_{\mathrm{B}} a_{\mathrm{B},x}, \tag{8.4.16}$$

in the $+\hat{\mathbf{j}}$ -direction:

$$N - m_{\mathrm{B}} g = 0. \tag{8.4.17}$$

Eq. (8.4.17) implies that $N = m_{\mathrm{B}} g$ and so the kinetic friction force acting on the block is

$$f_k = \mu_k N = \mu_k m_{\mathrm{B}} g. \tag{8.4.18}$$

We now substitute Eq. (8.4.18) into Eq. (8.4.16), which becomes

$$F_{\mathrm{L,B}} = \mu_k m_{\mathrm{B}} g + m_{\mathrm{B}} a_{\mathrm{B},x}, \tag{8.4.19}$$

Newton's Third Law for the block-rope interaction is given by

$$F_{\mathrm{L,B}} = F_{\mathrm{B,L}}. \tag{8.4.20}$$

We substitute Eq. (8.4.20) into Eq. (8.4.19) and then into Eq. (8.4.15) yielding

$$T(x) - (\mu_k m_{\mathrm{B}} g + m_{\mathrm{B}} a_{\mathrm{B},x}) = \frac{m}{d} x a_{\mathrm{L},x}. \tag{8.4.21}$$

Because the rope and block move together, the accelerations are equal which we denote by the symbol a

$$a \equiv a_{\mathrm{B},x} = a_{\mathrm{L},x}. \tag{8.4.22}$$

Then Eq. (8.4.21) becomes

$$T(x) = \mu_k m_{\mathrm{B}} g + (m_{\mathrm{B}} + \frac{m}{d} x) a. \tag{8.4.23}$$

Check our result: We could have used Eq. (8.4.14) to find the tension

$$T(x) = F_{\text{A,R}} - \frac{m}{d}(d - x)a. \tag{8.4.24}$$

The force diagram on the system consisting of the rope-block of the rope is shown in Figure 8.21.

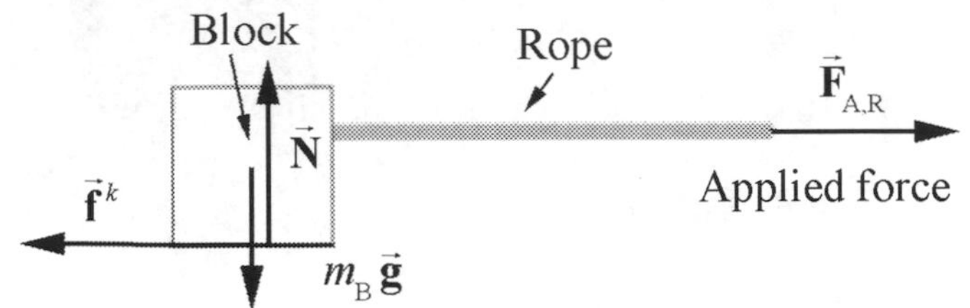

Figure 8.21 Force diagram on block-rope system

Therefore Newton's Second Law becomes

$$F_{\text{A,R}} - f_k = (m + m_{\text{B}})a. \tag{8.4.25}$$

Therefore substitute Eq. (8.4.25) into Eq. (8.4.24) yielding

$$T(x) = f_k + (m + m_{\text{B}})a - \frac{m_r}{d}(d - x)a = f_k + (m_{\text{B}} + \frac{m}{d}x)a. \tag{8.4.26}$$

in agreement with Eq. (8.4.23). We expect this result because the tension is accelerating both the left slice and the block and is opposed by the frictional force.

Example 8.4 Tension in a Suspended Rope

Suppose a uniform rope of mass M and length L is suspended from a ceiling (Figure 8.22). The magnitude of the acceleration due to gravity is g. (a) Find the tension in the rope at the upper end where the rope is fixed to the ceiling. (b) Find the tension in the rope as a function of the distance from the ceiling. (c) Find an equation for the rate of change of the tension with respect to distance from the ceiling in terms of M, L, and g. Be sure you show any free-body force diagrams or sketches that you plan to use.

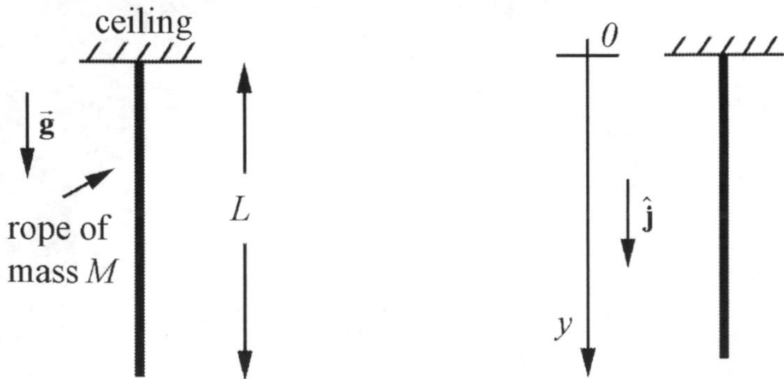

Figure 8.22 Rope suspended from ceiling **Figure 8.23** Coordinate system for suspended rope

Solution: (a) We begin by choosing a coordinate system with the origin at the ceiling and the positive y-direction pointing downward (Figure 8.23). In order to find the tension at then upper end of the rope, we draw a free-body diagram of the entire rope. The forces acting on the rope are the force at $y = 0$ holding the rope up, (we refer to that force as $T(y = 0)$, the tension at the upper end) and the gravitational force on the entire rope $Mg\,\hat{\mathbf{j}}$. The free-body force diagram is shown in Figure 8.24.

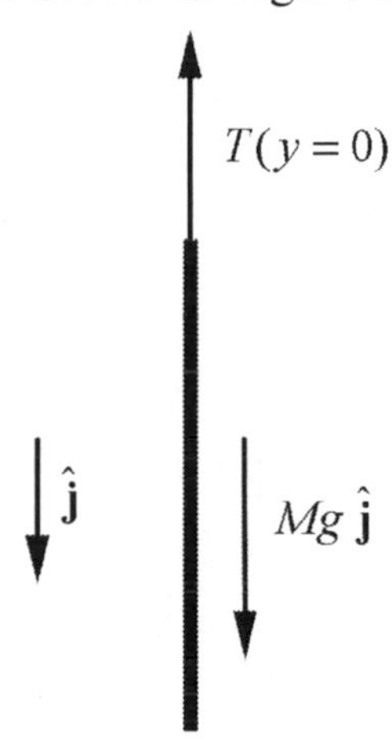

Figure 8.24 Force diagram on rope

We now apply Newton's Second Law noting that the acceleration is zero

$$Mg - T(y = 0) = 0 .$$

Thus we can solve for $T(y = 0)$, the tension at the upper end,

$$T(y = 0) = Mg .$$

(b) Recall that the tension at a point is the magnitude of the action-reaction pair of forces acting at that point. So we make an imaginary cut in the rope a distance y from the ceiling separating the rope into an upper and lower piece (Figure 8.25). We choose the upper piece as our system with mass $m = (M / L)y$. The forces acting on the upper rope are the gravitational force $mg\,\hat{\mathbf{j}} = (M / L)yg\hat{\mathbf{j}}$, the force at $y = 0$ holding the rope up, (we refer to that force as $T(y = 0)$, the tension at the upper end), and the tension at the point y, $T(y)$ that is pulling the upper piece down. The free-body force diagram is shown in Figure 8.26.

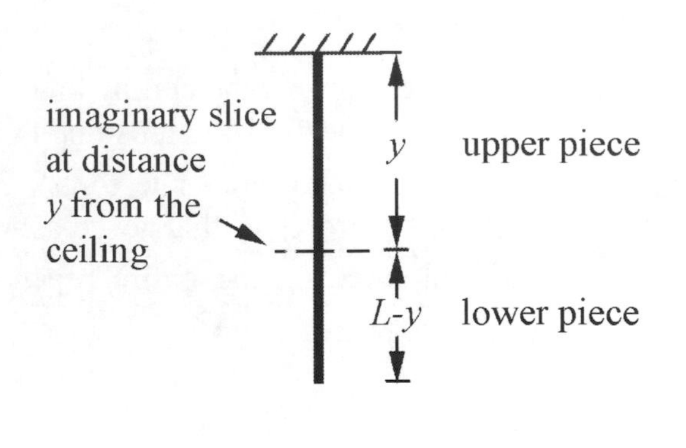

Figure 8.25 Imaginary slice separates rope into two pieces

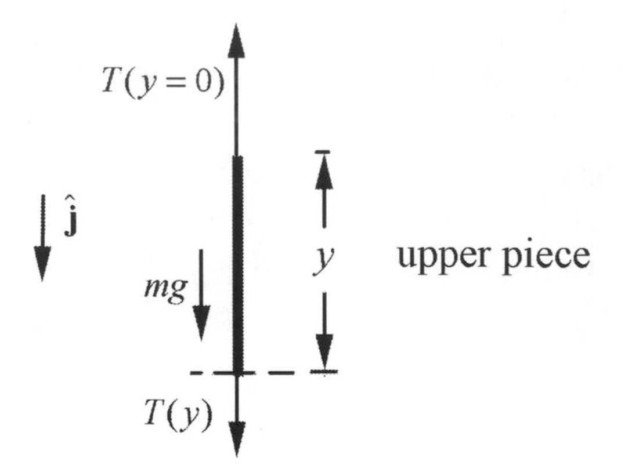

Figure 8.26 Free-body force diagram on upper piece of rope

We now apply Newton's Second Law noting that the acceleration is zero

$$mg + T(y) - T(y = 0) = 0.$$

Thus we can solve for $T(y)$, the tension at a distance y from the ceiling,

$$T(y) = T(y = 0) - mg.$$

Using our results for the mass of the upper piece and the tension at the upper end we have that

$$T(y) = Mg(1 - y / L) \tag{8.4.27}$$

As a check, we note that when $y = L$, the tension $T(y = L) = 0$ which is what we expect because there is no force acting at lower end of the rope.

(c) We can differentiate Eq. (8.4.27) with respect to y and find that

$$\frac{dT}{dy}(y) = -(M / L)g. \tag{8.4.28}$$

So the rate that the tension is changing is constant.

8.4.2 Continuous Systems and Newton's Second Law as a Differential Equations

We can determine the tension at a distance y from the ceiling in Example 8.4, by an alternative method, a technique that will generalize to many types of "continuous systems". We consider as our system an imaginary segment of the rope between the points y and $y + \Delta y$. This small element has length Δy and mass $\Delta m = (M / L)\Delta y$.

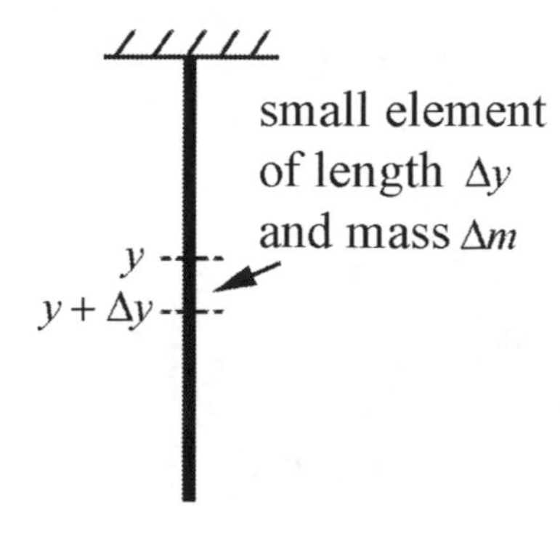

Figure 8.27 Rope with small element identified

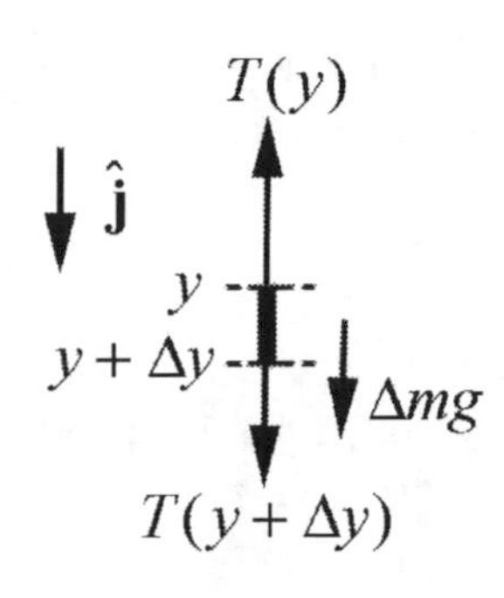

Figure 8.28 Infinitesimal slice of rope

The rope has now been divided into three pieces, an upper piece, the small element, and a lower piece (Figure 8.27). The forces acting on it are the tension $T(y)$ at y directed upward, (the force of the upper piece holding the element up), the tension $T(y+\Delta y)$ at $y+\Delta y$ directed downward (the force of the lower piece pulling the element down), and the gravitational force $\Delta mg\,\hat{\mathbf{j}} = (M/L)\Delta yg\hat{\mathbf{j}}$. The free-body force diagram is shown in Figure 8.28.

We now apply Newton's Second Law to the small element noting that the acceleration is zero

$$\Delta mg + T(y) - T(y+\Delta y) = 0 .$$

We now solve for the difference in the tension

$$T(y+\Delta y) - T(y) = -\Delta mg .$$

We now substitute our result for the mass of the element $\Delta m = (M/L)\Delta y$, and find that that

$$T(y+\Delta y) - T(y) = -(M/L)\Delta yg .$$

We now divide through by Δy yielding

$$\frac{T(y+\Delta y) - T(y)}{\Delta y} = -(M/L)g .$$

Now here comes the crucial step, the limiting argument. We consider the limit in which the length of the small element goes to zero, $\Delta y \to 0$.

$$\lim_{\Delta y \to 0} \frac{T(y+\Delta y) - T(y)}{\Delta y} = -(M/L)g .$$

Recall that the left hand side is the definition of the derivative of the tension with respect to y, and so we arrive at

$$\frac{dT}{dy}(y) = -(M / L)g$$

in agreement with Eq. (8.4.28).

We can solve this differential equation by a technique called ***separation of variables***. We rewrite the equation as

$$dT = -(M / L)gdy$$

and integrate both sides. Our integral will be a definite integral, the limits of the right hand side are from $y = 0$ to y, and the corresponding limits on the left hand side are from $T(y = 0)$ to $T(y)$:

$$\int_{T(y=0)}^{T(y)} dT = -(M / L)g\int_{y'=0}^{y'=y} dy'.$$

Notice that we have introduced a "dummy" integration variable y' to distinguish the integration variable from the endpoint of the integral y, the point that we are trying to find the tension, $T(y)$. We now integrate and find that

$$T(y) - T(y = 0) = -(M / L)gy .$$

We use our earlier result that $T(y = 0) = Mg$ and find that

$$T(y) = Mg(1 - y / L) .$$

in agreement with our earlier result.

8.5 Frictional Force as a Linear Function of Velocity

In many physical situations the force on an object will be modeled as depending on the object's velocity. We have already seen static and kinetic friction between surfaces modeled as being independent of the surfaces' relative velocity. Common experience (swimming, throwing a Frisbee) tells us that the frictional force between an object and a fluid can be a complicated function of velocity. Indeed, these complicated relations are an important part of such topics as aircraft design.

A reasonable model for the frictional force on an object m moving at low speeds through a viscous medium is

$$\vec{\mathbf{F}}_{\text{friction}} = -\gamma\, m\vec{\mathbf{v}} \qquad (8.5.1)$$

where γ is a constant that depends on the properties (density, viscosity) of the medium and the size and shape of the object. Note that γ has dimensions of inverse time,

$$\dim[\gamma] = \frac{\dim\left[\text{Force}\right]}{\dim[\text{mass}]\cdot\dim\left[\text{velocity}\right]} = \frac{\text{M}\cdot\text{L}\cdot\text{T}^{-2}}{\text{M}\cdot\text{L}\cdot\text{T}^{-1}} = \text{T}^{-1}. \tag{8.5.2}$$

The minus sign in Equation (8.5.1) indicates that the frictional force is directed against the object's velocity (relative to the fluid).

In a situation where $\vec{\mathbf{F}}_{\text{friction}}$ is the net force, Newton's Second Law becomes

$$-\gamma\, m\vec{\mathbf{v}} = m\vec{\mathbf{a}} \tag{8.5.3}$$

and so the acceleration is

$$\vec{\mathbf{a}} = -\gamma\,\vec{\mathbf{v}}\,. \tag{8.5.4}$$

The acceleration has no component perpendicular to the velocity, and in the absence of other forces will move in a straight line, but with varying speed. Denote the direction of this motion as the x-direction, so that Equation (8.5.4) becomes

$$a_x = \frac{d\,v_x}{dt} = -\gamma\,v_x\,. \tag{8.5.5}$$

Equation (8.5.5) is now a differential equation. For our purposes, we'll create an initial-condition problem by specifying that the initial x-component of velocity is $v(t=0) = v_{x0}$. The differential equation in (8.5.5) is also separable, in that the equation may be rewritten as

$$\frac{d\,v_x}{v_x} = -\gamma dt\;. \tag{8.5.6}$$

and each side can be separately integrated. The integration in this case is simple, leading to

$$\int_{v_{x0}}^{v_x(t)} \frac{d\,v_x}{v_x} = -\gamma\int_0^t dt\;. \tag{8.5.7}$$

The left hand side is

$$\text{LHS} = \ln(v_x)\Big|_{v_{x0}}^{v_x(t)} = \ln(v_x(t) - \ln(v_{x0}) = \ln(v_x(t)\,/\,v_{x0}), \tag{8.5.8}$$

and the right hand side is

$$\text{RHS} = -\gamma t\,. \tag{8.5.9}$$

Equating the two sides yields

$$\ln(v_x(t)/v_{x0}) = -\gamma t. \tag{8.5.10}$$

Exponentiate each side of the above equation yields

$$v_x(t)/v_{x0} = e^{-\gamma t}. \tag{8.5.11}$$

Thus the x-component of the velocity as a function of time is given by

$$v_x(t) = v_{x0} e^{-\gamma t} \tag{8.5.12}$$

A plot of v_x vs. t is shown in Figure 8.29 with initial conditions $v_{x0} = 10\ \text{m/s}$ and $\gamma = 0.5\ \text{N/J}$.

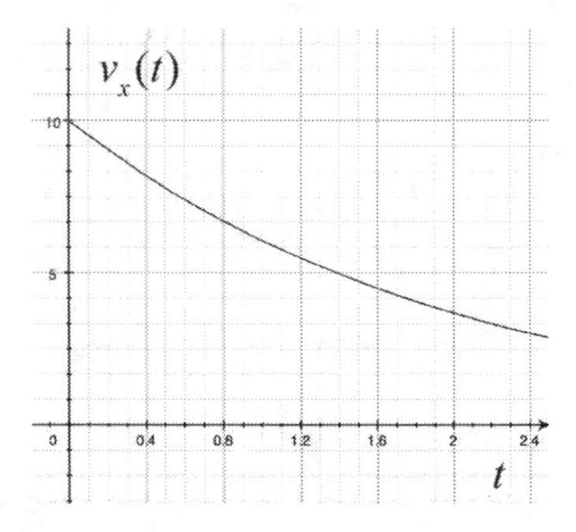

Figure 8.29 Plot of v_x vs. t for damping force $\vec{\mathbf{F}}_{\text{friction}} = -\gamma m \vec{\mathbf{v}}$

Example 8.6 An object moving the x-axis with an initial x-component of the velocity $v_x(t=0) = v_{x0}$ experiences a retarding frictional force whose magnitude is proportional to the square of the speed (a case known as ***Newtonian Damping***),

$$\left|\vec{\mathbf{F}}_{\text{friction}}\right| = \gamma m v^2 \tag{8.5.13}$$

Show that the x-component of the velocity of the object as a function of time is given by

$$v_x(t) = v_{x0} \frac{1}{1 + t/\tau}, \tag{8.5.14}$$

and find the constant τ.

Solution: Newton's Second Law can be written as

$$-\gamma v_x^2 = \frac{dv_x}{dt}. \tag{8.5.15}$$

Differentiating our possible solution yields

$$\frac{dv_x}{dt} = -v_{x0}\frac{1}{\tau}\frac{1}{\left(1+t/\tau\right)^2} = -\frac{1}{v_{x0}\tau}v_x^2 \tag{8.5.16}$$

Substituting into the Second Law yields

$$-\gamma v_x^2 = -\frac{1}{v_{x0}\tau}v_x^2. \tag{8.5.17}$$

Thus our function is a solution providing that

$$\tau = \frac{1}{v_{x0}\gamma}. \tag{8.5.18}$$

A plot of v_x vs. t is shown in Figure 8.30 with initial conditions $v_{x0} = 10\text{ m/s}$ and $\gamma = 0.5\text{ N/J}$.

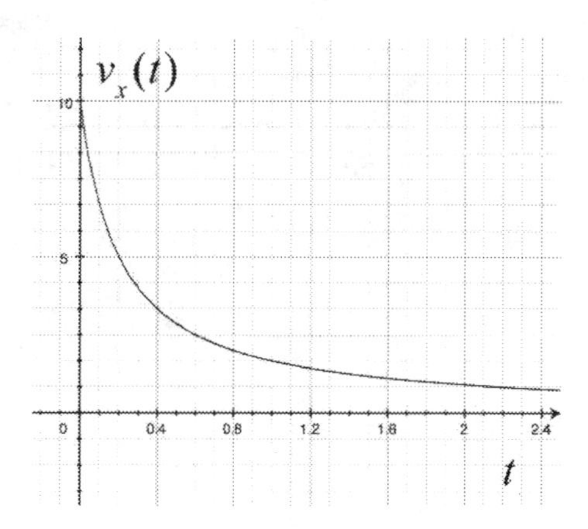

Figure 8.30 Plot of v_x vs. t for damping force $\left|\vec{\mathbf{F}}_{\text{friction}}\right| = \gamma m v^2$

Figure 8.31 shows the two x-component of the velocity functions Eq. (8.5.12) and Eq. (8.5.14) plotted on the same graph.

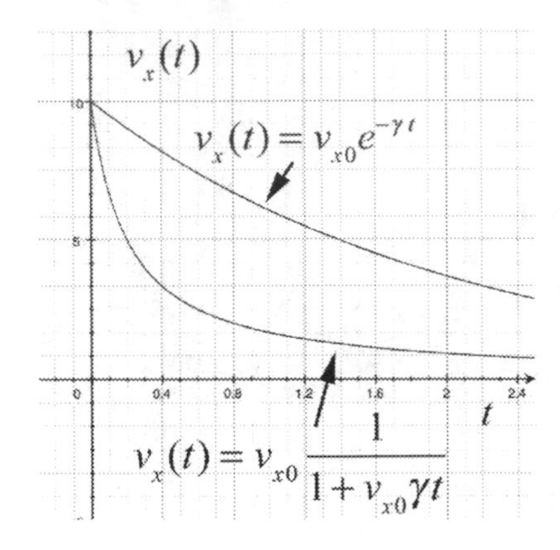

Figure 8.31 Plot of two different x-component of the velocity functions

8.6 Worked Examples

Example 8.6 Staircase

An object of given mass m starts with a given velocity v_0 and slides an unknown distance s along a floor and then off the top of a staircase (Figure 8.32). The goal of this problem is to find the distance s. The coefficient of kinetic friction between the object and the floor is given by μ_k. The object strikes at the far end of the third stair. Each stair has a given rise of h and a given run of d. Neglect air resistance and use g for the gravitational constant. (a) Briefly describe how you intend to model the motion of the object. What are the given quantities in this problem? (b) What is the distance s that the object slides along the floor? Express your answer in terms of the given quantities only.

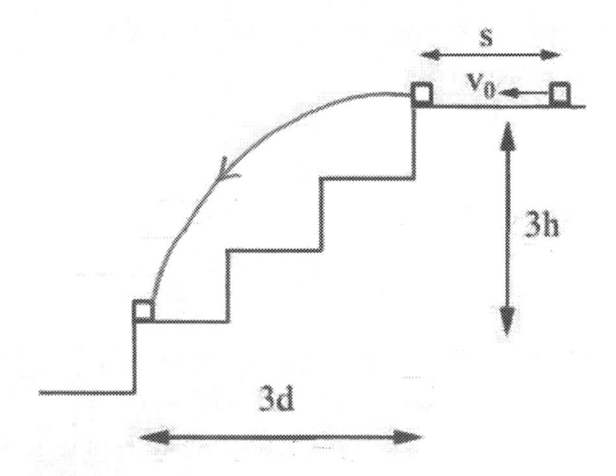

Figure 8.32 Object falling down a staircase

Solution: a) There are two distinct stages to the object's motion, the initial horizontal motion (the floor must be assumed horizontal) and the motion in free fall. The given final position of the object, at the far end of the third stair, will determine the horizontal component of the velocity at the instant the object left the top of the stair. This in turn determined the time the object decelerated, and the deceleration while on the floor determined the distance traveled on the floor. The given quantities are m, v_0, μ_k, g, h and d.

b) From the top of the stair to the far end of the third stair, the object is in free fall. Take the positive $\hat{\mathbf{i}}$-direction to be horizontal, directed to the left in the figure, take the positive $\hat{\mathbf{j}}$-direction to be vertical (up) and take the origin at the top of the stair, where the object first goes into free fall. The components of acceleration are $a_x = 0$, $a_y = -g$, the initial x-component of velocity will be denoted $v_{x,0}$, the initial y-component of velocity is $v_{y,0} = 0$, the initial x-position is $x_0 = 0$ and the initial y-position is $y_0 = 0$. The equations describing the object's motion as a function of time t are then

$$x(t) = x_0 + v_{x,0}\,t + \frac{1}{2}a_x t^2 = v_{x,0}\,t \qquad (8.6.1)$$

$$y(t) = y_0 + v_{y,0}\,t + \frac{1}{2}a_y t^2 = -\frac{1}{2}gt^2. \qquad (8.6.2)$$

It's crucial to see that in this notation and that given in the problem, $v_{x,0} < v_0$. In the above equations, Eq. (8.6.1) may be solved for t to give

$$t = \frac{x(t)}{v_{x,0}}. \qquad (8.6.3)$$

Substituting Eq. (8.6.3) into Eq. (8.6.2) and eliminating the variable t,

$$y(t) = -\frac{1}{2}g\frac{x^2(t)}{v_{x,0}^{\;2}}. \qquad (8.6.4)$$

Eq. (8.6.4) can now be solved for the square of the horizontal component of the velocity,

$$v_{x,0}^{\;2} = -\frac{1}{2}g\frac{x^2(t)}{y(t)}. \qquad (8.6.5)$$

At the far end of the third stair, $x = 3d$ and $y = -3h$; substitution into Eq. (8.6.5) gives

$$v_{x,0}^{\;2} = \frac{3gd^2}{2h}. \qquad (8.6.6)$$

For the horizontal motion, use the same coordinates with the origin at the edge of the landing. The forces on the object are gravity $m\vec{\mathbf{g}} = -mg\,\hat{\mathbf{j}}$, the normal force $\vec{\mathbf{N}} = N\,\hat{\mathbf{j}}$ and the kinetic frictional force $\vec{\mathbf{f}}_{\text{k}} = -f_{\text{k}}\,\hat{\mathbf{i}}$. The components of the vectors in Newton's Second Law, $\vec{\mathbf{F}} = m\vec{\mathbf{a}}$, are

$$\begin{aligned} -f_{\text{k}} &= m\,a_x \\ N - m\,g &= m\,a_y. \end{aligned} \tag{8.6.7}$$

The object does not move in the y-direction; $a_y = 0$ and thus from the second expression in (8.6.7), $N = m\,g$. The magnitude of the frictional force is then $f_{\text{k}} = \mu_{\text{k}} N = \mu_{\text{k}} mg$, and the first expression in (8.6.7) gives the x-component of acceleration as $a_x = -\mu_{\text{k}} g$. Since the acceleration is constant the x-component of the velocity is given by

$$v_x(t) = v_0 + a_x\,t\,, \tag{8.6.8}$$

where v_0 is the x-component of the velocity of the object when it just started sliding (this is a different from the x-component of the velocity when it just leaves the top landing, a quantity we denoted by earlier by $v_{x,0}$). The displacement is given by

$$x(t) - x_0 = v_0\,t + \frac{1}{2}a_x\,t^2. \tag{8.6.9}$$

Denote the time the block is decelerating as t_1. The initial conditions needed to solve the problem are the given v_0 (assuming we've "reset" our clocks appropriately), $v_x(t_1) = v_{x,0}$ as found above, $x_0 = -s$ and $x(t_1) = 0$. (Note that in the conditions for the initial and final position, we've kept the origin at the top of the stair. This is not necessary, but it works. The important matter is that the x-coordinate increases from right to left.) During this time, the block's x-component of velocity decreases from v_0 to $v_x(t_1) = v_{x,0}$ with constant acceleration $a_x = -\mu_{\text{k}} g$. Using the initial and final conditions, and the value of the acceleration, Eq. (8.6.9) becomes

$$s = v_0\,t_1 - \frac{1}{2}\mu_k g\,t_1^2 \tag{8.6.10}$$

and we can solve Eq. (8.6.8) for the time the block reaches the edge of the landing,

$$t_1 = \frac{v_{x,0} - v_0}{-\mu_{\text{k}} g} = \frac{v_0 - v_{x,0}}{\mu_{\text{k}} g}. \tag{8.6.11}$$

Substituting Eq. (8.6.11) into Eq. (8.6.10) yields

$$s = v_0\left(\frac{v_0 - v_{x,0}}{\mu_k g}\right) - \frac{1}{2}\mu_k g\left(\frac{v_0 - v_{x,0}}{\mu_k g}\right)^2 \tag{8.6.12}$$

and after some algebra, we can rewrite Eq. (8.6.12) as

$$s = v_0\left(\frac{v_0 - v_{x,0}}{\mu_k g}\right) - \frac{1}{2}\mu_k g\left(\frac{v_0 - v_{x,0}}{\mu_k g}\right)^2 = \frac{v_0^{\,2}}{2\mu_k g} - \frac{v_{x,0}^{\,2}}{2\mu_k g}. \tag{8.6.13}$$

Substituting Eq. (8.6.6) for $v_{x,0}^2$ into Eq. (8.6.13), yields the distance the object traveled on the landing,

$$s = \frac{v_0^2 - (3gd^2 / 2h)}{2\mu_k g}. \tag{8.6.14}$$

Example 8.7 Cart Moving on a Track

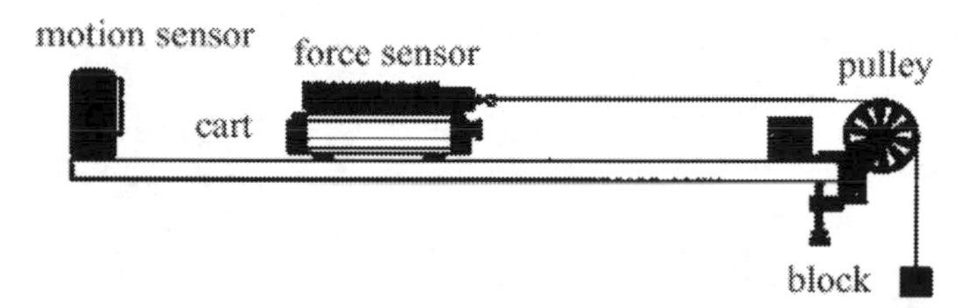

Figure 8.33 A falling block will accelerate a cart on a track via the pulling force of the string. The force sensor measures the tension in the string.

Consider a cart that free to move along a horizontal track (Figure 8.33). A force is applied to the cart via a string that is attached to a force sensor mounted on the cart, wrapped around a pulley and attached to a block on the other end. When the block is released the cart will begin to accelerate. The force sensor and cart together have a mass m_C, and the suspended block has mass m_B. You may neglect the small mass of the string and pulley, and assume the string is inextensible. The frictional force is modeled as a coefficient of kinetic friction μ_k between the cart and the track (almost all of the friction is in the wheel bearings, and the model is quite good). (a) What is the acceleration of the cart? (b) What is the tension in the string?

Solution: In general, we would like to draw free-body diagrams on all the individual objects (cart, sensor, pulley, rope, and block) but we can also choose a system consisting of two (or more) objects knowing that the forces of interaction between any two objects will cancel in pairs by Newton's Third Law. In this example, we shall choose the sensor/cart as one free-body, and the block as the other free-body. The free-body force diagram for the sensor/cart is shown in Figure 8.34.

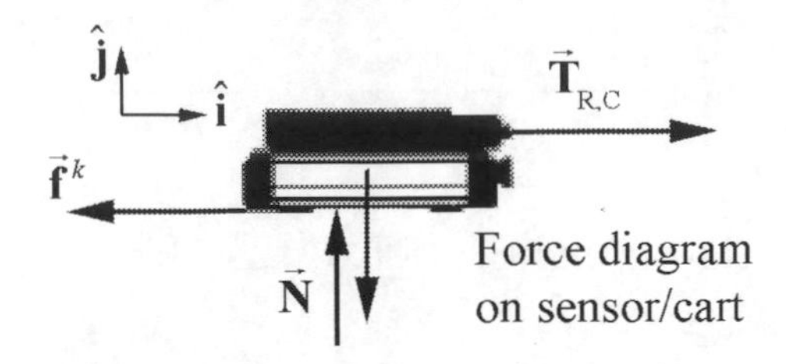

Figure 8.34 Force diagram on sensor/cart with a vector decomposition of the contact force into horizontal and vertical components

There are three forces acting on the sensor/cart: the gravitational force $m_C\vec{\mathbf{g}}$, the pulling force $\vec{\mathbf{T}}_{R,C}$ of the rope on the force sensor, and the contact force between the track and the cart. In Figure 8.34, we decompose the contact force into its two components, the kinetic frictional force $\vec{\mathbf{f}}_k = -f_k\hat{\mathbf{i}}$ and the normal force, $\vec{\mathbf{N}} = N\,\hat{\mathbf{j}}$.

The cart is only accelerating in the horizontal direction with $\vec{\mathbf{a}}_C = a_{C,x}\,\hat{\mathbf{i}}$, so the component of the force in the vertical direction must be zero, $a_{C,y} = 0$. We can now apply Newton's Second Law in the horizontal and vertical directions and find that

$$\hat{\mathbf{i}}:\ T_{R,C} - f_k = m_C a_{C,x} \tag{8.6.15}$$

$$\hat{\mathbf{j}}:\ \ N - m_C g = 0\,. \tag{8.6.16}$$

From Eq. (8.6.16), we conclude that the normal component is

$$N = m_C g\,. \tag{8.6.17}$$

We use Equation (8.6.17) for the normal force to find that the magnitude of the kinetic frictional force is

$$f_k = \mu_k N = m_C g\,. \tag{8.6.18}$$

Then Equation (8.6.15) becomes

$$T_{R,C} - \mu_k m_C g = m_C a_{C,x}\,. \tag{8.6.19}$$

The force diagram for the block is shown in Figure 8.35. The two forces acting on the block are the pulling force $\vec{\mathbf{T}}_{R,B}$ of the string and the gravitational force $m_B\vec{\mathbf{g}}$. We now apply Newton's Second Law to the block and find that

$$\hat{\mathbf{j}}: m_B g - T_{R,B} = m_B a_{B,y}\,. \tag{8.6.20}$$

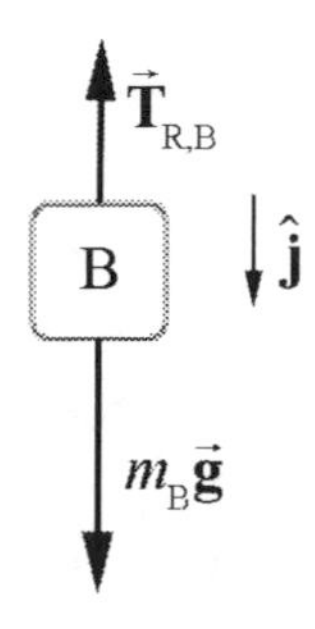

Figure 8.35 Forces acting on the block

In Equation (8.6.20), the symbol $a_{B,y}$ represents the component of the acceleration with sign determined by our choice of direction for the unit vector $\hat{\mathbf{j}}$. Note that we made a different choice of direction for the unit vector in the vertical direction in the free-body diagram for the block shown in Figure 8.34. Each free-body diagram has an independent set of unit vectors that define a sign convention for vector decomposition of the forces acting on the free-body and the acceleration of the free-body. In our example, with the unit vector pointing downwards in Figure 8.35, if we solve for the component of the acceleration and it is positive, then we know that the direction of the acceleration is downwards.

There is a second subtle way that signs are introduced with respect to the forces acting on a free-body. In our example, the force between the string and the block acting on the block points upwards, so in the vector decomposition of the forces acting on the block that appears on the left-hand side of Equation (8.6.20), this force has a minus sign and the quantity $\vec{\mathbf{T}}_{\mathrm{R,B}} = -T_{\mathrm{R,B}}\hat{\mathbf{j}}$ where $T_{\mathrm{R,B}}$ is assumed positive. However, if for some reason we were uncertain about the direction of the force between the string and the block acting on the block, and drew the arrow downwards, then when we solved the problem we would discover that $T_{\mathrm{R,B}}$ is negative, indicating that the force points in a direction opposite the direction of the arrow on the free-body diagram.

Our assumption that the mass of the rope and the mass of the pulley are negligible enables us to assert that the tension in the rope is uniform and equal in magnitude to the forces at each end of the rope,

$$T_{\mathrm{R,B}} = T_{\mathrm{R,C}} \equiv T\,. \tag{8.6.21}$$

We also assumed that the string is inextensible (does not stretch). This implies that the rope, block, and sensor/cart all have the same magnitude of acceleration,

$$a_{\mathrm{C},x} = a_{\mathrm{B},y} \equiv a\,. \tag{8.6.22}$$

Using Equations (8.6.21) and (8.6.22), we can now rewrite the equation of motion for the sensor/cart, Equation (8.6.19), as

$$T - \mu_{\mathrm{k}} m_{\mathrm{C}} g = m_{\mathrm{C}} a, \tag{8.6.23}$$

and the equation of motion (8.6.20) for the block as

$$m_{\mathrm{B}} g - T = m_{\mathrm{B}} a. \tag{8.6.24}$$

We have only two unknowns T and a, so we can now solve the two equations (8.6.23) and (8.6.24) simultaneously for the acceleration of the sensor/cart and the tension in the rope. We first solve Equation (8.6.23) for the tension

$$T = \mu_{\mathrm{k}} m_{\mathrm{C}} g + m_{\mathrm{C}} a \tag{8.6.25}$$

and then substitute Equation (8.6.25) into Equation (8.6.24) and find that

$$m_{\mathrm{B}} g - (\mu_{\mathrm{k}} m_{\mathrm{C}} g + m_{\mathrm{C}} a) = m_{\mathrm{B}} a. \tag{8.6.26}$$

We can now solve Equation (8.6.26) for the acceleration,

$$a = \frac{m_{\mathrm{B}} g - \mu_{\mathrm{k}} m_{\mathrm{C}} g}{m_{\mathrm{C}} + m_{\mathrm{B}}}. \tag{8.6.27}$$

Substitution of Equation (8.6.27) this into Equation (8.6.25) gives the tension in the string,

$$\begin{aligned} T &= \mu_{\mathrm{k}} m_{\mathrm{C}} g + m_{\mathrm{C}} a \\ &= \mu_{\mathrm{k}} m_{\mathrm{C}} g + m_{\mathrm{C}} \frac{m_{\mathrm{B}} g - \mu_{\mathrm{k}} m_{\mathrm{C}} g}{m_{\mathrm{C}} + m_{\mathrm{B}}} \\ &= (\mu_{\mathrm{k}} + 1) \frac{m_{\mathrm{C}} m_{\mathrm{B}}}{m_{\mathrm{C}} + m_{\mathrm{B}}} g. \end{aligned} \tag{8.6.28}$$

In this example, we applied Newton's Second Law to two objects, one a composite object consisting of the sensor and the cart, and the other the block. We analyzed the forces acting on each object and also any constraints imposed on the acceleration of each object. We used the force laws for kinetic friction and gravitation on each free-body system. The three equations of motion enable us to determine the forces that depend on the parameters in the example: the tension in the rope, the acceleration of the objects, and normal force between the cart and the table.

Example 8.8 Pulleys and Ropes Constraint Conditions

Consider the arrangement of pulleys and blocks shown in Figure 8.36. The pulleys are assumed massless and frictionless and the connecting strings are massless and unstretchable. Denote the respective masses of the blocks as m_1, m_2 and m_3. The upper pulley in the figure is free to rotate but its center of mass does not move. Both pulleys have the same radius R. (a) How are the accelerations of the objects related? (b) Draw force diagrams on each moving object. (c) Solve for the accelerations of the objects and the tensions in the ropes.

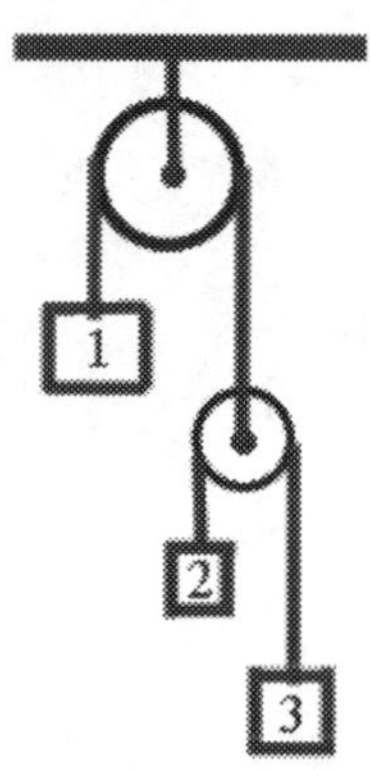

Figure 8.36 Constrained pulley system

Solution: (a) Choose an origin at the center of the upper pulley. Introduce coordinate functions for the three moving blocks, y_1, y_2 and y_3. Introduce a coordinate function y_P for the moving pulley (the pulley on the lower right in Figure 8.37). Choose downward for positive direction; the coordinate system is shown in the figure below then.

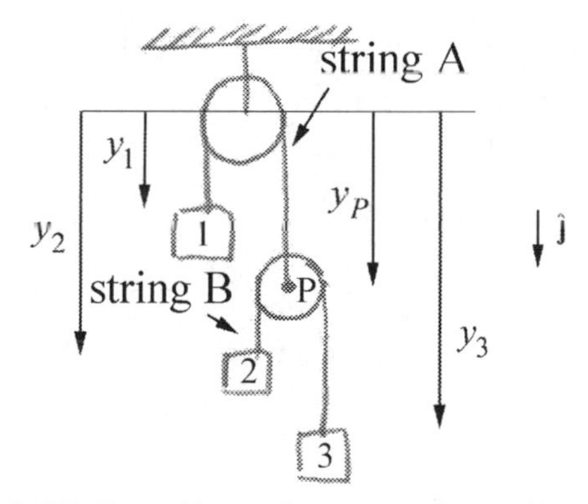

Figure 8.37 Coordinated system for pulley system

The length of string A is given by

$$l_A = y_1 + y_P + \pi R \tag{8.6.29}$$

where πR is the arc length of the rope that is in contact with the pulley. This length is constant, and so the second derivative with respect to time is zero,

$$0 = \frac{d^2 l_A}{dt^2} = \frac{d^2 y_1}{dt^2} + \frac{d^2 y_P}{dt^2} = a_{y,1} + a_{y,P}. \tag{8.6.30}$$

Thus block 1 and the moving pulley's components of acceleration are equal in magnitude but opposite in sign,

$$a_{y,P} = -a_{y,1}. \tag{8.6.31}$$

The length of string B is given by

$$l_B = (y_3 - y_P) + (y_2 - y_P) + \pi R = y_3 + y_2 - 2y_P + \pi R \tag{8.6.32}$$

where πR is the arc length of the rope that is in contact with the pulley. This length is also constant so the second derivative with respect to time is zero,

$$0 = \frac{d^2 l_B}{dt^2} = \frac{d^2 y_2}{dt^2} + \frac{d^2 y_3}{dt^2} - 2\frac{d^2 y_P}{dt^2} = a_{y,2} + a_{y,3} - 2a_{y,P}. \tag{8.6.33}$$

We can substitute Equation (8.6.31) for the pulley acceleration into Equation (8.6.33) yielding the *constraint relation* between the components of the acceleration of the three blocks,

$$0 = a_{y,2} + a_{y,3} + 2a_{y,1}. \tag{8.6.34}$$

b) Free-body Force diagrams: the forces acting on block 1 are: the gravitational force $m_1\vec{\mathbf{g}}$ and the pulling force $\vec{\mathbf{T}}_{1,r}$ of the string acting on the block 1. Since the string is assumed to be massless and the pulley is assumed to be massless and frictionless, the tension T_A in the string is uniform and equal in magnitude to the pulling force of the string on the block. The free-body diagram is shown in Figure 8.38.

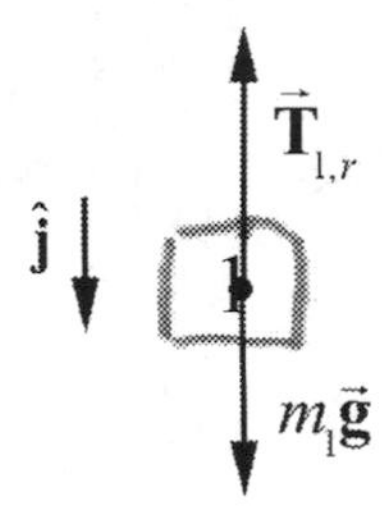

Figure 8.38 Free-body force diagram on block 1

Newton's Second Law applied to block 1 is then

$$\hat{\mathbf{j}}: \; m_1 g - T_A = m_1\, a_{y,1}. \tag{8.6.35}$$

The forces on the block 2 are the gravitational force $m_2\vec{\mathbf{g}}$ and the string holding the block, $\vec{\mathbf{T}}_{2,r}$, with magnitude T_B. The free-body diagram for the forces acting on block 2 is shown in Figure 8.39.

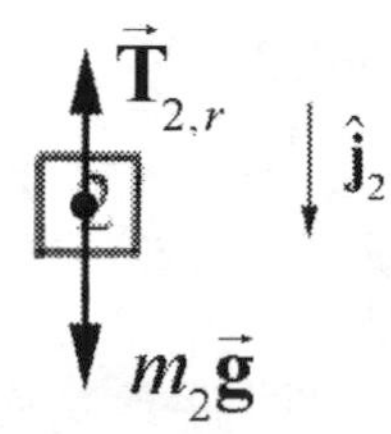

Figure 8.39 Free-body force diagram on block 2

Newton's second Law applied to block 2 is

$$\hat{\mathbf{j}}:\ m_2 g - T_B = m_2\, a_{y,2}\,. \tag{8.6.36}$$

The forces on the block 3 are the gravitational force $m_3\vec{\mathbf{g}}$ and the string holding the block, $\vec{\mathbf{T}}_{3,r}$, with magnitude T_B. The free-body diagram for the forces acting on block 3 is shown in Figure 8.40. Newton's second Law applied to block 3 is

$$\hat{\mathbf{j}}:\ m_3 g - T_B = m_3\, a_{y,3}\,. \tag{8.6.37}$$

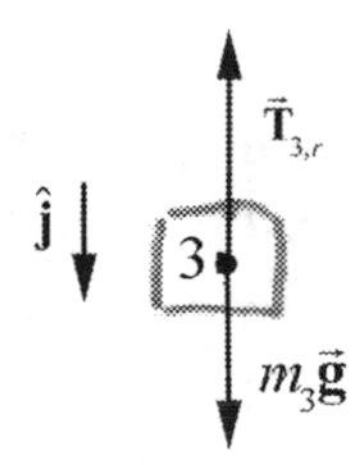

Figure 8.40 Free-body force diagram on block 3

The forces on the moving pulley are the gravitational force $m_P\,\vec{\mathbf{g}} = \vec{\mathbf{0}}$ (the pulley is assumed massless); string B pulls down on the pulley on each side with a force, $\vec{\mathbf{T}}_{P,B}$, which has magnitude T_B. String A holds the pulley up with a force $\vec{\mathbf{T}}_{P,A}$ with the magnitude T_A equal to the tension in string A. The free-body diagram for the forces acting on the moving pulley is shown in Figure 8.41.

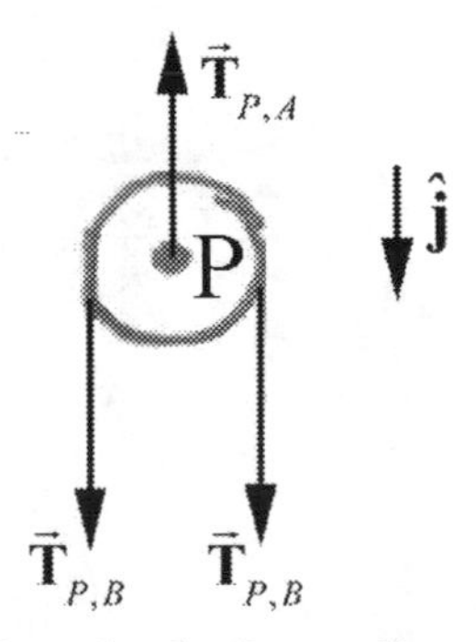

Figure 8.41 Free-body force diagram on pulley

Newton's second Law applied to the pulley is

$$\hat{\mathbf{j}}:\ 2T_B - T_A = m_P\, a_{y,P} = 0\,. \tag{8.6.38}$$

Because the pulley is assumed to be massless, we can use this last equation to determine the condition that the tension in the two strings must satisfy,

$$2T_B = T_A \tag{8.6.39}$$

We are now in position to determine the accelerations of the blocks and the tension in the two strings. We record the relevant equations as a summary.

$$0 = a_{y,2} + a_{y,3} + 2a_{y,1} \tag{8.6.40}$$

$$m_1 g - T_A = m_1\, a_{y,1} \tag{8.6.41}$$

$$m_2 g - T_B = m_2\, a_{y,2} \tag{8.6.42}$$

$$m_3 g - T_B = m_3\, a_{y,3} \tag{8.6.43}$$

$$2T_B = T_A\,. \tag{8.6.44}$$

There are five equations with five unknowns, so we can solve this system. We shall first use Equation (8.6.44) to eliminate the tension T_A in Equation (8.6.41), yielding

$$m_1 g - 2T_B = m_1\, a_{y,1}\,. \tag{8.6.45}$$

We now solve Equations (8.6.42), (8.6.43) and (8.6.45) for the accelerations,

$$a_{y,2} = g - \frac{T_B}{m_2} \tag{8.6.46}$$

$$a_{y,3} = g - \frac{T_B}{m_3} \tag{8.6.47}$$

$$a_{y,1} = g - \frac{2T_B}{m_1}\,. \tag{8.6.48}$$

We now substitute these results for the accelerations into the constraint equation, Equation (8.6.40),

$$0 = g - \frac{T_B}{m_2} + g - \frac{T_B}{m_3} + 2g - \frac{4T_B}{m_1} = 4g - T_B\left(\frac{1}{m_2} + \frac{1}{m_3} + \frac{4}{m_1}\right). \tag{8.6.49}$$

We can now solve this last equation for the tension in string B,

$$T_B = \frac{4g}{\left(\frac{1}{m_2} + \frac{1}{m_3} + \frac{4}{m_1}\right)} = \frac{4g\, m_1\, m_2\, m_3}{m_1\, m_3 + m_1 m_2 + 4 m_2\, m_3}. \tag{8.6.50}$$

From Equation (8.6.44), the tension in string A is

$$T_A = 2T_B = \frac{8g\, m_1\, m_2\, m_3}{m_1\, m_3 + m_1 m_2 + 4 m_2\, m_3}. \tag{8.6.51}$$

We find the acceleration of block 1 from Equation (8.6.48), using Equation (8.6.50) for the tension in string B,

$$a_{y,1} = g - \frac{2T_B}{m_1} = g - \frac{8g\, m_2\, m_3}{m_1\, m_3 + m_1\, m_2 + 4 m_2\, m_3} = g\frac{m_1\, m_3 + m_1\, m_2 - 4 m_2\, m_3}{m_1\, m_3 + m_1 m_2 + 4 m_2\, m_3}. \tag{8.6.52}$$

We find the acceleration of block 2 from Equation (8.6.46), using Equation (8.6.50) for the tension in string B,

$$a_{y,2} = g - \frac{T_B}{m_2} = g - \frac{4g\, m_1\, m_3}{m_1\, m_3 + m_1\, m_2 + 4 m_2\, m_3} = g\frac{-3m_1\, m_3 + m_1\, m_2 + 4 m_2\, m_3}{m_1\, m_3 + m_1\, m_2 + 4 m_2\, m_3}. \tag{8.6.53}$$

Similarly, we find the acceleration of block 3 from Equation (8.6.47), using Equation (8.6.50) for the tension in string B,

$$a_{y,3} = g - \frac{T_B}{m_3} = g - \frac{4g\, m_1\, m_2}{m_1\, m_3 + m_1\, m_2 + 4 m_2\, m_3} = g\frac{m_1\, m_3 - 3m_1\, m_2 + 4 m_2\, m_3}{m_1\, m_3 + m_1\, m_2 + 4 m_2\, m_3}. \tag{8.6.54}$$

As a check on our algebra we note that

$$2a_{1,y} + a_{2,y} + a_{3,y} =$$

$$2g\frac{m_1\, m_3 + m_1\, m_2 - 4 m_2\, m_3}{m_1\, m_3 + m_1\, m_2 + 4 m_2\, m_3} + g\frac{-3m_1\, m_3 + m_1\, m_2 + 4 m_2\, m_3}{m_1\, m_3 + m_1 m_2 + 4 m_2\, m_3} + g\frac{m_1\, m_3 - 3m_1\, m_2 + 4 m_2\, m_3}{m_1\, m_3 + m_1 m_2 + 4 m_2 m_3}$$

$$= 0.$$

Example 8.9 Accelerating Wedge

A 45° wedge is pushed along a table with constant acceleration $\vec{\mathbf{A}}$ according to an observer at rest with respect to the table. A block of mass m slides without friction down the wedge (Figure 8.42). Find its acceleration with respect to an observer at rest with respect to the table. Write down a plan for finding the magnitude of the acceleration of the block. Make sure you clearly state which concepts you plan to use to calculate any relevant physical quantities. Also clearly state any assumptions you make. Be sure you include any free-body force diagrams or sketches that you plan to use.

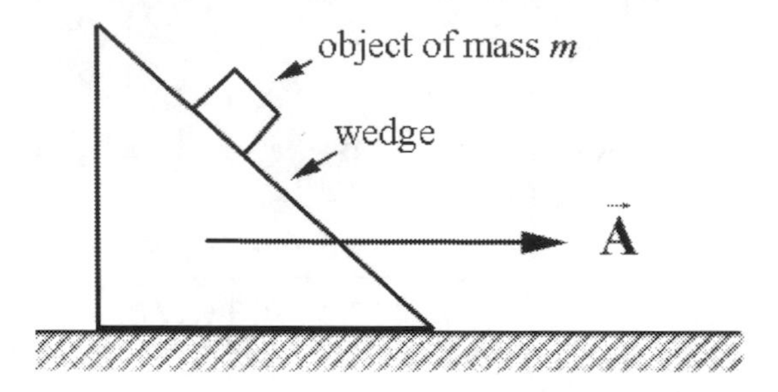

Figure 8.42 Block on accelerating wedge

Solution: Choose a coordinate system for the block and wedge as shown in Figure 8.43. Then $\vec{\mathbf{A}} = A_{x,w}\ \hat{\mathbf{i}}$ where $A_{x,w}$ is the x-component of the acceleration of the wedge.

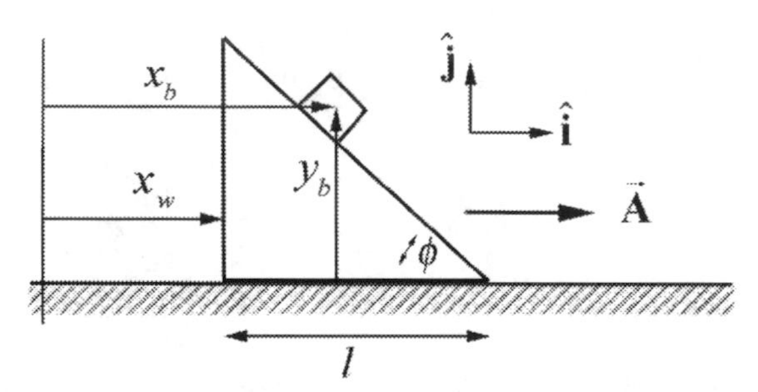

Figure 8.43 Coordinate system for block on accelerating wedge

We shall apply Newton's Second Law to the block sliding down the wedge. Because the wedge is accelerating, there is a constraint relation between the x- and y- components of the acceleration of the block. In order to find that constraint we choose a coordinate system for the wedge and block sliding down the wedge shown in the figure below. We shall find the constraint relationship between the components of the accelerations of the block and wedge by a geometric argument. From the figure above, we see that

$$\tan\phi = \frac{y_b}{l - (x_b - x_w)}. \tag{8.6.55}$$

Therefore

$$y_b = (l - (x_b - x_w))\tan\phi\,. \tag{8.6.56}$$

If we differentiate Eq. (8.6.56) twice with respect to time noting that

$$\frac{d^2 l}{dt^2} = 0 \tag{8.6.57}$$

we have that

$$\frac{d^2y_b}{dt^2} = -\left(\frac{d^2x_b}{dt^2} - \frac{d^2x_w}{dt^2}\right)\tan\phi \,. \tag{8.6.58}$$

Therefore

$$a_{b,y} = -(a_{b,x} - A_{x,w})\tan\phi \tag{8.6.59}$$

recalling that

$$A_{x,w} = \frac{d^2x_w}{dt^2} \,. \tag{8.6.60}$$

We now draw a free-body force diagram for the block (Figure 8.44). Newton's Second Law in the $\hat{\mathbf{i}}$ - direction becomes

$$N\sin\phi = ma_{b,x} \,. \tag{8.6.61}$$

and the $\hat{\mathbf{j}}$-direction becomes

$$N\cos\phi - mg = ma_{b,y} \tag{8.6.62}$$

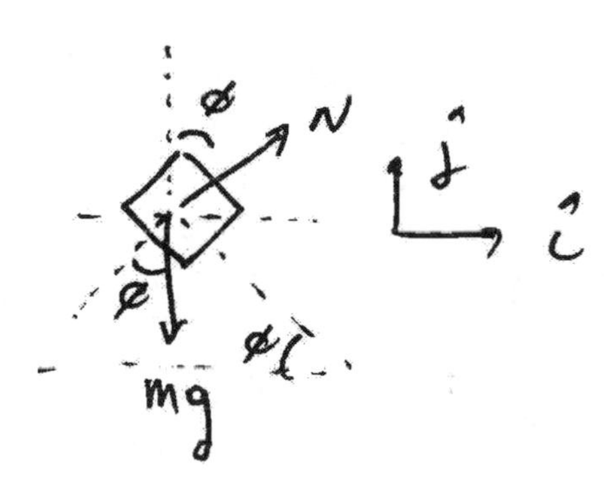

Figure 8.44 Free-body force diagram on block

We can solve for the normal force from Eq. (8.6.61):

$$N = \frac{ma_{b,x}}{\sin\phi} \tag{8.6.63}$$

We now substitute Eq. (8.6.59) and Eq. (8.6.63) into Eq. (8.6.62) yielding

$$\frac{ma_{b,x}}{\sin\phi}\cos\phi - mg = m(-(a_{b,x} - A_{w,x})\tan\phi) \,. \tag{8.6.64}$$

We now clean this up yielding

$$ma_{b,x}(\text{cotan}\,\phi + \tan\phi) = m(g + A_{w,x}\tan\phi) \tag{8.6.65}$$

Thus the x-component of the acceleration is then

$$a_{b,x} = \frac{g + A_{w,x} \tan\phi}{\text{cotan}\,\phi + \tan\phi} \tag{8.6.66}$$

From Eq. (8.6.59), the y-component of the acceleration is then

$$a_{b,y} = -(a_{b,x} - A_{w,x})\tan\phi = -\left(\frac{g + A_{w,x}\tan\phi}{\text{cotan}\,\phi + \tan\phi} - A_{w,x}\right)\tan\phi . \tag{8.6.67}$$

This simplifies to

$$a_{b,y} = \frac{A_{w,x} - g\tan\phi}{\text{cotan}\,\phi + \tan\phi} \tag{8.6.68}$$

When $\phi = 45^\circ$, $\text{cotan}\,45^\circ = \tan 45^\circ = 1$, and so Eq. (8.6.66) becomes

$$a_{b,x} = \frac{g + A_{w,x}}{2} \tag{8.6.69}$$

and Eq. (8.6.68) becomes

$$a_{b,y} = \frac{A - g}{2} . \tag{8.6.70}$$

The magnitude of the acceleration is then

$$a = \sqrt{a_{b,a}^2 + a_{b,y}^2} = \sqrt{\left(\frac{g + A_{w,x}}{2}\right)^2 + \left(\frac{A_{w,x} - g}{2}\right)^2} \tag{8.6.71}$$

$$a = \sqrt{\left(\frac{g^2 + A_{w,x}^2}{2}\right)} .$$

Example 8.10: Capstan

A device called a capstan is used aboard ships in order to control a rope that is under great tension. The rope is wrapped around a fixed drum of radius R, usually for several turns (Figure 8.45 shows about three fourths turn as seen from overhead). The load on the rope pulls it with a force T_A, and the sailor holds the other end of the rope with a much smaller force T_B. The coefficient of static friction between the rope and the drum is μ_s. The sailor is holding the rope so that it is just about to slip. Show that $T_B = T_A e^{-\mu_s \theta_{BA}}$, where θ_{BA} is the angle subtended by the rope on the drum.

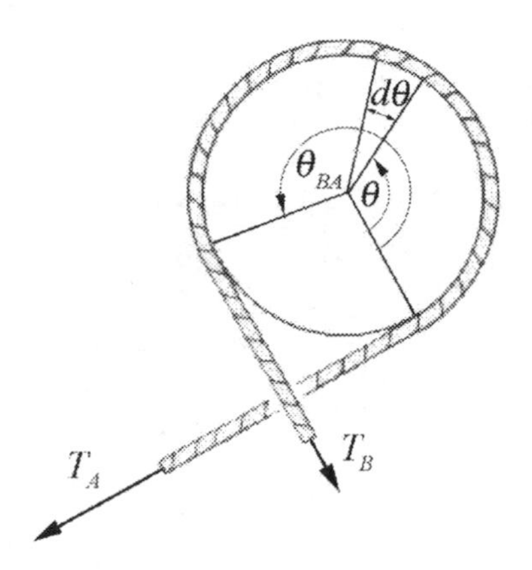

Figure 8.45 Capstan

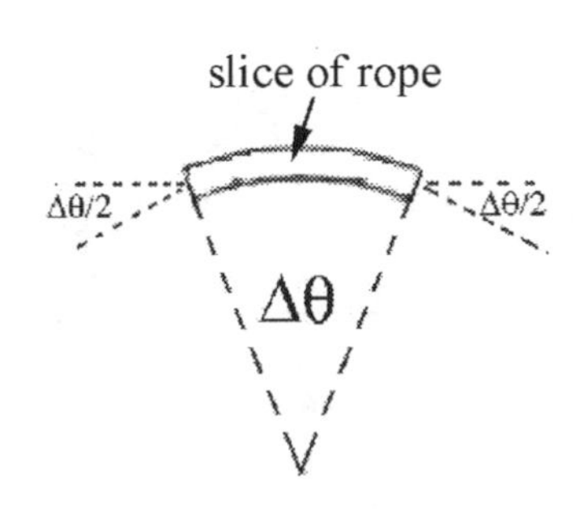

Figure 8.46 Small slice of rope

Solution: We begin by considering a small slice of rope of arc length $R\Delta\theta$, shown in the Figure 8.46. We choose unit vectors for the force diagram on this section of the rope and indicate them on Figure 8.47. The right edge of the slice is at angle θ and the left edge of the slice is at $\theta+\Delta\theta$. The angle edge end of the slice makes with the horizontal is $\Delta\theta/2$. There are four forces acting on the on this section of the rope. The forces are the normal force between the capstan and the rope pointing outward, a static frictional force and the tensions at either end of the slice. The rope is held at the just slipping point, so if the load exerts a greater force the rope will slip to the right. Therefore the direction of the static frictional force between the capstan and the rope, acting on the rope, points to the left. The tension on the right side of the slice is denoted by $T(\theta)\equiv T$, while the tension on the left side of the slice is denoted by $T(\theta+\Delta\theta)\equiv T+\Delta T$. Does the tension in this slice from the right side to the left, increase, remain the same, or decrease? The tension decreases because the load on the left side is less than the load on the right side. Note that $\Delta T<0$.

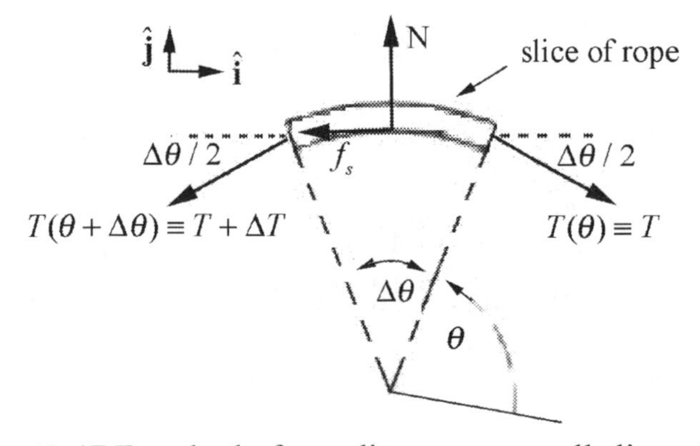

Figure 8.47 Free-body force diagram on small slice of rope

The vector decomposition of the forces is given by

$$\hat{\mathbf{i}}: T\cos(\Delta\theta/2) - f_s - (T+\Delta T)\cos(\Delta\theta/2) \tag{8.6.72}$$

$$\hat{\mathbf{j}}: -T\sin(\Delta\theta/2) + N - (T+\Delta T)\sin(\Delta\theta/2). \tag{8.6.73}$$

For small angles $\Delta\theta$, $\cos(\Delta\theta/2) \cong 1$ and $\sin(\Delta\theta/2) \cong \Delta\theta/2$. Using the small angle approximations, the vector decomposition of the forces in the x-direction (the $+\hat{\mathbf{i}}$-direction) becomes

$$\begin{aligned} T\cos(\Delta\theta/2) - f_s - (T+\Delta T)\cos(\Delta\theta/2) &\simeq T - f_s - (T+\Delta T) \\ &= -f_s - \Delta T \end{aligned} \tag{8.6.74}$$

By the static equilibrium condition the sum of the x-components of the forces is zero,

$$-f_s - \Delta T = 0. \tag{8.6.75}$$

The vector decomposition of the forces in the y-direction (the $+\hat{\mathbf{j}}$-direction) is approximately

$$\begin{aligned} -T\sin(\Delta\theta/2) + N - (T+\Delta T)\sin(\Delta\theta/2) &\simeq -T\Delta\theta/2 + N - (T+\Delta T)\Delta\theta/2 \\ &= -T\Delta\theta + N - \Delta T\Delta\theta/2 \end{aligned} \tag{8.6.76}$$

In the last equation above we can ignore the terms proportional to $\Delta T\,\Delta\theta$ because these are the product of two small quantities and hence are much smaller than the terms proportional to either ΔT or $\Delta\theta$. The vector decomposition in the y-direction becomes

$$-T\Delta\theta + N. \tag{8.6.77}$$

Static equilibrium implies that this sum of the y-components of the forces is zero,

$$-T\Delta\theta + N = 0. \tag{8.6.78}$$

We can solve this equation for the magnitude of the normal force

$$N = T\Delta\theta. \tag{8.6.79}$$

The just slipping condition is that the magnitude of the static friction attains its maximum value

$$f_s = (f_s)_{max} = \mu_s N. \tag{8.6.80}$$

We can now combine the Equations (8.6.75) and (8.6.80) to yield

$$\Delta T = -\mu_s N. \tag{8.6.81}$$

Now substitute the magnitude of the normal force, Equation (8.6.79), into Equation (8.6.81), yielding

$$-\mu_s T\Delta\theta - \Delta T = 0. \tag{8.6.82}$$

Finally, solve this equation for the ratio of the change in tension to the change in angle,

$$\frac{\Delta T}{\Delta\theta} = -\mu_s T. \tag{8.6.83}$$

The derivative of tension with respect to the angle θ is defined to be the limit

$$\frac{dT}{d\theta} \equiv \lim_{\Delta\theta \to 0} \frac{\Delta T}{\Delta\theta}, \tag{8.6.84}$$

and Equation (8.6.83) becomes

$$\frac{dT}{d\theta} = -\mu_s T. \tag{8.6.85}$$

This is an example of a first order linear differential equation that shows that the rate of change of tension with respect to the angle θ is proportional to the negative of the tension at that angle θ. This equation can be solved by integration using the technique of separation of variables. We first rewrite Equation (8.6.85) as

$$\frac{dT}{T} = -\mu_s\, d\theta. \tag{8.6.86}$$

Integrate both sides, noting that when $\theta = 0$, the tension is equal to force of the load T_A, and when angle $\theta = \theta_{A,B}$ the tension is equal to the force T_B the sailor applies to the rope,

$$\int_{T=T_A}^{T=T_B} \frac{dT}{T} = -\int_{\theta=0}^{\theta=\theta_{BA}} \mu_s\, d\theta. \tag{8.6.87}$$

The result of the integration is

$$\ln\left(\frac{T_B}{T_A}\right) = -\mu_s \theta_{BA}. \tag{8.6.88}$$

Note that the exponential of the natural logarithm

$$\exp\left(\ln\left(\frac{T_B}{T_A}\right)\right) = \frac{T_B}{T_A}, \tag{8.6.89}$$

so exponentiating both sides of Equation (8.6.88) yields

$$\frac{T_B}{T_A} = e^{-\mu_s \theta_{BA}} ; \qquad (8.6.90)$$

the tension decreases exponentially,

$$T_B = T_A e^{-\mu_s \theta_{BA}} , \qquad (8.6.91)$$

Because the tension decreases exponentially, the sailor need only apply a small force to prevent the rope from slipping.

Chapter 9 Circular Motion Dynamics

Chapter 9 Circular Motion Dynamics

I shall now recall to mind that the motion of the heavenly bodies is circular, since the motion appropriate to a sphere is rotation in a circle.[1]

Nicholas Copernicus

9.1 Introduction Newton's Second Law and Circular Motion

We have already shown that when an object moves in a circular orbit of radius r with angular velocity $\vec{\boldsymbol{\omega}}$, it is most convenient to choose polar coordinates to describe the position, velocity and acceleration vectors. In particular, the acceleration vector is given by

$$\vec{\mathbf{a}}(t) = -r\left(\frac{d\theta}{dt}\right)^2 \hat{\mathbf{r}}(t) + r\frac{d^2\theta}{dt^2}\hat{\boldsymbol{\theta}}(t). \tag{9.1.1}$$

Then Newton's Second Law, $\vec{\mathbf{F}} = m\vec{\mathbf{a}}$, can be decomposed into radial ($\hat{\mathbf{r}}$-) and tangential ($\hat{\boldsymbol{\theta}}$-) components

$$F_r = -mr\left(\frac{d\theta}{dt}\right)^2 \quad \text{(circular motion)}, \tag{9.1.2}$$

$$F_\theta = mr\frac{d^2\theta}{dt^2} \quad \text{(circular motion)}. \tag{9.1.3}$$

For the special case of uniform circular motion, $d^2\theta / dt^2 = 0$, and so the sum of the tangential components of the force acting on the object must therefore be zero,

$$F_\theta = 0 \quad \text{(uniform circular motion)}. \tag{9.1.4}$$

9.2 Universal Law of Gravitation and the Circular Orbit of the Moon

An important example of (approximate) circular motion is the orbit of the Moon around the Earth. We can approximately calculate the time T the Moon takes to complete one circle around the earth (a calculation of great importance to early lunar calendar systems, which became the basis for our current model.) Denote the distance from the moon to the center of the earth by $R_{\text{e, m}}$.

[1] *Dedicatory Letter to Pope Paul III.*

Because the Moon moves nearly in a circular orbit with angular speed $\omega = 2\pi / T$ it is accelerating towards the Earth. The radial component of the acceleration (centripetal acceleration) is

$$a_r = -\frac{4\pi^2 R_{\text{e, m}}}{T^2}. \tag{9.2.1}$$

According to Newton's Second Law, there must be a centripetal force acting on the Moon directed towards the center of the Earth that accounts for this inward acceleration.

9.2.1 Universal Law of Gravitation

Newton's Universal Law of Gravitation describes the gravitational force between two bodies 1 and 2 with masses m_1 and m_2 respectively. This force is a radial force (always pointing along the radial line connecting the masses) and the magnitude is proportional to the inverse square of the distance that separates the bodies. Then the force on object 2 due to the gravitational interaction between the bodies is given by,

$$\vec{\mathbf{F}}_{1,2} = -G\frac{m_1 m_2}{r_{1,2}^2}\hat{\mathbf{r}}_{1,2}, \tag{9.2.2}$$

where $r_{1,2}$ is the distance between the two bodies and $\hat{\mathbf{r}}_{1,2}$ is the unit vector located at the position of object 2 and pointing from object 1 towards object 2. The Universal Gravitation Constant is $G = 6.67\times10^{-11}\,\text{N}\cdot\text{m}^2\cdot\text{kg}^{-2}$. Figure 9.1 shows the direction of the forces on bodies 1 and 2 along with the unit vector $\hat{\mathbf{r}}_{1,2}$.

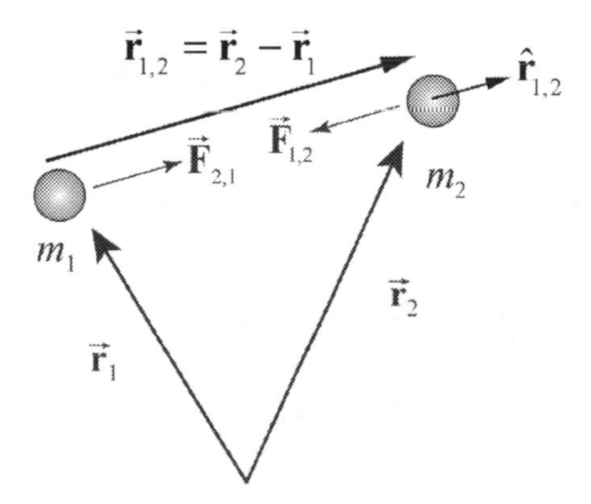

Figure 9.1 Gravitational force of interaction between two bodies

Newton realized that there were still some subtleties involved. First, why should the mass of the Earth act as if it were all placed at the center? Newton showed that for a perfect sphere with uniform mass distribution, all the mass may be assumed to be located at the center. (This calculation is difficult and can be found in Appendix 9A to this chapter.) We assume for the present calculation that the Earth and the Moon are perfect spheres with uniform mass distribution.

Second, does this gravitational force between the Earth and the Moon form an action-reaction Third Law pair? When Newton first explained the Moon's motion in 1666, he had still not formulated the Third Law, which accounted for the long delay in the publication of the *Principia*. The link between the concept of force and the concept of an action-reaction pair of forces was the last piece needed to solve the puzzle of the effect of gravity on planetary orbits. Once Newton realized that the gravitational force between any two bodies forms an action-reaction pair, and satisfies both the Universal Law of Gravitation and his newly formulated Third Law, he was able to solve the oldest and most important physics problem of his time, the motion of the planets.

The test for the Universal Law of Gravitation was performed through experimental observation of the motion of planets, which turned out to be resoundingly successful. For almost 200 years, Newton's Universal Law was in excellent agreement with observation. A sign of more complicated physics ahead, the first discrepancy only occurred when a slight deviation of the motion of Mercury was experimentally confirmed in 1882. The prediction of this deviation was the first success of Einstein's Theory of General Relativity (formulated in 1915).

We can apply this Universal Law of Gravitation to calculate the period of the Moon's orbit around the Earth. The mass of the Moon is $m_1 = 7.36 \times 10^{22}\ \text{kg}$ and the mass of the Earth is $m_2 = 5.98 \times 10^{24}\ \text{kg}$. The distance from the Earth to the Moon is $R_{\text{e, m}} = 3.82 \times 10^8\ \text{m}$. We show the force diagram in Figure 9.2.

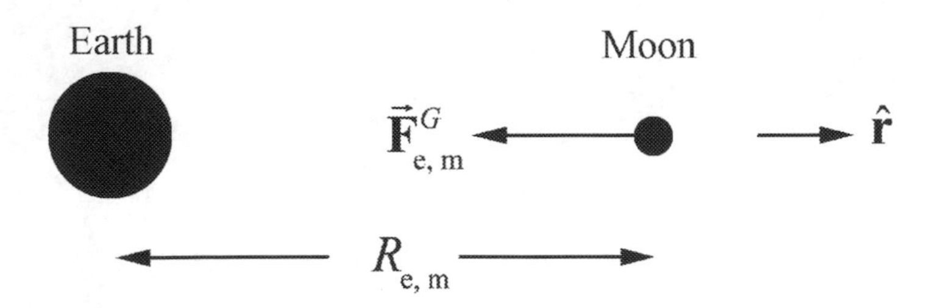

Figure 9.2 Gravitational force of moon

Newton's Second Law of motion for the radial direction becomes

$$-G\frac{m_1 m_2}{R_{\text{e,m}}^2} = -m_1 \frac{4\pi^2 R_{\text{e,m}}}{T^2}. \tag{9.2.3}$$

We can solve this equation for the period of the orbit,

$$T = \sqrt{\frac{4\pi^2 R_{\text{e,m}}^3}{G m_2}} \,. \tag{9.2.4}$$

Substitute the given values for the radius of the orbit, the mass of the earth, and the universal gravitational constant. The period of the orbit is

$$T = \sqrt{\frac{4\pi^2 (3.82\times10^8\ \text{m})^3}{(6.67\times10^{-11}\ \text{N}\cdot\text{m}^2\cdot\text{kg}^{-2})(5.98\times10^{24}\ \text{kg})}} = 2.35\times10^6\ \text{s} \,. \tag{9.2.5}$$

This period is given in days by

$$T = (2.35\times10^6\ \text{s})\left(\frac{1\ \text{day}}{8.64\times10^4\ \text{s}}\right) = 27.2\ \text{days}. \tag{9.2.6}$$

This period is called the *sidereal month* because it is the time that it takes for the Moon to return to a given position with respect to the stars.

The actual time T_1 between full moons, called the *synodic month* (the average period of the Moon's revolution with respect to the sun and is 29.53 days, it may range between 29.27 days and 29.83 days), is longer than the sidereal month because the Earth is traveling around the Sun. So for the next full moon the Moon must travel a little farther than one full circle around the Earth in order to be on the other side of the Earth from the Sun (Figure 9.3).

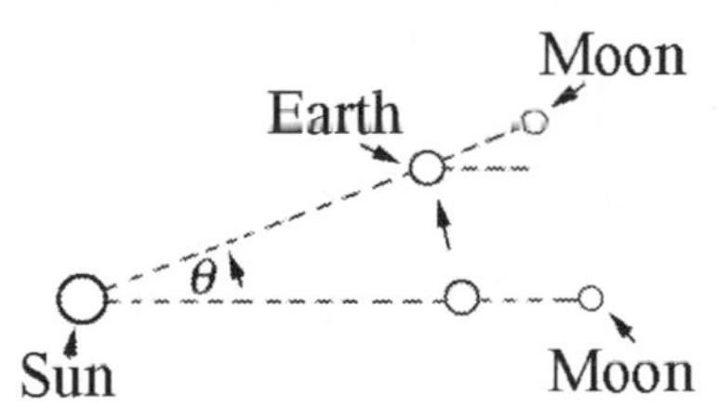

Figure 9.3: Orbital motion between full moons

Therefore the time T_1 between consecutive full moons is approximately $T_1 \simeq T + \Delta T$ where $\Delta T \simeq T/12 = 2.3\ \text{days}$. So $T_1 \simeq 29.5\ \text{days}$.

9.2.2 Kepler's Third Law and Circular Motion

The first thing that we notice from the above solution is that the period does not depend on the mass of the Moon. We also notice that the square of the period is proportional to the cube of the distance between the Earth and the Moon,

$$T^2 = \frac{4\pi^2 R_{\text{e,m}}^3}{G m_2}. \quad (9.2.7)$$

This is an example of Kepler's Third Law, of which Newton was aware. This confirmation was convincing evidence to Newton that his Universal Law of Gravitation was the correct mathematical description of the gravitational force law, even though he still could not explain what "caused" gravity.

9.3 Worked Examples Circular Motion

Example 9.1 Geosynchronous Orbit

A geostationary satellite goes around the earth once every 23 hours 56 minutes and 4 seconds, (a sidereal day, shorter than the noon-to-noon solar day of 24 hours) so that its position appears stationary with respect to a ground station. The mass of the earth is $m_e = 5.98\times10^{24}\ \text{kg}$. The mean radius of the earth is $R_e = 6.37\times10^{6}\ \text{m}$. The universal constant of gravitation is $G = 6.67\times10^{-11}\ \text{N}\cdot\text{m}^2\cdot\text{kg}^{-2}$. What is the radius of the orbit of a geostationary satellite? Approximately how many earth radii is this distance?

Solution: The satellite's motion can be modeled as uniform circular motion. The gravitational force between the earth and the satellite keeps the satellite moving in a circle (In Figure 9.4, the orbit is close to a scale drawing of the orbit). The acceleration of the satellite is directed towards the center of the circle, that is, along the radially inward direction.

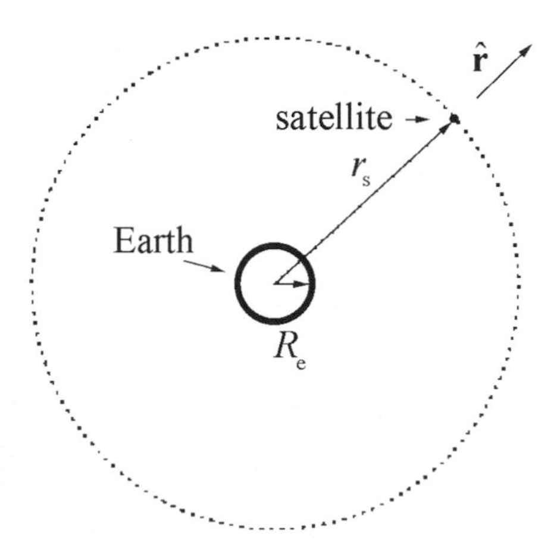

Figure 9.4 Geostationary satellite orbit (close to a scale drawing of orbit).

Choose the origin at the center of the earth, and the unit vector $\hat{\mathbf{r}}$ along the radial direction. This choice of coordinates makes sense in this problem since the direction of acceleration is along the radial direction.

Let $\vec{\mathbf{r}}$ be the position vector of the satellite. The magnitude of $\vec{\mathbf{r}}$ (we denote it as r_s) is the distance of the satellite from the center of the earth, and hence the radius of its circular orbit. Let ω be the angular velocity of the satellite, and the period is $T = 2\pi / \omega$. The acceleration is directed inward, with magnitude $r_s\omega^2$; in vector form,

$$\vec{\mathbf{a}} = -r_s\omega^2\hat{\mathbf{r}}. \tag{9.3.1}$$

Apply Newton's Second Law to the satellite for the radial component. The only force in this direction is the gravitational force due to the Earth,

$$\vec{\mathrm{F}}_{\mathrm{grav}} = -m_s\omega^2 r_s\,\hat{\mathbf{r}}. \tag{9.3.2}$$

The inward radial force on the satellite is the gravitational attraction of the earth,

$$-G\frac{m_s m_e}{r_s^2}\hat{\mathbf{r}} = -m_s\omega^2 r_s\,\hat{\mathbf{r}}. \tag{9.3.3}$$

Equating the $\hat{\mathbf{r}}$ components,

$$G\frac{m_s m_e}{r_s^2} = m_s\omega^2 r_s. \tag{9.3.4}$$

Solving for the radius of orbit of the satellite r_s,

$$r_s = \left(\frac{G\,m_e}{\omega^2}\right)^{1/3}. \tag{9.3.5}$$

The period T of the satellite's orbit in seconds is $86164\ \mathrm{s}$ and so the angular speed is

$$\omega = \frac{2\pi}{T} = \frac{2\pi}{86164\ \mathrm{s}} = 7.2921\times10^{-5}\ \mathrm{s}^{-1}. \tag{9.3.6}$$

Using the values of ω, G and m_e in Equation (9.3.5), we determine r_s,

$$r_s = 4.22\times10^{7}\ \mathrm{m} = 6.62\ R_e. \tag{9.3.7}$$

Example 9.2 Double Star System

Consider a double star system under the influence of gravitational force between the stars. Star 1 has mass m_1 and star 2 has mass m_2. Assume that each star undergoes uniform circular motion such that the stars are always a fixed distance s apart (rotating counterclockwise in Figure 9.5). What is the period of the orbit?

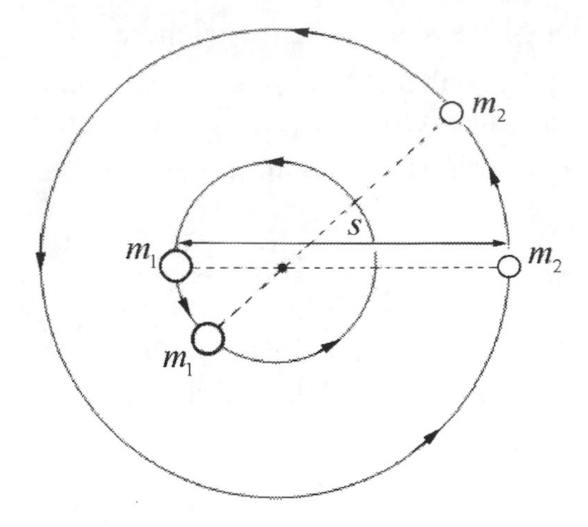

Figure 9.5 Two stars undergoing circular orbits about each other

Solution: Because the distance between the two stars doesn't change as orbit about each other there is a central point where the line connected the two objects intersect as the objects move as can be seen in the figure above. (We will see later on in the course that central point is the center of mass of the system.) Choose radial coordinates for each star with origin at that central point. Let $\hat{\mathbf{r}}_1$ be a unit vector at Star 1 pointing radially away from the center of mass. The position of object 1 is then $\vec{\mathbf{r}}_1 = r_1\,\hat{\mathbf{r}}_1$, where r_1 is the distance from the central point. Let $\hat{\mathbf{r}}_2$ be a unit vector at Star 2 pointing radially away from the center of mass. The position of object 2 is then $\vec{\mathbf{r}}_2 = r_2\,\hat{\mathbf{r}}_2$, where r_2 is the distance from the central point. Because the distance between the two stars is fixed we have that

$$s = r_1 + r_2 .$$

The coordinate system is shown in Figure 9.6

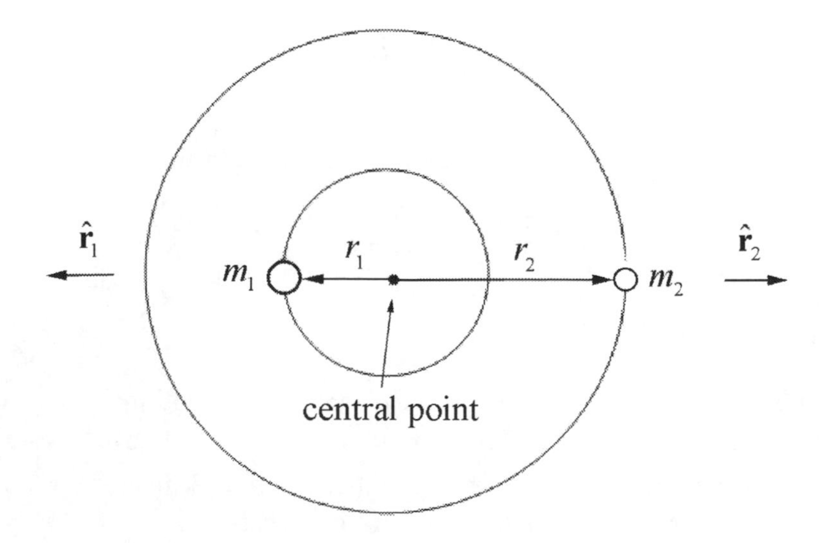

Figure 9.6 Coordinate system for double star orbits

The gravitational force on object 1 is then

$$\vec{\mathbf{F}}_{2,1} = -\frac{Gm_1m_2}{s^2}\hat{\mathbf{r}}_1 .$$

The gravitational force on object 2 is then

$$\vec{\mathbf{F}}_{1,2} = -\frac{Gm_1m_2}{s^2}\hat{\mathbf{r}}_2 .$$

The force diagrams on the two stars are shown in Figure 9.7.

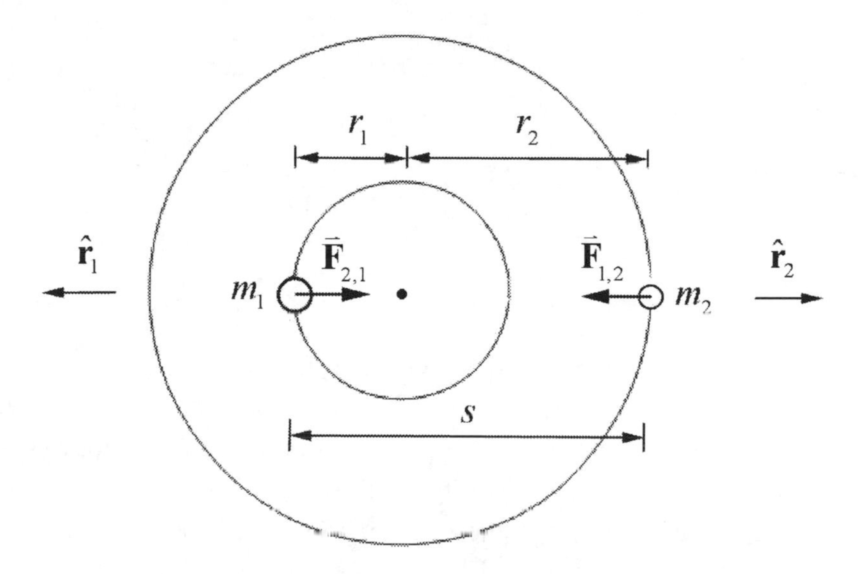

Figure 9.7 Force diagrams on objects 1 and 2

Let ω denote the magnitude of the angular velocity of each star about the central point. Then Newton's Second Law, $\vec{\mathbf{F}}_1 = m_1\vec{\mathbf{a}}_1$, for Star 1 in the radial direction $\hat{\mathbf{r}}_1$ is

$$-G\frac{m_1 m_2}{s^2} = -m_1 r_1 \omega^2 .$$

We can solve this for r_1,

$$r_1 = G\frac{m_2}{\omega^2 s^2} .$$

Newton's Second Law, $\vec{\mathbf{F}}_2 = m_2\vec{\mathbf{a}}_2$, for Star 2 in the radial direction $\hat{\mathbf{r}}_2$ is

$$-G\frac{m_1 m_2}{s^2} = -m_2 r_2 \omega^2 .$$

We can solve this for r_2,

$$r_2 = G\frac{m_1}{\omega^2 s^2}.$$

Because s, the distance between the stars, is constant

$$s = r_1 + r_2 = G\frac{m_2}{\omega^2 s^2} + G\frac{m_1}{\omega^2 s^2} = G\frac{(m_2 + m_1)}{\omega^2 s^2}.$$

Thus the magnitude of the angular velocity is

$$\omega = \left(G\frac{(m_2 + m_1)}{s^3} \right)^{1/2},$$

and the period is then

$$T = \frac{2\pi}{\omega} = \left(\frac{4\pi^2 s^3}{G(m_2 + m_1)} \right)^{1/2}. \tag{9.3.8}$$

Note that both masses appear in the above expression for the period unlike the expression for Kepler's Law for circular orbits. Eq. (9.2.7). The reason is that in the argument leading up to Eq. (9.2.7), we assumed that $m_1 << m_2$, this was equivalent to assuming that the central point was located at the center of the Earth. If we used Eq. (9.3.8) instead we would find that the orbital period for the circular motion of the Earth and moon about each other is

$$T = \sqrt{\frac{4\pi^2(3.82\times 10^8 \text{ m})^3}{(6.67\times 10^{-11} \text{ N}\cdot\text{m}^2\cdot\text{kg}^{-2})(5.98\times 10^{24} \text{ kg} + 7.36\times 10^{22} \text{ kg})}} = 2.33\times 10^6 \text{ s},$$

which is $1.43\times 10^4 \text{ s} = 0.17 \text{ d}$ shorter than our previous calculation.

Example 9.3 Rotating Objects

Two objects 1 and 2 of mass m_1 and m_2 are whirling around a shaft with a constant angular velocity ω. The first object is a distance d from the central axis, and the second object is a distance $2d$ from the axis (Figure 9.8). You may ignore the mass of the strings and neglect the effect of gravity. (a) What is the tension in the string between the inner object and the outer object? (b) What is the tension in the string between the shaft and the inner object?

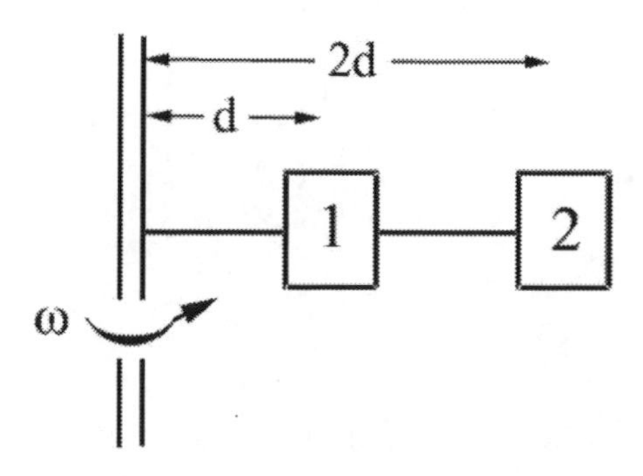

Figure 9.8 Objects attached to a rotating shaft

Solution: We begin by drawing separate force diagrams, Figure 9.9a for object 1 and Figure 9.9b for object 2.

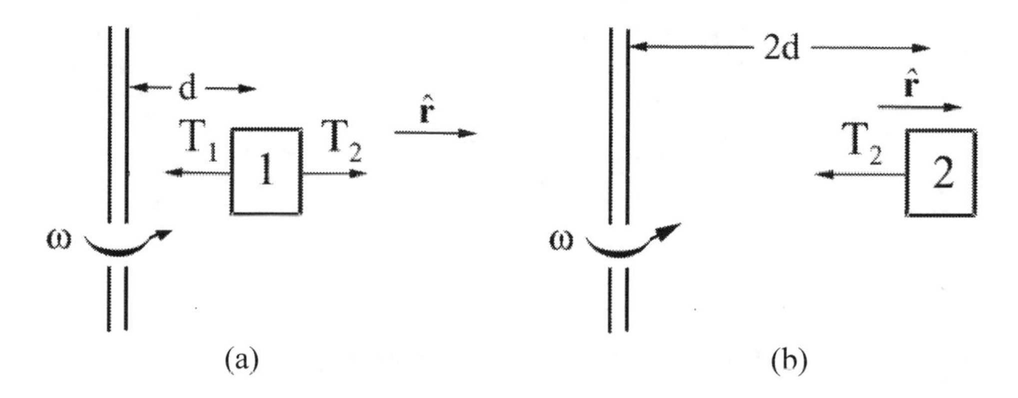

Figure 9.9 (a) and 9.9 (b) Free-body force diagrams for objects 1 and 2

Newton's Second Law, $\vec{\mathbf{F}}_1 = m_1 \vec{\mathbf{a}}_1$, for the inner object in the radial direction is

$$\hat{\mathbf{r}} : T_2 - T_1 = -m_1 d \omega^2 .$$

Newton's Second Law, $\vec{\mathbf{F}}_2 = m_2 \vec{\mathbf{a}}_2$, for the outer object in the radial direction is

$$\hat{\mathbf{r}} : -T_2 = -m_2 2d \omega^2 .$$

The tension in the string between the inner object and the outer object is therefore

$$T_2 = m_2 2d \omega^2 .$$

Using this result for T_2 in the force equation for the inner object yields

$$m_2 2d\omega^2 - T_1 = -m_1 d\omega^2,$$

which can be solved for the tension in the string between the shaft and the inner object

$$T_1 = d\omega^2(m_1 + 2m_2).$$

Example 9.4 Tension in a Rope

A uniform rope of mass m and length L is attached to shaft that is rotating at constant angular velocity ω. Find the tension in the rope as a function of distance from the shaft. You may ignore the effect of gravitation.

Solution: Divide the rope into small pieces of length Δr, each of mass $\Delta m = (m / L)\Delta r$. Consider the piece located a distance r from the shaft (Figure 9.10).

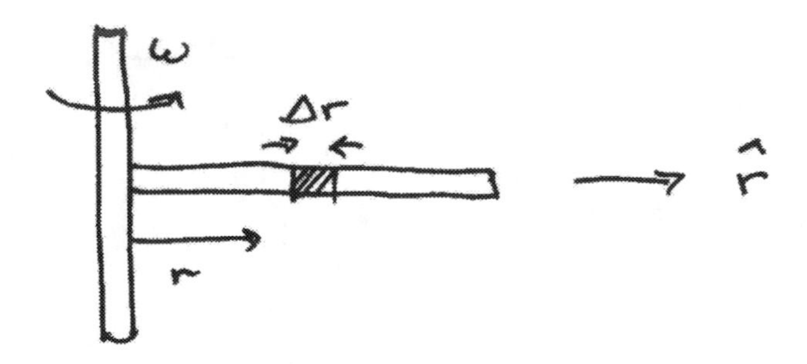

Figure 9.10 Small slice of rotating rope

The radial component of the force on that piece is the difference between the tensions evaluated at the sides of the piece, $F_r = T(r + \Delta r) - T(r)$, (Figure 9.11).

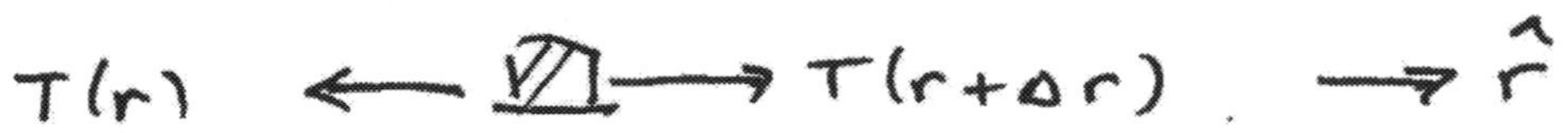

Figure 9.11 Free-body force diagram on small slice of rope

The piece is accelerating inward with a radial component $a_r = -r\omega^2$. Thus Newton's Second Law becomes

$$\begin{aligned} F_r &= -\Delta m \omega^2 r \\ T(r + \Delta r) - T(r) &= -(m / L)\Delta r\, r\omega^2. \end{aligned} \tag{9.3.9}$$

Denote the difference in the tension by $\Delta T = T(r + \Delta r) - T(r)$. After dividing through by Δr, Eq. (9.3.9) becomes

$$\frac{\Delta T}{\Delta r} = -(m / L)\, r\omega^2. \tag{9.3.10}$$

In the limit as $\Delta r \to 0$, Eq. (9.3.10) becomes a differential equation,

$$\frac{dT}{dr} = -(m / L)\omega^2 r . \tag{9.3.11}$$

From this, we see immediately that the tension decreases with increasing radius. We shall solve this equation by integration

$$\begin{aligned} T(r) - T(L) &= \int_{r'=L}^{r'=r} \frac{dT}{dr'} dr' = -(m\omega^2 / L) \int_L^r r' \, dr' \\ &= -(m\omega^2 / 2L)(r^2 - L^2) \\ &= (m\omega^2 / 2L)(L^2 - r^2). \end{aligned} \tag{9.3.12}$$

We use the fact that the tension, in the ideal case, will vanish at the end of the rope, $r = L$. Thus,

$$T(r) = (m\omega^2 / 2L)(L^2 - r^2). \tag{9.3.13}$$

This last expression shows the expected functional form, in that the tension is largest closest to the shaft, and vanishes at the end of the rope.

Example 9.5 Object Sliding in a Circular Orbit on the Inside of a Cone

Consider an object of mass m that slides without friction on the inside of a cone moving in a circular orbit with constant speed v_0. The cone makes an angle θ with respect to a vertical axis. The axis of the cone is vertical and gravity is directed downwards. The apex half-angle of the cone is θ as shown in Figure 9.12. Find the radius of the circular path and the time it takes to complete one circular orbit in terms of the given quantities and g.

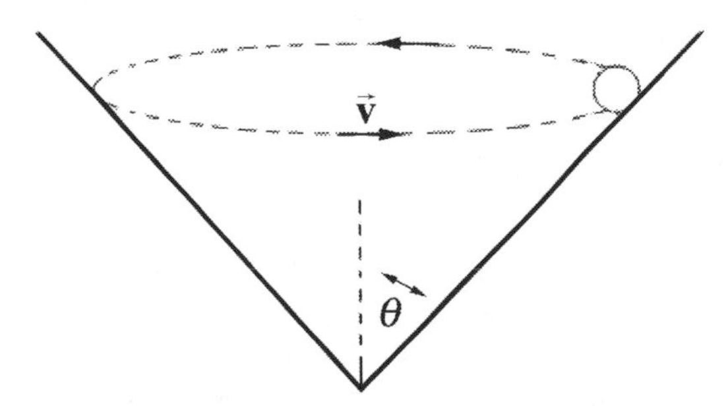

Figure 9.12 Object in a circular orbit on inside of a cone

Solution: Choose cylindrical coordinates as shown in the above figure. Choose unit vectors $\hat{\mathbf{r}}$ pointing in the radial outward direction and $\hat{\mathbf{k}}$ pointing upwards. The force diagram on the object is shown in Figure 9.13.

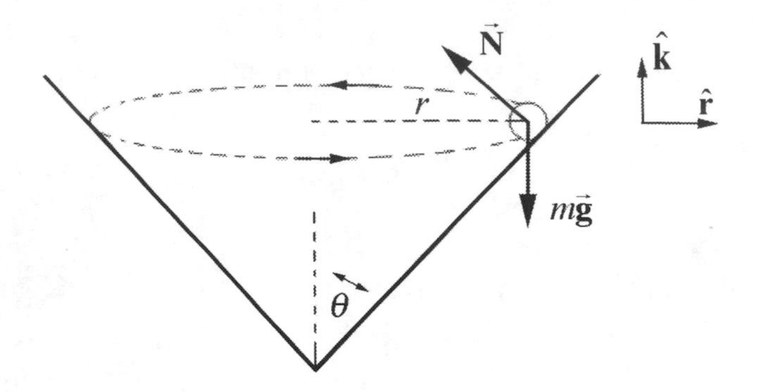

Figure 9.13 Free-body force diagram on object

The two forces acting on the object are the normal force of the wall on the object and the gravitational force. Then Newton's Second Law in the $\hat{\mathbf{r}}$-direction becomes

$$-N\cos\theta = \frac{-mv^2}{r}$$

and in the $\hat{\mathbf{k}}$-direction becomes

$$N\sin\theta - mg = 0.$$

These equations can be re-expressed as

$$N\cos\theta = m\frac{v^2}{r}$$

$$N\sin\theta = mg.$$

We can divide these two equations,

$$\frac{N\sin\theta}{N\cos\theta} = \frac{mg}{mv^2/r}$$

yielding

$$\tan\theta = \frac{rg}{v^2}.$$

This can be solved for the radius,

$$r = \frac{v^2}{g}\tan\theta.$$

The centripetal force in this problem is the vector component of the contact force that is pointing radially inwards,

$$F_{\text{cent}} = N\cos\theta = mg\cot\theta,$$

where $N\sin\theta = mg$ has been used to eliminate N in terms of m, g and θ. The radius is independent of the mass because the component of the normal force in the vertical direction must balance the gravitational force, and so the normal force is proportional to the mass.

Example 9.6 Coin on a Rotating Turntable

A coin of mass m (which you may treat as a point object) lies on a turntable, exactly at the rim, a distance R from the center. The turntable turns at constant angular speed ω and the coin rides without slipping. Suppose the coefficient of static friction between the turntable and the coin is given by μ. Let g be the gravitational constant. What is the maximum angular speed $\omega_{\max}$ such that the coin does not slip?

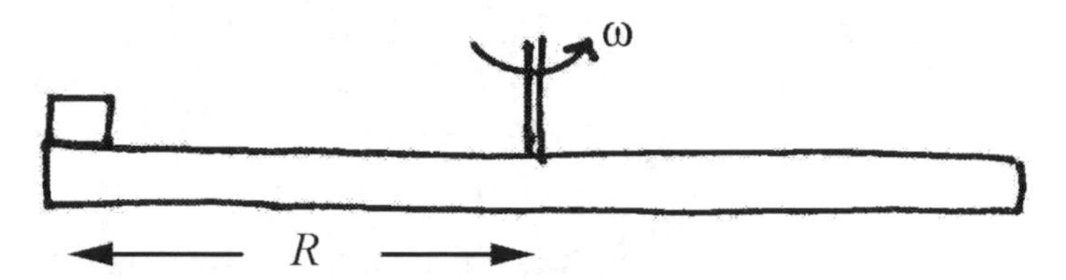

Figure 9.14 Coin on Rotating Turntable

Solution: The coin undergoes circular motion at constant speed so it is accelerating inward. The force inward is static friction and at the just slipping point it has reached its maximum value. We can use Newton's Second Law to find the maximum angular speed $\omega_{\max}$. We choose a polar coordinate system and the free-body force diagram is shown in the figure below.

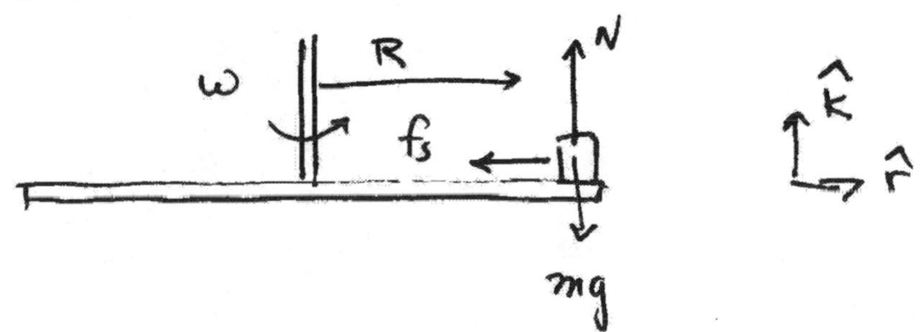

Figure 9.15 Free-body force diagram on coin

The contact force is given by

$$\vec{\mathbf{C}} = \vec{\mathbf{N}} + \vec{\mathbf{f}}_s = N\hat{\mathbf{k}} - f_s\hat{\mathbf{r}}. \tag{9.3.14}$$

The gravitational force is given by

$$\vec{\mathbf{F}}_{grav} = -mg\hat{\mathbf{k}}. \tag{9.3.15}$$

Newton's Second Law in the radial direction is given by

$$-f_s = -mR\omega^2. \tag{9.3.16}$$

Newton's Second Law, $F_z = ma_z$, in the z-direction, noting that the disc is static hence $a_z = 0$, is given by

$$N - mg = 0 . \tag{9.3.17}$$

Thus the normal force is

$$N = mg . \tag{9.3.18}$$

As ω increases, the static friction increases in magnitude until at $\omega = \omega_{\text{max}}$ and static friction reaches its maximum value (noting Eq. (9.3.18)).

$$(f_s)_{\text{max}} = \mu N = \mu mg . \tag{9.3.19}$$

At this value the disc slips. Thus substituting this value for the maximum static friction into Eq. (9.3.16) yields

$$\mu mg = mR\omega_{\text{max}}^2 . \tag{9.3.20}$$

We can now solve Eq. (9.3.20) for maximum angular speed ω_{max} such that the coin does not slip

$$\omega_{\text{max}} = \sqrt{\frac{\mu g}{R}} . \tag{9.3.21}$$

Appendix 9A The Gravitational Field of a Spherical Shell of Matter

When analyzing gravitational interactions between extended bodies we assumed that the when calculating the gravitational force between spherical objects with uniform mass distributions we could treat each sphere as a point-like mass located at the center of the sphere and then use the Universal Law of Gravitation to determine the force between the two point-like objects. We shall now justify that assumption. For simplicity we only need to consider the interaction between a spherical object and a point-like mass. We would like to determine the gravitational force on the point-like object of mass m_1 due to the gravitational interaction with a solid uniform sphere of mass m_2 and radius R. In order to determine the force law we shall first consider the interaction between the point-like object and a uniform spherical shell of mass m_s and radius R. We will show that:

1) The gravitational force acting on a point-like object of mass m_1 located a distance $r > R$ from the center of a uniform spherical shell of mass m_s and radius R is the same force that would arise if all the mass of the shell were placed at the center of the shell.

2) The gravitational force on an object of mass m_1 placed inside a spherical shell of matter is zero.

The force law summarizes these results:

$$\vec{\mathbf{F}}_{s,1}(r) = \begin{cases} -G\dfrac{m_s m_1}{r^2}\hat{\mathbf{r}}, & r > R \\ \vec{\mathbf{0}}, & r < R \end{cases},$$

where $\hat{\mathbf{r}}$ is the unit vector located at the position of the object and pointing radially away from the center of the shell.

For a uniform spherical distribution of matter, we can divide the sphere into thin shells. Then the force between the point-like object and each shell is the same as if all the mass of the shell were placed at the center of the shell. Then we add up all the contributions of the shells (integration), the spherical distribution can be treated as point-like object located at the center of the sphere justifying our assumption.

Thus it suffices to analyze the case of the spherical shell. We shall first divide the shell into small area elements and calculate the gravitational force on the point-like object due to one element of the shell and then add the forces due to all these elements via integration.

We begin by choosing a coordinate system. Choose our z-axis to be directed from the center of the sphere to the position of the object, at position $\vec{\mathbf{r}} = z\hat{\mathbf{k}}$, so that $z \geq 0$. (Figure 9A.1 shows the object lying outside the shell with $z > R$).

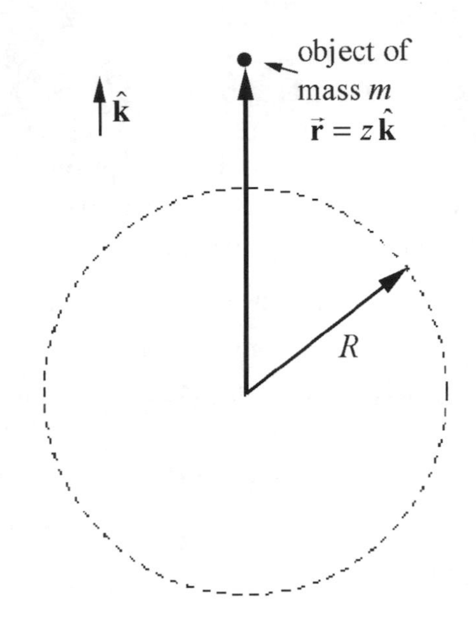

Figure 9A.1 Object lying outside shell with $z > R$.

Choose spherical coordinates as shown in Figure 9A.2.

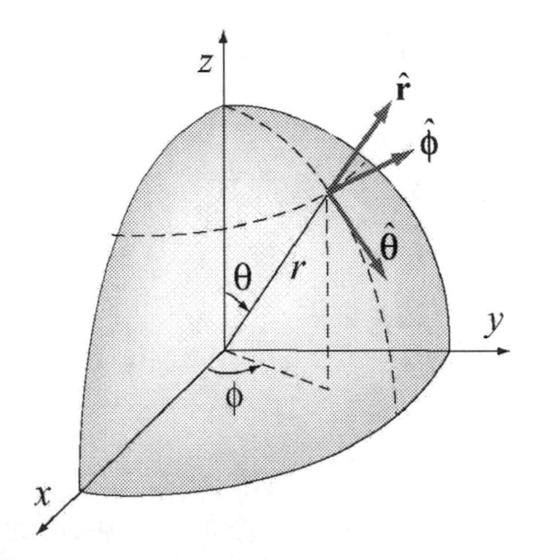

Figure 9A.2 Spherical coordinates

For a point on the surface of a sphere of radius $r = R$, the Cartesian coordinates are related to the spherical coordinates by

$$\begin{aligned} x &= R\sin\theta\cos\phi, \\ y &= R\sin\theta\sin\phi, \\ z &= R\cos\theta, \end{aligned} \qquad (9.A.1)$$

where $0 \le \theta \le \pi$ and $0 \le \phi \le 2\pi$.

Note that the angle θ in Figure 9A.2 and Equations (9.A.1) is not the same as that in plane polar coordinates or cylindrical coordinates. The angle θ is known as the *co-latitude*, the complement of the latitude. We now choose a small area element shown in Figure 9A.3.

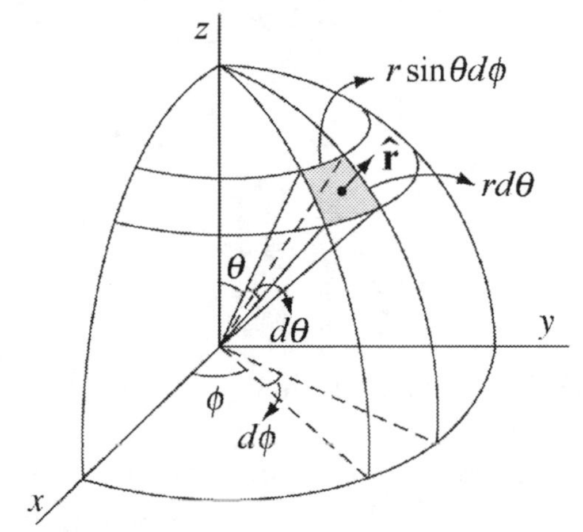

Figure 9A.3 Infinitesimal area element

The infinitesimal area element on the surface of the shell is given by

$$da = R\sin\theta d\theta d\phi .$$

Then the mass dm contained in that element is

$$dm = \sigma da = \sigma R^2 \sin\theta d\theta d\phi .$$

where σ is the surface mass density given by

$$\sigma = m_s / 4\pi R^2 .$$

The gravitational force $\vec{\mathbf{F}}_{dm,m_1}$ on the object of mass m_1 that lies outside the shell due to the infinitesimal piece of the shell (with mass dm) is shown in Figure 9A.4.

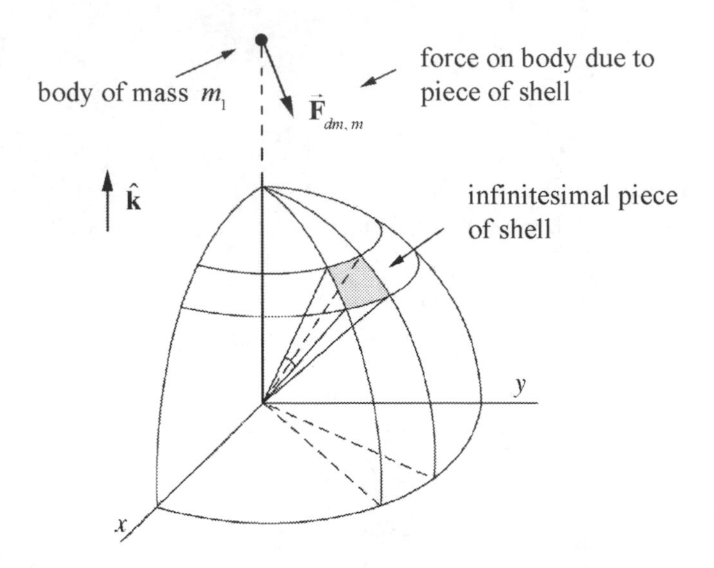

Figure 9A.4 Force on a point-like object due to piece of shell

The contribution from the piece with mass dm to the gravitational force on the object of mass m_1 that lies outside the shell has a component pointing in the negative $\hat{\mathbf{k}}$-direction and a component pointing radially away from the z-axis. By symmetry there is another mass element with the same differential mass $dm' = dm$ on the other side of the shell with same co-latitude θ but with ϕ replaced by $\phi \pm \pi$; this replacement changes the sign of x and y in Equations (9.A.1) but leaves z unchanged. This other mass element produces a gravitational force that exactly cancels the radial component of the force pointing away from the z-axis. Therefore the sum of the forces of these differential mass elements on the object has only a component in the negative $\hat{\mathbf{k}}$-direction (Figure 9A.5)

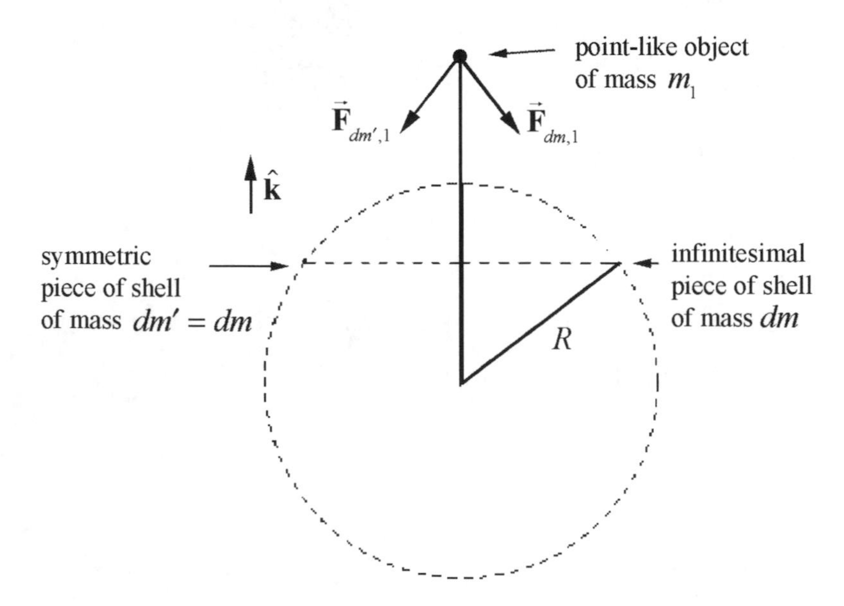

Figure 9A.5 Symmetric cancellation of components of force

Therefore we need only the z-component vector of the force due to the piece of the shell on the point-like object.

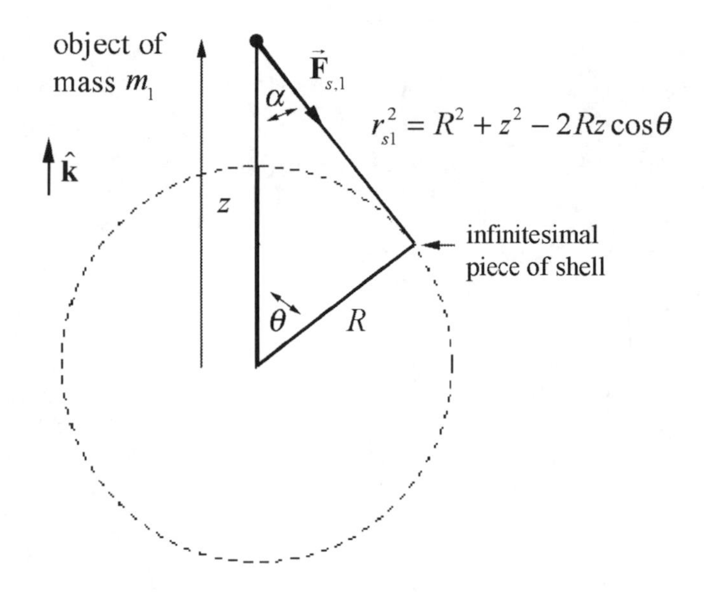

Figure 9A.6 Geometry for calculating the force due to piece of shell.

From the geometry of the set-up (Figure 9A.6) we see that

$$(d\vec{\mathbf{F}}_{s,1})_z \equiv dF_z\hat{\mathbf{k}} = -G\frac{m_1 dm}{r_{s1}^2}\cos\alpha\hat{\mathbf{k}}.$$

Thus

$$dF_z = -G\frac{m_1 dm}{r_{s1}^2}\cos\alpha = -\frac{Gm_s m_1}{4\pi}\frac{\cos\alpha\sin\theta d\theta d\phi}{r_{s1}^2}. \quad (9.A.2)$$

The integral of the force over the surface is then

$$F_z = -Gm_1\int_{\theta=0}^{\theta=\pi}\int_{\phi=0}^{\phi=2\pi}\frac{dm\cos\alpha}{r_{s1}^2} = -\frac{Gm_s m_1}{4\pi}\int_{\theta=0}^{\theta=\pi}\int_{\phi=0}^{\phi=2\pi}\frac{\cos\alpha\sin\theta d\theta d\phi}{r_{s1}^2}. \quad (9.A.3)$$

The ϕ-integral is straightforward yielding

$$F_z = -G\frac{m_1 dm}{r_{s1}^2}\cos\alpha = -\frac{Gm_s m_1}{2}\int_{\theta=0}^{\theta=\pi}\frac{\cos\alpha\sin\theta d\theta}{r_{s1}^2}. \quad (9.A.4)$$

From Figure 9A.6 we can use the law of cosines in two different ways

$$r_{s1}^2 = R^2 + z^2 - 2Rz\cos\theta$$
$$R^2 = z^2 + r_{s1}^2 - 2r_{s,1}z\cos\alpha. \tag{9.A.5}$$

Differentiating the first expression in (9.A.5), with R and z constant yields,

$$2r_{s,1}dr_{s,1} = 2Rz\sin\theta\, d\theta. \tag{9.A.6}$$

Hence

$$\sin\theta\, d\theta = \frac{r_{s,1}}{Rz}dr_{s,1}. \tag{9.A.7}$$

and from the second expression in (9.A.5) we have that

$$\cos\alpha = \frac{1}{2zr_{s,1}}\left[(z^2 - R^2) + r_{s1}^2\right]. \tag{9.A.8}$$

We now have everything we need in terms of $r_{s,1}$.

For the case when $z > R$, $r_{s,1}$ varies from $z + R$ to $z - R$. Substituting Equations (9.A.7) and (9.A.8) into Eq. (9.A.3) and using the limits for the definite integral yields

$$\begin{aligned} F_z &= -\frac{Gm_s m_1}{2}\int_{\theta=0}^{\theta=\pi}\frac{\cos\alpha\sin\theta}{r_{s,1}^2}d\theta \\ &= -\frac{Gm_s m_1}{2}\frac{1}{2z}\int_{z+R}^{z-R}\frac{1}{r_{s,1}}\left[\left(z^2 - R^2\right) + r_{s,1}^2\right]\frac{1}{r_{s,1}^2}\frac{r_{s,1}dr_{s,1}}{Rz} \\ &= -\frac{Gm_s m_1}{2}\frac{1}{2Rz^2}\left[\left(z^2 - R^2\right)\int_{z+R}^{z-R}\frac{dr_{s,1}}{r_{s,1}^2} + \int_{z+R}^{z-R}dr_{s,1}\right]. \end{aligned} \tag{9.A.9}$$

No tables should be needed for these; the result is

$$\begin{aligned} F_z &= -\frac{Gm_s m_1}{2}\frac{1}{2Rz^2}\left[-\frac{\left(z^2 - R^2\right)}{r_{s,1}} + r_{s,1}\right]_{z+R}^{z-R} \\ &= -\frac{Gm_s m_1}{2}\frac{1}{2Rz^2}\left[-(z+R) + (z-R) - 2R\right] \\ &= -\frac{Gm_s m_1}{z^2}. \end{aligned} \tag{9.A.10}$$

For the case when $z < R$, $r_{s,1}$ varies from $R+z$ to $R-z$. Then the integral is

$$\begin{aligned} F_z &= -\frac{Gm_s m_1}{2}\frac{1}{2Rz^2}\left[-\frac{(z^2 - R^2)}{r_{s,1}} + r_{s,1}\right]_{R+z}^{R-z} \\ &= -\frac{Gm_s m_1}{2}\frac{1}{2Rz^2}\left[+(R+z)+(z-R)-2z\right] \\ &= 0. \end{aligned} \tag{9.A.11}$$

So we have demonstrated the proposition that for a point-like object located on the z-axis a distance z from the center of a spherical shell, the gravitational force on the point like object is given by

$$\vec{\mathbf{F}}_{s,1}(r) = \begin{cases} -G\dfrac{m_s m_1}{z^2}\hat{\mathbf{k}}, & z > R \\ \vec{\mathbf{0}}, & z < R \end{cases}.$$

This proves the result that the gravitational force inside the shell is zero and the gravitational force outside the shell is equivalent to putting all the mass at the center of the shell.

Chapter 10 Momentum, System of Particles, and Conservation of Momentum

Chapter 10 Momentum, System of Particles, and Conservation of Momentum

Law II: The change of motion is proportional to the motive force impressed, and is made in the direction of the right line in which that force is impressed.

If any force generates a motion, a double force will generate double the motion, a triple force triple the motion, whether that force is impressed altogether and at once or gradually and successively. And this motion (being always directed the same way with the generating force), if the body moved before, is added or subtracted from the former motion, according as they directly conspire with or are directly contrary to each other; or obliquely joined, when they are oblique, so as to produce a new motion compounded from the determination of both. [1]

Isaac Newton *Principia*

10.1 Introduction

When we apply a force to an object, through pushing, pulling, hitting or otherwise, we are applying that force over a discrete interval of time, Δt. During this time interval, the applied force may be constant, or it may vary in magnitude or direction. Forces may also be applied continuously without interruption, such as the gravitational interaction between the earth and the moon. In this chapter we will investigate the relationship between forces and the time during which they are applied, and in the process learn about the quantity of momentum, the principle of conservation of momentum, and its use in solving a new set of problems in mechanics: collisions.

10.2 Momentum (Quantity of Motion) and Impulse

Newton defined the quantity of motion or the **momentum**, $\vec{\mathbf{p}}$, to be the product of the mass and the velocity

$$\vec{\mathbf{p}} = m\vec{\mathbf{v}} . \tag{10.2.1}$$

Momentum is a vector quantity, with direction and magnitude. The direction of momentum is the same as the direction of the velocity. The magnitude of the momentum is the product of the mass and the instantaneous speed.

Units: In the SI system of units, momentum has units of $[\mathrm{kg \cdot m \cdot s^{-1}}]$. There is no special name for this combination of units.

---

[1] Isaac Newton. *Mathematical Principles of Natural Philosophy.* Translated by Andrew Motte (1729). Revised by Florian Cajori. Berkeley: University of California Press, 1934. p. 13.

10.2.1 Average Force, Momentum, and Impulse

Suppose you are pushing a cart with a force that is non-uniform, but has an average value $\vec{\mathbf{F}}_{\text{ave}}$ during the time interval Δt. We can find the average acceleration according to Newton's Second Law,

$$\vec{\mathbf{F}}_{\text{ave}} = m\vec{\mathbf{a}}_{\text{ave}}. \tag{10.2.2}$$

Recall that the average acceleration is equal to the change in velocity $\Delta\vec{\mathbf{v}}$ divided by the time interval Δt,

$$\vec{\mathbf{a}}_{\text{ave}} = \frac{\Delta\vec{\mathbf{v}}}{\Delta t}. \tag{10.2.3}$$

Therefore Newton's Second Law can be recast as

$$\vec{\mathbf{F}}_{\text{ave}} = m\,\vec{\mathbf{a}}_{\text{ave}} = \frac{m\Delta\vec{\mathbf{v}}}{\Delta t}. \tag{10.2.4}$$

The change in momentum is the product of the mass and the change in velocity,

$$\Delta\vec{\mathbf{p}} = m\Delta\vec{\mathbf{v}}. \tag{10.2.5}$$

Newton's Second Law can be restated as follows: the product of the average force acting on an object and the time interval over which the force acts will produce a change in momentum of the object,

$$\vec{\mathbf{F}}_{\text{ave}}\Delta t = \Delta\vec{\mathbf{p}}. \tag{10.2.6}$$

This change in momentum is called the ***impulse***,

$$\vec{\mathbf{I}} = \vec{\mathbf{F}}_{\text{ave}}\Delta t = \Delta\vec{\mathbf{p}}. \tag{10.2.7}$$

Force is a vector quantity; impulse is obtained by multiplying a vector by a scalar, and so impulse is also a vector quantity. The SI units for impulse are $[\text{N}\cdot\text{s}] = [\text{kg}\cdot\text{m}\cdot\text{s}^{-1}]$, which are the same units as momentum.

10.2.2 Non-Constant Force and Impulse

Suppose you now let a cart roll down an inclined plane and bounce against a spring mounted at the bottom of the inclined plane (Figure 10.1). The spring is attached to a force sensor. The force between the spring and the cart is a non-constant force, $\vec{\mathbf{F}}(t)$, applied between times t_0 and t_f.

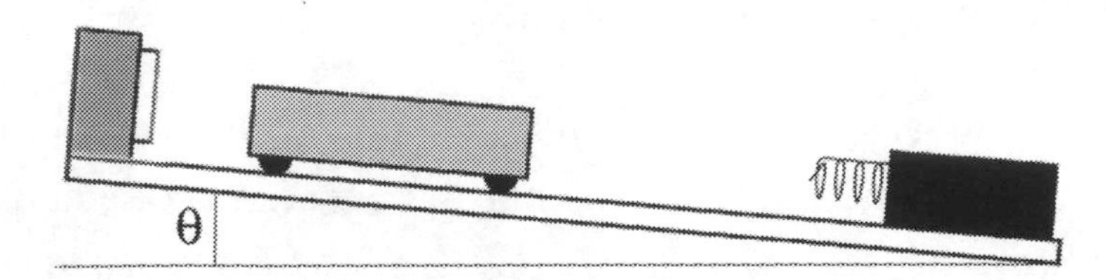

Figure 10.1: Cart sliding down inclined plane and colliding with a spring at the base and reversing motion

In Figure 10.2, we show a sample graph of force vs. time for the cart-spring system as measured by the force sensor during the time the spring is compressed by colliding with the cart.

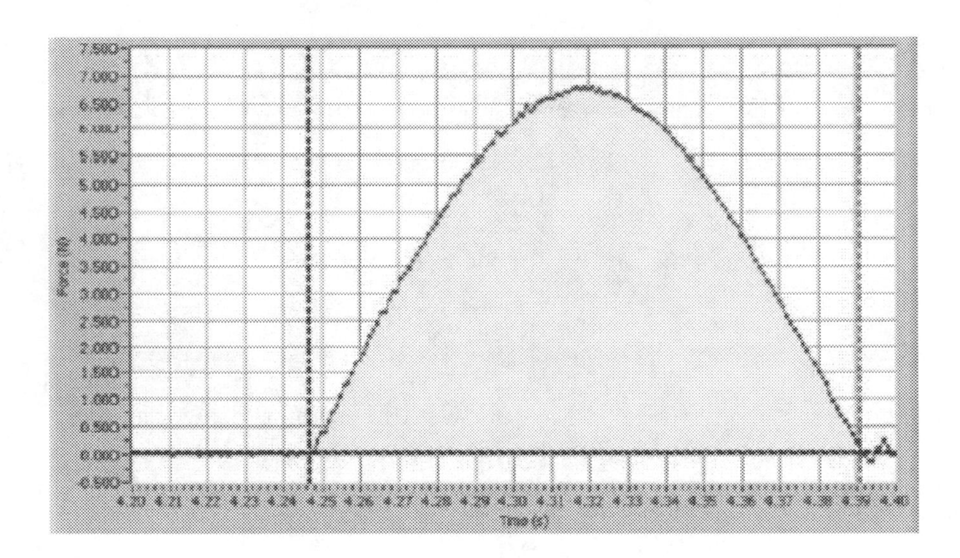

Figure 10.2 A graph of a non-constant force with respect to time.

If we divide up the time interval into N parts, then the impulse is approximately the vector sum of the impulse for each interval,

$$\vec{\mathbf{I}} \cong \sum_{i=1}^{i=N} \vec{\mathbf{F}}_i \, \Delta t_i . \tag{10.2.8}$$

The impulse $\vec{\mathbf{I}}$ is the limit of this sum as we take smaller and smaller intervals. This limit corresponds to the area under the force vs. time curve,

$$\vec{\mathbf{I}} = \lim_{\Delta t_i \to 0} \sum_{i=1}^{i=N} \vec{\mathbf{F}}_i \, \Delta t_i \equiv \int_{t=t_0}^{t=t_f} \vec{\mathbf{F}}(t)\, dt . \tag{10.2.9}$$

Because force is a vector quantity, the integral in Equation (10.2.9) is actually three integrals, one for each component of the force. Using Equation (10.2.7) in Equation (10.2.9) we see that

$$\Delta \vec{\mathbf{p}} = \vec{\mathbf{p}}(t_f) - \vec{\mathbf{p}}(t_0) = \int_{t=t_0}^{t=t_f} \vec{\mathbf{F}}(t)\,dt. \tag{10.2.10}$$

The Fundamental Theorem of Calculus, applied to vectors, then gives

$$\frac{d\vec{\mathbf{p}}}{dt}(t) = \vec{\mathbf{F}}(t) \tag{10.2.11}$$

for both constant and non-constant forces. Equation (10.2.11) is also obtained by taking the limit $\Delta t \to 0$ in Equation (10.2.4). In using either expression, it must be assumed that the mass of the object in question does not change during the interval Δt.

Example 10.1 Impulse for a Non-Constant Force

Suppose you push an object for a time $\Delta t = 1.0\text{ s}$ in the $+x$-direction. For the first half of the interval, you push with a force that increases linearly with time according to

$$\vec{\mathbf{F}}(t) = (bt)\hat{\mathbf{i}}, \quad 0 \le t \le 0.5\text{s} \ \text{ with } b = 2.0\times 10^1\ \text{N}\cdot\text{s}^{-1}. \tag{10.2.12}$$

Then for the second half of the interval, you push with a linearly decreasing force,

$$\vec{\mathbf{F}}(t) = (d - bt)\hat{\mathbf{i}}, \quad 0.5\text{s} \le t \le 1.0\text{s} \ \text{ with } d = 2.0\times 10^1\ \text{N} \tag{10.2.13}$$

The force vs. time graph is shown in Figure 10.3. What is the impulse applied to the object?

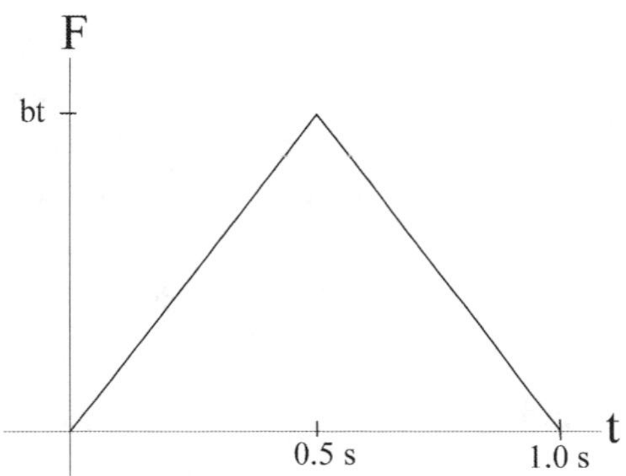

Figure 10.3 Graph of force vs. time

Solution: We can find the impulse by calculating the area under the force vs. time curve. Since the force vs. time graph consists of two triangles, the area under the curve is easy to calculate and is given by

$$
\begin{aligned}
\vec{\mathbf{I}} &= \left[\frac{1}{2}(b\Delta t)(\Delta t/2) + \frac{1}{2}(b\Delta t)(\Delta t/2)\right]\hat{\mathbf{i}} \\
&= \frac{1}{2}b(\Delta t)^2\,\hat{\mathbf{i}} = \frac{1}{2}(2.0\times 10^1\,\mathrm{N\cdot s^{-1}})(1.0\,\mathrm{s})^2\,\hat{\mathbf{i}} = (1.0\times 10^1\,\mathrm{N\cdot s})\hat{\mathbf{i}}.
\end{aligned} \tag{10.2.14}
$$

10.3 External and Internal Forces and the Change in Momentum of a System

So far we have restricted ourselves to considering how the momentum of an object changes under the action of a force. For example, if we analyze in detail the forces acting on the cart rolling down the inclined plane (Figure 10.4), we determine that there are three forces acting on the cart: the force $\vec{\mathbf{F}}_{\text{spring, cart}}$ the spring applies to the cart; the gravitational interaction $\vec{\mathbf{F}}_{\text{earth, cart}}$ between the cart and the earth; and the contact force $\vec{\mathbf{F}}_{\text{plane, cart}}$ between the inclined plane and the cart. If we define the cart as our *system*, then everything else acts as the *surroundings*. We illustrate this division of system and surroundings in Figure 10.4.

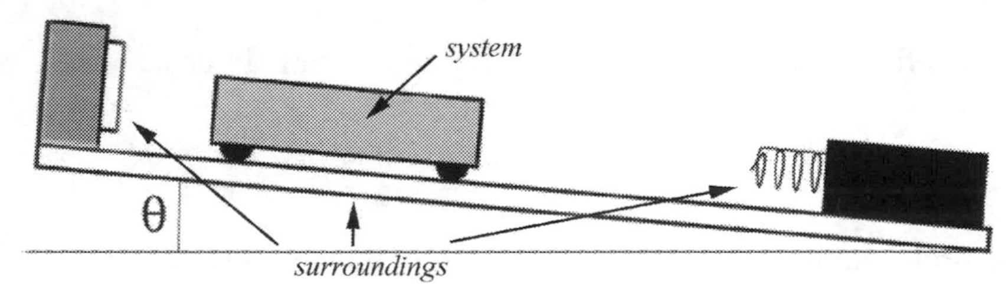

Figure 10.4 A diagram of a cart as a system and its surroundings

The forces acting on the cart are *external* forces. We refer to the vector sum of these external forces that are applied to the system (the cart) as the external force,

$$
\vec{\mathbf{F}}^{\text{ext}} = \vec{\mathbf{F}}_{\text{spring, cart}} + \vec{\mathbf{F}}_{\text{earth, cart}} + \vec{\mathbf{F}}_{\text{plane, cart}}\,. \tag{10.3.1}
$$

Then Newton's Second Law applied to the cart, in terms of impulse, is

$$
\Delta\vec{\mathbf{p}}_{\text{sys}} = \int_{t_0}^{t_f} \vec{\mathbf{F}}^{\text{ext}}\,dt \equiv \vec{\mathbf{I}}_{\text{sys}}. \tag{10.3.2}
$$

Let's extend our system to two interacting objects, for example the cart and the spring. The forces between the spring and cart are now *internal* forces. Both objects, the cart and the spring, experience these internal forces, which by Newton's Third Law are equal in magnitude and applied in opposite directions. So when we sum up the internal forces for the whole system, they cancel. Thus the sum of all the internal forces is always zero,

$$\vec{\mathbf{F}}^{\text{int}} = \vec{\mathbf{0}}. \tag{10.3.3}$$

External forces are still acting on our system; the gravitational force, the contact force between the inclined plane and the cart, and also a new external force, the force between the spring and the force sensor. The force acting on the system is the sum of the internal and the external forces. However, as we have shown, the internal forces cancel, so we have that

$$\vec{\mathbf{F}} = \vec{\mathbf{F}}^{\text{ext}} + \vec{\mathbf{F}}^{\text{int}} = \vec{\mathbf{F}}^{\text{ext}}. \tag{10.3.4}$$

10.4 System of Particles

Suppose we have a system of N particles labeled by the index $i = 1, 2, 3, \cdots, N$. The force on the ith particle is

$$\vec{\mathbf{F}}_i = \vec{\mathbf{F}}_i^{\text{ext}} + \sum_{j=1,\, j\neq i}^{j=N} \vec{\mathbf{F}}_{i,j}. \tag{10.4.1}$$

In this expression $\vec{\mathbf{F}}_{j,i}$ is the force on the ith particle due to the interaction between the ith and j^{th} particles. We sum over all j particles with $j \neq i$ since a particle cannot exert a force on itself (equivalently, we could define $\vec{\mathbf{F}}_{i,i} = \vec{\mathbf{0}}$), yielding the internal force acting on the ith particle,

$$\vec{\mathbf{F}}_i^{\text{int}} = \sum_{j=1,\, j\neq i}^{j=N} \vec{\mathbf{F}}_{j,i}. \tag{10.4.2}$$

The force acting on the system is the sum over all i particles of the force acting on each particle,

$$\vec{\mathbf{F}} = \sum_{i=1}^{i=N} \vec{\mathbf{F}}_i = \sum_{i=1}^{i=N} \vec{\mathbf{F}}_i^{\text{ext}} + \sum_{i=1}^{i=N} \sum_{j=1,\, j\neq i}^{j=N} \vec{\mathbf{F}}_{j,i} = \vec{\mathbf{F}}^{\text{ext}}. \tag{10.4.3}$$

Note that the double sum vanishes,

$$\sum_{i=1}^{i=N} \sum_{j=1,\, j\neq i}^{j=N} \vec{\mathbf{F}}_{j,i} = \vec{\mathbf{0}}, \tag{10.4.4}$$

because all internal forces cancel in pairs,

$$\vec{\mathbf{F}}_{j,i} + \vec{\mathbf{F}}_{i,j} = \vec{\mathbf{0}}. \tag{10.4.5}$$

The force on the ith particle is equal to the rate of change in momentum of the ith particle,

$$\vec{\mathbf{F}}_i = \frac{d\vec{\mathbf{p}}_i}{dt}. \qquad (10.4.6)$$

When can now substitute Equation (10.4.6) into Equation (10.4.3) and determine that that the external force is equal to the sum over all particles of the momentum change of each particle,

$$\vec{\mathbf{F}}^{\text{ext}} = \sum_{i=1}^{i=N} \frac{d\vec{\mathbf{p}}_i}{dt}. \qquad (10.4.7)$$

The momentum of the system is given by the sum

$$\vec{\mathbf{p}}_{\text{sys}} = \sum_{i=1}^{i=N} \vec{\mathbf{p}}_i\,; \qquad (10.4.8)$$

momenta add as vectors. We conclude that the external force causes the momentum of the system to change, and we thus restate and generalize Newton's Second Law for a system of objects as

$$\vec{\mathbf{F}}^{\text{ext}} = \frac{d\vec{\mathbf{p}}_{\text{sys}}}{dt}. \qquad (10.4.9)$$

In terms of impulse, this becomes the statement

$$\Delta\vec{\mathbf{p}}_{\text{sys}} = \int_{t_0}^{t_f} \vec{\mathbf{F}}^{\text{ext}}\,dt \equiv \vec{\mathbf{I}}. \qquad (10.4.10)$$

10.5 Center of Mass

Consider two point-like particles with masses m_1 and m_2. Choose a coordinate system with a choice of origin such that body 1 has position $\vec{\mathbf{r}}_1$ and body 2 has position $\vec{\mathbf{r}}_2$ (Figure 10.5).

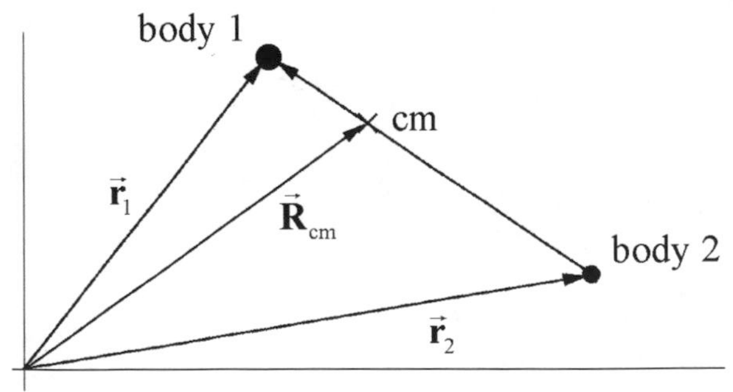

Figure 10.5 Center of mass coordinate system.

The center of mass vector, $\vec{\mathbf{R}}_{\text{cm}}$, of the two-body system is defined as

$$\vec{\mathbf{R}}_{\text{cm}} = \frac{m_1 \vec{\mathbf{r}}_1 + m_2 \vec{\mathbf{r}}_2}{m_1 + m_2}. \tag{10.5.1}$$

We shall now extend the concept of the center of mass to more general systems. Suppose we have a system of N particles labeled by the index $i = 1, 2, 3, \cdots, N$. Choose a coordinate system and denote the position of the ith particle as $\vec{\mathbf{r}}_i$. The mass of the system is given by the sum

$$m_{\text{sys}} = \sum_{i=1}^{i=N} m_i \tag{10.5.2}$$

and the position of the center of mass of the system of particles is given by

$$\vec{\mathbf{R}}_{\text{cm}} = \frac{1}{m_{\text{sys}}} \sum_{i=1}^{i=N} m_i \vec{\mathbf{r}}_i. \tag{10.5.3}$$

(For a continuous rigid body, each point-like particle has mass dm and is located at the position $\vec{\mathbf{r}}'$. The center of mass is then defined as an integral over the body,

$$\vec{\mathbf{R}}_{\text{cm}} = \frac{\int_{\text{body}} dm\, \vec{\mathbf{r}}'}{\int_{\text{body}} dm}. \tag{10.5.4}$$

Example 10.2 Center of Mass of the Earth-Moon System

The mean distance from the center of the earth to the center of the moon is $r_{em} = 3.84 \times 10^8\ \text{m}$. The mass of the earth is $m_e = 5.98 \times 10^{24}\ \text{kg}$ and the mass of the moon is $m_m = 7.34 \times 10^{22}\ \text{kg}$. The mean radius of the earth is $r_e = 6.37 \times 10^6\ \text{m}$. The mean radius of the moon is $r_m = 1.74 \times 10^6\ \text{m}$. Where is the location of the center of mass of the earth-moon system? Is it inside the earth's radius or outside?

Solution: The center of mass of the earth-moon system is defined to be

$$\vec{\mathbf{R}}_{cm} = \frac{1}{m_{\text{sys}}} \sum_{i=1}^{i=N} m_i \vec{\mathbf{r}}_i = \frac{1}{m_e + m_m}(m_e \vec{\mathbf{r}}_e + m_m \vec{\mathbf{r}}_m). \tag{10.5.5}$$

Choose an origin at the center of the earth and a unit vector $\hat{\mathbf{i}}$ pointing towards the moon, then $\vec{\mathbf{r}}_e = \vec{\mathbf{0}}$. The center of mass of the earth-moon system is then

$$\vec{\mathbf{R}}_{cm} = \frac{1}{m_e + m_m}(m_e \vec{\mathbf{r}}_e + m_m \vec{\mathbf{r}}_m) = \frac{m_m \vec{\mathbf{r}}_{em}}{m_e + m_m} = \frac{m_m r_{em}}{m_e + m_m}\hat{\mathbf{i}} \quad (10.5.6)$$

$$\vec{\mathbf{R}}_{cm} = \frac{(7.34\times 10^{22}\ \text{kg})(3.84\times 10^{8}\ \text{m})}{(5.98\times 10^{24}\ \text{kg} + 7.34\times 10^{22}\ \text{kg})}\hat{\mathbf{i}} = 4.66\times 10^{6}\ \text{m}\ \hat{\mathbf{i}} \quad (10.5.7)$$

The earth's mean radius is $r_e = 6.37\times 10^6\ \text{m}$ so the center of mass of the earth-moon system lies within the earth.

Example 10.3 Center of Mass of a Rod

A thin rod has length L and mass M. (a) Suppose the rod is uniform. Find the position of the center of mass with respect to the left end of the rod. (b) Now suppose the rod is not uniform but has a linear mass density that varies with the distance x from the left end according to

$$\lambda = \frac{\lambda_0}{L^2}x^2 \quad (10.5.8)$$

where λ_0 is a constant and has SI units $[\text{kg}\cdot\text{m}^{-1}]$. Find λ_0 and the position of the center of mass with respect to the left end of the rod.

Solution: (a) Choose a coordinate system with the rod aligned along the x-axis and the origin located at the left end of the rod. The center of mass of the rod can be found using the definition given in Eq. (10.5.4). In that expression dm is an infinitesimal mass element and $\vec{\mathbf{r}}$ is the vector from the origin to the mass element dm (Figure 10.6).

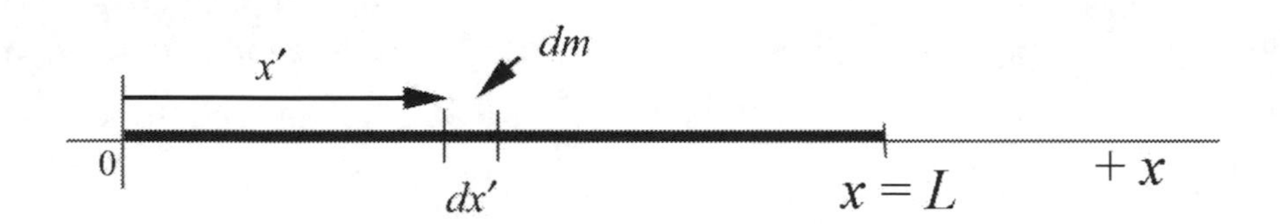

Figure 10.6 Infinitesimal mass element for rod

Choose an infinitesimal mass element dm located a distance x' from the origin. In this problem x' will be the integration variable. Let the length of the mass element be dx'. Then

$$dm = \frac{M}{L}dx' \quad (10.5.9)$$

The vector $\vec{\mathbf{r}} = x'\,\hat{\mathbf{i}}$. The center of mass is found by integration

$$\vec{\mathbf{R}}_{cm} = \frac{1}{M}\int_{\text{body}} \vec{\mathbf{r}}\,dm = \frac{1}{L}\int_{x'=0}^{x} x'dx'\,\hat{\mathbf{i}} = \frac{1}{2L}x'^2\Big|_{x'=0}^{x'=L}\hat{\mathbf{i}} = \frac{1}{2L}(L^2 - 0)\,\hat{\mathbf{i}} = \frac{L}{2}\hat{\mathbf{i}}. \quad (10.5.10)$$

(b) For a non-uniform rod, the mass element is found using Eq. (10.5.8)

$$dm = \lambda(x')dx' = \lambda = \frac{\lambda_0}{L^2} x'^2 dx' . \qquad (10.5.11)$$

The vector $\vec{\mathbf{r}} = x'\,\hat{\mathbf{i}}$. The mass is found by integrating the mass element over the length of the rod

$$M = \int_{\text{body}} dm = \int_{x'=0}^{x=L} \lambda(x')dx' = \frac{\lambda_0}{L^2} \int_{x'=0}^{x=L} x'^2 dx' = \frac{\lambda_0}{3L^2} x'^3 \Big|_{x'=0}^{x'=L} = \frac{\lambda_0}{3L^2}(L^3 - 0) = \frac{\lambda_0}{3} L . \qquad (10.5.12)$$

Therefore

$$\lambda_0 = \frac{3M}{L} \qquad (10.5.13)$$

The center of mass is again found by integration

$$\begin{aligned} \vec{\mathbf{R}}_{cm} &= \frac{1}{M} \int_{\text{body}} \vec{\mathbf{r}}\, dm = \frac{3}{\lambda_0 L} \int_{x'=0}^{x} \lambda(x') x'\, dx'\, \hat{\mathbf{i}} = \frac{3}{L^3} \int_{x'=0}^{x} x'^3\, dx'\, \hat{\mathbf{i}} \\ \vec{\mathbf{R}}_{cm} &= \frac{3}{4L^3} x'^4 \Big|_{x'=0}^{x'=L} \hat{\mathbf{i}} = \frac{3}{4L^3}(L^4 - 0)\,\hat{\mathbf{i}} = \frac{3}{4} L\,\hat{\mathbf{i}}. \end{aligned} \qquad (10.5.14)$$

.

10.6 Translational Motion of the Center of Mass

The velocity of the center of mass is found by differentiation,

$$\vec{\mathbf{V}}_{\text{cm}} = \frac{1}{m_{\text{sys}}} \sum_{i=1}^{i=N} m_i \vec{\mathbf{v}}_i = \frac{\vec{\mathbf{p}}_{\text{sys}}}{m_{\text{sys}}} . \qquad (10.6.1)$$

The momentum is then expressed in terms of the velocity of the center of mass by

$$\vec{\mathbf{p}}_{\text{sys}} = m_{\text{sys}} \vec{\mathbf{V}}_{\text{cm}} . \qquad (10.6.2)$$

We have already determined that the external force is equal to the change of the momentum of the system (Equation (10.4.9)). If we now substitute Equation (10.6.2) into Equation (10.4.9), and continue with our assumption of constant masses m_i, we have that

$$\vec{\mathbf{F}}^{\text{ext}} = \frac{d\vec{\mathbf{p}}_{\text{sys}}}{dt} = m_{\text{sys}} \frac{d\vec{\mathbf{V}}_{\text{cm}}}{dt} = m_{\text{sys}} \vec{\mathbf{A}}_{\text{cm}} , \qquad (10.6.3)$$

where $\vec{\mathbf{A}}_{cm}$, the derivative with respect to time of $\vec{\mathbf{V}}_{cm}$, is the acceleration of the center of mass. From Equation (10.6.3) we can conclude that in considering the linear motion of the center of mass, the sum of the external forces may be regarded as acting at the center of mass.

Example 10.4 Forces on a Baseball Bat

Suppose you push a baseball bat lying on a nearly frictionless table at the center of mass, position 2, with a force $\vec{\mathbf{F}}_{ext}$ (Figure 10.7). Will the acceleration of the center of mass be *greater than, equal to, or less than* if you push the bat with the same force at either end, positions 1 and 3

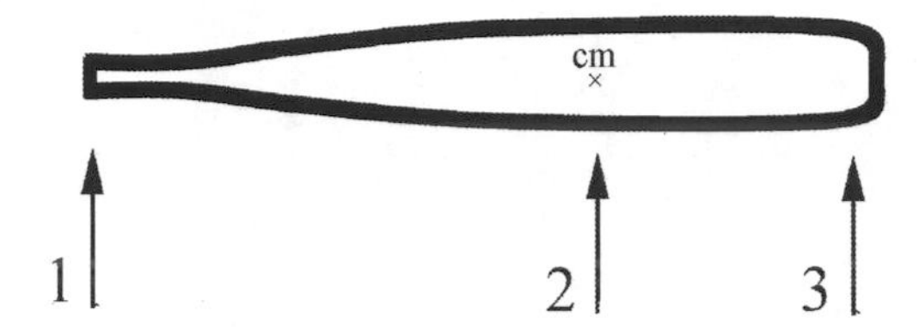

Figure 10.7 Forces acting on a baseball bat

Solution: The acceleration of the center of mass will be equal in the three cases. From our previous discussion, (Equation (10.6.3)), the acceleration of the center of mass is independent of where the force is applied. However, the bat undergoes a very different motion if we apply the force at one end or at the center of mass. When we apply the force at the center of mass all the particles in the baseball bat will undergo linear motion. When we push the bat at one end, the particles that make up the baseball bat will no longer undergo a linear motion even though the center of mass undergoes linear motion. In fact, each particle will rotate about the center of mass of the bat while the center of mass of the bat accelerates in the direction of the applied force.

10.7 Constancy of Momentum and Isolated Systems

Suppose we now completely isolate our system from the surroundings. When the external force acting on the system is zero,

$$\vec{\mathbf{F}}^{ext} = \vec{\mathbf{0}}. \tag{10.7.1}$$

the system is called an ***isolated system***. For an isolated system, the change in the momentum of the system is zero,

$$\Delta\vec{\mathbf{p}}_{sys} = \vec{\mathbf{0}} \quad \text{(isolated system)}, \tag{10.7.2}$$

therefore the momentum of the isolated system is constant. The initial momentum of our system is the sum of the initial momentum of the individual particles,

$$\vec{\mathbf{p}}_{\text{sys},i} = m_1\vec{\mathbf{v}}_{1,i} + m_2\vec{\mathbf{v}}_{2,i} + \cdots. \quad (10.7.3)$$

The final momentum is the sum of the final momentum of the individual particles,

$$\vec{\mathbf{p}}_{\text{sys},f} = m_1\vec{\mathbf{v}}_{1,f} + m_2\vec{\mathbf{v}}_{2,f} + \cdots. \quad (10.7.4)$$

Note that the right-hand-sides of Equations. (10.7.3) and (10.7.4) are vector sums.

When the external force on a system is zero, then the initial momentum of the system equals the final momentum of the system,

$$\vec{\mathbf{p}}_{\text{sys},i} = \vec{\mathbf{p}}_{\text{sys},f}. \quad (10.7.5)$$

10.8 Momentum Changes and Non-isolated Systems

Suppose the external force acting on the system is not zero,

$$\vec{\mathbf{F}}^{\text{ext}} \neq \vec{\mathbf{0}}. \quad (10.8.1)$$

and hence the system is not isolated. By Newton's Third Law, the sum of the force on the surroundings is equal in magnitude but opposite in direction to the external force acting on the system,

$$\vec{\mathbf{F}}^{\text{sur}} = -\vec{\mathbf{F}}^{\text{ext}}. \quad (10.8.2)$$

It's important to note that in Equation (10.8.2), all internal forces in the surroundings sum to zero. Thus the sum of the external force acting on the system and the force acting on the surroundings is zero,

$$\vec{\mathbf{F}}^{\text{sur}} + \vec{\mathbf{F}}^{\text{ext}} = \vec{\mathbf{0}}. \quad (10.8.3)$$

We have already found (Equation (10.4.9)) that the external force $\vec{\mathbf{F}}^{\text{ext}}$ acting on a system is equal to the rate of change of the momentum of the system. Similarly, the force on the surrounding is equal to the rate of change of the momentum of the surroundings. Therefore the momentum of both the system and surroundings is always conserved.

For a system and all of the surroundings that undergo any change of state, the change in the momentum of the system and its surroundings is zero,

$$\Delta\vec{\mathbf{p}}_{\text{sys}} + \Delta\vec{\mathbf{p}}_{\text{sur}} = \vec{\mathbf{0}}. \quad (10.8.4)$$

Equation (10.8.4) is referred to as the ***Principle of Conservation of Momentum.***

10.9 Worked Examples

10.9.1 Problem Solving Strategies

When solving problems involving changing momentum in a system, we shall employ our general problem solving strategy involving four basic steps:

1. Understand – get a conceptual grasp of the problem.
2. Devise a Plan - set up a procedure to obtain the desired solution.
3. Carry our your plan – solve the problem!
4. Look Back – check your solution and method of solution.

We shall develop a set of guiding ideas for the first two steps.

1. Understand – get a conceptual grasp of the problem

The first question you should ask is whether or not momentum is constant in some system that is changing its state after undergoing an interaction. First you must identify the objects that compose the system and how they are changing their state due to the interaction. As a guide, try to determine which objects change their momentum in the course of interaction. You must keep track of the momentum of these objects before and after any interaction. Second, momentum is a vector quantity so the question of whether momentum is constant or not must be answered in each relevant direction. In order to determine this, there are two important considerations. You should identify any external forces acting on the system. Remember that a non-zero external force will cause the momentum of the system to change, (Equation (10.4.9) above),

$$\vec{\mathbf{F}}^{\text{ext}} = \frac{d\vec{\mathbf{p}}_{\text{sys}}}{dt}. \qquad (10.9.1)$$

Equation (10.9.1) is a vector equation; if the external force in some direction is zero, then the change of momentum in that direction is zero. In some cases, external forces may act but the time interval during which the interaction takes place is so small that the impulse is small in magnitude compared to the momentum and might be negligible. Recall that the average external impulse changes the momentum of the system

$$\vec{\mathbf{I}} = \vec{\mathbf{F}}^{\text{ext}} \Delta t_{\text{int}} = \Delta\vec{\mathbf{p}}_{\text{sys}}. \qquad (10.9.2)$$

If the interaction time is small enough, the momentum of the system is constant, $\Delta\vec{\mathbf{p}} \to \vec{\mathbf{0}}$. If the momentum is not constant then you must apply either Equation (10.9.1) or Equation (10.9.2). If the momentum of the system is constant, then you can apply Equation (10.7.5),

$$\vec{\mathbf{p}}_{\text{sys},i} = \vec{\mathbf{p}}_{\text{sys},f}. \qquad (10.9.3)$$

If there is no net external force in some direction, for example the x-direction, the component of momentum is constant in that direction, and you must apply

$$p_{\text{sys},x,i} = p_{\text{sys},x,f} \tag{10.9.4}$$

2. Devise a Plan - set up a procedure to obtain the desired solution

Draw diagrams of all the elements of your system for the two states immediately before and after the system changes its state. Choose symbols to identify each mass and velocity in the system. Identify a set of positive directions and unit vectors for each state. Choose your symbols to correspond to the state and motion (this facilitates an easy interpretation, for example $(v_{x,i})_1$ represents the x-component of the velocity of object 1 in the initial state and $(v_{x,f})_1$ represents the x-component of the velocity of object 1 in the final state). Decide whether you are using components or magnitudes for your velocity symbols. Since momentum is a vector quantity, identify the initial and final vector components of the momentum. We shall refer to these diagrams as ***momentum flow diagrams***. Based on your model you can now write expressions for the initial and final momentum of your system. As an example in which two objects are moving only in the x-direction, the initial x-component of the momentum is

$$p_{\text{sys},x,i} = m_1(v_{x,i})_1 + m_2(v_{x,i})_2 + \cdots. \tag{10.9.5}$$

The final x-component of the momentum is

$$p_{\text{sys},x,f} = m_1(v_{x,f})_1 + m_2(v_{x,f})_2 + \cdots. \tag{10.9.6}$$

If the x-component of the momentum is constant then

$$p_{\text{sys},x,i} = p_{\text{sys},x,f}. \tag{10.9.7}$$

We can now substitute Equations (10.9.5) and (10.9.6) into Equation (10.9.7), yielding

$$m_1(v_{x,i})_1 + m_2(v_{x,i})_2 + \cdots = m_1(v_{x,f})_1 + m_2(v_{x,f})_2 + \cdots. \tag{10.9.8}$$

Equation (10.9.8) can now be used for any further analysis required by a particular problem. For example, you may have enough information to calculate the final velocities of the objects after the interaction. If so then carry out your plan and check your solution, especially dimensions or units and any relevant vector directions.

Example 10.5 Exploding Projectile

An instrument-carrying projectile of mass m_1 accidentally explodes at the top of its trajectory. The horizontal distance between launch point and the explosion is x_0. The projectile breaks into two pieces that fly apart horizontally. The larger piece, m_3, has three times the mass of the smaller piece, m_2. To the surprise of the scientist in charge, the smaller piece returns to earth at the launching station. Neglect air resistance and effects due to the earth's curvature. How far away, x_{3f}, from the original launching point does the larger piece land?

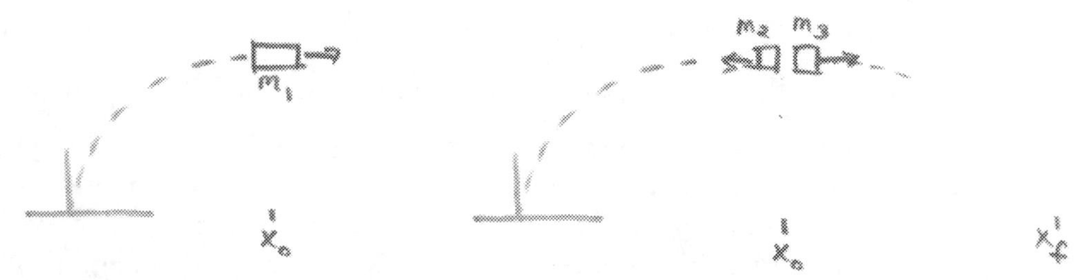

Figure 10.8 Exploding projectile trajectories

Solution: We can solve this problem two different ways. The easiest approach is utilizes the fact that the external force is the gravitational force and therefore the center of mass of the system follows a parabolic trajectory. From the information given in the problem $m_2 = m_1/4$ and $m_3 = 3m_1/4$. Thus when the two objects return to the ground the center of mass of the system has traveled a distance $R_{cm} = 2x_0$. We now use the definition of center of mass to find where the object with the greater mass hits the ground. Choose an origin at the starting point. The center of mass of the system is given by

$$\vec{\mathbf{R}}_{cm} = \frac{m_2\vec{\mathbf{r}}_2 + m_3\vec{\mathbf{r}}_3}{m_2 + m_3} .$$

So when the objects hit the ground $\vec{\mathbf{R}}_{cm} = 2x_0\ \hat{\mathbf{i}}$, the object with the smaller mass returns to the origin, $\vec{\mathbf{r}}_2 = \vec{\mathbf{0}}$, and the position vector of the other object is $\vec{\mathbf{r}}_3 = x_{3f}\ \hat{\mathbf{i}}$. So using the definition of the center of mass,

$$2x_0\ \hat{\mathbf{i}} = \frac{(3m_1/4)x_{3f}\ \hat{\mathbf{i}}}{m_1/4 + 3m_1/4} = \frac{(3m_1/4)x_{3f}\ \hat{\mathbf{i}}}{m_1} = \frac{3}{4}x_{3f}\ \hat{\mathbf{i}} .$$

Therefore

$$x_{3f} = \frac{8}{3}x_0 .$$

Note that the neither the vertical height above ground nor the gravitational acceleration g entered into our solution.

Alternatively, we can use conservation of momentum and kinematics to find the distance traveled. Since the smaller piece returns to the starting point after the collision, it must have the same speed v_0 as the projectile before the collision. Because the collision is instantaneous, the horizontal component of the momentum is constant during the collision. We can use this to determine the speed of the larger piece after the collision. The larger piece takes the same amount of time to return to the ground as the projectile originally takes to reach the top of the flight. We can therefore determine how far the larger piece traveled horizontally.

We begin by identifying various states in the problem.

Initial state, time t_0 : the projectile is launched.

State 1 time t_1: the projectile is at the top of its flight trajectory immediately before the explosion. The mass is m_1 and the speed of the projectile is v_1.

State 2 time t_2: immediately after the explosion, the projectile has broken into two pieces, one of mass m_2 moving backwards (in the $-x$-direction) with speed v_2 and the other of mass m_3 moving forward with speed v_3, (Figure 10.9).

State 3 time t_f: the two pieces strike the ground, one at the original launch site and the other at a distance x_f from the launch site, as indicated in the figure. The pieces take the same amount of time to reach the ground since they are falling from the same height and both have no velocity in the vertical direction immediately after the explosion.

Now we can pose some questions that may help us understand how to solve the problem. What is the speed of the projectile at the top of its flight just before the collision? What is the speed of the smaller piece just after the collision? What is the speed of the larger piece just after the collision?

The momentum flow diagram with state 1 as the initial state and state 2 as the final state is shown below (Figure 10.9). In the momentum flow diagram and analysis we shall use symbols that represent the magnitudes of the magnitudes x-components of the velocities and arrows to indicate the directions of the velocities; for example the symbol $v_1 \equiv |(v_x(t_1))_1|$ for the magnitude of the x-component of the velocity of the object before the explosion at time t_1, $v_2 \equiv |(v_x(t_2))_2|$ and $v_3 \equiv |(v_x(t_2))_3|$ for the magnitudes of the x-component of the velocity of objects 2 and 3 immediately after the collision at time t_2.

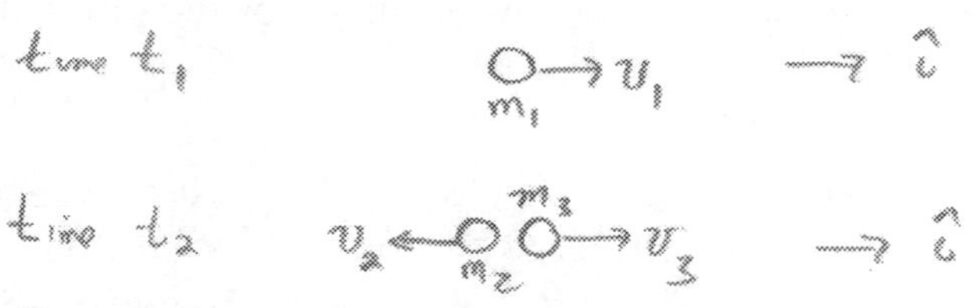

Figure 10.9 Momentum flow diagrams for the two middle states of the problem.

The initial momentum before the explosion is

$$p_{\text{sys},\,x,i} = p_{\text{sys},x}(t_1) = m_1\, v_1\,. \tag{10.9.9}$$

The momentum immediately after the explosion is

$$p_{\text{sys},\,x,i} = p_{\text{sys},x}(t_2) = -m_2\, v_2 + m_3\, v_3 \tag{10.9.10}$$

Note that in Equations (10.9.9) and (10.9.10) the signs of the terms are obtained directly from the momentum flow diagram, consistent with the use of magnitudes; we are told that the smaller piece moves in a direction opposite the original direction after the explosion.

During the duration of the explosion, impulse due to the external force, gravity in this case, may be neglected. The collision is considered to be instantaneous, and momentum is constant. In the horizontal direction, we have that

$$p_{\text{sys},x}(t_1) = p_{\text{sys},x}(t_2)\,. \tag{10.9.11}$$

If the collision were not instantaneous, then the masses would descend during the explosion, and the action of gravity would add downward velocity to the system. Equation (10.9.11) would still be valid, but our analysis of the motion between state 2 and state 3 would be affected. Substituting Equations (10.9.9) and (10.9.10) into Equation (10.9.11) yields

$$m_1\, v_1 = -m_2\, v_2 + m_3\, v_3\,. \tag{10.9.12}$$

The mass of the projectile is equal to the sum of the masses of the ejected pieces,

$$m_1 = m_2 + m_3\,. \tag{10.9.13}$$

The heavier fragment is three times the mass of the lighter piece, $m_3 = 3\,m_2$. Therefore

$$m_2 = (1/4)m_1, \quad m_3 = (3/4)\,m_1\,. \tag{10.9.14}$$

There are still two unknowns to consider, v_2 and v_3. However there is an additional piece of information. We know that the lighter object returns exactly to the starting position, which implies that $v_2 = v_1$ (we have already accounted for the change in direction by considering magnitudes, as discussed above.)

Recall from our study of projectile motion that the horizontal distance is given by $x_0 = v_1 t_1$, independent of the mass. The time that it takes the lighter mass to hit the ground is the same as the time it takes the original projectile to reach the top of its flight (neglecting air resistance). Therefore the speeds must be the same since the original projectile and the smaller fragment traveled the same distance. We can use the values for the respective masses (Equation (10.9.14)) in Equation (10.9.12), which becomes

$$m_1 v_1 = -\frac{1}{4} m_1 v_1 + \frac{3}{4} m_1 v_3 . \tag{10.9.15}$$

Equation (10.9.15) can now solved for the speed of the larger piece immediately after the collision,

$$v_3 = \frac{5}{3} v_1 . \tag{10.9.16}$$

The larger piece also takes the same amount of time t_1 to hit the ground as the smaller piece. Hence the larger piece travels a distance

$$x_3 = v_3 t_1 = \frac{5}{3} v_1 t_1 = \frac{5}{3} x_0 . \tag{10.9.17}$$

Therefore the total distance the larger piece traveled from the launching station is

$$x_f = x_0 + \frac{5}{3} x_0 = \frac{8}{3} x_0 , \tag{10.9.18}$$

in agreement with our previous approach.

Example 10.6 Landing Plane and Sandbag

A light plane of mass $1000\ \text{kg}$ makes an emergency landing on a short runway. With its engine off, it lands on the runway at a speed of $40\ \text{m}\cdot\text{s}^{-1}$. A hook on the plane snags a cable attached to a $120\ \text{kg}$ sandbag and drags the sandbag along. If the coefficient of friction between the sandbag and the runway is $\mu_k = 0.4$, and if the plane's brakes give an additional retarding force of magnitude $1400\ \text{N}$, how far does the plane go before it comes to a stop?

Solution: We shall assume that when the plane snags the sandbag, the collision is instantaneous so the momentum in the horizontal direction remains constant,

$$p_{x,i} = p_{x,1}. \tag{10.9.19}$$

We then know the speed of the plane and the sandbag immediately after the collision. After the collision, there are two external forces acting on the system of the plane and sandbag, the friction between the sandbag and the ground and the braking force of the runway on the plane. So we can use the Newton's Second Law to determine the acceleration and then one-dimensional kinematics to find the distance the plane traveled since we can determine the change in kinetic energy.

The momentum of the plane immediately before the collision is

$$\vec{\mathbf{p}}_i = m_p v_{p,i}\,\hat{\mathbf{i}} \tag{10.9.20}$$

The momentum of the plane and sandbag immediately after the collision is

$$\vec{\mathbf{p}}_1 = (m_p + m_s) v_{p,1}\,\hat{\mathbf{i}} \tag{10.9.21}$$

Because the x- component of the momentum is constant, we can substitute Eqs. (10.9.20) and (10.9.21) into Eq. (10.9.19) yielding

$$m_p v_{p,i} = (m_p + m_s) v_{p,1}. \tag{10.9.22}$$

The speed of the plane and sandbag immediately after the collision is

$$v_{p,1} = \frac{m_p v_{p,i}}{m_p + m_s} \tag{10.9.23}$$

The forces acting on the system consisting of the plane and the sandbag are the normal force on the sandbag,

$$\vec{\mathbf{N}}_{g,s} = N_{g,s}\hat{\mathbf{j}}, \tag{10.9.24}$$

the frictional force between the sandbag and the ground

$$\vec{\mathbf{f}}_k = -f_k\hat{\mathbf{i}} = -\mu_k N_{g,s}\hat{\mathbf{i}}, \tag{10.9.25}$$

the braking force on the plane

$$\vec{\mathbf{F}}_{g,p} = -F_{g,p}\hat{\mathbf{i}}, \tag{10.9.26}$$

and the gravitational force on the system,

$$(m_p + m_s)\vec{\mathbf{g}} = -(m_p + m_s)g\hat{\mathbf{j}}. \qquad (10.9.27)$$

Newton's Second Law in the $\hat{\mathbf{i}}$-direction becomes

$$-F_{g,p} - f_k = (m_p + m_s)a_x. \qquad (10.9.28)$$

If we just look at the vertical forces on the sandbag alone then Newton's Second Law in the $\hat{\mathbf{j}}$-direction becomes

$$N - m_s g = 0.$$

The frictional force on the sandbag is then

$$\vec{\mathbf{f}}_k = -\mu_k N_{g,s}\hat{\mathbf{i}} = -\mu_k m_s g\hat{\mathbf{i}}. \qquad (10.9.29)$$

Newton's Second Law in the $\hat{\mathbf{i}}$-direction becomes

$$-F_{g,p} - \mu_k m_s g = (m_p + m_s)a_x.$$

The x-component of the acceleration of the plane and the sand bag is then

$$a_x = \frac{-F_{g,p} - \mu_k m_s g}{m_p + m_s} \qquad (10.9.30)$$

We choose our origin at the location of the plane immediately after the collision, $x_p(0) = 0$. Set $t = 0$ immediately after the collision. The x-component of the velocity of the plane immediately after the collision is $v_{x,0} = v_{p,1}$. Set $t = t_f$ when the plane just comes to a stop. Because the acceleration is constant, the kinematic equations for the change in velocity is

$$v_{x,f}(t_f) - v_{p,1} = a_x t_f.$$

We can solve this equation for $t = t_f$, where $v_{x,f}(t_f) = 0$

$$t_f = -v_{p,1} / a_x t.$$

Then the position of the plane when it first comes to rest is

$$x_p(t_f) - x_p(0) = v_{p,1}t_f + \frac{1}{2}a_x t_f^2 = -\frac{1}{2}\frac{v_{p,1}^2}{a_x}. \qquad (10.9.31)$$

Then using $x_p(0) = 0$ and substituting Eq. (10.9.30) into Eq. (10.9.31) yields

$$x_p(t_f) = \frac{1}{2}\frac{(m_p + m_s)v_{p,1}^2}{(F_{g,p} + \mu_k m_s g)}. \quad (10.9.32)$$

We now use the condition from conservation of the momentum law during the collision, Eq. (10.9.23) in Eq. (10.9.32) yielding

$$x_p(t_f) = \frac{m_p^2 v_{p,i}^2}{2(m_p + m_s)(F_{g,p} + \mu_k m_s g)}. \quad (10.9.33)$$

Substituting the given values into Eq. (10.9.32) yields

$$x_p(t_f) = \frac{(1000\text{ kg})^2(40\text{ m}\cdot\text{s}^{-1})^2}{2(1000\text{ kg} + 120\text{ kg})(1400\text{ N} + (0.4)(120\text{ kg})(9.8\text{m}\cdot\text{s}^{-2}))} = 3.8\times 10^2\text{ m}. \quad (10.9.34)$$

Chapter 11 Reference Frames

Chapter 11 Reference Frames

Examples of this sort, together with the unsuccessful attempts to discover any motion of the earth relatively to the "light medium" suggest that the phenomena of electromagnetism as well as mechanics possess no properties corresponding to the idea of absolute rest. They suggest rather that, ..., the same laws of electrodynamics and optics will be valid for all frames of reference for which the equations of mechanics hold good. We will raise this conjecture (the purport of which will hereafter be called the "Principle of Relativity") to the status of a postulate, and also introduce another postulate, ..., namely that light is always propagated in empty space with a definite velocity c, which is independent of the state of motion of the emitting body. [1]

Albert Einstein

11.1 Introduction

Equation Chapter 11 Section 1

In order to describe physical events that occur in space and time such as the motion of bodies, we introduced a coordinate system. Its spatial and temporal coordinates can now specify a ***space-time event***. In particular, the position of a moving body can be described by space-time events specified by its space-time coordinates. You can place an *observer* at the origin of coordinate system. The coordinate system with your observer acts as a ***reference frame*** for describing the position, velocity, and acceleration of bodies. The position vector of the body depends on the choice of origin (location of your observer) but the displacement, velocity, and acceleration vectors are independent of the location of the observer.

You can always choose a second reference frame that is moving with respect to the first reference frame. Then the position, velocity and acceleration of bodies as seen by the different observers do depend on the relative motion of the two reference frames. The relative motion can be described in terms of the relative position, velocity, and acceleration of the observer at the origin, O, in reference frame S with respect to a second observer located at the origin, O', in reference frame S'.

11.2 Galilean Coordinate Transformations

Equation Section (Next)

Let the vector $\vec{\mathbf{R}}$ point from the origin of frame S to the origin of reference frame S'. Suppose an object is located at a point 1. Denote the position vector of the object with respect to origin of reference frame S by $\vec{\mathbf{r}}$. Denote the position vector of the object with respect to origin of reference frame S' by $\vec{\mathbf{r}}'$ (Figure 11.1).

---

[1] A. Einstein, *Zur Elektrodynamik begetter Körper*, (*On the Electrodynamics of Moving Bodies*), Ann. Physik, **17**, 891 (1905); translated by W. Perrett and G.B. Jeffrey, 19223, in *The Principle of Relativity*, Dover, New York.

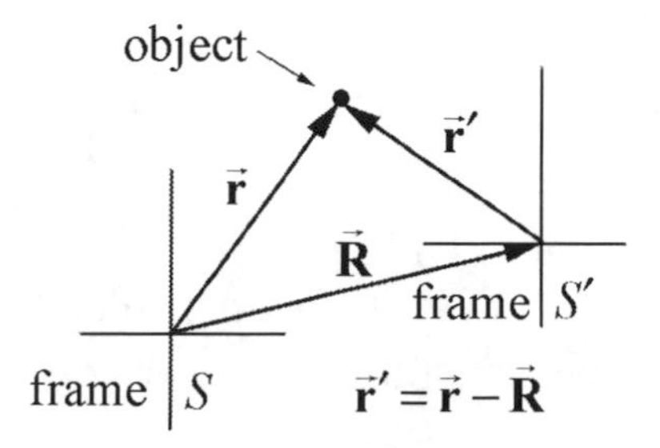

Figure 11.1 Two reference frames S and S'.

The position vectors are related by

$$\vec{\mathbf{r}}' = \vec{\mathbf{r}} - \vec{\mathbf{R}} \qquad (11.2.1)$$

These coordinate transformations are called the ***Galilean Coordinate Transformations.*** They enable the observer in frame S to predict the position vector in frame S', based only on the position vector in frame S and the relative position of the origins of the two frames.

The relative velocity between the two reference frames is given by the time derivative of the vector $\vec{\mathbf{R}}$, defined as the limit as of the displacement of the two origins divided by an interval of time, as the interval of time becomes infinitesimally small,

$$\vec{\mathbf{V}} = \frac{d\vec{\mathbf{R}}}{dt}. \qquad (11.2.2)$$

11.2.1 Relatively Inertial Reference Frames and the Principle of Relativity

If the relative velocity between the two reference frames is constant, then the relative acceleration between the two reference frames is zero,

$$\vec{\mathbf{A}} = \frac{d\vec{\mathbf{V}}}{dt} = \vec{\mathbf{0}}. \qquad (11.2.3)$$

When two reference frames are moving with a constant velocity relative to each other as above, the reference frames are called ***relatively inertial reference frames***.

We can reinterpret Newton's First Law

> *Law 1: Every body continues in its state of rest, or of uniform motion in a right line, unless it is compelled to change that state by forces impressed upon it.*

as the Principle of Relativity:

> *In relatively inertial reference frames, if there is no net force impressed on an object at rest in frame S, then there is also no net force impressed on the object in frame S'.*

11.3 Law of Addition of Velocities: Newtonian Mechanics

Equation Section (Next)

Suppose the object in Figure 11.1 is moving; then observers in different reference frames will measure different velocities. Denote the velocity of the object in frame S by $\vec{\mathbf{v}} = d\vec{\mathbf{r}}/dt$, and the velocity of the object in frame S' by $\vec{\mathbf{v}}' = d\vec{\mathbf{r}}'/dt'$. Since the derivative of the position is velocity, the velocities of the object in two different reference frames are related according to

$$\frac{d\vec{\mathbf{r}}'}{dt'} = \frac{d\vec{\mathbf{r}}}{dt} - \frac{d\vec{\mathbf{R}}}{dt}, \tag{11.3.1}$$

$$\vec{\mathbf{v}}' = \vec{\mathbf{v}} - \vec{\mathbf{V}}. \tag{11.3.2}$$

This is called the ***Law of Addition of Velocities.***

11.4 Worked Examples

Equation Section (Next)

Example 11.1 Relative Velocities of Two Moving Planes

An airplane A is traveling northeast with a speed of $v_{\text{A}} = 160\ \text{m}\cdot\text{s}^{-1}$. A second airplane B is traveling southeast with a speed of $v_{\text{B}} = 200\ \text{m}\cdot\text{s}^{-1}$. (a) Choose a coordinate system and write down an expression for the velocity of each airplane as vectors, $\vec{\mathbf{v}}_{\mathbf{A}}$ and $\vec{\mathbf{v}}_{B}$. Carefully use unit vectors to express your answer. (b) Sketch the vectors $\vec{\mathbf{v}}_{\mathbf{A}}$ and $\vec{\mathbf{v}}_{B}$ on your coordinate system. (c) Find a vector expression that expresses the velocity of aircraft A as seen from an observer flying in aircraft B. Calculate this vector. What is its magnitude and direction? Sketch it on your coordinate system.

Solution: From the information given in the problem we draw the velocity vectors of the airplanes as shown in Figure 11.2a.

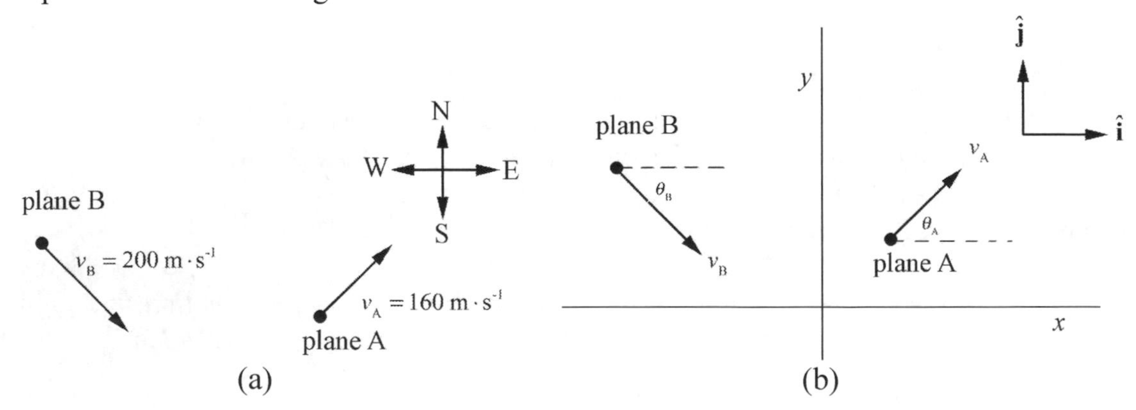

Figure 11.2 (a): Motion of two planes **Figure 11.2 (b):** Coordinate System

An observer at rest with respect to the ground defines a reference frame S. Choose a coordinate system shown in Figure 11.2b. According to this observer, airplane A is moving with velocity $\vec{\mathbf{v}}_A = v_A \cos\theta_A \hat{\mathbf{i}} + v_A \sin\theta_A \hat{\mathbf{j}}$, and airplane B is moving with velocity $\vec{\mathbf{v}}_B = v_B \cos\theta_B \hat{\mathbf{i}} + v_B \sin\theta_B \hat{\mathbf{j}}$. According to the information given in the problem airplane A flies northeast so $\theta_A = \pi/4$ and airplane B flies southeast east so $\theta_B = -\pi/4$. Thus $\vec{\mathbf{v}}_A = (80\sqrt{2}\ \text{m}\cdot\text{s}^{-1})\hat{\mathbf{i}} + (80\sqrt{2}\ \text{m}\cdot\text{s}^{-1})\hat{\mathbf{j}}$ and $\vec{\mathbf{v}}_B = (100\sqrt{2}\ \text{m}\cdot\text{s}^{-1})\hat{\mathbf{i}} - (100\sqrt{2}\ \text{m}\cdot\text{s}^{-1})\hat{\mathbf{j}}$

Consider a second observer moving along with airplane B, defining reference frame S'. What is the velocity of airplane A according to this observer moving in airplane B? The velocity of the observer moving along in airplane B with respect to an observer at rest on the ground is just the velocity of airplane B and is given by $\vec{\mathbf{V}} = \vec{\mathbf{v}}_B = v_B \cos\theta_B \hat{\mathbf{i}} + v_B \sin\theta_B \hat{\mathbf{j}}$. Using the Law of Addition of Velocities, Equation (11.3.2), the velocity of airplane A with respect to an observer moving along with Airplane B is given by

$$\begin{aligned}
\vec{\mathbf{v}}'_A &= \vec{\mathbf{v}}_A - \vec{\mathbf{V}} = (v_A \cos\theta_A \hat{\mathbf{i}} + v_A \sin\theta_A \hat{\mathbf{j}}) - (v_B \cos\theta_B \hat{\mathbf{i}} + v_B \sin\theta_B \hat{\mathbf{j}}) \\
&= (v_A \cos\theta_A - v_B \cos\theta_B)\hat{\mathbf{i}} + (v_A \sin\theta_A - v_B \sin\theta_B)\hat{\mathbf{j}} \\
&= ((80\sqrt{2}\ \text{m}\cdot\text{s}^{-1}) - (100\sqrt{2}\ \text{m}\cdot\text{s}^{-1}))\hat{\mathbf{i}} + ((80\sqrt{2}\ \text{m}\cdot\text{s}^{-1}) + (100\sqrt{2}\ \text{m}\cdot\text{s}^{-1}))\hat{\mathbf{j}}. \\
&= -(20\sqrt{2}\ \text{m}\cdot\text{s}^{-1})\hat{\mathbf{i}} + (180\sqrt{2}\ \text{m}\cdot\text{s}^{-1})\hat{\mathbf{j}} \\
&= v'_{Ax}\hat{\mathbf{i}} + v'_{Ay}\hat{\mathbf{j}}
\end{aligned} \tag{11.4.1}$$

Figure 11.3 shows the velocity of airplane A with respect to airplane B in reference frame S'.

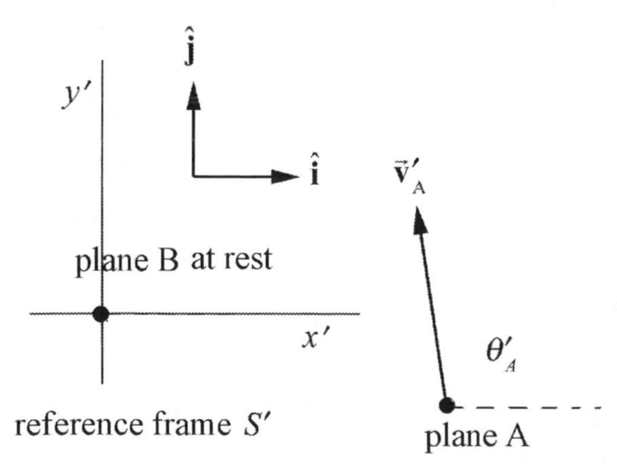

Figure 11.3 Airplane A as seen from observer in airplane B

The magnitude of velocity of airplane A as seen by an observer moving with airplane B is given by

$$\left|\vec{v}'_A\right| = (v'^{\,2}_{Ax} + v'^{\,2}_{Ay})^{1/2} = ((-20\sqrt{2}\ \text{m}\cdot\text{s}^{-1})^2 + (180\sqrt{2}\ \text{m}\cdot\text{s}^{-1})^2)^{1/2} = 256\ \text{m}\cdot\text{s}^{-1}. \qquad (11.4.2)$$

The angle of velocity of airplane A as seen by an observer moving with airplane B is given by,

$$\begin{aligned}\theta'_A &= \tan^{-1}(v'_{Ay} / v'_{Ax}) = \tan^{-1}((180\sqrt{2}\ \text{m}\cdot\text{s}^{-1}) / (-20\sqrt{2}\ \text{m}\cdot\text{s}^{-1})) \\ &= \tan^{-1}(-9) = 180^\circ - 83.7^\circ = 96.3^\circ\end{aligned}. \qquad (11.4.3)$$

Example 11.2 Relative Motion and Polar Coordinates

By relative velocity we mean velocity with respect to a specified coordinate system. (The term velocity, alone, is understood to be relative to the observer's coordinate system.) (a) A point is observed to have velocity $\vec{\mathbf{v}}_A$ relative to coordinate system A. What is its velocity relative to coordinate system B, which is displaced from system A by distance $\vec{\mathbf{R}}$? ($\vec{\mathbf{R}}$ can change in time.) (b) Particles a and b move in opposite directions around a circle with the magnitude of the angular velocity ω, as shown in Figure 11.4. At $t = 0$ they are both at the point $\vec{\mathbf{r}} = l\hat{\mathbf{j}}$, where l is the radius of the circle. Find the velocity of a relative to b.

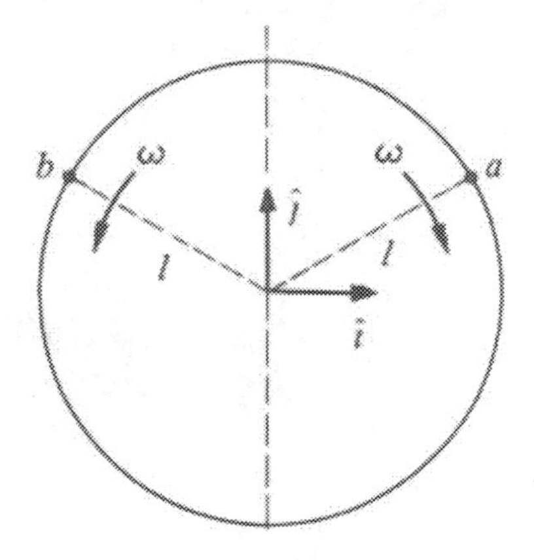

Figure 11.4 Particles a and b moving relative to each other

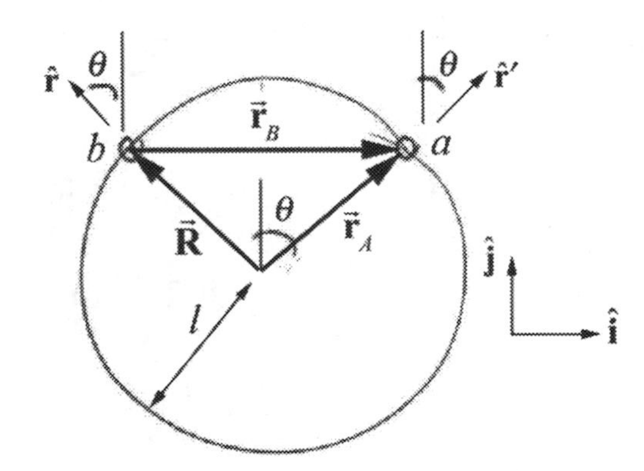

Figure 11.5 Particles a and b moving relative to each other

Solution: (a) The position vectors are related by

$$\vec{\mathbf{r}}_B = \vec{\mathbf{r}}_A - \vec{\mathbf{R}}. \qquad (11.4.4)$$

The velocities are related by the taking derivatives, (law of addition of velocities Eq. (11.3.2))

$$\vec{\mathbf{v}}_B = \vec{\mathbf{v}}_A - \vec{\mathbf{V}}. \qquad (11.4.5)$$

(b) Let's choose two reference frames; frame B is centered at particle b, and frame A is centered at the center of the circle in Figure 11.5. Then the relative position vector between the origins of the two frames is given by

$$\vec{\mathbf{R}} = l\,\hat{\mathbf{r}}\,. \tag{11.4.6}$$

The position vector of particle a relative to frame A is given by

$$\vec{\mathbf{r}}_A = l\,\hat{\mathbf{r}}'\,. \tag{11.4.7}$$

The position vector of particle b in frame B can be found by substituting Eqs. (11.4.7) and (11.4.6) into Eq. (11.4.4),

$$\vec{\mathbf{r}}_B = \vec{\mathbf{r}}_A - \vec{\mathbf{R}} = l\,\hat{\mathbf{r}}' - l\,\hat{\mathbf{r}}\,. \tag{11.4.8}$$

We can decompose each of the unit vectors $\hat{\mathbf{r}}$ and $\hat{\mathbf{r}}'$ with respect to the Cartesian unit vectors $\hat{\mathbf{i}}$ and $\hat{\mathbf{j}}$ (see Figure 11.5),

$$\hat{\mathbf{r}} = -\sin\theta\,\hat{\mathbf{i}} + \cos\theta\,\hat{\mathbf{j}} \tag{11.4.9}$$

$$\hat{\mathbf{r}}' = \sin\theta\,\hat{\mathbf{i}} + \cos\theta\,\hat{\mathbf{j}}\,. \tag{11.4.10}$$

Then Eq. (11.4.8) giving the position vector of particle b in frame B becomes

$$\vec{\mathbf{r}}_B = l\,\hat{\mathbf{r}}' - l\,\hat{\mathbf{r}} = l\,(\sin\theta\,\hat{\mathbf{i}} + \cos\theta\,\hat{\mathbf{j}}) - l\,(-\sin\theta\,\hat{\mathbf{i}} + \cos\theta\,\hat{\mathbf{j}}) = 2l\sin\theta\,\hat{\mathbf{i}}\,. \tag{11.4.11}$$

In order to find the velocity vector of particle a in frame B (i.e. with respect to particle b), differentiate Eq. (11.4.11)

$$\vec{\mathbf{v}}_B = \frac{d}{dt}(2l\sin\theta)\,\hat{\mathbf{i}} = (2l\cos\theta)\frac{d\theta}{dt}\,\hat{\mathbf{i}} = 2\omega l\cos\theta\,\hat{\mathbf{i}}\,. \tag{11.4.12}$$

Example 11.3 Recoil in Different Frames

A person of mass m_1 is standing on a cart of mass m_2. Assume that the cart is free to move on its wheels without friction. The person throws a ball of mass m_3 at an angle of θ with respect to the horizontal as measured by the person in the cart. The ball is thrown with a speed v_0 with respect to the cart (Figure 11.6). (a) What is the final velocity of the ball as seen by an observer fixed to the ground? (b) What is the final velocity of the cart as seen by an observer fixed to the ground? (c) With respect to the horizontal, what angle the fixed observer see the ball leave the cart?

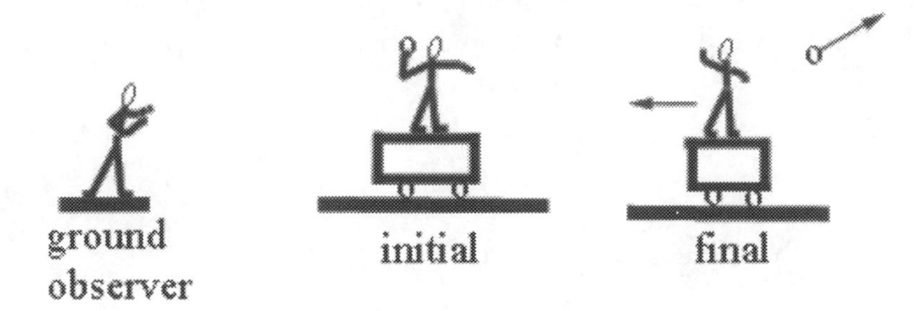

Figure 11.6 Recoil of a person on cart due to thrown ball

Solution: a), b) Our reference frame will be that fixed to the ground. We shall take as our initial state that before the ball is thrown (cart, ball, throwing person stationary) and our final state that after the ball is thrown. We are assuming that there is no friction, and so there are no external forces acting in the horizontal direction. The initial x-component of the total momentum is zero,

$$p_{x,0}^{\text{total}} = 0. \tag{11.4.13}$$

After the ball is thrown, the cart and person have a final momentum

$$\vec{\mathbf{p}}_{f,\text{cart}} = -(m_2 + m_1)v_{f,\text{cart}}\,\hat{\mathbf{i}} \tag{11.4.14}$$

as measured by the person on the ground, where $v_{f,\text{cart}}$ is the speed of the person and cart. (The person's center of mass will move with respect to the cart while the ball is being thrown, but since we're interested in velocities, not positions, we need only assume that the person is at rest with respect to the cart after the ball is thrown.)

The ball is thrown with a speed v_0 and at an angle θ with respect to the horizontal as measured by the person in the cart. Therefore the person in the cart throws the ball with velocity

$$\vec{\mathbf{v}}'_{f,\text{ball}} = v_0 \cos\theta\,\hat{\mathbf{i}} + v_0 \sin\theta\,\hat{\mathbf{j}}. \tag{11.4.15}$$

Because the cart is moving in the negative x-direction with speed $v_{f,\text{cart}}$ just as the ball leaves the person's hand, the x-component of the velocity of the ball as measured by an observer on the ground is given by

$$v_{xf,\text{ball}} = v_0 \cos\theta - v_{f,\text{cart}}. \tag{11.4.16}$$

The ball appears to have a smaller x-component of the velocity according to the observer on the ground. The velocity of the ball as measured by an observer on the ground is

$$\vec{\mathbf{v}}_{f,\text{ball}} = (v_0 \cos\theta - v_{f,\text{cart}})\,\hat{\mathbf{i}} + v_0 \sin\theta\,\hat{\mathbf{j}}. \tag{11.4.17}$$

The final momentum of the ball according to an observer on the ground is

$$\vec{\mathbf{p}}_{f,\text{ball}} = m_3\left[(v_0\cos\theta - v_{f,\text{cart}})\,\hat{\mathbf{i}} + v_0\sin\theta\,\hat{\mathbf{j}}\right]. \tag{11.4.18}$$

The momentum flow diagram is shown in (Figure 11.7).

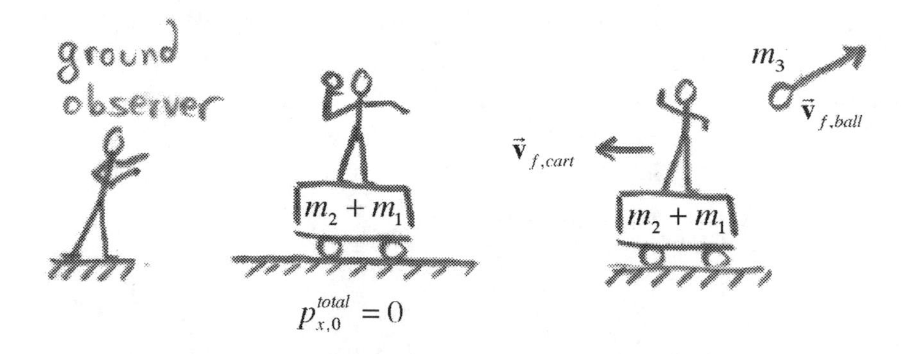

Figure 11.7 Momentum flow diagram for recoil

Because the x -component of the momentum of the system is constant, we have that

$$\begin{aligned} 0 &= (p_{x,f})_{\text{cart}} + (p_{x,f})_{\text{ball}} \\ &= -(m_2 + m_1)v_{f,\text{cart}} + m_3(v_0\cos\theta - v_{f,\text{cart}}). \end{aligned} \tag{11.4.19}$$

We can solve Equation (11.4.19) for the final speed and velocity of the cart as measured by an observer on the ground,

$$v_{f,\text{cart}} = \frac{m_3\, v_0 \cos\theta}{m_2 + m_1 + m_3}, \tag{11.4.20}$$

$$\vec{\mathbf{v}}_{f,\text{cart}} = v_{f,\text{cart}}\,\hat{\mathbf{i}} = \frac{m_3\, v_0\cos\theta}{m_2 + m_1 + m_3}\hat{\mathbf{i}}. \tag{11.4.21}$$

Note that the y-component of the momentum is not constant because as the person is throwing the ball he or she is pushing off the cart and the normal force with the ground exceeds the gravitational force so the net external force in the y-direction is non-zero.

Substituting Equation (11.4.20) into Equation (11.4.17) gives

$$\begin{aligned} \vec{\mathbf{v}}_{f,\text{ball}} &= (v_0\cos\theta - v_{f,\text{cart}})\,\hat{\mathbf{i}} + v_0\sin\theta\,\hat{\mathbf{j}} \\ &= \frac{m_1 + m_2}{m_1 + m_2 + m_3}(v_0\cos\theta\,\hat{\mathbf{i}} + v_0\sin\theta)\,\hat{\mathbf{j}}. \end{aligned} \tag{11.4.22}$$

As a check, note that in the limit $m_3 << m_1 + m_2$, $\vec{\mathbf{v}}_{f,\text{ball}}$ has speed v_0 and is directed at an angle θ above the horizontal; the fact that the much more massive person-cart combination is free to move doesn't affect the flight of the ball as seen by the fixed observer. Also note that in the unrealistic limit $m >> m_1 + m_2$ the ball is moving at a speed much smaller than v_0 as it leaves the cart.

c) The angle ϕ at which the ball is thrown as seen by the observer on the ground is given by

$$\begin{aligned}\phi &= \tan^{-1}\frac{(v_{f,\text{ball}})_y}{(v_{f,\text{ball}})_x} = \tan^{-1}\frac{v_0 \sin\theta}{\left[(m_1+m_2)/(m_1+m_2+m_3)\right]v_0\cos\theta} \\ &= \tan^{-1}\left[\tan\theta\frac{m_1+m_2+m_3}{m_1+m_2}\right].\end{aligned} \tag{11.4.23}$$

For arbitrary values for the masses, the above expression will not reduce to a simplified form. However, we can see that $\tan\phi > \tan\theta$ for arbitrary masses, and that in the limit $m_3 << m_1 + m_2$, $\phi \to \theta$ and in the unrealistic limit $m_3 >> m_1 + m_2$, $\phi \to \pi/2$. Can you explain this last odd prediction?

Chapter 12 Momentum and the Flow of Mass

Chapter 12 Momentum and the Flow of Mass

Even though the release was pulled, the rocket did not rise at first, but the flame came out, and there was a steady roar. After a number of seconds it rose, slowly until it cleared the flame, and then at express-train speed, curving over to the left, and striking the ice and snow, still going at a rapid rate. It looked almost magical as it rose, without any appreciably greater noise or flame, as if it said, "I've been here long enough; I think I'll be going somewhere else, if you don't mind." [1]

Robert Goddard

Preface The Challenger Flight

When the Rogers Commission in 1986 investigated the Challenger Flight disaster, a commission member, physicist Richard Feynman, made an extraordinary demonstration during the hearings.

"He (Feynman) also learned that rubber used to seal the solid rocket booster joints using O-rings, failed to expand when the temperature was at or below 32 degrees F (0 degrees C). The temperature at the time of the Challenger liftoff was 32 degrees F. Feynman now believed that he had the solution, but to test it, he dropped a piece of the O-ring material, squeezed with a C-clamp to simulate the actual conditions of the shuttle, into a glass of ice water. Ice, of course, is 32 degrees F. At this point one needs to understand exactly what role the O-rings play in the solid rocket booster (SRB) joints. When the material in the SRB start to heat up, it expands and pushes against the sides of the SRB. If there is an opening in a joint in the SRB, the gas tries to escape through that opening (think of it like water in a tea kettle escaping through the spout.) This leak in the Challenger's SRB was easily visible as a small flicker in a launch photo. This flicker turned into a flame and began heating the fuel tank, which then ruptured. When this happened, the fuel tank released liquid hydrogen into the atmosphere where it exploded. As Feynman explained, because the O-rings cannot expand in 32 degree weather, the gas finds gaps in the joints, which led to the explosion of the booster and then the shuttle itself."[2]

In the Report of the Presidential Commission on the Space Shuttle Challenger Accident (1986), *Appendix F - Personal observations on the reliability of the Shuttle*, Feynman wrote

The Challenger flight is an excellent example. ... The O-rings of the Solid Rocket Boosters were not designed to erode. Erosion was a clue that something was wrong. Erosion was not something from which safety can be inferred. There was no way, without

---

[1] describing the first rocket flight using liquid propellants at Aunt Effie's farm, 17 March 1926.

[2] http://www.fotuva.org/online/frameload.htm?/online/challenger.htm.

full understanding, that one could have confidence that conditions the next time might not produce erosion three times more severe than the time before. Nevertheless, officials fooled themselves into thinking they had such understanding and confidence, in spite of the peculiar variations from case to case. A mathematical model was made to calculate erosion. This was a model based not on physical understanding but on empirical curve fitting. To be more detailed, it was supposed a stream of hot gas impinged on the O-ring material, and the heat was determined at the point of stagnation (so far, with reasonable physical, thermodynamic laws). But to determine how much rubber eroded it was assumed this depended only on this heat by a formula suggested by data on a similar material. A logarithmic plot suggested a straight line, so it was supposed that the erosion varied as the .58 power of the heat, the .58 being determined by a nearest fit. At any rate, adjusting some other numbers, it was determined that the model agreed with the erosion (to depth of one-third the radius of the ring). There is nothing much so wrong with this as believing the answer! Uncertainties appear everywhere. How strong the gas stream might be was unpredictable, it depended on holes formed in the putty. Blow-by showed that the ring might fail even though not, or only partially eroded through. The empirical formula was known to be uncertain, for it did not go directly through the very data points by which it was determined. There were a cloud of points some twice above, and some twice below the fitted curve, so erosions twice predicted were reasonable from that cause alone. Similar uncertainties surrounded the other constants in the formula, etc., etc. When using a mathematical model careful attention must be given to uncertainties in the model. ...

In any event this has had very unfortunate consequences, the most serious of which is to encourage ordinary citizens to fly in such a dangerous machine, as if it had attained the safety of an ordinary airliner. The astronauts, like test pilots, should know their risks, and we honor them for their courage. Who can doubt that McAuliffe was equally a person of great courage, who was closer to an awareness of the true risk than NASA management would have us believe? Let us make recommendations to ensure that NASA officials deal in a world of reality in understanding technological weaknesses and imperfections well enough to be actively trying to eliminate them. .... For a successful technology, reality must take precedence over public relations, for nature cannot be fooled.[3]

12.1 Introduction

So far we have restricted ourselves to considering systems consisting of discrete objects or point-like objects that have fixed amounts of mass. We shall now consider systems in which material flows between the objects in the system, for example we shall consider coal falling from a hopper into a moving railroad car, sand leaking from railroad car fuel, grain moving forward into a railroad car, and fuel ejected from the back of a rocket, In each of these examples material is continuously flows into or out of an object. We have already shown that the total external force causes the momentum of a system to change,

---

[3] R. P. Feynman, *Appendix F - Personal observations on the reliability of the Shuttle*, Report of the PRESIDENTIAL COMMISSION on the Space Shuttle Challenger Accident (1986), http://history.nasa.gov/rogersrep/genindex.htm.

$$\vec{\mathbf{F}}_{\text{ext}}^{\text{total}} = \frac{d\vec{\mathbf{p}}_{\text{system}}}{dt}. \quad (12.1.1)$$

We shall analyze how the momentum of the constituent elements our system change over a time interval $[t, t + \Delta t]$, and then consider the limit as $\Delta t \to 0$. We can then explicit calculate the derivative on the right hand side of Eq. (12.1.1) and Eq. (12.1.1) becomes

$$\vec{\mathbf{F}}_{\text{ext}}^{\text{total}} = \frac{d\vec{\mathbf{p}}_{\text{system}}}{dt} = \lim_{\Delta t \to 0} \frac{\Delta\vec{\mathbf{p}}_{\text{system}}}{\Delta t} = \lim_{\Delta t \to 0} \frac{\vec{\mathbf{p}}_{\text{system}}(t + \Delta t) - \vec{\mathbf{p}}_{\text{system}}(t)}{\Delta t}. \quad (12.1.2)$$

We need to be very careful how we apply this generalized version of Newton's Second Law to systems in which mass flows between constituent objects. In particular, when we isolate elements as part of our system we must be careful to identify the mass Δm of the material that continuous flows in or out of an object that is part of our system during the time interval Δt under consideration.

We shall consider four categories of mass flow problems that are characterized by the momentum transfer of the material of mass Δm.

12.1.1 Transfer of Material into an Object, but no Transfer of Momentum

Consider for example rain falling vertically downward with speed u into cart of mass m moving forward with speed v. A small amount of falling rain Δm_r has no component of momentum in the direction of motion of the cart. There is a transfer of rain into the cart but no transfer of momentum in the direction of motion of the cart (Figure 12.1).

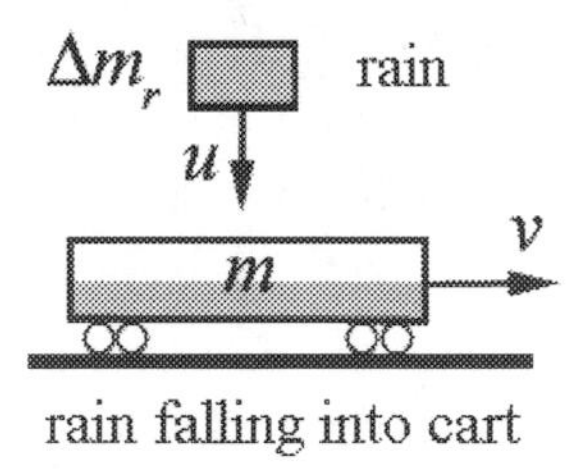

Figure 12.1 Transfer of rain mass into the cart but no transfer of momentum in direction of motion

12.1.2 Transfer of Material Out of an Object, but no Transfer of Momentum

The material continually leaves the object but it does not transport any momentum away from the object in the direction of motion of the object (Figure 12.2). Consider an ice skater gliding on ice at speed v holding a bag of sand that is leaking straight down with respect to the moving skater. The sand continually leaves the bag but it does not transport any momentum away from the bag in the direction of motion of the object. In Figure 12.2, sand of mass Δm_s leaves the bag.

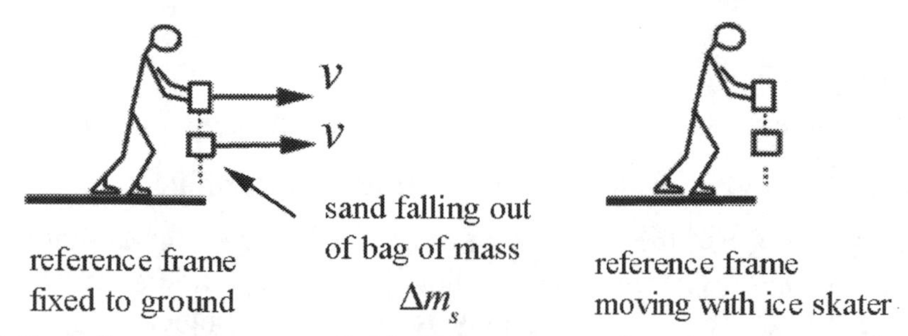

Figure 12.2 Transfer of mass out of object but no transfer of momentum in direction of motion

12.1.3 Transfer of Material Impulses Object Via Transfer of Momentum

Suppose a fire hose is used to put out a fire on a boat. The incoming water with speed u continually hits the boat impulsing it forward. Figure 12.3 shows a column of water of mass Δm_s approaching a boat of mass m_b that is moving forward with speed v.

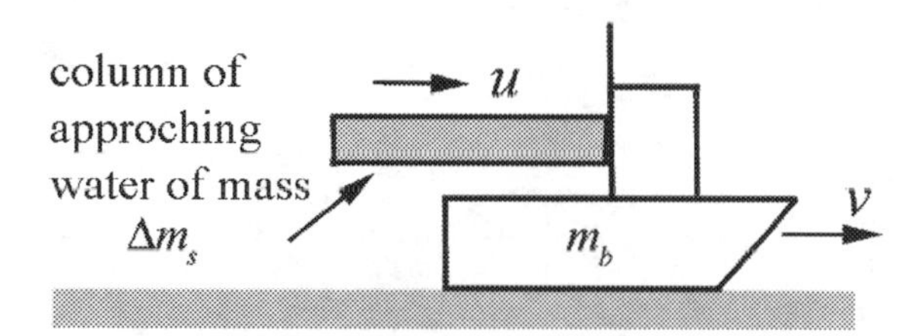

Figure 12.3 Transfer of mass provides impulse on object

12.1.4 Material Continually Ejected From Object results in Recoil of Object

When fuel of mass Δm_f is ejected from the back of a rocket with speed u relative to the rocket, the rocket of mass m_r recoils forward. Figure 12.4a shows the recoil of the rocket in the reference frame of the rocket. The rocket recoils forward with speed Δv_r. In a reference frame in which the rocket is moving forward with speed v_r, then the speed after recoil is $v_r + \Delta v_r$. The speed of the backwardly ejected fuel is $u - v_r$ (Figure 12.4b).

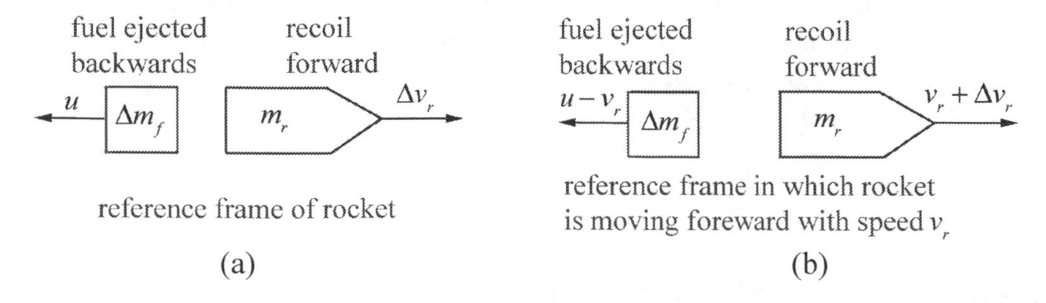

Figure 12.4 Transfer of mass out of rocket provides impulse on rocket in (a) reference frame of rocket, (b) reference frame in which rocket moves with speed v_r

We must carefully identify the momentum of the object and the material transferred at time t in order to determine $\vec{\mathbf{p}}_{\text{system}}(t)$. We must also identify the momentum of the object and the material transferred at time $t+\Delta t$ in order to determine $\vec{\mathbf{p}}_{\text{system}}(t+\Delta t)$ as well. Recall that when we defined the momentum of a system, we assumed that the mass of the system remain constant. Therefore we cannot ignore the momentum of the transferred material at time $t+\Delta t$ even though it may have left the object; it is still part of our system (or at time t even though it has not flowed into the object yet).

12.2 Worked Examples

Example 12.1 Filling a Coal Car

An empty coal car of mass m_0 starts from rest under an applied force of magnitude F. At the same time coal begins to run into the car at a steady rate b from a coal hopper at rest along the track (Figure 12.5). Find the speed when a mass m_c of coal has been transferred.

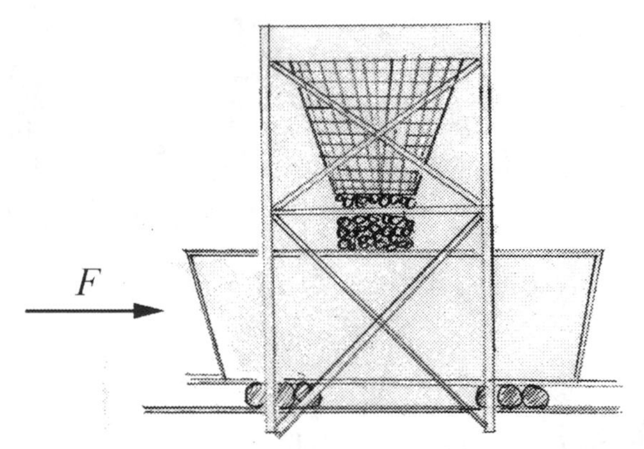

Figure 12.5 Filling a coal car

Solution: We shall analyze the momentum changes in the horizontal direction, which we call the x-direction. Because the falling coal does not have any horizontal velocity, the falling coal is not transferring any momentum in the x-direction to the coal car. So we shall take as our system the empty coal car and a mass m_c of coal that has been transferred. Our initial state at $t=0$ is when the coal car is empty and at rest before any coal has been transferred. The x-component of the momentum of this initial state is zero,

$$p_x(0)=0. \tag{12.1.3}$$

Our final state at $t=t_f$ is when all the coal of mass $m_c=bt_f$ has been transferred into the car that is now moving at speed v_f. The x-component of the momentum of this final state is

$$p_x(t_f) = (m_0 + m_c)v_f = (m_0 + bt_f)v_f . \quad (12.1.4)$$

There is an external constant force $F_x = F$ applied through the transfer. The momentum principle applied to the x-direction is

$$\int_0^{t_f} F_x dt = \Delta p_x = p_x(t_f) - p_x(0) . \quad (12.1.5)$$

Because the force is constant, the integral is simple and the momentum principle becomes

$$Ft_f = (m_0 + bt_f)v_f . \quad (12.1.6)$$

So the final speed is

$$v_f = \frac{Ft_f}{(m_0 + bt_f)} . \quad (12.1.7)$$

Example 12.2 Emptying a Freight Car

A freight car of mass m_c contains a mass of sand m_s. At $t = 0$ a constant horizontal force of magnitude F is applied in the direction of rolling and at the same time a port in the bottom is opened to let the sand flow out at the constant rate $b = dm_s / dt$. Find the speed of the freight car when all the sand is gone (Figure 12.6). Assume that the freight car is at rest at $t = 0$.

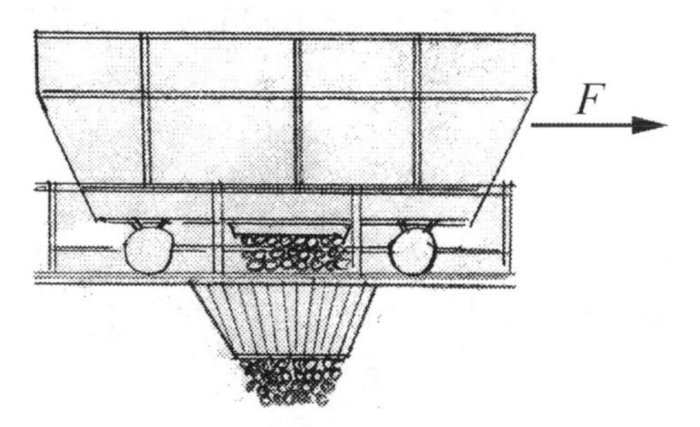

Figure 12.6 Emptying a freight car

Solution: Choose the positive x-direction to point in the direction that the car is moving. Let's take as our system the amount of sand of mass Δm_s that leaves the freight car during the time interval $[t, t + \Delta t]$, and the freight car and whatever sand is in it at time t. The momentum diagram for the system at time t is shown in Figure 12.7.

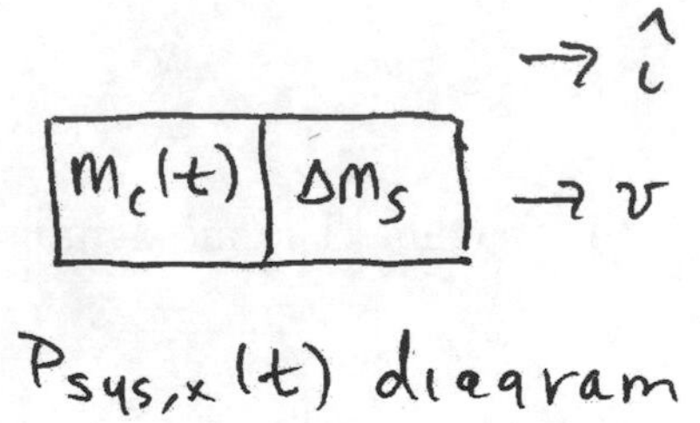

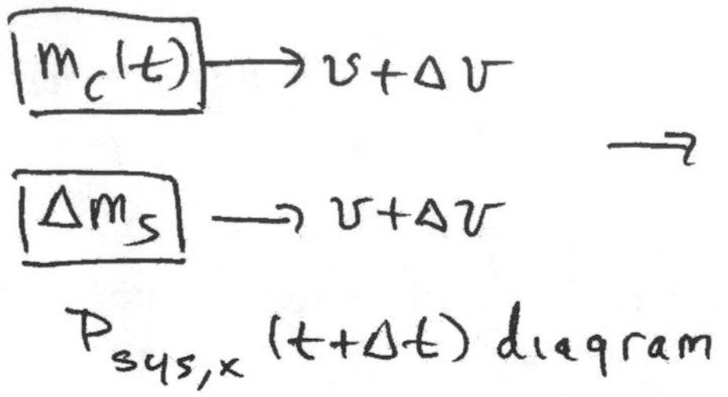

Figure 12.7 Momentum diagram at time t **Figure 12.8** Momentum diagram at time $t+\Delta t$

At the beginning of the interval the car and sand are moving with speed v so the x-component of the momentum at time t is given by

$$p_{sys,x}(t) = (\Delta m_s + m_c(t))v), \tag{12.1.8}$$

where $m_c(t)$ is the mass of the car and sand in it at time t. The momentum diagram for the system at time $t+\Delta t$ is shown in Figure 12.8.

Note that the sand that leaves the car is shown with speed $v+\Delta v$. This implies that all the sand leaves the car with the speed of the car at the end of the interval. This is an approximation. Because the sand leaves continuous, the speed will vary from v to $v+\Delta v$. As we will shortly see, the momentum of the sand $p_{s,x}(t+\Delta t) = \Delta m_s(v+\Delta v)$ is approximately $\Delta m_s v$ because the additional term $\Delta m_s \Delta v$ is a "second-order" term, the product of two infinitesimal terms and hence will vanish when we take the limit that $\Delta t \to 0$. Therefore the x-component of the momentum at time $t+\Delta t$ is given by

$$p_{sys,x}(t+\Delta t) = (\Delta m_s + m_c(t))(v+\Delta v). \tag{12.1.9}$$

We now construct a diagram that illustrates the changes in the x-component of momentum for each element of our system by subtracting Eq.(12.1.8) from Eq. (12.1.9), (Figure 12.9).

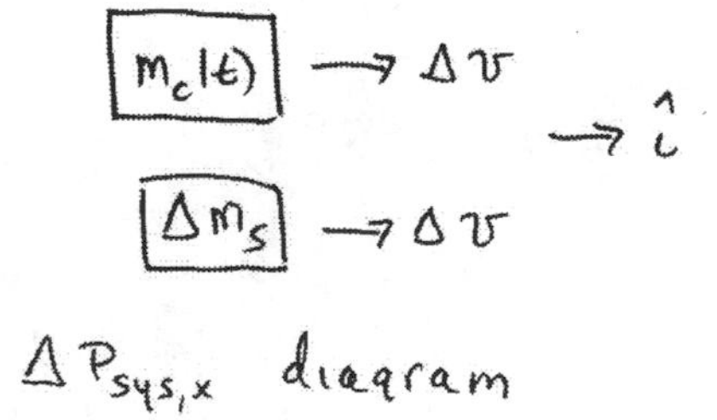

Figure 12.9 Change in momentum diagram

The change in the x-component of momentum of the system is then

$$\Delta p_{sys,x} = p_{sys,x}(t+\Delta t) - p_{sys,x}(t) = \Delta m_s \Delta v + m_c(t)\Delta v \simeq m_c(t)\Delta v . \quad (12.1.10)$$

In terms of the individual elements we have that $\Delta p_{sys,x} = \Delta p_{c,x} + \Delta p_{s,x}$ where

$$\Delta p_{c,x} = m_c(t)\Delta v , \quad (12.1.11)$$

$$\Delta p_{s,x} = \Delta m_s \Delta v . \quad (12.1.12)$$

Throughout the interval a constant force F is applied to the car so

$$F = \lim_{\Delta t \to 0} \frac{p_{sys,x}(t+\Delta t) - p_{sys,x}(t)}{\Delta t} = \lim_{\Delta t \to 0} \frac{\Delta p_{c,x}}{\Delta t} + \lim_{\Delta t \to 0} \frac{\Delta p_{s,x}}{\Delta t} . \quad (12.1.13)$$

From our analysis above, (Eqs. (12.1.11) and (12.1.12)). Eq. (12.1.13) becomes

$$F = \lim_{\Delta t \to 0} m_c(t)\frac{\Delta v}{\Delta t} + \lim_{\Delta t \to 0} \frac{\Delta m_s \Delta v}{\Delta t} . \quad (12.1.14)$$

The second term vanishes when we take $\Delta t \to 0$ because it is of second order in the infinitesimal quantities (in this case $\Delta m_s \Delta v$) and so when dividing by Δt the quantity is of first order and hence vanishes since both $\Delta m_s \to 0$ and $\Delta v \to 0$. Based on this elimination of the second order term, our diagram for the changes in the x-component of momentum for each element of our system simplifies to

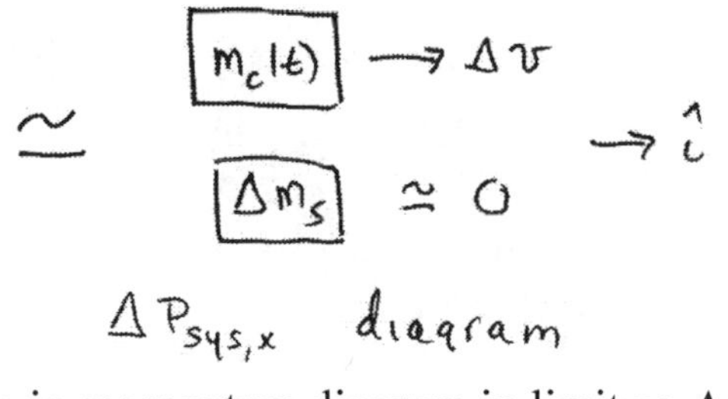

Figure 12.10 Change in momentum diagram in limit as $\Delta m_s \to 0$ and $\Delta v \to 0$

Eq. (12.1.14) becomes the differential equation

$$F = m_c(t)\lim_{\Delta t \to 0} \frac{\Delta v}{\Delta t} = m_c(t)\frac{dv}{dt} . \quad (12.1.15)$$

Denote by $m_{c,0} = m_c + m_s$ where m_c is the mass of the car and m_s is the mass of the sand in the car at $t = 0$, and $m_s(t) = bt$ is the mass of the sand that has left the car at time t,

$$m_s(t) = \int_0^t \frac{dm_s}{dt} dt = \int_0^t b dt = bt \,. \qquad (12.1.16)$$

Thus

$$m_c(t) = m_{c,0} - bt = m_c + m_s - bt \,. \qquad (12.1.17)$$

Using Eq. (12.1.17) we have

$$F = (m_c + m_s - bt)\frac{dv}{dt} \,. \qquad (12.1.18)$$

(b) We can integrate this equation through the separation of variable technique. Rewrite Eq. (12.1.18) as

$$dv = \frac{Fdt}{(m_c + m_s - bt)} \,. \qquad (12.1.19)$$

We can then integrate both sides of Eq. (12.1.19) with the limits as shown

$$\int_{v'=0}^{v'=v(t)} dv' = \int_{t'=0}^{t'=t} \frac{Fdt'}{m_c + m_s - bt'} \,. \qquad (12.1.20)$$

Integration yields the x-component of the velocity of the car as a function of time

$$v(t) = -\frac{F}{b}\ln(m_c + m_s - bt')\Big|_{t'=0}^{t'=t} = -\frac{F}{b}\ln\left(\frac{m_c + m_s - bt}{m_c + m_s}\right) = \frac{F}{b}\ln\left(\frac{m_c + m_s}{m_c + m_s - bt}\right). \qquad (12.1.21)$$

In writing Eq. (12.1.21), we used the property that $\ln(a) - \ln(b) = \ln(a/b)$ and consequently $\ln(a/b) = -\ln(b/a)$. Note that $m_c + m_s \geq m_c + m_s - bt$, so the term $\ln\left(\frac{m_c + m_s}{m_c + m_s - bt}\right) \geq 0$, and the x-component of the velocity of the car increases as we expect.

Example 12.3 Filling a Freight Car

Material is blown into cart A from cart B at a rate of b kilograms per second. The material leaves the chute vertically downward, so that it has the same horizontal velocity, u as cart B, (Figure 12.11). At the moment of interest, cart A has mass m_A and speed v. (a) Define the objects that will constitute your system. (b) Based on momentum flow diagrams, derive a differential equation for the velocity v.

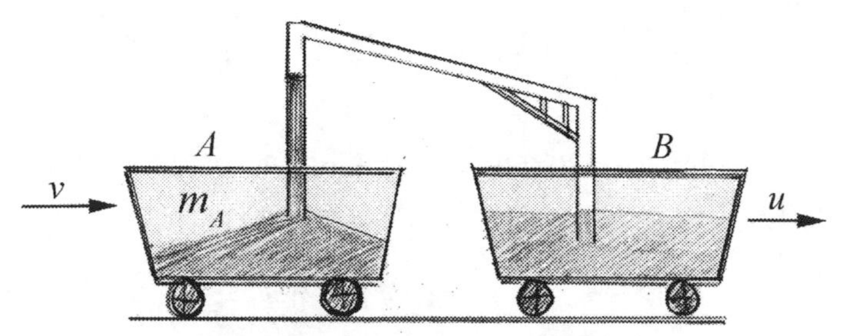

Figure 12.11 Filling a freight car

Solution: Choose positive x-direction to the right in the figure below. Define the system at time t to be the cart with whatever material is in it and the material blown into cart A during the time interval $[t, t+\Delta t]$. Denote the mass of the cart and material at time t by $m_A(t)$ and let Δm_g denote the material blown into cart A during the time interval $[t, t+\Delta t]$, with x-component of the velocity u. At time t, cart A is moving with x-component of the velocity v_A. At time $t+\Delta t$, Cart A is moving with x-component of the velocity $v_A + \Delta v_A$. The momentum diagram for time t is shown in Figure 12.12a and for $t+\Delta t$ is shown in Figure 12.12b.

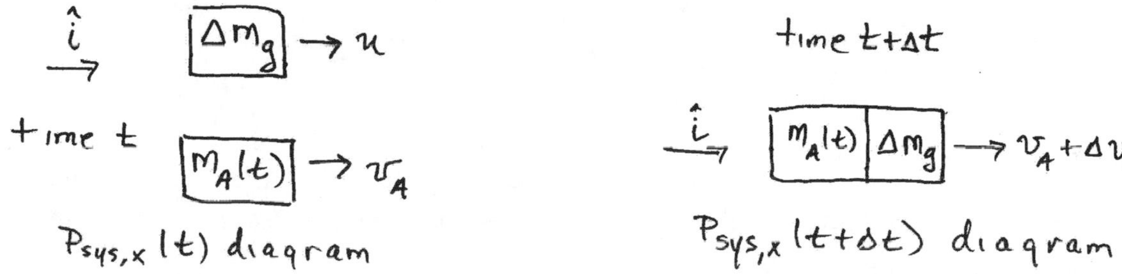

$\hat{i}$ → $\boxed{\Delta m_g} \rightarrow u$

time t $\boxed{m_A(t)} \rightarrow v_A$

$P_{sys,x}(t)$ diagram

(a)

time $t+\Delta t$

$\hat{i}$ → $\boxed{m_A(t)}\boxed{\Delta m_g} \rightarrow v_A + \Delta v_A$

$P_{sys,x}(t+\Delta t)$ diagram

(b)

Figure 12.12 Momentum diagram at (a) time t and, (b) $t+\Delta t$

The diagram representing the change in the x-component of the momentum is shown in Figure 12.13.

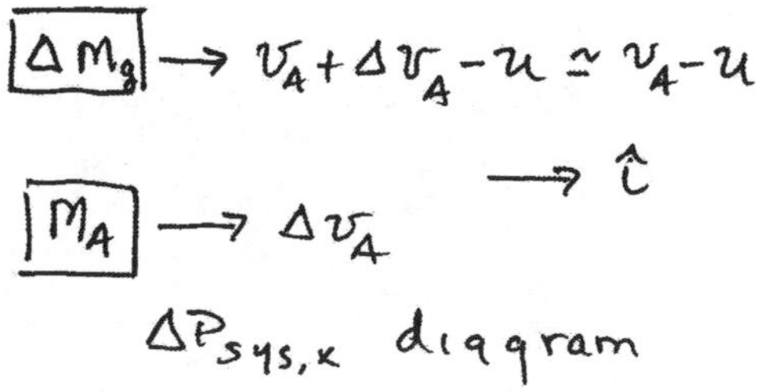

$\boxed{\Delta m_g} \rightarrow v_A + \Delta v_A - u \simeq v_A - u$

→ $\hat{i}$

$\boxed{m_A} \rightarrow \Delta v_A$

$\Delta P_{sys,x}$ diagram

Figure 12.13 Change in momentum diagram

There are no external forces in the x-direction acting on the system, $F_{ext,x} = 0$, so the momentum principle becomes

$$0 = \Delta p_{sys,x} = \Delta p_{A,x} + \Delta p_{g,x}. \tag{12.1.22}$$

From our diagram showing the change in the x-component of the momentum of the elements of the system, Eq. (12.1.22) becomes

$$0 = m_A \Delta v_A + \Delta m_g (v_A - u), \tag{12.1.23}$$

where we can ignore the contribution form the second order term $\Delta m \Delta v_A$. Because the cart's mass is increasing due to the material entering we have that

$$\Delta m_g = \Delta m_A. \tag{12.1.24}$$

and so Eq. (12.1.23) can now be written after taking limits as $\Delta t \to 0$

$$m_A dv_A = dm_A (u - v_A). \tag{12.1.25}$$

We can divide both sides of Eq. (12.1.25) by dt yielding

$$m_A \frac{dv_A}{dt} = \frac{dm_A}{dt}(u - v_A). \tag{12.1.26}$$

Rearranging terms and using the fact that the material is blown into the cart at a constant rate $b \equiv dm_A / dt$, we have that the rate of change of the x-component of the velocity of the cart is given by

$$\frac{dv_A}{dt} = \frac{b(u - v_A)}{m_A}. \tag{12.1.27}$$

We cannot directly integrate Eq. (12.1.27) with respect to dt because the mass of the cart is a function of time. In order to find the x-component of the velocity of the cart we need to know the relationship between the mass of the cart and the x-component of the velocity of the cart. There are two approaches. In the first approach we separate variables in Eq. (12.1.25)

$$\frac{dv_A}{u - v_A} = \frac{dm_A}{m_A}, \tag{12.1.28}$$

and then integrate

$$\int_{v'_A=0}^{v'_A=v_A(t)} \frac{dv'_A}{u - v'_A} = \int_{m'_A=m_{A,0}}^{m'_A=m_A(t)} \frac{dm'_A}{m'_A}, \tag{12.1.29}$$

where $m_{A,0}$ is the mass of the cart before any material has been blown in. After integration we have that

$$\ln\frac{u}{u - v_A(t)} = \ln\frac{m_A(t)}{m_{A,0}}. \quad (12.1.30)$$

Exponentiate both side gives

$$\frac{u}{u - v_A(t)} = \frac{m_A(t)}{m_{A,0}}. \quad (12.1.31)$$

We can solve this equation for the x-component of the velocity of the cart

$$v_A(t) = \frac{m_A(t) - m_{A,0}}{m_A(t)} u. \quad (12.1.32)$$

Because the material is blown into the cart at a constant rate $b \equiv dm_A / dt$, the mass of the cart as a function of time is given by

$$m_A(t) = m_{A,0} + bt. \quad (12.1.33)$$

Therefore substituting Eq. (12.1.33) into Eq. (12.1.32) yields the x-component of the velocity of the cart as a function of time

$$v_A(t) = \frac{bt}{m_{A,0} + bt} u. \quad (12.1.34)$$

In the second approach, we substitute Eq. (12.1.33) into Eq. (12.1.27) yielding

$$\frac{dv_A}{dt} = \frac{b(u - v_A)}{m_{A,0} + bt}. \quad (12.1.35)$$

We can now separate variables

$$\frac{dv_A}{u - v_A} = \frac{bdt}{m_{A,0} + bt}. \quad (12.1.36)$$

Now we can integrate

$$\int_{v_A'=0}^{v_A'=v_A(t)} \frac{dv_A'}{u - v_A'} = \int_{t'=0}^{t'=t} \frac{dt'}{m_{A,0} + bt'} \quad (12.1.37)$$

yielding

$$\ln\frac{u}{u - v_A(t)} = \ln\frac{m_{A,0} + bt}{m_{A,0}}. \quad (12.1.38)$$

Again exponentiating both sides yields

$$\frac{u}{u - v_A(t)} = \frac{m_{A,0} + bt}{m_{A,0}}. \tag{12.1.39}$$

and after some algebraic manipulation we can find the speed of the cart as a function of time

$$v_A(t) = \frac{bt}{m_{A,0} + bt} u. \tag{12.1.40}$$

in agreement with Eq. (12.1.34).

12.3 Rocket Propulsion

A rocket at time $t = 0$ is moving with speed $v_{r,0}$ in the positive x-direction. The rocket burns fuel that is then ejected backward with velocity $\vec{\mathbf{u}} = -u\hat{\mathbf{i}}$ relative to the rocket, where $u > 0$ is the relative speed of the ejected fuel. This exhaust velocity is independent of the velocity of the rocket. The rocket must exert a force to accelerate the ejected fuel backwards and therefore by Newton's Third law, the fuel exerts a force that is equal in magnitude but opposite in direction resulting in propelling the rocket forward. The rocket velocity is a function of time, $\vec{\mathbf{v}}_r(t) = v_{r,x}(t)\hat{\mathbf{i}}$, and the x-component increases at a rate $dv_{r,x} / dt$. Because fuel is leaving the rocket, the mass of the rocket is also a function of time, $m_r(t)$, and is decreasing at a rate dm_r / dt. We shall use the momentum principle, Eq. (12.1.2), to determine a differential equation that relates $dv_{r,x} / dt$, dm_r / dt, u, $v_{r,x}(t)$, and $F_{ext,x}$, an equation known as the rocket equation.

Let $t = t_i$ denote the instant the rocket begins to burn fuel and let $t = t_f$ denote the instant the rocket has finished burning fuel. At some arbitrary time t during this process, the rocket has velocity $\vec{\mathbf{v}}_r(t) = v_{r,x}(t)\hat{\mathbf{i}}$, with the mass of the rocket denoted by $m_r(t) \equiv m_r$. During the time interval $[t, t + \Delta t]$, with Δt taken to be a small interval (we shall eventually consider the limit that $\Delta t \to 0$), a small amount of fuel of mass Δm_f (in the limit that $\Delta t \to 0$, $\Delta m_f \to 0$) is ejected backwards with speed u relative to the rocket. The fuel was initially traveling at the speed of the rocket and so undergoes a change in momentum. The rocket recoils forward, undergoing a change in momentum. In order to keep track of all momentum changes, we define our system to be the rocket (including all the fuel that is not burned during the time interval Δt) and the small amount of fuel that is ejected during the interval Δt. At time t, the fuel has not yet been ejected so it is still inside the rocket. Figure 12.14 represents a momentum diagram at time t for our system relative to a fixed inertial reference frame in which the rocket at time t is moving with speed $v_{r,x}(t)$.

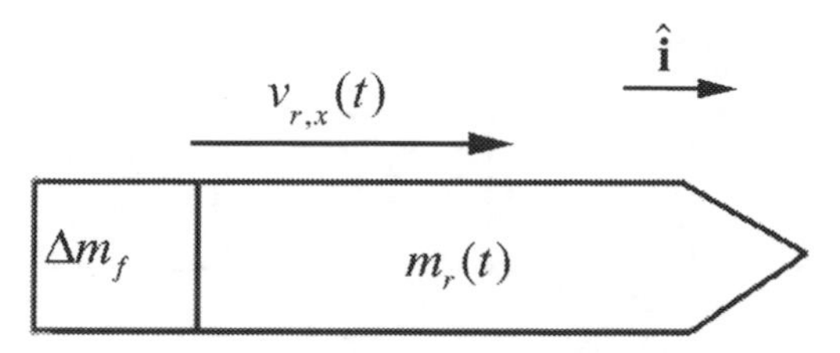

Figure 12.14 Momentum diagram for system at time t

The x-component of the momentum of the system at time t is therefore

$$p_{sys,x}(t) = (m_r(t) + \Delta m_f)v_{r,x}(t). \tag{12.1.41}$$

During the interval $[t, t + \Delta t]$ the fuel is ejected backwards relative to the rocket with speed u. The rocket recoils forward with an increased x-component of the velocity $v_{r,x}(t + \Delta t) = v_{r,x}(t) + \Delta v_{r,x}$, where $\Delta v_{r,x}$ represents the increase the rocket's x-component of the velocity. As usual let's assume that the fuel element, with mass Δm_f, has left the rocket at the end of the time interval, so that the x-component of the velocity of the fuel is $v_{f,x} = v_{r,x} + \Delta v_{r,x} - u$. The momentum diagram of the system at time $t + \Delta t$ is shown in Figure 12.15.

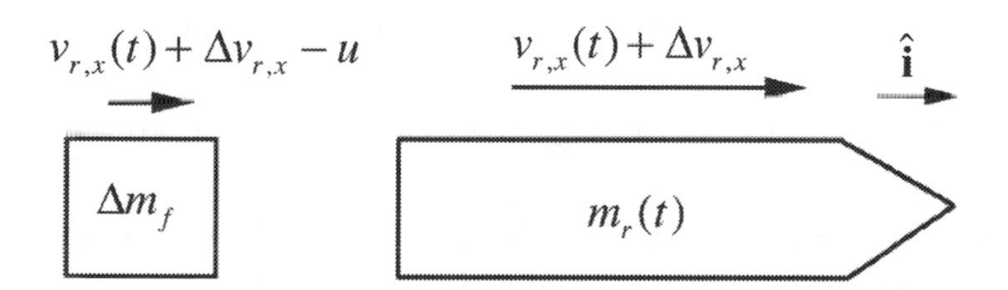

Figure 12.15 Momentum diagram for system at time $t + \Delta t$

The x-component of the momentum of the system at time $t + \Delta t$ is therefore

$$p_{sys,x}(t + \Delta t) = m_r(t)(v_{r,x}(t) + \Delta v_{r,x}) + \Delta m_f(v_{r,x}(t) + \Delta v_{r,x} - u). \tag{12.1.42}$$

In Figure 12.16, we show the diagram depicting the change in the x-component of the momentum of the system consisting of the ejected fuel and rocket.

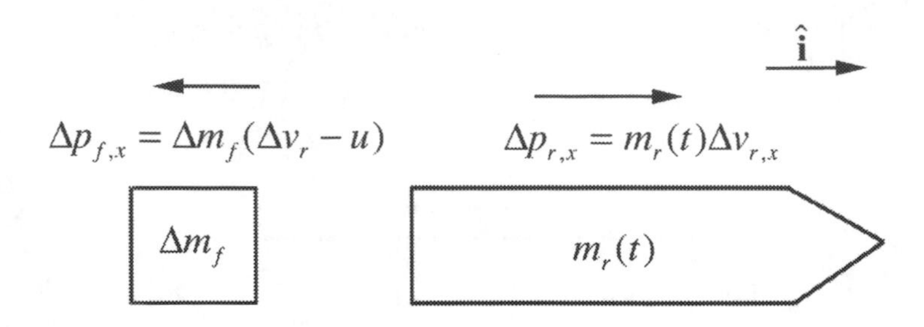

Figure 12.16 Change in momentum for system during time interval $[t, t+\Delta t]$

Therefore the change in the x-component of the momentum of the system is given by

$$\Delta p_{sys,x} = \Delta p_{r,x} + \Delta p_{f,x} = m_r(t)\Delta v_{r,x} + \Delta m_f(\Delta v_{r,x} - u). \qquad (12.1.43)$$

We again note that $\Delta p_{f,x} = \Delta m_f(\Delta v_{r,x} - u) \simeq -\Delta m_f u$, and we show the modified diagram for the change in the x-component of the momentum of the system in Figure 12.17.

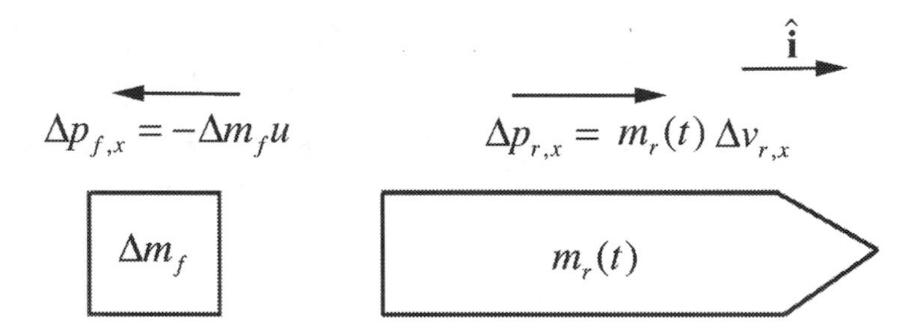

Figure 12.17 Modified change in momentum for system during time interval $[t, t+\Delta t]$

We can now apply Newton's Second Law in the form of the momentum principle (Eq. (12.1.2)), for the system consisting of the rocket and exhaust fuel,

$$F_{ext,x} = \lim_{\Delta t \to 0} \frac{p_{sys,x}(t+\Delta t) - p_{sys,x}(t)}{\Delta t} = \lim_{\Delta t \to 0} \frac{\Delta p_{sys,x}}{\Delta t} = \lim_{\Delta t \to 0} \frac{\Delta p_{r,x}}{\Delta t} + \lim_{\Delta t \to 0} \frac{\Delta p_{f,x}}{\Delta t}. \qquad (12.1.44)$$

From our diagram depicting the change in the x-component of the momentum of the system, we have that

$$F_{ext,x} = \lim_{\Delta t \to 0} \frac{m_r(t)\Delta v_{r,x}}{\Delta t} + \lim_{\Delta t \to 0} \frac{\Delta m_f(\Delta v_{r,x} - u)}{\Delta t}. \qquad (12.1.45)$$

We note that $\Delta m_f \Delta v_{r,x}$ is a second order differential, therefore

$$\lim_{\Delta t \to 0} \frac{\Delta m_f \Delta v_{r,x}}{\Delta t} = 0. \qquad (12.1.46)$$

We also note that

$$\frac{dv_{r,x}}{dt} \equiv \lim_{\Delta t \to 0} \frac{\Delta v_{r,x}}{\Delta t}, \tag{12.1.47}$$

and

$$\frac{dm_f}{dt} \equiv \lim_{\Delta t \to 0} \frac{\Delta m_f}{\Delta t}. \tag{12.1.48}$$

Therefore Eq. (12.1.45) becomes

$$F_{ext,x} = m_r(t)\frac{dv_{r,x}}{dt} - \frac{dm_f}{dt}u. \tag{12.1.49}$$

We can rewrite Eq. (12.1.49) as

$$F_{ext,x} + \frac{dm_f}{dt}u = m_r(t)\frac{dv_{r,x}}{dt}. \tag{12.1.50}$$

The second term on the left-hand-side of Eq. (12.1.50) is called the ***thrust***

$$F_{thrust,x} \equiv \frac{dm_f}{dt}u. \tag{12.1.51}$$

Note that this is not an extra force but the result of the forward recoil due to the ejection of the fuel. Because we are burning fuel at a positive rate $dm_f / dt > 0$ and the speed $u > 0$, the direction of the thrust is in the positive x-direction.

The rate of decrease of the mass of the rocket, dm_r / dt, is equal to the negative of the rate of increase of the exhaust fuel

$$\frac{dm_r}{dt} = -\frac{dm_f}{dt}. \tag{12.1.52}$$

Therefore substituting Eq. (12.1.52) into Eq. (12.1.49), we find that the differential equation describing the motion of the rocket and exhaust fuel is given by

$$F_{ext,x} - \frac{dm_r}{dt}u = m_r(t)\frac{dv_{r,x}}{dt}. \tag{12.1.53}$$

Eq. (12.1.53) is called the ***rocket equation***.

12.3.1 Rocket Equation in Gravity-free Space

We shall first consider the case in which there are no external forces acting on the system, then Eq. (12.1.53) becomes

$$-\frac{dm_r}{dt}u = m_r(t)\frac{dv_{r,x}}{dt}. \tag{12.1.54}$$

In order to solve this equation, we separate the variable quantities $v_{r,x}(t)$ and $m_r(t)$

$$\frac{dv_{r,x}}{dt} = -\frac{u}{m_r(t)}\frac{dm_r}{dt}. \tag{12.1.55}$$

We now multiply both sides by dt and integrate with respect to time between the initial time t_i when the ejection of the burned fuel began and the final time t_f when the process stopped.

$$\int_{t'=t_i}^{t'=t_f} \frac{dv_{r,x}}{dt'}dt' = -\int_{t'=t_i}^{t'=t_f} \frac{u}{m_r(t)}\frac{dm_r}{dt'}dt'. \tag{12.1.56}$$

We can rewrite the integrands and endpoints as

$$\int_{v'_{r,x}=v_{r,x,i}}^{v'_{r,x}=v_{r,x,f}} dv'_{r,x} = -\int_{m'_r=m_{r,i}}^{m'_r=m_{r,f}} \frac{u}{m'_r}dm'_r. \tag{12.1.57}$$

Performing the integration and substituting in the values at the endpoints gives

$$v_{r,x,f} - v_{r,x,i} = -u\ln\left(\frac{m_{r,f}}{m_{r,i}}\right). \tag{12.1.58}$$

Because the rocket is losing fuel, $m_{r,f} < m_{r,i}$, we can rewrite Eq. (12.1.58) as

$$v_{r,x,f} - v_{r,x,i} = u\ln\left(\frac{m_{r,i}}{m_{r,f}}\right). \tag{12.1.59}$$

We note $\ln(m_{r,i} / m_{r,f}) > 1$. Therefore $v_{r,x,f} > v_{r,x,i}$, as we expect.

After a slight rearrangement of Eq. (12.1.59), we have an expression for the velocity of the rocket as a function of the mass m_r of the rocket

$$v_{r,x,f} = v_{r,x,i} + u\ln\left(\frac{m_{r,i}}{m_{r,f}}\right). \tag{12.1.60}$$

Let's examine our result. First, let's suppose that all the fuel was burned and ejected. Then $m_{r,f} \equiv m_{r,d}$ is the final dry mass of the rocket (empty of fuel). The ratio

$$R = \frac{m_{r,i}}{m_{r,d}} \tag{12.1.61}$$

is the ratio of the initial mass of the rocket (including the mass of the fuel) to the final dry mass of the rocket (empty of fuel). The final velocity of the rocket is then

$$v_{r,x,f} = v_{r,x,i} + u \ln R \,. \tag{12.1.62}$$

This is why multistage rockets are used. You need a big container to store the fuel. Once all the fuel is burned in the first stage, the stage is disconnected from the rocket. During the next stage the dry mass of the rocket is much less and so R is larger than the single stage, so the next burn stage will produce a larger final speed then if the same amount of fuel were burned with just one stage (more dry mass of the rocket). In general rockets do not burn fuel at a constant rate but if we assume that the burning rate is constant where

$$b = \frac{dm_f}{dt} = -\frac{dm_r}{dt} \tag{12.1.63}$$

then we can integrate Eq. (12.1.63)

$$\int_{m'_r = m_{r,i}}^{m'_r = m_r(t)} dm'_r = -b \int_{t'=t_i}^{t'=t} dt' \tag{12.1.64}$$

and find an equation that describes how the mass of the rocket changes in time

$$m_r(t) = m_{r,i} - b(t - t_i) \,. \tag{12.1.65}$$

For this special case, if we set $t_f = t$ in Eq. (12.1.60), then the velocity of the rocket as a function of time is given by

$$v_{r,x,f} = v_{r,x,i} + u \ln\left(\frac{m_{r,i}}{m_{r,i} - bt}\right). \tag{12.1.66}$$

Example 12.4 Single-Stage Rocket

Before a rocket begins to burn fuel, the rocket has a mass of $m_{r,i} = 2.81 \times 10^7 \, \text{kg}$, of which the mass of the fuel is $m_{f,i} = 2.46 \times 10^7 \, \text{kg}$. The fuel is burned at a constant rate with total burn time is 510 s and ejected at a speed $u = 3000 \text{ m/s}$ relative to the rocket. If the rocket starts from rest in empty space, what is the final speed of the rocket after all the fuel has been burned?

Solution: The dry mass of the rocket is $m_{r,d} \equiv m_{r,i} - m_{f,i} = 0.35 \times 10^7 \,\text{kg}$, hence $R = m_{r,i} / m_{r,d} = 8.03$. The final speed of the rocket after all the fuel has burned is

$$v_{r,f} = \Delta v_r = u \ln R = 6250 \text{ m/s} \ . \tag{12.1.67}$$

Example 12.5 Two-Stage Rocket

Now suppose that the same rocket in Example 12.4 burns the fuel in two stages ejecting the fuel in each stage at the same relative speed. In stage one, the available fuel to burn is $m_{f,1,i} = 2.03 \times 10^7 \,\text{kg}$ with burn time $150 \,\text{s}$. Then the empty fuel tank and accessories from stage one are disconnected from the rest of the rocket. These disconnected parts have a mass $m = 1.4 \times 10^6 \,\text{kg}$. All the remaining fuel with mass is burned during the second stage with burn time of $360 \,\text{s}$. What is the final speed of the rocket after all the fuel has been burned?

Solution: The mass of the rocket after all the fuel in the first stage is burned is $m_{r,1,d} = m_{r,1,i} - m_{f,1,i} = 0.78 \times 10^7 \,\text{kg}$ and $R_1 = m_{r,1,i} / m_{r,1,d} = 3.60$. The change in speed after the first stage is complete is

$$\Delta v_{r,1} = u \ln R_1 = 3840 \text{ m/s} . \tag{12.1.68}$$

After the empty fuel tank and accessories from stage one are disconnected from the rest of the rocket, the remaining mass of the rocket is $m_{r,2,d} = 2.1 \times 10^6 \,\text{kg}$. The remaining fuel has mass $m_{f,2,i} = 4.3 \times 10^6 \,\text{kg}$. The mass of the rocket plus the unburned fuel at the beginning of the second stage is $m_{r,2,i} = 6.4 \times 10^6 \, kg$. Then $R_2 = m_{r,2,i} / m_{r,2,d} = 3.05$. Therefore the rocket increases its speed during the second stage by an amount

$$\Delta v_{r,2} = u \ln R_2 = 3340 \text{ m/s} . \tag{12.1.69}$$

The final speed of the rocket is the sum of the change in speeds due to each stage,

$$v_f = \Delta v_r = u \ln R_1 + u \ln R_2 = u \ln(R_1 R_2) = 7190 \text{ m/s} , \tag{12.1.70}$$

which is greater than if the fuel were burned in one stage. Plots of the speed of the rocket as a function time for both one-stage and two-stage burns are shown Figure 12.18.

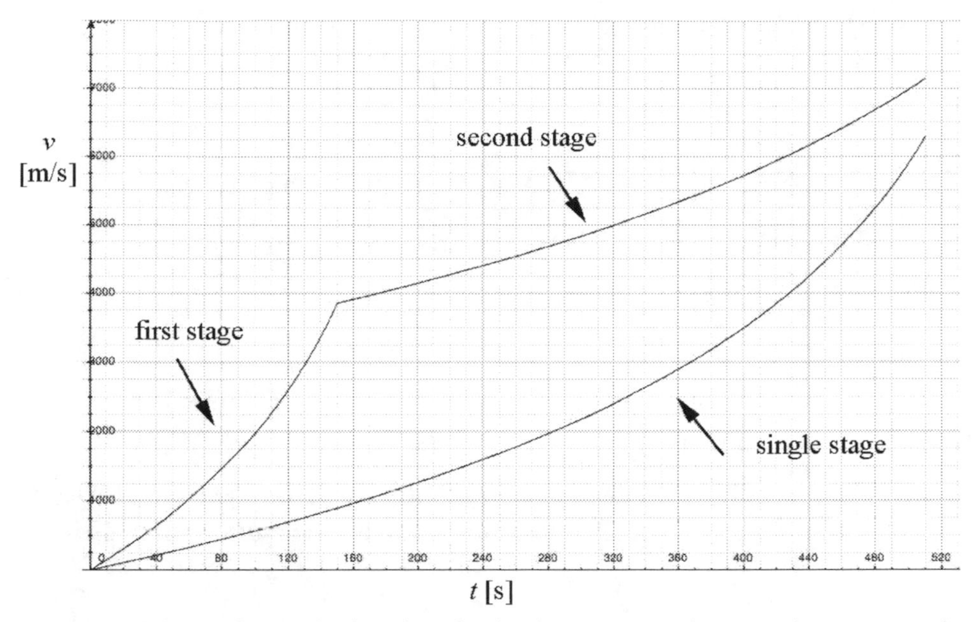

Figure 12.18 Plots of speed of rocket for both one-stage burn and two-stage burn

12.3.2 Rocket in a Constant Gravitational Field:

Now suppose that the rocket takes off from rest at time $t = 0$ in a constant gravitational field then the external force is

$$\vec{\mathbf{F}}_{\text{ext}}^{\text{total}} = m_r \vec{\mathbf{g}} . \tag{12.1.71}$$

Choose the positive x-axis in the upward direction then $F_{ext,x}(t) = -m_r(t)g$. Then the rocket equation (Eq. (12.1.53) becomes

$$-m_r(t)g - \frac{dm_r}{dt}u = m_r(t)\frac{dv_{r,x}}{dt} . \tag{12.1.72}$$

Multiply both sides of Eq. (12.1.72) by dt, and divide both sides by $m_r(t)$. Then Eq. (12.1.72) can be written as

$$dv_{r,x} = -gdt - \frac{dm_r}{m_r(t)}u . \tag{12.1.73}$$

We now integrate both sides

$$\int_{v_{r,x,i}=0}^{v_{r,x}(t)} dv'_{r,x} = -u \int_{m_{r,i}}^{m_r(t)} \frac{dm'_r}{m'_r} - g\int_0^t dt' , \tag{12.1.74}$$

where $m_{r,i}$ is the initial mass of the rocket and the fuel. Integration yields

$$v_{r,x}(t) = -u\ln\left(\frac{m_r(t)}{m_{r,i}}\right) - gt = u\ln\left(\frac{m_{r,i}}{m_r(t)}\right) - gt\,. \qquad (12.1.75)$$

After all the fuel is burned at $t = t_f$, the mass of the rocket is equal to the dry mass $m_{r,f} = m_{r,d}$ and so

$$v_{r,x}(t_f) = u\ln R - gt_f\,. \qquad (12.1.76)$$

The first term on the right hand side is independent of the burn time. However the second term depends on the burn time. The shorter the burn time, the smaller the negative contribution from the third turn, and hence the larger the final speed. So the rocket engine should burn the fuel as fast as possible in order to obtain the maximum possible speed.

Chapter 13 The Concept of Energy and Conservation of Energy

Chapter 13 Energy, Kinetic Energy, and Work

Acceleration of the expansion of the universe is one of the most exciting and significant discoveries in physics, with implications that could revolutionize theories of quantum physics, gravitation, and cosmology. With its revelation that close to the three-quarters of the energy density of the universe, given the name dark energy, is of a new, unknown origin and that its exotic gravitational "repulsion" will govern the fate of the universe, dark energy and the accelerating universe becomes a topic not just of great interest to research physicists but to science students at all levels. [1]

Eric Linder

13.1 The Concept of Energy and Conservation of Energy

The transformation of energy is a powerful concept that enables us to describe a vast number of processes:

Falling water releases stored *gravitational potential energy*, which can become the *kinetic energy* associated with a *coherent motion* of matter. The harnessed *mechanical energy* can be used to spin turbines and alternators, doing *work* to generate *electrical energy*, transmitted to consumers along power lines. When you use any electrical device, the electrical energy is transformed into other forms of energy. In a refrigerator, electrical energy is used to compress a gas into a liquid. During the compression, some of the internal energy of the gas is transferred to the *random motion* of molecules in the outside environment. The liquid flows from a high-pressure region into a low-pressure region where the liquid evaporates. During the evaporation, the liquid absorbs energy from the *random motion* of molecules inside of the refrigerator. The gas returns to the compressor.

"Human beings transform the stored chemical energy of food into various forms necessary for the maintenance of the functions of the various organ system, tissues and cells in the body."[2] A person can do work on their surroundings – for example, by pedaling a bicycle – and transfer energy to the surroundings in the form of increasing random motion of air molecules, by using this catabolic energy.

Burning gasoline in car engines converts *chemical energy*, stored in the molecular bonds of the constituent molecules of gasoline, into coherent (ordered) motion of the molecules that constitute a piston. With the use of gearing and tire/road friction, this motion is converted into kinetic energy of the car; the automobile moves.

---

[1] Eric Linder, *Resource Letter: Dark Energy and the Accelerating Universe*, Am.J.Phys.76: 197-204, 2008; p. 197.

[2] George B. Benedek and Felix M.H. Villars, *Physics with Illustrative Examples from Medicine and Biology, Volume 1: Mechanics,* Addison-Wesley, Reading, 1973, p. 115-6.

Stretching or compressing a spring stores *elastic potential energy* that can be released as kinetic energy.

The process of vision begins with stored *atomic energy* released as electromagnetic radiation (light), which is detected by exciting photoreceptors in the eye, releasing chemical energy.

When a proton fuses with deuterium (a hydrogen atom with a neutron and proton for a nucleus), helium-three is formed (with a nucleus of two protons and one neutron) along with radiant energy in the form of photons. The combined *internal energy* of the proton and deuterium are greater than the internal energy of the helium-three. This difference in internal energy is carried away by the photons as light energy.

There are many such processes in the manmade and natural worlds, involving different forms of energy: kinetic energy, gravitational energy, thermal energy, elastic energy, electrical energy, chemical energy, electromagnetic energy, nuclear energy and more. The total energy is always conserved in these processes, although different forms of energy are converted into others.

Any physical process can be characterized by two states, initial and final, between which energy transformations can occur. Each form of energy E_i, where "i" is an arbitrary label identifying one of the N forms of energy, may undergo a change during this transformation,

$$\Delta E_i \equiv E_{\text{final},i} - E_{\text{initial},i}\,. \tag{13.1.1}$$

Conservation of energy means that the sum of these changes is zero,

$$\Delta E_1 + \Delta E_2 + \cdots + \Delta E_N = \sum_{i=1}^{N} \Delta E_i = 0\,. \tag{13.1.2}$$

Two important points emerge from this idea. First, we are interested primarily in changes in energy and so we search for relations that describe how each form of energy changes. Second, we must account for all the ways energy can change. If we observe a process, and the sum of the changes in energy is not zero, either our expressions for energy are incorrect, or there is a new type of change of energy that we had not previously discovered. This is our first example of the importance of conservation laws in describing physical processes, as energy is a key quantity conserved in all physical processes. If we can quantify the changes of different forms of energy, we have a very powerful tool to understand nature.

We will begin our analysis of conservation of energy by considering processes involving only a few forms of changing energy. We will make assumptions that greatly simplify our description of these processes. At first we shall only consider processes acting on bodies in which the atoms move in a coherent fashion, ignoring processes in which energy is transferred into the random motion of atoms. Thus we will initially ignore the effects of friction. We shall then treat processes involving friction between rigid bodies. We will later return to processes in which there is an energy transfer resulting in an increase or decrease in random motion when we study the First Law of Thermodynamics.

Energy is always conserved but we often prefer to restrict our attention to a set of objects that we define to be our *system*. The rest of the universe acts as the *surroundings*. We illustrate this division of system and surroundings in Figure 13.1.

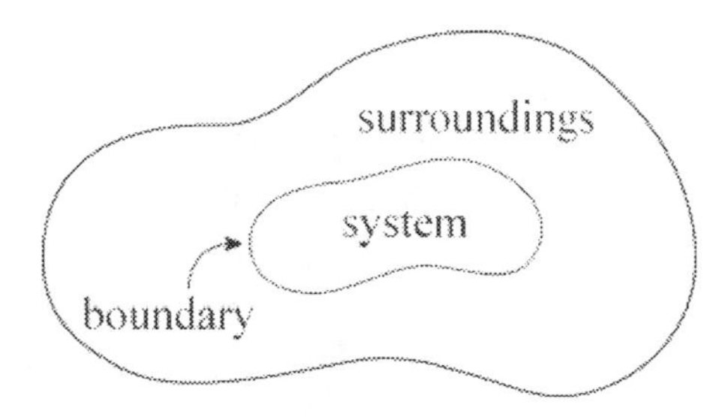

Figure 13.1 A diagram of a system and its surroundings with boundary

Because energy is conserved, any energy that leaves the system must cross through the boundary and enter the surroundings. Consider any physical process in which energy transformations occur that induces a transition between initial and final states.

When a system and its surroundings undergo a transition from an initial state to a final state, the change in energy is zero,

$$\Delta E = \Delta E_{\text{system}} + \Delta E_{\text{surroundings}} = 0 . \qquad (13.1.3)$$

Eq. (13.1.3) is called ***conservation of energy*** and is our operating definition for energy. We will sometime refer to Eq. (13.1.3) as the *energy principle*. In any physical application, we first identify our system and surroundings, and then attempt to quantify changes in energy. In order to do this, we need to identify every type of change of energy in every possible physical process. When there is no change in energy in the surroundings then the system is called a ***closed system***, and consequently the energy of a closed system is constant.

$$\Delta E_{\text{system}} = 0, \qquad (\text{closed system}) . \qquad (13.1.4)$$

If we add up all known changes in energy in the system and surroundings and do not arrive at a zero sum, we have an open scientific problem. By searching for the missing changes in energy, we may uncover some new physical phenomenon. Recently, one of the most exciting open problems in cosmology is the apparent acceleration of the expansion of the universe, which has been attributed to *dark energy* that resides in space itself, an energy type without a clearly known source.[3]

---

[3] http://www-supernova.lbl.gov/~evlinder/sci.html

13.2 Kinetic Energy

The first form of energy that we will study is an energy associated with the coherent motion of molecules that constitute a body of mass m; this energy is called the *kinetic energy* (from the Greek "*kinetikos,*" moving). Let us consider a car moving along a straight road (along which we will place the x-axis). For an observer at rest with respect to the ground, the car has velocity $\vec{\mathbf{v}} = v_x \hat{\mathbf{i}}$. The speed of the car is the magnitude of the velocity, $v \equiv |v_x|$.

> *The* ***kinetic energy*** *K of a non-rotating body of mass m moving with speed v is defined to be the positive scalar quantity*

$$K \equiv \frac{1}{2} m v^2 \tag{13.2.1}$$

The kinetic energy is proportional to the square of the speed. The SI units for kinetic energy are $[\mathrm{kg} \cdot \mathrm{m}^2 \cdot \mathrm{s}^{-2}]$. This combination of units is defined to be a joule and is denoted by $[\mathrm{J}]$, thus $1\,\mathrm{J} \equiv 1\,\mathrm{kg} \cdot \mathrm{m}^2 \cdot \mathrm{s}^{-2}$. (The SI unit of energy is named for James Prescott Joule.) The above definition of kinetic energy does not refer to any direction of motion, just the speed of the body.

Let's consider a case in which our car changes velocity. For our initial state, the car moves with an initial velocity $\vec{\mathbf{v}}_0 = v_{x,0}\,\hat{\mathbf{i}}$ along the x-axis. For the final state (at some later time), the car has changed its velocity and now moves with a final velocity $\vec{\mathbf{v}}_f = v_{x,f}\,\hat{\mathbf{i}}$. Therefore the change in the kinetic energy is

$$\Delta K = \frac{1}{2} m v_f^{\,2} - \frac{1}{2} m v_0^{\,2} \tag{13.2.2}$$

Example 13.1 Change in Kinetic Energy of a Car

Suppose car A increases its speed from 10 to 20 mph and car B increases its speed from 50 to 60 mph. Both cars have the same mass m. (a) What is the ratio of the change of kinetic energy of car B to the change of kinetic energy of car A? Which car has a greater change in kinetic energy? (b) What is the ratio of the change in kinetic energy of car B to car A as seen by an observer moving with the initial velocity of car A?

Solution: (a) The ratio of the change in kinetic energy of car B to car A is

$$\frac{\Delta K_B}{\Delta K_A} = \frac{\frac{1}{2}m(v_{B,f})^2 - \frac{1}{2}m(v_{B,0})^2}{\frac{1}{2}m(v_{A,f})^2 - \frac{1}{2}m(v_{A,0})^2} = \frac{(v_{B,f})^2 - (v_{B,0})^2}{(v_{A,f})^2 - (v_{A,0})^2}$$

$$= \frac{(60\text{ mph})^2 - (50\text{ mph})^2}{(10\text{ mph})^2} = 11/3.$$

Thus car B has a much greater increase in its kinetic energy than car A.

(b) Car A now increases its speed from rest to 10 mph and car B increases its speed from 40 to 50 mph. The ratio is now

$$\frac{\Delta K_B}{\Delta K_A} = \frac{\frac{1}{2}m(v_{B,f})^2 - \frac{1}{2}m(v_{B,0})^2}{\frac{1}{2}m(v_{A,f})^2 - \frac{1}{2}m(v_{A,0})^2} = \frac{(v_{B,f})^2 - (v_{B,0})^2}{(v_{A,f})^2 - (v_{A,0})^2}$$

$$= \frac{(50\text{ mph})^2 - (40\text{ mph})^2}{(10\text{ mph})^2} = 9.$$

The ratio is greater than that found in part a). Note that from the new reference frame both car A and car B have smaller increases in kinetic energy.

13.3 Kinematics and Kinetic Energy in One Dimension

13.3.1 Constant Accelerated Motion

Let's consider a constant accelerated motion of a *rigid body* in one dimension. We begin the discussion by treating our object as a point mass. Suppose at $t = 0$ the object has an initial x-component of the velocity given by $v_{x,i}$. If the acceleration is in the direction of the displacement of the body then the body will increase its speed. If the acceleration is opposite the direction of the displacement then the acceleration will decrease the body's speed. The displacement of the body is given by

$$\Delta x = v_{x,i}\,t + \frac{1}{2}a_x t^2. \tag{13.3.1}$$

The product of acceleration and the displacement is

$$a_x \Delta x = a_x (v_{x,i}\,t + \frac{1}{2}a_x t^2). \tag{13.3.2}$$

The acceleration is given by

$$a_x = \frac{\Delta v_x}{\Delta t} = \frac{(v_{x,f} - v_{x,i})}{t}. \quad (13.3.3)$$

Therefore

$$a_x \Delta x = \frac{(v_{x,f} - v_{x,i})}{t}\left(v_{x,0}\, t + \frac{1}{2}\frac{(v_{x,f} - v_{x,i})}{t} t^2 \right). \quad (13.3.4)$$

Equation (13.3.4) becomes

$$a_x \Delta x = (v_{x,f} - v_{x,i})(v_{x,i}) + \frac{1}{2}(v_{x,f} - v_{x,i})(v_{x,f} - v_{x,i}) = \frac{1}{2}v_{x,f}^2 - \frac{1}{2}v_{x,i}^2. \quad (13.3.5)$$

If we multiply each side of Equation (13.3.5) by the mass m of the object this kinematical result takes on an interesting interpretation for the motion of the object. We have

$$m a_x \Delta x = \frac{1}{2} m v_{x,f}^2 - m \frac{1}{2} v_{x,i}^2 = K_f - K_i. \quad (13.3.6)$$

Recall that for one-dimensional motion, Newton's Second Law is $F_x = m a_x$, for the motion considered here, Equation (13.3.6) becomes

$$F_x \Delta x = K_f - K_i. \quad (13.3.7)$$

13.3.2 Non-constant Accelerated Motion

If the acceleration is not constant, then we can divide the displacement into N intervals indexed by $j = 1$ to N. It will be convenient to denote the displacement intervals by Δx_j, the corresponding time intervals by Δt_j and the x-components of the velocities at the beginning and end of each interval as $v_{x,j-1}$ and $v_{x,j}$. Note that the x-component of the velocity at the beginning and end of the first interval $j = 1$,is then $v_{x,1} = v_{x,i}$ and the velocity at the end of the last interval, $j = N$ is $v_{x,N} = v_{x,j}$. Consider the sum of the products of the average acceleration $(a_{x,j})_{ave}$ and displacement Δx_j in each interval,

$$\sum_{j=1}^{j=N} (a_{x,j})_{\text{ave}} \Delta x_j. \quad (13.3.8)$$

The average acceleration over each interval is equal to

$$(a_{x,j})_{\text{ave}} = \frac{\Delta v_{x,j}}{\Delta t_j} = \frac{(v_{x,j+1} - v_{x,j})}{\Delta t_j}, \quad (13.3.9)$$

and so the contribution in each integral can be calculated as above and we have that

$$(a_{x,j})_{\text{ave}}\,\Delta x_j = \frac{1}{2}v_{x,j}{}^2 - \frac{1}{2}v_{x,j-1}{}^2. \tag{13.3.10}$$

When we sum over all the terms only the last and first terms survive, all the other terms cancel in pairs, and we have that

$$\sum_{j=1}^{j=N}(a_{x,j})_{\text{ave}}\,\Delta x_j = \frac{1}{2}v_{x,f}^2 - \frac{1}{2}v_{x,i}^{2\;2}. \tag{13.3.11}$$

In the limit as $N \to \infty$ *and* $\Delta x_j \to 0$ for all j (both conditions must be met!), the limit of the sum is the definition of the definite integral of the acceleration with respect to the position,

$$\lim_{\substack{N\to\infty \\ \Delta x_j\to 0}} \sum_{j=1}^{j=N}(a_{x,j})_{\text{ave}}\,\Delta x_j \equiv \int_{x=x_i}^{x=x_f} a_x(x)\,dx. \tag{13.3.12}$$

Therefore In the limit as $N \to \infty$ *and* $\Delta x_j \to 0$ for all j, with $v_{x,N} \to v_{x,f}$, Eq. (13.3.11) becomes

$$\int_{x=x_i}^{x=x_f} a_x(x)\,dx = \frac{1}{2}(v_{x,f}^2 - v_{x,i}^2) \tag{13.3.13}$$

This integral result is consequence of the definition that $a_x \equiv dv_x / dt$. Notice how Eq. (13.3.13) compares to the integral of acceleration with respect to time

$$\int_{t=t_i}^{t=t_f} a_x(t)\,dt = v_{x,f} - v_{x,i}. \tag{13.3.14}$$

Multiplying both sides of Eq. (13.3.13) by the mass m yields

$$\int_{x=x_i}^{x=x_f} ma_x(x)\,dx = \frac{1}{2}m(v_{x,f}^2 - v_{x,i}^2) = K_f - K_i. \tag{13.3.15}$$

When we introduce Newton's Second Law in the form $F_x = ma_x$, then Eq. (13.3.15) becomes

$$\int_{x=x_i}^{x=x_f} F_x(x)\,dx = K_f - K_i. \tag{13.3.16}$$

The integral of the x-component of the force with respect to displacement in Eq. (13.3.16) applies to the motion of a point-like object. For extended bodies, Eq. (13.3.16) applies to the center of mass motion because the external force on a rigid body causes the center of mass to accelerate.

13.4 Work done by Constant Forces

We will begin our discussion of the concept of work by analyzing the motion of an object in one dimension acted on by constant forces. Let's consider an example of this type of motion: pushing a cup forward with a constant force along a desktop. When the cup changes speed (and hence kinetic energy), the sum of the forces acting on the cup must be non-zero according to Newton's Second Law. There are three forces involved in this motion: the applied pushing force $\vec{\mathbf{F}}^a$; the contact force $\vec{\mathbf{C}} \equiv \vec{\mathbf{N}} + \vec{\mathbf{f}}_k$; and gravity $\vec{\mathbf{F}}^g = m\vec{\mathbf{g}}$. The force diagram on the cup is shown in Figure 13.2.

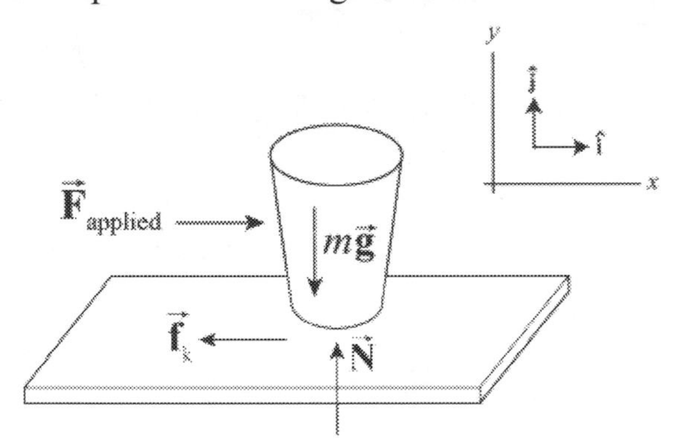

Figure 13.2 Force diagram for cup.

Let's choose our coordinate system so that the $+x$-direction is the direction of the forward motion of the cup. The pushing force can then be described by

$$\vec{\mathbf{F}}^a = F_x^a \,\hat{\mathbf{i}}. \tag{13.4.1}$$

Suppose a body moves from an initial point x_0 to a final point x_f so that the displacement of the point the force acts on is positive $\Delta x \equiv x_f - x_0 > 0$. The ***work done by a constant force*** $\vec{\mathbf{F}}^a = F_x^a \hat{\mathbf{i}}$ acting on the body is the product of the component of the force F_x^a and the displacement Δx,

$$W^a = F_x^a \Delta x. \tag{13.4.2}$$

Work is a scalar quantity; it is not a vector quantity. The SI unit for work is

$$[1\ \text{N}\cdot\text{m}] = [1\ \text{kg}\cdot\text{m}\cdot\text{s}^{-2}][1\ \text{m}] = [1\ \text{kg}\cdot\text{m}^2\cdot\text{s}^{-2}] = [1\ \text{J}]. \qquad (13.4.3)$$

Note that work has the same dimension and the same SI unit as kinetic energy. Because our applied force is along the direction of motion, both $F_x^a > 0$ and $\Delta x > 0$. The work done is just the product of the magnitude of the applied force and the distance through which that force acts and is positive. In the definition of work done by a force, the force can act at any point on the body. The displacement that appears in Equation (13.4.2) is not the displacement of the body but the *displacement of the point of application of the force*. For point-like objects, the displacement of the point of application of the force is equal to the displacement of the body. However for an extended body, we need to focus on where the force acts and whether or not that point of application undergoes any displacement in the direction of the force as the following example illustrates.

Example 13.2 Work Done by Static Fiction

Suppose you are initially standing and you start walking by pushing against the ground with your feet and your feet do not slip. How much work does the static frictional force do on you?

Solution: When you apply a contact force against the ground, the ground applies an equal and opposite contact force on you. The tangential component of this constant force is the force of static friction acting on you. Since your foot is at rest while you are pushing against the ground, there is no displacement of the point of application of this static frictional force. Therefore static friction does zero work on you while you are accelerating. You may be surprised by this result but if you think about energy transformation, chemical energy stored in your muscle cells is being transformed into kinetic energy of motion and thermal energy.

We can extend the concept of work to forces that oppose the motion, like friction. In our example of the moving cup, the kinetic frictional force is

$$\vec{\mathbf{F}}^f = f_x\,\hat{\mathbf{i}} = -\mu_k N\,\hat{\mathbf{i}} = -\mu_k mg\,\hat{\mathbf{i}}, \qquad (13.4.4)$$

where $N = mg$ from consideration of the $\hat{\mathbf{j}}$-components of force in Figure 13.2 and the model $f_k = \mu_k N$ for kinetic friction have been used.

Here the component of the force is in the opposite direction as the displacement. The work done by the frictional force is negative,

$$W^f = -\mu_k mg\Delta x. \qquad (13.4.5)$$

Since the gravitation force is perpendicular to the motion of the cup, the gravitational force has no component along the line of motion. Therefore the gravitation force does zero work on the cup when the cup is slid forward in the horizontal direction. The normal force is also perpendicular to the motion, and hence does no work.

We see that the pushing force does positive work, the frictional force does negative work, and the gravitation and normal force does zero work.

Example 13.3 Work Done by Force Applied in the Direction of Displacement

Push a cup of mass 0.2 kg along a horizontal table with a force of magnitude 2.0 N for a distance of 0.5 m. The coefficient of friction between the table and the cup is $\mu_k = 0.10$. Calculate the work done by the pushing force and the work done by the frictional force.

Solution: The work done by the pushing force is

$$W^a = F_x^a \Delta x = (2.0\ \text{N})(0.5\ \text{m}) = 1.0\ \text{J}. \qquad (13.4.6)$$

The work done by the frictional force is

$$W^f = -\mu_k mg\Delta x = -(0.1)(0.2\ \text{kg})(9.8\ \text{m}\cdot\text{s}^{-2})(0.5\ \text{m}) = -0.10\ \text{J}. \qquad (13.4.7)$$

Example 13.4 Work Done by Force Applied at an Angle to the Direction of Displacement

Suppose we push the cup in the previous example with a force of the same magnitude but at an angle $\theta = 30^o$ upwards with respect to the table. Calculate the work done by the pushing force. Calculate the work done by the kinetic frictional force.

Solution: The force diagram on the cup and coordinate system is shown in Figure 13.3.

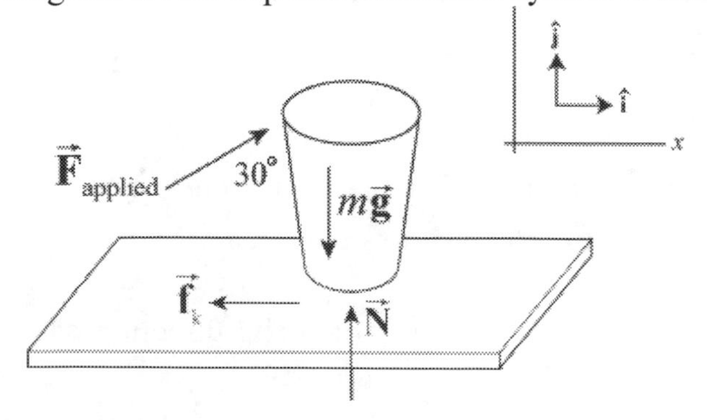

Figure 13.3 Force diagram on cup.

The x-component of the pushing force is now

$$F_x^a = F^a\cos(\theta) = (2.0\text{ N})(\cos(30^\circ)) = 1.7\text{ N}. \qquad (13.4.8)$$

The work done by the pushing force is

$$W^a = F_x^a\Delta x = (1.7\text{ N})(0.5\text{ m}) = 8.7\times10^{-1}\text{ J}. \qquad (13.4.9)$$

The kinetic frictional force is

$$\vec{\mathbf{F}}^f = -\mu_k N\,\hat{\mathbf{i}}. \qquad (13.4.10)$$

In this case, the magnitude of the normal force is not simply the same as the weight of the cup. We need to find the y-component of the applied force,

$$F_y^a = F^a\sin(\theta) = (2.0\text{ N})(\sin(30^o) = 1.0\text{ N}. \qquad (13.4.11)$$

To find the normal force, we apply Newton's Second Law in the y-direction,

$$F_y^a + N - mg = 0. \qquad (13.4.12)$$

Then the normal force is

$$N = mg - F_y^a = (0.2\text{ kg})(9.8\text{ m}\cdot\text{s}^{-2}) - (1.0\text{ N}) = 9.6\times10^{-1}\text{ N}. \qquad (13.4.13)$$

The work done by the kinetic frictional force is

$$W^f = -\mu_k N\Delta x = -(0.1)(9.6\times10^{-1}\text{ N})(0.5\text{ m}) = 4.8\times10^{\ 2}\text{ J}. \qquad (13.4.14)$$

Example 13.5 Work done by Gravity Near the Surface of the Earth

Consider a point-like body of mass m near the surface of the earth falling directly towards the center of the earth. The gravitation force between the body and the earth is nearly constant, $\vec{\mathbf{F}}_{grav} = m\vec{\mathbf{g}}$. Let's choose a coordinate system with the origin at the surface of the earth and the $+y$-direction pointing away from the center of the earth Suppose the body starts from an initial point y_0 and falls to a final point y_f closer to the earth. How much work does the gravitation force do on the body as it falls?

Solution: The displacement of the body is negative, $\Delta y \equiv y_f - y_0 < 0$. The gravitation force is given by

$$\vec{\mathbf{F}}^g = m\vec{\mathbf{g}} = F_y^g\,\hat{\mathbf{j}} = -mg\,\hat{\mathbf{j}}. \qquad (13.4.15)$$

The work done on the body is then

$$W^g = F^g_y \Delta y = -mg\Delta y . \qquad (13.4.16)$$

For a falling body, the displacement of the body is negative, $\Delta y \equiv y_f - y_0 < 0$; therefore the work done by gravity is positive, $W^g > 0$. The gravitation force is pointing in the same direction as the displacement of the falling object so the work should be positive.

When an object is rising while under the influence of a gravitation force, $\Delta y \equiv y_f - y_0 > 0$. The work done by the gravitation force for a rising body is negative, $W^g < 0$, because the gravitation force is pointing in the opposite direction from that in which the object is displaced.

It's important to note that the choice of the positive direction as being away from the center of the earth ("up") does not make a difference. If the downward direction were chosen positive, the falling body would have a positive displacement and the gravitational force as given in Equation (13.4.15) would have a positive downward component; the product $F^g_y \Delta y$ would still be positive.

13.5 Work done by Non-Constant Forces

Consider a body moving in the x-direction under the influence of a non-constant force in the x-direction, $\vec{\mathbf{F}} = F_x \,\hat{\mathbf{i}}$. The body moves from an initial position x_0 to a final position x_f. In order to calculate the work done by a non-constant force, we will divide up the displacement of the point of application of the force into a large number N of small displacements Δx_j where the index j marks the jth displacement and takes integer values from 1 to N. Let $(F_{x,j})_{ave}$ denote the average value of the x-component of the force in the displacement interval $[x_{j-1}, x_j]$. For the jth displacement interval we calculate the contribution to the work

$$\Delta W_j = (F_{x,j})_{\text{ave}} \Delta x_j \qquad (13.5.1)$$

This contribution is a scalar so we add up these scalar quantities to get the total work

$$W_N = \sum_{j=1}^{j=N} \Delta W_j = \sum_{j=1}^{j=N} (F_{x,j})_{\text{ave}} \Delta x_j . \qquad (13.5.2)$$

The sum in Equation (13.5.2) depends on the number of divisions N and the width of the intervals Δx_j. In order to define a quantity that is independent of the divisions, we take the limit as $N \to \infty$ and $|\Delta x_j| \to 0$ for all j. The work is then

$$W = \lim_{\substack{N\to\infty \\ |\Delta x_j|\to 0}} \sum_{j=1}^{j=N} (F_{x,j})_{\text{ave}} \Delta x_j = \int_{x=x_i}^{x=x_f} F_x(x)\,dx \qquad (13.5.3)$$

This last expression is the definition of the definite integral of the x-component of the force with respect to the parameter x. In Figure 13.5 we graph the x-component of the force as a function of the parameter x. The work integral is the area under this curve between $x = x_i$ and $x = x_f$.

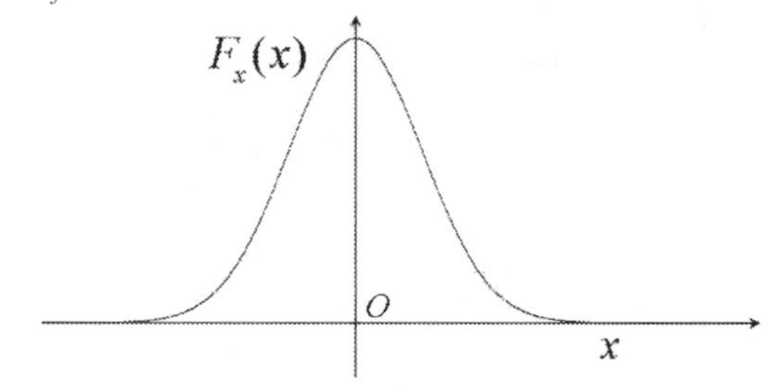

Figure 5 Plot of x-component of a sample force $F_x(x)$ as a function of x.

Example 13.6 Work done by the Spring Force

Connect one end of a spring to a body resting on a smooth (frictionless) table and fix the other end of the spring to a wall. Stretch the spring and release the spring-body system. How much work does the spring do on the body as a function of the stretched or compressed length of the spring?

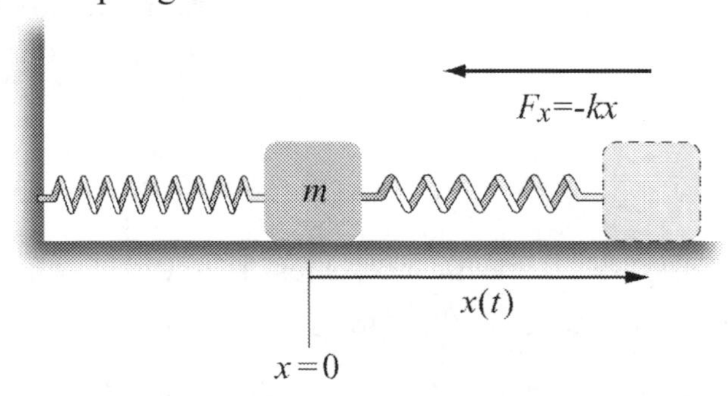

Figure 13.6 Equilibrium position and position at time t

Solution: We first begin by choosing a coordinate system with origin at the position of the body when the spring is at rest in the equilibrium position. We choose the $\hat{\mathbf{i}}$ unit vector to point in the direction the body moves when the spring is being stretched and the coordinate x to denote the position of the body with respect to the equilibrium position, as in Figure 13.6 (which indicates that in general the position x will be a function of time). The spring force on the body is given by

$$\vec{\mathbf{F}}^s = F_x^s\,\hat{\mathbf{i}} = -kx\,\hat{\mathbf{i}}. \tag{13.5.4}$$

In Figure 13.7 we show the graph of the x-component of the spring force as a function of x for both positive values of x corresponding to stretching, and negative values of x corresponding to compressing of the spring. Note that x_0 and x_f can be positive, zero, or negative.

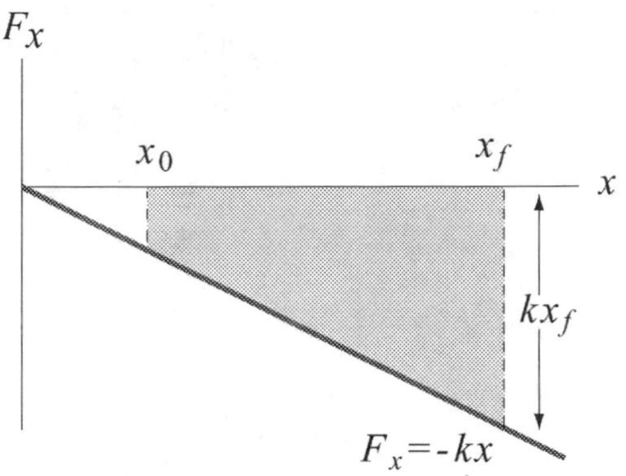

Figure 13.7 The x-component of the spring force as a function of x

The work done is just the area under the curve for the interval x_0 to x_f,

$$W = \int_{x=x_0}^{x=x_f} F_x^s\,dx = \int_{x=x_0}^{x=x_f} (-kx)\,dx. \tag{13.5.5}$$

This integral is straightforward; the work done by the spring force on the body is

$$W = \int_{x=x_0}^{x=x_f} (-kx)dx = -\frac{1}{2}k(x_f^{\,2} - x_0^{\,2}). \tag{13.5.6}$$

When the absolute value of the final distance is less than the absolute value of the initial distance, $\left|x_f\right| < \left|x_0\right|$, the work done by the spring force is positive. This means that if the spring is less stretched or compressed in the final state than in the initial state, the work done by the spring force is positive. The spring force does positive work on the body when the spring goes from a state of 'greater tension' to a state of 'lesser tension'.

13.6 Work-Kinetic Energy Theorem

There is a direct connection between the work done on a point-like object and the change in kinetic energy the point-like object undergoes. If the work done on the object is non-zero, this implies that an unbalanced force has acted on the object, and the object will have undergone acceleration. For an object undergoing one-dimensional motion the left hand side of Equation (13.3.16) is the work done on the object by the component of the sum of the forces in the direction of displacement,

$$W = \int_{x=x_i}^{x=x_f} F_x \, dx = \frac{1}{2}mv_f^{\,2} - \frac{1}{2}mv_i^{\,2} = K_f - K_i = \Delta K \tag{13.6.1}$$

When the work done on an object is positive, the object will increase its speed, and negative work done on an object causes a decrease in speed. When the work done is zero, the object will maintain a constant speed. In fact, the work-energy relationship is quite precise; the work done by the applied force on an object is identically equal to the change in kinetic energy of the object.

Example 13.7 Gravity and the Work-Energy Theorem

Suppose a ball of mass $m = 0.2\ \text{kg}$ starts from rest at a height $y_0 = 15\ \text{m}$ above the surface of the earth and falls down to a height $y_f = 5.0\ \text{m}$ above the surface of the earth. What is the change in the kinetic energy? Find the final velocity using the work-energy theorem.

Solution: As only one force acts on the ball, the change in kinetic energy is the work done by gravity,

$$\begin{aligned} W^g &= -mg(y_f - y_0) \\ &= (-2.0\times10^{-1}\ \text{kg})(9.8\ \text{m}\cdot\text{s}^{-2})(5\ \text{m} - 15\ \text{m}) = 2.0\times10^{1}\ \text{J}. \end{aligned} \tag{13.6.2}$$

The ball started from rest, $v_{y,0} = 0$. So the change in kinetic energy is

$$\Delta K = \frac{1}{2}mv_{y,f}^{\,2} - \frac{1}{2}mv_{y,0}^{\,2} = \frac{1}{2}mv_{y,f}^{\,2}. \tag{13.6.3}$$

We can solve Equation (13.6.3) for the final velocity using Equation (13.6.2)

$$v_{y,f} = \sqrt{\frac{2\Delta K}{m}} = \sqrt{\frac{2W^g}{m}} = \sqrt{\frac{2(2.0\times10^{1}\ \text{J})}{0.2\ \text{kg}}} = 1.4\times10^{1}\ \text{m}\cdot\text{s}^{-1}. \tag{13.6.4}$$

For the falling ball in a constant gravitation field, the positive work of the gravitation force on the body corresponds to an increasing kinetic energy and speed. For a rising body in the same field, the kinetic energy and hence the speed decrease since the work done is negative.

Example 13.7 Final Kinetic Energy of Moving Cup

A person pushes a cup of mass 0.2 kg along a horizontal table with a force of magnitude 2.0 N at an angle of 30° with respect to the horizontal for a distance of 0.5 m as in Example 13.4. The coefficient of friction between the table and the cup is $\mu_k = 0.1$. If the cup was initially at rest, what is the final kinetic energy of the cup after being pushed 0.5 m? What is the final speed of the cup?

Solution: The total work done on the cup is the sum of the work done by the pushing force and the work done by the frictional force, as given in Equations (13.4.9) and (13.4.14),

$$\begin{aligned} W &= W^a + W^f = (F_x^a - \mu_k N)(x_f - x_i) \\ &= (1.7\,\mathrm{N} - 9.6\times 10^{-2}\,\mathrm{N})(0.5\,\mathrm{m}) = 8.0\times 10^{-1}\,\mathrm{J} \end{aligned}. \tag{13.6.5}$$

The initial velocity is zero so the change in kinetic energy is just

$$\Delta K = \frac{1}{2} m v_{y,f}^{\,2} - \frac{1}{2} m v_{y,0}^{\,2} = \frac{1}{2} m v_{y,f}^{\,2}. \tag{13.6.6}$$

Thus the work-kinetic energy theorem, Eq.(13.6.1)), enables us to solve for the final kinetic energy,

$$K_f = \frac{1}{2} m v_f^{\,2} = \Delta K = W = 8.0\times 10^{-1}\,\mathrm{J}. \tag{13.6.7}$$

We can solve for the final speed,

$$v_{y,f} = \sqrt{\frac{2K_f}{m}} = \sqrt{\frac{2W}{m}} = \sqrt{\frac{2(8.0\times 10^{-1}\,\mathrm{J})}{0.2\,\mathrm{kg}}} = 2.9\,\mathrm{m}\cdot\mathrm{s}^{-1}. \tag{13.6.8}$$

13.7 Power Applied by a Constant Force

Suppose that an applied force $\vec{\mathbf{F}}^a$ acts on a body during a time interval Δt, and the displacement of the point of application of the force is in the x-direction by an amount Δx. The work done, ΔW^a, during this interval is

$$\Delta W^a = F_x^a\,\Delta x. \tag{13.7.1}$$

where F_x^a is the x-component of the applied force. (Equation (13.7.1) is the same as Equation (13.4.2).)

The ***average power*** of an applied force is defined to be the rate at which work is done,

$$P^a_{ave} = \frac{\Delta W^a}{\Delta t} = \frac{F^a_x \Delta x}{\Delta t} = F^a_x v_{\text{ave},x} . \tag{13.7.2}$$

The average power delivered to the body is equal to the component of the force in the direction of motion times the component of the average velocity of the body. Power is a scalar quantity and can be positive, zero, or negative depending on the sign of work. The SI units of power are called watts $[\text{W}]$ and $[1\ \text{W}] = [1\ \text{J} \cdot \text{s}^{-1}]$.

The ***instantaneous power*** at time t is defined to be the limit of the average power as the time interval $[t, t + \Delta t]$ approaches zero,

$$P^a = \lim_{\Delta t \to 0} \frac{\Delta W^a}{\Delta t} = \lim_{\Delta t \to 0} \frac{F^a_x \Delta x}{\Delta t} = F^a_x \left(\lim_{\Delta t \to 0} \frac{\Delta x}{\Delta t} \right) = F^a_x v_x . \tag{13.7.3}$$

The instantaneous power of a constant applied force is the product of the component of the force in the direction of motion and the instantaneous velocity of the moving object.

Example 13.8 Gravitational Power for a Falling Object

Suppose a ball of mass $m = 0.2\ \text{kg}$ starts from rest at a height $y_0 = 15\ \text{m}$ above the surface of the earth and falls down to a height $y_f = 5.0\ \text{m}$ above the surface of the earth. What is the average power exerted by the gravitation force? What is the instantaneous power when the ball is at a height $y_f = 5.0\ \text{m}$ above the surface of the Earth? Make a graph of power vs. time. You may ignore the effects of air resistance.

Solution: There are two ways to solve this problem. Both approaches require calculating the time interval Δt for the ball to fall. Set $t_0 = 0$ for the time the ball was released. We can solve for the time interval $\Delta t = t_f$ that it takes the ball to fall using the equation for a freely falling object that starts from rest,

$$y_f = y_0 - \frac{1}{2} g t_f^{\ 2} . \tag{13.7.4}$$

Thus the time interval for falling is

$$t_f = \sqrt{\frac{2}{g}(y_0 - y_f)} = \sqrt{\frac{2}{9.8\ \text{m} \cdot \text{s}^{-2}}(15\ \text{m} - 5\ \text{m})} = 1.4\ \text{s} . \tag{13.7.5}$$

First approach: we can calculate the work done by gravity,

$$W^g = -mg(y_f - y_0)$$
$$= (-2.0\times10^{-1}\text{ kg})(9.8\text{ m}\cdot\text{s}^{-2})(5\text{ m} - 15\text{ m}) = 2.0\times10^{1}\text{ J}. \tag{13.7.6}$$

Then the average power is

$$P^g_{\text{ave}} = \frac{\Delta W}{\Delta t} = \frac{2.0\times10^{1}\text{ J}}{1.4\text{ s}} = 1.4\times10^{1}\text{ W}. \tag{13.7.7}$$

Second Approach. We calculate the gravitation force and the average velocity. The gravitation force is

$$F^g_y = -mg = -(2.0\times10^{-1}\text{ kg})(9.8\text{ m}\cdot\text{s}^{-2}) = -2.0\text{ N}. \tag{13.7.8}$$

The average velocity is

$$v_{\text{ave},y} = \frac{\Delta y}{\Delta t} = \frac{5\text{ m} - 15\text{ m}}{1.4\text{ s}} = -7.0\text{ m}\cdot\text{s}^{-1}. \tag{13.7.9}$$

The average power is therefore

$$P^g_{\text{ave}} = F^g_y\, v_{\text{ave},y} = (-mg)v_{\text{ave},y}$$
$$= (-2.0\text{ N})(-7.0\text{ m}\cdot\text{s}^{-1}) = 1.4\times10^{1}\text{ W}. \tag{13.7.10}$$

In order to find the instantaneous power at any time, we need to find the instantaneous velocity at that time. The ball takes a time $t_f = 1.4\text{ s}$ to reach the height $y_f = 5.0\text{ m}$. The velocity at that height is given by

$$v_y = -gt_f = -(9.8\text{ m}\cdot\text{s}^{-2})(1.4\text{ s}) = -1.4\times10^{1}\text{ m}\cdot\text{s}^{-1}. \tag{13.7.11}$$

So the instantaneous power at time $t_f = 1.4\text{ s}$ is

$$P^g = F^g_y\, v_y = (-mg)(-gt_f) = mg^2 t_f$$
$$= (0.2\text{ kg})(9.8\text{ m}\cdot\text{s}^{-2})^2(1.4\text{ s}) = 2.7\times10^{1}\text{ W} \tag{13.7.12}$$

If this problem were done symbolically, the answers given in Equation (13.7.11) and Equation (13.7.12) would differ by a factor of two; the answers have been rounded to two significant figures.

The instantaneous power grows linearly with time. The graph of power vs. time is shown in Figure 13.8. From the figure, it should be seen that the instantaneous power at any time is twice the average power between $t = 0$ and that time.

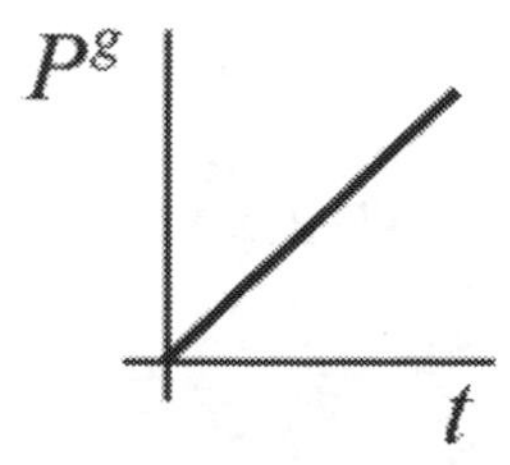

Figure 13.8 Graph of power vs. time

Example 13.9 Power Pushing a Cup

A person pushes a cup of mass 0.2 kg along a horizontal table with a force of magnitude 2.0 N at an angle of 30° with respect to the horizontal for a distance of 0.5 m, as in Example 13.4. The coefficient of friction between the table and the cup is $\mu_k = 0.1$. What is the average power of the pushing force? What is the average power of the kinetic frictional force?

Solution: We will use the results from Examples 13.4 and 13.7 but keeping extra significant figures in the intermediate calculations. The work done by the pushing force is

$$W^a = F_x^a(x_f - x_0) = (1.732\text{ N})(0.50\text{ m}) = 8.660\times 10^{-1}\text{ J}. \tag{13.7.13}$$

The final speed of the cup is $v_{x,f} = 2.860\text{ m}\cdot\text{s}^{-1}$. Assuming constant acceleration, the time during which the cup was pushed is

$$t_f = \frac{2(x_f - x_0)}{v_{x,f}} = 0.3496\text{s}. \tag{13.7.14}$$

The average power of the pushing force is then, with $\Delta t = t_f$,

$$P_{\text{ave}}^a = \frac{\Delta W^a}{\Delta t} = \frac{8.660\times 10^{-1}\text{ J}}{0.3496\text{ s}} = 2.340\text{ W}, \tag{13.7.15}$$

or 2.3W to two significant figures. The work done by the frictional force is

$$\begin{aligned} W^f &= f_{\text{k}}(x_f - x_0) \\ &= -\mu_{\text{k}} N(x_f - x_0) = -(9.6\times 10^{-2}\text{ N})(0.50\text{ m}) = -(4.8\times 10^{-2}\text{ J}). \end{aligned} \tag{13.7.16}$$

The average power of kinetic friction is

$$P^{f}_{\text{ave}} = \frac{\Delta W^{f}}{\Delta t} = \frac{-4.8\times 10^{-2}\ \text{J}}{0.3496\ \text{s}} = -1.4\times 10^{-1}\ \text{W}. \tag{13.7.17}$$

The time rate of change of the kinetic energy for a body of mass m moving in the x-direction is

$$\frac{dK}{dt} = \frac{d}{dt}\left(\frac{1}{2}m v_x^{\,2}\right) = m\frac{dv_x}{dt}v_x = m a_x v_x . \tag{13.7.18}$$

By Newton's Second Law, $F_x = m a_x$, and so Equation (13.7.18) becomes

$$\frac{dK}{dt} = F_x v_x = P . \tag{13.7.19}$$

The instantaneous power delivered to the body is equal to the time rate of change of the kinetic energy of the body.

13.8 Work and the Scalar Product

We shall introduce a vector operation, called the **scalar product** or "dot product" that takes any two vectors and generates a scalar quantity (a number). We shall see that the physical concept of work can be mathematically described by the scalar product between the force and the displacement vectors.

13.8.1 Scalar Product

Let $\vec{\mathbf{A}}$ and $\vec{\mathbf{B}}$ be two vectors. Because any two non-collinear vectors form a plane, we define the angle θ to be the angle between the vectors $\vec{\mathbf{A}}$ and $\vec{\mathbf{B}}$ as shown in Figure 13.9. Note that θ can vary from 0 to π.

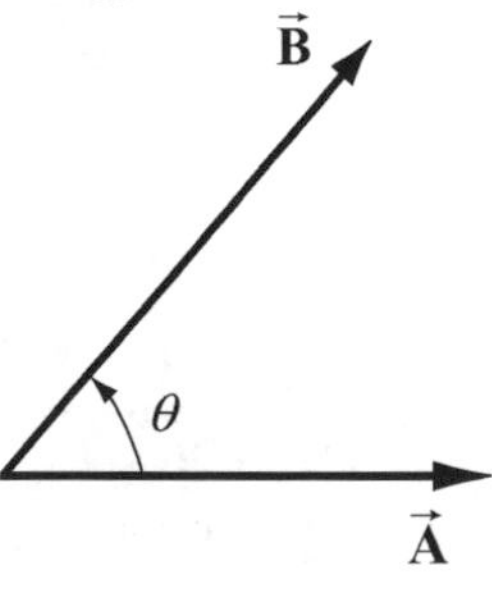

Figure 13.9 Scalar product geometry.

The ***scalar product*** $\vec{\mathbf{A}}\cdot\vec{\mathbf{B}}$ *of the vectors* $\vec{\mathbf{A}}$ *and* $\vec{\mathbf{B}}$ *is defined to be product of the magnitude of the vectors* $\vec{\mathbf{A}}$ *and* $\vec{\mathbf{B}}$ *with the cosine of the angle* θ *between the two vectors:*

$$\vec{\mathbf{A}}\cdot\vec{\mathbf{B}} = AB\cos(\theta), \tag{13.8.1}$$

where $A = |\vec{\mathbf{A}}|$ and $B = |\vec{\mathbf{B}}|$ represent the magnitude of $\vec{\mathbf{A}}$ and $\vec{\mathbf{B}}$ respectively. The scalar product can be positive, zero, or negative, depending on the value of $\cos\theta$. The scalar product is always a scalar quantity.

The angle formed by two vectors is therefore

$$\theta = \cos^{-1}\left(\frac{\vec{\mathbf{A}} \cdot \vec{\mathbf{B}}}{|\vec{\mathbf{A}}||\vec{\mathbf{B}}|}\right). \tag{13.8.2}$$

The magnitude of a vector $\vec{\mathbf{A}}$ is given by the square root of the scalar product of the vector $\vec{\mathbf{A}}$ with itself.

$$|\vec{\mathbf{A}}| = (\vec{\mathbf{A}} \cdot \vec{\mathbf{A}})^{1/2}. \tag{13.8.3}$$

We can give a geometric interpretation to the scalar product by writing the definition as

$$\vec{\mathbf{A}} \cdot \vec{\mathbf{B}} = (A\cos(\theta))\, B. \tag{13.8.4}$$

In this formulation, the term $A\cos\theta$ is the projection of the vector $\vec{\mathbf{B}}$ in the direction of the vector $\vec{\mathbf{B}}$. This projection is shown in Figure 13.10a. So the scalar product is the product of the projection of the length of $\vec{\mathbf{A}}$ in the direction of $\vec{\mathbf{B}}$ with the length of $\vec{\mathbf{B}}$. Note that we could also write the scalar product as

$$\vec{\mathbf{A}} \cdot \vec{\mathbf{B}} = A(B\cos(\theta)). \tag{13.8.5}$$

Now the term $B\cos(\theta)$ is the projection of the vector $\vec{\mathbf{B}}$ in the direction of the vector $\vec{\mathbf{A}}$ as shown in Figure 13.10b. From this perspective, the scalar product is the product of the projection of the length of $\vec{\mathbf{B}}$ in the direction of $\vec{\mathbf{A}}$ with the length of $\vec{\mathbf{A}}$.

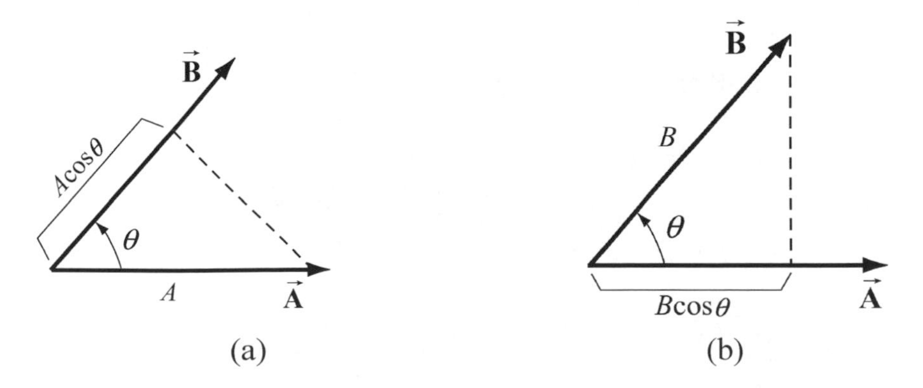

Figure 13.10 (a) and (b) Projection of vectors and the scalar product

From our definition of the scalar product we see that the scalar product of two vectors that are perpendicular to each other is zero since the angle between the vectors is $\pi/2$ and $\cos(\pi/2)=0$.

We can calculate the scalar product between two vectors in a Cartesian coordinates system as follows. Consider two vectors $\vec{\mathbf{A}} = A_x\,\hat{\mathbf{i}} + A_y\,\hat{\mathbf{j}} + A_z\,\hat{\mathbf{k}}$ and $\vec{\mathbf{B}} = B_x\,\hat{\mathbf{i}} + B_y\,\hat{\mathbf{j}} + B_z\,\hat{\mathbf{k}}$. Recall that

$$\begin{aligned}\hat{\mathbf{i}}\cdot\hat{\mathbf{i}} = \hat{\mathbf{j}}\cdot\hat{\mathbf{j}} = \hat{\mathbf{k}}\cdot\hat{\mathbf{k}} = 1\\ \hat{\mathbf{i}}\cdot\hat{\mathbf{j}} = \hat{\mathbf{j}}\cdot\hat{\mathbf{k}} = \hat{\mathbf{i}}\cdot\hat{\mathbf{k}} = 0.\end{aligned} \tag{13.8.6}$$

The scalar product between $\vec{\mathbf{A}}$ and $\vec{\mathbf{B}}$ is then

$$\vec{\mathbf{A}}\cdot\vec{\mathbf{B}} = A_x B_x + A_y B_y + A_z B_z\,. \tag{13.8.7}$$

The time derivative of the scalar product of two vectors is given by

$$\begin{aligned}&\frac{d}{dt}(\vec{\mathbf{A}}\cdot\vec{\mathbf{B}}) = \frac{d}{dt}(A_x B_x + A_y B_y + A_z B_z)\\ &= \frac{d}{dt}(A_x)B_x + \frac{d}{dt}(A_y)B_y + \frac{d}{dt}(A_z)B_z + A_x\frac{d}{dt}(B_x) + A_y\frac{d}{dt}(B_y) + A_z\frac{d}{dt}(B_z)\\ &= \left(\frac{d}{dt}\vec{\mathbf{A}}\right)\cdot\vec{\mathbf{B}} + \vec{\mathbf{A}}\cdot\left(\frac{d}{dt}\vec{\mathbf{B}}\right).\end{aligned} \tag{13.8.8}$$

In particular when $\vec{\mathbf{A}} = \vec{\mathbf{B}}$, then the time derivative of the square of the magnitude of the vector $\vec{\mathbf{A}}$ is given by

$$\frac{d}{dt}A^2 = \frac{d}{dt}(\vec{\mathbf{A}}\cdot\vec{\mathbf{A}}) = \left(\frac{d}{dt}\vec{\mathbf{A}}\right)\cdot\vec{\mathbf{A}} + \vec{\mathbf{A}}\cdot\left(\frac{d}{dt}\vec{\mathbf{A}}\right) = 2\left(\frac{d}{dt}\vec{\mathbf{A}}\right)\cdot\vec{\mathbf{A}}\,. \tag{13.8.9}$$

13.8.2 Kinetic Energy and the Scalar Product

For an object undergoing three-dimensional motion, the velocity of the object in Cartesian components is given by $\vec{\mathbf{v}} = v_x\hat{\mathbf{i}} + v_y\hat{\mathbf{j}} + v_z\hat{\mathbf{k}}$. Recall that the magnitude of a vector is given by the square root of the scalar product of the vector with itself,

$$A \equiv \left|\vec{\mathbf{A}}\right| \equiv (\vec{\mathbf{A}}\cdot\vec{\mathbf{A}})^{1/2} = (A_x^2 + A_y^2 + A_z^2)^{1/2}\,. \tag{13.8.10}$$

Therefore the square of the magnitude of the velocity is given by the expression

$$v^2 \equiv (\vec{\mathbf{v}} \cdot \vec{\mathbf{v}}) = v_x^2 + v_y^2 + v_z^2 . \tag{13.8.11}$$

Hence the kinetic energy of the object is given by

$$K = \frac{1}{2} m(\vec{\mathbf{v}} \cdot \vec{\mathbf{v}}) = \frac{1}{2} m(v_x^2 + v_y^2 + v_z^2) . \tag{13.8.12}$$

13.8.2 Work and the Scalar Product

Work is an important physical example of the mathematical operation of taking the scalar product between two vectors. Recall that when a constant force acts on a body and the point of application of the force undergoes a displacement along the x-axis, only the component of the force along that direction contributes to the work,

$$W = F_x \Delta x . \tag{13.8.13}$$

Suppose we are pulling a body along a horizontal surface with a force $\vec{\mathbf{F}}$. Choose coordinates such that horizontal direction is the x-axis and the force $\vec{\mathbf{F}}$ forms an angle β with the positive x-direction. In Figure 13.11 we show the force vector $\vec{\mathbf{F}} = F_x \, \hat{\mathbf{i}} + F_y \, \hat{\mathbf{j}}$ and the displacement vector of the point of application of the force $\Delta\vec{\mathbf{x}} = \Delta x \, \hat{\mathbf{i}}$. Note that $\Delta\vec{\mathbf{x}} = \Delta x \, \hat{\mathbf{i}}$ is the component of the displacement and hence can be greater, equal, or less than zero (but is shown as greater than zero in the figure for clarity). The scalar product between the force vector $\vec{\mathbf{F}}$ and the displacement vector $\Delta\vec{\mathbf{x}}$ is

$$\vec{\mathbf{F}} \cdot \Delta\vec{\mathbf{x}} = (F_x \, \hat{\mathbf{i}} + F_y \, \hat{\mathbf{j}}) \cdot (\Delta x \, \hat{\mathbf{i}}) = F_x \, \Delta x . \tag{13.8.14}$$

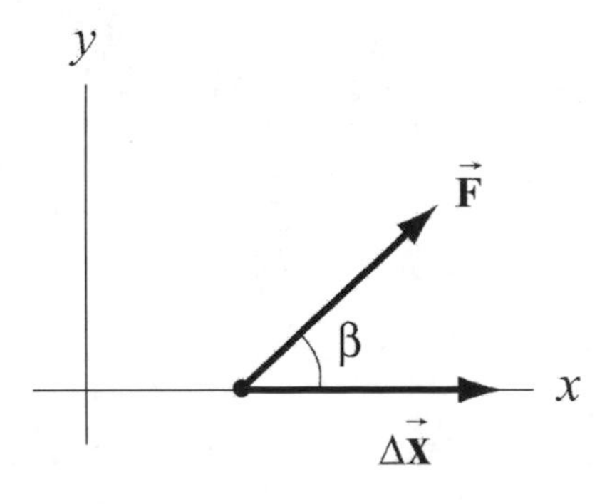

Figure 13.11 Force and displacement vectors

The work done by the force is then

$$\Delta W = \vec{\mathbf{F}} \cdot \Delta\vec{\mathbf{x}} . \tag{13.8.15}$$

In general, the angle β takes values within the range $-\pi \le \beta \le \pi$ (in Figure 13.11, $0 \le \beta \le \pi / 2$). Because the x-component of the force is $F_x = F\cos(\beta)$ where $F = |\vec{\mathbf{F}}|$ denotes the magnitude of $\vec{\mathbf{F}}$, the work done by the force is

$$W = \vec{\mathbf{F}} \cdot \Delta\vec{\mathbf{x}} = (F\cos(\beta))\Delta x\,. \tag{13.8.16}$$

Example 13.10 Object Sliding Down an Inclined Plane

An object of mass $m = 4.0\,kg$, starting from rest, slides down an inclined plane of length $l = 3.0\,m$. The plane is inclined by an angle of $\theta = 30^0$ to the ground. The coefficient of kinetic friction is $\mu_k = 0.2$. (a) What is the work done by each of the three forces while the object is sliding down the inclined plane? (b) For each force, is the work done by the force positive or negative? (c) What is the sum of the work done by the three forces? Is this positive or negative?

Solution: (a) and (b) Choose a coordinate system with the origin at the top of the inclined plane and the positive x-direction pointing down the inclined plane, and the positive y-direction pointing towards the upper right as shown in Figure 13.12. While the object is sliding down the inclined plane, three uniform forces act on the object, the gravitational force which points downward and has magnitude $F_g = mg$, the normal force N which is perpendicular to the surface of the inclined plane, and the frictional force which opposes the motion and is equal in magnitude to $f_k = \mu_k N$. A force diagram on the object is shown in Figure 13.13.

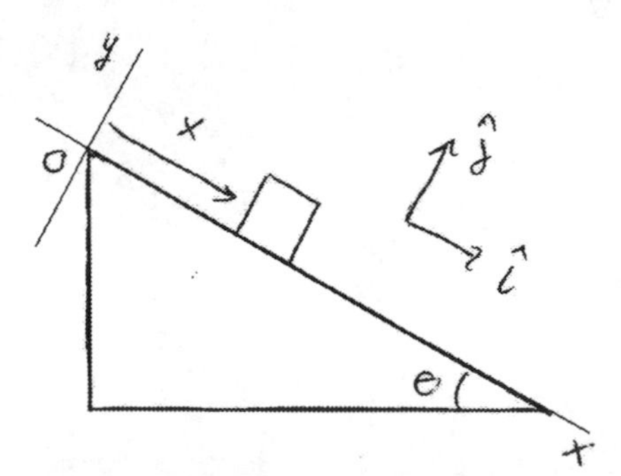

Figure 13.12 Coordinate system for object sliding down inclined plane

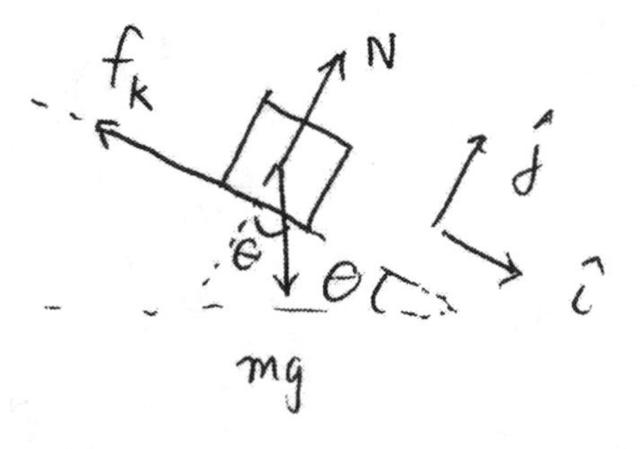

Figure 13.13 Free-body force diagram for object

In order to calculate the work we need to determine which forces have a component in the direction of the displacement. Only the component of the gravitational force along the positive x-direction $F_{gx} = mg\sin\theta$ and the frictional force are directed along the displacement and therefore contribute to the work. We need to use Newton's Second Law to determine the magnitudes of the normal force. Because the object is constrained to move along the positive x-direction, $a_y = 0$, Newton's Second Law in the $\hat{\mathbf{j}}$-direction $N - mg\cos\theta = 0$. Therefore $N = mg\cos\theta$ and the magnitude of the frictional force is $f_k = \mu_k mg\cos\theta$.

With our choice of coordinate system with the origin at the top of the inclined plane and the positive x-direction pointing down the inclined plane, the displacement of the object is given by the vector $\Delta\vec{\mathbf{r}} = \Delta x\,\hat{\mathbf{i}}$ (Figure 13.14).

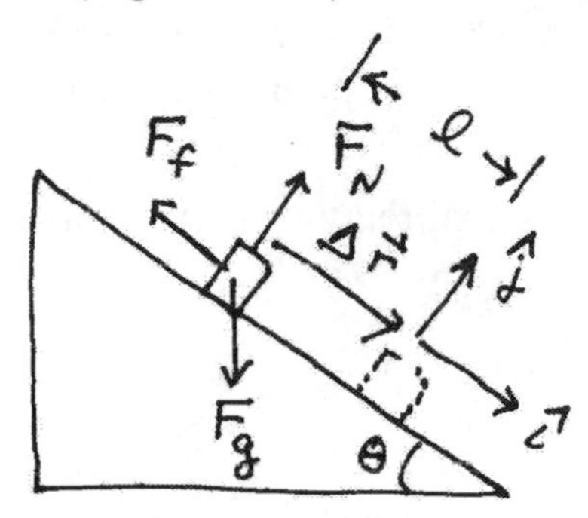

Figure 13.14 Force vectors and displacement vector for object

The vector decomposition of the three forces are $\vec{\mathbf{F}}^g = mg\sin\theta\,\hat{\mathbf{i}} - mg\cos\theta\hat{\mathbf{j}}$, $\vec{\mathbf{F}}^f = -\mu_k mg\cos\theta\hat{\mathbf{i}}$, and $\vec{\mathbf{F}}^N = mg\cos\theta\hat{\mathbf{j}}$. The work done by the normal force is zero because the normal force is perpendicular the displacement

$$W^N = \vec{\mathbf{F}}^N \cdot \Delta\vec{\mathbf{r}} = mg\cos\theta\hat{\mathbf{j}} \cdot l\,\hat{\mathbf{i}} = 0.$$

Then the work done by the frictional force is negative and given by

$$W^f = \vec{\mathbf{F}}^f \cdot \Delta\vec{\mathbf{r}} = -\mu_k mg\cos\theta\hat{\mathbf{i}} \cdot l\,\hat{\mathbf{i}} = -\mu_k mg\cos\theta l < 0.$$

Substituting in the appropriate values yields

$$W^f = -\mu_k mg\cos\theta l = -(0.2)(4.0\,\text{kg})(9.8\,\text{m}\cdot\text{s}^{-2})(3.0\,\text{m})(\cos(30^o)(3.0\,\text{m}) = -20.4\,\text{J}.$$

The work done by the gravitational force is positive and given by

$$W^g = \vec{\mathbf{F}}^g \cdot \Delta\vec{\mathbf{r}} = (mg\sin\theta\,\hat{\mathbf{i}} - mg\cos\theta\hat{\mathbf{j}}) \cdot l\,\hat{\mathbf{i}} = mgl\sin\theta > 0.$$

Substituting in the appropriate values yields

$$W^g = mgl\sin\theta = (4.0\,\text{kg})(9.8\,\text{m}\cdot\text{s}^{-2})(3.0\,\text{m})(\sin(30^o) = 58.8\,\text{J}.$$

(c) The scalar sum of the work done by the three forces is then

$$W = W^g + W^f = mgl(\sin\theta - \mu_k \cos\theta)$$
$$W = (4.0\,\text{kg})(9.8\,\text{m}\cdot\text{s}^{-2})(3.0\,\text{m})(\sin(30^o) - (0.2)(\cos(30^o)) = 38.4\ \text{J}.$$

13.9 Work done by a Non-Constant Force Along an Arbitrary Path

Suppose that a non-constant force $\vec{\mathbf{F}}$ acts on a point-like body of mass m while the body is moving on a three dimensional curved path. The position vector of the body at time t with respect to a choice of origin is $\vec{\mathbf{r}}(t)$. In Figure 13.15 we show the orbit of the body for a time interval $[t_i, t_f]$ moving from an initial position $\vec{\mathbf{r}}_i \equiv \vec{\mathbf{r}}(t = t_i)$ at time $t = t_i$ to a final position $\vec{\mathbf{r}}_f \equiv \vec{\mathbf{r}}(t = t_f)$ at time $t = t_f$.

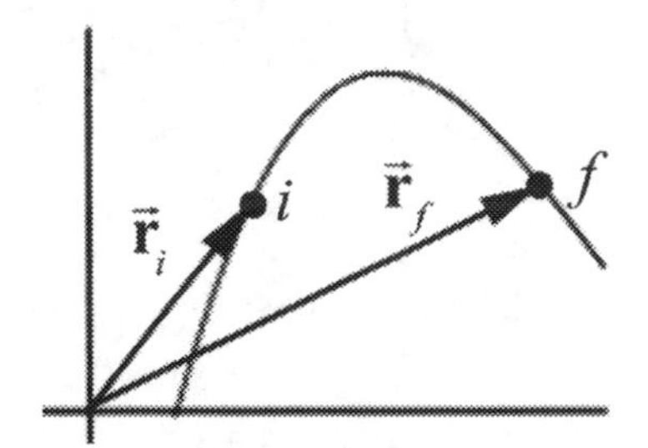

Figure 13.15 Path traced by the motion of a body.

We divide the time interval $[t_i, t_f]$ into N smaller intervals with $[t_{j-1}, t_j]$, $j = 1, \cdots, N$ with $t_N = t_f$. Consider two position vectors $\vec{\mathbf{r}}_j \equiv \vec{\mathbf{r}}(t = t_j)$ and $\vec{\mathbf{r}}_{j-1} \equiv \vec{\mathbf{r}}(t = t_{j-1})$ the displacement vector during the corresponding time interval as $\Delta\vec{\mathbf{r}}_j = \vec{\mathbf{r}}_j - \vec{\mathbf{r}}_{j-1}$. Let $\vec{\mathbf{F}}$ denote the force acting on the body during the interval $[t_{j-1}, t_j]$. The average force in this interval is $(\vec{\mathbf{F}}_j)_{\text{ave}}$ and the average work ΔW_j done by the force during the time interval $[t_{j-1}, t_j]$ is the scalar product between the average force vector and the displacement vector,

$$\Delta W_j = (\vec{\mathbf{F}}_j)_{\text{ave}} \cdot \Delta\vec{\mathbf{r}}_j. \qquad (13.8.17)$$

The force and the displacement vectors for the time interval $[t_{j-1}, t_j]$ are shown in Figure 13.16 (note that the subscript "ave" on $(\vec{\mathbf{F}}_j)_{\text{ave}}$ has been suppressed).

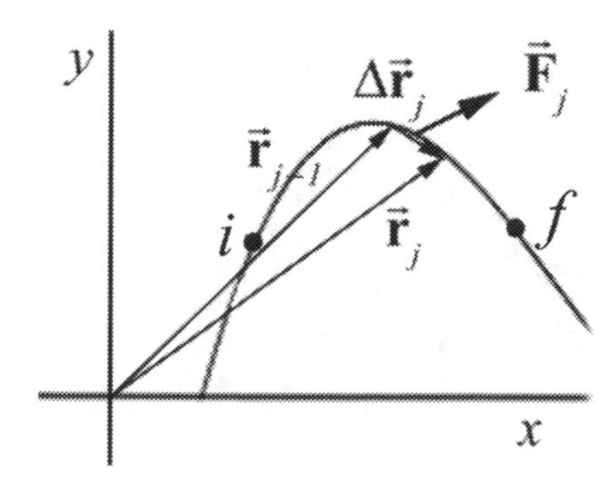

Figure 13.16 An infinitesimal work element.

We calculate the work by adding these scalar contributions to the work for each interval $[t_{j-1}, t_j]$, for $j = 1$ to N,

$$W_N = \sum_{j=1}^{j=N} \Delta W_j = \sum_{j=1}^{j=N} (\vec{\mathbf{F}}_j)_{\text{ave}} \cdot \Delta\vec{\mathbf{r}}_j . \tag{13.8.18}$$

We would like to define work in a manner that is independent of the way we divide the interval, so we take the limit as $N \to \infty$ and $\left|\Delta\vec{\mathbf{r}}_j\right| \to 0$ for all j. In this limit, as the intervals become smaller and smaller, the distinction between the average force and the actual force vanishes. Thus if this limit exists and is well defined, then the work done by the force is

$$W = \lim_{\substack{N\to\infty \\ \left|\Delta\vec{\mathbf{r}}_j\right|\to 0}} \sum_{j=1}^{j=N} (\vec{\mathbf{F}}_j)_{\text{ave}} \cdot \Delta\vec{\mathbf{r}}_j = \int_i^f \vec{\mathbf{F}} \cdot d\vec{\mathbf{r}} . \tag{13.8.19}$$

Notice that this summation involves adding scalar quantities. This limit is called the ***line integral*** of the force $\vec{\mathbf{F}}$. The symbol $d\vec{\mathbf{r}}$ is called the ***infinitesimal vector line element.*** At time t, $d\vec{\mathbf{r}}$ is tangent to the orbit of the body and is the limit of the displacement vector $\Delta\vec{\mathbf{r}} = \vec{\mathbf{r}}(t + \Delta t) - \vec{\mathbf{r}}(t)$ as Δt approaches zero. In this limit, the parameter t does not appear in the expression in Equation (13.8.19).

In general this line integral depends on the particular path the body takes between the initial position $\vec{\mathbf{r}}_i$ and the final position $\vec{\mathbf{r}}_f$, which matters when the force $\vec{\mathbf{F}}$ is non-constant in space, and when the contribution to the work can vary over different paths in space. We can represent the integral in Equation (13.8.19) explicitly in a coordinate system by specifying the infinitesimal vector line element $d\vec{\mathbf{r}}$ and then explicitly computing the scalar product.

13.9.1 Work Integral in Cartesian Coordinates

In Cartesian coordinates the line element is

$$d\vec{\mathbf{r}} = dx\,\hat{\mathbf{i}} + dy\,\hat{\mathbf{j}} + dz\,\hat{\mathbf{k}}, \tag{13.8.20}$$

where dx, dy, and dz represent arbitrary displacements in the $\hat{\mathbf{i}}$-, $\hat{\mathbf{j}}$-, and $\hat{\mathbf{k}}$-directions respectively as seen in Figure 13.17.

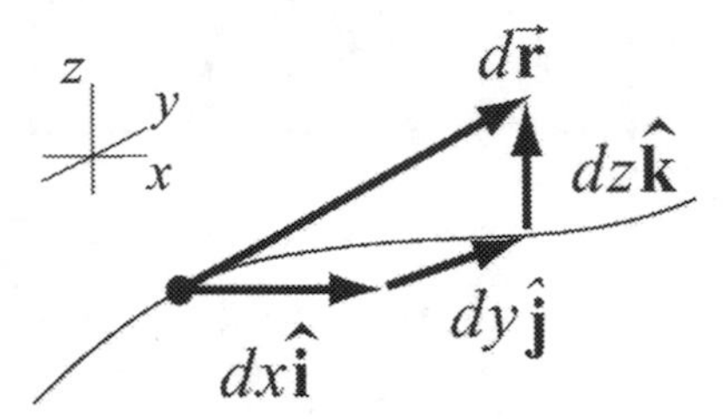

Figure 13.17 A line element in Cartesian coordinates.

The force vector can be represented in vector notation by

$$\vec{\mathbf{F}} = F_x\,\hat{\mathbf{i}} + F_y\,\hat{\mathbf{j}} + F_z\,\hat{\mathbf{k}}. \tag{13.8.21}$$

The infinitesimal work is the sum of the work done by the component of the force times the component of the displacement in each direction,

$$dW = F_x dx + F_y dy + F_z dz\,. \tag{13.8.22}$$

Eq. (13.8.22) is just the scalar product

$$\begin{aligned} dW &= \vec{\mathbf{F}}\cdot d\vec{\mathbf{r}} = (F_x\,\hat{\mathbf{i}} + F_y\,\hat{\mathbf{j}} + F_z\,\hat{\mathbf{k}})\cdot(dx\,\hat{\mathbf{i}} + dy\,\hat{\mathbf{j}} + dz\,\hat{\mathbf{k}}) \\ &= F_x dx + F_y dy + F_z dz \end{aligned}, \tag{13.8.23}$$

The work is

$$W = \int_{\vec{\mathbf{r}}=\vec{\mathbf{r}}_0}^{\vec{\mathbf{r}}=\vec{\mathbf{r}}_f} \vec{\mathbf{F}}\cdot d\vec{\mathbf{r}} = \int_{\vec{\mathbf{r}}=\vec{\mathbf{r}}_0}^{\vec{\mathbf{r}}=\vec{\mathbf{r}}_f} (F_x dx + F_y dy + F_z dz) = \int_{\vec{\mathbf{r}}=\vec{\mathbf{r}}_0}^{\vec{\mathbf{r}}=\vec{\mathbf{r}}_f} F_x dx + \int_{\vec{\mathbf{r}}=\vec{\mathbf{r}}_0}^{\vec{\mathbf{r}}=\vec{\mathbf{r}}_f} F_y dy + \int_{\vec{\mathbf{r}}=\vec{\mathbf{r}}_0}^{\vec{\mathbf{r}}=\vec{\mathbf{r}}_f} F_z dz\,. \tag{13.8.24}$$

13.9.2 Work Integral in Cylindrical Coordinates

In cylindrical coordinates the line element is

$$d\vec{\mathbf{r}} = dr\,\hat{\mathbf{r}} + rd\theta\,\hat{\boldsymbol{\theta}} + dz\,\hat{\mathbf{k}}, \tag{13.8.25}$$

where dr, $rd\theta$, and dz represent arbitrary displacements in the $\hat{\mathbf{r}}$-, $\hat{\boldsymbol{\theta}}$-, and $\hat{\mathbf{k}}$-directions respectively as seen in Figure 13.18.

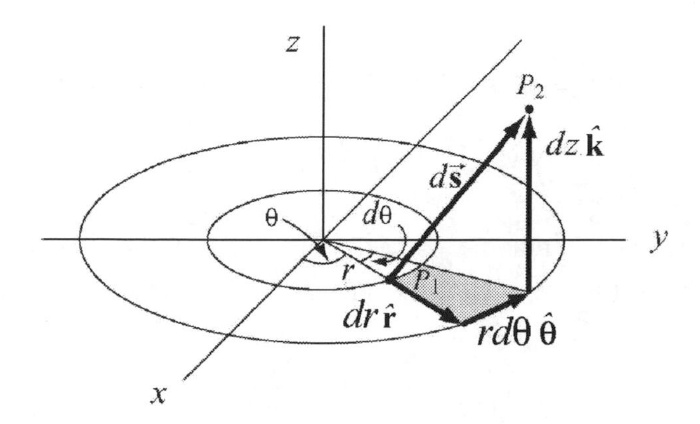

Figure 13.18 Displacement vector $d\vec{\mathbf{s}}$ between two points

The force vector can be represented in vector notation by

$$\vec{\mathbf{F}} = F_r\,\hat{\mathbf{r}} + F_\theta\,\hat{\boldsymbol{\theta}} + F_z\,\hat{\mathbf{k}}\,. \tag{13.8.26}$$

The infinitesimal work is the scalar product

$$\begin{aligned} dW &= \vec{\mathbf{F}}\cdot d\vec{\mathbf{r}} = (F_r\,\hat{\mathbf{r}} + F_\theta\,\hat{\boldsymbol{\theta}} + F_z\,\hat{\mathbf{k}})\cdot(dr\,\hat{\mathbf{r}} + rd\theta\,\hat{\boldsymbol{\theta}} + dz\,\hat{\mathbf{k}}) \\ &= F_r dr + F_\theta r d\theta + F_z dz. \end{aligned} \tag{13.8.27}$$

The work is

$$W = \int_{\vec{\mathbf{r}}=\vec{\mathbf{r}}_0}^{\vec{\mathbf{r}}=\vec{\mathbf{r}}_f} \vec{\mathbf{F}}\cdot d\vec{\mathbf{r}} = \int_{\vec{\mathbf{r}}=\vec{\mathbf{r}}_0}^{\vec{\mathbf{r}}=\vec{\mathbf{r}}_f} (F_r dr + F_\theta r d\theta + F_z dz) = \int_{\vec{\mathbf{r}}=\vec{\mathbf{r}}_0}^{\vec{\mathbf{r}}=\vec{\mathbf{r}}_f} F_r dr + \int_{\vec{\mathbf{r}}=\vec{\mathbf{r}}_0}^{\vec{\mathbf{r}}=\vec{\mathbf{r}}_f} F_\theta r d\theta + \int_{\vec{\mathbf{r}}=\vec{\mathbf{r}}_0}^{\vec{\mathbf{r}}=\vec{\mathbf{r}}_f} F_z dz\,. \tag{13.8.28}$$

13.10 Worked Examples

Example 13.11 Work Done in a Constant Gravitation Field

The work done in a uniform gravitation field is a fairly straightforward calculation when the body moves in the direction of the field. Suppose the body is moving under the influence of gravity, $\vec{\mathbf{F}} = -mg\,\hat{\mathbf{j}}$ along a parabolic curve. The body begins at the point (x_0, y_0) and ends at the point (x_f, y_f). What is the work done by the gravitation force on the body?

Solution: The infinitesimal line element $d\vec{\mathbf{r}}$ is therefore

$$d\vec{\mathbf{r}} = dx\,\hat{\mathbf{i}} + dy\,\hat{\mathbf{j}}\,. \tag{13.9.1}$$

The scalar product that appears in the line integral can now be calculated,

$$\vec{\mathbf{F}} \cdot d\vec{\mathbf{r}} = -mg\,\hat{\mathbf{j}} \cdot [dx\,\hat{\mathbf{i}} + dy\,\hat{\mathbf{j}}] = -mgdy\,. \tag{13.9.2}$$

This result is not surprising since the force is only in the y-direction. Therefore the only non-zero contribution to the work integral is in the y-direction, with the result that

$$W = \int_{\mathbf{r}_0}^{\mathbf{r}_f} \vec{\mathbf{F}} \cdot d\vec{\mathbf{r}} = \int_{y=y_0}^{y=y_f} F_y dy = \int_{y=y_0}^{y=y_f} -mgdy = -mg(y_f - y_0)\,. \tag{13.9.3}$$

In this case of a constant force, the work integral is independent of path.

Example 13.12 Hooke's Law Spring-Body System

Consider a spring-body system lying on a frictionless horizontal surface with one end of the spring fixed to a wall and the other end attached to a body of mass m (Figure 13.19). Calculate the work done by the spring force on body as the body moves from some initial position to some final position.

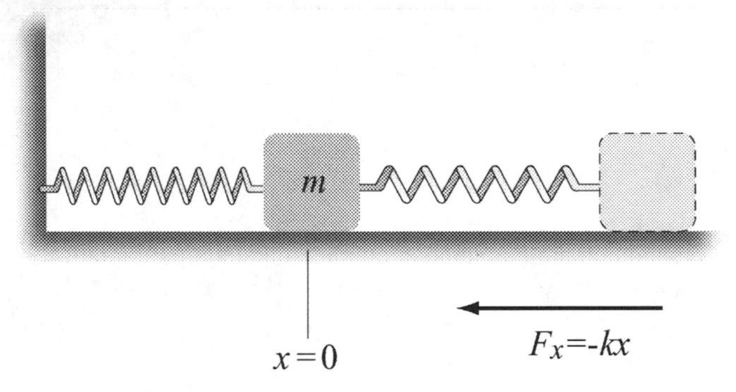

Figure 13.19 A spring-body system.

Solution: Choose the origin at the position of the center of the body when the spring is relaxed (the equilibrium position). Let x be the displacement of the body from the origin. We choose the $+\hat{\mathbf{i}}$ unit vector to point in the direction the body moves when the spring is being stretched (to the right of $x = 0$ in the figure). The spring force on the body is then given by

$$\vec{\mathbf{F}} = F_x\,\hat{\mathbf{i}} = -kx\,\hat{\mathbf{i}}\,. \tag{13.9.4}$$

The work done by the spring force on the mass is

$$W_{\text{spring}} = \int_{x=x_0}^{x=x_f} (-kx)\,dx = -\frac{1}{2}k(x_f^{\,2} - x_0^{\,2})\,. \tag{13.9.5}$$

Example 13.13 Work done by the Inverse Square Gravitation Force

Consider a body of mass m in moving in a fixed orbital plane about the sun. The mass of the sun is m_s. How much work does the gravitation interaction between the sun and the body done on the body during this motion?

Solution: Let's assume that the sun is fixed and choose a polar coordinate system with the origin at the center of the sun. Initially the body is at a distance r_0 from the center of the sun. In the final configuration the body has moved to a distance $r_f < r_0$ from the center of the sun. The infinitesimal displacement of the body is given by $d\vec{\mathbf{r}} = dr\,\hat{\mathbf{r}} + rd\theta\,\hat{\boldsymbol{\theta}}$. The gravitation force between the sun and the body is given by

$$\vec{\mathbf{F}}_{grav} = F_{grav}\ \hat{\mathbf{r}} = -\frac{Gm_s m}{r^2}\hat{\mathbf{r}}. \tag{13.9.6}$$

The infinitesimal work done work done by this gravitation force on the body is given by

$$dW = \vec{\mathbf{F}}_{grav} \cdot d\vec{\mathbf{r}} = (F_{grav,r}\ \hat{\mathbf{r}}) \cdot (dr\,\hat{\mathbf{r}} + rd\theta\,\hat{\boldsymbol{\theta}}) = F_{grav,r}\,dr\ . \tag{13.9.7}$$

Therefore the work done on the object as the object moves from r_i to r_f is given by the integral

$$W = \int_{r_i}^{r_f} \vec{\mathbf{F}}_{grav} \cdot d\vec{\mathbf{r}} = \int_{r_i}^{r_f} F_{grav,r}\,dr = \int_{r_i}^{r_f} \left(-\frac{Gm_{\text{sun}} m}{r^2}\right) dr\ . \tag{13.9.8}$$

Upon evaluation of this integral, we have for the work

$$W = \int_{r_i}^{r_f} \left(-\frac{Gm_{\text{sun}} m}{r^2}\right) dr = \left.\frac{Gm_{\text{sun}} m}{r}\right|_{r_i}^{r_f} = Gm_{\text{sun}} m \left(\frac{1}{r_f} - \frac{1}{r_i}\right). \tag{13.9.9}$$

Because the body has moved closer to the sun, $r_f < r_i$, hence $1/r_f > 1/r_i$. Thus the work done by gravitation force between the sun and the body, on the body is positive,

$$W = Gm_{\text{sun}} m \left(\frac{1}{r_f} - \frac{1}{r_i}\right) > 0 \tag{13.9.10}$$

We expect this result because the gravitation force points along the inward radial direction, so the scalar product and hence work of the force and the displacement is positive when the body moves closer to the sun. Also we expect that the sign of the work is the same for a body moving closer to the sun as a body falling towards the earth in a constant gravitation field, as seen in Example 4.7.1 above.

Example 13.14 Work Done by the Inverse Square Electrical Force

Let's consider two point-like bodies, body 1 and body 2, with charges q_1 and q_2 respectively interacting via the electric force alone. Body 1 is fixed in place while body 2 is free to move in an orbital plane. How much work does the electric force do on the body 2 during this motion?

Solution: The calculation in nearly identical to the calculation of work done by the gravitational inverse square force in Example 13.13. The most significant difference is that the electric force can be either attractive or repulsive while the gravitation force is always attractive. Once again we choose polar coordinates centered on body 2 in the plane of the orbit. Initially a distance r_0 separates the bodies and in the final state a distance r_f separates the bodies. The electric force between the bodies is given by

$$\vec{\mathbf{F}}_{\text{elec}} = F_{\text{elec}}\,\hat{\mathbf{r}} = F_{elec,r}\,\hat{\mathbf{r}} = \frac{1}{4\pi\varepsilon_0}\frac{q_1 q_2}{r^2}\,\hat{\mathbf{r}}. \tag{13.9.11}$$

The work done by this electric force on the body 2 is given by the integral

$$W = \int_{r_i}^{r_f} \vec{\mathbf{F}}_{elec} \cdot d\vec{\mathbf{r}} = \int_{r_i}^{r_f} F_{elec,r}\, dr = \frac{1}{4\pi\varepsilon_0}\int_{r_i}^{r_f} \frac{q_1 q_2}{r^2}\, dr. \tag{13.9.12}$$

Evaluating this integral, we have for the work done by the electric force

$$W = \int_{r_i}^{r_f} \frac{1}{4\pi\varepsilon_0}\frac{q_1 q_2}{r^2}\, dr = -\frac{1}{4\pi\varepsilon_0}\frac{q_1 q_2}{r^2}\bigg|_{r_i}^{r_f} = -\frac{1}{4\pi\varepsilon_0} q_1 q_2 \left(\frac{1}{r_f} - \frac{1}{r_i}\right). \tag{13.9.13}$$

If the charges have opposite signs, $q_1 q_2 < 0$, we expect that the body 2 will move closer to body 1 so $r_f < r_i$, and $1/r_f > 1/r_i$. From our result for the work, the work done by electrical force in moving body 2 is positive,

$$W = -\frac{1}{4\pi\varepsilon_0} q_1 q_2 \left(\frac{1}{r_f} - \frac{1}{r_i}\right) > 0. \tag{13.9.14}$$

Once again we see that bodies under the influence of electric forces only will naturally move in the directions in which the force does positive work. If the charges have the same sign, then $q_1 q_2 > 0$. They will repel with $r_f > r_i$ and $1/r_f < 1/r_i$. Thus the work is once again positive:

$$W = -\frac{1}{4\pi\varepsilon_0} q_1 q_2 \left(\frac{1}{r_f} - \frac{1}{r_i} \right) > 0. \qquad (13.9.15)$$

13.11 Work-Kinetic Energy Theorem in Three Dimensions

Recall our mathematical result that for one-dimensional motion

$$m\int_i^f a_x\,dx = m\int_i^f \frac{dv_x}{dt}\,dx = m\int_i^f dv_x \frac{dx}{dt} = m\int_i^f v_x\,dv_x = \frac{1}{2}mv_{x,f}^2 - \frac{1}{2}mv_{x,i}^2. \qquad (13.11.1)$$

Using Newton's Second Law in the form $F_x = ma_x$, we concluded that

$$\int_i^f F_x\,dx = \frac{1}{2}mv_{x,f}^2 - \frac{1}{2}mv_{x,i}^2. \qquad (13.11.2)$$

Eq. (13.11.2) generalizes to the y- and z-directions:

$$\int_i^f F_y\,dy = \frac{1}{2}mv_{y,f}^2 - \frac{1}{2}mv_{y,i}^2, \qquad (13.11.3)$$

$$\int_i^f F_z\,dz = \frac{1}{2}mv_{z,f}^2 - \frac{1}{2}mv_{z,i}^2. \qquad (13.11.4)$$

Adding Eqs. (13.11.2), (13.11.3), and (13.11.4) yields

$$\int_i^f (F_x\,dx + F_y\,dy + F_z\,dz) = \frac{1}{2}m(v_{x,f}^2 + v_{y,f}^2 + v_{z,f}^2) - \frac{1}{2}m(v_{x,i}^2 + v_{y,i}^2 + v_{z,i}^2). \qquad (13.11.5)$$

Recall (Eq. (13.8.24)) that the left hand side of Eq. (13.11.5) is the work done by the force $\vec{\mathbf{F}}$ on the object

$$W = \int_i^f dW = \int_i^f (F_x\,dx + F_y\,dy + F_z\,dz) = \int_i^f \vec{\mathbf{F}} \cdot d\vec{\mathbf{r}} \qquad (13.11.6)$$

The right hand side of Eq. (13.11.5) is the change in kinetic energy of the object

$$\Delta K \equiv K_f - K_i = \frac{1}{2}mv_f^{\,2} - \frac{1}{2}mv_0^{\,2} = \frac{1}{2}m(v_{x,f}^2 + v_{y,f}^2 + v_{z,f}^2) - \frac{1}{2}m(v_{x,i}^2 + v_{y,i}^2 + v_{z,i}^2). \; (13.11.7)$$

Therefore Eq. (13.11.5) is the three dimensional generalization of the work-kinetic energy theorem

$$\int_i^f \vec{\mathbf{F}} \cdot d\vec{\mathbf{r}} = K_f - K_i. \qquad (13.11.8)$$

When the work done on an object is positive, the object will increase its speed, and negative work done on an object causes a decrease in speed. When the work done is zero, the object will maintain a constant speed.

13.11.1 Instantaneous Power Applied by a Non-Constant Force for Three Dimensional Motion

Recall that for one-dimensional motion, the *instantaneous power* at time t is defined to be the limit of the average power as the time interval $[t, t+\Delta t]$ approaches zero,

$$P(t) = F_x^a(t) v_x(t). \qquad (13.11.9)$$

A more general result for the instantaneous power is found by using the expression for dW as given in Equation (13.8.23),

$$P = \frac{dW}{dt} = \frac{\vec{\mathbf{F}} \cdot d\vec{\mathbf{r}}}{dt} = \vec{\mathbf{F}} \cdot \vec{\mathbf{v}}. \qquad (13.11.10)$$

The time rate of change of the kinetic energy for a body of mass m is equal to the power,

$$\frac{dK}{dt} = \frac{1}{2} m \frac{d}{dt}\left(\vec{\mathbf{v}} \cdot \vec{\mathbf{v}}\right) = m \frac{d\vec{\mathbf{v}}}{dt} \cdot \vec{\mathbf{v}} = m\vec{\mathbf{a}} \cdot \vec{\mathbf{v}} = \vec{\mathbf{F}} \cdot \vec{\mathbf{v}} = P. \qquad (13.11.11)$$

where the we used Eq. (13.8.9), Newton's Second Law and Eq. (13.11.10).

Appendix 13A Work Done on a System of Two Particles

We shall show that the work done by an internal force in changing a system of two particles of masses m_1 and m_2 respectively from an initial state A to a final state B is equal to

$$W_c = \frac{1}{2}\mu(v_B^2 - v_A^2) \tag{13.A.1}$$

where v_B^2 is the square of the relative velocity in state B, v_A^2 is the square of the relative velocity in state A, and $\mu = m_1 m_2 / (m_1 + m_2)$.

Consider two bodies 1 and 2 and an interaction pair of forces shown in Figure 13A.1.

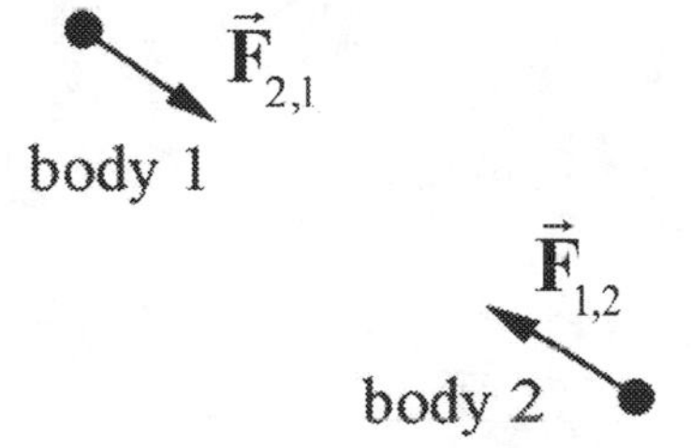

Figure 13A.1 System of two bodies interacting

We choose a coordinate system shown in Figure 13A.2.

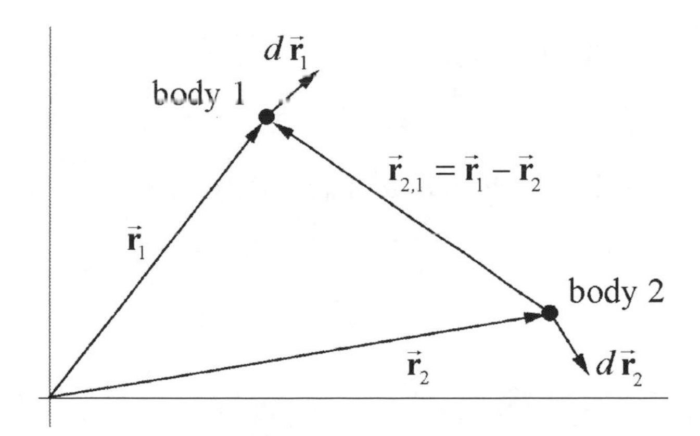

Figure 13A.2 Coordinate system for two-body interaction

Newton's Second Law applied to body 1 is

$$\vec{F}_{2,1} = m_1 \frac{d^2\vec{r}_1}{dt^2} \tag{13.A.2}$$

and applied to body 2 is

$$\vec{\mathbf{F}}_{1,2} = m_2 \frac{d^2\vec{\mathbf{r}}_2}{dt^2}. \tag{13.A.3}$$

Divide each side of Equation (13.A.2) by m_1,

$$\frac{\vec{\mathbf{F}}_{2,1}}{m_1} = \frac{d^2\vec{\mathbf{r}}_1}{dt^2} \tag{13.A.4}$$

and divide each side of Equation (13.A.3) by m_2,

$$\frac{\vec{\mathbf{F}}_{1,2}}{m_2} = \frac{d^2\vec{\mathbf{r}}_2}{dt^2}. \tag{13.A.5}$$

Subtract Equation (13.A.5) from Equation (13.A.4) yielding

$$\frac{\vec{\mathbf{F}}_{2,1}}{m_1} - \frac{\vec{\mathbf{F}}_{1,2}}{m_2} = \frac{d^2\vec{\mathbf{r}}_1}{dt^2} - \frac{d^2\vec{\mathbf{r}}_2}{dt^2} = \frac{d^2\vec{\mathbf{r}}_{2,1}}{dt^2}, \tag{13.A.6}$$

where $\vec{\mathbf{r}}_{2,1} = \vec{\mathbf{r}}_1 - \vec{\mathbf{r}}_2$. Use Newton's Third Law, $\vec{\mathbf{F}}_{2,1} = -\vec{\mathbf{F}}_{1,2}$ on the left hand side of Equation (13.A.6) to obtain

$$\vec{\mathbf{F}}_{2,1}\left(\frac{1}{m_1} + \frac{1}{m_2}\right) = \frac{d^2\vec{\mathbf{r}}_1}{dt^2} - \frac{d^2\vec{\mathbf{r}}_2}{dt^2} = \frac{d^2\vec{\mathbf{r}}_{2,1}}{dt^2}. \tag{13.A.7}$$

The quantity $d^2\vec{\mathbf{r}}_{1,2} / dt^2$ is the *relative acceleration* of body 1 with respect to body 2. Define

$$\frac{1}{\mu} \equiv \frac{1}{m_1} + \frac{1}{m_2}. \tag{13.A.8}$$

The quantity μ is known as the ***reduced mass*** of the system. Equation (13.A.7) now takes the form

$$\vec{\mathbf{F}}_{2,1} = \mu \frac{d^2\vec{\mathbf{r}}_{2,1}}{dt^2}. \tag{13.A.9}$$

The work done in the system in displacing the two masses from an initial state A to a final state B is given by

$$W = \int_A^B \vec{\mathbf{F}}_{2,1} \cdot d\vec{\mathbf{r}}_1 + \int_A^B \vec{\mathbf{F}}_{1,2} \cdot d\vec{\mathbf{r}}_2 . \tag{13.A.10}$$

Recall by the work energy theorem that the LHS is the work done on the system,

$$W = \int_A^B \vec{\mathbf{F}}_{2,1} \cdot d\vec{\mathbf{r}}_1 + \int_A^B \vec{\mathbf{F}}_{1,2} \cdot d\vec{\mathbf{r}}_2 = \Delta K . \tag{13.A.11}$$

From Newton's Third Law, the sum in Equation (13.A.10) becomes

$$W = \int_A^B \vec{\mathbf{F}}_{2,1} \cdot d\vec{\mathbf{r}}_1 - \int_A^B \vec{\mathbf{F}}_{2,1} \cdot d\vec{\mathbf{r}}_2 = \int_A^B \vec{\mathbf{F}}_{2,1} \cdot (d\vec{\mathbf{r}}_1 - d\vec{\mathbf{r}}_2) = \int_A^B \vec{\mathbf{F}}_{2,1} \cdot d\vec{\mathbf{r}}_{2,1} , \tag{13.A.12}$$

where $d\vec{\mathbf{r}}_{2,1}$ is the relative displacement of the two bodies. We can now substitute Newton's Second Law, Equation (13.A.9), for the relative acceleration into Equation (13.A.12),

$$W = \int_A^B \vec{\mathbf{F}}_{2,1} \cdot d\vec{\mathbf{r}}_{2,1} = \int_A^B \mu \frac{d^2\vec{\mathbf{r}}_{2,1}}{dt^2} \cdot d\vec{\mathbf{r}}_{2,1} = \mu \int_A^B \left(\frac{d^2\vec{\mathbf{r}}_{2,1}}{dt^2} \cdot \frac{d\vec{\mathbf{r}}_{2,1}}{dt} \right) dt , \tag{13.A.13}$$

where we have used the relation between the differential elements $d\vec{\mathbf{r}}_{2,1} = \dfrac{d\vec{\mathbf{r}}_{2,1}}{dt} dt$. The product rule for derivatives of the scalar product of a vector with itself is given for this case by

$$\frac{1}{2} \frac{d}{dt} \left(\frac{d\vec{\mathbf{r}}_{2,1}}{dt} \cdot \frac{d\vec{\mathbf{r}}_{2,1}}{dt} \right) = \frac{d^2\vec{\mathbf{r}}_{2,1}}{dt^2} \cdot \frac{d\vec{\mathbf{r}}_{2,1}}{dt} . \tag{13.A.14}$$

Substitute Equation (13.A.14) into Equation (13.A.13), which then becomes

$$W = \mu \int_A^B \frac{1}{2} \frac{d}{dt} \left(\frac{d\vec{\mathbf{r}}_{2,1}}{dt} \cdot \frac{d\vec{\mathbf{r}}_{2,1}}{dt} \right) dt . \tag{13.A.15}$$

Equation (13.A.15) is now the integral of an exact derivative, yielding

$$W = \frac{1}{2} \mu \left(\frac{d\vec{\mathbf{r}}_{2,1}}{dt} \cdot \frac{d\vec{\mathbf{r}}_{2,1}}{dt} \right) \Bigg|_A^B = \frac{1}{2} \mu (\vec{\mathbf{v}}_{2,1} \cdot \vec{\mathbf{v}}_{2,1}) \Big|_A^B = \frac{1}{2} \mu (v_B^2 - v_A^2) , \tag{13.A.16}$$

where $\vec{\mathbf{v}}_{2,1}$ is the *relative velocity* between the two bodies. It's important to note that in the above derivation had we exchanged the roles of body 1 and 2 i.e. $1 \to 2$ and $2 \to 1$, we would have obtained the identical result because

$$\begin{aligned}
\vec{\mathbf{F}}_{1,2} &= -\vec{\mathbf{F}}_{2,1} \\
\vec{\mathbf{r}}_{1,2} &= \vec{\mathbf{r}}_2 - \vec{\mathbf{r}}_1 = -\vec{\mathbf{r}}_{2,1} \\
d\vec{\mathbf{r}}_{1,2} &= d(\vec{\mathbf{r}}_2 - \vec{\mathbf{r}}_1) = -d\vec{\mathbf{r}}_{2,1} \\
\vec{\mathbf{v}}_{1,2} &= -\vec{\mathbf{v}}_{2,1}.
\end{aligned} \tag{13.A.17}$$

Equation (13.A.16) implies that the work done is the change in the kinetic energy of the system, which we can write in terms of the reduced mass and the change in the square of relative speed of the two objects

$$\Delta K = \frac{1}{2}\mu(v_B^2 - v_A^2). \tag{13.A.18}$$

Chapter 14 Potential Energy and Conservation of Energy

Chapter 14 Potential Energy and Conservation of Energy

There is a fact, or if you wish, a law, governing all natural phenomena that are known to date. There is no exception to this law — it is exact as far as we know. The law is called the conservation of energy. It states that there is a certain quantity, which we call energy that does not change in the manifold changes which nature undergoes. That is a most abstract idea, because it is a mathematical principle; it says that there is a numerical quantity, which does not change when something happens. It is not a description of a mechanism, or anything concrete; it is just a strange fact that we can calculate some number and when we finish watching nature go through her tricks and calculate the number again, it is the same. [1]

Richard Feynman

So far we have analyzed the motion of point-like objects under the action of forces using Newton's Laws of Motion. We shall now introduce the Principle of Conservation of Energy to study the change in energy of a system between its initial and final states. In particular we shall introduce the concept of potential energy to describe the effect of conservative internal forces acting on the constituent components of a system.

14.1 Conservation of Energy

Recall from Chapter 13.1, the principle of conservation of energy. When a system and its surroundings undergo a transition from an initial state to a final state, the change in energy is zero,

$$\Delta E = \Delta E_{\text{system}} + \Delta E_{\text{surroundings}} = 0 . \tag{14.1.1}$$

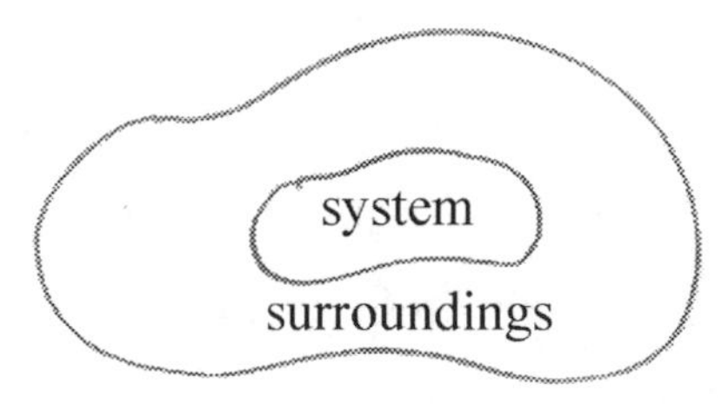

Figure 14.1 Diagram of a system and its surroundings

We shall study types of energy transformations due to interactions both inside and across the boundary of a system.

---

[1] Richard P. Feynman, Robert B. Leighton, and Matthew Sands, *The Feynman Lectures on Physics,* Vol. 1, p. 4.1.

14.2 Conservative and Non-Conservative Forces

Our first type of "energy accounting" involves *mechanical energy.* There are two types of mechanical energy, *kinetic energy* and *potential energy*. Our first task is to define what we mean by the change of the potential energy of a system.

We defined the work done by a force $\vec{\mathbf{F}}$, on an object, which moves along a path from an initial position $\vec{\mathbf{r}}_i$ to a final position $\vec{\mathbf{r}}_f$, as the integral of the component of the force tangent to the path with respect to the displacement of the point of contact of the force and the object,

$$W = \int_{\text{path}} \vec{\mathbf{F}} \cdot d\vec{\mathbf{r}} . \tag{14.2.1}$$

Does the work done on the object by the force depend on the path taken by the object?

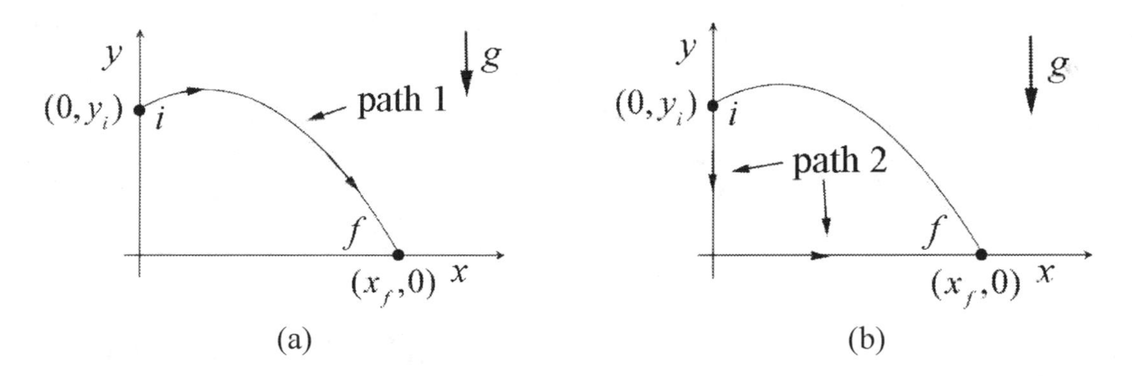

Figure 14.2 (a) and (b) Two different paths connecting the same initial and final points

First consider the motion of an object under the influence of a gravitational force near the surface of the earth. Let's consider two paths 1 and 2 shown in Figure 14.2. Both paths begin at the initial point $(x_i, y_i) = (0, y_i)$ and end at the final point $(x_f, y_f) = (x_f, 0)$. The gravitational force always points downward, so with our choice of coordinates, $\vec{\mathbf{F}} = -mg\,\hat{\mathbf{j}}$. The infinitesimal displacement along path 1 (Figure 14.2a) is given by $d\vec{\mathbf{r}}_1 = dx_1\,\hat{\mathbf{i}} + dy_1\,\hat{\mathbf{j}}$. The scalar product is then

$$\vec{\mathbf{F}} \cdot d\vec{\mathbf{r}}_1 = -mg\,\hat{\mathbf{j}} \cdot (dx_1\,\hat{\mathbf{i}} + dy_1\,\hat{\mathbf{j}}) = -mgdy_1 . \tag{14.2.2}$$

The work done by gravity along path 1 is the integral

$$W_1 = \int_{\text{path 1}} \vec{\mathbf{F}} \cdot d\vec{\mathbf{r}} = \int_{(0,y_i)}^{(x_f,0)} -mg\,dy_1 = -mg(0 - y_i) = mgy_i . \tag{14.2.3}$$

Path 2 consists of two legs (Figure 14.2b), leg A goes from the initial point $(0, y_i)$ to the origin $(0,0)$, and leg B goes from the origin $(0,0)$ to the final point $(x_f, 0)$. We shall calculate the work done along the two legs and then sum them up. The infinitesimal displacement along leg A is given by $d\vec{\mathbf{r}}_A = dy_A\,\hat{\mathbf{j}}$. The scalar product is then

$$\vec{\mathbf{F}} \cdot d\vec{\mathbf{r}}_A = -mg\,\hat{\mathbf{j}} \cdot dy_A\,\hat{\mathbf{j}} = -mgdy_A. \tag{14.2.4}$$

The work done by gravity along leg A is the integral

$$W_A = \int_{\text{leg } A} \vec{\mathbf{F}} \cdot d\vec{\mathbf{r}}_A = \int_{(0,y_i)}^{(0,0)} -mg\,dy_A = -mg(0 - y_i) = mgy_i. \tag{14.2.5}$$

The infinitesimal displacement along leg B is given by $d\vec{\mathbf{r}}_B = dx_B\,\hat{\mathbf{i}}$. The scalar product is then

$$\vec{\mathbf{F}} \cdot d\vec{\mathbf{r}}_B = -mg\,\hat{\mathbf{j}} \cdot dx_B\,\hat{\mathbf{i}} = 0. \tag{14.2.6}$$

Therefore the work done by gravity along leg B is zero, $W_B = 0$, which is no surprise because leg B is perpendicular to the direction of the gravitation force. Therefore the work done along path 2 is equal to the work along path 1,

$$W_2 = W_A + W_B = mgy_i = W_1. \tag{14.2.7}$$

Now consider the motion of an object on a surface with a kinetic frictional force between the object and the surface and denote the coefficient of kinetic friction by μ_k. Let's compare two paths from an initial point x_i to a final point x_f. The first path is a straight-line path. Along this path the work done is just

$$W^f = \int_{\text{path 1}} \vec{\mathbf{F}} \cdot d\vec{\mathbf{r}} = \int_{\text{path 1}} F_x\,dx = -\mu_k N s_1 = -\mu_k N\,\Delta x < 0, \tag{14.2.8}$$

where the length of the path is equal to the displacement, $s_1 = \Delta x$. Note that the fact that the kinetic frictional force is directed opposite to the displacement, which is reflected in the minus sign in Equation (14.2.8). The second path goes past x_f some distance and them comes back to x_f (Figure 14.3). Because the force of friction always opposes the motion, the work done by friction is negative,

$$W^f = \int_{\text{path 2}} \vec{\mathbf{F}} \cdot d\vec{\mathbf{r}} = \int_{\text{path 2}} F_x\,dx = -\mu_k N s_2 < 0. \tag{14.2.9}$$

The work depends on the total distance traveled s_2, and is greater than the displacement $s_2 > \Delta x$. The magnitude of the work done along the second path is greater than the magnitude of the work done along the first path.

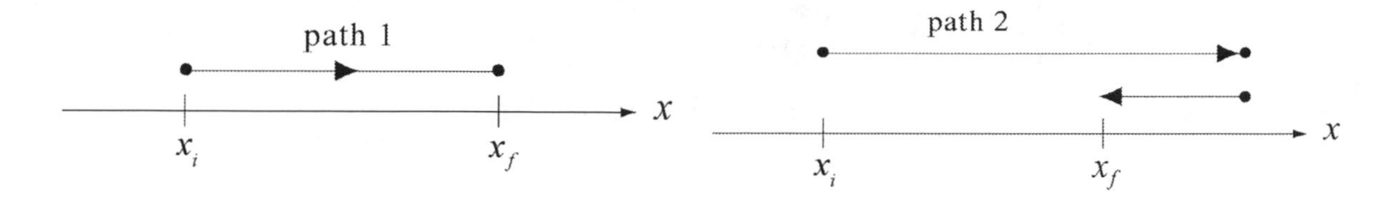

Figure 14.3 Two different paths from x_i to x_f.

These two examples typify two fundamentally different types of forces and their contribution to work. The work done by the gravitational force near the surface of the earth is independent of the path taken between the initial and final points. In the case of sliding friction, the work done depends on the path taken.

*Whenever the work done by a force in moving an object from an initial point to a final point is independent of the path, the force is called a **conservative force**.*

The work done by a conservative force $\vec{\mathbf{F}}_c$ in going around a closed path is zero. Consider the two paths shown in Figure 14.4 that form a closed path starting and ending at the point A with Cartesian coordinates $(1,0)$.

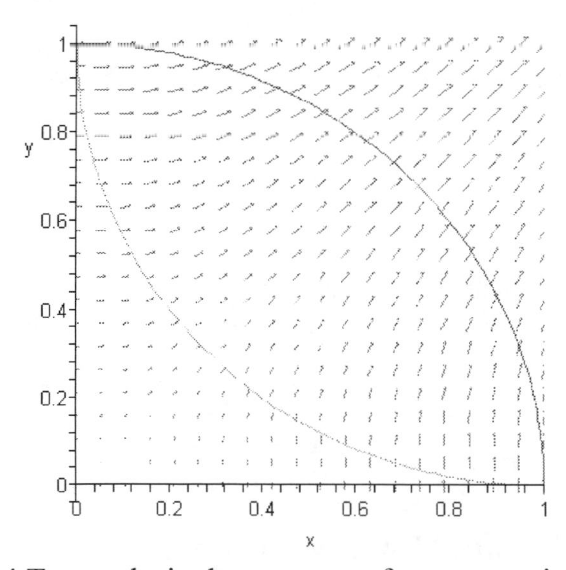

Figure 14.4 Two paths in the presence of a conservative force.

The work done along path 1 (the upper path in the figure, blue if viewed in color) from point A to point B with coordinates $(0,1)$ is given by

$$W_1 = \int_A^B \vec{\mathbf{F}}_c(1) \cdot d\vec{\mathbf{r}}_1 . \quad (14.2.10)$$

The work done along path 2 (the lower path, green in color) from B to A is given by

$$W_2 = \int_B^A \vec{\mathbf{F}}_c(2) \cdot d\vec{\mathbf{r}}_2 . \quad (14.2.11)$$

The work done around the closed path is just the sum of the work along paths 1 and 2,

$$W = W_1 + W_2 = \int_A^B \vec{\mathbf{F}}_c(1) \cdot d\vec{\mathbf{r}}_1 + \int_B^A \vec{\mathbf{F}}_c(2) \cdot d\vec{\mathbf{r}}_2 . \quad (14.2.12)$$

If we reverse the endpoints of path 2, then the integral changes sign,

$$W_2 = \int_B^A \vec{\mathbf{F}}_c(2) \cdot d\vec{\mathbf{r}}_2 = -\int_A^B \vec{\mathbf{F}}_c(2) \cdot d\vec{\mathbf{r}}_2 . \quad (14.2.13)$$

We can then substitute Equation (14.2.13) into Equation (14.2.12) to find that the work done around the closed path is

$$W = \int_A^B \vec{\mathbf{F}}_c(1) \cdot d\vec{\mathbf{r}}_1 - \int_A^B \vec{\mathbf{F}}_c(2) \cdot d\vec{\mathbf{r}}_2 . \quad (14.2.14)$$

Since the force is conservative, the work done between the points A to B is independent of the path, so

$$\int_A^B \vec{\mathbf{F}}_c(1) \cdot d\vec{\mathbf{r}}_1 = \int_A^B \vec{\mathbf{F}}_c(2) \cdot d\vec{\mathbf{r}}_2 . \quad (14.2.15)$$

We now use path independence of work for a conservative force (Equation (14.2.15) in Equation (14.2.14)) to conclude that the work done by a conservative force around a closed path is zero,

$$W = \oint_{\substack{\text{closed} \\ \text{path}}} \vec{\mathbf{F}}_c \cdot d\vec{\mathbf{r}} = 0 . \quad (14.2.16)$$

14.3 Changes in Potential Energies of a System

Consider an object near the surface of the earth as a system that is initially given a velocity directed upwards. Once the object is released, the gravitation force, acting as an external force, does a negative amount of work on the object, and the kinetic energy decreases until the object reaches its highest point, at which its kinetic energy is zero. The gravitational force then does positive work until the object returns to its initial starting point with a velocity directed downward. If we ignore any effects of air resistance, the descending object will then have the identical kinetic energy as when it was thrown. All the kinetic energy was completely recovered.

Now consider both the earth and the object as a system and assume that there are no other external forces acting on the system. Then the gravitational force is an internal conservative force, and does work on both the object and the earth during the motion. As the object moves upward, the kinetic energy of the system decreases, primarily because the object slows down, but there is also an imperceptible increase in the kinetic energy of the earth. The change in kinetic energy of the earth must also be included because the earth is part of the system. When the object returns to its original height (vertical distance from the surface of the earth), all the kinetic energy in the system is recovered, even though a very small amount has been transferred to the Earth.

If we included the air as part of the system, and the air resistance as a non-conservative internal force, then the kinetic energy lost due to the work done by the air resistance is not recoverable. This lost kinetic energy, which we have called thermal energy, is distributed as random kinetic energy in both the air molecules and the molecules that compose the object (and, to a smaller extent, the earth).

We shall define a new quantity, the change in the internal *potential energy* of the system, which measures the amount of lost kinetic energy that can be recovered during an interaction.

When only internal conservative forces act in a closed system, the sum of the changes of the kinetic and potential energies of the system is zero.

Consider a closed system, $\Delta E_{sys} = 0$, that consists of two objects with masses m_1 and m_2 respectively. Assume that there is only one conservative force (internal force) that is the source of the interaction between two objects. We denote the force on object 1 due to the interaction with object 2 by $\vec{\mathbf{F}}_{2,1}$ and the force on object 2 due to the interaction with object 1 by $\vec{\mathbf{F}}_{1,2}$. From Newton's Third Law,

$$\vec{\mathbf{F}}_{2,1} = -\vec{\mathbf{F}}_{1,2}. \qquad (14.3.1)$$

The forces acting on the objects are shown in Figure 14.5.

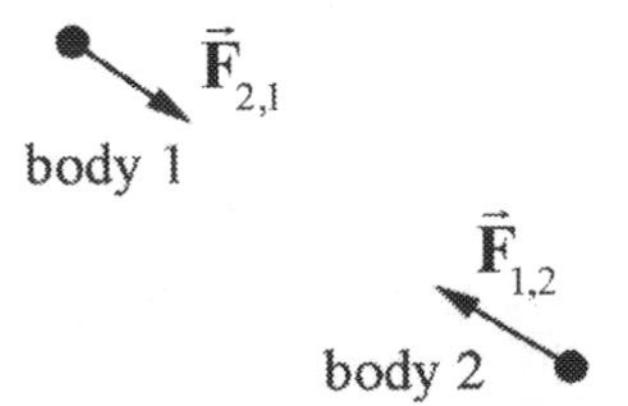

Figure 14.5 Internal forces acting on two objects

Choose a coordinate system (Figure 14.6) in which the position vector of object 1 is given by $\vec{\mathbf{r}}_1$ and the position vector of object 2 is given by $\vec{\mathbf{r}}_2$. The relative position of object 1 with respect to object 2 is given by $\vec{\mathbf{r}}_{2,1} = \vec{\mathbf{r}}_1 - \vec{\mathbf{r}}_2$. During the course of the interaction, object 1 is displaced by $d\vec{\mathbf{r}}_1$ and object 2 is displaced by $d\vec{\mathbf{r}}_2$, so the relative displacement of the two objects during the interaction is given by $d\vec{\mathbf{r}}_{2,1} = d\vec{\mathbf{r}}_1 - d\vec{\mathbf{r}}_2$.

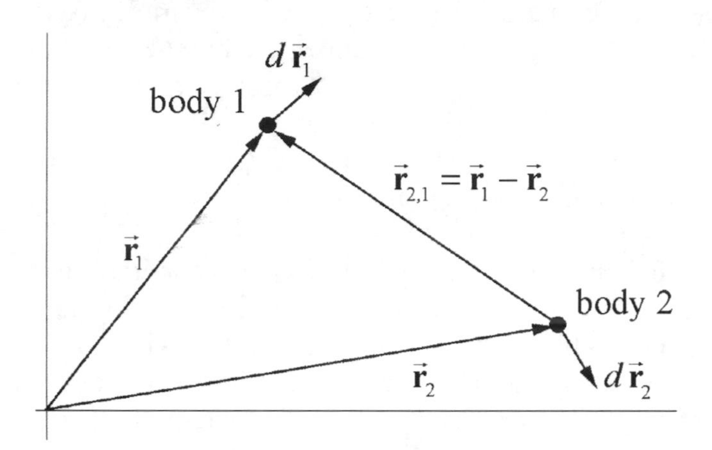

Figure 14.6 Coordinate system for two objects with relative position vector $\vec{\mathbf{r}}_{2,1} = \vec{\mathbf{r}}_1 - \vec{\mathbf{r}}_2$

Recall that the change in the kinetic energy of an object is equal to the work done by the forces in displacing the object. For two objects displaced from an initial state A to a final state B,

$$\Delta K_{\text{sys}} = \Delta K_1 + \Delta K_2 = W_{\text{c}} = \int_A^B \vec{\mathbf{F}}_{2,1} \cdot d\vec{\mathbf{r}}_1 + \int_A^B \vec{\mathbf{F}}_{1,2} \cdot d\vec{\mathbf{r}}_2 . \qquad (14.3.2)$$

(In Equation (14.3.2), the labels "A" and "B" refer to initial and final states, not paths.)

From Newton's Third Law, Equation (14.3.1), the sum in Equation (14.3.2) becomes

$$\Delta K_{\text{sys}} = W_c = \int_A^B \vec{\mathbf{F}}_{2,1} \cdot d\vec{\mathbf{r}}_1 - \int_A^B \vec{\mathbf{F}}_{2,1} \cdot d\vec{\mathbf{r}}_2 = \int_A^B \vec{\mathbf{F}}_{2,1} \cdot (d\vec{\mathbf{r}}_1 - d\vec{\mathbf{r}}_2) = \int_A^B \vec{\mathbf{F}}_{2,1} \cdot d\vec{\mathbf{r}}_{2,1} \qquad (14.3.3)$$

where $d\vec{\mathbf{r}}_{1,2} = d\vec{\mathbf{r}}_1 - d\vec{\mathbf{r}}_2$ is the relative displacement of the two objects. Note that since $\vec{\mathbf{F}}_{2,1} = -\vec{\mathbf{F}}_{1,2}$ and $d\vec{\mathbf{r}}_{2,1} = -d\vec{\mathbf{r}}_{1,2}$, $\int_A^B \vec{\mathbf{F}}_{2,1} \cdot d\vec{\mathbf{r}}_{2,1} = \int_A^B \vec{\mathbf{F}}_{1,2} \cdot d\vec{\mathbf{r}}_{1,2}$.

Consider a system consisting of two objects interacting through a conservative force. Let $\vec{\mathbf{F}}_{2,1}$ *denote the force on object 1 due to the interaction with object 2 and let* $d\vec{\mathbf{r}}_{2,1} = d\vec{\mathbf{r}}_1 - d\vec{\mathbf{r}}_2$ *be the relative displacement of the two objects. The* ***change in internal potential energy of the system*** *is defined to be the negative of the work done by the conservative force when the objects undergo a relative displacement from the initial state* A *to the final state* B *along any displacement that changes the initial state* A *to the final state* B,

$$\Delta U_{\text{sys}} = -W_{\text{c}} = -\int_A^B \vec{\mathbf{F}}_{2,1} \cdot d\vec{\mathbf{r}}_{2,1} = -\int_A^B \vec{\mathbf{F}}_{1,2} \cdot d\vec{\mathbf{r}}_{1,2}. \tag{14.3.4}$$

Our definition of potential energy only holds for conservative forces, because the work done by a conservative force does not depend on the path but only on the initial and final positions. Because the work done by the conservative force is equal to the change in kinetic energy, we have that

$$\Delta U_{\text{sys}} = -\Delta K_{\text{sys}}, \text{ (closed system with no non-conservative forces)}. \tag{14.3.5}$$

Recall that the work done by a conservative force in going around a closed path is zero (Equation (14.2.16)); therefore the change in kinetic energy when a system returns to its initial state is zero. This means that the kinetic energy is completely recoverable.

In the *Appendix 13A: Work Done on a System of Two Particles,* we showed that the work done by an internal force in changing a system of two particles of masses m_1 and m_2 respectively from an initial state A to a final state B is equal to

$$W = \frac{1}{2}\mu(v_B^2 - v_A^2) = \Delta K_{\text{sys}}, \tag{14.3.6}$$

where v_B^2 is the square of the relative velocity in state B, v_A^2 is the square of the relative velocity in state A, and $\mu = m_1 m_2 / (m_1 + m_2)$ is a quantity known as the *reduced mass* of the system.

14.3.1 Change in Potential Energy for Several Conservative Forces

When there are several internal conservative forces acting on the system we define a separate change in potential energy for the work done by each conservative force,

$$\Delta U_{\text{sys},i} = -W_{c,i} = -\int_A^B \vec{\mathbf{F}}_{\text{c},i} \cdot d\vec{\mathbf{r}}_i. \tag{14.3.7}$$

where $\vec{\mathbf{F}}_{c,i}$ is a conservative internal force and $d\vec{\mathbf{r}}_i$ a change in the relative positions of the objects on which $\vec{\mathbf{F}}_{c,i}$ when the system is changed from state A to state B. The work done is the sum of the work done by the individual conservative forces,

$$W_c = W_{c,1} + W_{c,2} + \cdots. \tag{14.3.8}$$

Hence, the sum of the changes in potential energies for the system is the sum

$$\Delta U_{sys} = \Delta U_{sys,1} + \Delta U_{sys,2} + \cdots. \tag{14.3.9}$$

Therefore the change in potential energy of the system is equal to the negative of the work done

$$\Delta U_{sys} = -W_c = -\sum_i \int_A^B \vec{\mathbf{F}}_{c,i} \cdot d\vec{\mathbf{r}}_i. \tag{14.3.10}$$

If the system is closed (external forces do no work), and there are no non-conservative internal forces then Eq. (14.3.5) holds.

14.4 Change in Potential Energy and Zero Point for Potential Energy

We already calculated the work done by different conservative forces: constant gravity near the surface of the earth, the spring force, and the universal gravitation force. We chose the system in each case so that the conservative force was an external force. In each case, there was no change of potential energy and the work done was equal to the change of kinetic energy,

$$W_{ext} = \Delta K_{sys}. \tag{14.4.1}$$

We now treat each of these conservative forces as internal forces and calculate the change in potential energy of the system according to our definition

$$\Delta U_{sys} = -W_c = -\int_A^B \vec{\mathbf{F}}_c \cdot d\vec{\mathbf{r}}. \tag{14.4.2}$$

We shall also choose a ***zero reference potential*** for the potential energy of the system, so that we can consider all changes in potential energy relative to this reference potential.

14.4.1 Change in Gravitational Potential Energy Near the Surface of the Earth

Let's consider the example of an object falling near the surface of the earth. Choose our system to consist of the earth and the object. The gravitational force is now an internal conservative force acting inside the system. The distance separating the object and the center of mass of the earth, and the velocities of the earth and the object specifies the initial and final states.

Let's choose a coordinate system with the origin on the surface of the earth and the $+y$-direction pointing away from the center of the earth. Because the displacement of the earth is negligible, we need only consider the displacement of the object in order to calculate the change in potential energy of the system.

Suppose the object starts at an initial height y_i above the surface of the earth and ends at final height y_f. The gravitational force on the object is given by $\vec{\mathbf{F}}^g = -mg\,\hat{\mathbf{j}}$, the displacement is given by $d\vec{\mathbf{r}} = dy\,\hat{\mathbf{j}}$, and the scalar product is given by $\vec{\mathbf{F}}^g \cdot d\vec{\mathbf{r}} = -mg\,\hat{\mathbf{j}} \cdot dy\hat{\mathbf{j}} = -mgdy$. The work done by the gravitational force on the object is then

$$W^g = \int_{y_i)}^{y_f} \vec{\mathbf{F}}^g \cdot d\vec{\mathbf{r}} = \int_{y_i)}^{y_f} -mg\,dy = -mg(y_f - y_i) \ . \tag{14.4.3}$$

The change in potential energy is then given by

$$\Delta U^g = -W^g = mg\,\Delta y = mg\,y_f - mg\,y_i \,. \tag{14.4.4}$$

We introduce a potential energy function U so that

$$\Delta U^g \equiv U_f^g - U_i^g \,. \tag{14.4.5}$$

Only differences in the function U^g have a physical meaning. We can choose a zero reference point for the potential energy anywhere we like. We have some flexibility to adapt our choice of zero for the potential energy to best fit a particular problem. Because the change in potential energy only depended on the displacement, Δy. In the above expression for the change of potential energy (Eq. (14.4.4)), let $y_f = y$ be an arbitrary point and $y_i = 0$ denote the surface of the earth. Choose the zero reference potential for the potential energy to be at the surface of the earth corresponding to our origin $y = 0$, with $U^g(0) = 0$. Then

$$\Delta U^g = U^g(y) - U^g(0) = U^g(y) \,. \tag{14.4.6}$$

Substitute $y_i = 0$, $y_f = y$ and Eq. (14.4.6) into Eq. (14.4.4) yielding a potential energy as a function of the height y above the surface of the earth,

$$U^g(y) = mgy, \text{ with } U^g(y = 0) = 0 \,. \tag{14.4.7}$$

14.4.2 Hooke's Law Spring-Object System

Consider a spring-object system lying on a frictionless horizontal surface with one end of the spring fixed to a wall and the other end attached to an object of mass m (Figure 14.7). The spring force is an internal conservative force. The wall exerts an external force on the spring-object system but since the point of contact of the wall with the spring undergoes no displacement, this external force does no work.

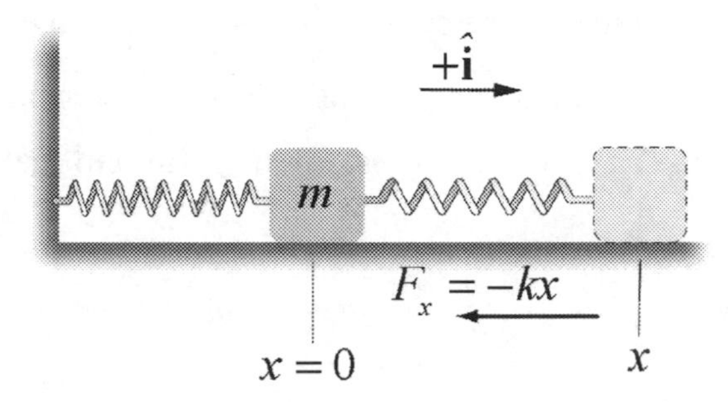

Figure 14.7 A spring-object system.

Choose the origin at the position of the center of the object when the spring is relaxed (the equilibrium position). Let x be the displacement of the object from the origin. We choose the $+\hat{\mathbf{i}}$ unit vector to point in the direction the object moves when the spring is being stretched (to the right of $x = 0$ in the figure). The spring force on a mass is then given by $\vec{\mathbf{F}}^s = F_x^s\,\hat{\mathbf{i}} = -kx\,\hat{\mathbf{i}}$. The displacement is $d\vec{\mathbf{r}} = dx\,\hat{\mathbf{i}}$. The scalar product is $\vec{\mathbf{F}} \cdot d\vec{\mathbf{r}} = -kx\,\hat{\mathbf{i}} \cdot dx\,\hat{\mathbf{i}} = -kx\,dx$. The work done by the spring force on the mass is

$$W^s = \int_{x=x_i}^{x=x_f} \vec{\mathbf{F}} \cdot d\vec{\mathbf{r}} = -\frac{1}{2} \int_{x=x_i}^{x=x_f} -\frac{1}{2}(-kx)\,dx = -\frac{1}{2}k({x_f}^2 - {x_i}^2). \tag{14.4.8}$$

We then define the change in potential energy in the spring-object system in moving the object from an initial position x_i from equilibrium to a final position x_f from equilibrium by

$$\Delta U^s \equiv U^s(x_f) - U^s(x_i) = -W^s = \frac{1}{2}k(x_f^2 - x_i^2). \tag{14.4.9}$$

Therefore an arbitrary stretch or compression of a spring-object system from equilibrium $x_i = 0$ to a final position $x_f = x$ changes the potential energy by

$$\Delta U^s = U^s(x_f) - U^s(0) = \frac{1}{2}kx^2. \tag{14.4.10}$$

For the spring-object system, there is an obvious choice of position where the potential energy is zero, the equilibrium position of the spring- object,

$$U^s(0) \equiv 0 . \qquad (14.4.11)$$

Then with this choice of zero reference potential, the potential energy as a function of the displacement x from the equilibrium position is given by

$$U^s(x) = \frac{1}{2}k x^2, \text{ with } U^s(0) \equiv 0 . \qquad (14.4.12)$$

14.4.3 Inverse Square Gravitation Force

Consider a system consisting of two objects of masses m_1 and m_2 that are separated by a center-to-center distance $r_{2,1}$. A coordinate system is shown in the Figure 14.8. The internal gravitational force on object 1 due to the interaction between the two objects is given by

$$\vec{\mathbf{F}}^G_{2,1} = -\frac{G m_1 m_2}{r_{2,1}^2}\hat{\mathbf{r}}_{2,1} . \qquad (14.4.13)$$

The displacement vector is given by $d\vec{\mathbf{r}}_{2,1} = dr_{2,1}\,\hat{\mathbf{r}}_{2,1}$. So the scalar product is

$$\vec{\mathbf{F}}^G_{2,1} \cdot d\vec{\mathbf{r}}_{2,1} = -\frac{G m_1 m_2}{r_{2,1}^2}\hat{\mathbf{r}}_{2,1} \cdot dr_{2,1}\,\hat{\mathbf{r}}_{2,1} = -\frac{G m_1 m_2}{r_{2,1}^2} dr_{2,1} . \qquad (14.4.14)$$

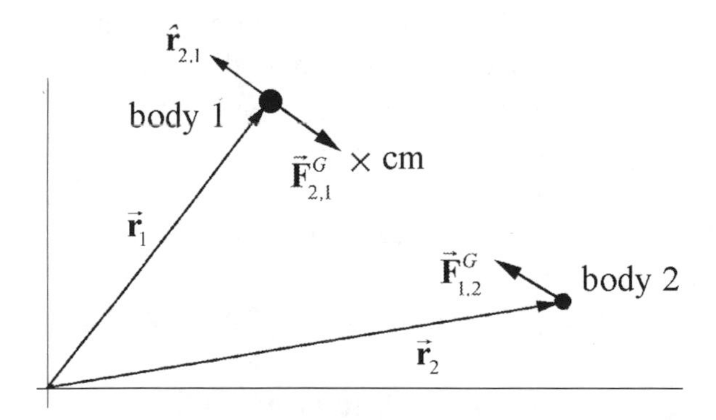

Figure 14.8 Gravitational interaction

Using our definition of potential energy (Eq. (14.3.4)), we have that the change in the gravitational potential energy of the system in moving the two objects from an initial position in which the center of mass of the two objects are a distance r_i apart to a final position in which the center of mass of the two objects are a distance r_f apart is given by

$$\Delta U^G = -\int_A^B \vec{\mathbf{F}}^G_{2,1} \cdot d\vec{\mathbf{r}}_{2,1} = -\int_{r_i}^{f} -\frac{G m_1 m_2}{r_{2,1}^2} dr_{2,1} = -\frac{G m_1 m_2}{r_{2,1}}\Bigg|_{r_i}^{r_f} = -\frac{G m_1 m_2}{r_f} + \frac{G m_1 m_2}{r_i}. \quad (14.4.15)$$

We now choose our reference point for the zero of the potential energy to be at infinity, $r_i = \infty$, with the choice that $U^G(\infty) \equiv 0$. By making this choice, the term $1/r$ in the expression for the change in potential energy vanishes when $r_i = \infty$. The gravitational potential energy as a function of the relative distance r between the two objects is given by

$$U^G(r) = -\frac{G m_1 m_2}{r}, \text{ with } U^G(\infty) \equiv 0. \quad (14.4.16)$$

14.5 Mechanical Energy and Conservation of Mechanical Energy

The total change in the ***mechanical energy*** *of the system is defined to be the sum of the changes of the kinetic and the potential energies,*

$$\Delta E_m = \Delta K_{\text{sys}} + \Delta U_{\text{sys}}. \quad (14.4.17)$$

For a closed system with only conservative internal forces, the total change in the mechanical energy is zero,

$$\Delta E_m = \Delta K_{\text{sys}} + \Delta U_{\text{sys}} = 0. \quad (14.4.18)$$

Equation (14.4.18) is the symbolic statement of what is called ***conservation of mechanical energy.*** Recall that the work done by a conservative force in going around a closed path is zero (Equation (14.2.16)), therefore both the changes in kinetic energy and potential energy are zero when a closed system with only conservative internal forces returns to its initial state. Throughout the process, the kinetic energy may change into internal potential energy but if the system returns to its initial state, the kinetic energy is completely recoverable. We shall refer to a closed system in which processes take place in which only conservative forces act as ***completely reversible processes***.

14.5.1 Change in Gravitational potential Energy Near the Surface of the Earth

Let's consider the example of an object of mass m_o falling near the surface of the earth (mass m_e). Choose our system to consist of the earth and the object. The gravitational force is now an internal conservative force acting inside the system. The initial and final states are specified by the distance separating the object and the center of mass of the earth, and the velocities of the earth and the object. The change in kinetic energy between the initial and final states for the system is

$$\Delta K_{\text{sys}} = \Delta K_e + \Delta K_o, \quad (14.4.19)$$

$$\Delta K_{\text{sys}} = \left(\frac{1}{2}m_{\text{e}}(v_{e,f})^2 - \frac{1}{2}m_{\text{e}}(v_{\text{e},\text{i}})^2\right) + \left(\frac{1}{2}m_o(v_{o,f})^2 - \frac{1}{2}m_o(v_{o,i})^2\right). \tag{14.4.20}$$

The change of kinetic energy of the earth due to the gravitational interaction between the earth and the object is negligible. The change in kinetic energy of the system is approximately equal to the change in kinetic energy of the object,

$$\Delta K_{\text{sys}} \cong \Delta K_o = \frac{1}{2}m_o(v_{o,f})^2 - \frac{1}{2}m_o(v_{o,i})^2. \tag{14.4.21}$$

We now define the mechanical energy function for the system

$$E_m = K + U^g = \frac{1}{2}m_o(v_b)^2 + m_o gy, \text{ with } U^g(0) = 0, \tag{14.4.22}$$

where K is the kinetic energy and U^g is the potential energy. The change in mechanical energy is then

$$\Delta E_m \equiv E_{m,f} - E_{m,i} = (K_f + U_f^g) - (K_i + U_i^g). \tag{14.4.23}$$

When the work done by the external forces is zero and there are no internal non-conservative forces, the total mechanical energy of the system is constant,

$$E_{m,f} = E_{m,i}, \tag{14.4.24}$$

or equivalently

$$(K_f + U_f) = (K_i + U_i). \tag{14.4.25}$$

14.6 Spring Force Energy Diagram

The spring force on an object is a restoring force $\vec{\mathbf{F}}^s = F_x^s\,\hat{\mathbf{i}} = -k\,x\,\hat{\mathbf{i}}$ where we choose a coordinate system with the equilibrium position at $x_i = 0$ and x is the amount the spring has been stretched $(x > 0)$ or compressed $(x < 0)$ from its equilibrium position. We calculate the potential energy difference Eq. (14.4.9) and found that

$$U^s(x) - U^s(x_i) = -\int_{x_i}^{x} F_x^s\,dx = \frac{1}{2}k(x^2 - x_i^2). \tag{14.5.1}$$

The first fundamental theorem of calculus states that

$$U(x) - U(x_i) = \int_{x'=x_i}^{x'=x} \frac{dU}{dx'}dx'. \tag{14.5.2}$$

Comparing Equation (14.5.1) with Equation (14.5.2) shows that the force is the negative derivative (with respect to position) of the potential energy,

$$F_x^s = -\frac{dU^s(x)}{dx}. \tag{14.5.3}$$

Choose the zero reference point for the potential energy to be at the equilibrium position, $U^s(0) \equiv 0$. Then the potential energy function becomes

$$U^s(x) = \frac{1}{2}kx^2. \tag{14.5.4}$$

From this, we obtain the spring force law as

$$F_x^s = -\frac{dU^s(x)}{dx} = -\frac{d}{dx}\left(\frac{1}{2}kx^2\right) = -kx. \tag{14.5.5}$$

In Figure 14.9 we plot the potential energy function $U^s(x)$ for the spring force as function of x with $U^s(0) \equiv 0$ (the units are arbitrary).

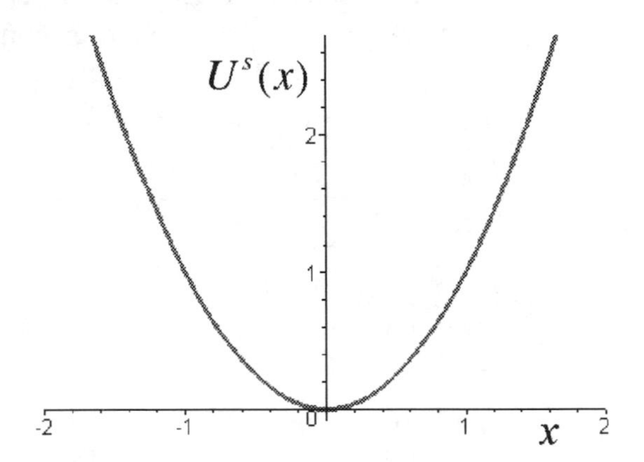

Figure 14.9 Graph of potential energy function as function of x for the spring.

The minimum of the potential energy function occurs at the point where the first derivative vanishes

$$\frac{dU^s(x)}{dx} = 0. \tag{14.5.6}$$

From Equation (14.5.4), the minimum occurs at $x = 0$,

$$0 = \frac{dU^s(x)}{dx} = kx. \tag{14.5.7}$$

Because the force is the negative derivative of the potential energy, and this derivative vanishes at the minimum, we have that the spring force is zero at the minimum $x = 0$ agreeing with our force law, $F^s_x\big|_{x=0} = -k\,x\big|_{x=0} = 0$.

The potential energy function has positive curvature in the neighborhood of a minimum equilibrium point. If the object is extended a small distance $x > 0$ away from equilibrium, the slope of the potential energy function is positive, $dU(x)/dx > 0$, hence the component of the force is negative because $F_x = -dU(x)/dx < 0$. Thus the object experiences a restoring force towards the minimum point of the potential. If the object is compresses with $x < 0$ then $dU(x)/dx < 0$, hence the component of the force is positive, $F_x = -dU(x)/dx > 0$, and the object again experiences a restoring force back towards the minimum of the potential energy as in Figure 14.10.

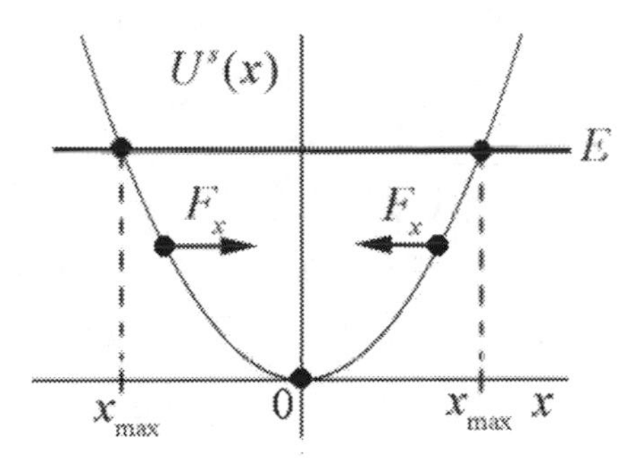

Figure 14.10 Stability diagram for the spring force.

The mechanical energy at any time is the sum of the kinetic energy $K(x)$ and the potential energy $U^s(x)$

$$E_m = K(x) + U^s(x). \tag{14.5.8}$$

Suppose our spring-object system has no loss of mechanical energy due to dissipative forces such as friction or air resistance. Both the kinetic energy and the potential energy are functions of the position of the object with respect to equilibrium. The energy is a constant of the motion and with our choice of $U^s(0) \equiv 0$, the energy can be either a positive value or zero. When the energy is zero, the object is at rest at the equilibrium position.

In Figure 14.10, we draw a straight horizontal line corresponding to a non-zero positive value for the energy E_m on the graph of potential energy as a function of x. The energy intersects the potential energy function at two points $\{-x_{\max}, x_{\max}\}$ with $x_{\max} > 0$. These points correspond to the maximum compression and maximum extension of the spring, which are called the ***turning points***. The kinetic energy is the difference between the energy and the potential energy,

$$K(x) = E_m - U^s(x). \quad (14.5.9)$$

At the turning points, where $E_m = U^s(x)$, the kinetic energy is zero. Regions where the kinetic energy is negative, $x < -x_{max}$ or $x > x_{max}$ are called the ***classically forbidden regions***, which the object can never reach if subject to the laws of classical mechanics. In quantum mechanics, with similar energy diagrams for quantum systems, there is a very small probability that the quantum object can be found in a classically forbidden region.

Example 14.1 Energy Diagram

The potential energy function for a particle of mass m, moving in the x-direction is given by

$$U(x) = -U_1\left(\left(\frac{x}{x_1}\right)^3 - \left(\frac{x}{x_1}\right)^2\right), \quad (14.5.10)$$

where U_1 and x_1 are positive constants and $U(0) = 0$. (a) Sketch $U(x)/U_1$ as a function of x/x_1. (b) Find the points where the force on the particle is zero. Classify them as stable or unstable. Calculate the value of $U(x)/U_1$ at these equilibrium points. (c) For energies E that lies in $0 < E < (4/27)U_1$ find an equation whose solution yields the turning points along the x-axis about which the particle will undergo periodic motion. (d) Suppose $E = (4/27)U_1$ and that the particle starts at $x = 0$ with speed v_0. Find v_0.

Solution: a) Figure 14.11 shows a graph of $U(x)$ vs. x, with the choice of values $x_1 = 1.5\text{ m}$, $U_1 = 27/4\text{ J}$, and $E = 0.2\text{ J}$.

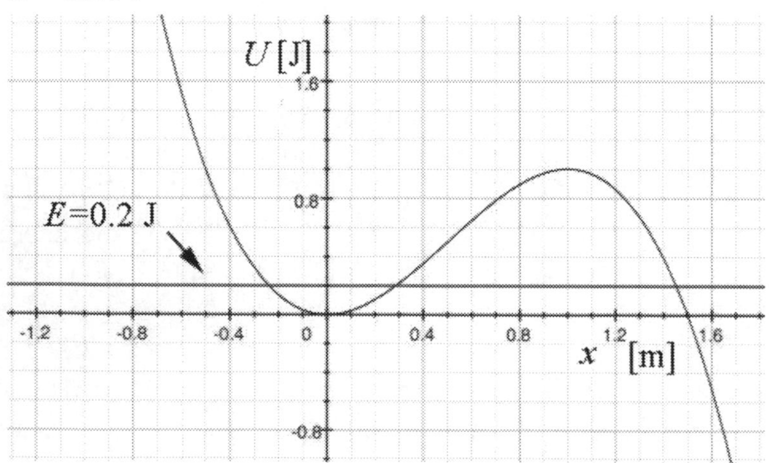

Figure 14.11 Energy diagram for Example 14.1

b) The force on the particle is zero at the minimum of the potential which occurs at

$$F_x(x) = -\frac{dU}{dx}(x) = U_1\left(\left(\frac{3}{x_1^{\,3}}\right)x^2 - \left(\frac{2}{x_1^{\,2}}\right)x\right) = 0 \qquad (14.5.11)$$

which becomes

$$x^2 = (2x_1 / 3)x\,. \qquad (14.5.12)$$

We can solve Eq. (14.5.12) for the extrema. This has two solutions

$$x = (2x_1 / 3) \;\;\text{and}\;\; x = 0\,. \qquad (14.5.13)$$

The second derivative is given by

$$\frac{d^2U}{dx^2}(x) = -U_1\left(\left(\frac{6}{x_1^{\,3}}\right)x - \left(\frac{2}{x_1^{\,2}}\right)\right). \qquad (14.5.14)$$

Evaluating the second derivative at $x = (2x_1 / 3)$ yields a negative quantity

$$\frac{d^2U}{dx^2}(x = (2x_1 / 3)) = -U_1\left(\left(\frac{6}{x_1^{\,3}}\right)\frac{2x_1}{3} - \left(\frac{2}{x_1^{\,2}}\right)\right) = -\frac{2U_1}{x_1^{\,2}} < 0\,, \qquad (14.5.15)$$

indicating the solution $x = (2x_1 / 3)$ represents a local maximum and hence is an unstable point. At $x = (2x_1 / 3)$, the potential energy is given by the value $U((2x_1 / 3)) = (4 / 27)U_1$. Evaluating the second derivative at $x = 0$ yields a positive quantity

$$\frac{d^2U}{dx^2}(x = 0) = -U_1\left(\left(\frac{6}{x_1^{\,3}}\right)0 - \left(\frac{2}{x_1^{\,2}}\right)\right) = \frac{2U_1}{x_1^{\,2}} > 0\,, \qquad (14.5.16)$$

indicating the solution $x = 0$ represents a local minimum and is a stable point. At the local minimum $x = 0$, the potential energy $U(0) = 0$.

c) Consider a fixed value of the energy of the particle within the range

$$U(0) = 0 < E < U(2x_1 / 3) = \frac{4U_1}{27}\,. \qquad (14.5.17)$$

If the particle at any time is found in the region $x_a < x < x_b < 2x_1 / 3$, where x_a and x_b are the turning points and are solutions to the equation

$$E = U(x) = -U_1\left(\left(\frac{x}{x_1}\right)^3 - \left(\frac{x}{x_1}\right)^2\right). \tag{14.5.18}$$

then the particle will undergo periodic motion between the values $x_a < x < x_b$. Within this region $x_a < x < x_b$, the kinetic energy is always positive because $K(x) = E - U(x)$. There is another solution x_c to Eq. (14.5.18) somewhere in the region $x_c > 2x_1/3$. If the particle at any time is in the region $x > x_c$ then it at any later time it is restricted to the region $x_c < x < +\infty$.

For $E > U(2x_1/3) = (4/27)U_1$, Eq. (14.5.18) has only one solution x_d. For all values of $x > x_d$, the kinetic energy is positive, which means that the particle can "escape" to infinity but can never enter the region $x < x_d$.

For $E < U(0) = 0$, the kinetic energy is negative for the range $-\infty < x < x_e$ where x_e satisfies Eq. (14.5.18) and therefore this region of space is forbidden.

(d) If the particle has speed v_0 at $x = 0$ where the potential energy is zero, $U(0) = 0$, the energy of the particle is constant and equal to kinetic energy

$$E = K(0) = \frac{1}{2}mv_0^2. \tag{14.5.19}$$

Therefore

$$(4/27)U_1 = \frac{1}{2}mv_0^2, \tag{14.5.20}$$

which we can solve for the speed

$$v_0 = \sqrt{8U_1/27m}. \tag{14.5.21}$$

14.7 Change of Mechanical Energy for a Closed System with Internal Non-conservative Forces

Consider a closed system (energy of the system is constant) that undergoes a transformation from an initial state to a final state by a prescribed set of changes.

> *Whenever the work done by a force in moving an object from an initial point to a final point depends on the path, the force is called a* ***non-conservative force****.*

Suppose the internal forces are both conservative and non-conservative. The work W done by the forces is a sum of the conservative work W_c, which is path-independent, and the non-conservative work W_{nc}, which is path-dependent,

$$W = W_c + W_{nc} . \tag{14.6.1}$$

The work done by the conservative forces is equal to the negative of the change in the potential energy

$$\Delta U = -W_c . \tag{14.6.2}$$

Substituting Equation (14.6.2) into Equation (14.6.1) yields

$$W = -\Delta U + W_{nc} . \tag{14.6.3}$$

The work done is equal to the change in the kinetic energy,

$$W = \Delta K . \tag{14.6.4}$$

Substituting Equation (14.6.4) into Equation (14.6.3) yields

$$\Delta K = -\Delta U + W_{nc} . \tag{14.6.5}$$

which we can rearrange as

$$W_{nc} = \Delta K + \Delta U . \tag{14.6.6}$$

We can now substitute Equation (14.6.4) into our expression for the change in the mechanical energy, Equation (14.4.17), with the result

$$W_{nc} = \Delta E_m . \tag{14.6.7}$$

The mechanical energy is no longer constant. The total change in energy of the system is zero,

$$\Delta E_{\text{system}} = \Delta E_m - W_{nc} = 0 . \tag{14.6.8}$$

Energy is conserved but some mechanical energy has been transferred into non-recoverable energy W_{nc}. We shall refer to processes in which there is non-zero non-recoverable energy as ***irreversible processes***.

14.7.1 Change of Mechanical Energy for a Non-closed System

When the system is no longer closed but in contact with its surroundings, the change in energy of the system is equal to the negative of the change in energy of the surroundings (Eq. (14.1.1)),

$$\Delta E_{\text{system}} = -\Delta E_{\text{surroundings}} \tag{14.6.9}$$

If the system is not isolated, the change in energy of the system can be the result of external work done by the surroundings on the system (which can be positive or negative)

$$W_{\text{ext}} = \int_A^B \vec{\mathbf{F}}_{\text{ext}} \cdot d\vec{\mathbf{r}} \,. \tag{14.6.10}$$

This work will result in the system undergoing ***coherent motion***. Note that $W_{\text{ext}} > 0$ if work is done on the system ($\Delta E_{\text{surroundings}} < 0$) and $W_{\text{ext}} < 0$ if the system does work on the surroundings ($\Delta E_{\text{surroundings}} > 0$). If the system is in thermal contact with the surroundings, then energy can flow into or out of the system. This energy flow due to thermal contact is often denoted by Q with the convention that $Q > 0$ if the energy flows into the system ($\Delta E_{\text{surroundings}} < 0$) and $Q < 0$ if the energy flows out of the system ($\Delta E_{\text{surroundings}} > 0$). Then Eq. (14.6.9) can be rewritten as

$$W^{\text{ext}} + Q = \Delta E_{\text{sys}} \tag{14.6.11}$$

Equation (14.6.11) is also called ***the first law of thermodynamics.***

This will result in either an increase or decrease in random thermal motion of the molecules inside the system, There may also be other forms of energy that enter the system, for example *radiative energy*.

Several questions naturally arise from this set of definitions and physical concepts. Is it possible to identify all the conservative forces and calculate the associated changes in potential energies? How do we account for non-conservative forces such as friction that act at the boundary of the system?

14.8 Dissipative Forces: Friction

Suppose we consider an object moving on a rough surface. As the object slides it slows down and stops. While the sliding occurs both the object and the surface increase in temperature. The increase in temperature is due to the molecules inside the materials increasing their kinetic energy. This random kinetic energy is called *thermal energy*. Kinetic energy associated with the coherent motion of the molecules of the object has been dissipated into kinetic energy associated with random motion of the molecules composing the object and surface.

If we define the system to be just the object, then the friction force acts as an external force on the system and results in the dissipation of energy into both the block and the surface. Without knowing further properties of the material we cannot determine the exact changes in the energy of the system.

Friction introduces a problem in that the point of contact is not well defined because the surface of contact is constantly deforming as the object moves along the surface. If we considered the object and the surface as the system, then the friction force is an internal force, and the decrease in the kinetic energy of the moving object ends up as an increase in the internal random kinetic energy of the constituent parts of the system. When there is dissipation at the boundary of the system, we need an additional model (thermal equation of state) for how the dissipated energy distributes itself among the constituent parts of the system.

14.8.1 Source Energy

Consider a person walking. The frictional force between the person and the ground does no work because the point of contact between the person's foot and the ground undergoes no displacement as the person applies a force against the ground, (there may be some slippage but that would be opposite the direction of motion of the person). However the kinetic energy of the object increases. Have we disproved the work-energy theorem? The answer is no! The chemical energy stored in the body tissue is converted to kinetic energy and thermal energy. Because the person-air-ground can be treated as a closed system, we have that

$$0 = \Delta E_{\text{sys}} = \Delta E_{\text{chemical}} + \Delta E_{\text{thermal}} + \Delta E_{\text{mechanical}}, \quad \text{(closed system)}. \qquad (14.7.1)$$

If we assume that there is no change in the potential energy of the system, then $\Delta E_{\text{mechanical}} = \Delta K$. Therefore some of the internal chemical energy has been transformed into thermal energy and the rest has changed into the kinetic energy of the system,

$$-\Delta E_{\text{chemical}} - \Delta E_{\text{thermal}} + \Delta K\,. \qquad (14.7.2)$$

14.9 Worked Examples

Example 14.2 Escape Velocity of Toro

The asteroid Toro, discovered in 1964, has a radius of about $R = 5.0\,\text{km}$ and a mass of about $m_t = 2.0\times10^{15}\,\text{kg}$. Let's assume that Toro is a perfectly uniform sphere. What is the escape velocity for an object of mass m on the surface of Toro? Could a person reach this speed (on earth) by running?

Solution: The only potential energy in this problem is the gravitational potential energy. We choose the zero point for the potential energy to be when the object and Toro are an infinite distance apart, $U^G(\infty) \equiv 0$. With this choice, the potential energy when the object and Toro are a finite distance r apart is given by

$$U^G(r) = -\frac{Gm_t\,m}{r} \qquad (14.8.1)$$

with $U^G(\infty) \equiv 0$. The expression ***escape velocity*** refers to the minimum speed necessary for an object to escape the gravitational interaction of the asteroid and move off to an infinite distance away. If the object has a speed less than the escape velocity, it will be unable to escape the gravitational force and must return to Toro. If the object has a speed greater than the escape velocity, it will have a non-zero kinetic energy at infinity. The condition for the escape velocity is that the object will have exactly zero kinetic energy at infinity.

We choose our initial state, at time t_i, when the object is at the surface of the asteroid with speed equal to the escape velocity. We choose our final state, at time t_f, to occur when the separation distance between the asteroid and the object is infinite.

The initial kinetic energy is $K_i = (1/2)m v_{\text{esc}}{}^2$. The initial potential energy is $U_i = -Gm_t m / R$, and so the initial mechanical energy is

$$E_i = K_i + U_i = \frac{1}{2} m v_{\text{esc}}^2 - \frac{Gm_t m}{R}. \tag{14.8.2}$$

The final kinetic energy is $K_f = 0$, because this is the condition that defines the escape velocity. The final potential energy is zero, $U_f = 0$ because we chose the zero point for potential energy at infinity. The final mechanical energy is then

$$E_f = K_f + U_f = 0. \tag{14.8.3}$$

There is no non-conservative work, so the change in mechanical energy is zero

$$0 = W_{\text{nc}} = \Delta E_m = E_f - E_i. \tag{14.8.4}$$

Therefore

$$0 = -\left(\frac{1}{2} m v_{\text{esc}}^2 - \frac{Gm_t m}{R}\right). \tag{14.8.5}$$

This can be solved for the escape velocity,

$$\begin{aligned} v_{\text{esc}} &= \sqrt{\frac{2Gm_t}{R}} \\ &= \sqrt{\frac{2(6.67 \times 10^{-11} \text{N} \cdot \text{m}^2 \cdot \text{kg}^{-2})(2.0 \times 10^{15} \text{kg})}{(5.0 \times 10^3 \text{m})}} = 7.3 \text{ m} \cdot \text{s}^{-1}. \end{aligned} \tag{14.8.6}$$

Considering that Olympic sprinters typically reach velocities of $12\ \text{m}\cdot\text{s}^{-1}$, this is an easy speed to attain by running on earth. It may be harder on Toro to generate the acceleration necessary to reach this speed by pushing off the ground, since any slight upward force will raise the runner's center of mass and it will take substantially more time than on earth to come back down for another push off the ground.

Example 14.3 Spring-Block-Loop-the-Loop

A small block of mass m is pushed against a spring with spring constant k and held in place with a catch. The spring is compressed an unknown distance x (Figure 14.12). When the catch is removed, the block leaves the spring and slides along a frictionless circular loop of radius r. When the block reaches the top of the loop, the force of the loop on the block (the normal force) is equal to twice the gravitational force on the mass. (a) Using conservation of energy, find the kinctic cnergy of the block at the top of the loop. (b) Using Newton's Second Law, derive the equation of motion for the block when it is at the top of the loop. Specifically, find the speed v_{top} in terms of the gravitation constant g and the loop radius r. (c) What distance was the spring compressed?

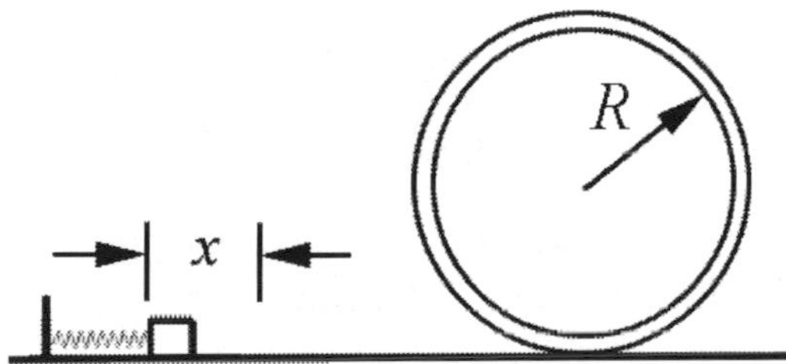

Figure 14.12 Initial state for spring-block-loop-the-loop system

Solution: a) Choose for the initial state the instant before the catch is released. The initial kinetic energy is $K_i = 0$. The initial potential energy is non-zero, $U_i = (1/2)kx^2$. The initial mechanical energy is then

$$E_i = K_i + U_i = \frac{1}{2}kx^2. \tag{14.8.7}$$

Choose for the final state the instant the block is at the top of the loop. The final kinetic energy is $K_f = (1/2)mv_{\text{top}}^2$; the block is in motion with speed v_{top}. The final potential energy is non-zero, $U_f = (mg)(2R)$. The final mechanical energy is then

$$E_f = K_f + U_f = 2mgR + \frac{1}{2}mv_{\text{top}}^2. \tag{14.8.8}$$

Because we are assuming the track is frictionless and neglecting air resistance, there is no non- conservative work. The change in mechanical energy is therefore zero,

$$0 = W_{\text{nc}} = \Delta E_m = E_f - E_i . \tag{14.8.9}$$

Mechanical energy is conserved, $E_f = E_i$, therefore

$$2mgR + \frac{1}{2} m v_{\text{top}}^2 = \frac{1}{2} k x^2 . \tag{14.8.10}$$

From Equation (14.8.10), the kinetic energy at the top of the loop is

$$\frac{1}{2} m v_{\text{top}}^2 = \frac{1}{2} k x^2 - 2mgR . \tag{14.8.11}$$

b) At the top of the loop, the forces on the block are the gravitational force of magnitude mg and the normal force of magnitude N, both directed down. Newton's Second Law in the radial direction, which is the downward direction, is

$$-mg - N = -\frac{m v_{\text{top}}^2}{R} . \tag{14.8.12}$$

In this problem, we are given that when the block reaches the top of the loop, the force of the loop on the block (the normal force, *downward* in this case) is equal to twice the weight of the block, $N = 2mg$. The Second Law, Eq. (14.8.12), then becomes

$$3mg = \frac{m v_{\text{top}}^2}{R} . \tag{14.8.13}$$

We can rewrite Equation (14.8.13) in terms of the kinetic energy as

$$\frac{3}{2} mg R = \frac{1}{2} m v_{\text{top}}^2 . \tag{14.8.14}$$

The speed at the top is therefore

$$v_{\text{top}} = \sqrt{3mg R} . \tag{14.8.15}$$

c) Combing Equations (14.8.11) and (14.8.14) yields

$$\frac{7}{2} mg R = \frac{1}{2} k x^2 . \tag{14.8.16}$$

Thus the initial displacement of the spring from equilibrium is

$$x = \sqrt{\frac{7mg\,R}{k}}. \qquad (14.8.17)$$

Example 14.4 Mass-Spring on a Rough Surface

A block of mass m slides along a horizontal table with speed v_0. At $x = 0$ it hits a spring with spring constant k and begins to experience a friction force. The coefficient of friction is variable and is given by $\mu = bx$, where b is a positive constant. Find the loss in mechanical energy when the block first momentarily comes to rest.

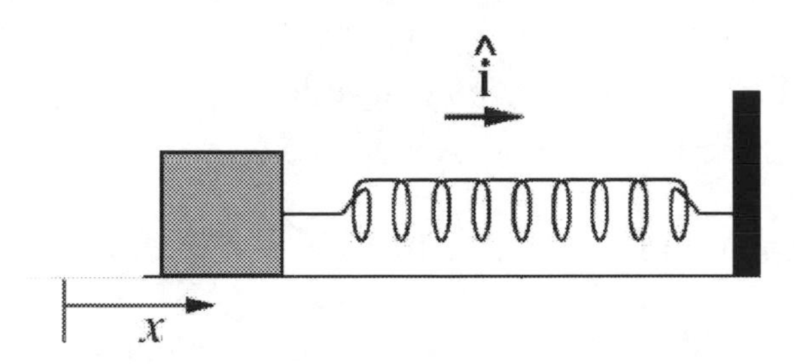

Figure 14.13 Spring-block system

Solution: From the model given for the frictional force, we could find the non-conservative work done, which is the same as the loss of mechanical energy, if we knew the position x_f where the block first comes to rest. The most direct (and easiest) way to find x_f is to use the work-energy theorem. The initial mechanical energy is $E_i = mv_i^2/2$ and the final mechanical energy is $E_f = k\,x_f^2/2$ (note that there is no potential energy term in E_i and no kinetic energy term in E_f). The difference between these two mechanical energies is the non-conservative work done by the frictional force,

$$\begin{aligned} W_{\text{nc}} &= \int_{x=0}^{x=x_f} F_{\text{nc}}\,dx = \int_{x=0}^{x=x_f} -F_{\text{friction}}\,dx = \int_{x=0}^{x=x_f} -\mu\,N\,dx \\ &= -\int_0^{x_f} b\,x\,mg\,dx = -\frac{1}{2}bmg\,x_f^2. \end{aligned} \qquad (14.8.18)$$

We then have that

$$\begin{aligned} W_{\text{nc}} &= \Delta E_m \\ W_{\text{nc}} &= E_f - E_i \\ -\frac{1}{2}bmg\,x_f^2 &= \frac{1}{2}k\,x_f^2 - \frac{1}{2}mv_i^2. \end{aligned} \qquad (14.8.19)$$

Solving the last of these equations for x_f^2 yields

$$x_f^2 = \frac{m v_0^2}{k + bmg}. \quad (14.8.20)$$

Substitute Eq. (14.8.20) into Eq. (14.8.18) gives the result that

$$W_{\text{nc}} = -\frac{bmg}{2}\frac{mv_0^2}{k+bmg} = -\frac{mv_0^2}{2}\left(1+\frac{k}{bmg}\right)^{-1}. \quad (14.8.21)$$

It is worth checking that the above result is dimensionally correct. From the model, the parameter b must have dimensions of inverse length (the coefficient of friction μ must be dimensionless), and so the product bmg has dimensions of force per length, as does the spring constant k; the result is dimensionally consistent.

Example 14.5 Cart-Spring on an Inclined Plane

An object of mass m slides down a plane that is inclined at an angle θ from the horizontal (Figure 14.14). The object starts out at rest. The center of mass of the cart is a distance d from an unstretched spring that lies at the bottom of the plane. Assume the spring is massless, and has a spring constant k. Assume the inclined plane to be frictionless. (a) How far will the spring compress when the mass first comes to rest? (b) Now assume that the inclined plane has a coefficient of kinetic friction μ_{k}. How far will the spring compress when the mass first comes to rest? The friction is primarily between the wheels and the bearings, not between the cart and the plane, but the friction force may be modeled by a coefficient of friction μ_{k}. (c) In case (b), how much energy has been lost to friction?

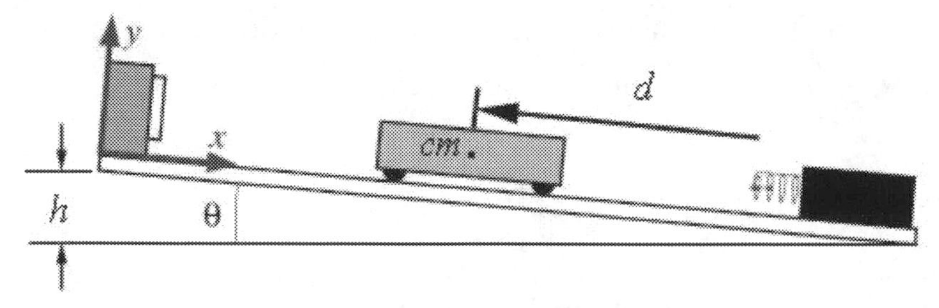

Figure 14.14 Cart on inclined plane

Solution: Let x denote the displacement of the spring from the equilibrium position. Choose the zero point for the gravitational potential energy $U^g(0) = 0$ not at the very bottom of the inclined plane, but at the location of the end of the unstretched spring. Choose the zero point for the spring potential energy where the spring is at its equilibrium position, $U^s(0) = 0$.

a) Choose for the initial state the instant the object is released (Figure 14.15). The initial kinetic energy is $K_i = 0$. The initial potential energy is non-zero, $U_i = mg\,d\sin\theta$. The initial mechanical energy is then

$$E_i = K_i + U_i = mg\,d\sin\theta \qquad (14.8.22)$$

Choose for the final state the instant when the object first comes to rest and the spring is compressed a distance x at the bottom of the inclined plane (Figure 14.16). The final kinetic energy is $K_f = 0$ since the mass is not in motion. The final potential energy is non-zero, $U_f = kx^2/2 - x\,mg\sin\theta$. Notice that the gravitational potential energy is negative because the object has dropped below the height of the zero point of gravitational potential energy.

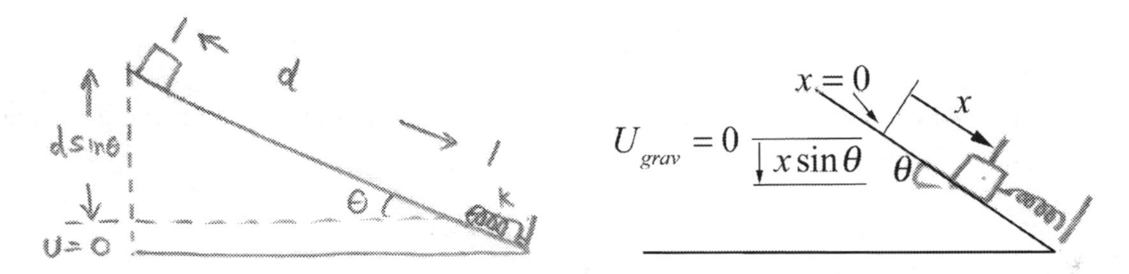

Figure 14.15 Initial state

Figure 14.16 Final state

The final mechanical energy is then

$$E_f = K_f + U_f = \frac{1}{2}kx^2 - x\,mg\sin\theta\,. \qquad (14.8.23)$$

Because we are assuming the track is frictionless and neglecting air resistance, there is no non- conservative work. The change in mechanical energy is therefore zero,

$$0 = W_{\mathrm{nc}} = \Delta E_m = E_f - E_i\,. \qquad (14.8.24)$$

Therefore

$$d\,mg\sin\theta = \frac{1}{2}kx^2 - x\,mg\sin\theta\,. \qquad (14.8.25)$$

This is a quadratic equation in x,

$$x^2 - \frac{2mg\sin\theta}{k}x - \frac{2d\,mg\sin\theta}{k} = 0\,. \qquad (14.8.26)$$

In the quadratic formula, we want the positive choice of square root for the solution to ensure a positive displacement of the spring from equilibrium,

$$x = \frac{mg\sin\theta}{k} + \left(\frac{m^2 g^2 \sin^2\theta}{k^2} + \frac{2d\, mg\sin\theta}{k}\right)^{1/2}$$
$$= \frac{mg}{k}(\sin\theta + \sqrt{1 + 2(k\, d / mg)\sin\theta}). \quad (14.8.27)$$

(What would the solution with the negative root represent?)

b) The effect of kinetic friction is that there is now a non-zero non-conservative work done on the object, which has moved a distance, $d + x$, given by

$$W_{\mathrm{nc}} = -f_{\mathrm{k}}(d + x) = -\mu_{\mathrm{k}} N(d + x) = -\mu_{\mathrm{k}} mg\cos\theta(d + x). \quad (14.8.28)$$

Note the normal force is found by using Newton's Second Law in the perpendicular direction to the inclined plane,

$$N - mg\cos\theta = 0. \quad (14.8.29)$$

The change in mechanical energy is therefore

$$W_{\mathrm{nc}} = \Delta E_m = E_f - E_i, \quad (14.8.30)$$

which becomes

$$-\mu_{\mathrm{k}} mg\cos\theta(d + x) = \left(\frac{1}{2}k x^2 - x\, mg\sin\theta\right) - d\, mg\sin\theta. \quad (14.8.31)$$

Equation (14.8.31) simplifies to

$$0 = \left(\frac{1}{2}k x^2 - x\, mg(\sin\theta - \mu_{\mathrm{k}}\cos\theta)\right) - d\, mg(\sin\theta - \mu_{\mathrm{k}}\cos\theta). \quad (14.8.32)$$

This is the same as Equation (14.8.25) above, but with $\sin\theta \to \sin\theta - \mu_k\cos\theta$. The maximum displacement of the spring is when there is friction is then

$$x = \frac{mg}{k}((\sin\theta - \mu_{\mathrm{k}}\cos\theta) + \sqrt{1 + 2(k\, d / mg)(\sin\theta - \mu_{\mathrm{k}}\cos\theta)}). \quad (14.8.33)$$

.

c) The energy lost to friction is given by $W_{\mathrm{nc}} = -\mu_{\mathrm{k}} mg\cos\theta(d + x)$, where x is given in part b).

Example 14.6 Object Sliding on a Sphere

A small point like object of mass m rests on top of a sphere of radius R. The object is released from the top of the sphere with a negligible speed and it slowly starts to slide (Figure 14.17). Let g denote the gravitation constant. (a) Determine the angle θ_1 with respect to the vertical at which the object will lose contact with the surface of the sphere. (b) What is the speed v_1 of the object at the instant it loses contact with the surface of the sphere.

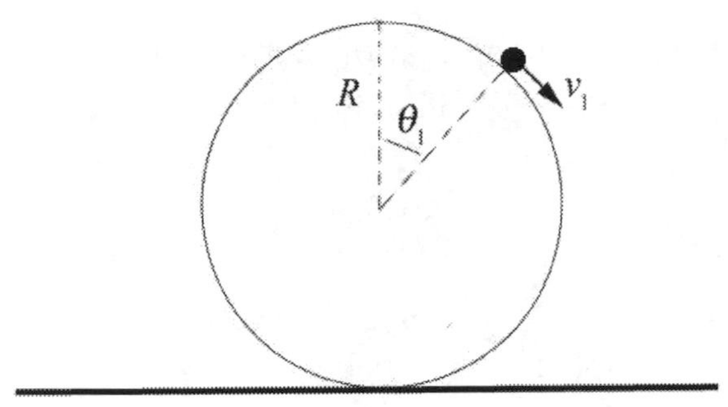

Figure 14.17 Object sliding on surface of sphere

Solution: We begin by identifying the forces acting on the object. There are two forces acting on the object, the gravitation and radial normal force that the sphere exerts on the particle that we denote by N. We draw a free-body force diagram for the object while it is sliding on the sphere. We choose polar coordinates as shown in Figure 14.18.

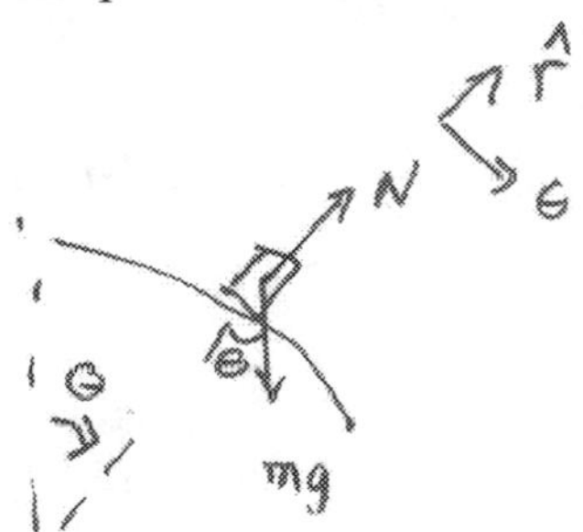

Figure 14.18 Free-body force diagram on object

The key constraint is that when the particle just leaves the surface the normal force is zero,

$$N(\theta_1) = 0, \tag{14.8.34}$$

where θ_1 denotes the angle with respect to the vertical at which the object will just lose contact with the surface of the sphere. Because the normal force is perpendicular to the displacement of the object, it does no work on the object and hence conservation of energy does not take into account the constraint on the motion imposed by the normal force. In order to analyze the effect of the normal force we must use the radial component of Newton's Second Law,

$$N - mg\cos\theta = -m\frac{v^2}{R}. \tag{14.8.35}$$

Then when the object just loses contact with the surface, Eqs. (14.8.34) and (14.8.35) require that

$$mg\cos\theta_1 = m\frac{v_1^{\,2}}{R}. \qquad (14.8.36)$$

where v_1 denotes the speed of the object at the instant it loses contact with the surface of the sphere. Note that the constrain condition Eq. (14.8.36) can be rewritten as

$$mgR\cos\theta_1 = mv_1^{\,2}. \qquad (14.8.37)$$

We can now apply conservation of energy. Choose the zero reference point $U = 0$ for potential energy to be the midpoint of the sphere.

Identify the initial state as the instant the object is released (Figure 14.19). We can neglect the very small initial kinetic energy needed to move the object away from the top of the sphere and so $K_i = 0$. The initial potential energy is non-zero, $U_i = mgR$. The initial mechanical energy is then

$$E_i = K_i + U_i = mgR. \qquad (14.8.38)$$

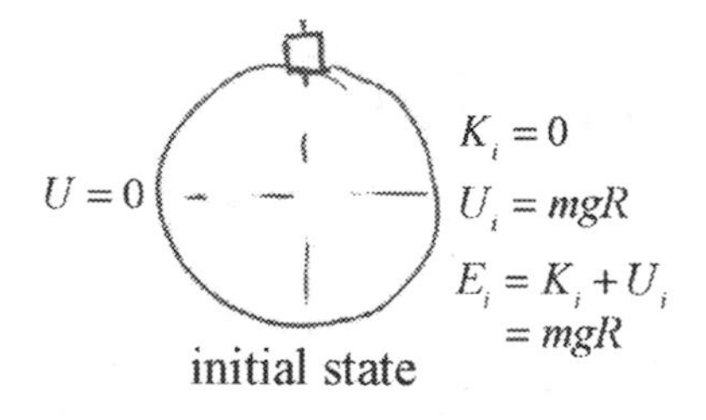

Figure 14.19 Initial state

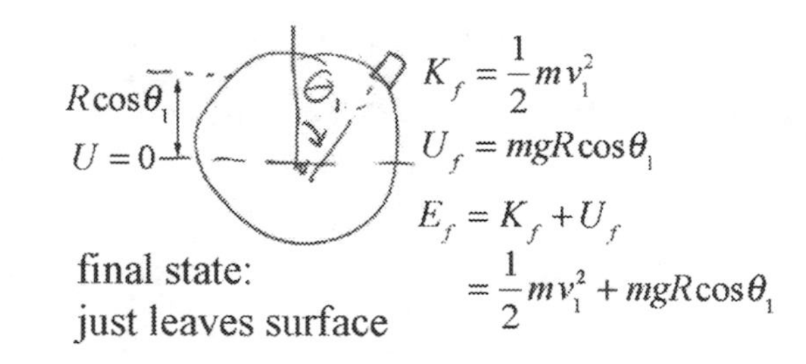

Figure 14.20 Final state

Choose for the final state the instant the object leaves the sphere (Figure 14.20). The final kinetic energy is $K_f = mv_1^2/2$; the object is in motion with speed v_1. The final potential energy is non-zero, $U_f = mgR\cos\theta_1$. The final mechanical energy is then

$$E_f = K_f + U_f = \frac{1}{2}mv_1^2 + mgR\cos\theta_1. \qquad (14.8.39)$$

Because we are assuming the contact surface is frictionless and neglecting air resistance, there is no non-conservative work. The change in mechanical energy is therefore zero,

$$0 = W_{\text{nc}} = \Delta E_m = E_f - E_i. \qquad (14.8.40)$$

Therefore

$$\frac{1}{2}mv_1^2 + mgR\cos\theta_1 = mgR. \qquad (14.8.41)$$

We now solve the constraint condition Eq. (14.8.37) into Eq. (14.8.41) yielding

$$\frac{1}{2} mgR\cos\theta_1 + mgR\cos\theta_1 = mgR. \qquad (14.8.42)$$

We can now solve for the angle at which the object just leaves the surface

$$\theta_1 = \cos^{-1}(2/3). \qquad (14.8.43)$$

We now substitute this result into Eq. (14.8.37) and solve for the speed

$$v_1 = \sqrt{2gR/3}. \qquad (14.8.44)$$

Chapter 15 Collision Theory

Chapter 15 Collision Theory

If it looks like a Higgs, swims like a Higgs, and quacks like a Higgs, then it probably is a Higgs[1]

Markus Klute

15.1 Introduction

When discussing conservation of momentum, we considered examples in which two objects collide and stick together, and either there are no external forces acting in some direction (or the collision was nearly instantaneous) so the component of the momentum of the system along that direction is constant. We shall now study collisions between objects in more detail. In particular we shall consider cases in which the objects do not stick together. The momentum along a certain direction may still be constant but the mechanical energy of the system may change. We will begin our analysis by considering two-particle collision. We introduce the concept of the relative velocity between two particles and show that it is independent of the choice of reference frame. We then show that the change in kinetic energy only depends on the change of the square of the relative velocity and therefore is also independent of the choice of reference frame. We will then study one- and two-dimensional collisions with zero change in potential energy. In particular we will characterize the types of collisions by the change in kinetic energy and analyze the possible outcomes of the collisions.

15.2 Reference Frames Relative and Velocities

We shall recall our definition of relative inertial reference frames. Let $\vec{\mathbf{R}}$ be the vector from the origin of frame S to the origin of reference frame S'. Denote the position vector of particle i with respect to the origin of reference frame S by $\vec{\mathbf{r}}_i$ and similarly, denote the position vector of particle i with respect to the origin of reference frame S' by $\vec{\mathbf{r}}'_i$ (Figure 15.1).

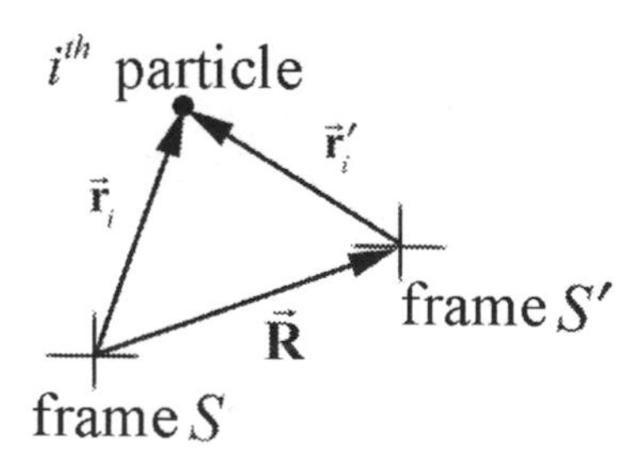

Figure 15.1 Position vector of ith particle in two reference frames.

---

[1] M.Klute, *Higgs Results from CMS,* The First Three Years of LHC 1st Workshop – Mainz, Ger., 2013.

The position vectors are related by

$$\vec{\mathbf{r}}_i = \vec{\mathbf{r}}'_i + \vec{\mathbf{R}} . \tag{15.2.1}$$

The relative velocity (call this the *boost velocity*) between the two reference frames is given by

$$\vec{\mathbf{V}} = \frac{d\vec{\mathbf{R}}}{dt} . \tag{15.2.2}$$

Assume the boost velocity between the two reference frames is constant. Then, the relative acceleration between the two reference frames is zero,

$$\vec{\mathbf{A}} = \frac{d\vec{\mathbf{V}}}{dt} = \vec{\mathbf{0}} . \tag{15.2.3}$$

When Eq. (15.2.3) is satisfied, the reference frames S and S' are called *relatively inertial reference frames.*

Suppose the ith particle in Figure 15.1 is moving; then observers in different reference frames will measure different velocities. Denote the velocity of ith particle in frame S by $\vec{\mathbf{v}}_i = d\vec{\mathbf{r}}_i / dt$, and the velocity of the same particle in frame S' by $\vec{\mathbf{v}}'_i = d\vec{\mathbf{r}}'_i / dt$. Taking derivative, the velocities of the particles in two different reference frames are related according to

$$\vec{\mathbf{v}}_i = \vec{\mathbf{v}}'_i + \vec{\mathbf{V}} . \tag{15.2.4}$$

15.2.1 Center of Mass Reference Frame

Let $\vec{\mathbf{R}}_{cm}$ be the vector from the origin of frame S to the center of mass of the system of particles, a point that we will choose as the origin of reference frame S_{cm}, called the ***center of mass reference frame***. Denote the position vector of particle i with respect to origin of reference frame S by $\vec{\mathbf{r}}_i$ and similarly, denote the position vector of particle i with respect to origin of reference frame S_{cm} by $\vec{\mathbf{r}}_{cm,i}$ (Figure 15.2).

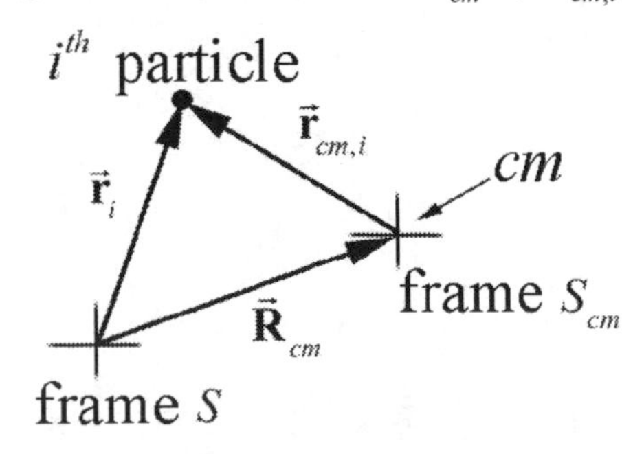

Figure 15.2 Position vector of ith particle in the center of mass reference frame.

The position vector of particle i in the center of mass frame is then given by

$$\vec{\mathbf{r}}_{cm,i} = \vec{\mathbf{r}}_i - \vec{\mathbf{R}}_{cm}. \tag{15.2.5}$$

The velocity of particle i in the center of mass reference frame is then given by

$$\vec{\mathbf{v}}_{cm,i} = \vec{\mathbf{v}}_i - \vec{\mathbf{V}}_{cm}. \tag{15.2.6}$$

There are many collision problems in which the center of mass reference frame is the most convenient reference frame to analyze the collision.

15.2.2 Relative Velocities

Consider two particles of masses m_1 and m_2 interacting via some force (Figure 15.3).

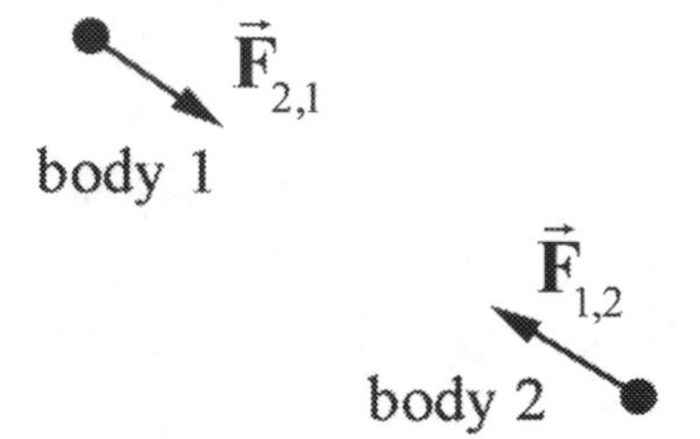

Figure 15.3 Two interacting particles

Choose a coordinate system (Figure 15.4) in which the position vector of body 1 is given by $\vec{\mathbf{r}}_1$ and the position vector of body 2 is given by $\vec{\mathbf{r}}_2$. The *relative position* of body 1 with respect to body 2 is given by $\vec{\mathbf{r}}_{1,2} = \vec{\mathbf{r}}_1 - \vec{\mathbf{r}}_2$.

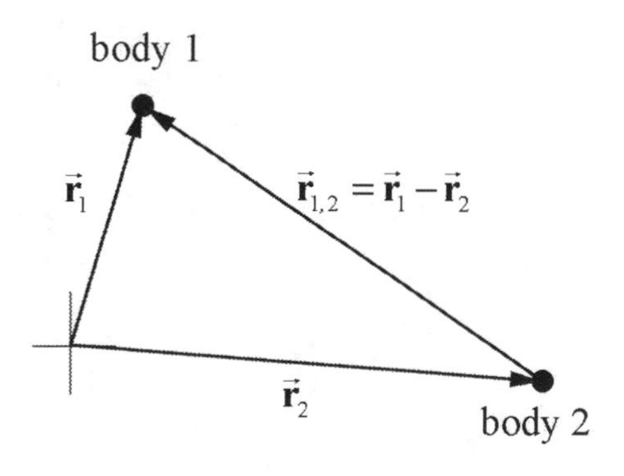

Figure 15.4 Coordinate system for two bodies.

During the course of the interaction, body 1 is displaced by $d\vec{\mathbf{r}}_1$ and body 2 is displaced by $d\vec{\mathbf{r}}_2$, so the ***relative displacement*** of the two bodies during the interaction is given by $d\vec{\mathbf{r}}_{1,2} = d\vec{\mathbf{r}}_1 - d\vec{\mathbf{r}}_2$. The ***relative velocity*** between the particles is

$$\vec{\mathbf{v}}_{1,2} = \frac{d\vec{\mathbf{r}}_{1,2}}{dt} = \frac{d\vec{\mathbf{r}}_1}{dt} - \frac{d\vec{\mathbf{r}}_2}{dt} = \vec{\mathbf{v}}_1 - \vec{\mathbf{v}}_2 . \qquad (15.2.7)$$

We shall now show that the relative velocity between the two particles is independent of the choice of reference frame providing that the reference frames are relatively inertial. The relative velocity $\vec{\mathbf{v}}'_{12}$ in reference frame S' can be determined from using Eq. (15.2.4) to express Eq. (15.2.7) in terms of the velocities in the reference frame S',

$$\vec{\mathbf{v}}_{1,2} = \vec{\mathbf{v}}_1 - \vec{\mathbf{v}}_2 = (\vec{\mathbf{v}}'_1 + \vec{\mathbf{V}}) - (\vec{\mathbf{v}}'_2 + \vec{\mathbf{V}}) = \vec{\mathbf{v}}'_1 - \vec{\mathbf{v}}'_2 = \vec{\mathbf{v}}'_{1,2} \qquad (15.2.8)$$

and is equal to the relative velocity in frame S.

For a two-particle interaction, the relative velocity between the two vectors is independent of the choice of relatively inertial reference frames.

We showed in Appendix 13A that when two particles of masses m_1 and m_2 interact, the change of kinetic energy between the final state B and the initial state A due to the interaction force is equal to

$$\Delta K = \frac{1}{2}\mu(v_B^2 - v_A^2), \qquad (15.2.9)$$

where $v_B \equiv \left|(\vec{\mathbf{v}}_{1,2})_B\right| = \left|(\vec{\mathbf{v}}_1)_B - (\vec{\mathbf{v}}_2)_B\right|$ denotes the relative speed in the state B and $v_A \equiv \left|(\vec{\mathbf{v}}_{1,2})_A\right| = \left|(\vec{\mathbf{v}}_1)_A - (\vec{\mathbf{v}}_2)_A\right|$ denotes the relative speed in state A, and $\mu = m_1 m_2 / (m_1 + m_2)$ is the reduced mass. (If the relative reference frames were non-inertial then Eq. (15.2.9) would not be valid in all frames.

Although kinetic energy is a reference dependent quantity, by expressing the change of kinetic energy in terms of the relative velocity, then

the change in kinetic energy is independent of the choice of relatively inertial reference frames.

15.3 Characterizing Collisions

In a collision, the ratio of the magnitudes of the initial and final relative velocities is called the coefficient of restitution and denoted by the symbol e,

$$e = \frac{v_B}{v_A} . \qquad (15.3.1)$$

If the magnitude of the relative velocity does not change during a collision, $e = 1$, then the change in kinetic energy is zero, (Eq. (15.2.9)). Collisions in which there is no change in kinetic energy are called ***elastic collisions***,

$$\Delta K = 0, \quad \textit{elastic collision}. \tag{15.3.2}$$

If the magnitude of the final relative velocity is less than the magnitude of the initial relative velocity, $e < 1$, then the change in kinetic energy is negative. Collisions in which the kinetic energy decreases are called ***inelastic collisions***,

$$\Delta K < 0, \quad \textit{inelastic collision}. \tag{15.3.3}$$

If the two objects stick together after the collision, then the relative final velocity is zero, $e = 0$. Such collisions are called ***totally inelastic***. The change in kinetic energy can be found from Eq. (15.2.9),

$$\Delta K = -\frac{1}{2}\mu\, v_A^2 = -\frac{1}{2}\frac{m_1 m_2}{m_1 + m_2} v_A^2, \quad \textit{totally inelastic collision}. \tag{15.3.4}$$

If the magnitude of the final relative velocity is greater than the magnitude of the initial relative velocity, $e > 1$, then the change in kinetic energy is positive. Collisions in which the kinetic energy increases are called ***superelastic collisions***,

$$\Delta K > 0, \quad \textit{superelastic collision}. \tag{15.3.5}$$

15.4 One-Dimensional Elastic Collision Between Two Objects

Consider a one-dimensional elastic collision between two objects moving in the x-direction. One object, with mass m_1 and initial x-component of the velocity $v_{1x,i}$, collides with an object of mass m_2 and initial x-component of the velocity $v_{2x,i}$. The scalar components $v_{1x,i}$ and $v_{1x,i}$ can be positive, negative or zero. No forces other than the interaction force between the objects act during the collision. After the collision, the final x-component of the velocities are $v_{1x,f}$ and $v_{2x,f}$. We call this reference frame the "laboratory reference frame".

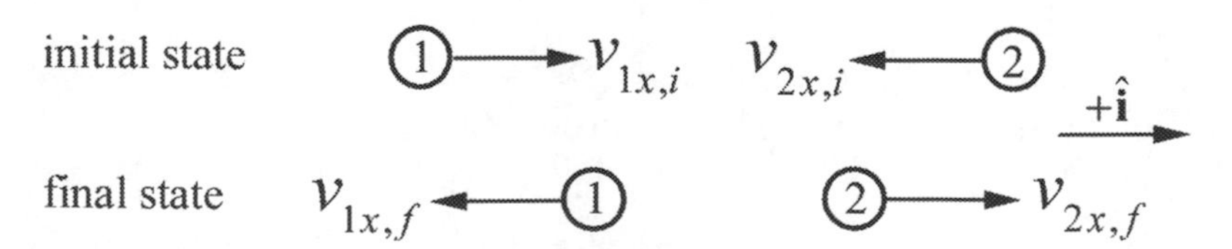

Figure 15.5 One-dimensional elastic collision, laboratory reference frame

For the collision depicted in Figure 15.5, $v_{1x,i} > 0$, $v_{2x,i} < 0$, $v_{1x,f} < 0$, and $v_{2x,f} > 0$. Because there are no external forces in the x-direction, momentum is constant in the x-direction. Equating the momentum components before and after the collision gives the relation

$$m_1 v_{1x,i} + m_2 v_{2x,0} = m_1 v_{1x,f} + m_2 v_{2x,f}. \qquad (15.3.6)$$

Because the collision is elastic, kinetic energy is constant. Equating the kinetic energy before and after the collision gives the relation

$$\frac{1}{2} m_1 v_{1x,i}^2 + \frac{1}{2} m_2 v_{2x,i}^2 = \frac{1}{2} m_1 v_{1x,f}^2 + \frac{1}{2} m_2 v_{2x,f}^2 \qquad (15.3.7)$$

Rewrite these Eqs. (15.3.6) and (15.3.7) as

$$m_1 (v_{1x,i} - v_{1x,f}) = m_2 (v_{2x,f} - v_{2x,i}) \qquad (15.3.8)$$

$$m_1 (v_{1x,i}^2 - v_{1x,f}^2) = m_2 (v_{2x,f}^2 - v_{2x,i}^2). \qquad (15.3.9)$$

Eq. (15.3.9) can be written as

$$m_1 (v_{1x,i} - v_{1x,f})(v_{1x,i} + v_{1x,f}) = m_2 (v_{2x,f} - v_{2x,i})(v_{2x,f} + v_{2x,i}). \qquad (15.3.10)$$

Divide Eq. (15.3.9) by Eq. (15.3.8), yielding

$$v_{1x,i} + v_{1x,f} = v_{2x,i} + v_{2x,f}. \qquad (15.3.11)$$

Eq. (15.3.11) may be rewritten as

$$v_{1x,i} - v_{2x,i} = v_{2x,f} - v_{1x,f}. \qquad (15.3.12)$$

Recall that the relative velocity between the two objects in state A is defined to be

$$\vec{\mathbf{v}}_A^{\text{rel}} = \vec{\mathbf{v}}_{1,A} - \vec{\mathbf{v}}_{2,A}. \qquad (15.3.13)$$

where we used the superscript "rel" to remind ourselves that the velocity is a relative velocity. Thus $v_{x,i}^{\text{rel}} = v_{1x,i} - v_{2x,i}$ is the initial x-component of the relative velocity, and $v_{x,f}^{\text{rel}} = v_{1x,f} - v_{2x,f}$ is the final x-component of the relative velocity. Therefore Eq. (15.3.12)
states that during the interaction the initial x-component of the relative velocity is equal to the negative of the final x-component of the relative velocity

$$v_{x,i}^{\text{rel}} = -v_{x,f}^{\text{rel}}. \tag{15.3.14}$$

Consequently the initial and final relative speeds are equal. We can now solve for the final x-component of the velocities, $v_{1x,f}$ and $v_{2x,f}$, as follows. Eq. (15.3.12) may be rewritten as

$$v_{2x,f} = v_{1x,f} + v_{1x,i} - v_{2x,i}. \tag{15.3.15}$$

Now substitute Eq. (15.3.15) into Eq. (15.3.6) yielding

$$m_1 v_{1x,i} + m_2 v_{2x,i} = m_1 v_{1x,f} + m_2 (v_{1x,f} + v_{1x,i} - v_{2x,i}). \tag{15.3.16}$$

Solving Eq. (15.3.16) for $v_{1x,f}$ involves some algebra and yields

$$v_{1x,f} = \frac{m_1 - m_2}{m_1 + m_2} v_{1x,i} + \frac{2m_2}{m_1 + m_2} v_{2x,i}. \tag{15.3.17}$$

To find $v_{2x,f}$, rewrite Eq. (15.3.12) as

$$v_{1x,f} = v_{2x,f} - v_{1x,i} + v_{2x,i}. \tag{15.3.18}$$

Now substitute Eq. (15.3.18) into Eq. (15.3.6) yielding

$$m_1 v_{1x,i} + m_2 v_{2x,i} = m_1 (v_{2x,f} - v_{1x,i} + v_{2x,i}) v_{1x,f} + m_2 v_{2x,f}. \tag{15.3.19}$$

We can solve Eq. (15.3.19) for $v_{2x,f}$ and determine that

$$v_{2x,f} = v_{2x,i} \frac{m_2 - m_1}{m_2 + m_1} + v_{1x,i} \frac{2m_1}{m_2 + m_1}. \tag{15.3.20}$$

Consider what happens in the limits $m_1 >> m_2$ in Eq. (15.3.17). Then

$$v_{1x,f} \to v_{1x,i} + \frac{2}{m_1} m_2 v_{2x,i}; \tag{15.3.21}$$

the more massive object's velocity component is only slightly changed by an amount proportional to the less massive object's x-component of momentum. Similarly, the less massive object's final velocity approaches

$$v_{2x,f} \to -v_{2x,i} + 2v_{1x,i} = v_{1x,i} + v_{1x,i} - v_{2x,i}. \tag{15.3.22}$$

We can rewrite this as

$$v_{2x,f} - v_{1x,i} = v_{1x,i} - v_{2x,i} = v_{x,i}^{\text{rel}}. \tag{15.3.23}$$

i.e. the less massive object "rebounds" with the same speed relative to the more massive object which barely changed its speed.

If the objects are identical, or have the same mass, Eqs. (15.3.17) and (15.3.20) become

$$v_{1x,f} = v_{2x,i}, \quad v_{2x,f} = v_{1x,i}; \tag{15.3.24}$$

the objects have exchanged x-components of velocities, and unless we could somehow distinguish the objects, we might not be able to tell if there was a collision at all.

15.4.1 One-Dimensional Collision Between Two Objects – Center of Mass Reference Frame

We analyzed the one-dimensional elastic collision (Figure 15.5) in Section 15.4 in the laboratory reference frame. Now let's view the collision from the center of mass (CM) frame. The x-component of velocity of the center of mass is

$$v_{x,\text{cm}} = \frac{m_1 v_{1x,i} + m_2 v_{2x,i}}{m_1 + m_2}. \tag{15.3.25}$$

With respect to the center of mass, the x-components of the velocities of the objects are

$$\begin{aligned} v'_{1x,i} &= v_{1x,i} - v_{x,\text{cm}} = (v_{1x,i} - v_{2x,i})\frac{m_2}{m_1 + m_2} \\ v'_{2x,i} &= v_{2x,i} - v_{x,\text{cm}} = (v_{2x,i} - v_{1x,i})\frac{m_1}{m_1 + m_2}. \end{aligned} \tag{15.3.26}$$

In the CM frame the momentum of the system is zero before the collision and hence the momentum of the system is zero after the collision. For an elastic collision, the only way for both momentum and kinetic energy to be the same before and after the collision is either the objects have the same velocity (a miss) or to reverse the direction of the velocities as shown in Figure 15.6.

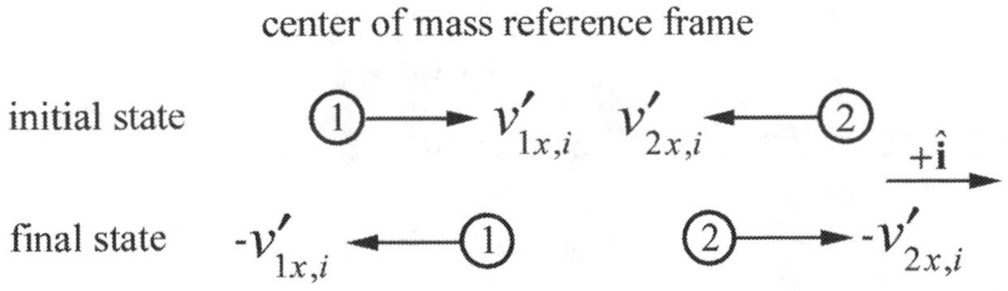

Figure 15.6 One-dimensional elastic collision in center of mass reference frame

In the CM frame, the final x-components of the velocities are

$$\begin{aligned} v'_{1x,f} &= -v'_{1x,i} = (v_{2x,i} - v_{1x,i})\frac{m_2}{m_1 + m_2} \\ v'_{2x,f} &= -v'_{2x,i} = (v_{2x,i} - v_{1x,i})\frac{m_1}{m_1 + m_2}. \end{aligned} \tag{15.3.27}$$

The final x-components of the velocities in the "laboratory frame" are then given by

$$\begin{aligned} v_{1x,f} &= v'_{1x,f} + v_{x,\text{cm}} \\ &= (v_{2x,i} - v_{1x,i})\frac{m_2}{m_1 + m_2} + \frac{m_1 v_{1x,i} + m_2 v_{2x,i}}{m_1 + m_2} \\ &= v_{1x,i}\frac{m_1 - m_2}{m_1 + m_2} + v_{2x,i}\frac{2m_2}{m_1 + m_2} \end{aligned} \tag{15.3.28}$$

as in Eq. (15.3.17) and a similar calculation reproduces Eq. (15.3.20).

15.5 Worked Examples

Example 15.1 Elastic One-Dimensional Collision Between Two Objects

Consider the elastic collision of two carts along a track; the incident cart 1 has mass m_1 and moves with initial speed $v_{1,i}$. The target cart has mass $m_2 = 2m_1$ and is initially at rest, $v_{2,i} = 0$. Immediately after the collision, the incident cart has final speed $v_{1,f}$ and the target cart has final speed $v_{2,f}$. Calculate the final x-component of the velocities of the carts as a function of the initial speed $v_{1,i}$.

Solution Draw a "momentum flow" diagram for the objects before (initial state) and after (final state) the collision (Figure 15.7).

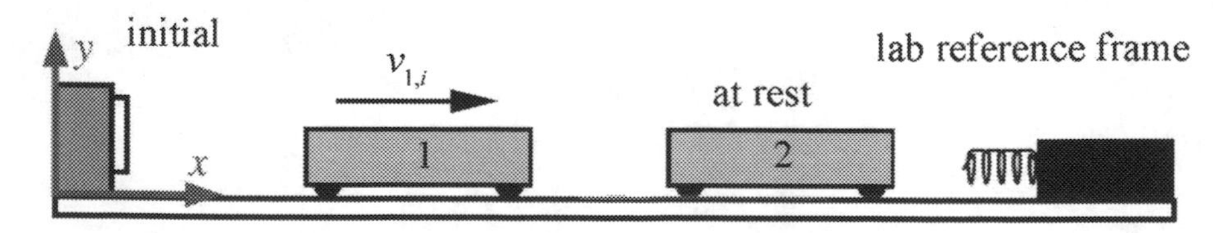

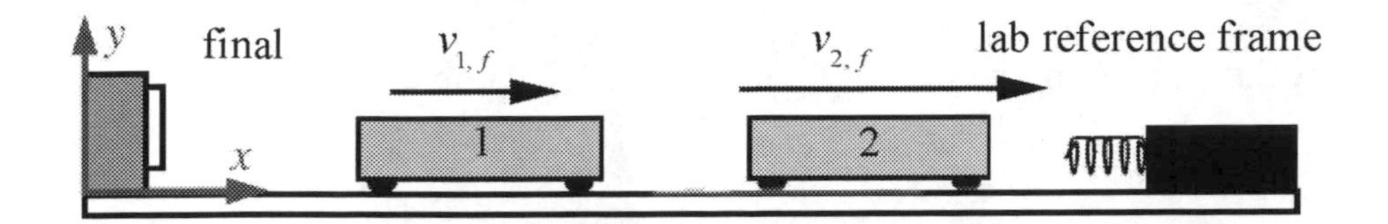

Figure 15.7 Momentum flow diagram for elastic one-dimensional collision

We can immediately use our results above with $m_2 = 2m_1$ and $v_{2,i} = 0$. The final x-component of velocity of cart 1 is given by Eq. (15.3.17), where we use $v_{1x,i} = v_{1,i}$

$$v_{1x,f} = -\frac{1}{3}v_{1,i}. \tag{15.3.29}$$

The final x-component of velocity of cart 2 is given by Eq. (15.3.20)

$$v_{2x,f} = \frac{2}{3}v_{1,i}. \tag{15.3.30}$$

Example 15.2 The Dissipation of Kinetic Energy in a Completely Inelastic Collision Between Two Objects

An incident object of mass m_1 and initial speed $v_{1,i}$ collides completely inelastically with an object of mass m_2 that is initially at rest. There are no external forces acting on the objects in the direction of the collision. Find $\Delta K / K_{\text{initial}} = (K_{\text{final}} - K_{\text{initial}}) / K_{\text{initial}}$.

Solution: In the absence of any net force on the system consisting of the two objects, the momentum after the collision will be the same as before the collision. After the collision the objects will move in the direction of the initial velocity of the incident object with a common speed v_f found from applying the momentum condition

$$\begin{aligned} m_1 v_{1,i} &= (m_1 + m_2)v_f \Rightarrow \\ v_f &= \frac{m_1}{m_1 + m_2}v_{1,i}. \end{aligned} \tag{15.3.31}$$

The initial relative speed is $v_i^{\text{rel}} = v_{1,i}$. The final relative velocity is zero because the objects stick together so using Eq. (15.2.9), the change in kinetic energy is

$$\Delta K = -\frac{1}{2}\mu(v_i^{rel})^2 = -\frac{1}{2}\frac{m_1 m_2}{m_1 + m_2}v_{1,i}^2. \tag{15.3.32}$$

The ratio of the change in kinetic energy to the initial kinetic energy is then

$$\Delta K / K_{\text{initial}} = -\frac{m_2}{m_1 + m_2}. \qquad (15.3.33)$$

As a check, we can calculate the change in kinetic energy via

$$\begin{aligned}
\Delta K = (K_f - K_i) &= \frac{1}{2}(m_1 + m_2)v_f^2 - \frac{1}{2}v_{1,i}^2 \\
&= \frac{1}{2}(m_1 + m_2)\left(\frac{m_1}{m_1 + m_2}\right)^2 v_{1,i}^2 - \frac{1}{2}v_{1,i}^2 \\
&= \left(\frac{m_1}{m_1 + m_2} - 1\right)\left(\frac{1}{2}m_1 v_{1,i}^2\right) = -\frac{1}{2}\frac{m_1 m_2}{m_1 + m_2}v_{1,i}^2.
\end{aligned} \qquad (15.3.34)$$

in agreement with Eq. (15.3.32).

Example 15.3 Elastic Two-Dimensional Collision

Object 1 with mass m_1 is initially moving with a speed $v_{1,i} = 3.0\,\text{m}\cdot\text{s}^{-1}$ and collides elastically with object 2 that has the same mass, $m_2 = m_1$, and is initially at rest. After the collision, object 1 moves with an unknown speed $v_{1,f}$ at an angle $\theta_{1,f} = 30^\circ$ with respect to its initial direction of motion and object 2 moves with an unknown speed $v_{2,f}$, at an unknown angle $\theta_{2,f}$ (as shown in the Figure 15.8). Find the final speeds of each of the objects and the angle $\theta_{2,f}$.

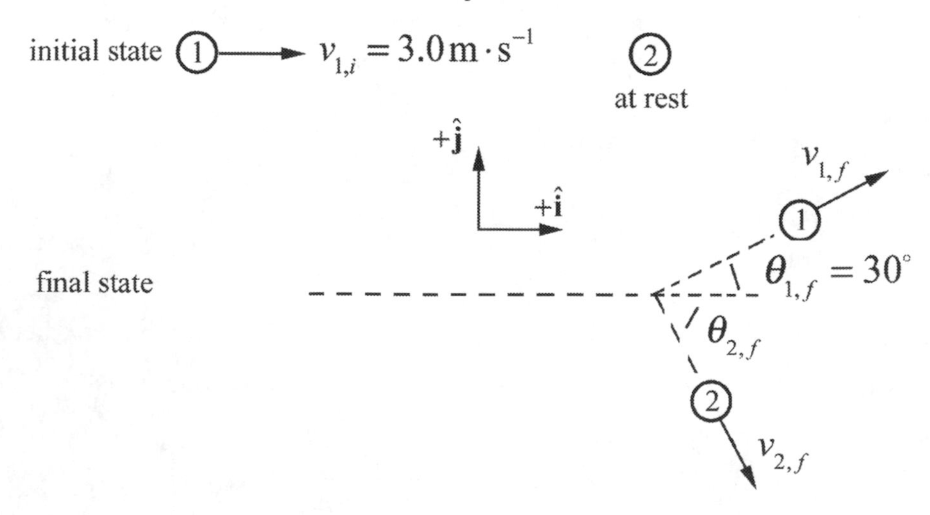

Figure 15.8 Momentum flow diagram for two-dimensional elastic collision

Solution: The components of the total momentum $\vec{\mathbf{p}}_i^{\text{sys}} = m_1\vec{\mathbf{v}}_{1,i} + m_2\vec{\mathbf{v}}_{2,i}$ in the initial state are given by

$$\begin{aligned} p_{x,i}^{\text{sys}} &= m_1 v_{1,i} \\ p_{y,i}^{\text{sys}} &= 0. \end{aligned} \tag{15.3.35}$$

The components of the momentum $\vec{\mathbf{p}}_f^{\text{sys}} = m_1\vec{\mathbf{v}}_{1,f} + m_2\vec{\mathbf{v}}_{2,f}$ in the final state are given by

$$\begin{aligned} p_{x,f}^{\text{sys}} &= m_1\, v_{1,f} \cos\theta_{1,f} + m_1\, v_{2,f} \cos\theta_{2,f} \\ p_{y,f}^{\text{sys}} &= m_1\, v_{1,f} \sin\theta_{1,f} - m_1\, v_{2,f} \sin\theta_{2,f}. \end{aligned} \tag{15.3.36}$$

There are no any external forces acting on the system, so each component of the total momentum remains constant during the collision,

$$p_{x,i}^{\text{sys}} = p_{x,f}^{\text{sys}} \tag{15.3.37}$$

$$p_{y,i}^{\text{sys}} = p_{y,f}^{\text{sys}}. \tag{15.3.38}$$

Eqs. (15.3.37) and (15.3.37) become

$$\begin{aligned} m_1\, v_{1,i} &= m_1\, v_{1,f} \cos\theta_{1,f} + m_1\, v_{2,f} \cos\theta_{2,f} \\ 0 &= m_1\, v_{1,f} \sin\theta_{1,f} - m_1\, v_{2,f} \sin\theta_{2,f}. \end{aligned} \tag{15.3.39}$$

The collision is elastic and therefore the system kinetic energy of is constant

$$K_i^{\text{sys}} = K_f^{\text{sys}}. \tag{15.3.40}$$

Using the given information, Eq. (15.3.40) becomes

$$\frac{1}{2} m_1 v_{1,i}^2 = \frac{1}{2} m_1 v_{1,f}^2 + \frac{1}{2} m_1 v_{2,f}^2. \tag{15.3.41}$$

We have three equations: two momentum equations and one energy equation, with three unknown quantities, $v_{1,f}$, $v_{2,f}$ and $\theta_{2,f}$ because we are given $v_{1,i} = 3.0\,\text{m}\cdot\text{s}^{-1}$ and $\theta_{1,f} = 30^o$. We first rewrite the expressions in Eq. (15.3.39), canceling the factor of m_1, as

$$\begin{aligned} v_{2,f} \cos\theta_{2,f} &= v_{1,i} - v_{1,f} \cos\theta_{1,f} \\ v_{2,f} \sin\theta_{2,f} &= v_{1,f} \sin\theta_{1,f}. \end{aligned} \tag{15.3.42}$$

We square each expressions in Eq. (15.3.42), yielding

$$v_{2,f}^2(\cos\theta_{2,f})^2 = v_{1,f}^2 - 2v_{1,i}v_{1,f}\cos\theta_{1,f} + v_{1,f}^2(\cos\theta_{1,f})^2 \quad (15.3.43)$$

$$v_{2,f}^2(\sin\theta_{2,f})^2 = v_{1,f}^2(\sin\theta_{1,f})^2 \quad (15.3.44)$$

We now add together Eqs. (15.3.43) and (15.3.44) yielding

$$v_{2,f}^2(\cos^2\theta_{2,f} + \sin^2\theta_{2,f}) = v_{1,f}^2 - 2v_{1,i}v_{1,f}\cos\theta_{1,f} + v_{1,f}^2(\cos\theta_{1,f} + \sin^2\theta_{1,f}). \quad (15.3.45)$$

We can use identity $\cos^2\theta + \sin^2\theta = 1$ to simplify Eq. (15.3.45), yielding

$$v_{2,f}^2 = v_{1,i}^2 - 2v_{1,i}v_{1,f}\cos\theta_{1,f} + v_{1,f}^2. \quad (15.3.46)$$

Substituting Eq. (15.3.46) into Eq. (15.3.41) yields

$$\frac{1}{2}m_1v_{1,i}^2 = \frac{1}{2}m_1v_{1,f}^2 + \frac{1}{2}m_1(v_{1,i}^2 - 2v_{1,i}v_{1,f}\cos\theta_{1,f} + v_{1,f}^2). \quad (15.3.47)$$

Eq. (15.3.47) simplifies to

$$0 = 2v_{1,f}^2 - 2v_{1,i}v_{1,f}\cos\theta_{1,f}, \quad (15.3.48)$$

which may be solved for the final speed of object 1,

$$v_{1,f} = v_{1,i}\cos\theta_{1,f} = (3.0\,\text{m}\cdot\text{s}^{-1})\cos 30^\circ = 2.6\,\text{m}\cdot\text{s}^{-1}. \quad (15.3.49)$$

Divide the expressions in Eq. (15.3.42), yielding

$$\frac{v_{2,f}\sin\theta_{2,f}}{v_{2,f}\cos\theta_{2,f}} = \frac{v_{1,f}\sin\theta_{1,f}}{v_{1,i} - v_{1,f}\cos\theta_{1,f}}. \quad (15.3.50)$$

Eq. (15.3.50) simplifies to

$$\tan\theta_{2,f} = \frac{v_{1,f}\sin\theta_{1,f}}{v_{1,i} - v_{1,f}\cos\theta_{1,f}}. \quad (15.3.51)$$

Thus object 2 moves at an angle

$$\theta_{2,f} = \tan^{-1}\left(\frac{v_{1,f}\sin\theta_{1,f}}{v_{1,i} - v_{1,f}\cos\theta_{1,f}}\right)$$

$$\theta_{2,f} = \tan^{-1}\left(\frac{(2.6\,\text{m}\cdot\text{s}^{-1})\sin 30^\circ}{3.0\,\text{m}\cdot\text{s}^{-1} - (2.6\,\text{m}\cdot\text{s}^{-1})\cos 30^\circ}\right) \tag{15.3.52}$$

$$= 60^\circ.$$

The above results for $v_{1,f}$ and $\theta_{2,f}$ may be substituted into either of the expressions in Eq. (15.3.42), or Eq. (15.3.41), to find $v_{2,f} = 1.5\text{m}\cdot\text{s}^{-1}$.

Before going on, the fact that $\theta_{1,f} + \theta_{2,f} = 90^\circ$, that is, the objects move away from the collision point at right angles, is not a coincidence. A vector derivation is presented below. We can see this result algebraically from the above result. Using the result of Eq. (15.3.49), $v_{1,f} = v_{1,i}\cos\theta_{1,f}$, in Eq. (15.3.51) yields

$$\tan\theta_{2,f} = \frac{\cos\theta_{1,f}\sin\theta_{1,f}}{1 - \cos\theta_{1,f}^2} = \cot\theta_{1,f}; \tag{15.3.53}$$

the angles $\theta_{1,f}$ and $\theta_{2,f}$ are complements. Eq. (15.3.48) also has the solution $v_{2,f} = 0$, which would correspond to the incident particle missing the target completely.

Example 15.4 Equal Mass Particles in a Two-Dimensional Elastic Collision Emerge at Right Angles

Show that the equal mass particles emerge from the collision at right angles by making explicit use of the fact that momentum is a vector quantity.

Solution: There are no external forces acting on the two objects during the collision (the collision forces are all internal), therefore momentum is constant

$$\vec{\mathbf{p}}_i^{\text{sys}} = \vec{\mathbf{p}}_f^{\text{sys}}, \tag{15.3.54}$$

which becomes

$$m_1\vec{\mathbf{v}}_{1,i} = m_1\vec{\mathbf{v}}_{1,f} + m_1\vec{\mathbf{v}}_{2,f}. \tag{15.3.55}$$

Eq. (15.3.55) simplifies to

$$\vec{\mathbf{v}}_{1,i} = \vec{\mathbf{v}}_{1,f} + \vec{\mathbf{v}}_{2,f}. \tag{15.3.56}$$

Recall the vector identity that the square of the speed is given by the dot product $\vec{\mathbf{v}}\cdot\vec{\mathbf{v}} = v^2$. With this identity in mind, we take the dot product of each side of Eq. (15.3.56) with itself,

$$\vec{\mathbf{v}}_{1,i} \cdot \vec{\mathbf{v}}_{1,i} = (\vec{\mathbf{v}}_{1,f} + \vec{\mathbf{v}}_{2,f}) \cdot (\vec{\mathbf{v}}_{1,f} + \vec{\mathbf{v}}_{2,f}) \\ = \vec{\mathbf{v}}_{1,f} \cdot \vec{\mathbf{v}}_{1,f} + 2\vec{\mathbf{v}}_{1,f} \cdot \vec{\mathbf{v}}_{2,f} + \vec{\mathbf{v}}_{2,f} \cdot \vec{\mathbf{v}}_{2,f}. \qquad (15.3.57)$$

This becomes

$$v_{1,i}^2 = v_{1,f}^2 + 2\vec{\mathbf{v}}_{1,f} \cdot \vec{\mathbf{v}}_{2,f} + v_{2,f}^2. \qquad (15.3.58)$$

Recall that kinetic energy is the same before and after an elastic collision, and the masses of the two objects are equal, so Eq. (15.3.41) simplifies to

$$v_{1,i}^2 = v_{1,f}^2 + v_{2,f}^2. \qquad (15.3.59)$$

Comparing Eq. (15.3.58) with Eq. (15.3.59), we see that

$$\vec{\mathbf{v}}_{1,f} \cdot \vec{\mathbf{v}}_{2,f} = 0 \ . \qquad (15.3.60)$$

The dot product of two nonzero vectors is zero when the two vectors are at right angles to each other justifying our claim that the collision particles emerge at right angles to each other.

Example 15.5 Bouncing Superballs

Two superballs are dropped from a height h_i above the ground, one on top of the other. Ball 1 is on top and has mass m_1, and ball 2 is underneath and has mass m_2 with $m_2 >> m_1$. Assume that there is no loss of kinetic energy during all collisions. Ball 2 first collides with the ground and rebounds. Then, as ball 2 starts to move upward, it collides with the ball 1 which is still moving downwards. How high will ball 1 rebound in the air? Hint: consider this collision as seen by an observer moving upward with the same speed as the ball 2 has after it collides with ground (Figure 15.9). What speed does ball 1 have in this reference frame after it collides with the ball 2?

Solution: The system consists of the two balls and the earth. There are five special states for this motion shown in the figure above.

Initial state: the balls are released from rest at a height h_i above the ground.

State A: the balls just reach the ground with speed $v_{1,A} = v_{2,A} = v_A = \sqrt{2gh_i}$.

State B: ball 2 has collided with the ground and reversed direction with the same speed, $v_{2,B} = v_A$, and ball 1 is still moving down with speed $v_{1,B} = v_A$.

State C: because we are assuming that $m_2 >> m_1$, ball 2 does not change speed as a result of the collision so it is still moving upward with speed $v_{2,C} = v_A$. As a result of the collision, ball 1 moves upward with speed $v_{1,C}$.

Final State: ball 1 reaches a maximum height $h_f = v_b^2 / 2g$ above the ground.

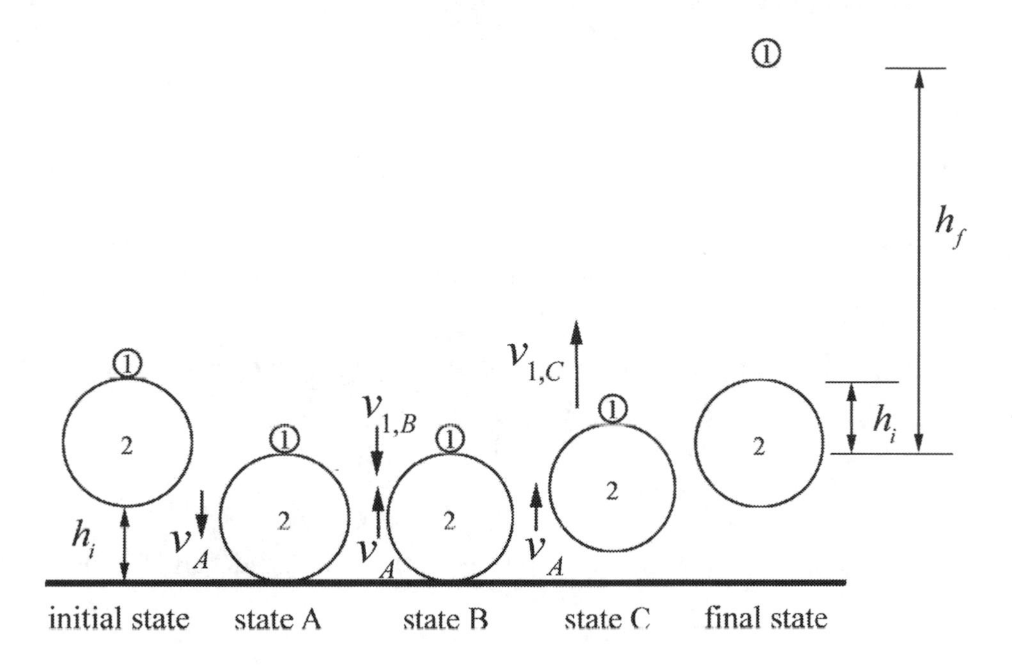

Figure 15.9 States in superball collisions

Choice of Reference Frame: For states B and C, the collision is best analyzed from the reference frame of an observer moving upward with speed v_A, the speed of ball 2 just after it rebounded from the ground. In this frame ball 1 is moving downward with a speed $v'_{1,B}$ that is twice the speed seen by an observer at rest on the ground (lab reference frame).

$$v'_{1,B} = 2v_A \tag{15.3.61}$$

The mass of ball 2 is much larger than the mass of ball 1, $m_2 >> m_1$. This enables us to consider the collision (between states B and C) to be equivalent to ball 1 bouncing off a hard wall, while ball 2 experiences virtually no recoil. Hence ball 2 remains at rest, $v'_{2,C} = 0$, in the reference frame moving upwards with speed v_A with respect to observer at rest on ground. Before the collision, ball 1 has speed $v'_{1,B} = 2v_a$. Because we assumed the collision is elastic there is no loss of kinetic energy during the collision, therefore ball 1 changes direction but maintains the same speed,

$$v'_{1,C} = 2v_A. \tag{15.3.62}$$

However, according to an observer at rest on the ground, after the collision ball 1 is moving upwards with speed

$$v_{1,C} = 2v_A + v_A = 3v_A . \tag{15.3.63}$$

While rebounding, the mechanical energy of the smaller superball is constant hence between state C and final state,

$$\Delta K + \Delta U = 0 . \tag{15.3.64}$$

The change in kinetic energy is

$$\Delta K = -\frac{1}{2} m_1 (3v_A)^2 . \tag{15.3.65}$$

The change in potential energy is

$$\Delta U = m_1 g h_f . \tag{15.3.66}$$

So the condition that mechanical energy is constant (Eq. (15.3.64)) is now

$$-\frac{1}{2} m_1 (3v_A)^2 + m_1 g h_f = 0 . \tag{15.3.67}$$

Recall that we can also use the fact that the mechanical energy doesn't change between the initial state and state A. Therefore

$$m_1 g h_i = \frac{1}{2} m_1 (v_A)^2 . \tag{15.3.68}$$

Substitute the expression for the kinetic energy in Eq. (15.3.68) into Eq. (15.3.67) yielding

$$m_1 g h_f = 9 m_1 g h_i . \tag{15.3.69}$$

Thus ball 1 reaches a maximum height

$$h_f = 9 h_i . \tag{15.3.70}$$

Chapter 16 Two Dimensional Rotational Kinematics

Chapter 16 Two Dimensional Rotational Kinematics

Most galaxies exhibit rising rotational velocities at the largest measured velocity; only for the very largest galaxies are the rotation curves flat. Thus the smallest SC's (i.e. lowest luminosity) exhibit the same lack of Keplerian velocity decrease at large R as do the high-luminosity spirals. The form for the rotation curves implies that the mass is not centrally condensed, but that significant mass is located at large R. The integral mass is increasing at least as fast as R. The mass is not converging to a limiting mass at the edge of the optical image. The conclusion is inescapable than non-luminous matter exists beyond the optical galaxy.[1]

Vera Rubin

16.1 Introduction

The physical objects that we encounter in the world consist of collections of atoms that are bound together to form systems of particles. When forces are applied, the shape of the body may be stretched or compressed like a spring, or sheared like jello. In some systems the constituent particles are very loosely bound to each other as in fluids and gasses, and the distances between the constituent particles will vary. We shall begin by restricting ourselves to an ideal category of objects, rigid bodies, which do not stretch, compress, or shear.

A body is called a ***rigid body*** if the distance between any two points in the body does not change in time. Rigid bodies, unlike point masses, can have forces applied at different points in the body. Let's start by considering the simplest example of rigid body motion, rotation about a fixed axis.

16.2 Fixed Axis Rotation: Rotational Kinematics

16.2.1 Fixed Axis Rotation

A simple example of rotation about a fixed axis is the motion of a compact disc in a CD player, which is driven by a motor inside the player. In a simplified model of this motion, the motor produces angular acceleration, causing the disc to spin. As the disc is set in motion, resistive forces oppose the motion until the disc no longer has any angular acceleration, and the disc now spins at a constant angular velocity. Throughout this process, the CD rotates about an axis passing through the center of the disc, and is perpendicular to the plane of the disc (see Figure 16.1). This type of motion is called ***fixed-axis rotation***.

[1]V.C. Rubin, W.K. Jr. Ford, N Thonnard, *Rotational properties of 21 SC galaxies with a large range of luminosities and radii, from NGC 4605 /R = 4kpc/ to UGC 2885 /R = 122 kpc/*, Astrophysical Journal, Part 1, vol. 238, June 1, 1980, p. 471-487.

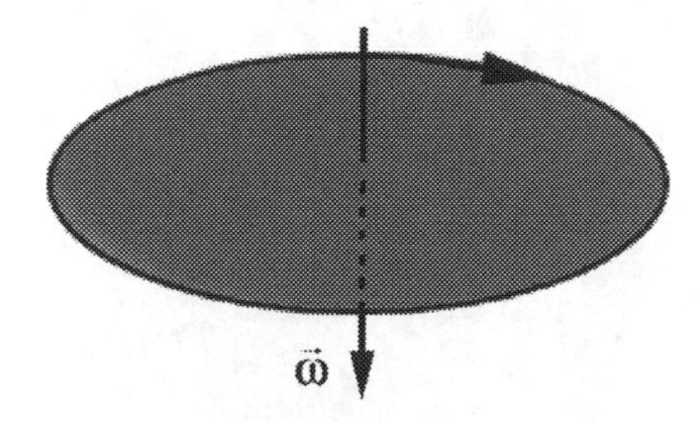

Figure 16.1 Rotation of a compact disc about a fixed axis.

When we ride a bicycle forward, the wheels rotate about an axis passing through the center of each wheel and perpendicular to the plane of the wheel (Figure 16.2). As long as the bicycle does not turn, this axis keeps pointing in the same direction. This motion is more complicated than our spinning CD because the wheel is both moving (translating) with some center of mass velocity, $\vec{\mathbf{v}}_{cm}$, and rotating with an angular speed ω.

Figure 16.2 Fixed axis rotation and center of mass translation for a bicycle wheel.

When we turn the bicycle's handlebars, we change the bike's trajectory and the axis of rotation of each wheel changes direction. Other examples of non-fixed axis rotation are the motion of a spinning top, or a gyroscope, or even the change in the direction of the earth's rotation axis. This type of motion is much harder to analyze, so we will restrict ourselves in this chapter to considering fixed axis rotation, with or without translation.

16.2.2 Angular Velocity and Angular Acceleration

For a rigid body undergoing fixed-axis rotation, we can divide the body up into small volume elements with mass Δm_i. Each of these volume elements is moving in a circle of radius r_i about the axis of rotation (Figure 16.3).

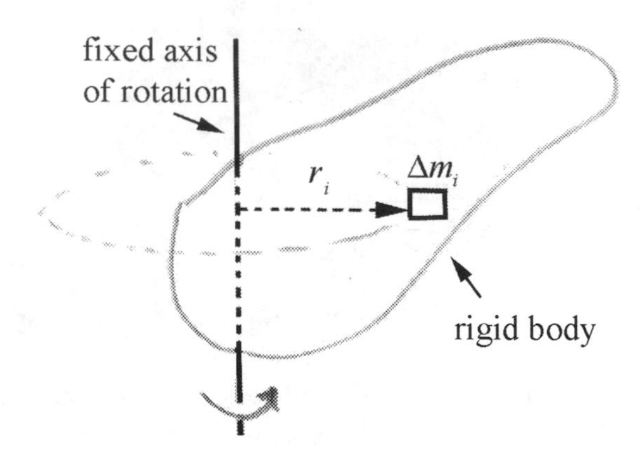

Figure 16.3 Coordinate system for fixed-axis rotation.

We will adopt the notation implied in Figure 16.3, and denote the vector from the axis to the point where the mass element is located as $\vec{\mathbf{r}}_i$, with magnitude $r_i = |\vec{\mathbf{r}}_i|$. Suppose the fixed axis of rotation is the z-axis. Introduce a right-handed coordinate system for an angle θ in the plane of rotation and the choice of the positive z-direction perpendicular to that plane of rotation. Recall our definition of the angular velocity vector. The angular velocity vector is directed along the z-axis with z-component equal to the time derivative of the angle θ,

$$\vec{\boldsymbol{\omega}} = \frac{d\theta}{dt}\hat{\mathbf{k}} = \omega_z \hat{\mathbf{k}}. \tag{16.1.1}$$

The angular velocity vector for the mass element undergoing fixed axis rotation with $\omega_z > 0$ is shown in Figure 16.4. Because the body is rigid, all the mass elements will have the same angular velocity $\vec{\boldsymbol{\omega}}$ and hence the same angular acceleration $\vec{\alpha}$. If the bodies did not have the same angular velocity, the mass elements would "catch up to" or "pass" each other, precluded by the rigid-body assumption.

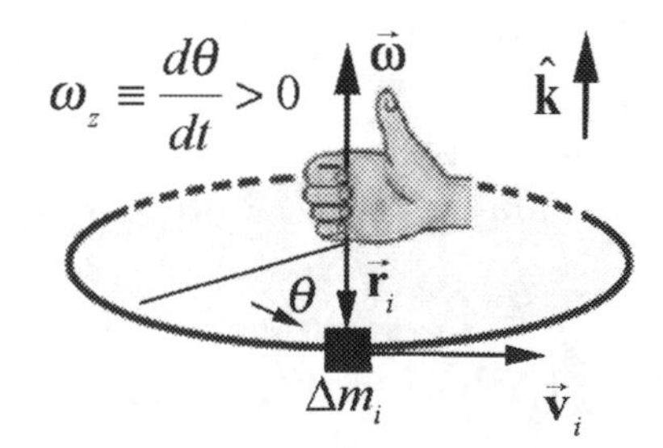

Figure 16.4 Angular velocity vector for a mass element for fixed axis rotation

In a similar fashion, all points in the rigid body have the same angular acceleration,

$$\vec{\boldsymbol{\alpha}} = \frac{d^2\theta}{dt^2}\hat{\mathbf{k}} = \alpha_z \hat{\mathbf{k}}. \tag{16.1.2}$$

The angular acceleration vector is shown in Figure 16.5.

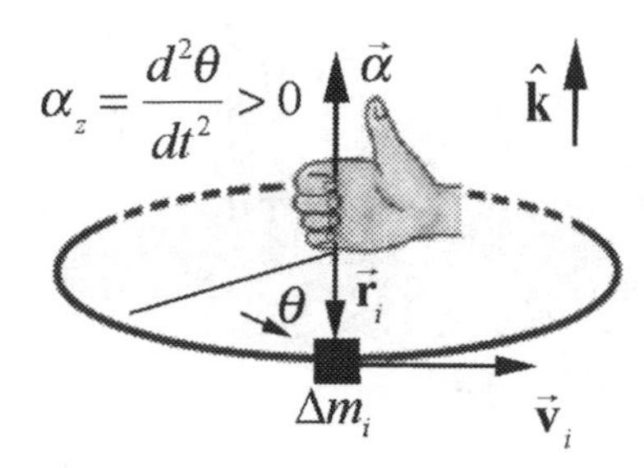

Figure 16.5 Angular acceleration vector for a rigid body rotating about the z-axis

16.2.3 Sign Convention: Angular Velocity and Angular Acceleration

For rotational problems we shall always choose a right-handed cylindrical coordinate system. If the positive z-axis points up, then we choose θ to be increasing in the counterclockwise direction as shown in Figures 16.4 and 16.5. If the rigid body rotates in the counterclockwise direction, then the z-component of the angular velocity is positive, $\omega_z = d\theta / dt > 0$. The angular velocity vector then points in the $+\hat{\mathbf{k}}$-direction as shown in Figure 16.4. If the rigid body rotates in the clockwise direction, then the z-component of the angular velocity angular velocity is negative, $\omega_z = d\theta / dt < 0$. The angular velocity vector then points in the $-\hat{\mathbf{k}}$-direction.

If the rigid body *increases* its rate of rotation in the counterclockwise (positive) direction then the z-component of the angular acceleration is positive, $\alpha_z = d^2\theta/dt^2 - d\omega_z / dt > 0$. The angular acceleration vector then points in the $+\hat{\mathbf{k}}$-direction as shown in Figure 16.5. If the rigid body *decreases* its rate of rotation in the counterclockwise (positive) direction then the z-component of the angular acceleration is negative, $\alpha_z = d^2\theta / dt^2 = d\omega_z / dt < 0$. The angular acceleration vector then points in the $-\hat{\mathbf{k}}$-direction. To phrase this more generally, if $\vec{\alpha}$ and $\vec{\boldsymbol{\omega}}$ point in the same direction, the body is speeding up, if in opposite directions, the body is slowing down. This general result is independent of the choice of positive direction of rotation. Note that in Figure 16.1, the CD has the angular velocity vector points downward (in the $-\hat{\mathbf{k}}$-direction).

16.2.4 Tangential Velocity and Tangential Acceleration

Because the small element of mass, Δm_i, is moving in a circle of radius r_i with angular velocity $\vec{\boldsymbol{\omega}} = \omega_z \hat{\mathbf{k}}$, the element has a tangential velocity component

$$v_{\theta,i} = r_i \omega_z . \tag{16.1.3}$$

If the magnitude of the tangential velocity is changing, the mass element undergoes a tangential acceleration given by

$$a_{\theta,i} = r_i \alpha_z . \tag{16.1.4}$$

Recall that the mass element is always accelerating inward with radial component given by

$$a_{r,i} = -\frac{v_{\theta,i}^2}{r_i} = -r_i \omega_z^2 . \tag{16.1.5}$$

Example 16.1 Turntable

A turntable is a uniform disc of mass $1.2\ \mathrm{kg}$ and a radius $1.3\times10^1\ \mathrm{cm}$. The turntable is spinning initially in a counterclockwise direction when seen from above at a constant rate of $f_0 = 33\ \mathrm{cycles}\cdot\mathrm{min}^{-1}$ ($33\ \mathrm{rpm}$). The motor is turned off and the turntable slows to a stop in $8.0\ \mathrm{s}$. Assume that the angular acceleration is constant. (a) What is the initial angular velocity of the turntable? (b) What is the angular acceleration of the turntable?

Solution: (a) Choose a coordinate system shown in Figure 16.6.

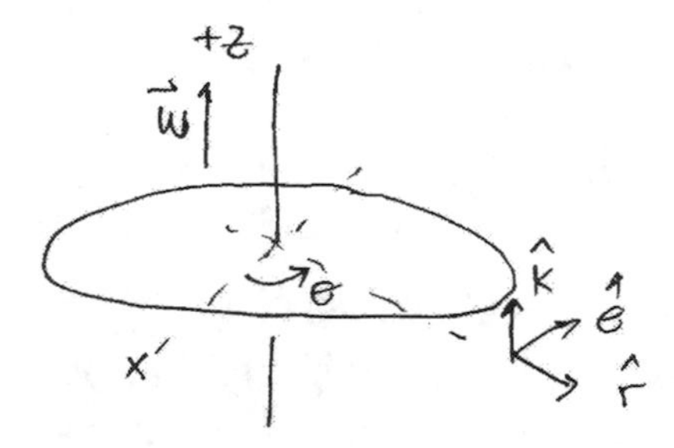

Figure 16.6 Coordinate system for turntable

Initially, the disc is spinning with a frequency

$$f_0 = \left(33\,\frac{\mathrm{cycles}}{\mathrm{min}}\right)\left(\frac{1\,\mathrm{min}}{60\ \mathrm{s}}\right) = 0.55\ \mathrm{cycles}\cdot\mathrm{s}^{-1} = 0.55\ \mathrm{Hz}, \tag{16.1.6}$$

so the initial angular velocity has magnitude

$$\omega_0 = 2\pi f_0 = \left(2\pi\,\frac{\mathrm{radian}}{\mathrm{cycle}}\right)\left(0.55\,\frac{\mathrm{cycles}}{\mathrm{s}}\right) = 3.5\ \mathrm{rad}\cdot\mathrm{s}^{-1} . \tag{16.1.7}$$

The angular velocity vector points in the $+\hat{\mathbf{k}}$-direction as shown above.

(b) The final angular velocity is zero, so the component of the angular acceleration is

$$\alpha_z = \frac{\Delta\omega_z}{\Delta t} = \frac{\omega_f - \omega_0}{t_f - t_0} = \frac{-3.5 \text{ rad} \cdot \text{s}^{-1}}{8.0 \text{ s}} = -4.3 \times 10^{-1} \text{ rad} \cdot \text{s}^{-2}. \tag{16.1.8}$$

The z-component of the angular acceleration is negative, the disc is slowing down and so the angular acceleration vector then points in the $-\hat{\mathbf{k}}$-direction as shown in Figure 16.7.

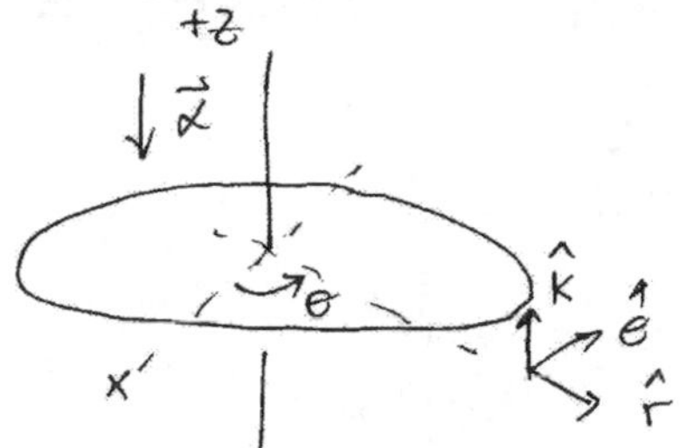

Figure 16.7 Angular acceleration vector for turntable

16.3 Rotational Kinetic Energy and Moment of Inertia

16.3.1 Rotational Kinetic Energy and Moment of Inertia

We have already defined translational kinetic energy for a point object as $K = (1/2)mv^2$; we now define the rotational kinetic energy for a rigid body about its center of mass.

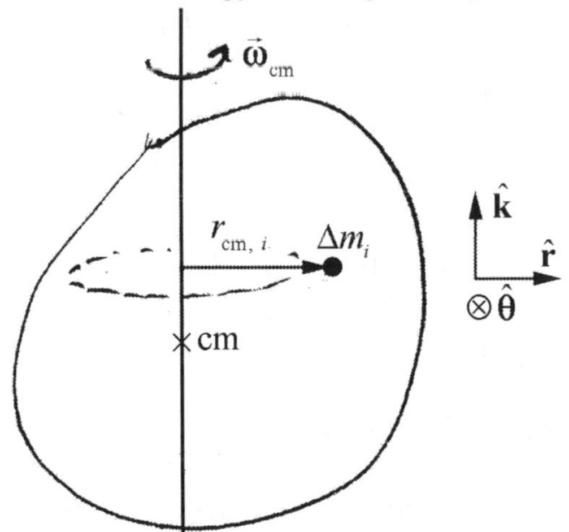

Figure 16.8 Volume element undergoing fixed-axis rotation about the z-axis that passes through the center of mass.

Choose the z-axis to lie along the axis of rotation passing through the center of mass. As in Section 16.2.2, divide the body into volume elements of mass Δm_i (Figure 16.8). Each individual mass element Δm_i undergoes circular motion about the center of mass with z-

component of angular velocity ω_{cm} in a circle of radius $r_{\text{cm},\,i}$. Therefore the velocity of each element is given by $\vec{\mathbf{v}}_{\text{cm},i} = r_{\text{cm},i}\,\omega_{\text{cm}}\,\hat{\boldsymbol{\theta}}$. The rotational kinetic energy is then

$$K_{\text{cm},\,i} = \frac{1}{2}\Delta m_i\, v_{\text{cm},\,i}^2 = \frac{1}{2}\Delta m_i r_{\text{cm},\,i}^2\,\omega_{\text{cm}}^2 . \qquad (16.2.1)$$

We now add up the kinetic energy for all the mass elements,

$$\begin{aligned} K_{\text{cm}} &= \lim_{\substack{i\to\infty \\ \Delta m_i \to 0}} \sum_{i=1}^{i=N} K_{\text{cm},\,i} = \lim_{\substack{i\to\infty \\ \Delta m_i \to 0}} \sum_{i=1}^{i=N} \left(\sum_i \frac{1}{2}\Delta m_i r_{\text{cm},\,i}^2 \right)\omega_{\text{cm}}^2 \\ &= \left(\frac{1}{2}\int_{\text{body}} dm\, r_{\text{dm}}^2 \right)\omega_{\text{cm}}^2 , \end{aligned} \qquad (16.2.2)$$

where dm is an infinitesimal mass element undergoing a circular orbit of radius r_{dm} about the axis passing through the center of mass.

The quantity

$$I_{cm} = \int_{\text{body}} dm\, r_{dm}^2 . \qquad (16.2.3)$$

is called the ***moment of inertia*** *of the rigid body about a fixed axis passing through the center of mass, and is a physical property of the body. The SI units for moment of inertia are* $\left[\text{kg}\cdot\text{m}^2\right]$.

Thus

$$K_{\text{cm}} = \left(\frac{1}{2}\int_{\text{body}} dm\, r_{\text{dm}}^2 \right)\omega_{\text{cm}}^2 \equiv \frac{1}{2} I_{cm}\,\omega_{\text{cm}}^2 . \qquad (16.2.4)$$

16.3.2 Moment of Inertia of a Rod of Uniform Mass Density

Consider a thin uniform rod of length L and mass m. In this problem, we will calculate the moment of inertia about an axis perpendicular to the rod that passes through the center of mass of the rod. A sketch of the rod, volume element, and axis is shown in Figure 16.9. Choose Cartesian coordinates, with the origin at the center of mass of the rod, which is midway between the endpoints since the rod is uniform. Choose the x-axis to lie along the length of the rod, with the positive x-direction to the right, as in the figure.

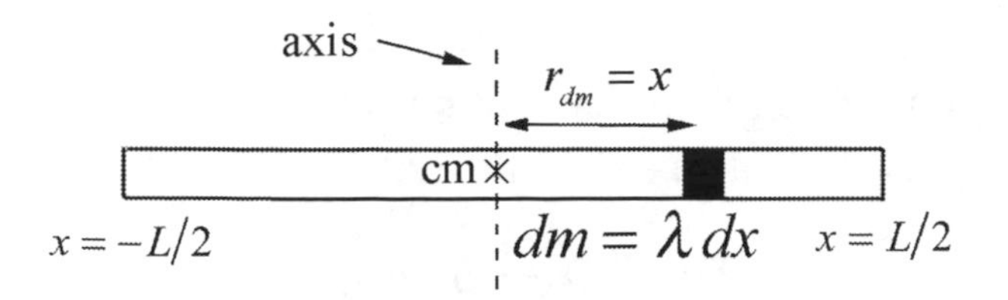

Figure 16.9 Moment of inertia of a uniform rod about center of mass.

Identify an infinitesimal mass element $dm = \lambda\, dx$, located at a displacement x from the center of the rod, where the mass per unit length $\lambda = m / L$ is a constant, as we have assumed the rod to be uniform. When the rod rotates about an axis perpendicular to the rod that passes through the center of mass of the rod, the element traces out a circle of radius $r_{dm} = x$. We add together the contributions from each infinitesimal element as we go from $x = -L/2$ to $x = L/2$. The integral is then

$$\begin{aligned} I_{\text{cm}} &= \int_{\text{body}} r_{dm}^2\, dm = \lambda \int_{-L/2}^{L/2} (x^2)\, dx = \lambda \frac{x^3}{3}\Bigg|_{-L/2}^{L/2} \\ &= \frac{m}{L}\frac{(L/2)^3}{3} - \frac{m}{L}\frac{(-L/2)^3}{3} = \frac{1}{12} m L^2 . \end{aligned} \tag{16.2.5}$$

By using a constant mass per unit length along the rod, we need not consider variations in the mass density in any direction other than the x- axis. We also assume that the width is the rod is negligible. (Technically we should treat the rod as a cylinder or a rectangle in the x-y plane if the axis is along the z- axis. The calculation of the moment of inertia in these cases would be more complicated.)

Example 16.2 Moment of Inertia of a Uniform Disc

A thin uniform disc of mass M and radius R is mounted on an axle passing through the center of the disc, perpendicular to the plane of the disc. Calculate the moment of inertia about an axis that passes perpendicular to the disc through the center of mass of the disc

Solution: As a starting point, consider the contribution to the moment of inertia from the mass element dm show in Figure 16.10. Let r denote the distance form the center of mass of the disc to the mass element.

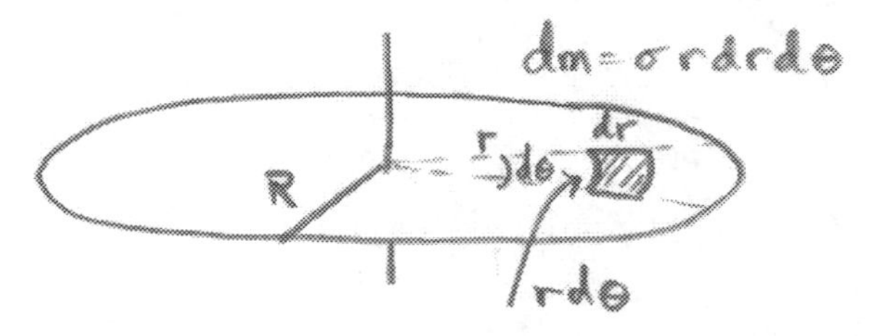

Figure 16.10 Infinitesimal mass element and coordinate system for disc.

Choose cylindrical coordinates with the coordinates (r,θ) in the plane and the z-axis perpendicular to the plane. The area element

$$da = r\,dr\,d\theta \tag{16.2.6}$$

may be thought of as the product of arc length $r\,d\theta$ and the radial width dr. Since the disc is uniform, the mass per unit area is a constant,

$$\sigma = \frac{dm}{da} = \frac{m_{\text{total}}}{\text{Area}} = \frac{M}{\pi R^2}. \tag{16.2.7}$$

Therefore the mass in the infinitesimal area element as given in Equation (16.2.6), a distance r from the axis of rotation, is given by

$$dm = \sigma\, r\,dr\,d\theta = \frac{M}{\pi R^2} r\,dr\,d\theta\,. \tag{16.2.8}$$

When the disc rotates, the mass element traces out a circle of radius $r_{dm} = r$; that is, the distance from the center is the perpendicular distance from the axis of rotation. The moment of inertia integral is now an integral in two dimensions; the angle θ varies from $\theta = 0$ to $\theta = 2\pi$, and the radial coordinate r varies from $r = 0$ to $r = R$. Thus the limits of the integral are

$$I_{\text{cm}} = \int_{\text{body}} r_{dm}^2\, dm = \frac{M}{\pi R^2}\int_{r=0}^{r=R}\int_{\theta=0}^{\theta=2\pi} r^3\, d\theta\, dr\,. \tag{16.2.9}$$

The integral can now be explicitly calculated by first integrating the θ-coordinate

$$I_{\text{cm}} = \frac{M}{\pi R^2}\int_{r=0}^{r=R}\left(\int_{\theta=0}^{\theta=2\pi} d\theta\right) r^3 dr = \frac{M}{\pi R^2}\int_{r=0}^{r=R} 2\pi r^3 dr = \frac{2M}{R^2}\int_{r=0}^{r=R} r^3 dr \tag{16.2.10}$$

and then integrating the r-coordinate,

$$I_{\text{cm}} = \frac{2M}{R^2}\int_{r=0}^{r=R} r^3 dr = \frac{2M}{R^2}\frac{r^4}{4}\bigg|_{r=0}^{r=R} = \frac{2M}{R^2}\frac{R^4}{4} = \frac{1}{2}MR^2\,. \tag{16.2.11}$$

Remark: Instead of taking the area element as a small patch $da = r\,dr\,d\theta$, choose a ring of radius r and width dr. Then the area of this ring is given by

$$da_{\text{ring}} = \pi(r+dr)^2 - \pi r^2 = \pi r^2 + 2\pi r\,dr + \pi(dr)^2 - \pi r^2 = 2\pi r\,dr + \pi(dr)^2\,. \tag{16.2.12}$$

In the limit that $dr \to 0$, the term proportional to $(dr)^2$ can be ignored and the area is $da = 2\pi r dr$. This equivalent to first integrating the $d\theta$ variable

$$da_{\text{ring}} = r\,dr\left(\int_{\theta=0}^{\theta=2\pi} d\theta\right) = 2\pi r\,dr\,. \tag{16.2.13}$$

Then the mass element is

$$dm_{\text{ring}} = \sigma da_{\text{ring}} = \frac{M}{\pi R^2} 2\pi r\,dr\,. \tag{16.2.14}$$

The moment of inertia integral is just an integral in the variable r,

$$I_{\text{cm}} = \int_{\text{body}} (r_\perp)^2\,dm = \frac{2\pi M}{\pi R^2}\int_{r=0}^{r=R} r^3 dr = \frac{1}{2}MR^2\,. \tag{16.2.15}$$

16.3.3 Parallel Axis Theorem

Consider a rigid body of mass m undergoing fixed-axis rotation. Consider two parallel axes. The first axis passes through the center of mass of the body, and the moment of inertia about this first axis is I_{cm}. The second axis passes through some other point S in the body. Let $d_{S,\text{cm}}$ denote the perpendicular distance between the two parallel axes (Figure 16.11).

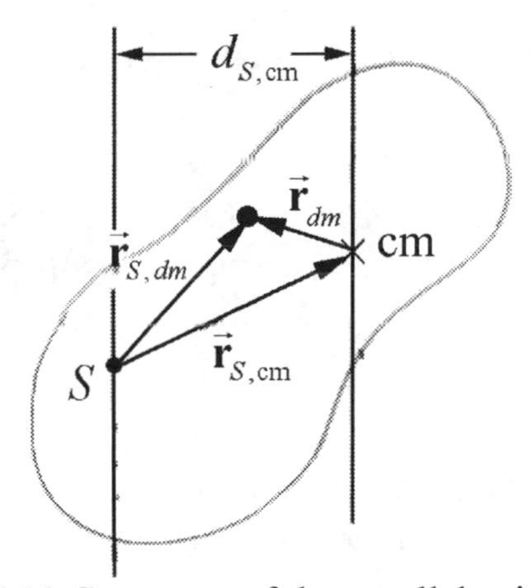

Figure 16.11 Geometry of the parallel axis theorem.

Then the moment of inertia I_S about an axis passing through a point S is related to I_{cm} by

$$I_S = I_{\text{cm}} + m\,d_{S,\text{cm}}^2\,. \tag{16.2.16}$$

16.3.4 Parallel Axis Theorem Applied to a Uniform Rod

Let point S be the left end of the rod of Figure 16.9. Then the distance from the center of mass to the end of the rod is $d_{S,\text{cm}} = L/2$. The moment of inertia $I_S = I_{\text{end}}$ about an axis passing through the endpoint is related to the moment of inertia about an axis passing through the center of mass, $I_{\text{cm}} = (1/12)\,m\,L^2$, according to Equation (16.2.16),

$$I_S = \frac{1}{12} m\,L^2 + \frac{1}{4} m\,L^2 = \frac{1}{3} m\,L^2 . \tag{16.2.17}$$

In this case it's easy and useful to check by direct calculation. Use Equation (16.2.5) but with the limits changed to $x' = 0$ and $x' = L$, where $x' = x + L/2$,

$$\begin{aligned} I_{\text{end}} &= \int_{\text{body}} r_\perp^{\,2}\, dm = \lambda \int_0^L x'^2\, dx' \\ &= \lambda \frac{x'^3}{3}\bigg|_0^L = \frac{m}{L}\frac{(L)^3}{3} - \frac{m}{L}\frac{(0)^3}{3} = \frac{1}{3} m\,L^2 . \end{aligned} \tag{16.2.18}$$

Example 16.3 Rotational Kinetic Energy of Disk

A disk with mass M and radius R is spinning with angular speed ω about an axis that passes through the rim of the disk perpendicular to its plane. The moment of inertia about cm is $I_{cm} = (1/2)mR^2$. What is the kinetic energy of the disk?

Solution: The parallel axis theorem states the moment of inertia about an axis passing perpendicular to the plane of the disc and passing through a point on the edge of the disc is equal to

$$I_{edge} = I_{cm} + mR^2 . \tag{16.2.19}$$

The moment of inertia about an axis passing perpendicular to the plane of the disc and passing through the center of mass of the disc is equal to $I_{cm} = (1/2)mR^2$. Therefore

$$I_{edge} = (3/2)mR^2 . \tag{16.2.20}$$

The kinetic energy is then

$$K = (1/2) I_{edge}\,\omega^2 = (3/4)mR^2\omega^2 . \tag{16.2.21}$$

16.4 Conservation of Energy for Fixed Axis Rotation

Consider a closed system ($\Delta E_{system} = 0$) under action of only conservative internal forces. Then the change in the mechanical energy of the system is zero

$$\Delta E_m = \Delta U + \Delta K = (U_f + K_f) - (U_i + K_i) = 0 . \qquad (16.3.1)$$

For fixed axis rotation with a component of angular velocity ω about the fixed axis, the change in kinetic energy is given by

$$\Delta K \equiv K_f - K_i = \frac{1}{2} I_S \omega_f^2 - \frac{1}{2} I_S \omega_i^2 , \qquad (16.3.2)$$

where S is a point that lies on the fixed axis. Then conservation of energy implies that

$$U_f + \frac{1}{2} I_S \omega_f^2 = U_i + \frac{1}{2} I_S \omega_i^2 \qquad (16.3.3)$$

Example 16.4 Energy and Pulley System

A wheel in the shape of a uniform disk of radius R and mass m_p is mounted on a frictionless horizontal axis. The wheel has moment of inertia about the center of mass $I_{cm} = (1/2) m_p R^2$. A massless cord is wrapped around the wheel and one end of the cord is attached to an object of mass m_2 that can slide up or down a frictionless inclined plane. The other end of the cord is attached to a second object of mass m_1 that hangs over the edge of the inclined plane. The plane is inclined from the horizontal by an angle θ (Figure 16.12). Once the objects are released from rest, the cord moves without slipping around the disk. Calculate the speed of block 2 as a function of distance that it moves down the inclined plane using energy techniques. Assume there are no energy losses due to friction and that the rope does not slip around the pulley

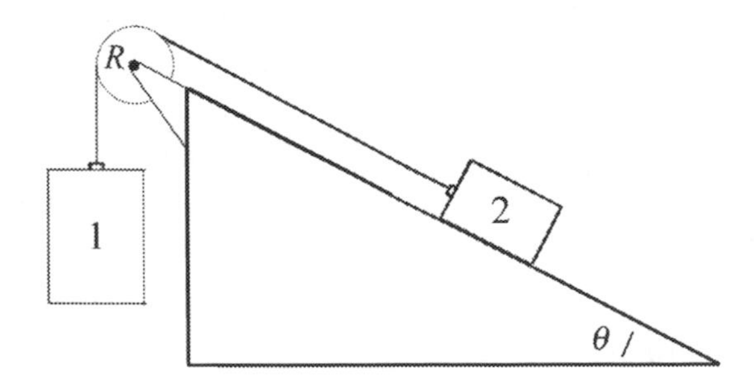

Figure 16.12 Pulley and blocks

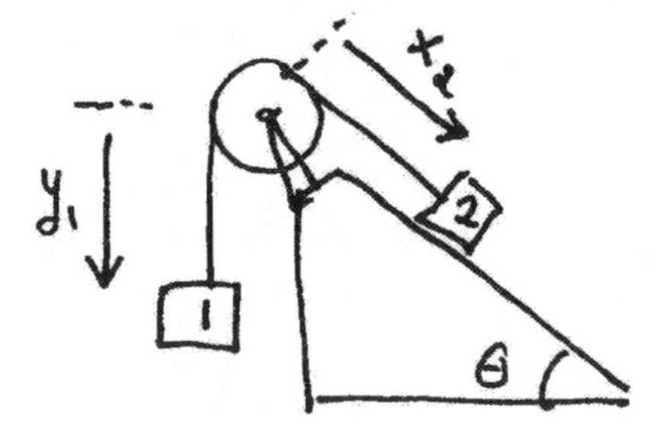

Figure 16.13 Coordinate system for pulley and blocks

Solution: Define a coordinate system as shown in Figure 16.13. Choose the zero for the gravitational potential energy at a height equal to the center of the pulley. In Figure 16.14 illustrates the energy diagrams for the initial state and a dynamic state at an arbitrary time when the blocks are sliding.

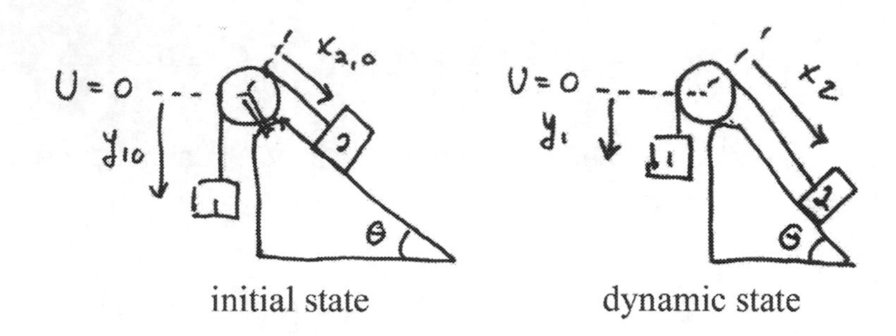

Figure 16.14 Energy diagrams for initial state and dynamic state at arbitrary time

Then the initial mechanical energy is

$$E_i = U_i = -m_1 g y_{1,i} - m_2 g x_{2,i} \sin\theta . \tag{16.3.4}$$

The mechanical energy, when block 2 has moved a distance

$$d = x_2 - x_{2,i} \tag{16.3.5}$$

is given by

$$E = U + K = -m_1 g y_1 - m_2 g x_2 \sin\theta + \frac{1}{2} m_1 v_1^{\,2} + \frac{1}{2} m_2 v_2^{\,2} + \frac{1}{2} I_P \omega^2 . \tag{16.3.6}$$

The rope connects the two blocks, and so the blocks move at the same speed

$$v \equiv v_1 = v_2 . \tag{16.3.7}$$

The rope does not slip on the pulley; therefore as the rope moves around the pulley the tangential speed of the rope is equal to the speed of the blocks

$$v_{\tan} = R\omega = v . \tag{16.3.8}$$

Eq. (16.3.6) can now be simplified

$$E = U + K = -m_1 g y_1 - m_2 g x_2 \sin\theta + \frac{1}{2}\left(m_1 + m_2 + \frac{I_P}{R^2} \right) v^2 . \tag{16.3.9}$$

Because we have assumed that there is no loss of mechanical energy, we can set $E_i = E$ and find that

$$-m_1 g y_{1,i} - m_2 g x_{2,i} \sin\theta = -m_1 g y_1 - m_2 g x_2 \sin\theta + \frac{1}{2}\left(m_1 + m_2 + \frac{I_P}{R^2} \right) v^2, \quad (16.3.10)$$

which simplifies to

$$-m_1 g (y_{1,0} - y_1) + m_2 g (x_2 - x_{2,0}) \sin\theta = \frac{1}{2}\left(m_1 + m_2 + \frac{I_P}{R^2} \right) v^2. \quad (16.3.11)$$

We finally note that the movement of block 1 and block 2 are constrained by the relationship

$$d = x_2 - x_{2,i} = y_{1,i} - y_1. \quad (16.3.12)$$

Then Eq. (16.3.11) becomes

$$gd(-m_1 + m_2 \sin\theta) = \frac{1}{2}\left(m_1 + m_2 + \frac{I_P}{R^2} \right) v^2. \quad (16.3.13)$$

We can now solve for the speed as a function of distance $d = x_2 - x_{2,i}$ that block 2 has traveled down the incline plane

$$v = \sqrt{\frac{2gd(-m_1 + m_2 \sin\theta)}{\left(m_1 + m_2 + (I_P / R^2)\right)}}. \quad (16.3.14)$$

If we assume that the moment of inertial of the pulley is $I_{\text{cm}} = (1/2) m_{\text{p}} R^2$, then the speed becomes

$$v = \sqrt{\frac{2gd(-m_1 + m_2 \sin\theta)}{\left(m_1 + m_2 + (1/2) m_P\right)}}. \quad (16.3.15)$$

Example 16.5 Physical Pendulum

A physical pendulum consists of a uniform rod of mass m_1 pivoted at one end about the point S. The rod has length l_1 and moment of inertia I_1 about the pivot point. A disc of mass m_2 and radius r_2 with moment of inertia I_{cm} about its center of mass is rigidly attached a distance l_2 from the pivot point. The pendulum is initially displaced to an angle θ_i and then released from rest. (a) What is the moment of inertia of the physical pendulum about the pivot point S? (b) How far from the pivot point is the center of mass of the system? (c) What is the angular speed of the pendulum when the pendulum is at the bottom of its swing?

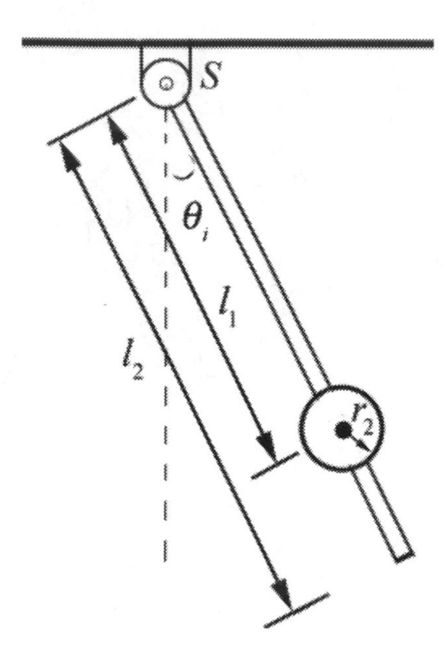

Figure 16.15 Rod and with fixed disc pivoted about the point S

Solution: a) The moment of inertia about the pivot point is the sum of the moment of inertia of the rod, given as I_1, and the moment of inertia of the disc about the pivot point. The moment of inertia of the disc about the pivot point is found from the parallel axis theorem,

$$I_{\text{disc}} = I_{\text{cm}} + m_2 l_2^2. \quad (16.3.16)$$

The moment of inertia of the system consisting of the rod and disc about the pivot point S is then

$$I_S = I_1 + I_{\text{disc}} = I_1 + I_{\text{cm}} + m_2 l_2^2. \quad (16.3.17)$$

The center of mass of the system is located a distance from the pivot point

$$l_{\text{cm}} = \frac{m_1(l_1/2) + m_2 l_2}{m_1 + m_2}. \quad (16.3.18)$$

b) We can use conservation of mechanical energy, to find the angular speed of the pendulum at the bottom of its swing. Take the zero point of gravitational potential energy to be the point where the bottom of the rod is at its lowest point, that is, $\theta = 0$. The initial state energy diagram for the rod is shown in Figure 16.16a and the initial state energy diagram for the disc is shown in Figure 16.16b.

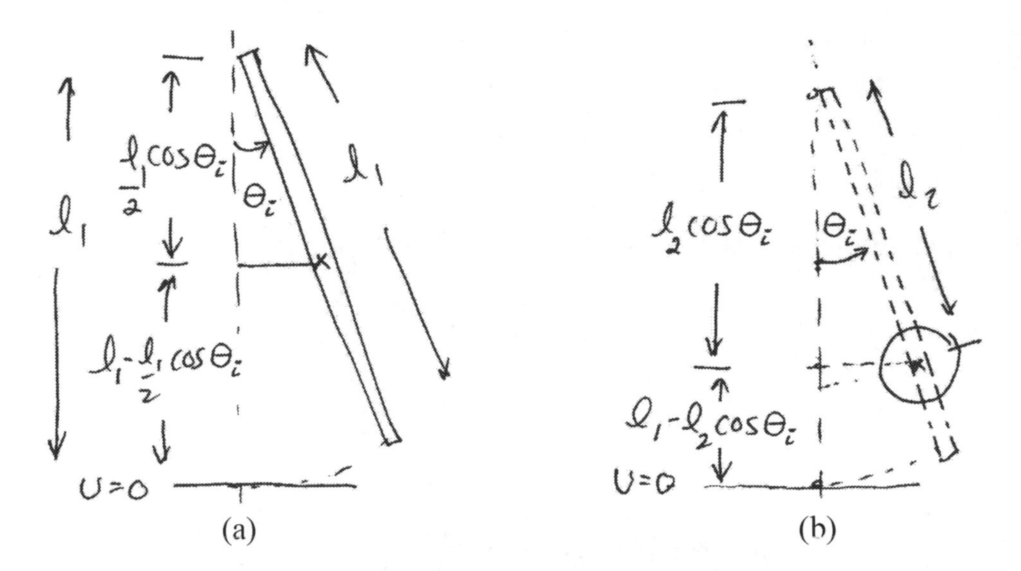

Figure 16.16 (a) Initial state energy diagram for rod (b) Initial state energy diagram for disc

The initial mechanical energy is then

$$E_i = U_i = m_1 g(l_1 - \frac{l_1}{2}\cos\theta_i) + m_2 g(l_1 - l_2\cos\theta_i), \tag{16.3.19}$$

At the bottom of the swing, $\theta_f = 0$, and the system has angular velocity ω_f. The mechanical energy at the bottom of the swing is

$$E_f = U_f + K_f = m_1 g\frac{l_1}{2} + m_2 g(l_1 - l_2) + \frac{1}{2}I_S\omega_f^2, \tag{16.3.20}$$

with I_S as found in Equation (16.3.17). There are no non-conservative forces acting, so the mechanical energy is constant therefore equating the expressions in (16.3.19) and (16.3.20) we get that

$$m_1 g(l_1 - \frac{l_1}{2}\cos\theta_i) + m_2 g(l_1 - l_2\cos\theta_i) = m_1 g\frac{l_1}{2} + m_2 g(l_1 - l_2) + \frac{1}{2}I_S\omega_f^2, \tag{16.3.21}$$

This simplifies to

$$\left(\frac{m_1 l_1}{2} + m_2 l_2\right)g(1-\cos\theta_i) = \frac{1}{2}I_S\omega_f^2, \tag{16.3.22}$$

We now solve for ω_f (taking the positive square root to insure that we are calculating angular speed)

$$\omega_f = \sqrt{\frac{2\left(\frac{m_1 l_1}{2} + m_2 l_2\right) g(1-\cos\theta_i)}{I_S}}, \tag{16.3.23}$$

Finally we substitute in Eq.(16.3.17) in to Eq. (16.3.23) and find

$$\omega_f = \sqrt{\frac{2\left(\frac{m_1 l_1}{2} + m_2 l_2\right) g(1-\cos\theta_i)}{I_1 + I_{\text{cm}} + m_2 l_2^2}}. \tag{16.3.24}$$

Note that we can rewrite Eq. (16.3.22), using Eq. (16.3.18) for the distance between the center of mass and the pivot point, to get

$$(m_1 + m_2) l_{cm} g(1-\cos\theta_i) = \frac{1}{2} I_S \omega_f^2, \tag{16.3.25}$$

We can interpret this equation as follows. Treat the system as a point particle of mass $m_1 + m_2$ located at the center of mass l_{cm}. Take the zero point of gravitational potential energy to be the point where the center of mass is at its lowest point, that is, $\theta = 0$. Then

$$E_i = (m_1 + m_2) l_{cm} g(1-\cos\theta_i), \tag{16.3.26}$$

$$E_f = \frac{1}{2} I_S \omega_f^2. \tag{16.3.27}$$

Thus conservation of energy reproduces Eq. (16.3.25).

Appendix 16A: Proof of the Parallel Axis Theorem

Identify an infinitesimal volume element of mass dm. The vector from the point S to the mass element is $\vec{\mathbf{r}}_{S,dm}$, the vector from the center of mass to the mass element is $\vec{\mathbf{r}}_{dm}$, and the vector from the point S to the center of mass is $\vec{\mathbf{r}}_{S,\text{cm}}$.

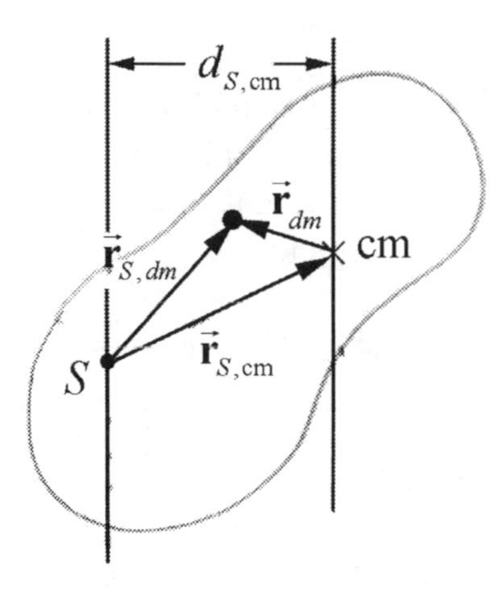

Figure 16A.1 Geometry of the parallel axis theorem.

From Figure 16A.1, we see that

$$\vec{\mathbf{r}}_{S,dm} = \vec{\mathbf{r}}_{S,\text{cm}} + \vec{\mathbf{r}}_{dm}. \tag{16.A.1}$$

The notation gets complicated at this point. The vector $\vec{\mathbf{r}}_{dm}$ has a component vector $\vec{\mathbf{r}}_{\parallel,dm}$ parallel to the axis through the center of mass and a component vector $\vec{\mathbf{r}}_{\perp,dm}$ perpendicular to the axis through the center of mass. The magnitude of the perpendicular component vector is

$$\left|\vec{\mathbf{r}}_{\text{cm},\perp,dm}\right| = r_{\perp,dm}. \tag{16.A.2}$$

The vector $\vec{\mathbf{r}}_{S,dm}$ has a component vector $\vec{\mathbf{r}}_{S,\parallel,dm}$ parallel to the axis through the point S and a component vector $\vec{\mathbf{r}}_{S,\perp,dm}$ perpendicular to the axis through the point S. The magnitude of the perpendicular component vector is

$$\left|\vec{\mathbf{r}}_{S,\perp,dm}\right| = r_{S,\perp,dm}. \tag{16.A.3}$$

The vector $\vec{\mathbf{r}}_{S,\text{cm}}$ has a component vector $\vec{\mathbf{r}}_{S,\parallel,\text{cm}}$ parallel to *both* axes and a perpendicular component vector $\vec{\mathbf{r}}_{S,\perp,\text{cm}}$ that is perpendicular to *both* axes (the axes are parallel, of course). The magnitude of the perpendicular component vector is

$$\left|\vec{\mathbf{r}}_{S,\perp,\text{cm}}\right| = d_{S,\text{cm}}. \tag{16.A.4}$$

Equation (16.A.1) is now expressed as two equations,

$$\begin{aligned} \vec{\mathbf{r}}_{S,\perp,dm} &= \vec{\mathbf{r}}_{S,\perp,\text{cm}} + \vec{\mathbf{r}}_{\perp,dm} \\ \vec{\mathbf{r}}_{S,\|,dm} &= \vec{\mathbf{r}}_{S,\|,\text{cm}} + \vec{\mathbf{r}}_{\|,dm}. \end{aligned} \tag{16.A.5}$$

At this point, note that if we had simply decided that the two parallel axes are parallel to the z-direction, we could have saved some steps and perhaps spared some of the notation with the triple subscripts. However, we want a more general result, one valid for cases where the axes are not fixed, or when different objects in the same problem have different axes. For example, consider the turning bicycle, for which the two wheel axes will not be parallel, or a spinning top that *precesses* (wobbles). Such cases will be considered in later on, and we will show the general case of the parallel axis theorem in anticipation of use for more general situations.

The moment of inertia about the point S is

$$I_S = \int_{\text{body}} dm (r_{S,\perp,dm})^2 . \tag{16.A.6}$$

From (16.A.5) we have

$$\begin{aligned} (r_{S,\perp,dm})^2 &= \vec{\mathbf{r}}_{S,\perp,dm} \cdot \vec{\mathbf{r}}_{S,\perp,dm} \\ &= (\vec{\mathbf{r}}_{S,\perp,\text{cm}} + \vec{\mathbf{r}}_{\perp,dm}) \cdot (\vec{\mathbf{r}}_{S,\perp,\text{cm}} + \vec{\mathbf{r}}_{\perp,dm}) \\ &= d_{S,\text{cm}}^2 + (r_{\perp,dm})^2 + 2\vec{\mathbf{r}}_{S,\perp,\text{cm}} \cdot \vec{\mathbf{r}}_{\perp,dm}. \end{aligned} \tag{16.A.7}$$

Thus we have for the moment of inertia about S,

$$I_S = \int_{\text{body}} dm\, d_{S,\text{cm}}^2 + \int_{\text{body}} dm (r_{\perp,dm})^2 + 2 \int_{\text{body}} dm (\vec{\mathbf{r}}_{S,\perp,\text{cm}} \cdot \vec{\mathbf{r}}_{\perp,dm}) . \tag{16.A.8}$$

In the first integral in Equation (16.A.8), $r_{S,\perp,\text{cm}} = d_{S,\text{cm}}$ is the distance between the parallel axes and is a constant. Therefore we can rewrite the integral as

$$d_{S,\text{cm}}^2 \int_{\text{body}} dm = m\, d_{S,\text{cm}}^2 . \tag{16.A.9}$$

The second term in Equation (16.A.8) is the moment of inertia about the axis through the center of mass,

$$I_{\text{cm}} = \int_{\text{body}} dm\, (r_{\perp,dm})^2 . \tag{16.A.10}$$

The third integral in Equation (16.A.8) is zero. To see this, note that the term $\vec{\mathbf{r}}_{S,\perp,\text{cm}}$ is a constant and may be taken out of the integral,

$$2\int_{\text{body}} dm\,(\vec{\mathbf{r}}_{S,\perp,\text{cm}}\cdot\vec{\mathbf{r}}_{\perp,dm}) = \vec{\mathbf{r}}_{S,\perp,\text{cm}}\cdot 2\int_{\text{body}} dm\,\vec{\mathbf{r}}_{\perp,dm} \qquad (16.\text{A}.11)$$

The integral $\int_{\text{body}} dm\,\vec{\mathbf{r}}_{\perp,dm}$ is the perpendicular component of the position of the center of mass with respect to the center of mass, and hence $\vec{\mathbf{0}}$, with the result that

$$2\int_{\text{body}} dm\,(\vec{\mathbf{r}}_{S,\perp,\text{cm}}\cdot\vec{\mathbf{r}}_{\perp,dm}) = 0\,. \qquad (16.\text{A}.12)$$

Thus, the moment of inertia about S is just the sum of the first two integrals in Equation (16.A.8)

$$I_S = I_{\text{cm}} + m d_{S,\text{cm}}^2\,, \qquad (16.\text{A}.13)$$

proving the parallel axis theorem.

Chapter 17 Two Dimensional Rotational Dynamics

Chapter 17 Two Dimensional Rotational Dynamics

torque, n.

a. The twisting or rotary force in a piece of mechanism (as a measurable quantity); the moment of a system of forces producing rotation.

Oxford English Dictionary

17.1 Introduction

A body is called a *rigid body* if the distance between any two points in the body does not change in time. Rigid bodies, unlike point masses, can have forces applied at different points in the body. For most objects, treating as a rigid body is an idealization, but a very good one. In addition to forces applied at points, forces may be distributed over the entire body. Forces that are distributed over a body are difficult to analyze; however, for example, we regularly experience the effect of the gravitational force on bodies. Based on our experience observing the effect of the gravitational force on rigid bodies, we shall demonstrate that the gravitational force can be concentrated at a point in the rigid body called the *center of gravity*, which for small bodies (so that $\vec{\mathbf{g}}$ may be taken as constant within the body) is identical to the *center of mass* of the body.

Let's consider a rigid rod thrown in the air (Figure 17.1) so that the rod is spinning as its center of mass moves with velocity $\vec{\mathbf{v}}_{\text{cm}}$. We have explored the physics of translational motion; now, we wish to investigate the properties of rotational motion exhibited in the rod's motion, beginning with the notion that every particle is rotating about the center of mass with the same angular (rotational) velocity.

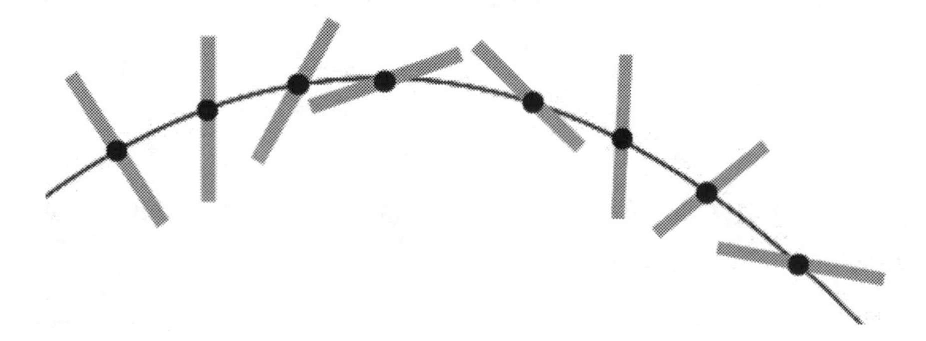

Figure 17.1 The center of mass of a thrown rigid rod follows a parabolic trajectory while the rod rotates about the center of mass.

We can use Newton's Second Law to predict how the center of mass will move. Because the only external force on the rod is the gravitational force (neglecting the action of air resistance), the center of mass of the body will move in a parabolic trajectory.

How was the rod induced to rotate? In order to spin the rod, we applied a torque with our fingers and wrist to one end of the rod as the rod was released. The applied torque is proportional to the angular acceleration. The constant of proportionality is the moment of inertia. When external forces and torques are present, the motion of a rigid body can be extremely complicated while it is translating and rotating in space.

In order to describe the relationship between torque, moment of inertia, and angular acceleration, we will introduce a new vector operation called the ***vector product*** also know as the "cross product" that takes any two vectors and generates a new vector. The vector product is a type of "multiplication" law that turns our vector space (law for addition of vectors) into a vector algebra (a vector algebra is a vector space with an additional rule for multiplication of vectors).

17.2 Vector Product (Cross Product)

Let $\vec{\mathbf{A}}$ and $\vec{\mathbf{B}}$ be two vectors. Because any two non-parallel vectors form a plane, we denote the angle θ to be the angle between the vectors $\vec{\mathbf{A}}$ and $\vec{\mathbf{B}}$ as shown in Figure 17.2. The ***magnitude of the vector product*** $\vec{\mathbf{A}}\times\vec{\mathbf{B}}$ *of the vectors $\vec{\mathbf{A}}$ and $\vec{\mathbf{B}}$ is defined to be product of the magnitude of the vectors $\vec{\mathbf{A}}$ and $\vec{\mathbf{B}}$ with the sine of the angle θ between the two vectors,*

$$\left|\vec{\mathbf{A}}\times\vec{\mathbf{B}}\right| = \left|\vec{\mathbf{A}}\right|\left|\vec{\mathbf{B}}\right|\sin(\theta). \qquad (17.2.1)$$

The angle θ between the vectors is limited to the values $0 \le \theta \le \pi$ ensuring that $\sin(\theta) \ge 0$.

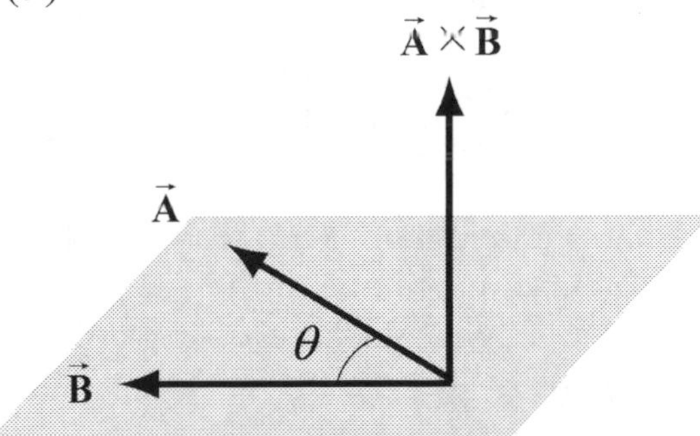

Figure 17.2 Vector product geometry.

*The direction of the vector product is defined as follows. The vectors $\vec{\mathbf{A}}$ and $\vec{\mathbf{B}}$ form a plane. Consider the direction perpendicular to this plane. There are two possibilities: we shall choose one of these two (the one shown in Figure 17.2) for the direction of the vector product $\vec{\mathbf{A}}\times\vec{\mathbf{B}}$ using a convention that is commonly called the "**right-hand rule**".*

17.2.1 Right-hand Rule for the Direction of Vector Product

The first step is to redraw the vectors $\vec{\mathbf{A}}$ and $\vec{\mathbf{B}}$ so that the tails are touching. Then draw an arc starting from the vector $\vec{\mathbf{A}}$ and finishing on the vector $\vec{\mathbf{B}}$. Curl your right fingers the same way as the arc. Your right thumb points in the direction of the vector product $\vec{\mathbf{A}}\times\vec{\mathbf{B}}$ (Figure 17.3).

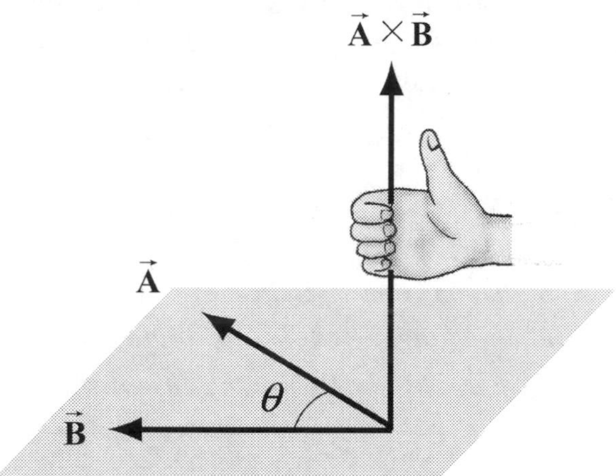

Figure 17.3 Right-Hand Rule.

You should remember that the direction of the vector product $\vec{\mathbf{A}}\times\vec{\mathbf{B}}$ is perpendicular to the plane formed by $\vec{\mathbf{A}}$ and $\vec{\mathbf{B}}$. We can give a geometric interpretation to the magnitude of the vector product by writing the magnitude as

$$\left|\vec{\mathbf{A}}\times\vec{\mathbf{B}}\right| = \left|\vec{\mathbf{A}}\right|\left(\left|\vec{\mathbf{B}}\right|\sin\theta\right). \qquad (17.2.2)$$

The vectors $\vec{\mathbf{A}}$ and $\vec{\mathbf{B}}$ form a parallelogram. The area of the parallelogram is equal to the height times the base, which is the magnitude of the vector product. In Figure 17.4, two different representations of the height and base of a parallelogram are illustrated. As depicted in Figure 17.4a, the term $\left|\vec{\mathbf{B}}\right|\sin\theta$ is the projection of the vector $\vec{\mathbf{B}}$ in the direction perpendicular to the vector $\vec{\mathbf{B}}$. We could also write the magnitude of the vector product as

$$\left|\vec{\mathbf{A}}\times\vec{\mathbf{B}}\right| = \left(\left|\vec{\mathbf{A}}\right|\sin\theta\right)\left|\vec{\mathbf{B}}\right|. \qquad (17.2.3)$$

The term $\left|\vec{\mathbf{A}}\right|\sin\theta$ is the projection of the vector $\vec{\mathbf{A}}$ in the direction perpendicular to the vector $\vec{\mathbf{B}}$ as shown in Figure 17.4(b). The vector product of two vectors that are parallel (or anti-parallel) to each other is zero because the angle between the vectors is 0 (or π) and $\sin(0)=0$ (or $\sin(\pi)=0$). Geometrically, two parallel vectors do not have a unique component perpendicular to their common direction.

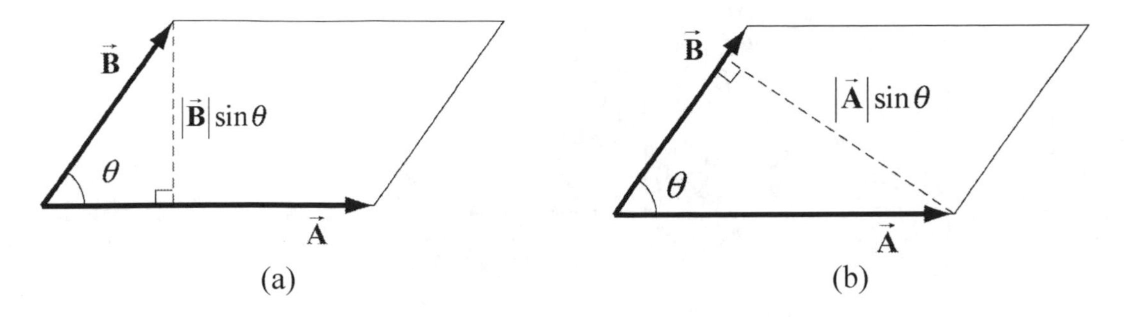

Figure 17.4 Projection of (a) $\vec{\mathbf{B}}$ perpendicular to $\vec{\mathbf{A}}$, (b) of $\vec{\mathbf{A}}$ perpendicular to $\vec{\mathbf{B}}$

17.2.2 Properties of the Vector Product

(1) The vector product is anti-commutative because changing the order of the vectors changes the direction of the vector product by the right hand rule:

$$\vec{\mathbf{A}}\times\vec{\mathbf{B}} = -\vec{\mathbf{B}}\times\vec{\mathbf{A}}\,. \tag{17.2.4}$$

(2) The vector product between a vector $c\vec{\mathbf{A}}$ where c is a scalar and a vector $\vec{\mathbf{B}}$ is

$$c\vec{\mathbf{A}}\times\vec{\mathbf{B}} = c(\vec{\mathbf{A}}\times\vec{\mathbf{B}})\,. \tag{17.2.5}$$

Similarly,

$$\vec{\mathbf{A}}\times c\vec{\mathbf{B}} = c(\vec{\mathbf{A}}\times\vec{\mathbf{B}})\,. \tag{17.2.6}$$

(3) The vector product between the sum of two vectors $\vec{\mathbf{A}}$ and $\vec{\mathbf{B}}$ with a vector $\vec{\mathbf{C}}$ is

$$(\vec{\mathbf{A}}+\vec{\mathbf{B}})\times\vec{\mathbf{C}} = \vec{\mathbf{A}}\times\vec{\mathbf{C}}+\vec{\mathbf{B}}\times\vec{\mathbf{C}} \tag{17.2.7}$$

Similarly,

$$\vec{\mathbf{A}}\times(\vec{\mathbf{B}}+\vec{\mathbf{C}}) = \vec{\mathbf{A}}\times\vec{\mathbf{B}}+\vec{\mathbf{A}}\times\vec{\mathbf{C}}\,. \tag{17.2.8}$$

17.2.3 Vector Decomposition and the Vector Product: Cartesian Coordinates

We first calculate that the magnitude of vector product of the unit vectors $\hat{\mathbf{i}}$ and $\hat{\mathbf{j}}$:

$$|\,\hat{\mathbf{i}}\times\hat{\mathbf{j}}\,| = |\,\hat{\mathbf{i}}\,||\,\hat{\mathbf{j}}\,|\sin(\pi/2) = 1, \tag{17.2.9}$$

because the unit vectors have magnitude $|\,\hat{\mathbf{i}}\,| = |\,\hat{\mathbf{j}}\,| = 1$ and $\sin(\pi/2) = 1$. By the right hand rule, the direction of $\hat{\mathbf{i}}\times\hat{\mathbf{j}}$ is in the $+\hat{\mathbf{k}}$ as shown in Figure 17.5. Thus $\hat{\mathbf{i}}\times\hat{\mathbf{j}} = \hat{\mathbf{k}}$.

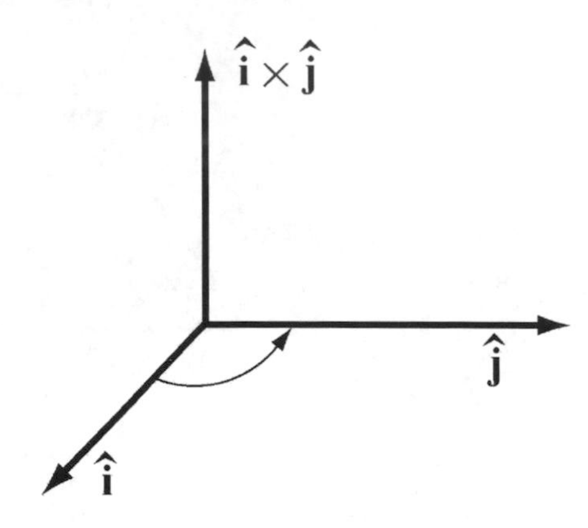

Figure 17.5 Vector product of $\hat{\mathbf{i}}\times\hat{\mathbf{j}}$

We note that the same rule applies for the unit vectors in the y and z directions,

$$\hat{\mathbf{j}}\times\hat{\mathbf{k}} = \hat{\mathbf{i}}, \quad \hat{\mathbf{k}}\times\hat{\mathbf{i}} = \hat{\mathbf{j}}. \tag{17.2.10}$$

By the anti-commutatively property (1) of the vector product,

$$\hat{\mathbf{j}}\times\hat{\mathbf{i}} = -\hat{\mathbf{k}}, \quad \hat{\mathbf{i}}\times\hat{\mathbf{k}} = -\hat{\mathbf{j}} \tag{17.2.11}$$

The vector product of the unit vector $\hat{\mathbf{i}}$ with itself is zero because the two unit vectors are parallel to each other, ($\sin(0) = 0$),

$$|\hat{\mathbf{i}}\times\hat{\mathbf{i}}| = |\hat{\mathbf{i}}||\hat{\mathbf{i}}|\sin(0) = 0. \tag{17.2.12}$$

The vector product of the unit vector $\hat{\mathbf{j}}$ with itself and the unit vector $\hat{\mathbf{k}}$ with itself are also zero for the same reason,

$$\left|\hat{\mathbf{j}}\times\hat{\mathbf{j}}\right| = 0, \quad \left|\hat{\mathbf{k}}\times\hat{\mathbf{k}}\right| = 0. \tag{17.2.13}$$

With these properties in mind we can now develop an algebraic expression for the vector product in terms of components. Let's choose a Cartesian coordinate system with the vector $\vec{\mathbf{B}}$ pointing along the positive x-axis with positive x-component B_x. Then the vectors $\vec{\mathbf{A}}$ and $\vec{\mathbf{B}}$ can be written as

$$\vec{\mathbf{A}} = A_x\,\hat{\mathbf{i}} + A_y\,\hat{\mathbf{j}} + A_z\,\hat{\mathbf{k}} \tag{17.2.14}$$

$$\vec{\mathbf{B}} = B_x\,\hat{\mathbf{i}}, \tag{17.2.15}$$

respectively. The vector product in vector components is

$$\vec{\mathbf{A}}\times\vec{\mathbf{B}} = (A_x\,\hat{\mathbf{i}} + A_y\,\hat{\mathbf{j}} + A_z\,\hat{\mathbf{k}})\times B_x\,\hat{\mathbf{i}}. \tag{17.2.16}$$

This becomes,

$$\begin{aligned}\vec{\mathbf{A}}\times\vec{\mathbf{B}} &= (A_x\,\hat{\mathbf{i}}\times B_x\,\hat{\mathbf{i}})+(A_y\,\hat{\mathbf{j}}\times B_x\,\hat{\mathbf{i}})+(A_z\,\hat{\mathbf{k}}\times B_x\,\hat{\mathbf{i}}) \\ &= A_xB_x\,(\hat{\mathbf{i}}\times\hat{\mathbf{i}})+A_yB_x\,(\hat{\mathbf{j}}\times\hat{\mathbf{i}})+A_zB_x\,(\hat{\mathbf{k}}\times\hat{\mathbf{i}}) \quad . \\ &= -A_yB_x\,\hat{\mathbf{k}}+A_zB_x\,\hat{\mathbf{j}}\end{aligned} \tag{17.2.17}$$

The vector component expression for the vector product easily generalizes for arbitrary vectors

$$\vec{\mathbf{A}} = A_x\,\hat{\mathbf{i}}+A_y\,\hat{\mathbf{j}}+A_z\,\hat{\mathbf{k}} \tag{17.2.18}$$

$$\vec{\mathbf{B}} = B_x\,\hat{\mathbf{i}}+B_y\,\hat{\mathbf{j}}+B_z\,\hat{\mathbf{k}}, \tag{17.2.19}$$

to yield

$$\vec{\mathbf{A}}\times\vec{\mathbf{B}} = (A_yB_z-A_zB_y)\,\hat{\mathbf{i}}+(A_zB_x-A_xB_z)\,\hat{\mathbf{j}}+(A_xB_y-A_yB_x)\,\hat{\mathbf{k}}\,. \tag{17.2.20}$$

17.2.4 Vector Decomposition and the Vector Product: Cylindrical Coordinates

Recall the cylindrical coordinate system, which we show in Figure 17.6. We have chosen two directions, radial and tangential in the plane, and a perpendicular direction to the plane.

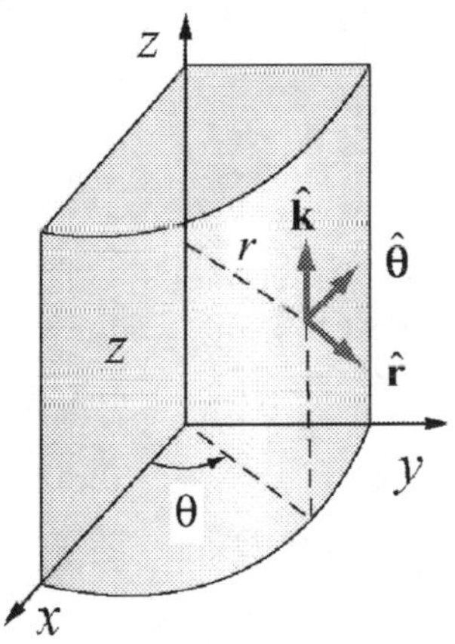

Figure 17.6 Cylindrical coordinates

The unit vectors are at right angles to each other and so using the right hand rule, the vector product of the unit vectors are given by the relations

$$\hat{\mathbf{r}}\times\hat{\boldsymbol{\theta}} = \hat{\mathbf{k}} \tag{17.2.21}$$

$$\hat{\boldsymbol{\theta}}\times\hat{\mathbf{k}} = \hat{\mathbf{r}} \tag{17.2.22}$$

$$\hat{\mathbf{k}}\times\hat{\mathbf{r}} = \hat{\boldsymbol{\theta}}\,. \tag{17.2.23}$$

Because the vector product satisfies $\vec{\mathbf{A}}\times\vec{\mathbf{B}} = -\vec{\mathbf{B}}\times\vec{\mathbf{A}}$, we also have that

$$\hat{\boldsymbol{\theta}}\times\hat{\mathbf{r}} = -\hat{\mathbf{k}} \tag{17.2.24}$$

$$\hat{\mathbf{k}}\times\hat{\boldsymbol{\theta}} = -\hat{\mathbf{r}} \qquad (17.2.25)$$

$$\hat{\mathbf{r}}\times\hat{\mathbf{k}} = -\hat{\boldsymbol{\theta}}. \qquad (17.2.26)$$

Finally

$$\hat{\mathbf{r}}\times\hat{\mathbf{r}} = \hat{\boldsymbol{\theta}}\times\hat{\boldsymbol{\theta}} = \hat{\mathbf{k}}\times\hat{\mathbf{k}} = \vec{\mathbf{0}}. \qquad (17.2.27)$$

Example 17.1 Vector Products

Given two vectors, $\vec{\mathbf{A}} = 2\,\hat{\mathbf{i}} + -3\,\hat{\mathbf{j}} + 7\,\hat{\mathbf{k}}$ and $\vec{\mathbf{B}} = 5\hat{\mathbf{i}} + \hat{\mathbf{j}} + 2\hat{\mathbf{k}}$, find $\vec{\mathbf{A}}\times\vec{\mathbf{B}}$.

Solution:

$$\begin{aligned}\vec{\mathbf{A}}\times\vec{\mathbf{B}} &= (A_yB_z - A_zB_y)\,\hat{\mathbf{i}} + (A_zB_x - A_xB_z)\,\hat{\mathbf{j}} + (A_xB_y - A_yB_x)\,\hat{\mathbf{k}} \\ &= ((-3)(2)-(7)(1))\,\hat{\mathbf{i}} + ((7)(5)-(2)(2))\,\hat{\mathbf{j}} + ((2)(1)-(-3)(5))\,\hat{\mathbf{k}} \\ &= -13\,\hat{\mathbf{i}} + 31\,\hat{\mathbf{j}} + 17\,\hat{\mathbf{k}}.\end{aligned}$$

Example 17.2 Law of Sines

For the triangle shown in Figure 17.7a, prove the law of sines, $\left|\vec{\mathbf{A}}\right| / \sin\alpha = \left|\vec{\mathbf{B}}\right| / \sin\beta = \left|\vec{\mathbf{C}}\right| / \sin\gamma$, using the vector product.

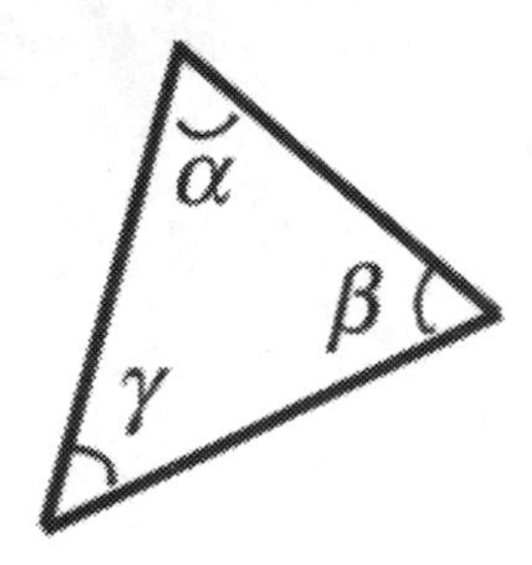

Figure 17.7 (a) Example 17.2

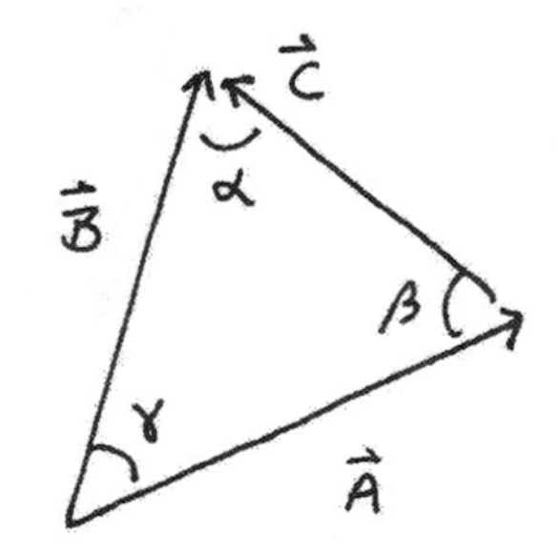

Figure 17.7 (b) Vector analysis

Solution: Consider the area of a triangle formed by three vectors $\vec{\mathbf{A}}$, $\vec{\mathbf{B}}$, and $\vec{\mathbf{C}}$, where $\vec{\mathbf{A}}+\vec{\mathbf{B}}+\vec{\mathbf{C}}=0$ (Figure 17.7b). Because $\vec{\mathbf{A}}+\vec{\mathbf{B}}+\vec{\mathbf{C}}=0$, we have that $0=\vec{\mathbf{A}}\times(\vec{\mathbf{A}}+\vec{\mathbf{B}}+\vec{\mathbf{C}})=\vec{\mathbf{A}}\times\vec{\mathbf{B}}+\vec{\mathbf{A}}\times\vec{\mathbf{C}}$. Thus $\vec{\mathbf{A}}\times\vec{\mathbf{B}}=-\vec{\mathbf{A}}\times\vec{\mathbf{C}}$ or $\left|\vec{\mathbf{A}}\times\vec{\mathbf{B}}\right|=\left|\vec{\mathbf{A}}\times\vec{\mathbf{C}}\right|$. From Figure 17.7b we see that $\left|\vec{\mathbf{A}}\times\vec{\mathbf{B}}\right|=\left|\vec{\mathbf{A}}\right|\left|\vec{\mathbf{B}}\right|\sin\gamma$ and $\left|\vec{\mathbf{A}}\times\vec{\mathbf{C}}\right|=\left|\vec{\mathbf{A}}\right|\left|\vec{\mathbf{C}}\right|\sin\beta$. Therefore $\left|\vec{\mathbf{A}}\right|\left|\vec{\mathbf{B}}\right|\sin\gamma=\left|\vec{\mathbf{A}}\right|\left|\vec{\mathbf{C}}\right|\sin\beta$, and hence $\left|\vec{\mathbf{B}}\right|/\sin\beta=\left|\vec{\mathbf{C}}\right|/\sin\gamma$. A similar argument shows that $\left|\vec{\mathbf{B}}\right|/\sin\beta=\left|\vec{\mathbf{A}}\right|/\sin\alpha$ proving the law of sines.

Example 17.3 Unit Normal

Find a unit vector perpendicular to $\vec{\mathbf{A}} = \hat{\mathbf{i}} + \hat{\mathbf{j}} - \hat{\mathbf{k}}$ and $\vec{\mathbf{B}} = -2\hat{\mathbf{i}} - \hat{\mathbf{j}} + 3\hat{\mathbf{k}}$.

Solution: The vector product $\vec{\mathbf{A}}\times\vec{\mathbf{B}}$ is perpendicular to both $\vec{\mathbf{A}}$ and $\vec{\mathbf{B}}$. Therefore the unit vectors $\hat{\mathbf{n}}=\pm\vec{\mathbf{A}}\times\vec{\mathbf{B}}/\left|\vec{\mathbf{A}}\times\vec{\mathbf{B}}\right|$ are perpendicular to both $\vec{\mathbf{A}}$ and $\vec{\mathbf{B}}$. We first calculate

$$\begin{aligned}\vec{\mathbf{A}}\times\vec{\mathbf{B}}&=(A_yB_z-A_zB_y)\,\hat{\mathbf{i}}+(A_zB_x-A_xB_z)\,\hat{\mathbf{j}}+(A_xB_y-A_yB_x)\,\hat{\mathbf{k}}\\&=((1)(3)-(-1)(-1))\,\hat{\mathbf{i}}+((-1)(2)-(1)(3))\,\hat{\mathbf{j}}+((1)(-1)-(1)(2))\,\hat{\mathbf{k}}\\&=2\,\hat{\mathbf{i}}-5\,\hat{\mathbf{j}}-3\,\hat{\mathbf{k}}.\end{aligned}$$

We now calculate the magnitude

$$\left|\vec{\mathbf{A}}\times\vec{\mathbf{B}}\right|=(2^2+5^2+3^2)^{1/2}=(38)^{1/2}\ .$$

Therefore the perpendicular unit vectors are

$$\hat{\mathbf{n}}=\pm\vec{\mathbf{A}}\times\vec{\mathbf{B}}/\left|\vec{\mathbf{A}}\times\vec{\mathbf{B}}\right|=\pm(2\,\hat{\mathbf{i}}-5\,\hat{\mathbf{j}}-3\,\hat{\mathbf{k}})/(38)^{1/2}.$$

Example 17.4 Volume of Parallelepiped

Show that the volume of a parallelepiped with edges formed by the vectors $\vec{\mathbf{A}}$, $\vec{\mathbf{B}}$, and $\vec{\mathbf{C}}$ is given by $\vec{\mathbf{A}}\cdot(\vec{\mathbf{B}}\times\vec{\mathbf{C}})$.

Solution: The volume of a parallelepiped is given by area of the base times height. If the base is formed by the vectors $\vec{\mathbf{B}}$ and $\vec{\mathbf{C}}$, then the area of the base is given by the magnitude of $\vec{\mathbf{B}}\times\vec{\mathbf{C}}$. The vector $\vec{\mathbf{B}}\times\vec{\mathbf{C}}=\left|\vec{\mathbf{B}}\times\vec{\mathbf{C}}\right|\hat{\mathbf{n}}$ where $\hat{\mathbf{n}}$ is a unit vector perpendicular to the base (Figure 17.8).

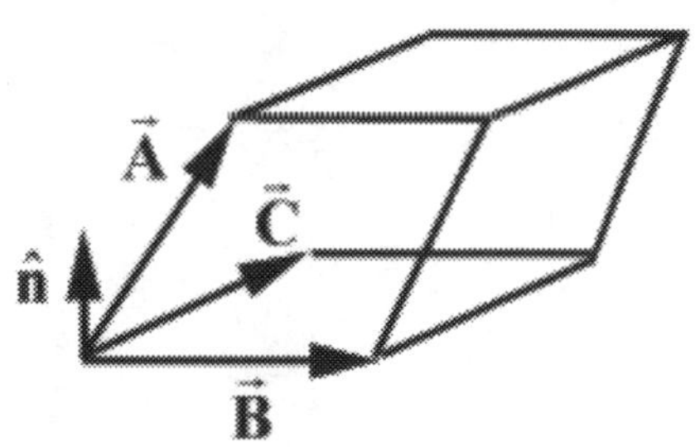

Figure 17.8 Example 17.4

The projection of the vector $\vec{\mathbf{A}}$ along the direction $\hat{\mathbf{n}}$ gives the height of the parallelepiped. This projection is given by taking the dot product of $\vec{\mathbf{A}}$ with a unit vector and is equal to $\vec{\mathbf{A}}\cdot\hat{\mathbf{n}}=height$. Therefore

$$\vec{\mathbf{A}}\cdot(\vec{\mathbf{B}}\times\vec{\mathbf{C}})=\vec{\mathbf{A}}\cdot(\left|\vec{\mathbf{B}}\times\vec{\mathbf{C}}\right|)\hat{\mathbf{n}}=(\left|\vec{\mathbf{B}}\times\vec{\mathbf{C}}\right|)\vec{\mathbf{A}}\cdot\hat{\mathbf{n}}=(area)(height)=(volume).$$

Example 17.5 Vector Decomposition

Let $\vec{\mathbf{A}}$ be an arbitrary vector and let $\hat{\mathbf{n}}$ be a unit vector in some fixed direction. Show that $\vec{\mathbf{A}} = (\vec{\mathbf{A}} \cdot \hat{\mathbf{n}})\hat{\mathbf{n}} + (\hat{\mathbf{n}} \times \vec{\mathbf{A}}) \times \hat{\mathbf{n}}$.

Solution: Let $\vec{\mathbf{A}} = A_{\|}\hat{\mathbf{n}} + A_{\perp}\hat{\mathbf{e}}$ where $A_{\|}$ is the component $\vec{\mathbf{A}}$ in the direction of $\hat{\mathbf{n}}$, $\hat{\mathbf{e}}$ is the direction of the projection of $\vec{\mathbf{A}}$ in a plane perpendicular to $\hat{\mathbf{n}}$, and $A_{\perp}$ is the component of $\vec{\mathbf{A}}$ in the direction of $\hat{\mathbf{e}}$. Because $\hat{\mathbf{e}} \cdot \hat{\mathbf{n}} = 0$, we have that $\vec{\mathbf{A}} \cdot \hat{\mathbf{n}} = A_{\|}$. Note that

$$\hat{\mathbf{n}} \times \vec{\mathbf{A}} = \hat{\mathbf{n}} \times (A_{\|}\hat{\mathbf{n}} + A_{\perp}\hat{\mathbf{e}}) = \hat{\mathbf{n}} \times A_{\perp}\hat{\mathbf{e}} = A_{\perp}(\hat{\mathbf{n}} \times \hat{\mathbf{e}}).$$

The unit vector $\hat{\mathbf{n}} \times \hat{\mathbf{e}}$ lies in the plane perpendicular to $\hat{\mathbf{n}}$ and is also perpendicular to $\hat{\mathbf{e}}$. Therefore $(\hat{\mathbf{n}} \times \hat{\mathbf{e}}) \times \hat{\mathbf{n}}$ is also a unit vector that is parallel to $\hat{\mathbf{e}}$ (by the right hand rule. So $(\hat{\mathbf{n}} \times \vec{\mathbf{A}}) \times \hat{\mathbf{n}} = A_{\perp}\hat{\mathbf{e}}$. Thus

$$\vec{\mathbf{A}} = A_{\|}\hat{\mathbf{n}} + A_{\perp}\hat{\mathbf{e}} = (\vec{\mathbf{A}} \cdot \hat{\mathbf{n}})\hat{\mathbf{n}} + (\hat{\mathbf{n}} \times \vec{\mathbf{A}}) \times \hat{\mathbf{n}}.$$

17.3 Torque

17.3.1 Definition of Torque about a Point

In order to understand the dynamics of a rotating rigid body we will introduce a new quantity, the torque. Let a force $\vec{\mathbf{F}}_P$ with magnitude $F = |\vec{\mathbf{F}}_P|$ act at a point P. Let $\vec{\mathbf{r}}_{S,P}$ be the vector from the point S to a point P, with magnitude $r = |\vec{\mathbf{r}}_{S,P}|$. The angle between the vectors $\vec{\mathbf{r}}_{S,P}$ and $\vec{\mathbf{F}}_P$ is θ with $[0 \le \theta \le \pi]$ (Figure 17.9).

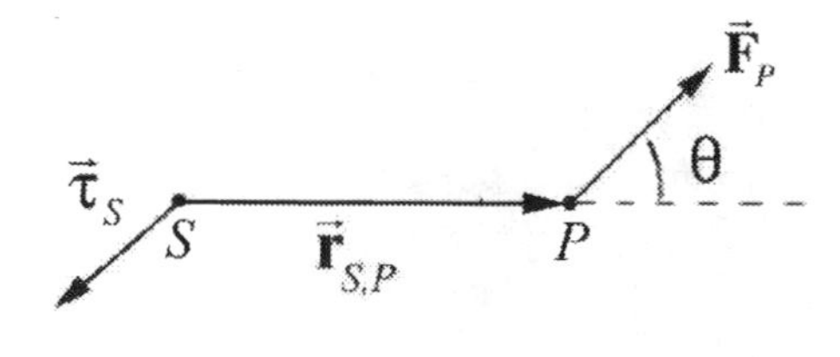

Figure 17.9 Torque about a point S due to a force acting at a point P

The torque about a point S due to force $\vec{\mathbf{F}}_P$ acting at P, is defined by

$$\vec{\boldsymbol{\tau}}_S = \vec{\mathbf{r}}_{S,P} \times \vec{\mathbf{F}}_P. \qquad (17.2.28)$$

The magnitude of the torque about a point S due to force $\vec{\mathbf{F}}_P$ acting at P, is given by

$$\tau_S \equiv \left|\vec{\boldsymbol{\tau}}_S\right| = r\,F\sin\theta . \qquad (17.2.29)$$

The SI units for torque are $[\mathrm{N}\cdot\mathrm{m}]$. The direction of the torque is perpendicular to the plane formed by the vectors $\vec{\mathbf{r}}_{S,P}$ and $\vec{\mathbf{F}}_P$ (for $[0<\theta<\pi]$), and by definition points in the direction of the unit normal vector to the plane $\hat{\mathbf{n}}_{RHR}$ as shown in Figure 17.10.

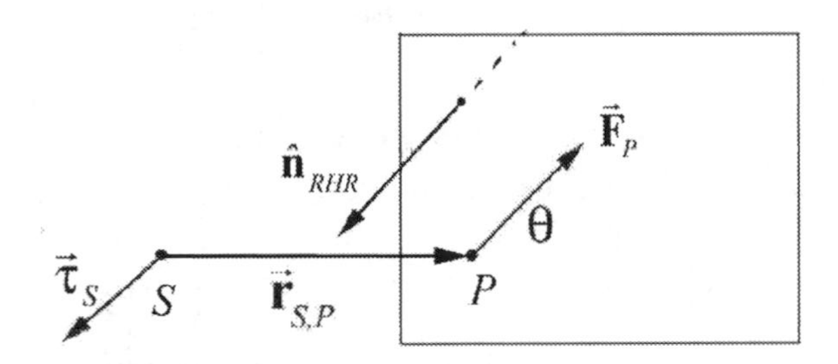

Figure 17.10 Vector direction for the torque

Figure 17.11 shows the two different ways of defining height and base for a parallelogram defined by the vectors $\vec{\mathbf{r}}_{S,P}$ and $\vec{\mathbf{F}}_P$.

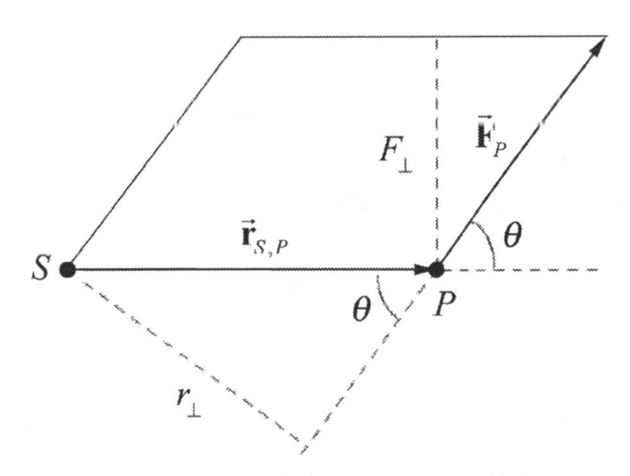

Figure 17.11 Area of the torque parallelogram.

Let $r_\perp = r\sin\theta$ and let $F_\perp = F\sin\theta$ be the component of the force $\vec{\mathbf{F}}_P$ that is perpendicular to the line passing from the point S to P. (Recall the angle θ has a range of values $0 \le \theta \le \pi$ so both $r_\perp \ge 0$ and $F_\perp \ge 0$.) Then the area of the parallelogram defined by $\vec{\mathbf{r}}_{S,P}$ and $\vec{\mathbf{F}}_P$ is given by

$$\text{Area} = \tau_S = r_\perp F = r\,F_\perp = r\,F\sin\theta . \qquad (17.2.30)$$

We can interpret the quantity $r_\perp$ as follows.

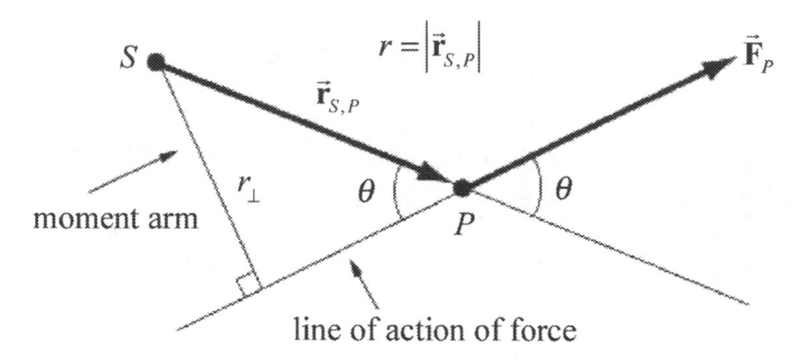

Figure 17.12 The moment arm about the point S and line of action of force passing through the point P

We begin by drawing the ***line of action of the force*** $\vec{\mathbf{F}}_P$. This is a straight line passing through P, parallel to the direction of the force $\vec{\mathbf{F}}_P$. Draw a perpendicular to this line of action that passes through the point S (Figure 17.12). The length of this perpendicular, $r_\perp = r\sin\theta$, is called ***the moment arm about the point S of the force*** $\vec{\mathbf{F}}_P$.

You should keep in mind three important properties of torque:

1. The torque is zero if the vectors $\vec{\mathbf{r}}_{S,P}$ and $\vec{\mathbf{F}}_P$ are parallel $(\theta = 0)$ or anti-parallel $(\theta = \pi)$.

2. Torque is a vector whose direction and magnitude depend on the choice of a point S about which the torque is calculated.

3. The direction of torque is perpendicular to the plane formed by the two vectors, $\vec{\mathbf{F}}_P$ and $r = |\vec{\mathbf{r}}_{S,P}|$ (the vector from the point S to a point P).

17.3.2 Alternative Approach to Assigning a Sign Convention for Torque

In the case where all of the forces $\vec{\mathbf{F}}_i$ and position vectors $\vec{\mathbf{r}}_{i,P}$ are coplanar (or zero), we can, instead of referring to the direction of torque, assign a purely algebraic positive or negative sign to torque according to the following convention. We note that the arc in Figure 17.13a circles in counterclockwise direction. (Figures 17.13a and 17.13b use the simplifying assumption, for the purpose of the figure only, that the two vectors in question, $\vec{\mathbf{F}}_P$ and $\vec{\mathbf{r}}_{S,P}$ are perpendicular. The point S about which torques are calculated is not shown.)

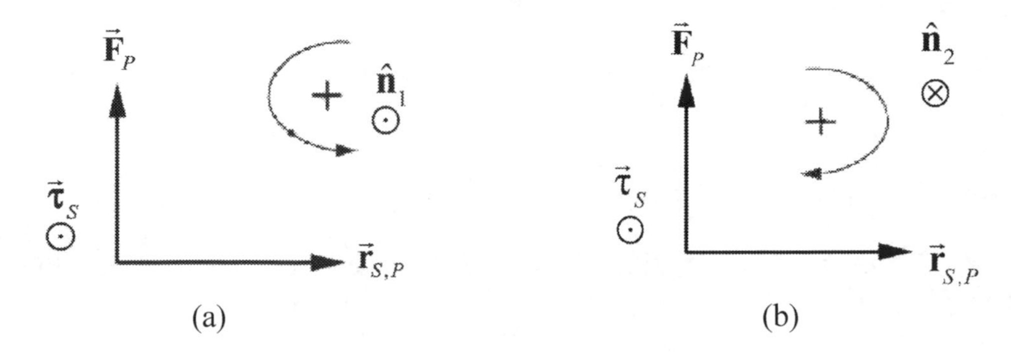

Figure 17.13 (a) Positive torque out of plane, **(b)** positive torque into plane

We can associate with this counterclockwise orientation a unit normal vector $\hat{\mathbf{n}}$ according to the right-hand rule: curl your right hand fingers in the counterclockwise direction and your right thumb will then point in the $\hat{\mathbf{n}}_1$ direction (Figure 17.13a). The arc in Figure 17.13b circles in the clockwise direction, and we associate this orientation with the unit normal $\hat{\mathbf{n}}_2$.

It's important to note that the terms "clockwise" and "counterclockwise" might be different for different observers. For instance, if the plane containing $\vec{\mathbf{F}}_P$ and $\vec{\mathbf{r}}_{S,P}$ is horizontal, an observer above the plane and an observer below the plane would disagree on the two terms. For a vertical plane, the directions that two observers on opposite sides of the plane would be mirror images of each other, and so again the observers would disagree.

1. Suppose we choose counterclockwise as positive. Then we assign a positive sign for the component of the torque when the torque is in the same direction as the unit normal $\hat{\mathbf{n}}_1$, i.e. $\vec{\boldsymbol{\tau}}_S = \vec{\mathbf{r}}_{S,P} \times \vec{\mathbf{F}}_P = +|\vec{\mathbf{r}}_{S,P}||\vec{\mathbf{F}}_P|\hat{\mathbf{n}}_l$, (Figure 17.13a).

2. Suppose we choose clockwise as positive. Then we assign a negative sign for the component of the torque in Figure 17.13b because the torque is directed opposite to the unit normal $\hat{\mathbf{n}}_2$, i.e. $\vec{\boldsymbol{\tau}}_S = \vec{\mathbf{r}}_{S,P} \times \vec{\mathbf{F}}_P = -|\vec{\mathbf{r}}_{S,P}||\vec{\mathbf{F}}_P|\hat{\mathbf{n}}_2$.

Example 17.6 Torque and Vector Product

Consider two vectors $\vec{\mathbf{r}}_{P,F} = x\hat{\mathbf{i}}$ with $x > 0$ and $\vec{\mathbf{F}} = F_x\hat{\mathbf{i}} + F_z\hat{\mathbf{k}}$ with $F_x > 0$ and $F_z > 0$. Calculate the torque $\vec{\mathbf{r}}_{P,F} \times \vec{\mathbf{F}}$.

Solution: We calculate the vector product noting that in a right handed choice of unit vectors, $\hat{\mathbf{i}} \times \hat{\mathbf{i}} = \vec{\mathbf{0}}$ and $\hat{\mathbf{i}} \times \hat{\mathbf{k}} = -\hat{\mathbf{j}}$,

$$\vec{\mathbf{r}}_{P,F} \times \vec{\mathbf{F}} = x\hat{\mathbf{i}} \times (F_x\hat{\mathbf{i}} + F_z\hat{\mathbf{k}}) = (x\hat{\mathbf{i}} \times F_x\hat{\mathbf{i}}) + (x\hat{\mathbf{i}} \times F_z\hat{\mathbf{k}}) = -xF_z\,\hat{\mathbf{j}}.$$

Because $x > 0$ and $F_z > 0$, the direction of the vector product is in the negative y-direction.

Example 17.7 Calculating Torque

In Figure 17.14, a force of magnitude F is applied to one end of a lever of length L. What is the magnitude and direction of the torque about the point S?

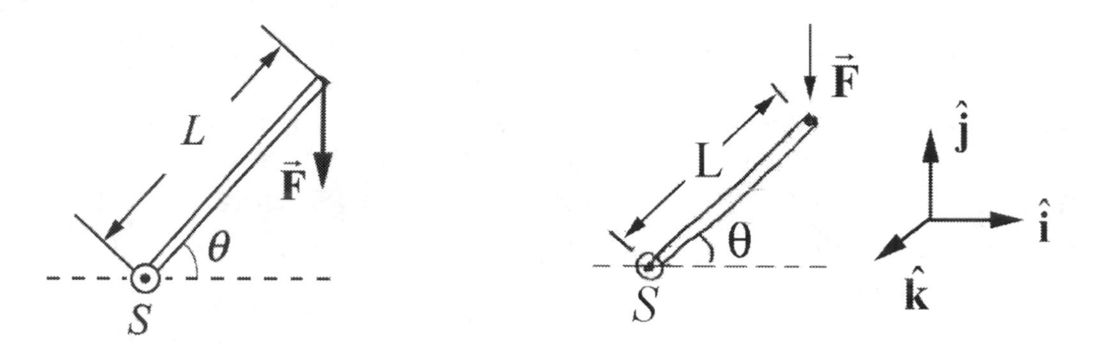

Figure 17.14 Example 17.7 **Figure 17.15** Coordinate system

Solution: Choose units vectors such that $\hat{\mathbf{i}} \times \hat{\mathbf{j}} = \hat{\mathbf{k}}$, with $\hat{\mathbf{i}}$ pointing to the right and $\hat{\mathbf{j}}$ pointing up (Figure 17.15). The torque about the point S is given by $\vec{\boldsymbol{\tau}}_S = \vec{\mathbf{r}}_{S,F} \times \vec{\mathbf{F}}$, where $\vec{\mathbf{r}}_{SF} = L\cos\theta\hat{\mathbf{i}} + L\sin\theta\hat{\mathbf{j}}$ and $\vec{\mathbf{F}} = -F\hat{\mathbf{j}}$ then

$$\vec{\boldsymbol{\tau}}_S = (L\cos\theta\,\hat{\mathbf{i}} + L\sin\theta\,\hat{\mathbf{j}}) \times -F\,\hat{\mathbf{j}} = -FL\cos\theta\,\hat{\mathbf{k}}.$$

Example 17.8 Torque and the Ankle

A person of mass m is crouching with their weight evenly distributed on both tiptoes. The free-body force diagram on the skeletal part of the foot is shown in Figure 17.16. The normal force $\vec{\mathbf{N}}$ acts at the contact point between the foot and the ground. In this position, the tibia acts on the foot at the point S with a force $\vec{\mathbf{F}}$ of an unknown magnitude $F = |\vec{\mathbf{F}}|$ and makes an unknown angle β with the vertical. This force acts on the ankle a horizontal distance s from the point where the foot contacts the floor. The Achilles tendon also acts on the foot and is under considerable tension with magnitude $T \equiv |\vec{\mathbf{T}}|$ and acts at an angle α with the horizontal as shown in the figure. The tendon acts on the ankle a horizontal distance b from the point S where the tibia acts on the foot. You may ignore the weight of the foot. Let g be the gravitational constant. Compute the torque about the point S due to (a) the tendon force on the foot; (b) the force of the tibia on the foot; (c) the normal force of the floor on the foot.

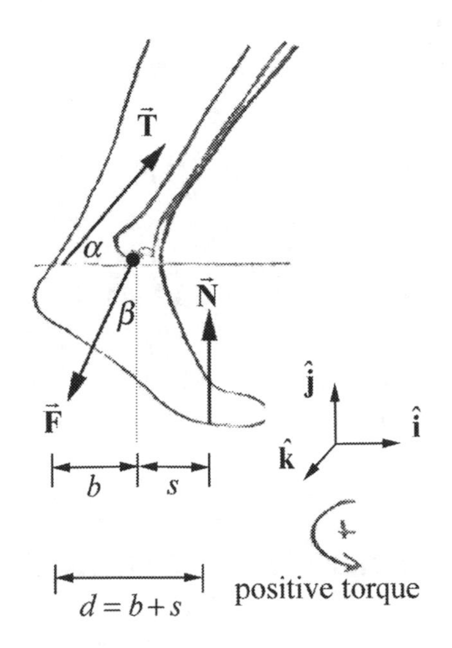

Figure 17.16 Force diagram and coordinate system for ankle

Solution: (a) We shall first calculate the torque due to the force of the Achilles tendon on the ankle. The tendon force has the vector decomposition $\vec{\mathbf{T}} = T\cos\alpha\hat{\mathbf{i}} + T\sin\alpha\hat{\mathbf{j}}$.

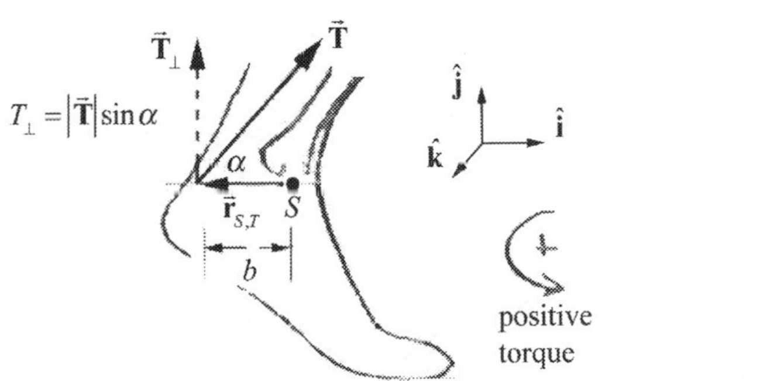

Figure 17.17 Torque diagram for tendon force on ankle

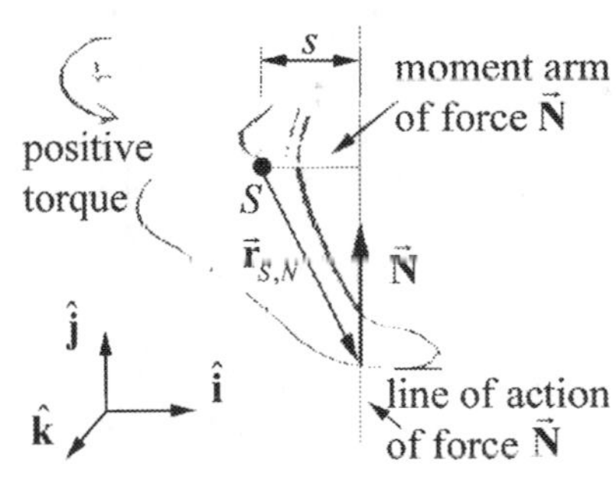

Figure 17.18 Torque diagram for normal force on ankle $\vec{\mathbf{r}}_{S,N}$

The vector from the point S to the point of action of the force is given by $\vec{\mathbf{r}}_{S,T} = -b\hat{\mathbf{i}}$ (Figure 17.17). Therefore the torque due to the force of the tendon $\vec{\mathbf{T}}$ on the ankle about the point S is then

$$\vec{\tau}_{S,T} = \vec{\mathbf{r}}_{S,T} \times \vec{\mathbf{T}} = -b\hat{\mathbf{i}} \times (T\cos\alpha\hat{\mathbf{i}} + T\sin\alpha\hat{\mathbf{j}}) = -bT\sin\alpha\hat{\mathbf{k}}.$$

(b) The torque diagram for the normal force is shown in Figure 17.18. The vector from the point S to the point where the normal force acts on the foot is given by

$\vec{\mathbf{r}}_{S,N} = (s\hat{\mathbf{i}} - h\hat{\mathbf{j}})$. Because the weight is evenly distributed on the two feet, the normal force on one foot is equal to half the weight, or $N = (1/2)mg$. The normal force is therefore given by $\vec{\mathbf{N}} = N\,\hat{\mathbf{j}} = (1/2)mg\,\hat{\mathbf{j}}$. Therefore the torque of the normal force about the point S is

$$\vec{\tau}_{S,N} = \vec{\mathbf{r}}_{S,N} \times N\,\hat{\mathbf{j}} = (s\hat{\mathbf{i}} - h\hat{\mathbf{j}}) \times N\,\hat{\mathbf{j}} = s\,N\,\hat{\mathbf{k}} = (1/2)\,s\,mg\,\hat{\mathbf{k}}.$$

(c) The force $\vec{\mathbf{F}}$ that the tibia exerts on the ankle will make no contribution to the torque about this point S since the tibia force acts at the point S and therefore the vector $\vec{\mathbf{r}}_{S,F} = \vec{\mathbf{0}}$.

17.4 Torque, Angular Acceleration, and Moment of Inertia

17.4.1 Torque Equation for Fixed Axis Rotation

For fixed-axis rotation, there is a direct relation between the component of the torque along the axis of rotation and angular acceleration. Consider the forces that act on the rotating body. Generally, the forces on different volume elements will be different, and so we will denote the force on the volume element of mass Δm_i by $\vec{\mathbf{F}}_i$. Choose the z-axis to lie along the axis of rotation. Divide the body into volume elements of mass Δm_i. Let the point S denote a specific point along the axis of rotation (Figure 17.19). Each volume element undergoes a tangential acceleration as the volume element moves in a circular orbit of radius $r_i = |\vec{\mathbf{r}}_i|$ about the fixed axis.

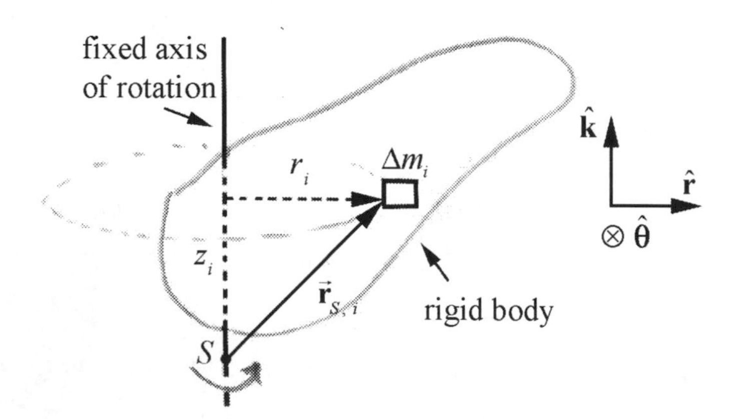

Figure 17.19: Volume element undergoing fixed-axis rotation about the z-axis.

The vector from the point S to the volume element is given by

$$\vec{\mathbf{r}}_{S,i} = z_i\,\hat{\mathbf{k}} + \vec{\mathbf{r}}_i = z_i\,\hat{\mathbf{k}} + r_i\,\hat{\mathbf{r}} \qquad (17.3.1)$$

where z_i is the distance along the axis of rotation between the point S and the volume element. The torque about S due to the force $\vec{\mathbf{F}}_i$ acting on the volume element is given by

$$\vec{\boldsymbol{\tau}}_{S,i} = \vec{\mathbf{r}}_{S,i} \times \vec{\mathbf{F}}_i . \tag{17.3.2}$$

Substituting Eq. (17.3.1) into Eq. (17.3.2) gives

$$\vec{\boldsymbol{\tau}}_{S,i} = (z_i\,\hat{\mathbf{k}} + r_i\,\hat{\mathbf{r}}) \times \vec{\mathbf{F}}_i . \tag{17.3.3}$$

For fixed-axis rotation, we are interested in the z-component of the torque, which must be the term

$$(\vec{\boldsymbol{\tau}}_{S,i})_z = (r_i\,\hat{\mathbf{r}} \times \vec{\mathbf{F}}_i)_z \tag{17.3.4}$$

because the vector product $z_i\,\hat{\mathbf{k}} \times \vec{\mathbf{F}}_i$ must be directed perpendicular to the plane formed by the vectors $\hat{\mathbf{k}}$ and $\vec{\mathbf{F}}_i$, hence perpendicular to the z-axis. The force acting on the volume element has components

$$\vec{\mathbf{F}}_i = F_{r,i}\,\hat{\mathbf{r}} + F_{\theta,i}\,\hat{\boldsymbol{\theta}} + F_{z,i}\,\hat{\mathbf{k}} . \tag{17.3.5}$$

The z-component $F_{z,i}$ of the force cannot contribute a torque in the z-direction, and so substituting Eq. (17.3.5) into Eq. (17.3.4) yields

$$(\vec{\boldsymbol{\tau}}_{S,i})_z = (r_i\,\hat{\mathbf{r}} \times (F_{r,i}\,\hat{\mathbf{r}} + F_{\theta,i}\,\hat{\boldsymbol{\theta}}))_z . \tag{17.3.6}$$

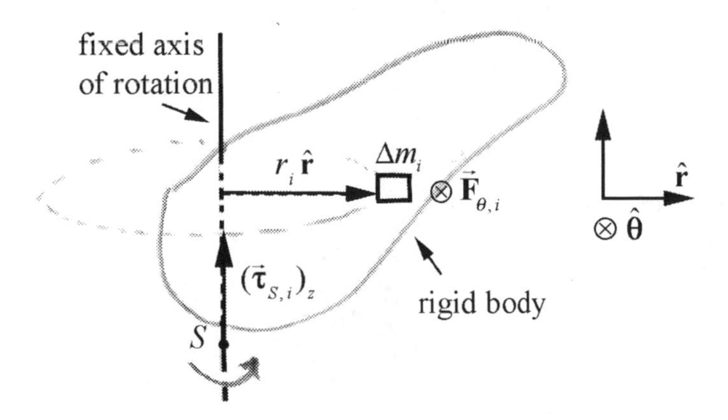

Figure 17.20 Tangential force acting on a volume element.

The radial force does not contribute to the torque about the z-axis, since

$$r_i\,\hat{\mathbf{r}} \times F_{r,i}\,\hat{\mathbf{r}} = \vec{\mathbf{0}} . \tag{17.3.7}$$

So, we are interested in the contribution due to torque about the z-axis due to the tangential component of the force on the volume element (Figure 17.20). The component of the torque about the z-axis is given by

$$(\vec{\boldsymbol{\tau}}_{S,i})_z = (r_i \hat{\mathbf{r}} \times F_{\theta,i} \hat{\boldsymbol{\theta}})_z = r_i F_{\theta,i} . \tag{17.3.8}$$

The z-component of the torque is directed upwards in Figure 17.20, where $F_{\theta,i}$ is positive (the tangential force is directed counterclockwise, as in the figure). Applying Newton's Second Law in the tangential direction,

$$F_{\theta,i} = \Delta m_i a_{\theta,i} . \tag{17.3.9}$$

Using our kinematics result that the tangential acceleration is $a_{\theta,i} = r_i \alpha_z$, where α_z is the z-component of angular acceleration, we have that

$$F_{\theta,i} = \Delta m_i r_i \alpha_z . \tag{17.3.10}$$

From Eq. (17.3.8), the component of the torque about the z-axis is then given by

$$(\vec{\boldsymbol{\tau}}_{S,i})_z = r_i F_{\theta,i} = \Delta m_i r_i^2 \alpha_z . \tag{17.3.11}$$

The component of the torque about the z-axis is the summation of the torques on all the volume elements,

$$\begin{aligned}(\vec{\boldsymbol{\tau}}_S)_z &= \sum_{i=1}^{i=N} (\vec{\boldsymbol{\tau}}_{S,i})_z = \sum_{i=1}^{i=N} r_{\perp,i} F_{\theta,i} \\ &= \sum_{i=1}^{i=N} \Delta m_i r_i^2 \alpha_z .\end{aligned} \tag{17.3.12}$$

Because each element has the same z-component of angular acceleration, α_z, the summation becomes

$$(\vec{\boldsymbol{\tau}}_S)_z = \left(\sum_{i=1}^{i=N} \Delta m_i r_i^2 \right) \alpha_z . \tag{17.3.13}$$

Recalling our definition of the moment of inertia, (Chapter 16.3) the z-component of the torque is proportional to the z-component of angular acceleration,

$$\tau_{S,z} = I_S \alpha_z , \tag{17.3.14}$$

and the moment of inertia, I_S, is the constant of proportionality. The torque about the point S is the sum of the external torques and the internal torques

$$\vec{\boldsymbol{\tau}}_S = \vec{\boldsymbol{\tau}}_S^{\text{ext}} + \vec{\boldsymbol{\tau}}_S^{\text{int}}. \tag{17.3.15}$$

The external torque about the point S is the sum of the torques due to the net external force acting on each element

$$\vec{\boldsymbol{\tau}}_S^{\text{ext}} = \sum_{i=1}^{i=N} \vec{\boldsymbol{\tau}}_{S,i}^{\text{ext}} = \sum_{i=1}^{i=N} \vec{\mathbf{r}}_{S,i} \times \vec{\mathbf{F}}_i^{\text{ext}}. \tag{17.3.16}$$

The internal torque arise from the torques due to the internal forces acting between pairs of elements

$$\vec{\boldsymbol{\tau}}_S^{\text{int}} = \sum_{i=1}^{N} \vec{\boldsymbol{\tau}}_{S,j}^{\text{int}} = \sum_{i=1}^{i=N} \sum_{\substack{j=1 \\ j \neq i}}^{j=N} \vec{\boldsymbol{\tau}}_{S,j,i}^{\text{int}} = \sum_{i=1}^{i=N} \sum_{\substack{j=1 \\ j \neq i}}^{j=N} \vec{\mathbf{r}}_{S,i} \times \vec{\mathbf{F}}_{j,i}. \tag{17.3.17}$$

We know by Newton's Third Law that the internal forces cancel in pairs, $\vec{\mathbf{F}}_{j,i} = -\vec{\mathbf{F}}_{i,j}$, and hence the sum of the internal forces is zero

$$\vec{\mathbf{0}} = \sum_{i=1}^{i=N} \sum_{\substack{j=1 \\ j \neq i}}^{j=N} \vec{\mathbf{F}}_{j,i}. \tag{17.3.18}$$

Does the same statement hold about pairs of internal torques? Consider the sum of internal torqucs arising from the interaction between the ith and j^{th} particles

$$\vec{\boldsymbol{\tau}}_{S,j,i}^{\text{int}} + \vec{\boldsymbol{\tau}}_{S,i,j}^{\text{int}} = \vec{\mathbf{r}}_{S,i} \times \vec{\mathbf{F}}_{j,i} + \vec{\mathbf{r}}_{S,j} \times \vec{\mathbf{F}}_{i,j}. \tag{17.3.19}$$

By the Newton's Third Law this sum becomes

$$\vec{\boldsymbol{\tau}}_{S,j,i}^{\text{int}} + \vec{\boldsymbol{\tau}}_{S,i,j}^{\text{int}} = (\vec{\mathbf{r}}_{S,i} - \vec{\mathbf{r}}_{S,j}) \times \vec{\mathbf{F}}_{j,i}. \tag{17.3.20}$$

In the Figure 17.21, the vector $\vec{\mathbf{r}}_{S,i} - \vec{\mathbf{r}}_{S,j}$ points from the jth element to the ith element. If the internal forces between a pair of particles are directed along the line joining the two particles then the torque due to the internal forces cancel in pairs.

$$\vec{\boldsymbol{\tau}}_{S,j,i}^{\text{int}} + \vec{\boldsymbol{\tau}}_{S,i,j}^{\text{int}} = (\vec{\mathbf{r}}_{S,i} - \vec{\mathbf{r}}_{S,j}) \times \vec{\mathbf{F}}_{j,i} = \vec{\mathbf{0}}. \tag{17.3.21}$$

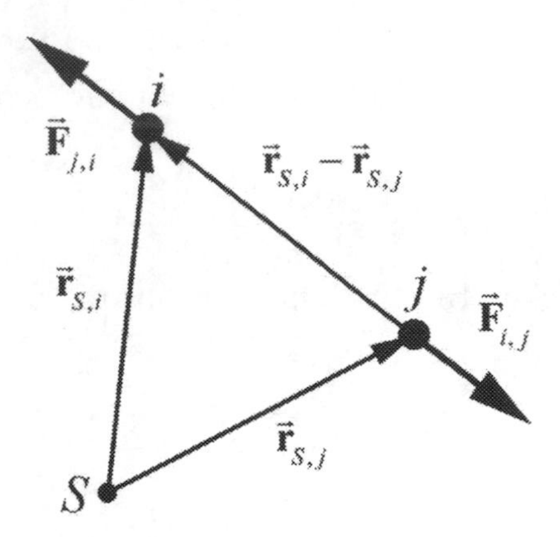

Figure 17.21 The internal force is directed along the line connecting the ith and j^{th} particles

This is a stronger version of Newton's Third Law than we have so far since we have added the additional requirement regarding the direction of all the internal forces between pairs of particles. With this assumption, the torque is just due to the external forces

$$\vec{\boldsymbol{\tau}}_S = \vec{\boldsymbol{\tau}}_S^{\text{ext}} . \tag{17.3.22}$$

Thus Eq. (17.3.14) becomes

$$(\tau_S^{\text{ext}})_z = I_S \alpha_z , \tag{17.3.23}$$

This is very similar to Newton's Second Law: the total force is proportional to the acceleration,

$$\vec{\mathbf{F}} = m\vec{\mathbf{a}} . \tag{17.3.24}$$

where the mass, m, is the constant of proportionality.

17.4.2 Torque Acts at the Center of Gravity

Suppose a rigid body in static equilibrium consists of N particles labeled by the index $i = 1, 2, 3, ..., N$. Choose a coordinate system with a choice of origin O such that mass m_i has position $\vec{\mathbf{r}}_i$. Each point particle experiences a gravitational force $\vec{\mathbf{F}}_{\text{gravity},i} = m_i \vec{\mathbf{g}}$. The total torque about the origin is then zero (static equilibrium condition),

$$\vec{\boldsymbol{\tau}}_O = \sum_{i=1}^{i=N} \vec{\boldsymbol{\tau}}_{O,i} = \sum_{i=1}^{i=N} \vec{\mathbf{r}}_i \times \vec{\mathbf{F}}_{\text{gravity},i} = \sum_{i=1}^{i=N} \vec{\mathbf{r}}_i \times m_i \vec{\mathbf{g}} = \vec{\mathbf{0}} . \tag{17.3.25}$$

If the gravitational acceleration $\vec{\mathbf{g}}$ is assumed constant, we can rearrange the summation in Eq. (17.3.25) by pulling the constant vector $\vec{\mathbf{g}}$ out of the summation ($\vec{\mathbf{g}}$ appears in each term in the summation),

$$\vec{\boldsymbol{\tau}}_O = \sum_{i=1}^{i=N} \vec{\mathbf{r}}_i \times m_i \vec{\mathbf{g}} = \left(\sum_{i=1}^{i=N} m_i \vec{\mathbf{r}}_i \right) \times \vec{\mathbf{g}} = \vec{\mathbf{0}}. \qquad (17.3.26)$$

We now use our definition of the center of the center of mass, Eq. (10.5.3), to rewrite Eq. (17.3.26) as

$$\vec{\boldsymbol{\tau}}_O = \sum_{i=1}^{i=N} \vec{\mathbf{r}}_i \times m_i \vec{\mathbf{g}} = M_{\mathrm{T}} \vec{\mathbf{R}}_{\mathrm{cm}} \times \vec{\mathbf{g}} = \vec{\mathbf{R}}_{\mathrm{cm}} \times M_{\mathrm{T}} \vec{\mathbf{g}} = \vec{\mathbf{0}}. \qquad (17.3.27)$$

Thus the torque due to the gravitational force acting on each point-like particle is equivalent to the torque due to the gravitational force acting on a point-like particle of mass M_{T} located at a point in the body called the ***center of gravity***, which is equal to the center of mass of the body in the typical case in which the gravitational acceleration $\vec{\mathbf{g}}$ is constant throughout the body.

Example 17.9 Turntable

The turntable in Example 16.1, of mass $1.2\ \mathrm{kg}$ and radius $1.3\times10^{1}\ \mathrm{cm}$, has a moment of inertia $I_S = 1.01\times10^{-2}\ \mathrm{kg}\cdot\mathrm{m}^2$ about an axis through the center of the turntable and perpendicular to the turntable. The turntable is spinning at an initial constant frequency $f_i = 33\,\mathrm{cycles}\cdot\mathrm{min}^{-1}$. The motor is turned off and the turntable slows to a stop in $8.0\ \mathrm{s}$ due to frictional torque. Assume that the angular acceleration is constant. What is the magnitude of the frictional torque acting on the turntable?

Solution: We have already calculated the angular acceleration of the turntable in Example 16.1, where we found that

$$\alpha_z = \frac{\Delta\omega_z}{\Delta t} = \frac{\omega_f - \omega_i}{t_f - t_i} = \frac{-3.5\ \mathrm{rad}\cdot\mathrm{s}^{-1}}{8.0\ \mathrm{s}} = -4.3\times10^{-1}\ \mathrm{rad}\cdot\mathrm{s}^{-2} \qquad (17.3.28)$$

and so the magnitude of the frictional torque is

$$\begin{aligned} \left|\tau_z^{\mathrm{fric}}\right| &= I_S \left|\alpha_z\right| = (1.01\times10^{-2}\ \mathrm{kg}\cdot\mathrm{m}^2)(4.3\times10^{-1}\ \mathrm{rad}\cdot\mathrm{s}^{-2}) \\ &= 4.3\times10^{-3}\ \mathrm{N}\cdot\mathrm{m}. \end{aligned} \qquad (17.3.29)$$

Example 17.10 Pulley and blocks

A pulley of mass m_{p}, radius R, and moment of inertia about its center of mass I_{cm}, is attached to the edge of a table. An inextensible string of negligible mass is wrapped around the pulley and attached on one end to block 1 that hangs over the edge of the table (Figure 17.22). The other end of the string is attached to block 2 that slides along a table.

The coefficient of sliding friction between the table and the block 2 is μ_k. Block 1 has mass m_1 and block 2 has mass m_2, with $m_1 > \mu_k m_2$. At time $t = 0$, the blocks are released from rest and the string does not slip around the pulley. At time $t = t_1$, block 1 hits the ground. Let g denote the gravitational constant. (a) Find the magnitude of the acceleration of each block. (b) How far did the block 1 fall before hitting the ground?

Figure 17.22 Example 17.10 **Figure 17.23** Torque diagram for pulley

Solution: The torque diagram for the pulley is shown in the figure below where we choose $\hat{\mathbf{k}}$ pointing into the page. Note that the tensions in the string on either side of the pulley are not equal. The reason is that the pulley is massive. To understand why, remember that the difference in the magnitudes of the torques due to the tension on either side of the pulley is equal to the moment of inertia times the magnitude of the angular acceleration, which is non-zero for a massive pulley. So the tensions cannot be equal. From our torque diagram, the torque about the point O at the center of the pulley is given by

$$\vec{\boldsymbol{\tau}}_O = \vec{\mathbf{r}}_{O,1} \times \vec{\mathbf{T}}_1 + \vec{\mathbf{r}}_{O,2} \times \vec{\mathbf{T}}_2 = R(T_1 - T_2)\hat{\mathbf{k}}. \quad (17.3.30)$$

Therefore the torque equation (17.3.23) becomes

$$R(T_1 - T_2) = I_z \alpha_z. \quad (17.3.31)$$

The free body force diagrams on the two blocks are shown in Figure 17.23.

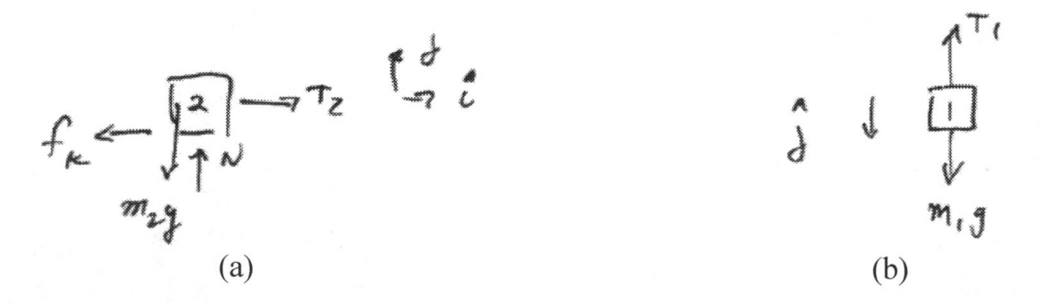

Figure 17.23 Free-body force diagrams on (a) block 2, (b) block 1

Newton's Second Law on block 1 yields

$$m_1 g - T_1 = m_1 a_{y1}. \tag{17.3.32}$$

Newton's Second Law on block 2 in the $\hat{\mathbf{j}}$ direction yields

$$N - m_2 g = 0. \tag{17.3.33}$$

Newton's Second Law on block 2 in the $\hat{\mathbf{i}}$ direction yields

$$T_2 - f_k = m_2 a_{x2}. \tag{17.3.34}$$

The kinetic friction force is given by

$$f_k = \mu_k N = \mu_k m_2 g \tag{17.3.35}$$

Therefore Eq. (17.3.34) becomes

$$T_2 - \mu_k m_2 g = m_2 a_{x2}. \tag{17.3.36}$$

Block 1 and block 2 are constrained to have the same acceleration so

$$a \equiv a_{x1} = a_{x2}. \tag{17.3.37}$$

We can solve Eqs. (17.3.32) and (17.3.36) for the two tensions yielding

$$T_1 = m_1 g - m_1 a, \tag{17.3.38}$$

$$T_2 = \mu_k m_2 g + m_2 a. \tag{17.3.39}$$

At point on the rim of the pulley has a tangential acceleration that is equal to the acceleration of the blocks so

$$a = a_\theta = R\alpha_z. \tag{17.3.40}$$

The torque equation (Eq. (17.3.31)) then becomes

$$T_1 - T_2 = \frac{I_z}{R^2} a. \tag{17.3.41}$$

Substituting Eqs. (17.3.38) and (17.3.39) into Eq. (17.3.41) yields

$$m_1 g - m_1 a - (\mu_k m_2 g + m_2 a) = \frac{I_z}{R^2} a, \tag{17.3.42}$$

which we can now solve for the accelerations of the blocks

$$a = \frac{m_1 g - \mu_k m_2 g}{m_1 + m_2 + I_z / R^2}. \quad (17.3.43)$$

Block 1 hits the ground at time t_1, therefore it traveled a distance

$$y_1 = \frac{1}{2}\left(\frac{m_1 g - \mu_k m_2 g}{m_1 + m_2 + I_z / R^2}\right) t_1^{\,2}. \quad (17.3.44)$$

Example 17.11 Experimental Method for Determining Moment of Inertia

A steel washer is mounted on a cylindrical rotor of radius $r = 12.7\ \text{mm}$. A massless string, with an object of mass $m = 0.055\ \text{kg}$ attached to the other end, is wrapped around the side of the rotor and passes over a massless pulley (Figure 17.24). Assume that there is a constant frictional torque about the axis of the rotor. The object is released and falls. As the object falls, the rotor undergoes an angular acceleration of magnitude α_1. After the string detaches from the rotor, the rotor coasts to a stop with an angular acceleration of magnitude α_2. Let $g = 9.8\ \text{m}\cdot\text{s}^{-2}$ denote the gravitational constant. Based on the data in the Figure 17.25, what is the moment of inertia I_R of the rotor assembly (including the washer) about the rotation axis?

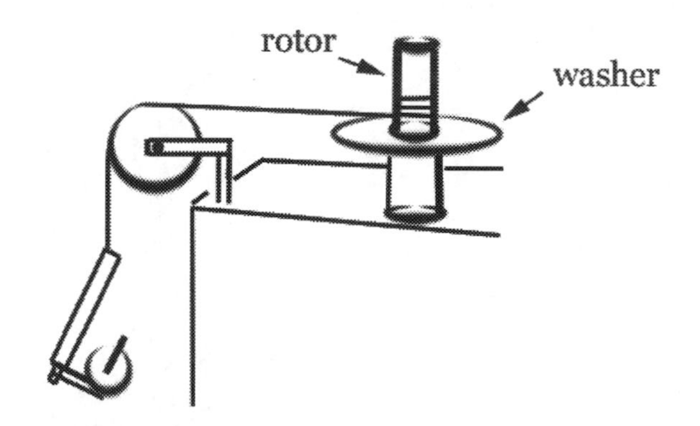

Figure 17.24 Steel washer, rotor, pulley, and hanging object

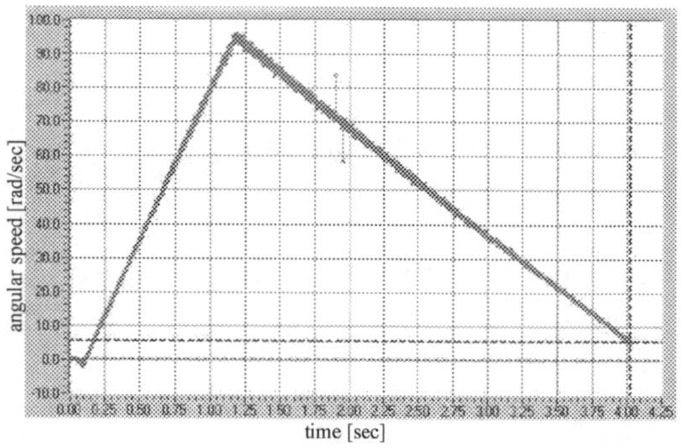

Figure 17.26 Graph of angular speed vs. time for falling object

Solution: We begin by drawing a force-torque diagram (Figure 17.26a) for the rotor and a free-body diagram for hanger (Figure 17.26b). (The choice of positive directions are indicated on the figures.) The frictional torque on the rotor is then given by $\vec{\boldsymbol{\tau}}_f = -\tau_f \hat{\mathbf{k}}$ where we use τ_f as the magnitude of the frictional torque. The torque about the center of the rotor due to the tension in the string is given by $\vec{\boldsymbol{\tau}}_T = rT\,\hat{\mathbf{k}}$ where r is the radius of the rotor. The angular acceleration of the rotor is given by $\vec{\boldsymbol{\alpha}}_1 = \alpha_1\,\hat{\mathbf{k}}$ and we expect that $\alpha_1 > 0$ because the rotor is speeding up.

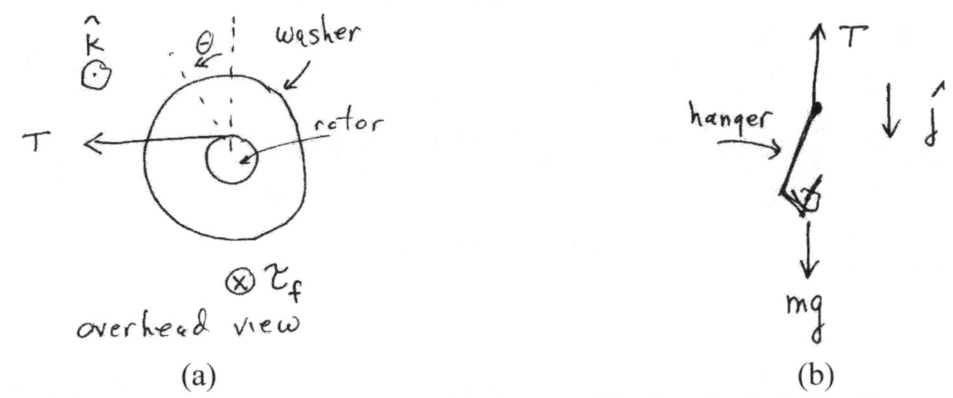

Figure 17.26 (a) Force-torque diagram on rotor and (b) free-body force diagram on hanging object

While the hanger is falling, the rotor-washer combination has a net torque due to the tension in the string and the frictional torque, and using the rotational equation of motion,

$$Tr - \tau_f = I_R \alpha_1 . \tag{17.4.1}$$

We apply Newton's Second Law to the hanger and find that

$$mg - T = ma_1 = m\alpha_1 r , \tag{17.4.2}$$

where $a_1 = r\alpha_1$ has been used to express the linear acceleration of the falling hanger to the angular acceleration of the rotor; that is, the string does not stretch. Before proceeding, it might be illustrative to multiply Eq. (17.4.2) by r and add to Eq. (17.4.1) to obtain

$$mgr - \tau_f = (I_R + mr^2)\alpha_1 . \tag{17.4.3}$$

Eq. (17.4.3) contains the unknown frictional torque, and this torque is determined by considering the slowing of the rotor/washer after the string has detached.

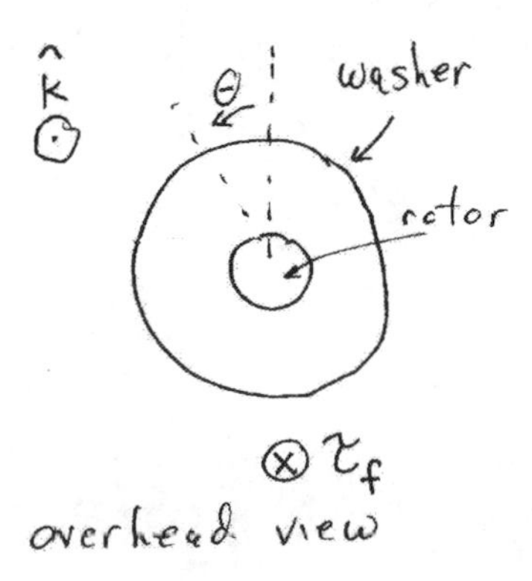

Figure 17.27 Torque diagram on rotor when string has detached

The torque on the system is just this frictional torque (Figure 17.27), and so

$$-\tau_f = I_R \alpha_2 \tag{17.4.4}$$

Note that in Eq. (17.4.4), $\tau_f > 0$ and $\alpha_2 < 0$. Subtracting Eq. (17.4.4) from Eq. (17.4.3) eliminates τ_f,

$$mgr = mr^2\alpha_1 + I_R(\alpha_1 - \alpha_2). \tag{17.4.5}$$

We can now solve for I_R yielding

$$I_R = \frac{mr(g - r\alpha_1)}{\alpha_1 - \alpha_2}. \tag{17.4.6}$$

For a numerical result, we use the data collected during a trial run resulting in the graph of angular speed vs. time for the falling object shown in Figure 17.25. The values for α_1 and α_2 can be determined by calculating the slope of the two straight lines in Figure 17.28 yielding

$$\alpha_1 = (96\,\text{rad}\cdot\text{s}^{-1})/(1.15\,\text{s}) = 83\ \text{rad}\cdot\text{s}^{-2},$$
$$\alpha_2 = -(89\,\text{rad}\cdot\text{s}^{-1})/(2.85\,\text{s}) = -31\,\text{rad}\cdot\text{s}^{-2}.$$

Inserting these values into Eq. (17.4.6) yields

$$I_R = 5.3\times10^{-5}\,\text{kg}\cdot\text{m}^2. \tag{17.4.7}$$

17.5 Torque and Rotational Work

When a constant torque $\tau_{s,z}$ is applied to an object, and the object rotates through an angle $\Delta\theta$ about a fixed z-axis through the center of mass, then the torque does an amount of work $\Delta W = \tau_{S,z}\Delta\theta$ on the object. By extension of the linear work-energy theorem, the amount of work done is equal to the change in the rotational kinetic energy of the object,

$$W_{\text{rot}} = \frac{1}{2}I_{\text{cm}}\omega_f^2 - \frac{1}{2}I_{\text{cm}}\omega_i^2 = K_{\text{rot},f} - K_{\text{rot},i}. \tag{17.4.8}$$

The rate of doing this work is the rotational power exerted by the torque,

$$P_{\text{rot}} \equiv \frac{dW_{\text{rot}}}{dt} = \lim_{\Delta t\to 0}\frac{\Delta W_{\text{rot}}}{\Delta t} = \tau_{S,z}\frac{d\theta}{dt} = \tau_{S,z}\omega_z. \tag{17.4.9}$$

17.5.1 Rotational Work

Consider a rigid body rotating about an axis. Each small element of mass Δm_i in the rigid body is moving in a circle of radius $(r_{S,i})_\perp$ about the axis of rotation passing through the point S. Each mass element undergoes a small angular displacement $\Delta\theta$ under the action of a tangential force, $\vec{\mathbf{F}}_{\theta,i} = F_{\theta,i}\,\hat{\boldsymbol{\theta}}$, where $\hat{\boldsymbol{\theta}}$ is the unit vector pointing in the tangential direction (Figure 17.20). The element will then have an associated displacement vector for this motion, $\Delta\vec{\mathbf{r}}_{S,i} = r_i\Delta\theta\,\hat{\boldsymbol{\theta}}$ and the work done by the tangential force is

$$\Delta W_{\text{rot},i} = \vec{\mathbf{F}}_{\theta,i}\cdot\Delta\vec{\mathbf{r}}_{S,i} = (F_{\theta,i}\,\hat{\boldsymbol{\theta}})\cdot(r_i\Delta\theta\,\hat{\boldsymbol{\theta}}) = r_iF_{\theta,i}\Delta\theta\,. \qquad (17.4.10)$$

Recall the result of Eq. (17.3.8) that the component of the torque (in the direction along the axis of rotation) about S due to the tangential force, $\vec{\mathbf{F}}_{\theta,i}$, acting on the mass element Δm_i is

$$(\tau_{S,i})_z = r_iF_{\theta,i}\,, \qquad (17.4.11)$$

and so Eq. (17.4.10) becomes

$$\Delta W_{\text{rot},\,i} = (\tau_{S,i})_z\Delta\theta\,. \qquad (17.4.12)$$

Summing over all the elements yields

$$W_{\text{rot}} = \sum_i \Delta W_{\text{rot},\,i} = \big((\tau_{S,i})_z\big)\Delta\theta = \tau_{S,z}\Delta\theta\,, \qquad (17.4.13)$$

the rotational work is the product of the torque and the angular displacement. In the limit of small angles, $\Delta\theta \to d\theta$, $\Delta W_{\text{rot}} \to dW_{\text{rot}}$ and the differential rotational work is

$$dW_{\text{rot}} = \tau_{S,z}d\theta\,. \qquad (17.4.14)$$

We can integrate this amount of rotational work as the angle coordinate of the rigid body changes from some initial value $\theta = \theta_i$ to some final value $\theta = \theta_f$,

$$W_{\text{rot}} = \int dW_{\text{rot}} = \int_{\theta_i}^{\theta_f} \tau_{S,z}\,d\theta\,. \qquad (17.4.15)$$

17.5.2 Rotational Work-Kinetic Energy Theorem

We will now show that the rotational work is equal to the change in rotational kinetic energy. We begin by substituting our result from Eq. (17.3.14) into Eq. (17.4.14) for the infinitesimal rotational work,

$$dW_{\text{rot}} = I_S\alpha_z\,d\theta\,. \qquad (17.4.16)$$

Recall that the rate of change of angular velocity is equal to the angular acceleration, $\alpha_z \equiv d\omega_z / dt$ and that the angular velocity is $\omega_z \equiv d\theta / dt$. Note that in the limit of small displacements,

$$\frac{d\omega_z}{dt} d\theta = d\omega_z \frac{d\theta}{dt} = d\omega_z \omega_z . \tag{17.4.17}$$

Therefore the infinitesimal rotational work is

$$dW_{\text{rot}} = I_S \alpha_z \, d\theta = I_S \frac{d\omega_z}{dt} d\theta = I_S \, d\omega_z \frac{d\theta}{dt} = I_S \, d\omega_z \, \omega_z . \tag{17.4.18}$$

We can integrate this amount of rotational work as the angular velocity of the rigid body changes from some initial value $\omega_z = \omega_{z,i}$ to some final value $\omega_z = \omega_{z,f}$,

$$W_{\text{rot}} = \int dW_{\text{rot}} = \int_{\omega_{z,i}}^{\omega_{z,f}} I_S \, d\omega_z \, \omega_z = \frac{1}{2} I_S \omega_{z,f}^2 - \frac{1}{2} I_S \omega_{z,i}^2 . \tag{17.4.19}$$

When a rigid body is rotating about a fixed axis passing through a point S in the body, there is both rotation and translation about the center of mass unless S is the center of mass. If we choose the point S in the above equation for the rotational work to be the center of mass, then

$$W_{\text{rot}} = \frac{1}{2} I_{\text{cm}} \omega_{\text{cm},f}^2 - \frac{1}{2} I_{\text{cm}} \omega_{\text{cm},i}^2 = K_{\text{rot},f} - K_{\text{rot},i} \equiv \Delta K_{\text{rot}} . \tag{17.4.20}$$

Note that because the z-component of the angular velocity of the center of mass appears as a square, we can just use its magnitude in Eq. (17.4.20).

17.5.3 Rotational Power

The rotational power is defined as the rate of doing rotational work,

$$P_{\text{rot}} \equiv \frac{dW_{\text{rot}}}{dt} . \tag{17.4.21}$$

We can use our result for the infinitesimal work to find that the rotational power is the product of the applied torque with the angular velocity of the rigid body,

$$P_{\text{rot}} \equiv \frac{dW_{\text{rot}}}{dt} = \tau_{S,z} \frac{d\theta}{dt} = \tau_{S,z} \omega_z . \tag{17.4.22}$$

Example 17.12 Work Done by Frictional Torque

A steel washer is mounted on the shaft of a small motor. The moment of inertia of the motor and washer is I_0. The washer is set into motion. When it reaches an initial angular velocity ω_0, at $t = 0$, the power to the motor is shut off, and the washer slows down during a time interval $\Delta t_1 = t_a$ until it reaches an angular velocity of ω_a at time t_a. At that instant, a second steel washer with a moment of inertia I_w is dropped on top of the first washer. Assume that the second washer is only in contact with the first washer. The collision takes place over a time $\Delta t_{\text{int}} = t_b - t_a$ after which the two washers and rotor rotate with angular speed ω_b. Assume the frictional torque on the axle (magnitude τ_f) is independent of speed, and remains the same when the second washer is dropped. (a) What angle does the rotor rotate through during the collision? (b) What is the work done by the friction torque from the bearings during the collision? (c) Write down an equation for conservation of energy. Can you solve this equation for ω_b? (d) What is the average rate that work is being done by the friction torque during the collision?

Solution: We begin by solving for the frictional torque during the first stage of motion when the rotor is slowing down. We choose a coordinate system shown in Figure 17.29.

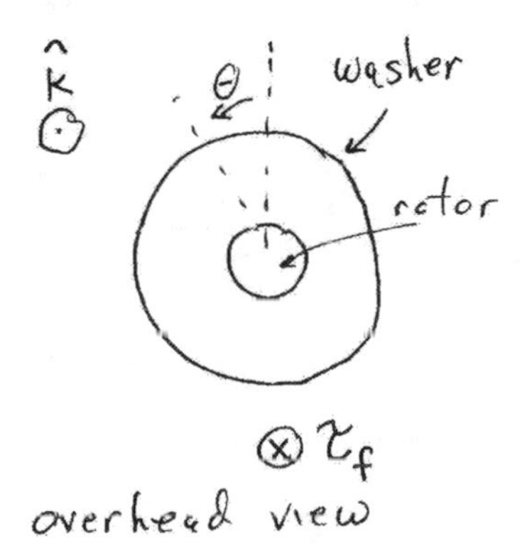

Figure 17.29 Coordinate system for Example 17.12

The component of average angular acceleration is given by

$$\alpha_1 = \frac{\omega_a - \omega_0}{t_a} < 0.$$

We can use the rotational equation of motion, and find that the frictional torque satisfies

$$-\tau_f = I_0 \left(\frac{\omega_a - \omega_0}{\Delta t_1} \right).$$

During the collision, the component of the average angular acceleration of the rotor is given by

$$\alpha_2 = \frac{\omega_b - \omega_a}{(\Delta t_{\text{int}})} < 0 .$$

The angle the rotor rotates through during the collision is (analogous to linear motion with constant acceleration)

$$\Delta\theta_2 = \omega_a \Delta t_{\text{int}} + \frac{1}{2}\alpha_2 \Delta t_{\text{int}}{}^2 = \omega_a \Delta t_{\text{int}} + \frac{1}{2}\left(\frac{\omega_b - \omega_a}{\Delta t_{\text{int}}}\right)\Delta t_{\text{int}}{}^2 = \frac{1}{2}(\omega_b + \omega_a)\Delta t_{\text{int}} > 0 .$$

The non-conservative work done by the bearing friction during the collision is

$$W_{f,b} = -\tau_f \Delta\theta_{rotor} = -\tau_f \frac{1}{2}(\omega_a + \omega_b)\Delta t_{\text{int}} .$$

Using our result for the frictional torque, the work done by the bearing friction during the collision is

$$W_{f,b} = \frac{1}{2} I_0 \left(\frac{\omega_a - \omega_0}{\Delta t_1}\right)(\omega_a + \omega_b)\Delta t_{\text{int}} < 0 .$$

The negative work is consistent with the fact that the kinetic energy of the rotor is decreasing as the rotor is slowing down. Using the work energy theorem during the collision the kinetic energy of the rotor has deceased by

$$W_{f,b} = \frac{1}{2}(I_0 + I_w)\omega_b^2 - \frac{1}{2} I_0 \omega_a^2 .$$

Using our result for the work, we have that

$$\frac{1}{2} I_0 \left(\frac{\omega_a - \omega_0}{\Delta t_1}\right)(\omega_a + \omega_b)\Delta t_{\text{int}} = \frac{1}{2}(I_0 + I_w)\omega_b^2 - \frac{1}{2} I_0 \omega_a^2 .$$

This is a quadratic equation for the angular speed ω_b of the rotor and washer immediately after the collision that we can in principle solve. However remember that we assumed that the frictional torque is independent of the speed of the rotor. Hence the best practice would be to measure ω_0, ω_a, ω_b, Δt_1, Δt_{int}, I_0, and I_w and then determine how closely our model agrees with conservation of energy. The rate of work done by the frictional torque is given by

$$P_f = \frac{W_{f,b}}{\Delta t_{\text{int}}} = \frac{1}{2} I_0 \left(\frac{\omega_a - \omega_0}{\Delta t_1}\right)(\omega_a + \omega_b) < 0 .$$

Chapter 18 Static Equilibrium

Chapter 18 Static Equilibrium

The proof of the correctness of a new rule can be attained by the repeated application of it, the frequent comparison with experience, the putting of it to the test under the most diverse circumstances. This process, would in the natural course of events, be carried out in time. The discoverer, however hastens to reach his goal more quickly. He compares the results that flow from his rule with all the experiences with which he is familiar, with all older rules, repeatedly tested in times gone by, and watches to see if he does not light on contradictions. In this procedure, the greatest credit is, as it should be, conceded to the oldest and most familiar experiences, the most thoroughly tested rules. Our instinctive experiences, those generalizations that are made involuntarily, by the irresistible force of the innumerable facts that press upon us, enjoy a peculiar authority; and this is perfectly warranted by the consideration that it is precisely the elimination of subjective caprice and of individual error that is the object aimed at.[1]

Ernst Mach

18.1 Introduction Static Equilibrium

When the vector sum of the forces acting on a point-like object is zero then the object will continue in its state of rest, or of uniform motion in a straight line. If the object is in uniform motion we can always change reference frames so that the object will be at rest. We showed that for a collection of point-like objects the sum of the external forces may be regarded as acting at the center of mass. So if that sum is zero the center of mass will continue in its state of rest, or of uniform motion in a straight line. We introduced the idea of a rigid body, and again showed that in addition to the fact that the sum of the external forces may be regarded as acting at the center of mass, forces like the gravitational force that acts at every point in the body may be treated as acting at the center of mass. However for an extended rigid body it matters where the force is applied because even though the sum of the forces on the body may be zero, a non-zero sum of torques on the body may still produce angular acceleration. In particular for fixed axis rotation, the torque along the axis of rotation on the object is proportional to the angular acceleration. It is possible that sum of the torques may be zero on a body that is not constrained to rotate about a fixed axis and the body may still undergo rotation. We will restrict ourselves to the special case in which in an inertial reference frame both the center of mass of the body is at rest and the body does not undergo any rotation, a condition that is called ***static equilibrium of an extended object.***

The two sufficient and necessary conditions for a rigid body to be in static equilibrium are:

---

[1] Ernst Mach, *The Science of Mechanics: A Critical and Historical Account of Its Development*, translated by Thomas J. McCormack, Sixth Edition with Revisions through the Ninth German Edition, Open Court Publishing, Illinois.

(1) The sum of the forces acting on the rigid body is zero,

$$\vec{\mathbf{F}} = \vec{\mathbf{F}}_1 + \vec{\mathbf{F}}_2 + \cdots = \vec{\mathbf{0}}. \qquad (18.1.1)$$

(2) The vector sum of the torques about any point S in a rigid body is zero,

$$\vec{\boldsymbol{\tau}}_S = \vec{\boldsymbol{\tau}}_{S,1} + \vec{\boldsymbol{\tau}}_{S,2} + \cdots = \vec{\mathbf{0}}. \qquad (18.1.2)$$

18.2 Lever Law

Let's consider a uniform rigid beam of mass m_b balanced on a pivot near the center of mass of the beam. We place two objects 1 and 2 of masses m_1 and m_2 on the beam, at distances d_1 and d_2 respectively from the pivot, so that the beam is static (that is, the beam is not rotating. See Figure 18.1.) We shall neglect the thickness of the beam and take the pivot point to be the center of mass.

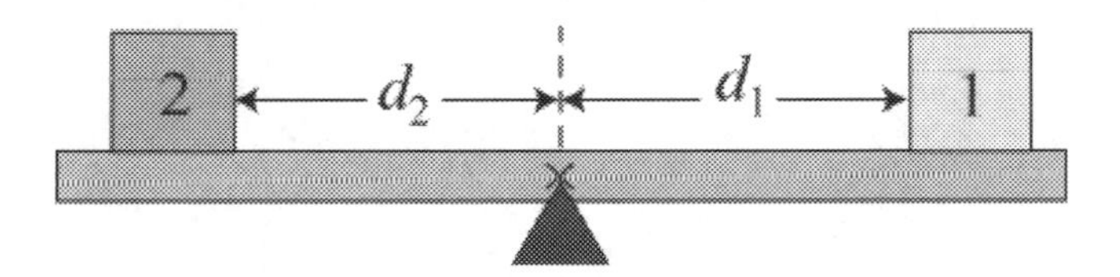

Figure 18.1 Pivoted Lever

Let's consider the forces acting on the beam. The earth attracts the beam downward. This gravitational force acts on every atom in the beam, but we can summarize its action by stating that the gravitational force $m_b\vec{\mathbf{g}}$ is concentrated at a point in the beam called the *center of gravity* of the beam, which is identical to the center of mass of the uniform beam. There is also a contact force $\vec{\mathbf{F}}_{\text{pivot}}$ between the pivot and the beam, acting upwards on the beam at the pivot point. The objects 1 and 2 exert normal forces downwards on the beam, $\vec{\mathbf{N}}_{1,b} \equiv \vec{\mathbf{N}}_1$, and $\vec{\mathbf{N}}_{2,b} \equiv \vec{\mathbf{N}}_2$, with magnitudes N_1, and N_2, respectively. Note that the normal forces are not the gravitational forces acting on the objects, but contact forces between the beam and the objects. (In this case, they are mathematically the same, due to the horizontal configuration of the beam and the fact that all objects are in static equilibrium.) The distances d_1 and d_2 are called the *moment arms* with respect to the pivot point for the forces $\vec{\mathbf{N}}_1$ and $\vec{\mathbf{N}}_2$, respectively. The force diagram on the beam is shown in Figure 18.2. Note that the pivot force $\vec{\mathbf{F}}_{\text{pivot}}$ and the force of gravity $m_b\vec{\mathbf{g}}$ each has a zero moment arm about the pivot point.

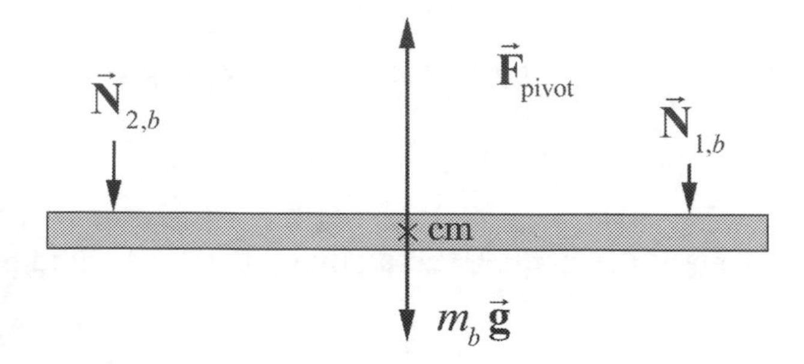

Figure 18.2 Free-body diagram on beam

Because we assume the beam is not moving, the sum of the forces in the vertical direction acting on the beam is therefore zero,

$$F_{\text{pivot}} - m_b g - N_1 - N_2 = 0. \qquad (18.2.1)$$

The force diagrams on the objects are shown in Figure 18.3. Note the magnitude of the normal forces on the objects are also N_1 and N_2 since these are each part of an action-reaction pair, $\vec{\mathbf{N}}_{1,b} = -\vec{\mathbf{N}}_{b,1}$, and $\vec{\mathbf{N}}_{2,b} = -\vec{\mathbf{N}}_{b,2}$.

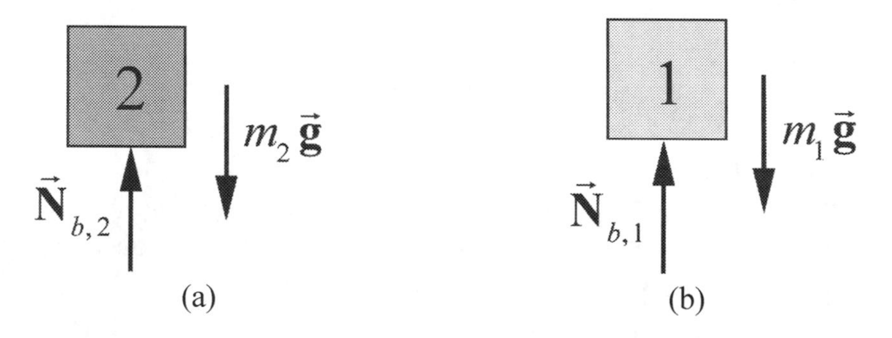

Figure 18.3 Free-body force diagrams for each body.

The condition that the forces sum to zero is not sufficient to completely predict the motion of the beam. All we can deduce is that the center of mass of the system is at rest (or moving with a uniform velocity). In order for the beam not to rotate the sum of the torques about any point must be zero. In particular the sum of the torques about the pivot point must be zero. Because the moment arm of the gravitational force and the pivot force is zero, only the two normal forces produce a torque on the beam. If we choose out of the page as positive direction for the torque (or equivalently counterclockwise rotations are positive) then the condition that the sum of the torques about the pivot point is zero becomes

$$d_2 N_2 - d_1 N_1 = 0. \qquad (18.2.2)$$

The magnitudes of the two torques about the pivot point are equal, a condition known as the lever law.

Lever Law: *A beam of length l is balanced on a pivot point that is placed directly beneath the center of mass of the beam. The beam will not undergo rotation if the product of the normal force with the moment arm to the pivot is the same for each body,*

$$d_1 N_1 = d_2 N_2 . \tag{18.2.3}$$

Example 18.1 Lever Law

Suppose a uniform beam of length $l = 1.0\ \text{m}$ and mass $m_B = 2.0\ \text{kg}$ is balanced on a pivot point, placed directly beneath the center of the beam. We place body 1 with mass $m_1 = 0.3\ \text{kg}$ a distance $d_1 = 0.4\ \text{m}$ to the right of the pivot point, and a second body 2 with $m_2 = 0.6\ \text{kg}$ a distance d_2 to the left of the pivot point, such that the beam neither translates nor rotates. (a) What is the force $\vec{\mathbf{F}}_{\text{pivot}}$ that the pivot exerts on the beam? (b) What is the distance d_2 that maintains static equilibrium?

Solution: a) By Newton's Third Law, the beam exerts equal and opposite normal forces of magnitude N_1 on body 1, and N_2 on body 2. The condition for force equilibrium applied separately to the two bodies yields

$$N_1 - m_1\, g = 0 , \tag{18.2.4}$$

$$N_2 - m_2\, g = 0 . \tag{18.2.5}$$

Thus the total force acting on the beam is zero,

$$F_{\text{pivot}} - (m_b + m_1 + m_2) g = 0 , \tag{18.2.6}$$

and the pivot force is

$$\begin{aligned} F_{\text{pivot}} &= (m_b + m_1 + m_2) g \\ &= (2.0\ \text{kg} + 0.3\ \text{kg} + 0.6\ \text{kg})(9.8\ \text{m} \cdot \text{s}^{-2}) = 2.8 \times 10^1\ \text{N} . \end{aligned} \tag{18.2.7}$$

b) We can compute the distance d_2 from the Lever Law,

$$d_2 = \frac{d_1 N_1}{N_2} = \frac{d_1 m_1 g}{m_2 g} = \frac{d_1 m_1}{m_2} = \frac{(0.4\ \text{m})(0.3\ \text{kg})}{0.6\ \text{kg}} = 0.2\ \text{m} . \tag{18.2.8}$$

18.3 Generalized Lever Law

We can extend the Lever Law to the case in which two external forces $\vec{\mathbf{F}}_1$ and $\vec{\mathbf{F}}_2$ are acting on the pivoted beam at angles θ_1 and θ_2 with respect to the horizontal as shown in the Figure 18.4. Throughout this discussion the angles will be limited to the range $[0 \le \theta_1, \theta_2 \le \pi]$. We shall again neglect the thickness of the beam and take the pivot point to be the center of mass.

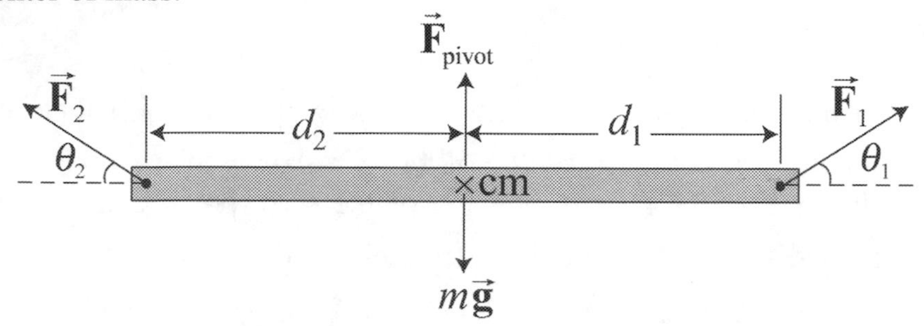

Figure 18.4 Forces acting at angles to a pivoted beam.

The forces $\vec{\mathbf{F}}_1$ and $\vec{\mathbf{F}}_2$ can be decomposed into separate vectors components respectively $(\vec{\mathbf{F}}_{1,\parallel}, \vec{\mathbf{F}}_{1,\perp})$ and $(\vec{\mathbf{F}}_{2,\parallel}, \vec{\mathbf{F}}_{2,\perp})$, where $\vec{\mathbf{F}}_{1,\parallel}$ and $\vec{\mathbf{F}}_{2,\parallel}$ are the horizontal vector projections of the two forces with respect to the direction formed by the length of the beam, and $\vec{\mathbf{F}}_{1,\perp}$ and $\vec{\mathbf{F}}_{2,\perp}$ are the perpendicular vector projections respectively to the beam (Figure 18.5), with

$$\vec{\mathbf{F}}_1 = \vec{\mathbf{F}}_{1,\parallel} + \vec{\mathbf{F}}_{1,\perp}, \quad (18.3.1)$$

$$\vec{\mathbf{F}}_2 = \vec{\mathbf{F}}_{2,\parallel} + \vec{\mathbf{F}}_{2,\perp}. \quad (18.3.2)$$

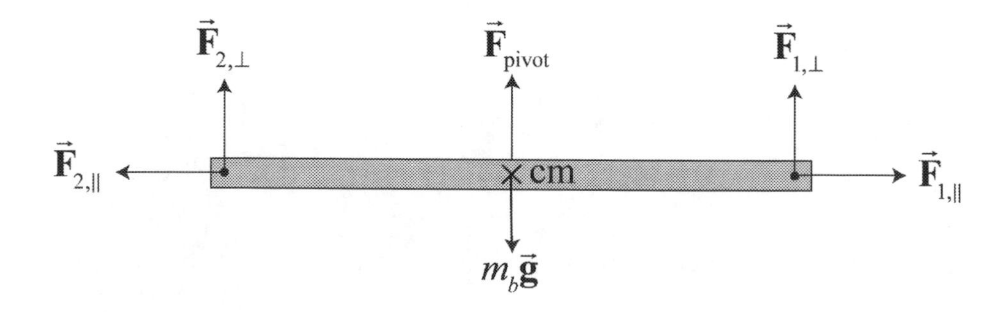

Figure 18.5 Vector decomposition of forces.

The horizontal components of the forces are

$$F_{1,\parallel} = F_1 \cos\theta_1, \quad (18.3.3)$$

$$F_{2,\parallel} = -F_2 \cos\theta_2, \quad (18.3.4)$$

where our choice of positive horizontal direction is to the right. Neither horizontal force component contributes to possible rotational motion of the beam. The sum of these horizontal forces must be zero,

$$F_1 \cos\theta_1 - F_2 \cos\theta_2 = 0. \qquad (18.3.5)$$

The perpendicular component forces are

$$F_{1,\perp} = F_1 \sin\theta_1, \qquad (18.3.6)$$

$$F_{2,\perp} = F_2 \sin\theta_2, \qquad (18.3.7)$$

where the positive vertical direction is upwards. The perpendicular components of the forces must also sum to zero,

$$F_{\text{pivot}} - m_b g + F_1 \sin\theta_1 + F_2 \sin\theta_2 = 0. \qquad (18.3.8)$$

Only the vertical components $F_{1,\perp}$ and $F_{2,\perp}$ of the external forces are involved in the lever law (but the horizontal components must balance, as in Equation (18.3.5), for equilibrium). Then the Lever Law can be extended as follows.

Generalized Lever Law *A beam of length l is balanced on a pivot point that is placed directly beneath the center of mass of the beam. Suppose a force $\vec{\mathbf{F}}_1$ acts on the beam a distance d_1 to the right of the pivot point. A second force $\vec{\mathbf{F}}_2$ acts on the beam a distance d_2 to the left of the pivot point. The beam will remain in static equilibrium if the following two conditions are satisfied:*

1) The total force on the beam is zero,

2) The product of the magnitude of the perpendicular component of the force with the distance to the pivot is the same for each force,

$$d_1 |F_{1,\perp}| = d_2 |F_{2,\perp}|. \qquad (18.3.9)$$

The Generalized Lever Law can be stated in an equivalent form,

$$d_1 F_1 \sin\theta_1 = d_2 F_2 \sin\theta_2. \qquad (18.3.10)$$

We shall now show that the generalized lever law can be reinterpreted as the statement that the vector sum of the torques about the pivot point S is zero when there are just two forces $\vec{\mathbf{F}}_1$ and $\vec{\mathbf{F}}_2$ acting on our beam as shown in Figure 18.6.

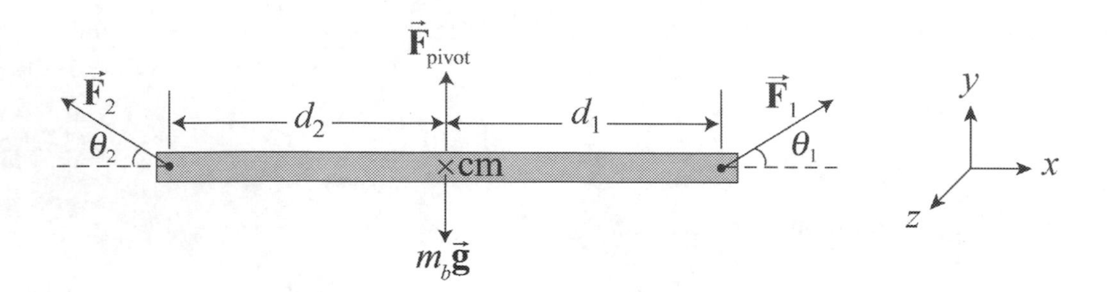

Figure 18.6 Force and torque diagram.

Let's choose the positive z-direction to point out of the plane of the page then torque pointing out of the page will have a positive z-component of torque (counterclockwise rotations are positive). From our definition of torque about the pivot point, the magnitude of torque due to force $\vec{\mathbf{F}}_1$ is given by

$$\tau_{S,1} = d_1 F_1 \sin\theta_1 . \tag{18.3.11}$$

From the right hand rule this is out of the page (in the counterclockwise direction) so the component of the torque is positive, hence,

$$(\tau_{S,1})_z = d_1 F_1 \sin\theta_1 . \tag{18.3.12}$$

The torque due to $\vec{\mathbf{F}}_2$ about the pivot point is into the page (the clockwise direction) and the component of the torque is negative and given by

$$(\tau_{S,2})_z = -d_2 F_2 \sin\theta_2 . \tag{18.3.13}$$

The z-component of the torque is the sum of the z-components of the individual torques and is zero,

$$(\tau_{S,\text{total}})_z = (\tau_{S,1})_z + (\tau_{S,2})_z = d_1 F_1 \sin\theta_1 - d_2 F_2 \sin\theta_2 = 0 , \tag{18.3.14}$$

which is equivalent to the Generalized Lever Law, Equation (18.3.10),

$$d_1 F_1 \sin\theta_1 = d_2 F_2 \sin\theta_2 .$$

18.4 Worked Examples

Example 18.2 Suspended Rod

A uniform rod of length $l = 2.0\text{ m}$ and mass $m = 4.0\text{ kg}$ is hinged to a wall at one end and suspended from the wall by a cable that is attached to the other end of the rod at an angle of $\beta = 30^o$ to the rod (see Figure 18.7). Assume the cable has zero mass. There is a contact force at the pivot on the rod. The magnitude and direction of this force is unknown. One of the most difficult parts of these types of problems is to introduce an angle for the pivot force and then solve for that angle if possible. In this problem you will solve for the magnitude of the tension in the cable and the direction and magnitude of the pivot force. (a) What is the tension in the cable? (b) What angle does the pivot force make with the beam? (c) What is the magnitude of the pivot force?

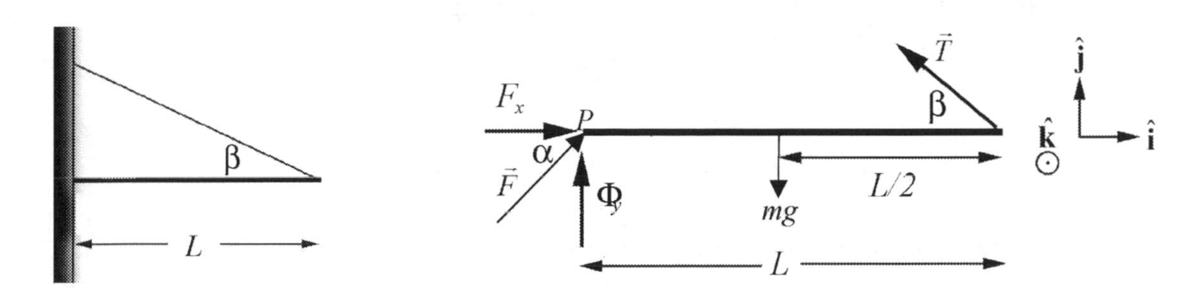

Figure 18.7 Example 18.2 **Figure 18.8** Force and torque diagram.

Solution: a) The force diagram is shown in Figure 18.8. Take the positive $\hat{\mathbf{i}}$-direction to be to the right in the figure above, and take the positive $\hat{\mathbf{j}}$-direction to be vertically upward. The forces on the rod are: the gravitational force $m\vec{\mathbf{g}} = -m\,g\,\hat{\mathbf{j}}$, acting at the center of the rod; the force that the cable exerts on the rod, $\vec{\mathbf{T}} = T(-\cos\beta\,\hat{\mathbf{i}} + \sin\beta\,\hat{\mathbf{j}})$, acting at the right end of the rod; and the pivot force $\vec{\mathbf{F}}_{\text{pivot}} = F(\cos\alpha\,\hat{\mathbf{i}} + \sin\alpha\,\hat{\mathbf{j}})$, acting at the left end of the rod. If $0 < \alpha < \pi/2$, the pivot force is directed up and to the right in the figure. If $0 > \alpha > -\pi/2$, the pivot force is directed down and to the right. We have no reason, at this point, to expect that α will be in either of the quadrants, but it must be in one or their other.

For static equilibrium, the sum of the forces must be zero, and hence the sums of the components of the forces must be zero,

$$\begin{aligned} 0 &= -T\cos\beta + F\cos\alpha \\ 0 &= -m\,g + T\sin\beta + F\sin\alpha. \end{aligned} \tag{18.4.1}$$

With respect to the pivot point, and taking positive torques to be counterclockwise, the gravitational force exerts a negative torque of magnitude $m\,g(l/2)$ and the cable exerts a positive torque of magnitude $T\,l\sin\beta$. The pivot force exerts no torque about the pivot. Setting the sum of the torques equal to zero then gives

$$\begin{aligned} 0 &= T\,l\sin\beta - m\,g(l/2) \\ T &= \frac{m\,g}{2\sin\beta}. \end{aligned} \tag{18.4.2}$$

This result has many features we would expect; proportional to the weight of the rod and inversely proportional to the sine of the angle made by the cable with respect to the horizontal. Inserting numerical values gives

$$T = \frac{mg}{2\sin\beta} = \frac{(4.0\,\mathrm{kg})(9.8\,\mathrm{m\cdot s^{-2}})}{2\sin 30^\circ} = 39.2\,\mathrm{N}. \qquad (18.4.3)$$

There are many ways to find the angle α. Substituting Eq. (18.4.2) for the tension into both force equations in Eq. (18.4.1) yields

$$\begin{aligned} F\cos\alpha &= T\cos\beta = (mg/2)\cot\beta \\ F\sin\alpha &= mg - T\sin\beta = mg/2. \end{aligned} \qquad (18.4.4)$$

In Eq. (18.4.4), dividing one equation by the other, we see that $\tan\alpha = \tan\beta$, $\alpha = \beta$.

The horizontal forces on the rod must cancel. The tension force and the pivot force act with the same angle (but in opposite horizontal directions) and hence must have the same magnitude,

$$F = T = 39.2\,\mathrm{N}. \qquad (18.4.5)$$

As an alternative, if we had not done the previous parts, we could find torques about the point where the cable is attached to the wall. The cable exerts no torque about this point and the y-component of the pivot force exerts no torque as well. The moment arm of the x-component of the pivot force is $l\tan\beta$ and the moment arm of the weight is $l/2$. Equating the magnitudes of these two torques gives

$$F\cos\alpha\, l\tan\beta = mg\frac{l}{2},$$

equivalent to the first equation in Eq. (18.4.4). Similarly, evaluating torques about the right end of the rod, the cable exerts no torques and the x-component of the pivot force exerts no torque. The moment arm of the y-component of the pivot force is l and the moment arm of the weight is $l/2$. Equating the magnitudes of these two torques gives

$$F\sin\alpha\, l = mg\frac{l}{2},$$

reproducing the second equation in Eq. (18.4.4). The point of this alternative solution is to show that choosing a different origin (or even more than one origin) in order to remove an unknown force from the torques equations might give a desired result more directly.

Example 18.3 Person Standing on a Hill

A person is standing on a hill that is sloped at an angle of α with respect to the horizontal (Figure 18.9). The person's legs are separated by a distance d, with one foot uphill and one downhill. The center of mass of the person is at a distance h above the ground, perpendicular to the hillside, midway between the person's feet. Assume that the coefficient of static friction between the person's feet and the hill is sufficiently large that the person will not slip. (a) What is the magnitude of the normal force on each foot? (b) How far must the feet be apart so that the normal force on the upper foot is just zero? This is the moment when the person starts to rotate and fall over.

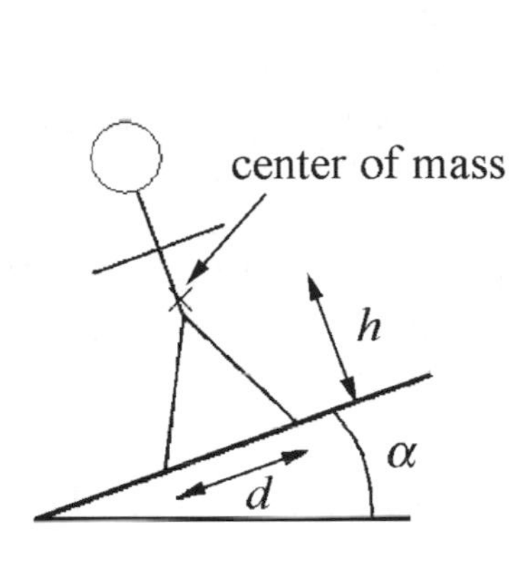

Figure 18.9 Person standing on hill

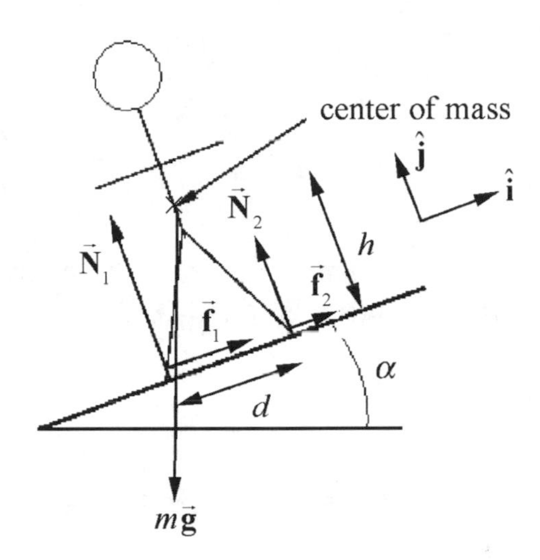

Figure 18.10 Free-body force diagram for person standing on hill

Solution: The force diagram on the person is shown in Figure 18.10. Note that the contact forces have been decomposed into components perpendicular and parallel to the hillside. A choice of unit vectors and positive direction for torque is also shown. Applying Newton's Second Law to the two components of the net force,

$$\hat{\mathbf{j}}: \; N_1 + N_2 - mg\cos\alpha = 0 \tag{18.4.6}$$

$$\hat{\mathbf{i}}: \; f_1 + f_2 - mg\sin\alpha = 0. \tag{18.4.7}$$

These two equations imply that

$$N_1 + N_2 = mg\cos\alpha \tag{18.4.8}$$

$$f_1 + f_2 = mg\sin\alpha. \tag{18.4.9}$$

Evaluating torques about the center of mass,

$$h(f_1 + f_2) + (N_2 - N_1)\frac{d}{2} = 0. \tag{18.4.10}$$

Equation (18.4.10) can be rewritten as

$$N_1 - N_2 = \frac{2h(f_1 + f_2)}{d}. \quad (18.4.11)$$

Substitution of Equation (18.4.9) into Equation (18.4.11) yields

$$N_1 - N_2 = \frac{2h(mg\sin\alpha)}{d}. \quad (18.4.12)$$

We can solve for N_1 by adding Equations (18.4.8) and (18.4.12), and then dividing by 2, yielding

$$N_1 = \frac{1}{2}mg\cos\alpha + \frac{h(mg\sin\alpha)}{d} = mg\left(\frac{1}{2}\cos\alpha + \frac{h}{d}\sin\alpha\right). \quad (18.4.13)$$

Similarly, we can solve for N_2 by subtracting Equation (18.4.12) from Equation (18.4.8) and dividing by 2, yielding

$$N_2 = mg\left(\frac{1}{2}\cos\alpha - \frac{h}{d}\sin\alpha\right). \quad (18.4.14)$$

The normal force N_2 as given in Equation (18.4.14) vanishes when

$$\frac{1}{2}\cos\alpha = \frac{h}{d}\sin\alpha, \quad (18.4.15)$$

which can be solved for the minimum distance between the legs,

$$d = 2h(\tan\alpha). \quad (18.4.16)$$

In the above figures, $\alpha = 20°$, $2\tan\alpha = 0.73$, and the stick-figure person is very close to tipping over. It should be noted that no specific model for the frictional force was used, that is, no coefficient of static friction entered the problem. The two frictional forces f_1 and f_2 were not determined separately; only their sum entered the above calculations.

Example 18.4 The Knee

A man of mass $m = 70\,\text{kg}$ is about to start a race. Assume the runner's weight is equally distributed on both legs. The patellar ligament in the knee is attached to the upper tibia and runs over the kneecap. When the knee is bent, a tensile force, $\vec{\mathbf{T}}$, that the ligament exerts on the upper tibia, is directed at an angle of $\theta = 40°$ with respect to the horizontal. The femur exerts a force $\vec{\mathbf{F}}$ on the upper tibia. The angle, α, that this force makes with the vertical will vary and is one of the unknowns to solve for. Assume that the ligament is connected a distance, $d = 3.8\,\text{cm}$, directly below the contact point of the femur on the tibia. The contact point between the foot and the ground is a distance $s = 3.6 \times 10^{1}\,\text{cm}$ from the vertical line passing through contact point of the femur on the tibia. The center of mass of the lower leg lies a distance $x = 1.8 \times 10^{1}\,\text{cm}$ from this same vertical line. Suppose the mass m_{L} of the lower leg is a 1/10 of the mass of the body (Figure 18.11). (a) Find the magnitude T of the force $\vec{\mathbf{T}}$ of the patellar ligament on the tibia. (b) Find the direction (the angle α) of the force $\vec{\mathbf{F}}$ of the femur on the tibia. (c) Find the magnitude F of the force $\vec{\mathbf{F}}$ of the femur on the tibia.

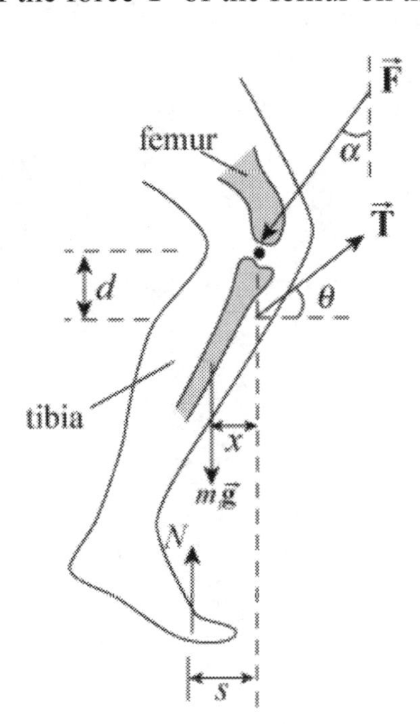

Figure 18.11 Example 18.4

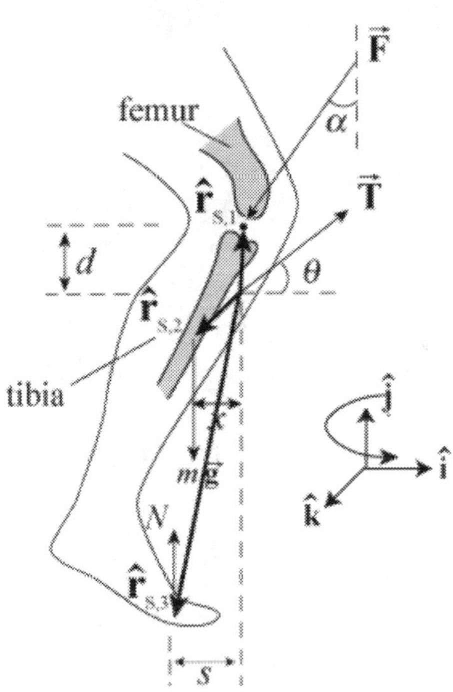

Figure 18.12 Torque-force diagram for knee

Solutions: a) Choose the unit vector $\hat{\mathbf{i}}$ to be directed horizontally to the right and $\hat{\mathbf{j}}$ directed vertically upwards. The first condition for static equilibrium, Eq. (18.1.1), that the sum of the forces is zero becomes

$$\hat{\mathbf{i}}: -F\sin\alpha + T\cos\theta = 0. \tag{18.4.17}$$

$$\hat{\mathbf{j}}: N - F\cos\alpha + T\sin\theta - (1/10)mg = 0. \tag{18.4.18}$$

Because the weight is evenly distributed on the two feet, the normal force on one foot is equal to half the weight, or

$$N = (1/2)mg\,; \tag{18.4.19}$$

Equation (18.4.18) becomes

$$\hat{\mathbf{j}}: (1/2)mg - F\cos\alpha + T\sin\theta - (1/10)mg = 0$$
$$(2/5)mg - F\cos\alpha + T\sin\theta = 0. \qquad (18.4.20)$$

The torque-force diagram on the knee is shown in Figure 18.12. Choose the point of action of the ligament on the tibia as the point S about which to compute torques. Note that the tensile force, $\vec{\mathbf{T}}$, that the ligament exerts on the upper tibia will make no contribution to the torque about this point S. This may help slightly in doing the calculations. Choose counterclockwise as the positive direction for the torque; this is the positive $\hat{\mathbf{k}}$ - direction. Then the torque due to the force $\vec{\mathbf{F}}$ of the femur on the tibia is

$$\vec{\boldsymbol{\tau}}_{S,1} = \vec{\mathbf{r}}_{S,1} \times \vec{\mathbf{F}} = d\,\hat{\mathbf{j}} \times (-F\sin\alpha\,\hat{\mathbf{i}} - F\cos\alpha\,\hat{\mathbf{j}}) = d\,F\sin\alpha\,\hat{\mathbf{k}}. \qquad (18.4.21)$$

The torque due to the mass of the leg is

$$\vec{\boldsymbol{\tau}}_{S,2} = \vec{\mathbf{r}}_{S,2} \times (-mg/10)\,\hat{\mathbf{j}} = (-x\,\hat{\mathbf{i}} - y_L\,\hat{\mathbf{j}}) \times (-mg/10)\,\hat{\mathbf{j}} = (1/10)x\,mg\,\hat{\mathbf{k}}. \qquad (18.4.22)$$

The torque due to the normal force of the ground is

$$\vec{\boldsymbol{\tau}}_{S,3} = \vec{\mathbf{r}}_{S,3} \times N\,\hat{\mathbf{j}} = (-s\,\hat{\mathbf{i}} - y_N\,\hat{\mathbf{j}}) \times N\,\hat{\mathbf{j}} = -s\,N\,\hat{\mathbf{k}} = -(1/2)s\,mg\,\hat{\mathbf{k}}. \qquad (18.4.23)$$

(In Equations (18.4.22) and (18.4.23), y_L and y_N are the vertical displacements of the point where the weight of the leg and the normal force with respect to the point S; as can be seen, these quantities do not enter directly into the calculations.) The condition that the sum of the torques about the point S vanishes, Eq. (18.1.2),

$$\vec{\boldsymbol{\tau}}_{S,\,\text{total}} = \vec{\boldsymbol{\tau}}_{S,1} + \vec{\boldsymbol{\tau}}_{S,2} + \vec{\boldsymbol{\tau}}_{S,3} = \vec{\mathbf{0}}, \qquad (18.4.24)$$

becomes

$$d\,F\sin\alpha\,\hat{\mathbf{k}} + (1/10)x\,mg\,\hat{\mathbf{k}} - (1/2)s\,mg\,\hat{\mathbf{k}} = \vec{\mathbf{0}}. \qquad (18.4.25)$$

The three equations in the three unknowns are summarized below:

$$\begin{aligned} -F\sin\alpha + T\cos\theta &= 0 \\ (2/5)mg - F\cos\alpha + T\sin\theta &= 0 \\ d\,F\sin\alpha + (1/10)x\,mg - (1/2)s\,mg &= 0. \end{aligned} \qquad (18.4.26)$$

The horizontal force equation, the first in (18.4.26), implies that

$$F\sin\alpha = T\cos\theta. \qquad (18.4.27)$$

Substituting this into the torque equation, the third equation of (18.4.26), yields

$$d\,T\cos\theta + (1/10)x\,mg - s(1/2)mg = 0\,. \tag{18.4.28}$$

Note that Equation (18.4.28) is the equation that would have been obtained if we had chosen the contact point between the tibia and the femur as the point about which to determine torques. Had we chosen this point, we would have saved one minor algebraic step. We can solve this Equation (18.4.28) for the magnitude T of the force $\vec{\mathbf{T}}$ of the patellar ligament on the tibia,

$$T = \frac{s(1/2)mg - (1/10)x\,mg}{d\cos\theta}. \tag{18.4.29}$$

Inserting numerical values into Equation (18.4.29),

$$\begin{aligned} T &= (70\,\text{kg})(9.8\,\text{m}\cdot\text{s}^{-2})\frac{(3.6\times10^{-1}\text{m})(1/2) - (1/10)(1.8\times10^{-1}\text{m})}{(3.8\times10^{-2}\text{m})\cos(40°)} \\ &= 3.8\times10^{3}\,\text{N}. \end{aligned} \tag{18.4.30}$$

b) We can now solve for the direction α of the force $\vec{\mathbf{F}}$ of the femur on the tibia as follows. Rewrite the two force equations in (18.4.26) as

$$\begin{aligned} F\cos\alpha &= (2/5)mg + T\sin\theta \\ F\sin\alpha &= T\cos\theta. \end{aligned} \tag{18.4.31}$$

Dividing these equations yields

$$\frac{F\cos\alpha}{F\sin\alpha} = \text{cotan}\,\alpha = \frac{(2/5)mg + T\sin\theta}{T\cos\theta}, \tag{18.4.32}$$

And so

$$\begin{aligned} \alpha &= \text{cotan}^{-1}\left(\frac{(2/5)mg + T\sin\theta}{T\cos\theta}\right) \\ \alpha &= \text{cotan}^{-1}\left(\frac{(2/5)(70\,\text{kg})(9.8\,\text{m}\cdot\text{s}^{-2}) + (3.4\times10^{3}\,\text{N})\sin(40°)}{(3.4\times10^{3}\text{N})\cos(40°)}\right) = 47°. \end{aligned} \tag{18.4.33}$$

c) We can now use the horizontal force equation to calculate the magnitude F of the force of the femur $\vec{\mathbf{F}}$ on the tibia from Equation (18.4.27),

$$F = \frac{(3.8\times10^{3}\,\text{N})\cos(40°)}{\sin(47°)} = 4.0\times10^{3}\,\text{N}\,. \tag{18.4.34}$$

Note you can find a symbolic expression for α that did not involve the intermediate numerical calculation of the tension. This is rather complicated algebraically; basically, the last two equations in (18.4.26) are solved for F and T in terms of α, θ and the other variables (Cramer's Rule is suggested) and the results substituted into the first of (18.4.26). The resulting expression is

$$\begin{aligned}\cot\alpha &= \frac{(s/2 - x/10)\sin(40^\circ) + ((2d/5)\cos(40^\circ))}{(s/2 - x/10)\cos(40^\circ)} \\ &= \tan(40^\circ) + \frac{2d/5}{s/2 - x/10}\end{aligned} \tag{18.4.35}$$

which leads to the same numerical result, $\alpha = 47^\circ$.

Appendix 18A The Torques About any Two Points are Equal for a Body in Static Equilibrium

When the net force on a body is zero, the torques about any two points are equal. To show this, consider any two points A and B. Choose a coordinate system with origin O and denote the constant vector from A to B by $\vec{\mathbf{r}}_{A,B}$. Suppose a force $\vec{\mathbf{F}}_i$ is acting at the point $\vec{\mathbf{r}}_{O,i}$. The vector from the point A to the point where the force acts is denoted by $\vec{\mathbf{r}}_{A,i}$, and the vectors from the point B to the point where the force acts is denoted by $\vec{\mathbf{r}}_{B,i}$.

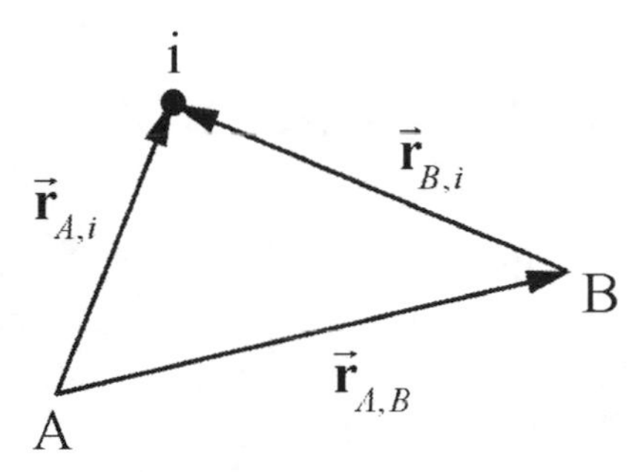

Figure 18A.1 Location of body i with respect to the points A and B.

In Figure 18A.1, the position vectors satisfy

$$\vec{\mathbf{r}}_{A,i} = \vec{\mathbf{r}}_{A,B} + \vec{\mathbf{r}}_{B,i}. \tag{18.A.1}$$

The sum of the torques about the point A is given by

$$\vec{\boldsymbol{\tau}}_{A} = \sum_{i=1}^{i=N} \vec{\mathbf{r}}_{A,i} \times \vec{\mathbf{F}}_i. \tag{18.A.2}$$

The sum of the torques about the point B is given by

$$\vec{\boldsymbol{\tau}}_{B} = \sum_{i=1}^{i=N} \vec{\mathbf{r}}_{B,i} \times \vec{\mathbf{F}}_i. \tag{18.A.3}$$

We can now substitute Equation (18.A.1) into Equation (18.A.2) and find that

$$\vec{\boldsymbol{\tau}}_{A} = \sum_{i=1}^{i=N} \vec{\mathbf{r}}_{A,i} \times \vec{\mathbf{F}}_i = \sum_{i=1}^{i=N} (\vec{\mathbf{r}}_{A,B} + \vec{\mathbf{r}}_{B,i}) \times \vec{\mathbf{F}}_i = \sum_{i=1}^{i=N} \vec{\mathbf{r}}_{A,B} \times \vec{\mathbf{F}}_i + \sum_{i=1}^{i=N} \vec{\mathbf{r}}_{B,i} \times \vec{\mathbf{F}}_i. \tag{18.A.4}$$

In the next-to-last term in Equation (18.A.4), the vector $\vec{\mathbf{r}}_{A,B}$ is constant and so may be taken outside the summation,

$$\sum_{i=1}^{i=N} \vec{\mathbf{r}}_{A,B} \times \vec{\mathbf{F}}_i = \vec{\mathbf{r}}_{A,B} \times \sum_{i=1}^{i=N} \vec{\mathbf{F}}_i . \tag{18.A.5}$$

We are assuming that there is no net force on the body, and so the sum of the forces on the body is zero,

$$\sum_{i=1}^{i=N} \vec{\mathbf{F}}_i = \vec{\mathbf{0}} . \tag{18.A.6}$$

Therefore the torque about point A, Equation (18.A.2), becomes

$$\vec{\boldsymbol{\tau}}_A = \sum_{i=1}^{i=N} \vec{\mathbf{r}}_{B,i} \times \vec{\mathbf{F}}_i = \vec{\boldsymbol{\tau}}_B . \tag{18.A.7}$$

For static equilibrium problems, the result of Equation (18.A.7) tells us that it does not matter which point we use to determine torques. In fact, note that the position of the chosen origin did not affect the result at all. Choosing the point about which to calculate torques (variously called "A", "B", "S" or sometimes "O") so that unknown forces do not exert torques about that point may often greatly simplify calculations.

Chapter 19 Angular Momentum

Chapter 19 Angular Momentum

The situation, in brief, is that newtonian physics is incapable of predicting conservation of angular momentum, but no isolated system has yet been encountered experimentally for which angular momentum is not conserved. We conclude that conservation of angular momentum is an independent physical law, and until a contradiction is observed, our physical understanding must be guided by it. [1]

Dan Kleppner

19.1 Introduction

When we consider a system of objects, we have shown that the external force, acting at the center of mass of the system, is equal to the time derivative of the total momentum of the system,

$$\vec{\mathbf{F}}^{\text{ext}} = \frac{d\vec{\mathbf{p}}_{\text{sys}}}{dt}. \tag{0.0.1}$$

We now introduce the rotational analog of Equation (0.0.1). We will first introduce the concept of angular momentum for a point-like particle of mass m with linear momentum $\vec{\mathbf{p}}$ about a point S, defined by the equation

$$\vec{\mathbf{L}}_S = \vec{\mathbf{r}}_S \times \vec{\mathbf{p}}, \tag{0.0.2}$$

where $\vec{\mathbf{r}}_S$ is the vector from the point S to the particle. We will show in this chapter that the torque about the point S acting on the particle is equal to the rate of change of the angular momentum about the point S of the particle,

$$\vec{\boldsymbol{\tau}}_S = \frac{d\vec{\mathbf{L}}_S}{dt}. \tag{0.0.3}$$

Equation (0.0.3) generalizes to any body undergoing rotation.

We shall concern ourselves first with the special case of rigid body undergoing fixed axis rotation about the z-axis with angular velocity $\vec{\boldsymbol{\omega}} = \omega_z \hat{\mathbf{k}}$. We divide up the rigid body into N elements labeled by the index i, $i = 1,2,\ldots N$, the ith element having mass m_i and position vector $\vec{\mathbf{r}}_{S,i}$. The rigid body has a moment of inertia I_S about some point S on the fixed axis, (often taken to be the *z*-axis, but not always) which rotates with angular velocity $\vec{\boldsymbol{\omega}}$ about this axis. The angular momentum is then the vector sum of the individual angular momenta,

[1] Kleppner, Daniel, An Introduction to Mechanics (1973), p. 307.

$$\vec{\mathbf{L}}_S = \sum_{i=1}^{i=N} \vec{\mathbf{L}}_{S,i} = \sum_{i=1}^{i=N} \vec{\mathbf{r}}_{S,i} \times \vec{\mathbf{p}}_i \tag{0.0.4}$$

When the rotation axis is the z-axis the z-component of the angular momentum, $L_{S,z}$, about the point S is then given by

$$L_{S,z} = I_S \, \omega_z \,. \tag{0.0.5}$$

We shall show that the z-component of the torque about the point S, $\tau_{S,z}$, is then the time derivative of the z-component of angular momentum about the point S,

$$\tau_{S,z} = \frac{dL_{S,z}}{dt} = I_S \frac{d\omega_z}{dt} = I_S \, \alpha_z \,. \tag{0.0.6}$$

19.2 Angular Momentum about a Point for a Particle

19.2.1 Angular Momentum for a Point Particle

Consider a point-like particle of mass m moving with a velocity $\vec{\mathbf{v}}$ (Figure 19.1).

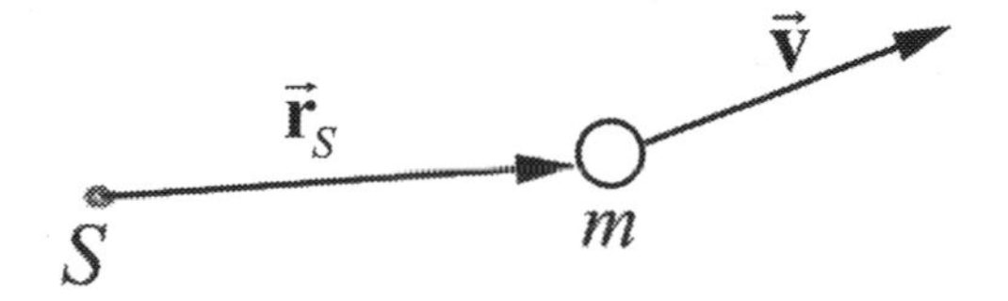

Figure 19.1 A point-like particle and its angular momentum about S.

The linear momentum of the particle is $\vec{\mathbf{p}} = m\vec{\mathbf{v}}$. Consider a point S located anywhere in space. Let $\vec{\mathbf{r}}_S$ denote the vector from the point S to the location of the object.

> *Define the angular momentum $\vec{\mathbf{L}}_S$ about the point S of a point-like particle as the vector product of the vector from the point S to the location of the object with the momentum of the particle,*

$$\vec{\mathbf{L}}_S = \vec{\mathbf{r}}_S \times \vec{\mathbf{p}} \,. \tag{0.1.1}$$

The derived SI units for angular momentum are $[\mathrm{kg} \cdot \mathrm{m}^2 \cdot \mathrm{s}^{-1}] = [\mathrm{N} \cdot \mathrm{m} \cdot \mathrm{s}] = [\mathrm{J} \cdot \mathrm{s}]$. There is no special name for this set of units.

Because angular momentum is defined as a vector, we begin by studying its magnitude and direction. The magnitude of the angular momentum about S is given by

$$\left|\vec{\mathbf{L}}_S\right| = \left|\vec{\mathbf{r}}_S\right|\left|\vec{\mathbf{p}}\right|\sin\theta\,, \qquad (0.1.2)$$

where θ is the angle between the vectors and $\vec{\mathbf{p}}$, and lies within the range $[0 \le \theta \le \pi]$ (Figure 19.2). Analogous to the magnitude of torque, there are two ways to determine the magnitude of the angular momentum about S.

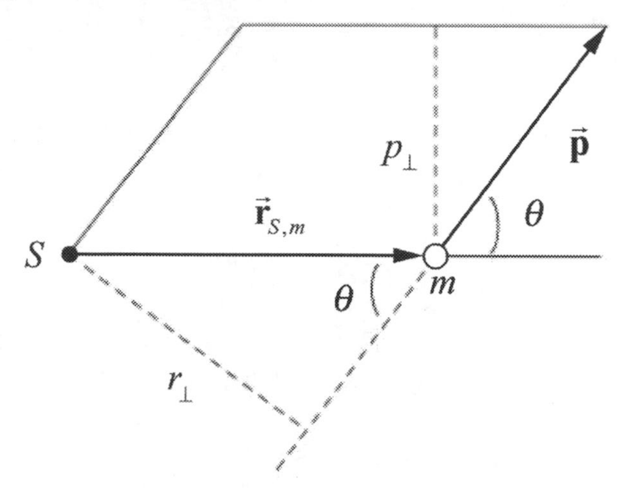

Figure 19.2 Vector diagram for angular momentum.

Define the *moment arm*, $r_\perp$, as the perpendicular distance from the point S to the line defined by the direction of the momentum. Then

$$r_\perp = \left|\vec{\mathbf{r}}_S\right|\sin\theta\,. \qquad (0.1.3)$$

Hence the magnitude of the angular momentum is the product of the moment arm with the magnitude of the momentum,

$$\left|\vec{\mathbf{L}}_S\right| = r_\perp\left|\vec{\mathbf{p}}\right|. \qquad (0.1.4)$$

Alternatively, define the *perpendicular momentum*, $p_\perp$, to be the magnitude of the component of the momentum perpendicular to the line defined by the direction of the vector $\vec{\mathbf{r}}_S$. Thus

$$p_\perp = \left|\vec{\mathbf{p}}\right|\sin\theta\,. \qquad (0.1.5)$$

We can think of the magnitude of the angular momentum as the product of the distance from S to the particle with the perpendicular momentum,

$$\left|\vec{\mathbf{L}}_S\right| = \left|\vec{\mathbf{r}}_S\right| p_\perp\,. \qquad (0.1.6)$$

19.2.2 Right-Hand-Rule for the Direction of the Angular Momentum

We shall define the direction of the angular momentum about the point S by a right hand rule. Draw the vectors $\vec{\mathbf{r}}_S$ and $\vec{\mathbf{p}}$ so their tails are touching. Then draw an arc starting from the vector $\vec{\mathbf{r}}_S$ and finishing on the vector $\vec{\mathbf{p}}$. (There are two such arcs; choose the shorter one.) This arc is either in the clockwise or counterclockwise direction. Curl the fingers of your right hand in the same direction as the arc. Your right thumb points in the direction of the angular momentum.

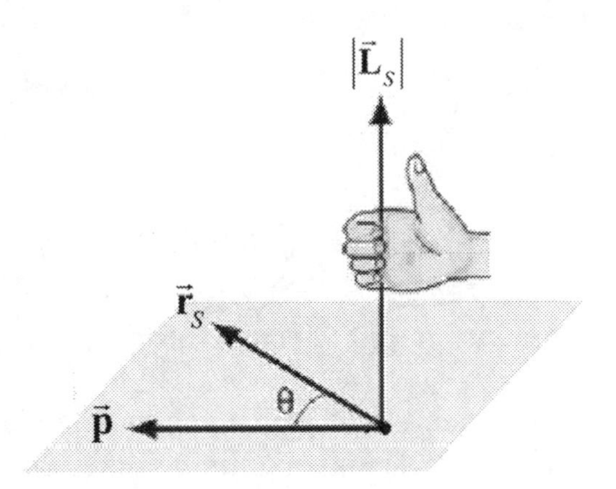

Figure 19.3 The right hand rule.

Remember that, as in all vector products, the direction of the angular momentum is perpendicular to the plane formed by $\vec{\mathbf{r}}_S$ and $\vec{\mathbf{p}}$.

Example 19.1 Angular Momentum: Constant Velocity

A particle of mass $m = 2.0\ \text{kg}$ moves as shown in Figure 19.4 with a uniform velocity $\vec{\mathbf{v}} = 3.0\ \text{m}\cdot\text{s}^{-1}\ \hat{\mathbf{i}} + 3.0\ \text{m}\cdot\text{s}^{-1}\ \hat{\mathbf{j}}$. At time t, the particle passes through the point $(2.0\ \text{m}, 3.0\ \text{m})$. Find the direction and the magnitude of the angular momentum about the origin (point O) at time t.

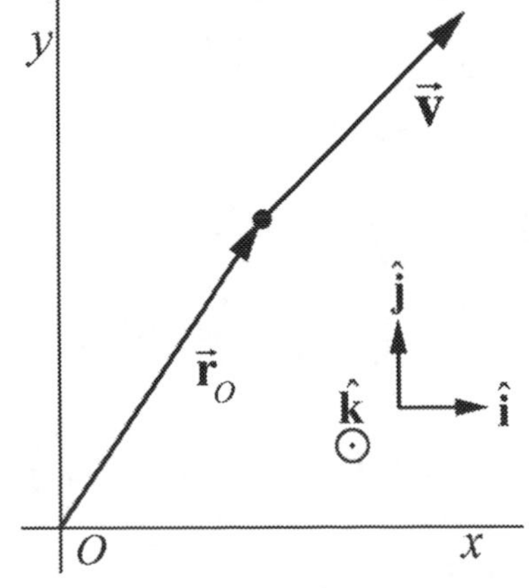

Figure 19.4 Example 19.4

Solution: Choose Cartesian coordinates with unit vectors shown in the figure above. The vector from the origin O to the location of the particle is $\vec{\mathbf{r}}_O = 2.0\,\text{m}\,\hat{\mathbf{i}} + 3.0\,\text{m}\,\hat{\mathbf{j}}$. The angular momentum vector $\vec{\mathbf{L}}_O$ of the particle about the origin O is given by:

$$\begin{aligned}
\vec{\mathbf{L}}_O &= \vec{\mathbf{r}}_O \times \vec{\mathbf{p}} = \vec{\mathbf{r}}_O \times m\vec{\mathbf{v}} \\
&= (2.0\text{m}\,\hat{\mathbf{i}} + 3.0\text{m}\,\hat{\mathbf{j}}) \times (2\text{kg})(3.0\text{m}\cdot\text{s}^{-1}\hat{\mathbf{i}} + 3.0\text{m}\cdot\text{s}^{-1}\hat{\mathbf{j}}) \\
&= 0 + 12\text{kg}\cdot\text{m}^2\cdot\text{s}^{-1}\hat{\mathbf{k}} - 18\text{kg}\cdot\text{m}^2\cdot\text{s}^{-1}(-\hat{\mathbf{k}}) + \vec{\mathbf{0}} \\
&= -6\text{kg}\cdot\text{m}^2\cdot\text{s}^{-1}\hat{\mathbf{k}}.
\end{aligned}$$

In the above, the relations $\vec{\mathbf{i}} \times \vec{\mathbf{j}} = \vec{\mathbf{k}}$, $\vec{\mathbf{j}} \times \vec{\mathbf{i}} = -\vec{\mathbf{k}}$, $\vec{\mathbf{i}} \times \vec{\mathbf{i}} = \vec{\mathbf{j}} \times \vec{\mathbf{j}} = \vec{\mathbf{0}}$ were used.

Example 19.2 Angular Momentum and Circular Motion

A particle of mass m moves in a circle of radius r at an angular speed ω about the z-axis in a plane parallel to the x-y plane passing through the origin O (Figure 19.5). Find the magnitude and the direction of the angular momentum $\vec{\mathbf{L}}_O$ relative to the origin.

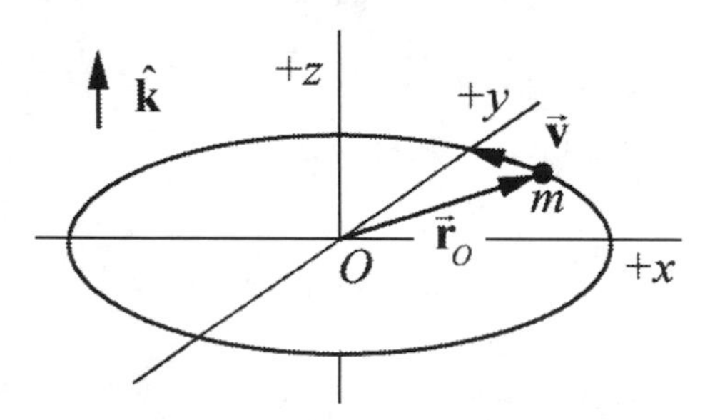

Figure 19.5 Example 19.2

Solution: The velocity of the particle is given by $\vec{\mathbf{v}} = r\omega\,\hat{\boldsymbol{\theta}}$. The vector from the center of the circle (the point O) to the object is given by $\vec{\mathbf{r}}_O = r\,\hat{\mathbf{r}}$. The angular momentum about the center of the circle is the vector product

$$\vec{\mathbf{L}}_O = \vec{\mathbf{r}}_O \times \vec{\mathbf{p}} = \vec{\mathbf{r}}_O \times m\vec{\mathbf{v}} = rmv\,\hat{\mathbf{k}} = rmr\omega\,\hat{\mathbf{k}} = mr^2\omega\,\hat{\mathbf{k}}.$$

The magnitude is $\left|\vec{\mathbf{L}}_O\right| = mr^2\omega$, and the direction is in the $+\hat{\mathbf{k}}$-direction.

Example 19.3 Angular Momentum About a Point along Central Axis for Circular Motion

A particle of mass m moves in a circle of radius r at an angular speed ω about the z-axis in a plane parallel to but a distance h above the x-y plane (Figure 19.6). (a) Find the magnitude and the direction of the angular momentum $\vec{\mathbf{L}}_O$ relative to the origin O. (b) Find the z-component of $\vec{\mathbf{L}}_O$.

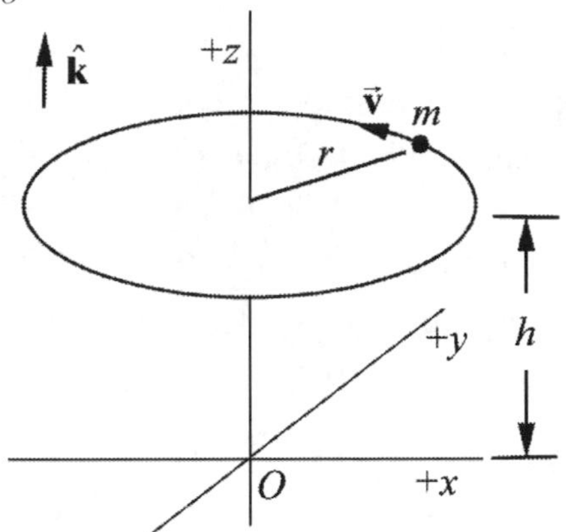

Figure 19.6 Example 19.3

Solution: We begin by making a geometric argument. Suppose the particle has coordinates (x, y, h). The angular momentum about the origin O is defined as

$$\vec{\mathbf{L}}_O = \vec{\mathbf{r}}_O \times m\vec{\mathbf{v}}\,. \quad (0.1.7)$$

The vectors $\vec{\mathbf{r}}_O$ and $\vec{\mathbf{v}}$ are perpendicular to each other so the angular momentum is perpendicular to the plane formed by those two vectors. The speed of the particle is $v = r\omega$. Suppose the vector $\vec{\mathbf{r}}_O$ forms an angle ϕ with the z-axis. Then $\vec{\mathbf{L}}_O$ forms an angle $90^\circ - \phi$ with respect to the z-axis or an angle ϕ with respect to the x-y plane as shown in Figure 19.7.

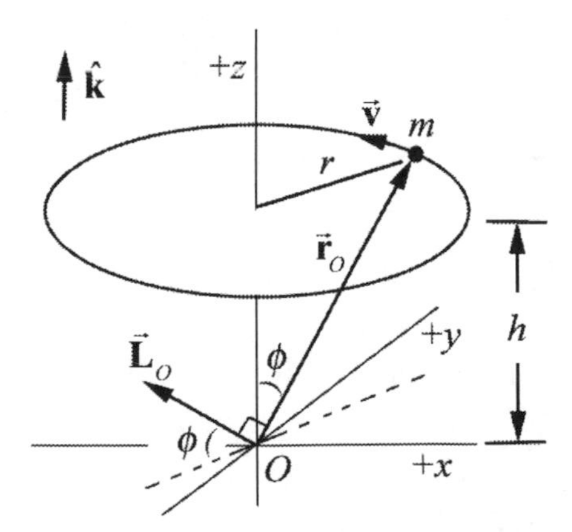

Figure 19.7 Direction of $\vec{\mathbf{L}}_O$

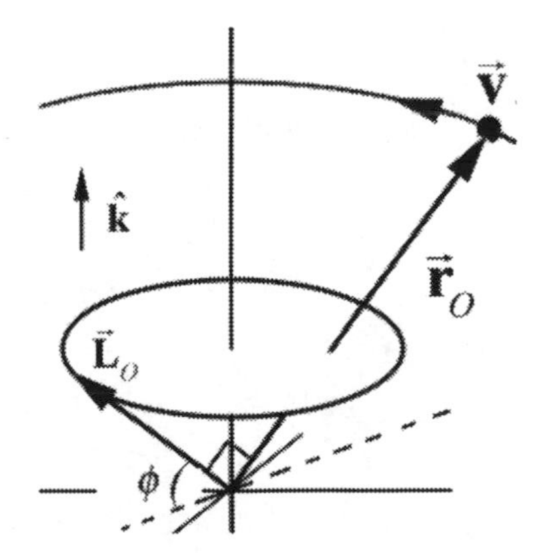

Figure 19.8 Direction of $\vec{\mathbf{L}}_O$ sweeps out a cone

The magnitude of $\vec{\mathbf{L}}_O$ is

$$\left|\vec{\mathbf{L}}_O\right| = \left|\vec{\mathbf{r}}_O\right| m \left|\vec{\mathbf{v}}\right| = m(h^2 + (x^2 + y^2))^{1/2} r\omega . \qquad (0.1.8)$$

The magnitude of $\vec{\mathbf{L}}_O$ is constant, but its direction is changing as the particle moves in a circular orbit about the z-axis, sweeping out a cone as shown in Figure 19.8. We draw the vector $\vec{\mathbf{L}}_O$ at the origin because it is defined at that point.

We shall now explicitly calculate the vector product. Determining the vector product using polar coordinates is the easiest way to calculate $\vec{\mathbf{L}}_O = \vec{\mathbf{r}}_O \times m\vec{\mathbf{v}}$. We begin by writing the two vectors that appear in Eq. (0.1.7) in polar coordinates. We start with the vector from the origin to the location of the moving object, $\vec{\mathbf{r}}_O = x\hat{\mathbf{i}} + y\hat{\mathbf{j}} + h\hat{\mathbf{k}} = r\hat{\mathbf{r}} + h\hat{\mathbf{k}}$ where $r = (x^2 + y^2)^{1/2}$. The velocity vector is tangent to the circular orbit so $\vec{\mathbf{v}} = v\,\hat{\boldsymbol{\theta}} = r\omega\,\hat{\boldsymbol{\theta}}$. Using the fact that $\hat{\mathbf{r}} \times \hat{\boldsymbol{\theta}} = \hat{\mathbf{k}}$ and $\hat{\mathbf{k}} \times \hat{\boldsymbol{\theta}} = -\hat{\mathbf{r}}$, the angular momentum about the origin $\vec{\mathbf{L}}_O$ is

$$\vec{\mathbf{L}}_O = \vec{\mathbf{r}}_O \times m\vec{\mathbf{v}} = (r\hat{\mathbf{r}} + h\hat{\mathbf{k}}) \times mr\omega\hat{\boldsymbol{\theta}} = rmr\omega\hat{\mathbf{k}} - hmr\omega\hat{\mathbf{r}} . \qquad (0.1.9)$$

The magnitude of $\vec{\mathbf{L}}_O$ is given by

$$\left|\vec{\mathbf{L}}_O\right| = ((rmr\omega)^2 + (hmr\omega)^2)^{1/2} = m(h^2 + r^2)^{1/2} r\omega = m(h^2 + (x^2 + y^2))^{1/2} r\omega . \qquad (0.1.10)$$

Agreeing with our geometric argument. In Figure 19.9, denote the angle $\vec{\mathbf{L}}_O$ forms with respect to the x-y plane by β. Then

$$\tan\beta = -\frac{L_{0z}}{L_{0r}} = \frac{r}{h} = \tan\phi , \qquad (0.1.11)$$

so $\beta = \phi$ also agreeing with our geometric argument.

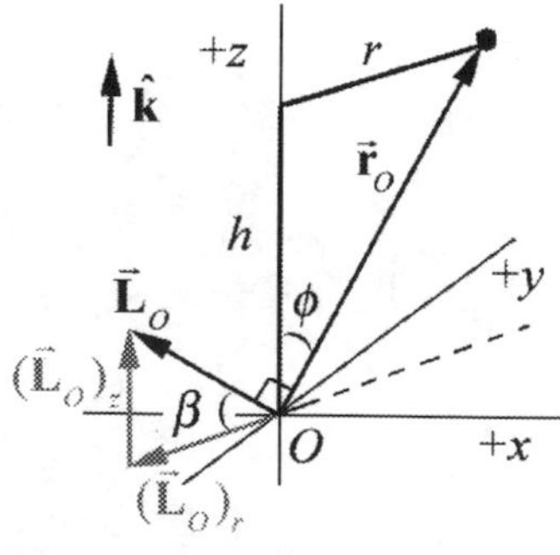

Figure 19.9 Geometry for components of $\vec{\mathbf{L}}_O$

The important point to keep in mind regarding this calculation is that for any point along the z-axis not at the center of the circular orbit of a single particle, the angular momentum about that point does not point along the z-axis but it is has a non-zero component in the x-y plane (or in the $\hat{\mathbf{r}}$-direction if you use polar coordinates). However, the z-component of the angular momentum about any point along the z-axis is independent of the location of the point along the axis.

19.3 Torque and the Time Derivative of Angular Momentum about a Point for a Particle

We will now show that the torque about a point S is equal to the time derivative of the angular momentum about S,

$$\vec{\boldsymbol{\tau}}_S = \frac{d\vec{\mathbf{L}}_S}{dt}. \tag{0.2.1}$$

Take the time derivative of the angular momentum about S,

$$\frac{d\vec{\mathbf{L}}_S}{dt} = \frac{d}{dt}\left(\vec{\mathbf{r}}_S \times \vec{\mathbf{p}}\right). \tag{0.2.2}$$

In this equation we are taking the time derivative of a vector product of two vectors. There are two important facts that will help us simplify this expression. First, the time derivative of the vector product of two vectors satisfies the product rule,

$$\frac{d\vec{\mathbf{L}}_S}{dt} = \frac{d}{dt}(\vec{\mathbf{r}}_S \times \vec{\mathbf{p}}) = \left(\left(\frac{d\vec{\mathbf{r}}_S}{dt}_S\right) \times \vec{\mathbf{p}}\right) + \left(\vec{\mathbf{r}}_S \times \left(\frac{d\vec{\mathbf{p}}}{dt}\right)\right). \tag{0.2.3}$$

Second, the first term on the right hand side vanishes,

$$\frac{d\vec{\mathbf{r}}_S}{dt} \times \vec{\mathbf{p}} = \vec{\mathbf{v}} \times m\vec{\mathbf{v}} = \vec{\mathbf{0}}. \tag{0.2.4}$$

The rate of angular momentum change about the point S is then

$$\frac{d\vec{\mathbf{L}}_S}{dt} = \vec{\mathbf{r}}_S \times \frac{d\vec{\mathbf{p}}}{dt}. \tag{0.2.5}$$

From Newton's Second Law, the force on the particle is equal to the derivative of the linear momentum,

$$\vec{\mathbf{F}} = \frac{d\vec{\mathbf{p}}}{dt}. \tag{0.2.6}$$

Therefore the rate of change in time of angular momentum about the point S is

$$\frac{d\vec{\mathbf{L}}_S}{dt} = \vec{\mathbf{r}}_S \times \vec{\mathbf{F}}. \qquad (0.2.7)$$

Recall that the torque about the point S due to the force $\vec{\mathbf{F}}$ acting on the particle is

$$\vec{\boldsymbol{\tau}}_S = \vec{\mathbf{r}}_S \times \vec{\mathbf{F}}. \qquad (0.2.8)$$

Combining the expressions in (0.2.7) and (0.2.8), it is readily seen that the torque about the point S is equal to the rate of change of angular momentum about the point S,

$$\vec{\boldsymbol{\tau}}_S = \frac{d\vec{\mathbf{L}}_S}{dt}. \qquad (0.2.9)$$

19.4 Conservation of Angular Momentum about a Point

So far we have introduced two conservation principles, showing that energy is constant for closed systems (no change in energy in the surroundings) and linear momentum is constant isolated system. The change in mechanical energy of a closed system is

$$W_{nc} = \Delta E_m = \Delta K + \Delta U, \qquad \text{(closed system)}. \qquad (0.2.10)$$

If the non-conservative work done in the system is zero, then the mechanical energy is constant,

$$0 = W_{\text{nc}} = \Delta E_{\text{mechanical}} = \Delta K + \Delta U, \quad \text{(closed system)}. \qquad (0.2.11)$$

The conservation of linear momentum arises from Newton's Second Law applied to systems,

$$\vec{\mathbf{F}}^{\text{ext}} = \sum_{i=1}^{N} \frac{d}{dt}\vec{\mathbf{p}}_i = \frac{d}{dt}\vec{\mathbf{p}}_{\text{sys}} \qquad (0.2.12)$$

Thus if the external force in any direction is zero, then the component of the momentum of the system in that direction is a constant. For example, if there are no external forces in the x- and y-directions then

$$\begin{aligned}\vec{\mathbf{0}} &= (\vec{\mathbf{F}}^{\text{ext}})_x = \frac{d}{dt}(\vec{\mathbf{p}}_{\text{sys}})_x \\ \vec{\mathbf{0}} &= (\vec{\mathbf{F}}^{\text{ext}})_y = \frac{d}{dt}(\vec{\mathbf{p}}_{\text{sys}})_y.\end{aligned} \qquad (0.2.13)$$

We can now use our relation between torque about a point S and the change of the angular momentum about S, Eq. (0.2.9), to introduce a new conservation law. Suppose we can find a point S such that torque about the point S is zero,

$$\vec{\mathbf{0}} = \vec{\boldsymbol{\tau}}_S = \frac{d\vec{\mathbf{L}}_S}{dt}, \qquad (0.2.14)$$

then the angular momentum about the point S is a constant vector, and so the change in angular momentum is zero,

$$\Delta\vec{\mathbf{L}}_S \equiv \vec{\mathbf{L}}_{S,f} - \vec{\mathbf{L}}_{S,i} = \vec{\mathbf{0}}. \qquad (0.2.15)$$

Thus when the torque about a point S is zero, the final angular momentum about S is equal to the initial angular momentum,

$$\vec{\mathbf{L}}_{S,f} = \vec{\mathbf{L}}_{S,i}. \qquad (0.2.16)$$

Example 19.4 Meteor Flyby of Earth

A meteor of mass $m = 2.1 \times 10^{13}\ \text{kg}$ is approaching earth as shown in Figure 19.10. The distance h is called the *impact parameter*. The radius of the earth is $r_e = 6.37 \times 10^6\ \text{m}$. The mass of the earth is $m_e = 5.98 \times 10^{24}\ \text{kg}$. Suppose the meteor has an initial speed of $v_0 = 1.0 \times 10^1\ \text{m} \cdot \text{s}^{-1}$. Assume that the meteor started very far away from the earth. Suppose the meteor just grazes the earth. You may ignore all other gravitational forces except the earth. Find the impact parameter h.

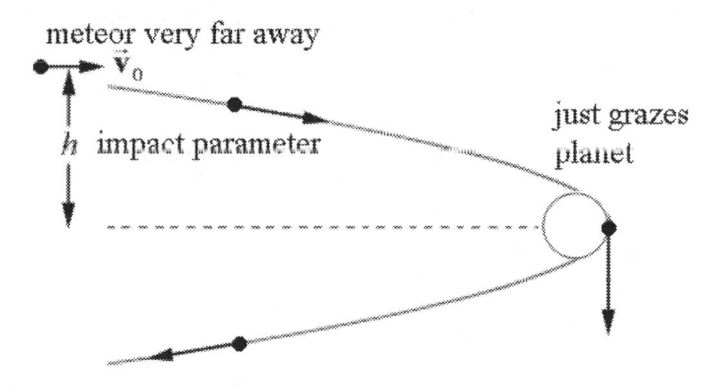

Figure 19.10 Meteor flyby of earth

Solution: The free-body force diagrams when the meteor is very far away and when the meteor just grazes the earth are shown in Figure 19.11. Denote the center of the earth by S. The force on the meteor is given by

$$\vec{\mathbf{F}} = -\frac{Gm_e m}{|\vec{r}|^2}\hat{\mathbf{r}} \qquad (0.2.17)$$

where $\hat{\mathbf{r}}$ is a unit vector pointing radially away from the center of the earth, and r is the distance from the center of the earth to the meteor. The torque on the meteor is given by

$\vec{\tau}_S = \vec{\mathbf{r}}_{S,F} \times \vec{\mathbf{F}}$, where $\vec{\mathbf{r}}_{S,F} = r\,\hat{\mathbf{r}}$ is the vector from the point S to the position of the meteor. Because the force and the position vector are collinear, the vector product vanishes and hence the torque on the meteor vanishes about S.

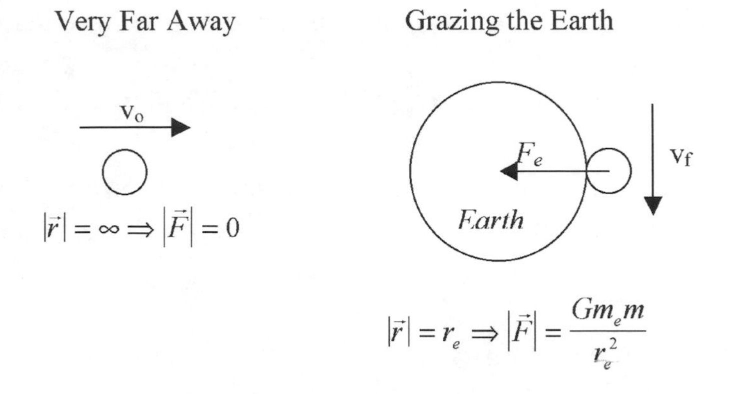

Figure 19.11 Free-body force diagrams for meteor

Choose Cartesian coordinates as shown in Figure 19.12. The initial angular momentum about the center of the earth is

$$(\vec{\mathbf{L}}_S)_i = \vec{\mathbf{r}}_{S,i} \times \vec{\mathbf{p}}_0, \tag{0.2.18}$$

where the vector from the center of the earth to the meteor is $\vec{\mathbf{r}}_{S,i} = -x_i\,\hat{\mathbf{i}} + h\,\hat{\mathbf{j}}$ (we can choose some arbitrary x_i for the initial x-component of position), and the momentum is $\vec{\mathbf{p}}_i = mv_i\hat{\mathbf{i}}$. Then the initial angular momentum is

$$(\vec{\mathbf{L}}_S)_i = \vec{\mathbf{r}}_{S,0} \times \vec{\mathbf{p}}_i = (-x_i\,\hat{\mathbf{i}} + h\,\hat{\mathbf{j}}) \times mv_i\hat{\mathbf{i}} = -mv_i h\,\hat{\mathbf{k}} \tag{0.2.19}$$

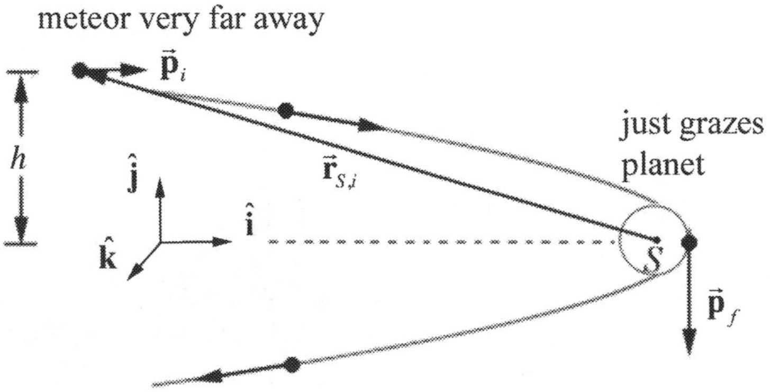

Figure 19.12 Momentum diagram for meteor

The final angular momentum about the center of the earth is

$$(\vec{\mathbf{L}}_S)_f = \vec{\mathbf{r}}_{S,f} \times \vec{\mathbf{p}}_f, \tag{0.2.20}$$

where the vector from the center of the earth to the meteor is $\vec{\mathbf{r}}_{S,f} = r_e\,\hat{\mathbf{i}}$ since the meteor is then just grazing the surface of earth, and the momentum is $\vec{\mathbf{p}}_f = -mv_f\,\hat{\mathbf{j}}$. Therefore

$$(\vec{\mathbf{L}}_S)_f = \vec{\mathbf{r}}_{S,f} \times \vec{\mathbf{p}}_f = r_e\,\hat{\mathbf{i}} \times (-mv_f\,\hat{\mathbf{j}}) = -mr_e v_f\,\hat{\mathbf{k}}. \tag{0.2.21}$$

Because the angular momentum about the center of the earth is constant throughout the motion

$$(\vec{\mathbf{L}}_S)_i = (\vec{\mathbf{L}}_S)_f, \tag{0.2.22}$$

which implies that

$$-mv_i h\,\hat{\mathbf{k}} = -mr_e v_f\,\hat{\mathbf{k}} \Rightarrow v_f = \frac{v_i h}{r_e}. \tag{0.2.23}$$

The mechanical energy is constant and with our choice of zero for potential energy when the meteor is very far away, the energy condition becomes

$$\frac{1}{2}mv_i^2 = \frac{1}{2}mv_f^2 - \frac{Gm_e m}{r_e}. \tag{0.2.24}$$

Therefore

$$v_f^2 = v_i^2 + \frac{2Gm_e}{r_e} \tag{0.2.25}$$

Substituting v_f from part (d) and solving for h, we have that

$$h = r_e\sqrt{1 + \frac{2Gm_e}{r_e v_i^2}} \tag{0.2.26}$$

On substituting the values we have,

$$h = 1117.4\,r_e = 7.12 \times 10^9\,\mathrm{m}. \tag{0.2.27}$$

19.5 Angular Impulse and Change in Angular Momentum

If there is a total applied torque $\vec{\boldsymbol{\tau}}_S$ about a point S over an interval of time $\Delta t = t_f - t_i$, then the torque applies an ***angular impulse*** about a point S, given by

$$\vec{\mathbf{J}}_S = \int_{t_i}^{t_f} \vec{\boldsymbol{\tau}}_S\, dt. \tag{0.3.1}$$

Because $\vec{\boldsymbol{\tau}}_S = d\vec{\mathbf{L}}_S^{\text{total}} / dt$, the angular impulse about S is equal to the change in angular momentum about S,

$$\vec{\mathbf{J}}_S = \int_{t_i}^{t_f} \vec{\boldsymbol{\tau}}_S \, dt = \int_{t_i}^{t_f} \frac{d\vec{\mathbf{L}}_S}{dt} dt = \Delta\vec{\mathbf{L}}_S = \vec{\mathbf{L}}_{S,f} - \vec{\mathbf{L}}_{S,i} . \qquad (0.3.2)$$

This result is the rotational analog to linear impulse, which is equal to the change in momentum,

$$\vec{\mathbf{I}} = \int_{t_i}^{t_f} \vec{\mathbf{F}} \, dt = \int_{t_i}^{t_f} \frac{d\vec{\mathbf{p}}}{dt} dt = \Delta\vec{\mathbf{p}} = \vec{\mathbf{p}}_f - \vec{\mathbf{p}}_i . \qquad (0.3.3)$$

19.6 Angular Momentum of a System of Particles

We now calculate the angular momentum about the point S associated with a system of N point particles. Label each individual particle by the index j, $j = 1, 2, \cdots, N$. Let the jth particle have mass m_j and velocity $\vec{\mathbf{v}}_j$. The momentum of an individual particle is then $\vec{\mathbf{p}}_j = m_j \vec{\mathbf{v}}_j$. Let $\vec{\mathbf{r}}_{S,j}$ be the vector from the point S to the jth particle, and let θ_j be the angle between the vectors $\vec{\mathbf{r}}_{S,j}$ and $\vec{\mathbf{p}}_j$ (Figure 19.13).

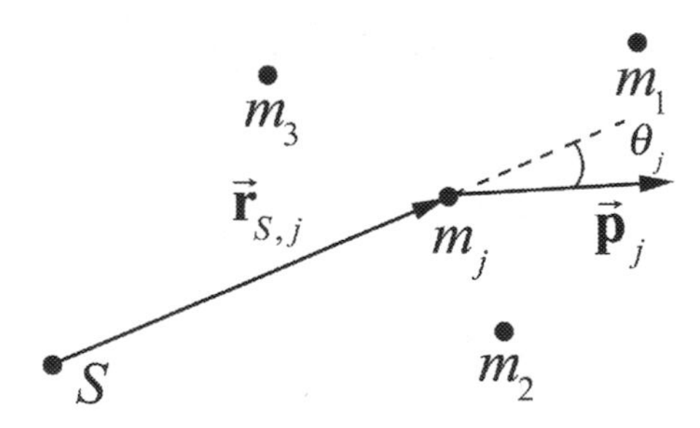

Figure 19.13 System of particles

The angular momentum $\vec{\mathbf{L}}_{S,j}$ of the jth particle is

$$\vec{\mathbf{L}}_{S,j} = \vec{\mathbf{r}}_{S,j} \times \vec{\mathbf{p}}_j . \qquad (0.4.1)$$

The angular momentum for the system of particles is the vector sum of the individual angular momenta,

$$\vec{\mathbf{L}}_S^{\text{sys}} = \sum_{j=1}^{j=N} \vec{\mathbf{L}}_{S,j} = \sum_{j=1}^{j=N} \vec{\mathbf{r}}_{S,j} \times \vec{\mathbf{p}}_j . \qquad (0.4.2)$$

The change in the angular momentum of the system of particles about a point S is given by

$$\frac{d\vec{\mathbf{L}}_S^{\text{sys}}}{dt} = \frac{d}{dt} \sum_{j=1}^{j=N} \vec{\mathbf{L}}_{S,j} = \sum_{j=1}^{j=N} \left(\frac{d\vec{\mathbf{r}}_{S,j}}{dt} \times \vec{\mathbf{p}}_j + \vec{\mathbf{r}}_{S,j} \times \frac{d\vec{\mathbf{p}}_j}{dt} \right) . \qquad (0.4.3)$$

Because the velocity of the jth particle is $\vec{\mathbf{v}}_{S,j} = d\vec{\mathbf{r}}_{S,j} / dt$, the first term in the parentheses vanishes (the cross product of a vector with itself is zero because they are parallel to each other)

$$\frac{d\vec{\mathbf{r}}_{S,j}}{dt} \times \vec{\mathbf{p}}_j = \vec{\mathbf{v}}_{S,j} \times m_j \vec{\mathbf{v}}_{S,j} = 0 . \tag{0.4.4}$$

Substitute Eq. (0.4.4) and $\vec{\mathbf{F}}_j = d\vec{\mathbf{p}}_j / dt$ into Eq. (0.4.3) yielding

$$\frac{d\vec{\mathbf{L}}_S^{\text{sys}}}{dt} = \sum_{j=1}^{j=N} \left(\vec{\mathbf{r}}_{S,j} \times \frac{d\vec{\mathbf{p}}_j}{dt} \right) = \sum_{j=1}^{j=N} \left(\vec{\mathbf{r}}_{S,j} \times \vec{\mathbf{F}}_j \right). \tag{0.4.5}$$

Because

$$\sum_{j=1}^{j=N} \left(\vec{\mathbf{r}}_{S,j} \times \vec{\mathbf{F}}_j \right) = \sum_{j=1}^{j=N} \vec{\boldsymbol{\tau}}_{S,j} = \vec{\boldsymbol{\tau}}_S^{\text{ ext}} + \vec{\boldsymbol{\tau}}_S^{\text{ int}} \tag{0.4.6}$$

We have already shown in Chapter 17.4 that when we assume all internal forces are directed along the line connecting the two interacting objects then the internal torque about the point S is zero,

$$\vec{\boldsymbol{\tau}}_S^{\text{int}} = \vec{\mathbf{0}} . \tag{0.4.7}$$

Eq. (0.4.6) simplifies to

$$\sum_{j=1}^{j=N} \left(\vec{\mathbf{r}}_{S,j} \times \vec{\mathbf{F}}_j \right) = \sum_{j=1}^{j=N} \vec{\boldsymbol{\tau}}_{S,j} = \vec{\boldsymbol{\tau}}_S^{\text{ ext}} . \tag{0.4.8}$$

Therefore Eq. (0.4.5) becomes

$$\vec{\boldsymbol{\tau}}_S^{\text{ext}} = \frac{d\vec{\mathbf{L}}_S^{\text{sys}}}{dt} . \tag{0.4.9}$$

The external torque about the point S is equal to the time derivative of the angular momentum of the system about that point.

Example 19.5 Angular Momentum of Two Particles undergoing Circular Motion

Two identical particles of mass m move in a circle of radius r, 180° out of phase at an angular speed ω about the z-axis in a plane parallel to but a distance h above the x-y plane (Figure 19.14). Find the magnitude and the direction of the angular momentum $\vec{\mathbf{L}}_O$ relative to the origin.

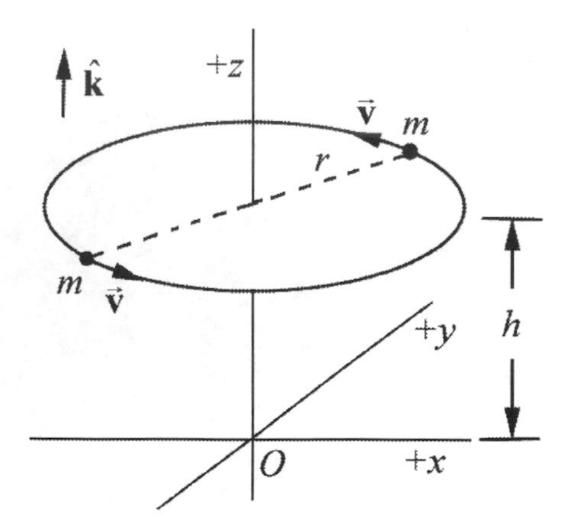

Figure 19.14 Example 19.5

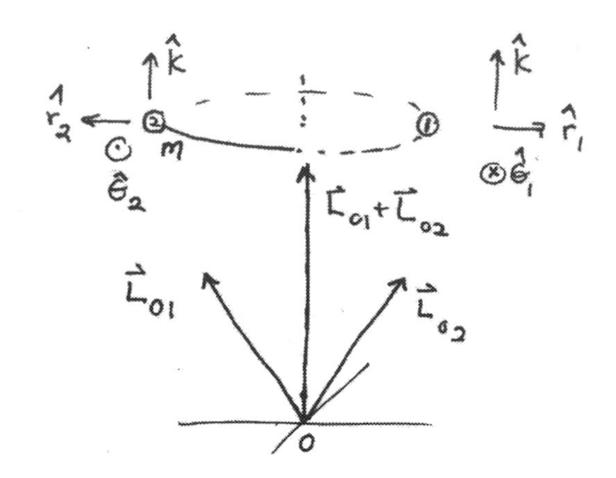

Figure 19.15 Angular momentum of each particle about origin and sum

Solution: The angular momentum about the origin is the sum of the contributions from each object. Since they have the same mass, the angular momentum vectors are shown in Figure 19.15. The components that lie in the x-y plane cancel leaving only a non-zero z-component,

$$\vec{\mathbf{L}}_O = \vec{\mathbf{L}}_{O,1} + \vec{\mathbf{L}}_{O,2} = 2mr^2\omega\hat{\mathbf{k}}. \quad (0.4.10)$$

If you explicitly calculate the cross product in polar coordinates you must be careful because the units vectors $\hat{\mathbf{r}}$ and $\hat{\boldsymbol{\theta}}$ at the position of objects 1 and 2 are different. If we set $\vec{\mathbf{r}}_{O,1} = r\hat{\mathbf{r}}_1 + h\hat{\mathbf{k}}$ and $\vec{\mathbf{v}}_1 = r\omega\,\hat{\boldsymbol{\theta}}_1$ such that $\hat{\mathbf{r}}_1 \times \hat{\boldsymbol{\theta}}_1 = \hat{\mathbf{k}}$ and similarly set $\vec{\mathbf{r}}_{O,2} = r\hat{\mathbf{r}}_2 + h\hat{\mathbf{k}}$ and $\vec{\mathbf{v}}_2 = r\omega\,\hat{\boldsymbol{\theta}}_2$ such that $\hat{\mathbf{r}}_2 \times \hat{\boldsymbol{\theta}}_2 = \hat{\mathbf{k}}$ then $\hat{\mathbf{r}}_1 = -\hat{\mathbf{r}}_2$ and $\hat{\boldsymbol{\theta}}_1 = -\hat{\boldsymbol{\theta}}_2$. With this in mind we can compute

$$\begin{aligned}
\vec{\mathbf{L}}_0 &= \vec{\mathbf{L}}_{0,1} + \vec{\mathbf{L}}_{0,2} = \vec{\mathbf{r}}_{0,1} \times m\vec{\mathbf{v}}_1 + \vec{\mathbf{r}}_{0,2} \times m\vec{\mathbf{v}}_2 \\
&= (r\hat{\mathbf{r}}_1 + h\hat{\mathbf{k}}) \times mr\omega\,\hat{\boldsymbol{\theta}}_1 + (r\hat{\mathbf{r}}_2 + h\hat{\mathbf{k}}) \times mr\omega\,\hat{\boldsymbol{\theta}}_2 \\
&= 2mr^2\omega\hat{\mathbf{k}} + hmr\omega(-\hat{\mathbf{r}}_1 - \hat{\mathbf{r}}_2) = 2mr^2\omega\hat{\mathbf{k}} + hmr\omega(-\hat{\mathbf{r}}_1 + \hat{\mathbf{r}}_1) = 2mr^2\omega\hat{\mathbf{k}}.
\end{aligned} \quad (0.4.11)$$

The important point about this example is that the two objects are symmetrically distributed with respect to the z-axis (opposite sides of the circular orbit). Therefore the angular momentum about any point S along the z-axis has the same value $\vec{\mathbf{L}}_S = 2mr^2\omega\hat{\mathbf{k}}$, which is constant in magnitude and points in the $+z$-direction for the motion shown in Figure 19.14.

Example 19.6 Angular Momentum of a System of Particles about Different Points

Consider a system of N particles, and two points A and B (Figure 19.6). The angular momentum of the jth particle about the point A is given by

$$\vec{\mathbf{L}}_{A,j} = \vec{\mathbf{r}}_{A,j} \times m_j \vec{\mathbf{v}}_j . \tag{0.4.12}$$

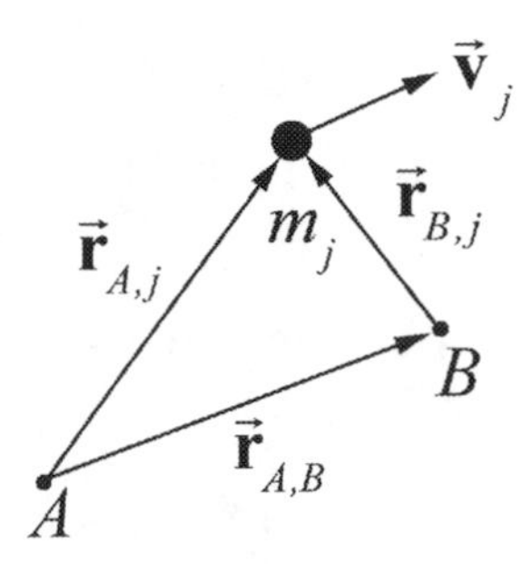

Figure 19.16 Vector triangle relating position of object and points A and B

The angular momentum of the system of particles about the point A is given by the sum

$$\vec{\mathbf{L}}_A = \sum_{j=1}^{N} \vec{\mathbf{L}}_{A,j} = \sum_{j=1}^{N} \vec{\mathbf{r}}_{A,j} \times m_j \vec{\mathbf{v}}_j \tag{0.4.13}$$

The angular momentum about the point B can be calculated in a similar way and is given by

$$\vec{\mathbf{L}}_B = \sum_{j=1}^{N} \vec{\mathbf{L}}_{B,j} = \sum_{j=1}^{N} \vec{\mathbf{r}}_{B,j} \times m_j \vec{\mathbf{v}}_j . \tag{0.4.14}$$

From Figure 19.16, the vectors

$$\vec{\mathbf{r}}_{A,j} = \vec{\mathbf{r}}_{B,j} + \vec{\mathbf{r}}_{A,B} . \tag{0.4.15}$$

We can substitute Eq. (0.4.15) into Eq. (0.4.13) yielding

$$\vec{\mathbf{L}}_A = \sum_{j=1}^{N} (\vec{\mathbf{r}}_{B,j} + \vec{\mathbf{r}}_{A,B}) \times m_j \vec{\mathbf{v}}_j = \sum_{j=1}^{N} \vec{\mathbf{r}}_{B,j} \times m_j \vec{\mathbf{v}}_j + \sum_{j=1}^{N} \vec{\mathbf{r}}_{A,B} \times m_j \vec{\mathbf{v}}_j . \tag{0.4.16}$$

The first term in Eq. (0.4.16) is the angular momentum about the point B. The vector $\vec{\mathbf{r}}_{A,B}$ is a constant and so can be pulled out of the sum in the second term, and Eq. (0.4.16) becomes

$$\vec{\mathbf{L}}_A = \vec{\mathbf{L}}_B + \vec{\mathbf{r}}_{A,B} \times \sum_{j=1}^{N} m_j \vec{\mathbf{v}}_j \tag{0.4.17}$$

The sum in the second term is the momentum of the system

$$\vec{\mathbf{p}}_{\text{sys}} = \sum_{j=1}^{N} m_j \vec{\mathbf{v}}_j \,. \tag{0.4.18}$$

Therefore the angular momentum about the points A and B are related by

$$\vec{\mathbf{L}}_A = \vec{\mathbf{L}}_B + \vec{\mathbf{r}}_{A,B} \times \vec{\mathbf{p}}_{\text{sys}} \tag{0.4.19}$$

Thus if the momentum of the system is zero, the angular momentum is the same about any point.

$$\vec{\mathbf{L}}_A = \vec{\mathbf{L}}_B, \qquad (\vec{\mathbf{p}}_{\text{sys}} = \vec{\mathbf{0}}) \,. \tag{0.4.20}$$

In particular, the momentum of a system of particles is zero by definition in the center of mass reference frame because in that reference frame $\vec{\mathbf{p}}_{\text{sys}} = \vec{\mathbf{0}}$. Hence the angular momentum is the same about any point in the center of mass reference frame.

19.7 Angular Momentum and Torque for Fixed Axis Rotation

We have shown that, for fixed axis rotation, the component of torque that causes the angular velocity to change is the rotational analog of Newton's Second Law,

$$\vec{\boldsymbol{\tau}}_S^{\text{ext}} = I_S \vec{\boldsymbol{\alpha}} \,. \tag{0.4.21}$$

We shall now see that this is a special case of the more general result

$$\vec{\boldsymbol{\tau}}_S^{\text{ext}} = \frac{d}{dt} \vec{\mathbf{L}}_S^{\text{sys}} \,. \tag{0.4.22}$$

Consider a rigid body rotating about a fixed axis passing through the point S and take the fixed axis of rotation to be the z-axis. Recall that all the points in the rigid body rotate about the z-axis with the same angular velocity $\vec{\boldsymbol{\omega}} \equiv (d\theta / dt)\hat{\mathbf{k}} \equiv \omega_z \hat{\mathbf{k}}$. In a similar fashion, all points in the rigid body have the same angular acceleration, $\vec{\boldsymbol{\alpha}} \equiv (d^2\theta / dt^2)\,\hat{\mathbf{k}} \equiv \alpha_z \hat{\mathbf{k}}$. The angular momentum is a vector, and will have a component along the direction of the fixed z-axis. Let the point S lie somewhere along the z-axis.

As before, the body is divided into individual elements. We calculate the contribution of each element to the angular momentum about the point S, and then sum over all the elements. The summation will become an integral for a continuous body.

Each individual element has a mass Δm_i and is moving in a circle of radius $r_{S,\perp,i}$ about the axis of rotation. Let $\vec{\mathbf{r}}_{S,i}$ be the vector from the point S to the element. The velocity of the element, $\vec{\mathbf{v}}_i$, is tangent to this circle (Figure 19.17).

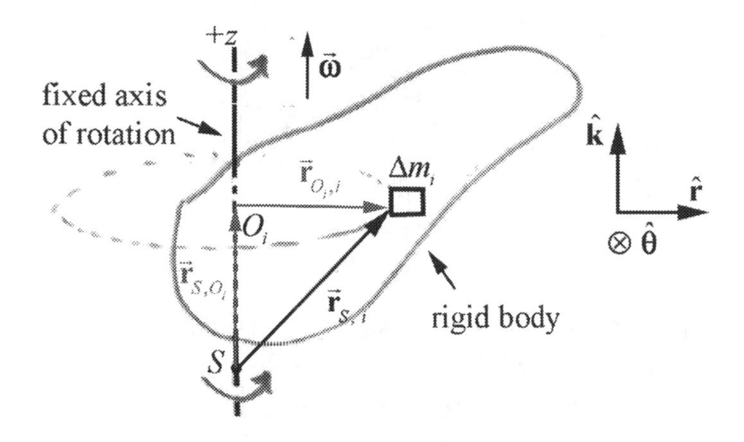

Figure 19.17 Geometry of instantaneous rotation.

The angular momentum of the ith element about the point S is given by

$$\vec{\mathbf{L}}_{S,i} = \vec{\mathbf{r}}_{S,i} \times \vec{\mathbf{p}}_i = \vec{\mathbf{r}}_{S,i} \times \Delta m_i \, \vec{\mathbf{v}}_i . \tag{0.4.23}$$

Let the point O_i denote the center of the circular orbit of the element. Define a vector $\vec{\mathbf{r}}_{O_i,i}$ from the point O_i to the element. Let $\vec{\mathbf{r}}_{S,O_i}$ denote the vector from the point S to the point O_i. Then $\dot{\mathbf{r}}_{S,i} = \dot{\mathbf{r}}_{S,O_i} + \dot{\mathbf{r}}_{O_i,i}$ (Figure 19.17). Because we are interested in the perpendicular and parallel components of $\vec{\mathbf{r}}_{S,i}$ with respect to the axis of rotation, denote the perpendicular component by $\vec{\mathbf{r}}_{O_i,i} \equiv \vec{\mathbf{r}}_{S,\perp,i}$ and the parallel component by $\vec{\mathbf{r}}_{S,O_i} \equiv \vec{\mathbf{r}}_{S,\parallel,i}$. The three vectors are related by

$$\vec{\mathbf{r}}_{S,i} = \vec{\mathbf{r}}_{S,\parallel,i} + \vec{\mathbf{r}}_{S,\perp,i} . \tag{0.4.24}$$

The angular momentum about S is then

$$\begin{aligned} \vec{\mathbf{L}}_{S,i} &= \vec{\mathbf{r}}_{S,i} \times \Delta m_i \, \vec{\mathbf{v}}_i = (\vec{\mathbf{r}}_{S,\parallel,i} + \vec{\mathbf{r}}_{S,\perp,i}) \times \Delta m_i \vec{\mathbf{v}}_i \\ &= (\vec{\mathbf{r}}_{S,\parallel,i} \times \Delta m_i \vec{\mathbf{v}}_i) + (\vec{\mathbf{r}}_{S,\perp,i} \times \Delta m_i \, \vec{\mathbf{v}}_i). \end{aligned} \tag{0.4.25}$$

In the last expression in Equation (0.4.25), the first term has a direction that is perpendicular to the z-axis; the direction of a vector product of two vectors is always perpendicular to the direction of either vector. Because $\vec{\mathbf{r}}_{S,\parallel,i}$ is in the z-direction, the first term in the last expression in Equation (0.4.25) has no component along the z-axis.

Therefore the z-component of the angular momentum about the point S, $(L_{S,i})_z$, arises entirely from the second term, $\vec{\mathbf{r}}_{S,\perp,i} \times \Delta m_i \vec{\mathbf{v}}_i$. The vectors $\vec{\mathbf{r}}_{S,\perp,i}$ and $\vec{\mathbf{v}}_i$ are perpendicular, as shown in Figure 19.18.

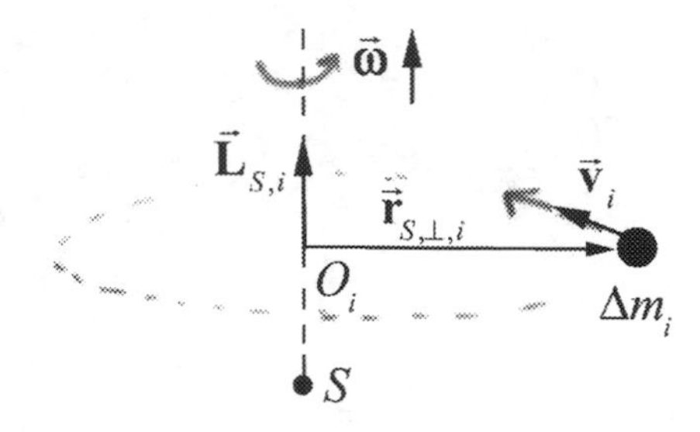

Figure 19.18 The z-component of angular momentum.

Therefore the z-component of the angular momentum about S is just the product of the radius of the circle, $r_{S,\perp,i}$, and the tangential component $\Delta m_i v_i$ of the momentum,

$$(L_{S,i})_z = r_{S,\perp,i}\,\Delta m_i v_i \,. \tag{0.4.26}$$

For a rigid body, all elements have the same z-component of the angular velocity, $\omega_z = d\theta / dt$, and the tangential velocity is

$$v_i = r_{S,\perp,i}\,\omega_z \,. \tag{0.4.27}$$

The expression in Equation (0.4.26) for the z-component of the momentum about S is then

$$(L_{S,i})_z = r_{S,\perp,i}\,\Delta m_i v_i = \Delta m_i (r_{S,\perp,i})^2 \omega_z \,. \tag{0.4.28}$$

The z-component of the angular momentum of the system about S is the summation over all the elements,

$$L_{S,z}^{\text{sys}} = \sum_i (L_{S,i})_z = \sum_i \Delta m_i (r_{S,\perp,i})^2 \omega_z \,. \tag{0.4.29}$$

For a continuous mass distribution the summation becomes an integral over the body,

$$L_{S,z}^{\text{sys}} = \int_{\text{body}} dm\,(r_{dm})^2 \omega_z \,, \tag{0.4.30}$$

where r_{dm} is the distance form the fixed z-axis to the infinitesimal element of mass dm. The moment of inertia of a rigid body about a fixed z-axis passing through a point S is given by an integral over the body

$$I_S = \int_{\text{body}} dm\,(r_{dm})^2 . \qquad (0.4.31)$$

Thus the z-component of the angular momentum about S for a fixed axis that passes through S in the z-direction is proportional to the z-component of the angular velocity, ω_z,

$$L^{\text{sys}}_{S,z} = I_S\,\omega_z . \qquad (0.4.32)$$

For fixed axis rotation, our result that torque about a point is equal to the time derivative of the angular momentum about that point,

$$\vec{\boldsymbol{\tau}}^{\text{ext}}_S = \frac{d}{dt}\vec{\mathbf{L}}^{\text{sys}}_S , \qquad (0.4.33)$$

can now be resolved in the z-direction,

$$\tau^{\text{ext}}_{S,z} = \frac{dL^{\text{sys}}_{S,z}}{dt} = \frac{d}{dt}(I_S\,\omega_z) = I_S\frac{d\omega_z}{dt} = I_S\frac{d^2\theta}{dt^2} = I_S\,\alpha_z , \qquad (0.4.34)$$

in agreement with our earlier result that the z-component of torque about the point S is equal to the product of moment of inertia about I_S, and the z-component of the angular acceleration, α_z.

Example 19.6 Circular Ring

A circular ring of radius r, and mass m is rotating about the z-axis in a plane parallel to but a distance h above the x-y plane. The z-component of the angular velocity is ω_z (Figure 19.19). Find the magnitude and the direction of the angular momentum $\vec{\mathbf{L}}_S$ along at any point S on the central z-axis.

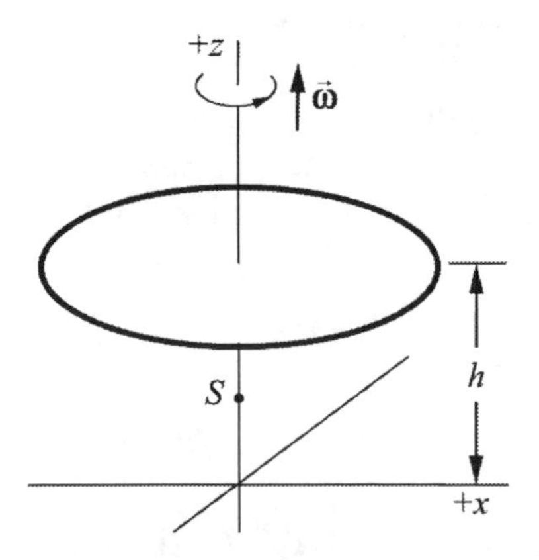

Figure 19.19 Example 19.6

Solution: Use the same symmetry argument as we did in Example 19.5. The ring can be thought of as made up of pairs of point like objects on opposite sides of the ring each of mass Δm. Each pair has a non-zero z-component of the angular momentum taken about any point S along the z-axis, $\vec{\mathbf{L}}_{S,pair} = \vec{\mathbf{L}}_{S,1} + \vec{\mathbf{L}}_{S,2} = 2\Delta m r^2 \omega_z \hat{\mathbf{k}}$. So summing up over all the pairs gives

$$\vec{\mathbf{L}}_S = m r^2 \omega_z \hat{\mathbf{k}}. \tag{0.4.35}$$

Recall that the moment of inertia of a ring is given by

$$I_S = \int_{\text{body}} dm\,(r_{dm})^2 = m r^2. \tag{0.4.36}$$

For the symmetric ring, the angular momentum about S points in the direction of the angular velocity and is equal to

$$\vec{\mathbf{L}}_S = I_S \omega_z \hat{\mathbf{k}} \tag{0.4.37}$$

19.8 Principle of Conservation of Angular Momentum

Consider a system of particles. We begin with the result that we derived in Section 19.7 that the torque about a point S is equal to the time derivative of the angular momentum about that point S,

$$\vec{\boldsymbol{\tau}}_S^{\text{ext}} = \frac{d\vec{\mathbf{L}}_S^{\text{sys}}}{dt}. \tag{0.4.38}$$

With this assumption, the torque due to the external forces is equal to the rate of change of the angular momentum

$$\vec{\boldsymbol{\tau}}_S^{\text{ext}} = \frac{d\vec{\mathbf{L}}_S^{\text{sys}}}{dt}. \tag{0.4.39}$$

Principle of Conservation of Angular Momentum

If the external torque acting on a system is zero, then the angular momentum of the system is constant. So for any change of state of the system the change in angular momentum is zero

$$\Delta\vec{\mathbf{L}}_S^{\text{sys}} \equiv (\vec{\mathbf{L}}_S^{\text{sys}})_f - (\vec{\mathbf{L}}_S^{\text{sys}})_i = \vec{\mathbf{0}}. \tag{0.4.40}$$

Equivalently the angular momentum is constant

$$(\vec{\mathbf{L}}_S^{\text{sys}})_f = (\vec{\mathbf{L}}_S^{\text{sys}})_i. \tag{0.4.41}$$

So far no isolated system has been encountered such that the angular momentum is not constant so our assumption that internal torques cancel is pairs can be taken as an experimental observation.

Example 19.7 Collision Between Pivoted Rod and Object

An object of mass m and speed v_0 strikes a rigid uniform rod of length l and mass m_r that is hanging by a frictionless pivot from the ceiling. Immediately after striking the rod, the object continues forward but its speed decreases to $v_0/2$ (Figure 19.20). The moment of inertia of the rod about its center of mass is $I_{cm} = (1/12)m_r l^2$. Gravity acts with acceleration g downward. (a) For what value of v_0 will the rod just touch the ceiling on its first swing? You may express your answer in terms of g, m_r, m, and l. (b) For what ratio m_r / mO will the collision be elastic?

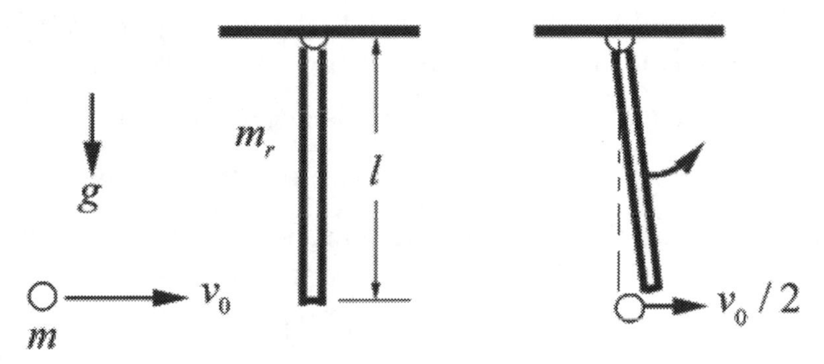

Figure 19.20 Example 19.7

Solution: We begin by identifying our system, which consists of the object and the uniform rod. We identify three states; state 1: immediately before the collision, state 2: immediately after the collision, and state 3: the instant the rod touches the ceiling when the final angular speed is zero. We would like to know if any of our fundamental quantities: momentum, energy, and angular momentum, are constant during these state changes, state 1 $\rightarrow$ state 2, state 2 $\rightarrow$ state 3.

We start with state 1 $\rightarrow$ state 2. The pivot force holding the rod to the ceiling is an external force acting at the pivot point S. There is also the gravitational force acting on at the center of mass of the rod and on the object. There are also internal forces due to the collision of the rod and the object at point A (Figure 19.21).

The external force means that momentum is not constant. The point of action of the external pivot force is fixed and so does no work. However, we do not know whether or not the collision is elastic and so we cannot assume that mechanical energy is constant. If we choose the pivot point S as the point in which to calculate torque, then the torque about the pivot is

$$\vec{\boldsymbol{\tau}}_S^{\text{sys}} = \vec{\mathbf{r}}_{S,S} \times \vec{\mathbf{F}}_{pivot} + \vec{\mathbf{r}}_{S,A} \times \vec{\mathbf{F}}_{r,o} + \vec{\mathbf{r}}_{S,A} \times \vec{\mathbf{F}}_{o,r} + \vec{\mathbf{r}}_{S,cm} \times \vec{\mathbf{F}}_{g,r} . \qquad (0.4.42)$$

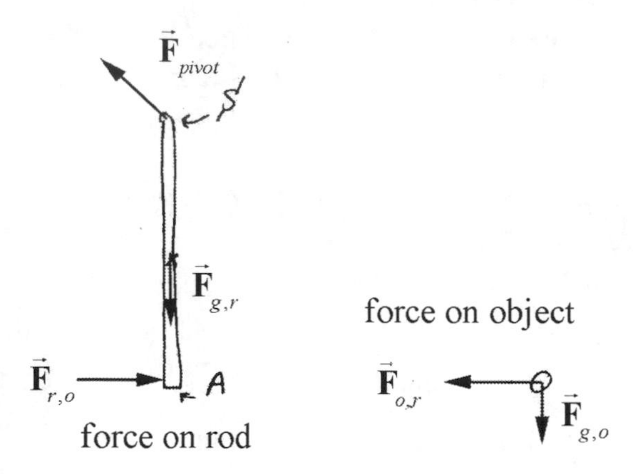

Figure 19.21 Free-body force diagrams on elements of system

The external pivot force does not contribute any torque because $\vec{\mathbf{r}}_{S,S} = \vec{\mathbf{0}}$. The internal forces between the rod and the object are equal in magnitude and opposite in direction, $\vec{\mathbf{F}}_{r,o} = -\vec{\mathbf{F}}_{o,r}$ (Newton's Third Law), and so their contributions to the torque add to zero, $\vec{\mathbf{r}}_{S,A} \times \vec{\mathbf{F}}_{o,r} + \vec{\mathbf{r}}_{S,A} \times \vec{\mathbf{F}}_{r,o} = \vec{\mathbf{0}}$. If the collision is instantaneous then the gravitational force is parallel to $\vec{\mathbf{r}}_{S,cm}$ and so $\vec{\mathbf{r}}_{S,cm} \times \vec{\mathbf{F}}_{r,g} = \vec{\mathbf{0}}$. Therefore the torque on the system about the pivot point is zero, $\vec{\boldsymbol{\tau}}_S^{\text{sys}} = \vec{\mathbf{0}}$. Thus the angular momentum about the pivot point is constant,

$$(\vec{\mathbf{L}}_S^{\text{sys}})_1 = (\vec{\mathbf{L}}_S^{\text{sys}})_2 \,. \tag{0.4.43}$$

In order to calculate the angular momentum we draw a diagram showing the momentum of the object and the angular speed of the rod in (Figure 19.22).

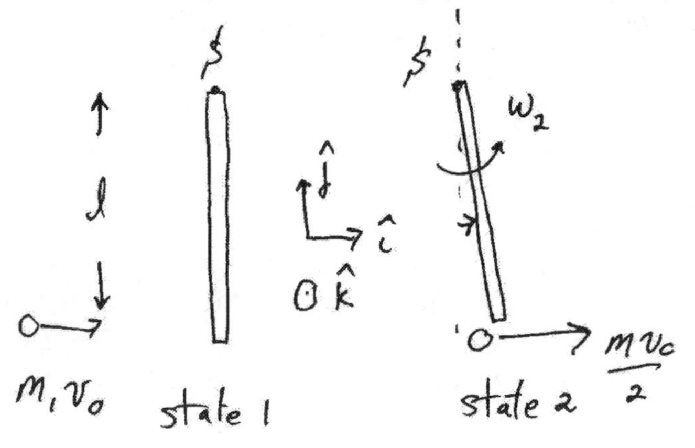

Figure 19.22 Angular momentum diagram

The angular momentum about S immediately before the collision is

$$(\vec{\mathbf{L}}_S^{\text{sys}})_1 = \vec{\mathbf{r}}_{S,0} \times m_1 \vec{\mathbf{v}}_0 = l(-\hat{\mathbf{j}}) \times m_1 v_0 \hat{\mathbf{i}} = l m_1 v_0 \hat{\mathbf{k}} \,.$$

The angular momentum about S immediately after the collision is

$$(\vec{\mathbf{L}}_S^{\text{sys}})_2 = \vec{\mathbf{r}}_{S,0} \times m_1 \vec{\mathbf{v}}_0 / 2 + I_s \vec{\boldsymbol{\omega}}_2 = l(-\hat{\mathbf{j}}) \times m_1 (v_0 / 2)\hat{\mathbf{i}} = \frac{l m_1 v_0}{2}\hat{\mathbf{k}} + I_s \omega_2 \hat{\mathbf{k}} .$$

Therefore the condition that the angular momentum about S is constant during the collision becomes

$$l m_1 v_0 \hat{\mathbf{k}} = \frac{l m_1 v_0}{2}\hat{\mathbf{k}} + I_s \omega_2 \hat{\mathbf{k}} .$$

We can solve for the angular speed immediately after the collision

$$\omega_2 = \frac{l m_1 v_0}{2 I_s} .$$

By the parallel axis theorem the moment of inertial of a uniform rod about the pivot point is

$$I_S = m(l/2)^2 + I_{cm} = (1/4) m_r l^2 + (1/12) m_r l^2 = (1/3) m_r l^2 . \qquad (0.4.44)$$

Therefore the angular speed immediately after the collision is

$$\omega_2 = \frac{3 m_1 v_0}{2 m_r l} . \qquad (0.4.45)$$

For the transition state 2 $\rightarrow$ state 3, we know that the gravitational force is conservative and the pivot force does no work so mechanical energy is constant.

$$E_{m,2} = E_{m,3}$$

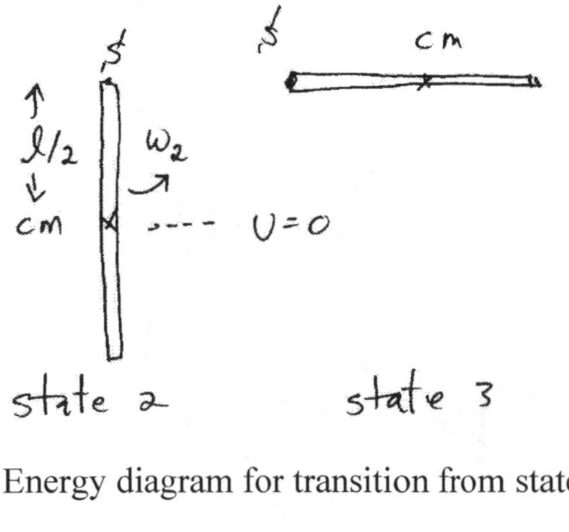

Figure 19.23 Energy diagram for transition from state 2 to state 3.

We draw an energy diagram in Figure 19.23, with a choice of zero for the potential energy at the center of mass. We only show the rod because the object undergoes no energy transformation during the transition state 2 $\to$ state 3. The mechanical energy immediately after the collision is

$$E_{m,2} = \frac{1}{2} I_S \omega_2^2 + \frac{1}{2} m_1 (v_0 / 2)^2 .$$

Using our results for the moment of inertia I_S (Eq. (0.4.44)) and ω_2 (Eq. (0.4.45)), we have that

$$E_{m,2} = \frac{1}{2}(1/3) m_r l^2 \left(\frac{3 m_1 v_0}{2 m_r l} \right)^2 + \frac{1}{2} m_1 (v_0 / 2)^2 = \frac{3 m_1^{\,2} v_0^{\,2}}{8 m_r} + \frac{1}{2} m_1 (v_0 / 2)^2 . \quad (0.4.46)$$

The mechanical energy when the rod just reaches the ceiling when the final angular speed is zero is then

$$E_{m,3} = m_r g (l/2) + \frac{1}{2} m_1 (v_0 / 2)^2 .$$

Then the condition that the mechanical energy is constant becomes

$$\frac{3 m_1^{\,2} v_0^{\,2}}{8 m_r} + \frac{1}{2} m_1 (v_0 / 2)^2 = m_r g (l/2) + \frac{1}{2} m_1 (v_0 / 2)^2 . \quad (0.4.47)$$

We can now solve Eq. (0.4.47) for the initial speed of the object

$$v_0 = \frac{m_r}{m_1} \sqrt{\frac{4 g l}{3}} . \quad (0.4.48)$$

We now return to the transition state 1 $\to$ state 2 and determine the constraint on the mass ratio in order for the collision to be elastic. The mechanical energy before the collision is

$$E_{m,1} = \frac{1}{2} m_1 v_0^{\,2} . \quad (0.4.49)$$

If we impose the condition that the collision is elastic then

$$E_{m,1} = E_{m,2} . \quad (0.4.50)$$

Substituting Eqs. (0.4.46) and (0.4.49) into Eq. (0.4.50) yields

$$\frac{1}{2}m_1 v_0^2 = \frac{3m_1^2 v_0^2}{8m_r} + \frac{1}{2}m_1 (v_0/2)^2 .$$

This simplifies to

$$\frac{3}{8}m_1 v_0^2 = \frac{3m_1^2 v_0^2}{8m_r}$$

Hence we can solve for the mass ratio necessary to insure that the collision is elastic if the final speed of the object is half it's initial speed

$$\frac{m_r}{m_1} = 1 . \qquad (0.4.51)$$

Notice that the mass ratio is independent of the initial speed of the object.

19.9 External Impulse and Change in Angular Momentum

Define the ***external impulse*** *about a point* S *applied as the integral of the external torque about* S

$$\vec{\mathbf{J}}_S^{\text{ext}} \equiv \int_{t_i}^{t_f} \vec{\boldsymbol{\tau}}_S^{\text{ext}} \, dt . \qquad (0.4.52)$$

Then the external impulse about S is equal to the change in angular momentum

$$\vec{\mathbf{J}}_S^{\text{ext}} \equiv \int_{t_i}^{t_f} \vec{\boldsymbol{\tau}}_S^{\text{ext}} \, dt = \int_{t_i}^{t_f} \frac{d\vec{\mathbf{L}}_S^{\text{sys}}}{dt} dt = \vec{\mathbf{L}}_{S,f}^{\text{sys}} - \vec{\mathbf{L}}_{S,i}^{\text{sys}} . \qquad (0.4.53)$$

Notice that this is the rotational analog to our statement about impulse and momentum,

$$\vec{\mathbf{I}}_S^{\text{ext}} \equiv \int_{t_i}^{t_f} \vec{\mathbf{F}}^{\text{ext}} \, dt = \int_{t_i}^{t_f} \frac{d\vec{\mathbf{p}}_{\text{sys}}}{dt} dt = \vec{\mathbf{p}}_{\text{sys},f} - \vec{\mathbf{p}}_{\text{sys},i} . \qquad (0.4.54)$$

Example 19.8 Angular Impulse on Steel Washer

A steel washer is mounted on the shaft of a small motor. The moment of inertia of the motor and washer is I_0. The washer is set into motion. When it reaches an initial angular speed ω_0, at $t = 0$, the power to the motor is shut off, and the washer slows down until it reaches an angular speed of ω_a at time t_a. At that instant, a second steel washer with a moment of inertia I_w is dropped on top of the first washer. Assume that the second washer is only in contact with the first washer. The collision takes place over a time

$\Delta t_{\text{int}} = t_b - t_a$. Assume the frictional torque on the axle is independent of speed, and remains the same when the second washer is dropped. The two washers continue to slow down during the time interval $\Delta t_2 = t_f - t_b$ until they stop at time $t = t_f$. (a) What is the angular acceleration while the washer and motor are slowing down during the interval $\Delta t_1 = t_a$? (b) Suppose the collision is nearly instantaneous, $\Delta t_{\text{int}} = (t_b - t_a) \simeq 0$. What is the angular speed ω_b of the two washers immediately after the collision is finished (when the washers rotate together)?

Now suppose the collision is not instantaneous but that the frictional torque is independent of the speed of the rotor. (c) What is the angular impulse during the collision? (d) What is the angular velocity ω_b of the two washers immediately after the collision is finished (when the washers rotate together)? (e) What is the angular deceleration α_2 after the collision?

Solution: a) The angular acceleration of the motor and washer from the instant when the power is shut off until the second washer was dropped is given by

$$\alpha_1 = \frac{\omega_a - \omega_0}{\Delta t_1} < 0. \tag{0.4.55}$$

(b) If the collision is nearly instantaneous, then there is no angular impulse and therefore the z-component of the angular momentum about the rotation axis of the motor remains constant

$$0 = \Delta L_z = L_{f,z} - L_{0,z} = (I_0 + I_{\text{w}})\omega_b - I_0\omega_a. \tag{0.4.56}$$

We can solve Eq. (0.4.56) for the angular speed ω_b of the two washers immediately after the collision is finished

$$\omega_b = \frac{I_0}{I_0 + I_{\text{w}}}\omega_a. \tag{0.4.57}$$

(c) The angular acceleration found in part a) is due to the frictional torque in the motor.

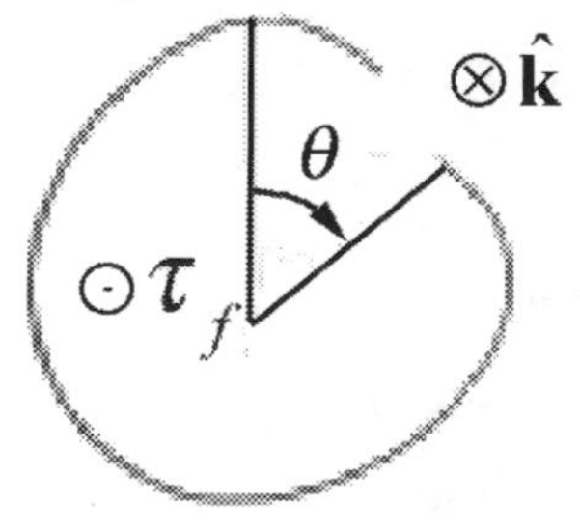

Figure 19.24 Frictional torque in the motor

Let $\vec{\boldsymbol{\tau}}_f = -\tau_f \hat{\mathbf{k}}$ where τ_f is the magnitude of the frictional torque (Figure 19.24) then

$$-\tau_f = I_0 \alpha_1 = \frac{I_0(\omega_a - \omega_0)}{\Delta t_1}. \tag{0.4.58}$$

During the collision with the second washer, the frictional torque exerts an angular impulse (pointing along the z-axis in the figure),

$$J_z = -\int_{t_a}^{t_b} \tau_f \, dt = -\tau_f \Delta t_{\text{int}} = I_0(\omega_a - \omega_0)\frac{\Delta t_{\text{int}}}{\Delta t_1}. \tag{0.4.59}$$

(d) The z-component of the angular momentum about the rotation axis of the motor changes during the collision,

$$\Delta L_z = L_{f,z} - L_{0,z} = (I_0 + I_{\text{w}})\omega_b - I_0 \omega_a. \tag{0.4.60}$$

The change in the z-component of the angular momentum is equal to the z-component of the angular impulse

$$J_z = \Delta L_z. \tag{0.4.61}$$

Thus, equating the expressions in Equations (0.4.59) and (0.4.60), yields

$$I_0(\omega_a - \omega_0)\left(\frac{\Delta t_{\text{int}}}{\Delta t_1}\right) = (I_0 + I_{\text{w}})\omega_b - (I_0)\omega_a. \tag{0.4.62}$$

Solve Equation (0.4.62) for the angular velocity immediately after the collision,

$$\omega_b = \frac{I_0}{(I_0 + I_{\text{w}})}\left((\omega_a - \omega_0)\left(\frac{\Delta t_{\text{int}}}{\Delta t_1}\right) + \omega_a\right). \tag{0.4.63}$$

If there were no frictional torque, then the first term in the brackets would vanish, and the second term of Eq. (0.4.63) would be the only contribution to the final angular speed.

(e) The final angular acceleration α_2 is given by

$$\alpha_2 = \frac{0 - \omega_b}{\Delta t_2} = -\frac{I_0}{(I_0 + I_{\text{w}})\Delta t_2}\left((\omega_a - \omega_0)\left(\frac{\Delta t_{\text{int}}}{\Delta t_1}\right) + \omega_a\right). \tag{0.4.64}$$

Chapter 20 Rigid Body: Translation and Rotational Motion Kinematics for Fixed Axis Rotation

Chapter 20 Rigid Body: Translation and Rotational Motion Kinematics for Fixed Axis Rotation

Hence I feel no shame in asserting that this whole region engirdled by the moon, and the center of the earth, traverse this grand circle amid the rest of the planets in an annual revolution around the sun. Near the sun is the center of the universe. Moreover, since the sun remains stationary, whatever appears as a motion of the sun is really due rather to the motion of the earth.[1]

Copernicus

20.1 Introduction

The general motion of a rigid body of mass m consists of a translation of the center of mass with velocity $\vec{\mathbf{V}}_{cm}$ and a rotation about the center of mass with all elements of the rigid body rotating with the same angular velocity $\vec{\boldsymbol{\omega}}_{\text{cm}}$. We prove this result in Appendix A. Figure 20.1 shows the center of mass of a thrown rigid rod follows a parabolic trajectory while the rod rotates about the center of mass.

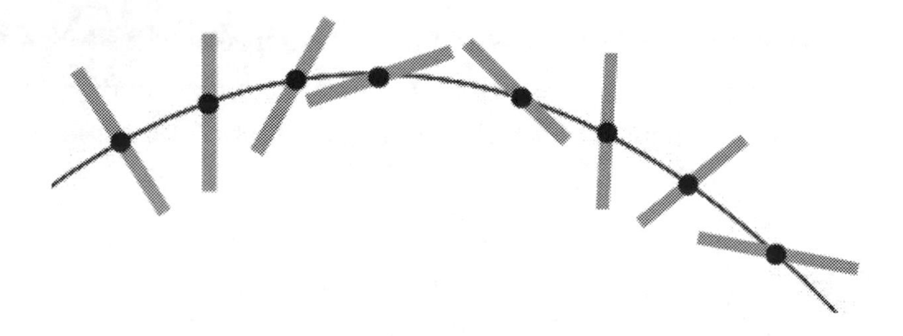

Figure 20.1 The center of mass of a thrown rigid rod follows a parabolic trajectory while the rod rotates about the center of mass.

20.2 Constrained Motion: Translation and Rotation

We shall encounter many examples of a rolling object whose motion is constrained. For example we will study the motion of an object rolling along a level or inclined surface and the motion of a yo-yo unwinding and winding along a string. We will examine the constraint conditions between the translational quantities that describe the motion of the center of mass, displacement, velocity and acceleration, and the rotational quantities that describe the motion about the center of mass, angular displacement, angular velocity and angular acceleration. We begin with a discussion about the rotation and translation of a rolling wheel.

---

[1]Nicolaus Copernicus, *De revolutionibus orbium coelestium* (*On the Revolutions of the Celestial Spheres*), Book 1 Chapter 10.

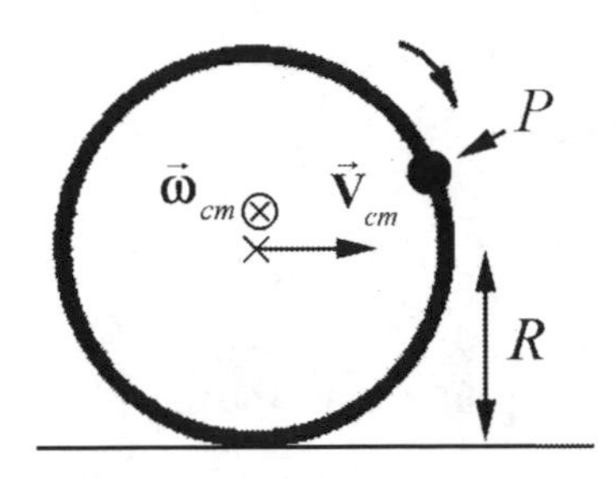

Figure 20.2 Rolling Wheel

Consider a wheel of radius R is rolling in a straight line (Figure 20.2). The center of mass of the wheel is moving in a straight line at a constant velocity $\vec{\mathbf{V}}_{cm}$. Let's analyze the motion of a point P on the rim of the wheel.

Let $\vec{\mathbf{v}}_P$ denote the velocity of a point P on the rim of the wheel with respect to reference frame O at rest with respect to the ground (Figure 20.3a). Let $\vec{\mathbf{v}}'_P$ denote the velocity of the point P on the rim with respect to the center of mass reference frame O_{cm} moving with velocity $\vec{\mathbf{V}}_{cm}$ with respect to at O (Figure 20.3b). (You should review the definition of the center of mass reference frame in Chapter 15.2.1.) We can use the law of addition of velocities (Eq.15.2.4) to relate these three velocities,

$$\vec{\mathbf{v}}_P = \vec{\mathbf{v}}'_P + \vec{\mathbf{V}}_{cm} . \tag{20.2.1}$$

Let's choose Cartesian coordinates for the translation motion and polar coordinates for the motion about the center of mass as shown in Figure 20.3.

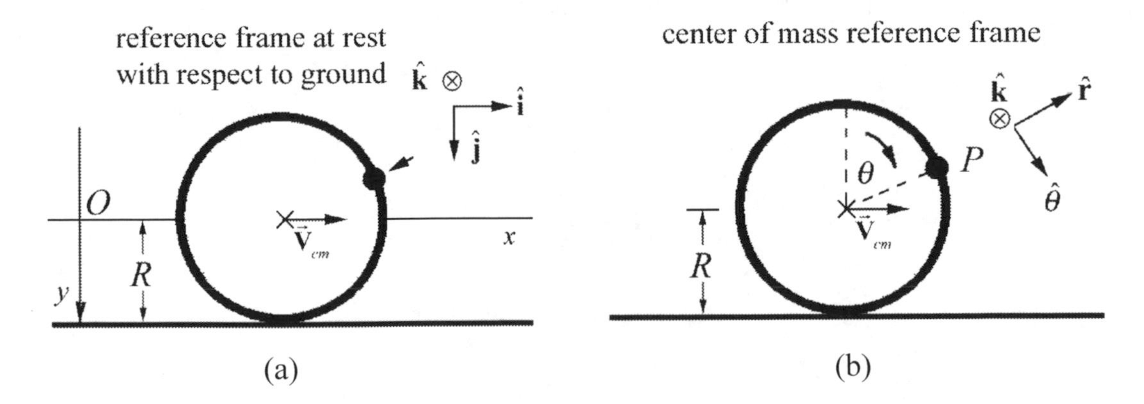

Figure 20.3 (a) reference frame fixed to ground, (b) center of mass reference frame

The center of mass velocity in the reference frame fixed to the ground is given by

$$\vec{\mathbf{V}}_{cm} = V_{cm}\,\hat{\mathbf{i}} . \tag{20.2.2}$$

where V_{cm} is the speed of the center of mass. The position of the center of mass in the reference frame fixed to the ground is given by

$$\vec{\mathbf{R}}_{\text{cm}}(t) = (X_{\text{cm},0} + V_{\text{cm}}t)\hat{\mathbf{i}} \ , \tag{20.2.3}$$

where $X_{\text{cm},0}$ is the initial x-component of the center of mass at $t = 0$. The angular velocity of the wheel in the center of mass reference frame is given by

$$\vec{\boldsymbol{\omega}}_{\text{cm}} = \omega_{\text{cm}}\hat{\mathbf{k}} . \tag{20.2.4}$$

where ω_{cm} is the angular speed. The point P on the rim is undergoing uniform circular motion with the velocity in the center of mass reference frame given by

$$\vec{\mathbf{v}}'_P = R\omega_{\text{cm}}\hat{\boldsymbol{\theta}} . \tag{20.2.5}$$

If we want to use the law of addition of velocities then we should express $\vec{\mathbf{v}}'_P = R\omega_{\text{cm}}\hat{\boldsymbol{\theta}}$ in Cartesian coordinates. Assume that at $t = 0$, $\theta(t = 0) = 0$ i.e. the point P is at the top of the wheel at $t = 0$. Then the unit vectors in polar coordinates satisfy (Figure 20.4)

$$\begin{aligned} \hat{\mathbf{r}} &= \sin\theta\hat{\mathbf{i}} - \cos\theta\hat{\mathbf{j}} \\ \hat{\boldsymbol{\theta}} &= \cos\theta\hat{\mathbf{i}} + \sin\theta\hat{\mathbf{j}} \end{aligned} . \tag{20.2.6}$$

Therefore the velocity of the point P on the rim in the center of mass reference frame is given by

$$\vec{\mathbf{v}}'_P = R\omega_{\text{cm}}\hat{\boldsymbol{\theta}} = R\omega_{\text{cm}}(\cos\theta\hat{\mathbf{i}} - \sin\theta\hat{\mathbf{j}}) . \tag{20.2.7}$$

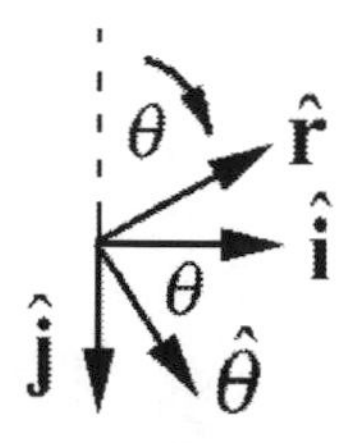

Figure 20.4 Unit vectors

Now substitute Eqs. (20.2.2) and (20.2.7) into Eq. (20.2.1) for the velocity of a point P on the rim in the reference frame fixed to the ground

$$\begin{aligned} \vec{\mathbf{v}}_P &= R\omega_{\text{cm}}(\cos\theta\hat{\mathbf{i}} + \sin\theta\hat{\mathbf{j}}) + V_{\text{cm}}\,\hat{\mathbf{i}} \\ &= (V_{\text{cm}} + R\omega_{\text{cm}}\cos\theta)\hat{\mathbf{i}} + R\omega_{\text{cm}}\sin\theta\hat{\mathbf{j}} \end{aligned} . \tag{20.2.8}$$

The point P is in contact with the ground when $\theta = \pi$. At that instant the velocity of a point P on the rim in the reference frame fixed to the ground is

$$\vec{\mathbf{v}}_P(\theta = \pi) = (V_{\text{cm}} - R\omega_{\text{cm}})\hat{\mathbf{i}}. \qquad (20.2.9)$$

What velocity does the observer at rest on the ground measure for the point on the rim when that point is in contact with the ground? In order to understand the relationship between V_{cm} and ω_{cm}, we consider the displacement of the center of mass for a small time interval Δt (Figure 20.5).

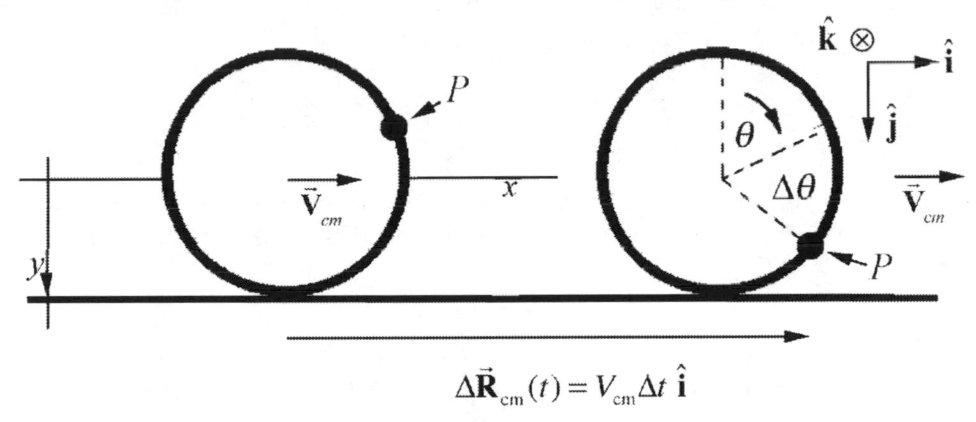

Figure 20.5 Displacement of center of mass in ground reference frame.

From Eq. (20.2.3) the x-component of the displacement of the center of mass is

$$\Delta X_{\text{cm}} = V_{\text{cm}} \Delta t . \qquad (20.2.10)$$

The point P on the rim in the center of mass reference frame is undergoing circular motion (Figure 20.6).

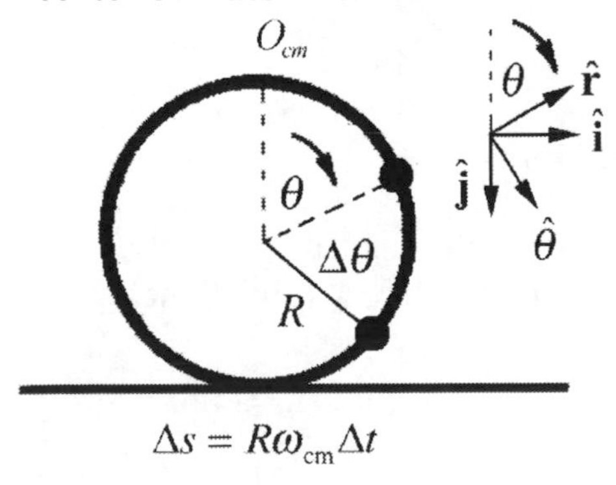

Figure 20.6: Small displacement of point on rim in center of mass reference frame.

In the center of mass reference frame, the magnitude of the tangential displacement is given by the arc length subtended by the angular displacement $\Delta\theta = \omega_{cm}\Delta t$,

$$\Delta s = R\Delta\theta = R\omega_{cm}\Delta t \,. \qquad (20.2.11)$$

Case 1: if the x-component of the displacement of the center of mass is equal to the arc length subtended by $\Delta\theta$, then the wheel is rolling without slipping or skidding, ***rolling without slipping*** for short, along the surface with

$$\Delta X_{cm} = \Delta s \,. \qquad (20.2.12)$$

Substitute Eq. (20.2.10) and Eq. (20.2.11) into Eq. (20.2.12) and divide through by Δt. Then the rolling without slipping condition becomes

$$V_{cm} = R\omega_{cm}, \qquad \text{(rolling without slipping)}. \qquad (20.2.13)$$

Case 2: if the x-component of the displacement of the center of mass is greater than the arc length subtended by $\Delta\theta$, then the wheel is ***skidding*** along the surface with

$$\Delta X_{cm} > \Delta s \,. \qquad (20.2.14)$$

Substitute Eqs. (20.2.10) and (20.2.11) into Eq. (20.2.14) and divide through by Δt, then

$$V_{cm} > R\omega_{cm}, \qquad \text{(skidding)}. \qquad (20.2.15)$$

Case 3: if the x-component of the displacement of the center of mass is less than the arc length subtended by $\Delta\theta$, then the wheel is slipping along the surface with

$$\Delta X_{cm} < \Delta s \,. \qquad (20.2.16)$$

Arguing as above the slipping condition becomes

$$V_{cm} < R\omega_{cm}, \qquad \text{(slipping)}. \qquad (20.2.17)$$

20.2.1 Rolling without slipping

When a wheel is rolling without slipping, the velocity of a point P on the rim is zero when it is in contact with the ground. In Eq. (20.2.9) set $\theta = \pi$,

$$\vec{\mathbf{v}}_P(\theta = \pi) = (V_{cm} - R\omega_{cm})\hat{\mathbf{i}} = (R\omega_{cm} - R\omega_{cm})\hat{\mathbf{i}} = \vec{\mathbf{0}} \,. \qquad (20.2.18)$$

This makes sense because the velocity of the point P on the rim in the center of mass reference frame when it is in contact with the ground points in the opposite direction as the translational motion of the center of mass of the wheel. The two velocities have the same magnitude so the vector sum is zero. The observer at rest on the ground sees the contact point on the rim at rest relative to the ground.

Thus any frictional force acting between the tire and the ground on the wheel is static friction because the two surfaces are instantaneously at rest with respect to each other. Recall that the direction of the static frictional force depends on the other forces acting on the wheel.

Example 20.1 Bicycle Wheel Rolling Without Slipping

Consider a bicycle wheel of radius R that is rolling in a straight line without slipping. The velocity of the center of mass in a reference frame fixed to the ground is given by velocity $\vec{\mathbf{V}}_{\text{cm}}$. A bead is fixed to a spoke a distance b from the center of the wheel (Figure 20.7). (a) Find the position, velocity, and acceleration of the bead as a function of time in the center of mass reference frame. (b) Find the position, velocity, and acceleration of the bead as a function of time as seen in a reference frame fixed to the ground.

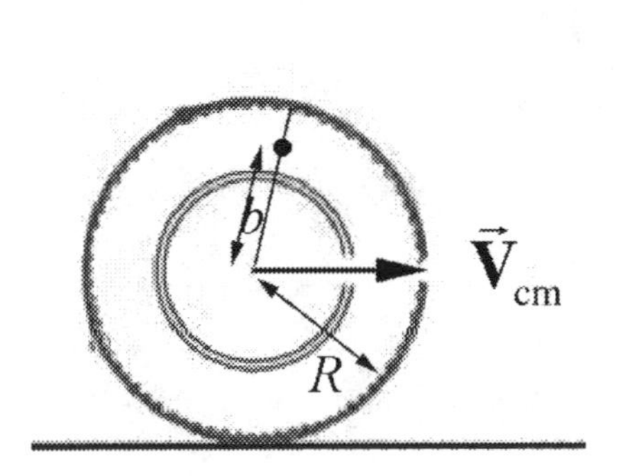

Figure 20.7 Example 20.1

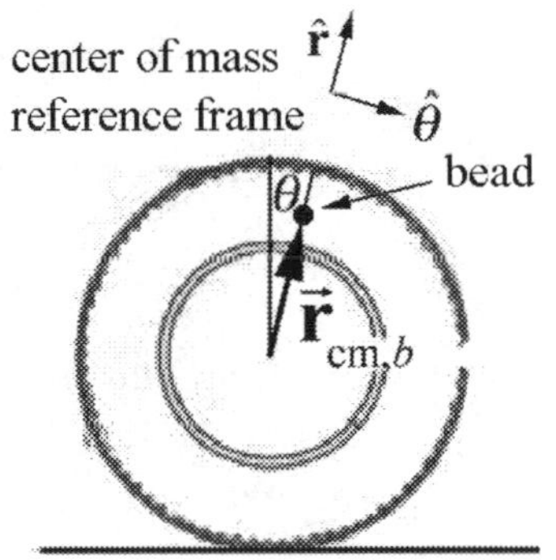

Figure 20.8 Coordinate system for bead in center of mass reference frame

Solution: a) Choose the center of mass reference frame with an origin at the center of the wheel, and moving with the wheel. Choose polar coordinates (Figure 20.8). The z-component of the angular velocity $\omega_{\text{cm}} = d\theta / dt > 0$. Then the bead is moving uniformly in a circle of radius $r = b$ with the position, velocity, and acceleration given by

$$\vec{\mathbf{r}}'_b = b\,\hat{\mathbf{r}}, \quad \vec{\mathbf{v}}'_b = b\omega_{\text{cm}}\,\hat{\theta}, \quad \vec{\mathbf{a}}'_b = -b\omega_{\text{cm}}^2\,\hat{\mathbf{r}}\ . \qquad (20.2.19)$$

Because the wheel is rolling without slipping, the velocity of a point on the rim of the wheel has speed $v'_P = R\omega_{\text{cm}}$. This is equal to the speed of the center of mass of the wheel V_{cm}, thus

$$V_{cm} = R\omega_{\text{cm}}\ . \qquad (20.2.20)$$

Note that at $t = 0$, the angle $\theta = \theta_0 = 0$. So the angle grows in time as

$$\theta(t) = \omega_{\text{cm}} t = (V_{\text{cm}} / R)t \,. \qquad (20.2.21)$$

The velocity and acceleration of the bead with respect to the center of the wheel are then

$$\vec{\mathbf{v}}'_b = \frac{bV_{cm}}{R}\hat{\theta}, \quad \vec{\mathbf{a}}'_b = -\frac{bV_{\text{cm}}^2}{R^2}\hat{\mathbf{r}} \,. \qquad (20.2.22)$$

b) Define a second reference frame fixed to the ground with choice of origin, Cartesian coordinates and unit vectors as shown in Figure 20.9.

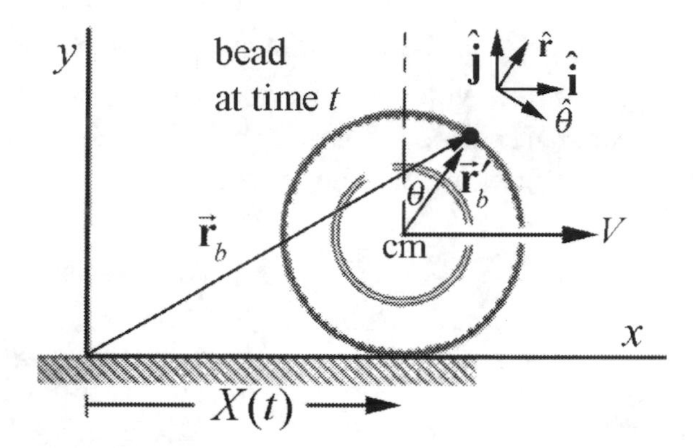

Figure 20.9 Coordinates of bead in reference frame fixed to ground

Then the position vector of the center of mass in the reference frame fixed to the ground is given by

$$\vec{\mathbf{R}}_{cm}(t) = X_{cm}\,\hat{\mathbf{i}} + R\,\hat{\mathbf{j}} = V_{cm} t\,\hat{\mathbf{i}} + R\,\hat{\mathbf{j}} \,. \qquad (20.2.23)$$

The relative velocity of the two frames is the derivative

$$\vec{\mathbf{V}}_{\text{cm}} = \frac{d\vec{\mathbf{R}}_{\text{cm}}}{dt} = \frac{dX_{\text{cm}}}{dt}\,\hat{\mathbf{i}} = V_{\text{cm}}\,\hat{\mathbf{i}} \,. \qquad (20.2.24)$$

Because the center of the wheel is moving at a uniform speed the relative acceleration of the two frames is zero,

$$\vec{\mathbf{A}}_{\text{cm}} = \frac{d\vec{\mathbf{V}}_{\text{cm}}}{dt} = \vec{\mathbf{0}} \,. \qquad (20.2.25)$$

Define the position, velocity, and acceleration in this frame (with respect to the ground) by

$$\vec{\mathbf{r}}_b(t) = x_b(t)\,\hat{\mathbf{i}} + y_b(t)\,\hat{\mathbf{j}}, \quad \vec{\mathbf{v}}_b(t) = v_{b,x}(t)\,\hat{\mathbf{i}} + v_{b,y}(t)\,\hat{\mathbf{j}}, \quad \vec{\mathbf{a}}(t) = a_{b,x}(t)\,\hat{\mathbf{i}} + a_{b,y}(t)\,\hat{\mathbf{j}} \,. \qquad (20.2.26)$$

Then the position vectors are related by

$$\vec{\mathbf{r}}_b(t) = \vec{\mathbf{R}}_{\text{cm}}(t) + \vec{\mathbf{r}}'_b(t). \tag{20.2.27}$$

In order to add these vectors we need to decompose the position vector in the center of mass reference frame into Cartesian components,

$$\vec{\mathbf{r}}'_b(t) = b\,\hat{\mathbf{r}}(t) = b\sin\theta(t)\,\hat{\mathbf{i}} + b\cos\theta(t)\,\hat{\mathbf{j}}. \tag{20.2.28}$$

Then using the relation $\theta(t) = (V_{\text{cm}} / R)t$, Eq. (20.2.28) becomes

$$\begin{aligned}\vec{\mathbf{r}}_b(t) &= \vec{\mathbf{R}}_{\text{cm}}(t) + \vec{\mathbf{r}}'_b(t) = (V_{\text{cm}}t\,\hat{\mathbf{i}} + R\,\hat{\mathbf{j}}) + (b\sin\theta(t)\,\hat{\mathbf{i}} + b\cos\theta(t)\,\hat{\mathbf{j}}) \\ &= \left(V_{\text{cm}}t + b\sin((V_{\text{cm}} / R)t)\right)\hat{\mathbf{i}} + \left(R + b\cos((V_{\text{cm}} / R)t)\right)\hat{\mathbf{j}}\end{aligned}. \tag{20.2.29}$$

Thus the position components of the bead with respect to the reference frame fixed to the ground are given by

$$x_b(t) = V_{\text{cm}}t + b\sin((V_{\text{cm}} / R)t) \tag{20.2.30}$$

$$y_b(t) = R + b\cos((V_{\text{cm}} / R)t)\ . \tag{20.2.31}$$

A plot of the y-component vs. the x-component of the position of the bead in the reference frame fixed to the ground is shown in Figure 20.10 below using the values $V_{\text{cm}} = 5\ \text{m}\cdot\text{s}^{-1}$, $R = 0.25\ \text{m}$, and $b = 0.125\ \text{m}$. This path is called a ***cycloid***. We can differentiate the position vector in the reference frame fixed to the ground to find the velocity of the bead

$$\vec{\mathbf{v}}_b(t) = \frac{d\vec{\mathbf{r}}_b}{dt}(t) = \frac{d}{dt}(V_{\text{cm}}t + b\sin((V_{\text{cm}} / R)t))\ \hat{\mathbf{i}} + \frac{d}{dt}(R + b\cos((V_{\text{cm}} / R)t)\,)\hat{\mathbf{j}}, \tag{20.2.32}$$

$$\vec{\mathbf{v}}_b(t) = (V_{\text{cm}} + (b / R)V\cos((V_{\text{cm}} / R)t))\,\hat{\mathbf{i}} - ((b / R)V_{\text{cm}}\sin((V_{\text{cm}} / R)t)\,)\hat{\mathbf{j}}. \tag{20.2.33}$$

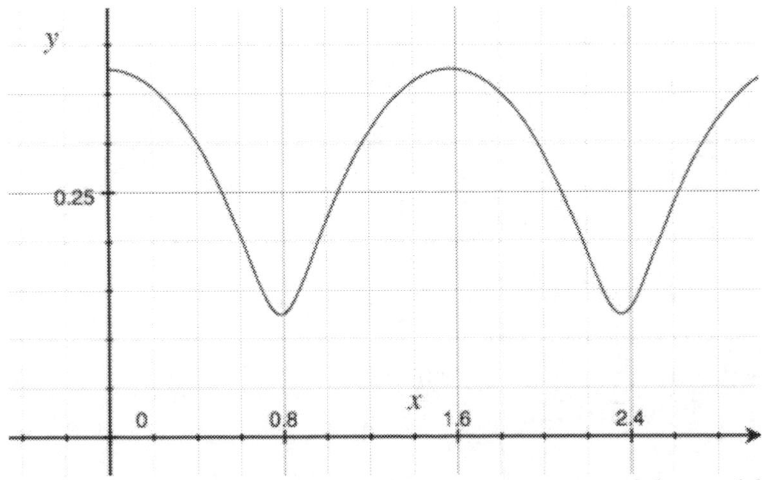

Figure 20.10 Plot of the y-component vs. the x-component of the position of the bead

Alternatively, we can decompose the velocity of the bead in the center of mass reference frame into Cartesian coordinates

$$\vec{\mathbf{v}}'_b(t) = (b / R)V_{\text{cm}} (\cos((V_{\text{cm}} / R)t)\,\hat{\mathbf{i}} - \sin((V_{\text{cm}} / R)t)\,\hat{\mathbf{j}}). \qquad (20.2.34)$$

The law of addition of velocities is then

$$\vec{\mathbf{v}}_b(t) = \vec{\mathbf{V}}_{\text{cm}} + \vec{\mathbf{v}}'_b(t), \qquad (20.2.35)$$

$$\vec{\mathbf{v}}_b(t) = V_{\text{cm}}\,\hat{\mathbf{i}} + (b / R)V_{\text{cm}} (\cos((V_{\text{cm}} / R)t)\,\hat{\mathbf{i}} - \sin((V_{\text{cm}} / R)t)\,\hat{\mathbf{j}}), \qquad (20.2.36)$$

$$\vec{\mathbf{v}}_b(t) = (V_{\text{cm}} + (b / R)V_{\text{cm}} \cos((V_{\text{cm}} / R)t))\,\hat{\mathbf{i}} - (b / R)\sin((V_{\text{cm}} / R)t)\,\hat{\mathbf{j}}, \qquad (20.2.37)$$

in agreement with our previous result. The acceleration is the same in either frame so

$$\vec{\mathbf{a}}_b(t) = \vec{\mathbf{a}}'_b = -(b / R^2)V_{\text{cm}}^2\,\hat{\mathbf{r}} = -(b / R^2)V_{\text{cm}}^2 (\sin((V_{\text{cm}} / R)t)\,\hat{\mathbf{i}} + \cos((V_{\text{cm}} / R)t)\,\hat{\mathbf{j}}). \qquad (20.2.38)$$

Example 20.2 Cylinder Rolling Without Slipping Down an Inclined Plane

A uniform cylinder of outer radius R and mass M with moment of inertia about the center of mass $I_{\text{cm}} = (1/2)M\,R^2$ starts from rest and rolls without slipping down an incline tilted at an angle β from the horizontal. The center of mass of the cylinder has dropped a vertical distance h when it reaches the bottom of the incline. Let g denote the gravitational constant. What is the relation between the component of the acceleration of the center of mass in the direction down the inclined plane and the component of the angular acceleration into the page of Figure 20.11?

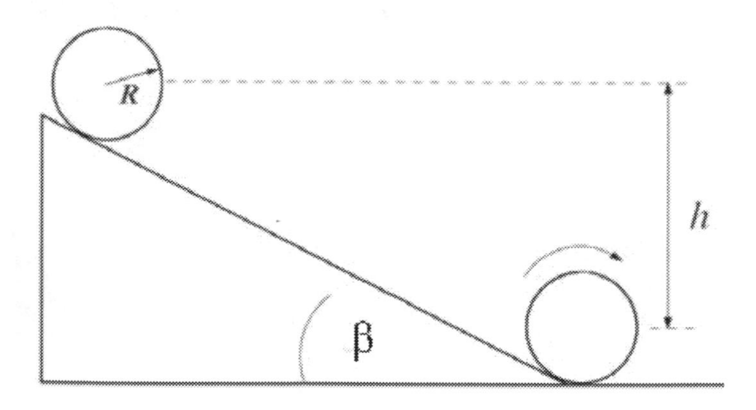

Figure 20.11 Example 20.2

Solution: We begin by choosing a coordinate system for the translational and rotational motion as shown in Figure 20.12.

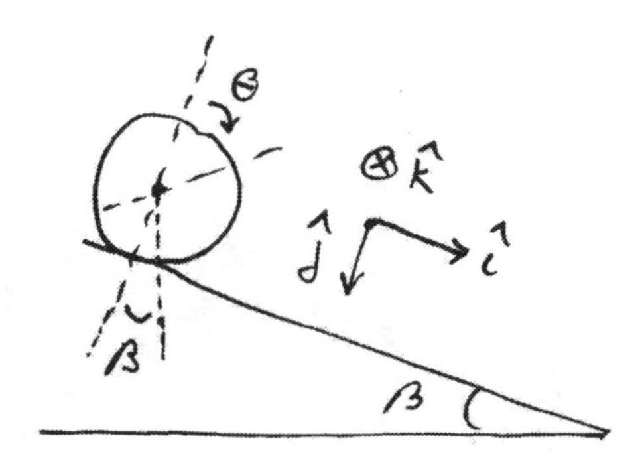

Figure 20.12 Coordinate system for rolling cylinder

For a time interval Δt, the displacement of the center of mass is given by $\Delta \vec{\mathbf{R}}_{cm}(t) = \Delta X_{cm}\,\hat{\mathbf{i}}$. The arc length due to the angular displacement of a point on the rim during the time interval Δt is given by $\Delta s = R\Delta\theta$. The rolling without slipping condition is

$$\Delta X_{cm} = R\Delta\theta \ .$$

If we divide both sides by Δt and take the limit as $\Delta t \to 0$ then the rolling without slipping condition show that the x-component of the center of mass velocity is equal to the magnitude of the tangential component of the velocity of a point on the rim

$$V_{\text{cm}} = \lim_{\Delta t \to 0} \frac{\Delta X_{\text{cm}}}{\Delta t} = \lim_{\Delta t \to 0} R\frac{\Delta\theta}{\Delta t} = R\omega_{\text{cm}} \ .$$

Similarly if we differentiate both sides of the above equation, we find a relation between the x-component of the center of mass acceleration is equal to the magnitude of the tangential component of the acceleration of a point on the rim

$$A_{\text{cm}} = \frac{dV_{\text{cm}}}{dt} = R\frac{d\omega_{\text{cm}}}{dt} = R\alpha_{\text{cm}} \ .$$

Example 20.3 Falling Yo-Yo

A Yo-Yo of mass m has an axle of radius b and a spool of radius R (Figure 20.13a). Its moment of inertia about the center of mass can be taken to be $I = (1/2)mR^2$ (the thickness of the string can be neglected). The Yo-Yo is released from rest. What is the relation between the angular acceleration about the center of mass and the linear acceleration of the center of mass?

Solution: Choose coordinates as shown in Figure 20.13b.

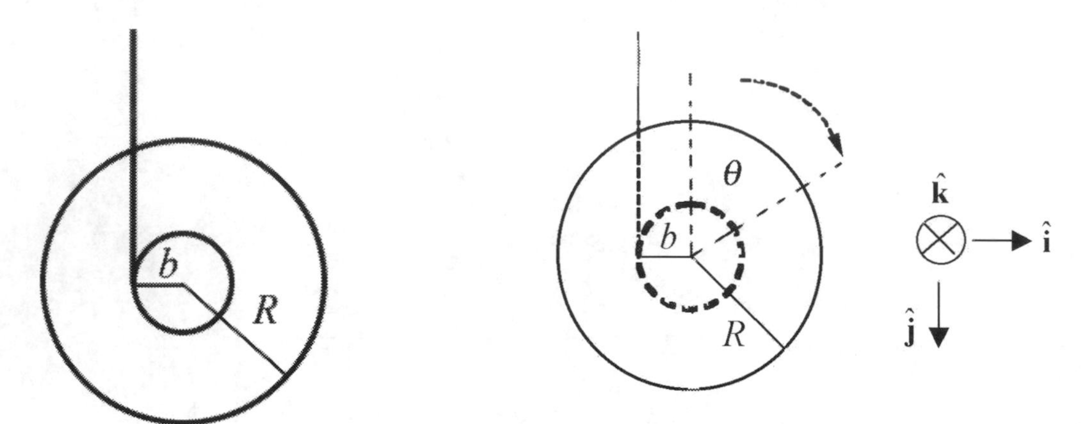

Figure 20.13a Example 20.3 **Figure 20.13b** Coordinate system for Yo-Yo

Consider a point on the rim of the axle at a distance $r = b$ from the center of mass. As the yo-yo falls, the arc length $\Delta s = b\Delta\theta$ subtended by the rotation of this point is equal to length of string that has unraveled, an amount Δl. In a time interval Δt, $b\Delta\theta = \Delta l$. Therefore $b\Delta\theta / \Delta t = \Delta l / \Delta t$. Taking limits, noting that, $V_{\mathrm{cm},y} = dl / dt$, we have that $b\omega_{\mathrm{cm}} = V_{\mathrm{cm},y}$. Differentiating a second time yields $b\alpha_{\mathrm{cm}} = A_{\mathrm{cm},y}$.

Example 20.4 Unwinding Drum

Drum A of mass m and radius R is suspended from a drum B also of mass m and radius R, which is free to rotate about its axis. The suspension is in the form of a massless metal tape wound around the outside of each drum, and free to unwind (Figure 20.14). Gravity acts with acceleration g downwards. Both drums are initially at rest. Find the initial acceleration of drum A, assuming that it moves straight down.

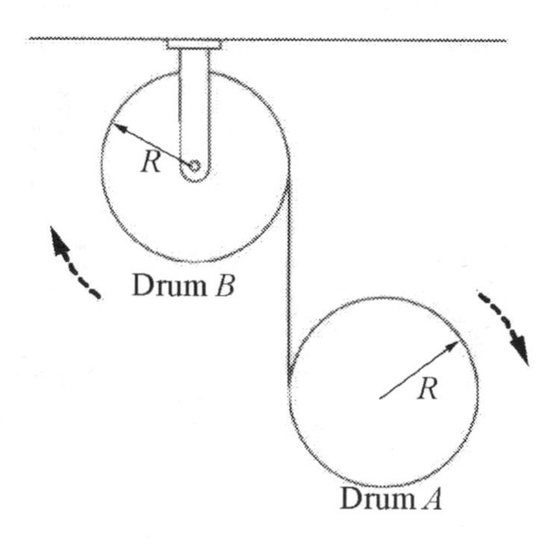

Figure 20.14 Example 20.4

Solution: The key to solving this problem is to determine the relation between the three kinematic quantities α_A, α_B, and a_A, the angular accelerations of the two drums and the linear acceleration of drum A. Choose the positive y-axis pointing downward with the origin at the center of drum B. After a time interval Δt, the center of drum A has undergone a displacement Δy. An amount of tape $\Delta l_A = R\Delta\theta_A$ has unraveled from drum A, and an amount of tape $\Delta l_B = R\Delta\theta_B$ has unraveled from drum B. Therefore the displacement of the center of drum A is equal to the total amount of tape that has unwound from the two drums, $\Delta y = \Delta l_A + \Delta l_B = R\Delta\theta_A + R\Delta\theta_B$. Dividing through by Δt and taking the limit as $\Delta t \to 0$ yields

$$\frac{dy}{dt} = R\frac{d\theta_A}{dt} + R\frac{d\theta_B}{dt}.$$

Differentiating a second time yields the desired relation between the angular accelerations of the two drums and the linear acceleration of drum A,

$$\frac{d^2y}{dt^2} = R\frac{d^2\theta_A}{dt^2} + R\frac{d^2\theta_B}{dt^2}$$

$$a_{A,y} = R\alpha_A + R\alpha_B.$$

20.3 Angular Momentum for a System of Particles Undergoing Translational and Rotational

We shall now show that the angular momentum of a body about a point S can be decomposed into two vector parts, the angular momentum of the center of mass (treated as a point particle) about the point S, and the angular momentum of the rotational motion about the center of mass.

Consider a system of N particles located at the points labeled $i = 1, 2, \cdots, N$. The angular momentum about the point S is the sum

$$\vec{\mathbf{L}}_S^{\text{total}} = \sum_{i=1}^{N} \vec{\mathbf{L}}_{S,i} = (\sum_{i=1}^{N} \vec{\mathbf{r}}_{S,i} \times m_i \vec{\mathbf{v}}_i), \tag{20.3.1}$$

where $\vec{\mathbf{r}}_{S,i}$ is the vector from the point S to the ith particle (Figure 20.15) satisfying

$$\vec{\mathbf{r}}_{S,i} = \vec{\mathbf{r}}_{S,cm} + \vec{\mathbf{r}}_{\text{cm},i}, \tag{20.3.2}$$

$$\vec{\mathbf{v}}_{S,i} = \vec{\mathbf{V}}_{cm} + \vec{\mathbf{v}}_{\text{cm},i}, \tag{20.3.3}$$

where $\vec{\mathbf{v}}_{S,cm} = \vec{\mathbf{V}}_{cm}$. We can now substitute both Eqs. (20.3.2) and (20.3.3) into Eq. (20.3.1) yielding

$$\vec{\mathbf{L}}_S^{\text{total}} = \sum_{i=1}^{N} (\vec{\mathbf{r}}_{S,cm} + \vec{\mathbf{r}}_{\text{cm},i}) \times m_i (\vec{\mathbf{V}}_{cm} + \vec{\mathbf{v}}_{\text{cm},i}). \tag{20.3.4}$$

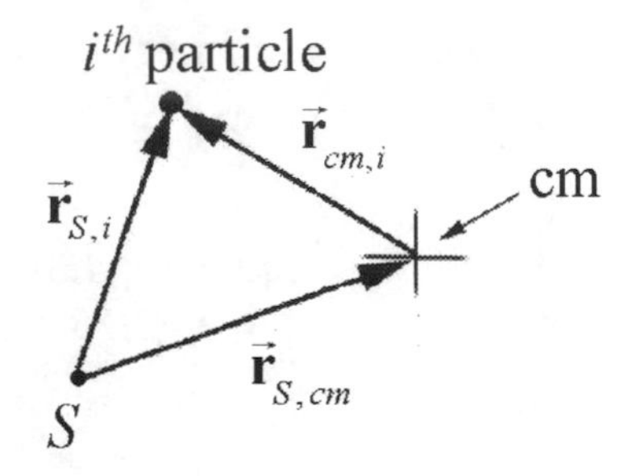

Figure 20.15 Vector Triangle

When we expand the expression in Equation (20.3.4), we have four terms,

$$\begin{aligned}\vec{\mathbf{L}}_S^{\text{total}} &= \sum_{i=1}^{N}(\vec{\mathbf{r}}_{S,cm} \times m_i \vec{\mathbf{v}}_{\text{cm},i}) + \sum_{i=1}^{N}(\vec{\mathbf{r}}_{S,cm} \times m_i \vec{\mathbf{V}}_{cm}) \\ &+ \sum_{i=1}^{N}(\vec{\mathbf{r}}_{\text{cm},i} \times m_i \vec{\mathbf{v}}_{\text{cm},i}) + \sum_{i=1}^{N}(\vec{\mathbf{r}}_{\text{cm},i} \times m_i \vec{\mathbf{V}}_{cm}).\end{aligned} \tag{20.3.5}$$

The vector $\vec{\mathbf{r}}_{S,cm}$ is a constant vector that depends only on the location of the center of mass and not on the location of the ith particle. Therefore in the first term in the above equation, $\vec{\mathbf{r}}_{S,cm}$ can be taken outside the summation. Similarly, in the second term the velocity of the center of mass $\vec{\mathbf{V}}_{cm}$ is the same for each term in the summation, and may be taken outside the summation,

$$\begin{aligned}\vec{\mathbf{L}}_S^{\text{total}} &= \vec{\mathbf{r}}_{S,cm} \times \left(\sum_{i=1}^{N} m_i \vec{\mathbf{v}}_{\text{cm},i}\right) + \vec{\mathbf{r}}_{S,cm} \times \left(\sum_{i=1}^{N} m_i\right) \vec{\mathbf{V}}_{cm} \\ &+ \sum_{i=1}^{N}(\vec{\mathbf{r}}_{\text{cm},i} \times m_i \vec{\mathbf{v}}_{\text{cm},i}) + \left(\sum_{i=1}^{N} m_i \vec{\mathbf{r}}_{\text{cm},i}\right) \times \vec{\mathbf{V}}_{cm}.\end{aligned} \tag{20.3.6}$$

The first and third terms in Eq. (20.3.6) are both zero due to the fact that

$$\begin{aligned}\sum_{i=1}^{N} m_i \vec{\mathbf{r}}_{\text{cm},i} &= 0 \\ \sum_{i=1}^{N} m_i \vec{\mathbf{v}}_{\text{cm},i} &= 0.\end{aligned} \tag{20.3.7}$$

We first show that $\sum_{i=1}^{N} m_i \vec{\mathbf{r}}_{\text{cm},i}$ is zero. We begin by using Eq. (20.3.2),

$$\begin{aligned}\sum_{i=1}^{N}(m_i\vec{\mathbf{r}}_{\text{cm},i}) &= \sum_{i=1}^{N}(m_i(\vec{\mathbf{r}}_i - \vec{\mathbf{r}}_{S,cm})) \\ &= \sum_{i=1}^{N}m_i\vec{\mathbf{r}}_i - \left(\sum_{i=1}^{N}(m_i)\right)\vec{\mathbf{r}}_{S,cm} = \sum_{i=1}^{N}m_i\vec{\mathbf{r}}_i - m^{total}\vec{\mathbf{r}}_{S,cm}.\end{aligned} \tag{20.3.8}$$

Substitute the definition of the center of mass (Eq. 10.5.3) into Eq. (20.3.8) yielding

$$\sum_{i=1}^{N}(m_i\vec{\mathbf{r}}_{\text{cm},i}) = \sum_{i=1}^{N}m_i\vec{\mathbf{r}}_i - m^{total}\frac{1}{m^{total}}\sum_{i=1}^{N}m_i\vec{\mathbf{r}}_i = \vec{0}. \tag{20.3.9}$$

The vanishing of $\sum_{i=1}^{N}m_i\vec{\mathbf{v}}_{\text{cm},i} = 0$ follows directly from the definition of the center of mass frame, that the momentum in the center of mass is zero. Equivalently the derivative of Eq. (20.3.9) is zero. We could also simply calculate and find that

$$\begin{aligned}\sum_i m_i\vec{\mathbf{v}}_{\text{cm},i} &= \sum_i m_i(\vec{\mathbf{v}}_i - \vec{\mathbf{V}}_{\text{cm}}) \\ &= \sum_i m_i\vec{\mathbf{v}}_i - \vec{\mathbf{V}}_{\text{cm}}\sum_i m_i \\ &= m^{\text{total}}\vec{\mathbf{V}}_{\text{cm}} - \vec{\mathbf{V}}_{\text{cm}}m^{\text{total}} \\ &= \vec{\mathbf{0}}.\end{aligned} \tag{20.3.10}$$

We can now simplify Eq. (20.3.6) for the angular momentum about the point S using the fact that, $m_T = \sum_{i=1}^{N}m_i$, and $\vec{\mathbf{p}}_{\text{sys}} = m_T\vec{\mathbf{V}}_{cm}$ (in reference frame O):

$$\vec{\mathbf{L}}_S^{\text{total}} = \vec{\mathbf{r}}_{S,cm}\times\vec{\mathbf{p}}^{\text{sys}} + \sum_{i=1}^{N}(\vec{\mathbf{r}}_{cm,i}\times m_i\vec{\mathbf{v}}_{cm,i}). \tag{20.3.11}$$

Consider the first term in Equation (20.3.11), $\vec{\mathbf{r}}_{S,\text{cm}}\times\vec{\mathbf{p}}_{\text{sys}}$; the vector $\vec{\mathbf{r}}_{S,\text{cm}}$ is the vector from the point S to the center of mass. If we treat the system as a point-like particle of mass m_T located at the center of mass, then the momentum of this point-like particle is $\vec{\mathbf{p}}_{\text{sys}} = m_T\vec{\mathbf{V}}_{cm}$. Thus the first term is the angular momentum about the point S of this "point-like particle", which is called the ***orbital angular momentum*** about S,

$$\vec{\mathbf{L}}_S^{\text{orbital}} = \vec{\mathbf{r}}_{S,\text{cm}}\times\vec{\mathbf{p}}_{\text{sys}}. \tag{20.3.12}$$

for the system of particles.

Consider the second term in Equation (20.3.11), $\sum_{i=1}^{N}(\vec{\mathbf{r}}_{cm,i} \times m_i \vec{\mathbf{v}}_{cm,i})$; the quantity inside the summation is the angular momentum of the ith particle with respect to the origin in the center of mass reference frame O_{cm} (recall the origin in the center of mass reference frame is the center of mass of the system),

$$\vec{\mathbf{L}}_{cm,i} = \vec{\mathbf{r}}_{cm,i} \times m_i \vec{\mathbf{v}}_{cm,i} . \tag{20.3.13}$$

Hence the total angular momentum of the system with respect to the center of mass in the center of mass reference frame is given by

$$\vec{\mathbf{L}}_{\text{cm}}^{\text{spin}} = \sum_{i=1}^{N} \vec{\mathbf{L}}_{cm,i} = \sum_{i=1}^{N}(\vec{\mathbf{r}}_{cm,i} \times m_i \vec{\mathbf{v}}_{cm,i}) . \tag{20.3.14}$$

a vector quantity we call the ***spin angular momentum***. Thus we see that the total angular momentum about the point S is the sum of these two terms,

$$\vec{\mathbf{L}}_{S}^{\text{total}} = \vec{\mathbf{L}}_{S}^{\text{orbital}} + \vec{\mathbf{L}}_{\text{cm}}^{\text{spin}} . \tag{20.3.15}$$

This decomposition of angular momentum into a piece associated with the translational motion of the center of mass and a second piece associated with the rotational motion about the center of mass in the center of mass reference frame is the key conceptual foundation for what follows.

Example 20.5 Earth's Motion Around the Sun

The earth, of mass $m_{\text{e}} = 5.97 \times 10^{24}$ kg and (mean) radius $R_{\text{e}} = 6.38 \times 10^{6}$ m, moves in a nearly circular orbit of radius $r_{\text{s,e}} = 1.50 \times 10^{11}$ m around the sun with a period $T_{\text{orbit}} = 365.25$ days, and spins about its axis in a period $T_{\text{spin}} = 23$ hr 56 min, the axis inclined to the normal to the plane of its orbit around the sun by $23.5°$ (in Figure 20.16, the relative size of the earth and sun, and the radius and shape of the orbit are not representative of the actual quantities).

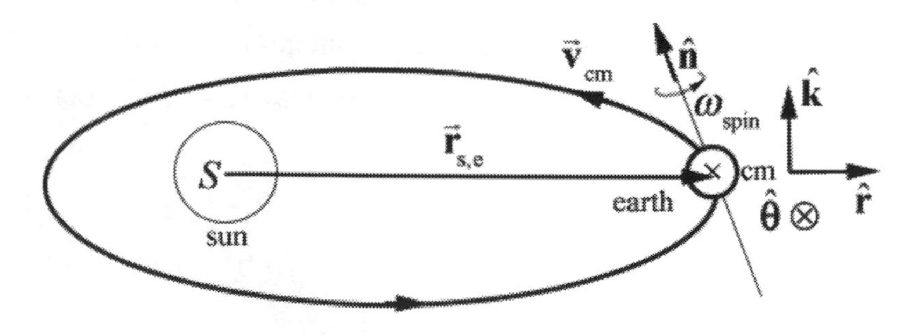

Figure 20.16 Example 20.5

If we approximate the earth as a uniform sphere, then the moment of inertia of the earth about its center of mass is

$$I_{\text{cm}} = \frac{2}{5} m_{\text{e}} R_{\text{e}}^2 . \quad (20.3.16)$$

If we choose the point S to be at the center of the sun, and assume the orbit is circular, then the orbital angular momentum is

$$\vec{\mathbf{L}}_S^{\text{orbital}} = \vec{\mathbf{r}}_{S,\text{cm}} \times \vec{\mathbf{p}}_{\text{sys}} = r_{\text{s,e}}\, \hat{\mathbf{r}} \times m_{\text{e}} v_{\text{cm}}\, \hat{\boldsymbol{\theta}} = r_{\text{s,e}} m_{\text{e}} v_{\text{cm}}\, \hat{\mathbf{k}} . \quad (20.3.17)$$

The velocity of the center of mass of the earth about the sun is related to the orbital angular velocity by

$$v_{\text{cm}} = r_{\text{s,e}}\, \omega_{\text{orbit}} , \quad (20.3.18)$$

where the orbital angular speed is

$$\begin{aligned} \omega_{\text{orbit}} &= \frac{2\pi}{T_{\text{orbit}}} = \frac{2\pi}{(365.25\ \text{d})(8.640 \times 10^4\ \text{s} \cdot \text{d}^{-1})} \\ &= 1.991 \times 10^{-7}\ \text{rad} \cdot \text{s}^{-1}. \end{aligned} \quad (20.3.19)$$

The orbital angular momentum about S is then

$$\begin{aligned} \vec{\mathbf{L}}_S^{\text{orbital}} &= m_{\text{e}} r_{\text{s,e}}^2\, \omega_{\text{orbit}}\, \hat{\mathbf{k}} \\ &= (5.97 \times 10^{24}\ \text{kg})(1.50 \times 10^{11}\ \text{m})^2 (1.991 \times 10^{-7}\ \text{rad} \cdot \text{s}^{-1})\, \hat{\mathbf{k}} \\ &= (2.68 \times 10^{40}\ \text{kg} \cdot \text{m}^2 \cdot \text{s}^{-1})\, \hat{\mathbf{k}}. \end{aligned} \quad (20.3.20)$$

The spin angular momentum is given by

$$\vec{\mathbf{L}}_{\text{cm}}^{\text{spin}} = I_{\text{cm}}\, \vec{\boldsymbol{\omega}}_{\text{spin}} = \frac{2}{5} m_{\text{e}} R_{\text{e}}^2\, \omega_{\text{spin}}\, \hat{\mathbf{n}} , \quad (20.3.21)$$

where $\hat{\mathbf{n}}$ is a unit normal pointing along the axis of rotation of the earth and

$$\omega_{\text{spin}} = \frac{2\pi}{T_{\text{spin}}} = \frac{2\pi}{8.616 \times 10^4\ \text{s}} = 7.293 \times 10^{-5}\ \text{rad} \cdot \text{s}^{-1} . \quad (20.3.22)$$

The spin angular momentum is then

$$\begin{aligned} \vec{\mathbf{L}}_{\text{cm}}^{\text{spin}} &= \frac{2}{5}(5.97 \times 10^{24}\ \text{kg})(6.38 \times 10^6\ \text{m})^2 (7.293 \times 10^{-5}\ \text{rad} \cdot \text{s}^{-1})\, \hat{\mathbf{n}} \\ &= (7.10 \times 10^{33}\ \text{kg} \cdot \text{m}^2 \cdot \text{s}^{-1})\, \hat{\mathbf{n}}. \end{aligned} \quad (20.3.23)$$

The ratio of the magnitudes of the orbital angular momentum about S to the spin angular momentum is greater than a million,

$$\frac{L_S^{\text{orbital}}}{L_{\text{cm}}^{\text{spin}}} = \frac{m_{\text{e}} r_{\text{s,e}}^2 \omega_{\text{orbit}}}{(2/5) m_{\text{e}} R_{\text{e}}^2 \omega_{\text{spin}}} = \frac{5}{2} \frac{r_{\text{s,e}}^2}{R_{\text{e}}^2} \frac{T_{\text{spin}}}{T_{\text{orbit}}} = 3.77 \times 10^6 , \tag{20.3.24}$$

as this ratio is proportional to the square of the ratio of the distance to the sun to the radius of the earth. The angular momentum about S is then

$$\vec{\mathbf{L}}_S^{\text{total}} = m_{\text{e}} r_{\text{s,e}}^2 \omega_{\text{orbit}} \, \hat{\mathbf{k}} + \frac{2}{5} m_{\text{e}} R_{\text{e}}^2 \omega_{\text{spin}} \, \hat{\mathbf{n}} . \tag{20.3.25}$$

The orbit and spin periods are known to far more precision than the average values used for the earth's orbit radius and mean radius. Two different values have been used for one "day;" in converting the orbit period from days to seconds, the value for the *solar day*, $T_{\text{solar}} = 86,400\,\text{s}$ was used. In converting the earth's spin angular frequency, the *sidereal day*, $T_{\text{sidereal}} = T_{\text{spin}} = 86,160\,\text{s}$ was used. The two periods, the solar day from noon to noon and the sidereal day from the difference between the times that a fixed star is at the same place in the sky, do differ in the third significant figure.

20.4 Kinetic Energy of a System of Particles

Consider a system of particles. The ith particle has mass m_i and velocity $\vec{\mathbf{v}}_i$ with respect to a reference frame O. The kinetic energy of the system of particles is given by

$$\begin{aligned} K &= \sum_i \frac{1}{2} m_i v_i^2 = \frac{1}{2} \sum_i m_i \vec{\mathbf{v}}_i \cdot \vec{\mathbf{v}}_i \\ &= \frac{1}{2} \sum_i m_i (\vec{\mathbf{v}}_{\text{cm},i} + \vec{\mathbf{V}}_{\text{cm}}) \cdot (\vec{\mathbf{v}}_{\text{cm},i} + \vec{\mathbf{V}}_{\text{cm}}). \end{aligned} \tag{20.4.1}$$

where Equation 15.2.6 has been used to express $\vec{\mathbf{v}}_i$ in terms of $\vec{\mathbf{v}}_{cm,i}$ and $\vec{\mathbf{V}}_{\text{cm}}$. Expanding the last dot product in Equation (20.4.1),

$$\begin{aligned} K &= \frac{1}{2} \sum_i m_i (\vec{\mathbf{v}}_{\text{cm},i} \cdot \vec{\mathbf{v}}_{i,\text{rel}} + \vec{\mathbf{V}}_{\text{cm}} \cdot \vec{\mathbf{V}}_{\text{cm}} + 2 \vec{\mathbf{v}}_{\text{cm},i} \cdot \vec{\mathbf{V}}_{\text{cm}}) \\ &= \frac{1}{2} \sum_i m_i (\vec{\mathbf{v}}_{\text{cm},i} \cdot \vec{\mathbf{v}}_{i,\text{rel}}) + \frac{1}{2} \sum_i m_i (\vec{\mathbf{V}}_{\text{cm}} \cdot \vec{\mathbf{V}}_{\text{cm}}) + \sum_i m_i \vec{\mathbf{v}}_{\text{cm},i} \cdot \vec{\mathbf{V}}_{\text{cm}} \\ &= \sum_i \frac{1}{2} m_i v_{\text{cm},i}^2 + \frac{1}{2} \sum_i m_i V_{\text{cm}}^2 + \left(\sum_i m \vec{\mathbf{v}}_{\text{cm},i} \right) \cdot \vec{\mathbf{V}}_{\text{cm}} . \end{aligned} \tag{20.4.2}$$

The last term in the third equation in (20.4.2) vanishes as we showed in Eq. (20.3.7). Then Equation (20.4.2) reduces to

$$\begin{aligned} K &= \sum_i \frac{1}{2} m_i v_{\text{cm},i}^2 + \frac{1}{2} \sum_i m_i V_{\text{cm}}^2 \\ &= \sum_i \frac{1}{2} m_i v_{\text{cm},i}^2 + \frac{1}{2} m^{\text{total}} V_{\text{cm}}^2 . \end{aligned} \tag{20.4.3}$$

We interpret the first term as the kinetic energy of the center of mass motion in reference frame O and the second term as the sum of the individual kinetic energies of the particles of the system in the center of mass reference frame O_{cm}.

At this point, it's important to note that no assumption was made regarding the mass elements being constituents of a rigid body. Equation (20.4.3) is valid for a rigid body, a gas, a firecracker (but K is certainly not the same before and after detonation), and the sixteen pool balls after the break, or any collection of objects for which the center of mass can be determined.

20.5 Rotational Kinetic Energy for a Rigid Body Undergoing Fixed Axis Rotation

The rotational kinetic energy for the rigid body, using $\vec{\mathbf{v}}_{\text{cm},i} = (r_{\text{cm},i})_{\perp} \omega_{\text{cm}} \hat{\boldsymbol{\theta}}$, simplifies to

$$K_{\text{rot}} = \frac{1}{2} I_{\text{cm}} \omega_{\text{cm}}^2 . \tag{20.5.1}$$

Therefore the total kinetic energy of a translating and rotating rigid body is

$$K_{\text{total}} = K_{\text{trans}} + K_{\text{rot}} = \frac{1}{2} m V_{\text{cm}}^2 + \frac{1}{2} I_{\text{cm}} \omega_{\text{cm}}^2 . \tag{20.5.2}$$

Appendix 20A Chasles's Theorem: Rotation and Translation of a Rigid Body

We now return to our description of the translating and rotating rod that we first considered when we began our discussion of rigid bodies. We shall now show that the motion of any rigid body consists of a translation of the center of mass and rotation about the center of mass.

We shall demonstrate this for a rigid body by dividing up the rigid body into point-like constituents. Consider two point-like constituents with masses m_1 and m_2. Choose a coordinate system with a choice of origin such that body 1 has position $\vec{\mathbf{r}}_1$ and body 2 has position $\vec{\mathbf{r}}_2$ (Figure 20A.1). The relative position vector is given by

$$\vec{\mathbf{r}}_{1,2} = \vec{\mathbf{r}}_1 - \vec{\mathbf{r}}_2 . \tag{20.A.1}$$

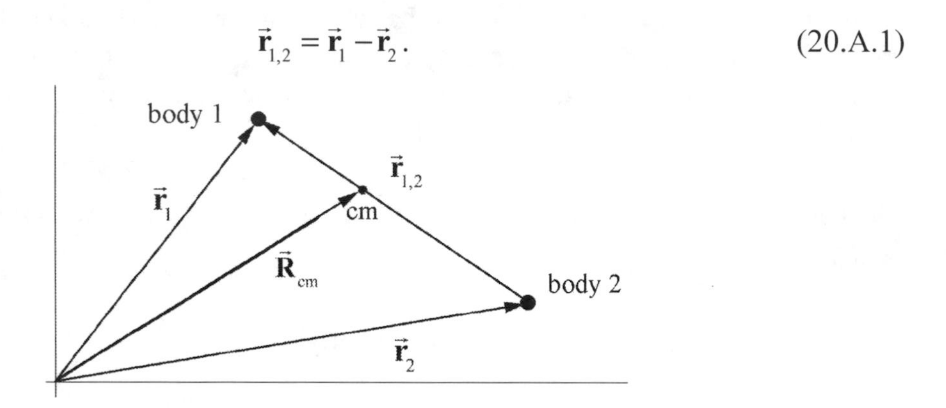

Figure 20A.1 Two-body coordinate system.

Recall we defined the center of mass vector, $\vec{\mathbf{R}}_{\text{cm}}$, of the two-body system as

$$\vec{\mathbf{R}}_{cm} = \frac{m_1 \vec{\mathbf{r}}_1 + m_2 \vec{\mathbf{r}}_2}{m_1 + m_2} . \tag{20.A.2}$$

In Figure 20A.2 we show the center of mass coordinate system.

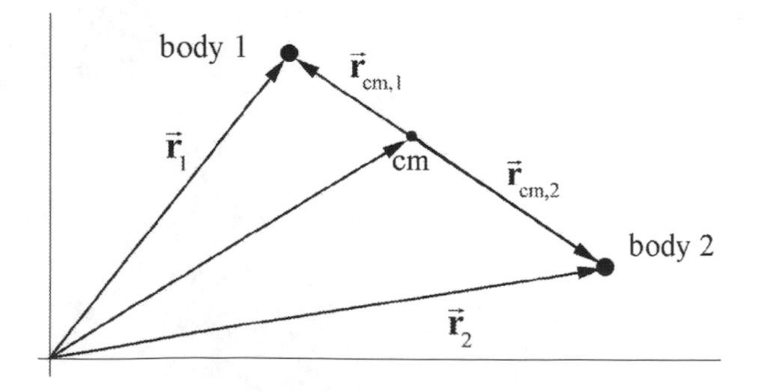

Figure 20A.2: Center of mass reference frame

The position vector of the object 1 with respect to the center of mass is given by

$$\vec{\mathbf{r}}_{\text{cm},1} = \vec{\mathbf{r}}_1 - \vec{\mathbf{R}}_{\text{cm}} = \vec{\mathbf{r}}_1 - \frac{m_1\vec{\mathbf{r}}_1 + m_2\vec{\mathbf{r}}_2}{m_1 + m_2} = \frac{m_2}{m_1 + m_2}(\vec{\mathbf{r}}_1 - \vec{\mathbf{r}}_2) = \frac{\mu}{m_1}\vec{\mathbf{r}}_{1,2}, \qquad (20.A.3)$$

where

$$\mu = \frac{m_1 m_2}{m_1 + m_2}, \qquad (20.A.4)$$

is the reduced mass. In addition, the relative position vector between the two objects is independent of the choice of reference frame,

$$\vec{\mathbf{r}}_{12} = \vec{\mathbf{r}}_1 - \vec{\mathbf{r}}_2 = (\vec{\mathbf{r}}_{\text{cm},1} + \vec{\mathbf{R}}_{\text{cm}}) - (\vec{\mathbf{r}}_{\text{cm},2} + \vec{\mathbf{R}}_{\text{cm}}) = \vec{\mathbf{r}}_{\text{cm},1} - \vec{\mathbf{r}}_{\text{cm},2} = \vec{\mathbf{r}}_{\text{cm},1,2}. \qquad (20.A.5)$$

Because the center of mass is at the origin in the center of mass reference frame,

$$\frac{m_1\vec{\mathbf{r}}_{\text{cm},1} + m_2\vec{\mathbf{r}}_{\text{cm},2}}{m_1 + m_2} = \vec{\mathbf{0}}. \qquad (20.A.6)$$

Therefore

$$m_1\vec{\mathbf{r}}_{\text{cm},1} = -m_2\vec{\mathbf{r}}_{\text{cm},2} \qquad (20.A.7)$$

$$m_1\left|\vec{\mathbf{r}}_{\text{cm},1}\right| = m_2\left|\vec{\mathbf{r}}_{\text{cm},2}\right|. \qquad (20.A.8)$$

The displacement of object 1 about the center of mass is given by taking the derivative of Eq. (20.A.3),

$$d\vec{\mathbf{r}}_{\text{cm},1} = \frac{\mu}{m_1}d\vec{\mathbf{r}}_{1,2}. \qquad (20.A.9)$$

A similar calculation for the position of object 2 with respect to the center of mass yields for the position and displacement with respect to the center of mass

$$\vec{\mathbf{r}}_{\text{cm},2} = \vec{\mathbf{r}}_2 - \vec{\mathbf{R}}_{\text{cm}} = -\frac{\mu}{m_2}\vec{\mathbf{r}}_{1,2}, \qquad (20.A.10)$$

$$d\vec{\mathbf{r}}_{\text{cm},2} = -\frac{\mu}{m_2}d\vec{\mathbf{r}}_{1,2}. \qquad (20.A.11)$$

Let $i = 1,2$. An arbitrary displacement of the ith object is given respectively by

$$d\vec{\mathbf{r}}_i = d\vec{\mathbf{r}}_{\text{cm},i} + d\vec{\mathbf{R}}_{\text{cm}}, \qquad (20.A.12)$$

which is the sum of a displacement about the center of mass $d\vec{\mathbf{r}}_{\text{cm},i}$ and a displacement of the center of mass $d\vec{\mathbf{R}}_{\text{cm}}$. The displacement of objects 1 and 2 are constrained by the condition that the distance between the objects must remain constant since the body is rigid. In particular, the distance between objects 1 and 2 is given by

$$\left|\vec{\mathbf{r}}_{1,2}\right|^2 = (\vec{\mathbf{r}}_1 - \vec{\mathbf{r}}_2)\cdot(\vec{\mathbf{r}}_1 - \vec{\mathbf{r}}_2). \qquad (20.A.13)$$

Because this distance is constant we can differentiate Eq. (20.A.13), yielding the ***rigid body condition*** that

$$0 = 2(\vec{\mathbf{r}}_1 - \vec{\mathbf{r}}_2)\cdot(d\vec{\mathbf{r}}_1 - d\vec{\mathbf{r}}_2) = 2\vec{\mathbf{r}}_{1,2}\cdot d\vec{\mathbf{r}}_{1,2} \qquad (20.A.14)$$

20A.1. Translation of the Center of Mass

The condition (Eq. (20.A.14)) can be satisfied if the relative displacement vector between the two objects is zero,

$$d\vec{\mathbf{r}}_{1,2} = d\vec{\mathbf{r}}_1 - d\vec{\mathbf{r}}_2 = \vec{\mathbf{0}}. \qquad (20.A.15)$$

This implies, using, Eq. (20.A.9) and Eq. (20.A.11), that the displacement with respect to the center of mass is zero,

$$d\vec{\mathbf{r}}_{\text{cm},1} = d\vec{\mathbf{r}}_{\text{cm},2} = \vec{\mathbf{0}}. \qquad (20.A.16)$$

Thus by Eq. (20.A.12), the displacement of each object is equal to the displacement of the center of mass,

$$d\vec{\mathbf{r}}_i = d\vec{\mathbf{R}}_{\text{cm}}, \qquad (20.A.17)$$

which means that the body is undergoing pure translation.

20A.2 Rotation about the Center of Mass

Now suppose that $d\vec{\mathbf{r}}_{1,2} = d\vec{\mathbf{r}}_1 - d\vec{\mathbf{r}}_2 \neq \vec{\mathbf{0}}$. The rigid body condition can be expressed in terms of the center of mass coordinates. Using Eq. (20.A.9), the rigid body condition (Eq. (20.A.14)) becomes

$$0 = 2\frac{\mu}{m_1}\vec{\mathbf{r}}_{1,2}\cdot d\vec{\mathbf{r}}_{\text{cm},1}. \qquad (20.A.18)$$

Because the relative position vector between the two objects is independent of the choice of reference frame (Eq. (20.A.5)), the rigid body condition Eq. (20.A.14) in the center of mass reference frame is then given by

$$0 = 2\vec{\mathbf{r}}_{\text{cm},1,2}\cdot d\vec{\mathbf{r}}_{\text{cm},1}. \qquad (20.A.19)$$

This condition is satisfied if the relative displacement is perpendicular to the line passing through the center of mass,

$$\vec{\mathbf{r}}_{\text{cm},1,2} \perp d\vec{\mathbf{r}}_{\text{cm},1}. \tag{20.A.20}$$

By a similar argument, $\vec{\mathbf{r}}_{\text{cm},1,2} \perp d\vec{\mathbf{r}}_{\text{cm},2}$. In order for these displacements to correspond to a rotation about the center of mass, the displacements must have the same angular displacement.

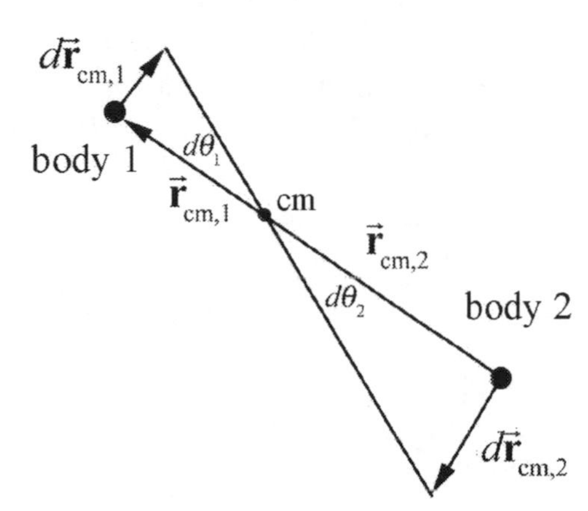

Figure 20A.3 Infinitesimal angular displacements in the center of mass reference frame

In Figure 20A.3, the infinitesimal angular displacement of each object is given by

$$d\theta_1 = \frac{\left|d\vec{\mathbf{r}}_{\text{cm},1}\right|}{\left|\vec{\mathbf{r}}_{\text{cm},1}\right|}, \tag{20.A.21}$$

$$d\theta_2 = \frac{\left|d\vec{\mathbf{r}}_{\text{cm},2}\right|}{\left|\vec{\mathbf{r}}_{\text{cm},2}\right|}. \tag{20.A.22}$$

From Eq. (20.A.9) and Eq. (20.A.11), we can rewrite Eqs. (20.A.21) and (20.A.22) as

$$d\theta_1 = \frac{\mu}{m_1}\frac{\left|d\vec{\mathbf{r}}_{1,2}\right|}{\left|\vec{\mathbf{r}}_{\text{cm},1}\right|}, \tag{20.A.23}$$

$$d\theta_2 = \frac{\mu}{m_2}\frac{\left|d\vec{\mathbf{r}}_{1,2}\right|}{\left|\vec{\mathbf{r}}_{\text{cm},2}\right|}. \tag{20.A.24}$$

Recall that in the center of mass reference frame $m_1\left|\vec{\mathbf{r}}_{\text{cm},1}\right| = m_2\left|\vec{\mathbf{r}}_{\text{cm},2}\right|$ (Eq. (20.A.8)) and hence the angular displacements are equal,

$$d\theta_1 = d\theta_2 = d\theta \,. \qquad (20.A.25)$$

Therefore the displacement of the ith object $d\vec{\mathbf{r}}_i$ differs from the displacement of the center of mass $d\vec{\mathbf{R}}_{cm}$ by a vector that corresponds to an infinitesimal rotation in the center of mass reference frame

$$d\vec{\mathbf{r}}_{cm,i} = d\vec{\mathbf{r}}_i - d\vec{\mathbf{R}}_{cm} \,. \qquad (20.A.26)$$

We have shown that the displacement of a rigid body is the vector sum of the displacement of the center of mass (translation of the center of mass) and an infinitesimal rotation about the center of mass.

Chapter 21 Rigid Body Dynamics: Rotation and Translation about a Fixed Axis

Chapter 21 Rigid Body Dynamics: Rotation and Translation about a Fixed Axis

Accordingly, we find Euler and D'Alembert devoting their talent and their patience to the establishment of the laws of rotation of the solid bodies. Lagrange has incorporated his own analysis of the problem with his general treatment of mechanics, and since his time M. Poinsôt has brought the subject under the power of a more searching analysis than that of the calculus, in which ideas take the place of symbols, and intelligent propositions supersede equations. [1]

James Clerk Maxwell

21.1 Introduction

We shall analyze the motion of systems of particles and rigid bodies that are undergoing translational and rotational motion about a fixed direction. Because the body is translating, the axis of rotation is no longer fixed in space. We shall describe the motion by a translation of the center of mass and a rotation about the center of mass. By choosing a reference frame moving with the center of mass, we can analyze the rotational motion separately and discover that the torque about the center of mass is equal to the change in the angular momentum about the center of mass. For a rigid body undergoing fixed axis rotation about the center of mass, our rotational equation of motion is similar to one we have already encountered for fixed axis rotation, $\vec{\boldsymbol{\tau}}_{\text{cm}}^{\text{ext}} = d\vec{\mathbf{L}}_{\text{cm}}^{\text{spin}} / dt$.

21.2 Translational Equation of Motion

We shall think about the system of particles as follows. We treat the whole system as a single point-like particle of mass m_T located at the center of mass moving with the velocity of the center of mass $\vec{\mathbf{V}}_{cm}$. The external force acting on the system acts at the center of mass and from our earlier result (Eq. 10.4.9) we have that

$$\vec{\mathbf{F}}^{\text{ext}} = \frac{d\vec{\mathbf{p}}_{\text{sys}}}{dt} = \frac{d}{dt}(m_T \vec{\mathbf{V}}_{cm}). \tag{22.2.1}$$

21.3 Translational and Rotational Equations of Motion

For a system of particles, the torque about a point S can be written as

[1] J. C. Maxwell on Louis Poinsôt (1777-1859) in 'On a Dynamical Top' (1857). In W. D. Niven (ed.), The Scientific Papers of James Clerk Maxwell (1890), Vol. 1, 248.

$$\vec{\boldsymbol{\tau}}_S^{\text{ext}} = \sum_{i=1}^{N} (\vec{\mathbf{r}}_i \times \vec{\mathbf{F}}_i) . \tag{22.3.1}$$

where we have assumed that all internal torques cancel in pairs. Let choose the point S to be the origin of the reference frame O, then $\vec{\mathbf{r}}_{S,\text{cm}} = \vec{\mathbf{R}}_{cm}$ (Figure 21.1). (You may want to recall the main properties of the center of mass reference frame by reviewing Chapter 15.2.1.)

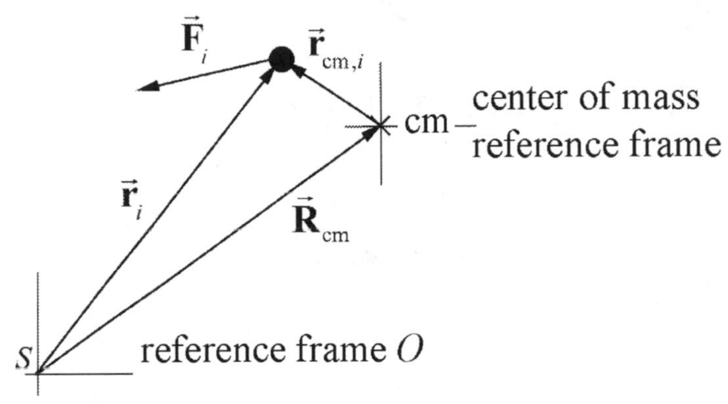

Figure 21.1 Torque diagram for center of mass reference frame

We can now apply $\vec{\mathbf{r}}_i = \vec{\mathbf{R}}_{cm} + \vec{\mathbf{r}}_{\text{cm},i}$ to Eq. (22.3.1) yielding

$$\vec{\boldsymbol{\tau}}_S^{\text{ext}} = \sum_{i=1}^{N} ((\vec{\mathbf{r}}_{S,\text{cm}} + \vec{\mathbf{r}}_{\text{cm},i}) \times \vec{\mathbf{F}}_i) = \sum_{i=1}^{N} (\vec{\mathbf{r}}_{S,\text{cm}} \times \vec{\mathbf{F}}_i) + \sum_{i=1}^{N} (\vec{\mathbf{r}}_{\text{cm},i} \times \vec{\mathbf{F}}_i) . \tag{22.3.2}$$

The term

$$\vec{\boldsymbol{\tau}}_{S,cm}^{\text{ext}} = \vec{\mathbf{r}}_{S,\text{cm}} \times \vec{\mathbf{F}}^{\text{ext}} \tag{22.3.3}$$

in Eq. (22.3.2) corresponds to the external torque about the point S where all the external forces act at the center of mass (Figure 21.2).

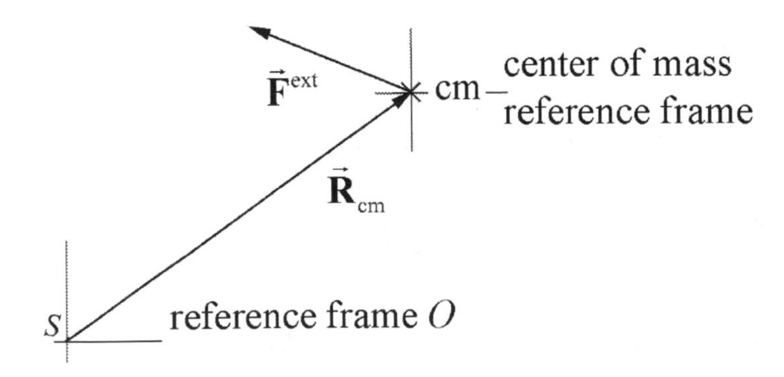

Figure 21.2 Torque diagram for "point-like" particle located at center of mass

The term,

$$\vec{\boldsymbol{\tau}}_{\text{cm}}^{\text{ext}} = \sum_{i=1}^{N} (\vec{\mathbf{r}}_{\text{cm},i} \times \vec{\mathbf{F}}_i) . \qquad (22.3.4)$$

is the sum of the torques on the individual particles in the center of mass reference frame. If we assume that all internal torques cancel in pairs, then

$$\vec{\boldsymbol{\tau}}_{\text{cm}}^{\text{ext}} = \sum_{i=1}^{N} (\vec{\mathbf{r}}_{\text{cm},i} \times \vec{\mathbf{F}}_i^{\text{ext}}) . \qquad (22.3.5)$$

We conclude that the external torque about the point S can be decomposed into two pieces,

$$\vec{\boldsymbol{\tau}}_S^{\text{ext}} = \vec{\boldsymbol{\tau}}_{S,cm}^{\text{ext}} + \vec{\boldsymbol{\tau}}_{\text{cm}}^{\text{ext}} . \qquad (22.3.6)$$

We showed in Chapter 20.3 that

$$\vec{\mathbf{L}}_S^{\text{sys}} = \vec{\mathbf{r}}_{S,\text{cm}} \times \vec{\mathbf{p}}^{\text{sys}} + \sum_{i=1}^{N} (\vec{\mathbf{r}}_{cm,i} \times m_i \vec{\mathbf{v}}_{cm,i}), \qquad (22.3.7)$$

where the first term in Eq. (22.3.7) is the orbital angular momentum of the center of mass about the point S

$$\vec{\mathbf{L}}_S^{\text{orbital}} = \vec{\mathbf{r}}_{S,\text{cm}} \times \vec{\mathbf{p}}^{\text{sys}} , \qquad (22.3.8)$$

and the second term in Eq. (22.3.7) is the spin angular momentum about the center of mass (independent of the point S)

$$\vec{\mathbf{L}}_S^{\text{spin}} = \sum_{i=1}^{N} (\vec{\mathbf{r}}_{cm,i} \times m_i \vec{\mathbf{v}}_{cm,i}) . \qquad (22.3.9)$$

The angular momentum about the point S can therefore be decomposed into two terms

$$\vec{\mathbf{L}}_S^{\text{sys}} = \vec{\mathbf{L}}_S^{\text{orbital}} + \vec{\mathbf{L}}_S^{\text{spin}} . \qquad (22.3.10)$$

Recall that that we have previously shown that it is always true that

$$\vec{\boldsymbol{\tau}}_S^{\text{ext}} = \frac{d\vec{\mathbf{L}}_S^{\text{sys}}}{dt} . \qquad (22.3.11)$$

Therefore we can therefore substitute Eq. (22.3.6) on the LHS of Eq. (22.3.11) and substitute Eq. (22.3.10) on the RHS of Eq. (22.3.11) yielding as

$$\vec{\boldsymbol{\tau}}^{\text{ext}}_{S,cm} + \vec{\boldsymbol{\tau}}^{\text{ext}}_{\text{cm}} = \frac{d\vec{\mathbf{L}}^{\text{orbital}}_{S}}{dt} + \frac{d\vec{\mathbf{L}}^{\text{spin}}_{S}}{dt}. \tag{22.3.12}$$

We shall now show that Eq. (22.3.12) can also be decomposed in two separate conditions. We begin by analyzing the first term on the RHS of Eq. (22.3.12). We differentiate Eq. (22.3.8) and find that

$$\frac{d\vec{\mathbf{L}}^{\text{orbital}}_{S}}{dt} = \frac{d}{dt}(\vec{\mathbf{r}}_{S,\text{cm}} \times \vec{\mathbf{p}}^{\text{sys}}). \tag{22.3.13}$$

We apply the vector identity

$$\frac{d}{dt}(\vec{\mathbf{A}} \times \vec{\mathbf{B}}) = \frac{d\vec{\mathbf{A}}}{dt} \times \vec{\mathbf{B}} + \vec{\mathbf{A}} \times \frac{d\vec{\mathbf{B}}}{dt}, \tag{22.3.14}$$

to Eq. (22.3.13) yielding

$$\frac{d\vec{\mathbf{L}}^{\text{orbital}}_{S}}{dt} = \frac{d\vec{\mathbf{r}}_{S,\text{cm}}}{dt} \times \vec{\mathbf{p}}_{\text{sys}} + \vec{\mathbf{r}}_{S,\text{cm}} \times \frac{d\vec{\mathbf{p}}_{\text{sys}}}{dt}. \tag{22.3.15}$$

The first term in Eq. (22.3.21) is zero because

$$\frac{d\vec{\mathbf{r}}_{S,\text{cm}}}{dt} \times \vec{\mathbf{p}}_{\text{sys}} = \vec{\mathbf{V}}_{cm} \times m^{\text{total}} \vec{\mathbf{V}}_{cm} = \vec{\mathbf{0}}. \tag{22.3.16}$$

Therefore the time derivative of the orbital angular momentum about a point S, Eq. (22.3.15), becomes

$$\frac{d\vec{\mathbf{L}}^{\text{orbital}}_{S}}{dt} = \vec{\mathbf{r}}_{S,\text{cm}} \times \frac{d\vec{\mathbf{p}}_{\text{sys}}}{dt}. \tag{22.3.17}$$

In Eq. (22.3.17), the time derivative of the momentum of the system is the external force,

$$\vec{\mathbf{F}}^{\text{ext}} = \frac{d\vec{\mathbf{p}}_{\text{sys}}}{dt}. \tag{22.3.18}$$

The expression in Eq. (22.3.17) then becomes the first of our relations

$$\frac{d\vec{\mathbf{L}}^{\text{orbital}}_{S}}{dt} = \vec{\mathbf{r}}_{S,\text{cm}} \times \vec{\mathbf{F}}^{\text{ext}} = \vec{\boldsymbol{\tau}}^{\text{ext}}_{S,cm}. \tag{22.3.19}$$

Thus the time derivative of the orbital angular momentum about the point S is equal to the external torque about the point S where all the external forces act at the center of mass, (we treat the system as a point-like particle located at the center of mass).

We now consider the second term on the RHS of Eq. (22.3.12), the time derivative of the spin angular momentum about the center of mass. We differentiate Eq. (22.3.9),

$$\frac{d\vec{\mathbf{L}}_S^{\text{spin}}}{dt} = \frac{d}{dt}\sum_{i=1}^{N}(\vec{\mathbf{r}}_{cm,i} \times m_i \vec{\mathbf{v}}_{cm,i}). \tag{22.3.20}$$

We again use the product rule for taking the time derivatives of a vector product (Eq. (22.3.14)). Then Eq. (22.3.20) the becomes

$$\frac{d\vec{\mathbf{L}}_S^{\text{spin}}}{dt} = \sum_{i=1}^{N}\left(\frac{d\vec{\mathbf{r}}_{cm,i}}{dt} \times m_i \vec{\mathbf{v}}_{cm,i}\right) + \sum_{i=1}^{N}\left(\vec{\mathbf{r}}_{cm,i} \times \frac{d}{dt}(m_i \vec{\mathbf{v}}_{cm,i})\right). \tag{22.3.21}$$

The first term in Eq. (22.3.21) is zero because

$$\sum_{i=1}^{N}\left(\frac{d\vec{\mathbf{r}}_{cm,i}}{dt} \times m_i \vec{\mathbf{v}}_{cm,i}\right) = \sum_{i=1}^{N}(\vec{\mathbf{v}}_{cm,i} \times m_i \vec{\mathbf{v}}_{cm,i}) = \vec{\mathbf{0}}. \tag{22.3.22}$$

Therefore the time derivative of the spin angular momentum about the center of mass, Eq. (22.3.21), becomes

$$\frac{d\vec{\mathbf{L}}_S^{\text{spin}}}{dt} = \sum_{i=1}^{N}\left(\vec{\mathbf{r}}_{cm,i} \times \frac{d}{dt}(m_i \vec{\mathbf{v}}_{cm,i})\right). \tag{22.3.23}$$

The force, acting on an element of mass m_i, is

$$\vec{\mathbf{F}}_i = \frac{d}{dt}(m_i \vec{\mathbf{v}}_{cm,i}). \tag{22.3.24}$$

The expression in Eq. (22.3.23) then becomes

$$\frac{d\vec{\mathbf{L}}_S^{\text{spin}}}{dt} = \sum_{i=1}^{N}(\vec{\mathbf{r}}_{cm,i} \times \vec{\mathbf{F}}_i). \tag{22.3.25}$$

The term, $\sum_{i=1}^{N}(\vec{\mathbf{r}}_{\text{cm},i} \times \vec{\mathbf{F}}_i)$, is the sum of the torques on the individual particles in the center of mass reference frame. If we again assume that all internal torques cancel in pairs, Eq. (22.3.25) may be expressed as

$$\frac{d\vec{\mathbf{L}}_S^{\text{spin}}}{dt} = \sum_{i=1}^{N}(\vec{\mathbf{r}}_{cm,i} \times \vec{\mathbf{F}}_i^{\text{ext}}) = \sum_{i=1}^{N}\vec{\boldsymbol{\tau}}_{\text{cm},i}^{\text{ext}} = \vec{\boldsymbol{\tau}}_{\text{cm}}^{\text{ext}}, \tag{22.3.26}$$

which is the second of our two relations.

21.3.1 Summary

For a system of particles, there are two conditions that always hold (Eqs. (22.3.19) and (22.3.26)) when we calculate the torque about a point S; we treat the system as a point-like particle located at the center of mass of the system. All the external forces $\vec{\mathbf{F}}^{\text{ext}}$ act at the center of mass. We calculate the orbital angular momentum of the center of mass and determine its time derivative and then apply

$$\vec{\boldsymbol{\tau}}_{S,\text{cm}}^{\text{ext}} = \vec{\mathbf{r}}_{S,\text{cm}} \times \vec{\mathbf{F}}^{\text{ext}} = \frac{d\vec{\mathbf{L}}_{S}^{\text{orbital}}}{dt}. \tag{22.3.27}$$

In addition, we calculate the torque about the center of mass due to all the forces acting on the particles in the center of mass reference frame. We calculate the time derivative of the angular momentum of the system with respect to the center of mass in the center of mass reference frame and then apply

$$\vec{\boldsymbol{\tau}}_{\text{cm}}^{\text{ext}} = \sum_{i=1}^{N}(\vec{\mathbf{r}}_{cm,i} \times \vec{\mathbf{F}}_{i}^{\text{ext}}) = \frac{d\vec{\mathbf{L}}_{\text{cm}}^{\text{spin}}}{dt}. \tag{22.3.28}$$

21.4 Translation and Rotation of a Rigid Body Undergoing Fixed Axis Rotation

For the special case of rigid body of mass m, we showed that with respect to a reference frame in which the center of mass of the rigid body is moving with velocity $\vec{\mathbf{V}}_{cm}$, all elements of the rigid body are rotating about the center of mass with the same angular velocity $\vec{\boldsymbol{\omega}}_{\text{cm}}$. For the rigid body of mass m and momentum $\vec{\mathbf{p}} = m\vec{\mathbf{V}}_{cm}$, the translational equation of motion is still given by Eq. (22.2.1), which we repeat in the form

$$\vec{\mathbf{F}}^{\text{ext}} = m\vec{\mathbf{A}}_{cm}. \tag{22.4.1}$$

For fixed axis rotation, choose the z-axis as the axis of rotation that passes through the center of mass of the rigid body. We have already seen in our discussion of angular momentum of a rigid body that the angular momentum does not necessary point in the same direction as the angular velocity. However we can take the z-component of Eq. (22.3.28)

$$\tau_{cm,z}^{\text{ext}} = \frac{dL_{cm,z}^{\text{spin}}}{dt}. \tag{22.4.2}$$

For a rigid body rotating about the center of mass with $\vec{\boldsymbol{\omega}}_{\text{cm}} = \omega_{\text{cm},z}\hat{\mathbf{k}}$, the z-component of angular momentum about the center of mass is

$$L^{\text{spin}}_{cm,z} = I_{\text{cm}} \omega_{\text{cm},z} \,. \quad (22.4.3)$$

The z-component of the rotational equation of motion about the center of mass is

$$\tau^{\text{ext}}_{cm,z} = I_{\text{cm}} \frac{d\omega_{\text{cm},z}}{dt} = I_{\text{cm}} \alpha_{\text{cm},z} \,. \quad (22.4.4)$$

21.5 Work-Energy Theorem

For a rigid body, we can also consider the work-energy theorem separately for the translational motion and the rotational motion. Once again treat the rigid body as a point-like particle moving with velocity $\vec{\mathbf{V}}_{cm}$ in reference frame O. We can use the same technique that we used when treating point particles to show that the work done by the external forces is equal to the change in kinetic energy

$$\begin{aligned} W^{\text{ext}}_{\text{trans}} &= \int_i^f \vec{\mathbf{F}}^{\text{ext}} \cdot d\vec{\mathbf{r}} = \int_i^f \frac{d(m\vec{\mathbf{V}}_{cm})}{dt} \cdot d\vec{\mathbf{R}}_{cm} = m\int_i^f \frac{d(\vec{\mathbf{V}}_{cm})}{dt} \cdot \vec{\mathbf{V}}_{cm}\, dt \\ &= \frac{1}{2} m \int_i^f d(\vec{\mathbf{V}}_{cm} \cdot \vec{\mathbf{V}}_{cm}) = \frac{1}{2} m V^2_{\text{cm},\,f} - \frac{1}{2} m V^2_{\text{cm},\,i} = \Delta K_{\text{trans}} . \end{aligned} \quad (22.5.1)$$

For the rotational motion we go to the center of mass reference frame and we determine the rotational work done i.e. the integral of the z-component of the torque about the center of mass with respect to $d\theta$ as we did for fixed axis rotational work. Then

$$\begin{aligned} \int_i^f (\vec{\boldsymbol{\tau}}^{\text{ext}}_{\text{cm}})_z \, d\theta &= \int_i^f I_{\text{cm}} \frac{d\omega_{\text{cm},z}}{dt} d\theta = \int_i^f I_{\text{cm}}\, d\omega_{\text{cm},z} \frac{d\theta}{dt} = \int_i^f I_{\text{cm}}\, d\omega_{\text{cm},z} \omega_{\text{cm},z} \\ &= \frac{1}{2} I_{\text{cm}} \omega^2_{\text{cm},\,f} - \frac{1}{2} I_{\text{cm}} \omega^2_{\text{cm},\,i} = \Delta K_{\text{rot}} \end{aligned} \,. \quad (22.5.2)$$

In Eq. (22.5.2) we expressed our result in terms of the angular speed ω_{cm} because it appears as a square. Therefore we can combine these two separate results, Eqs. (22.5.1) and (22.5.2), and determine the work-energy theorem for a rotating and translating rigid body that undergoes fixed axis rotation about the center of mass.

$$\begin{aligned} W &= \left(\frac{1}{2} m V^2_{\text{cm,f}} + \frac{1}{2} I_{\text{cm}} \omega^2_{\text{cm},f}\right) - \left(\frac{1}{2} m V^2_{\text{cm,f}} + \frac{1}{2} I_{\text{cm}} \omega^2_{\text{cm},i}\right) \\ &= \Delta K_{\text{trans}} + \Delta K_{\text{rot}} = \Delta K. \end{aligned} \quad (22.5.3)$$

Equations (22.4.1), (22.4.4), and (22.5.3) are principles that we shall employ to analyze the motion of a rigid bodies undergoing translation and fixed axis rotation about the center of mass.

21.6 Worked Examples

Example 21.1 Angular Impulse

Two point-like objects are located at the points A and B, of respective masses $M_A = 2M$, and $M_B = M$, as shown in the figure below. The two objects are initially oriented along the y-axis and connected by a rod of negligible mass of length D, forming a rigid body. A force of magnitude $F = |\vec{\mathbf{F}}|$ along the x direction is applied to the object at B at $t = 0$ for a short time interval Δt, (Figure 21.3). Neglect gravity. Give all your answers in terms of M and D as needed. (a) What is the magnitude of the angular velocity of the system after the collision?

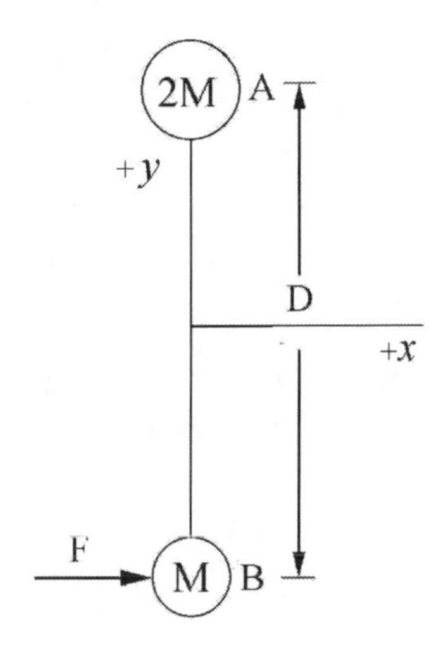

Figure 21.3 Example 21.1

Solutions: An impulse of magnitude $F\Delta t$ is applied in the $+x$ direction, and the center of mass of the system will move in this direction. The two masses will rotate about the center of mass, counterclockwise in the figure. Before the force is applied we can calculate the position of the center of mass (Figure 21.4a),

$$\vec{\mathbf{R}}_{\text{cm}} = \frac{M_A\vec{\mathbf{r}}_A + M_B\vec{\mathbf{r}}_B}{M_A + M_B} = \frac{2M(D/2)\hat{\mathbf{j}} + M(D/2)(-\hat{\mathbf{j}})}{3M} = (D/6)\hat{\mathbf{j}}. \qquad (22.6.1)$$

The center of mass is a distance (2/3)D from the object at B and is a distance (1/3)D from the object at A.

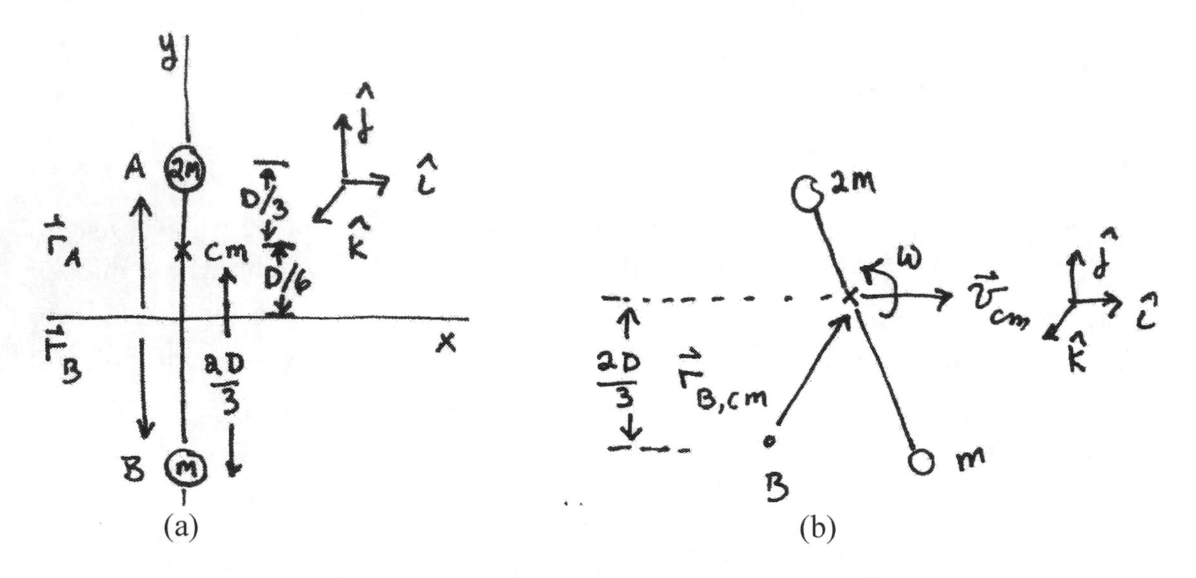

Figure 21.4 (a) Center of mass of system, (b) Angular momentum about point B

Because $F\Delta t\hat{\mathbf{i}} = 3M\vec{\mathbf{V}}_{\text{cm}}$, the magnitude of the velocity of the center of mass is then $F\Delta t / 3\text{M}$ and the direction is in the positive $\hat{\mathbf{i}}$-direction. Because the force is applied at the point B, there is no torque about the point B, hence the angular momentum is constant about the point B. The initial angular momentum about the point B is zero. The angular momentum about the point B (Figure 21.4b) after the impulse is applied is the sum of two terms,

$$\begin{aligned}\vec{\mathbf{0}} &= \vec{\mathbf{L}}_{B,f} = \vec{\mathbf{r}}_{B,f} \times 3M\vec{\mathbf{V}}_{\text{cm}} + \vec{\mathbf{L}}_{\text{cm}} = (2D/3)\hat{\mathbf{j}} \times F\Delta t\,\hat{\mathbf{i}} + \vec{\mathbf{L}}_{\text{cm}} \\ \vec{\mathbf{0}} &= (2DF\Delta t/3)(-\hat{\mathbf{k}}) + \vec{\mathbf{L}}_{\text{cm}}.\end{aligned} \qquad (22.6.2)$$

The angular momentum about the center of mass is given by

$$.\vec{\mathbf{L}}_{cm} = I_{cm}\omega\,\hat{\mathbf{k}} = (2M(D/3)^2 + M(2D/3)^2)\omega\,\hat{\mathbf{k}} = (2/3)MD^2\omega\,\hat{\mathbf{k}}. \qquad (22.6.3)$$

Thus the angular about the point B after the impulse is applied is

$$\vec{\mathbf{0}} = (2DF\Delta t/3)(-\hat{\mathbf{k}}) + (2/3)MD^2\omega\,\hat{\mathbf{k}} \qquad (22.6.4)$$

We can solve this Eq. (22.6.4) for the angular speed

$$\omega = \frac{F\Delta t}{MD}. \qquad (22.6.5)$$

Example 21.2 Person on a railroad car moving in a circle

A person of mass M is standing on a railroad car, which is rounding an unbanked turn of radius R at a speed v. His center of mass is at a height of L above the car midway between his feet, which are separated by a distance of d. The man is facing the direction of motion (Figure 21.5). What is the magnitude of the normal force on each foot?

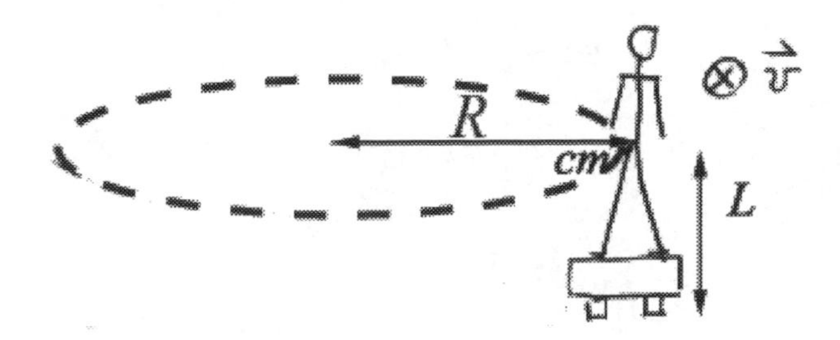

Figure 21.5 Example 21.2

Solution: We begin by choosing a cylindrical coordinate system and drawing a free-body force diagram, shown in Figure 21.6.

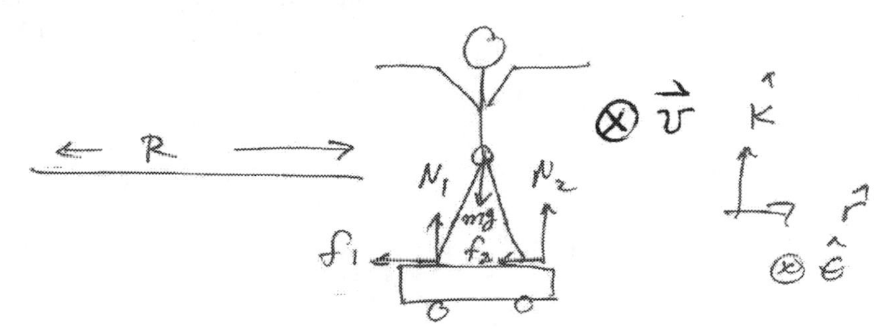

Figure 21.6 Coordinate system for Example 21.2

We decompose the contact force between the foot closest to the center of the circular motion and the ground into a tangential component corresponding to static friction $\vec{\mathbf{f}}_1$ and a perpendicular component, $\vec{\mathbf{N}}_1$. In a similar fashion we decompose the contact force between the foot furthest from the center of the circular motion and the ground into a tangential component corresponding to static friction $\vec{\mathbf{f}}_2$ and a perpendicular component, $\vec{\mathbf{N}}_2$. We do not assume that the static friction has its maximum magnitude nor do we assume that $\vec{\mathbf{f}}_1 = \vec{\mathbf{f}}_2$ or $\vec{\mathbf{N}}_1 = \vec{\mathbf{N}}_2$. The gravitational force acts at the center of mass.

We shall use our two dynamical equations of motion, Eq. (22.4.1) for translational motion and Eq. (22.4.4) for rotational motion about the center of mass noting that we are considering the special case that $\vec{\boldsymbol{\alpha}}_{cm} = 0$ because the object is not rotating about the center of mass. In order to apply Eq. (22.4.1), we treat the person as a point-like particle located at the center of mass and all the external forces act at this point. The radial component of Newton's Second Law (Eq. (22.4.1) is given by

$$\hat{\mathbf{r}} : -f_1 - f_2 = -m\frac{v^2}{R}. \qquad (22.6.6)$$

The vertical component of Newton's Second Law is given by

$$\hat{\mathbf{k}}: N_1 + N_2 - mg = 0. \tag{22.6.7}$$

The rotational equation of motion (Eq. (22.4.4) is

$$\vec{\boldsymbol{\tau}}_{\text{cm}}^{\text{total}} = 0. \tag{22.6.8}$$

We begin our calculation of the torques about the center of mass by noting that the gravitational force does not contribute to the torque because it is acting at the center of mass. We draw a torque diagram in Figure 21.7a showing the location of the point of application of the forces, the point we are computing the torque about (which in this case is the center of mass), and the vector $\vec{\mathbf{r}}_{\text{cm},1}$ from the point we are computing the torque about to the point of application of the forces.

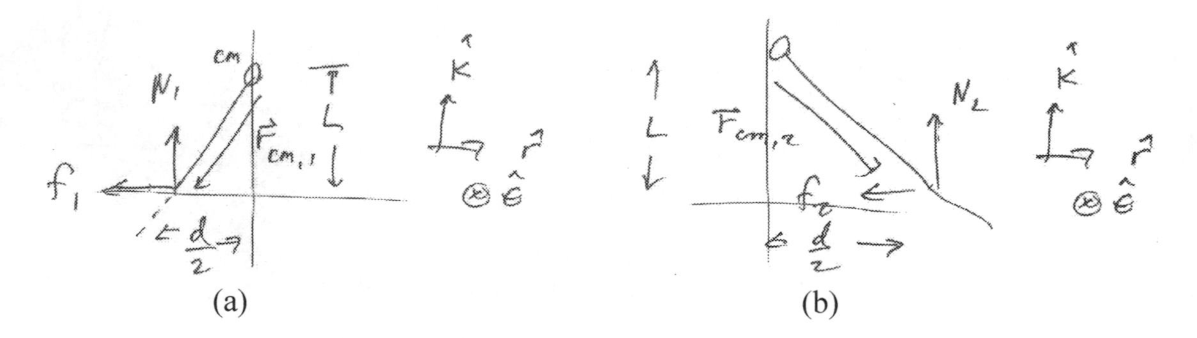

Figure 21.7 Torque diagram for (a) inner foot, (b) outer foot

The torque on the inner foot is given by

$$\vec{\boldsymbol{\tau}}_{\text{cm},1} = \vec{\mathbf{r}}_{\text{cm},1} \times (\vec{\mathbf{f}}_1 + \vec{\mathbf{N}}_1) = \left(-\frac{d}{2}\hat{\mathbf{r}} - L\hat{\mathbf{k}}\right) \times (-f_1\hat{\mathbf{r}} + N_1\hat{\mathbf{k}}) = \left(\frac{d}{2}N_1 + Lf_1\right)\hat{\boldsymbol{\theta}}. \tag{22.6.9}$$

We draw a similar torque diagram (Figure 21.7b) for the forces applied to the outer foot. The torque on the outer foot is given by

$$\vec{\boldsymbol{\tau}}_{\text{cm},2} = \vec{\mathbf{r}}_{\text{cm},2} \times (\vec{\mathbf{f}}_2 + \vec{\mathbf{N}}_2) = \left(+\frac{d}{2}\hat{\mathbf{r}} - L\hat{\mathbf{k}}\right) \times (-f_2\hat{\mathbf{r}} + N_2\hat{\mathbf{k}}) = \left(-\frac{d}{2}N_2 + Lf_2\right)\hat{\boldsymbol{\theta}}. \tag{22.6.10}$$

Notice that the forces $\vec{\mathbf{f}}_1$, $\vec{\mathbf{N}}_1$, and $\vec{\mathbf{f}}_2$ all contribute torques about the center of mass in the positive $\hat{\boldsymbol{\theta}}$-direction while $\vec{\mathbf{N}}_2$ contributes a torque about the center of mass in the negative $\hat{\boldsymbol{\theta}}$-direction. According to Eq. (22.6.8) the sum of these torques about the center of mass must be zero. Therefore

$$\vec{\boldsymbol{\tau}}_{\text{cm}}^{\text{ext}} = \vec{\boldsymbol{\tau}}_{\text{cm},1} + \vec{\boldsymbol{\tau}}_{\text{cm},2} = \left(\frac{d}{2}N_1 + Lf_1\right)\hat{\boldsymbol{\theta}} + \left(-\frac{d}{2}N_2 + Lf_2\right)\hat{\boldsymbol{\theta}}$$
$$= \left(\frac{d}{2}(N_1 - N_2) + L(f_1 + f_2)\right)\hat{\boldsymbol{\theta}} = \vec{\mathbf{0}}. \qquad (22.6.11)$$

Notice that the magnitudes of the two frictional forces appear together as a sum in Eqs. (22.6.11) and (22.6.6). We now can solve Eq. (22.6.6) for $f_1 + f_2$ and substitute the result into Eq. (22.6.11) yielding the condition that

$$\frac{d}{2}(N_1 - N_2) + Lm\frac{v^2}{R} = 0. \qquad (22.6.12)$$

We can rewrite this Eq. as

$$N_2 - N_1 = \frac{2Lmv^2}{dR}. \qquad (22.6.13)$$

We also rewrite Eq. (22.6.7) in the form

$$N_2 + N_1 = mg. \qquad (22.6.14)$$

We now can solve for N_2 by adding together Eqs. (22.6.13) and (22.6.14), and then divide by two,

$$N_2 = \frac{1}{2}\left(\frac{2Lmv^2}{dR} + mg\right). \qquad (22.6.15)$$

We now can solve for N_1 by subtracting Eqs. (22.6.13) from (22.6.14), and then divide by two,

$$N_1 = \frac{1}{2}\left(mg - \frac{2Lmv^2}{dR}\right). \qquad (22.6.16)$$

Check the result: we see that the normal force acting on the outer foot is greater in magnitude than the normal force acting on the inner foot. We expect this result because as we increase the speed v, we find that at a maximum speed v_{max}, the normal force on the inner foot goes to zero and we start to rotate in the positive $\hat{\boldsymbol{\theta}}$-direction, tipping outward. We can find this maximum speed by setting $N_1 = 0$ in Eq. (22.6.16) resulting in

$$v_{\text{max}} = \sqrt{\frac{gdR}{2L}}. \qquad (22.6.17)$$

Example 21.3 Torque, Rotation and Translation: Yo-Yo

A Yo-Yo of mass m has an axle of radius b and a spool of radius R. Its moment of inertia about the center can be taken to be $I_{cm} = (1/2)mR^2$ and the thickness of the string can be neglected (Figure 21.8). The Yo-Yo is released from rest. You will need to assume that the center of mass of the Yo-Yo descends vertically, and that the string is vertical as it unwinds. (a) What is the tension in the cord as the Yo-Yo descends? (b) What is the magnitude of the angular acceleration as the yo-yo descends and the magnitude of the linear acceleration? (c) Find the magnitude of the angular velocity of the Yo-Yo when it reaches the bottom of the string, when a length l of the string has unwound.

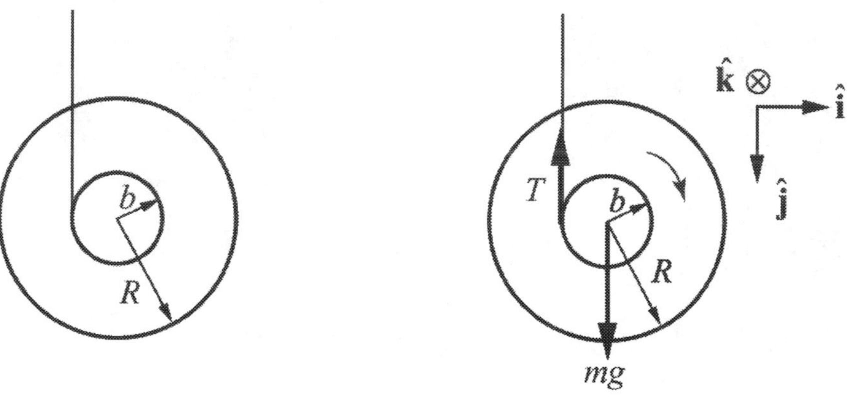

Figure 21.8 Example 21.3 **Figure 21.9** Torque diagram for Yo-Yo

Solutions: a) as the Yo-Yo descends it rotates clockwise in Figure 21.9. The torque about the center of mass of the Yo-Yo is due to the tension and increases the magnitude of the angular velocity. The direction of the torque is into the page in Figure 21.9 (positive z-direction). Use the right-hand rule to check this, or use the vector product definition of torque,

$$\vec{\tau}_{\text{cm}} = \vec{\mathbf{r}}_{\text{cm},T} \times \vec{\mathbf{T}}. \tag{22.6.18}$$

About the center of mass, $\vec{\mathbf{r}}_{\text{cm},T} = -b\,\hat{\mathbf{i}}$ and $\vec{\mathbf{T}} = -T\,\hat{\mathbf{j}}$, so the torque is

$$\vec{\tau}_{\text{cm}} = \vec{\mathbf{r}}_{\text{cm},T} \times \vec{\mathbf{T}} = (-b\,\hat{\mathbf{i}}) \times (-T\,\hat{\mathbf{j}}) = bT\,\hat{\mathbf{k}}. \tag{22.6.19}$$

Apply Newton's Second Law in the $\hat{\mathbf{j}}$-direction,

$$mg - T = ma_y. \tag{22.6.20}$$

Apply the rotational equation of motion for the Yo-Yo,

$$bT = I_{\text{cm}}\alpha_z, \tag{22.6.21}$$

where α_z is the z-component of the angular acceleration. The z-component of the angular acceleration and the y-component of the linear acceleration are related by the constraint condition

$$a_y = b\alpha_z, \tag{22.6.22}$$

where b is the axle radius of the Yo-Yo. Substitute Eq. (22.6.22) into (22.6.20) yielding

$$mg - T = mb\alpha_z. \tag{22.6.23}$$

Now solve Eq. (22.6.21) for α_z and substitute the result into Eq.(22.6.23),

$$mg - T = \frac{mb^2 T}{I_{\text{cm}}}. \tag{22.6.24}$$

Solve Eq. (22.6.24) for the tension T,

$$T = \frac{mg}{\left(1 + \dfrac{mb^2}{I_{\text{cm}}}\right)} = \frac{mg}{\left(1 + \dfrac{mb^2}{(1/2)mR^2}\right)} = \frac{mg}{\left(1 + \dfrac{2b^2}{R^2}\right)}. \tag{22.6.25}$$

b) Substitute Eq. (22.6.25) into Eq. (22.6.21) to determine the z-component of the angular acceleration,

$$\alpha_z = \frac{bT}{I_{\text{cm}}} = \frac{2bg}{(R^2 + 2b^2)}. \tag{22.6.26}$$

Using the constraint condition Eq. (22.6.22), we determine the y-component of linear acceleration

$$a_y = b\alpha_z = \frac{2b^2 g}{(R^2 + 2b^2)} = \frac{g}{1 + R^2/2b^2}. \tag{22.6.27}$$

Note that both quantities $a_z > 0$ and $\alpha_z > 0$, so Eqs. (22.6.26) and (22.6.27) are the magnitudes of the respective quantities. For a typical Yo-Yo, the acceleration is much less than that of an object in free fall.

c) Use conservation of energy to determine the magnitude of the angular velocity of the Yo-Yo when it reaches the bottom of the string. As in Figure 21.9, choose the downward vertical direction as the positive $\hat{\mathbf{j}}$-direction and let $y = 0$ designate the location of the center of mass of the Yo-Yo when the string is completely wound. Choose $U(y=0)=0$ for the zero reference potential energy. This choice of direction and reference means that the gravitational potential energy will be negative and decreasing while the Yo-Yo descends. For this case, the gravitational potential energy is

$$U = -mg\,y\,. \qquad (22.6.28)$$

The Yo-Yo is not yet moving downward or rotating, and the center of mass is located at $y=0$ so the mechanical energy in the initial state, when the Yo-Yo is completely wound, is zero

$$E_i = U(y=0) = 0\,. \qquad (22.6.29)$$

Denote the linear speed of the Yo-Yo as v_f and its angular speed as ω_f (at the point $y=l$). The constraint condition between v_f and ω_f is given by

$$v_f = b\omega_f\,, \qquad (22.6.30)$$

consistent with Eq. (22.6.22). The kinetic energy is the sum of translational and rotational kinetic energy, where we have used $I_{\text{cm}} = (1/2)mR^2$, and so mechanical energy in the final state, when the Yo-Yo is completely unwound, is

$$\begin{aligned} E_f = K_f + U_f &= \frac{1}{2}mv_f^2 + \frac{1}{2}I_{\text{cm}}\omega_f^2 - mgl \\ &= \frac{1}{2}mb^2\omega_f^2 + \frac{1}{4}mR^2\omega_f^2 - mgl. \end{aligned} \qquad (22.6.31)$$

There are no external forces doing work on the system (neglect air resistance), so

$$0 = E_f = E_i\,. \qquad (22.6.32)$$

Thus

$$\left(\frac{1}{2}mb^2 + \frac{1}{4}mR^2\right)\omega_f^2 = mgl\,. \qquad (22.6.33)$$

Solving for ω_f,

$$\omega_f = \sqrt{\frac{4gl}{(2b^2 + R^2)}}\,. \qquad (22.6.34)$$

We may also use kinematics to determine the final angular velocity by solving for the time interval Δt that it takes for the Yo-Yo to travel a distance l at the constant acceleration found in Eq. (22.6.27)),

$$\Delta t = \sqrt{2l / a_y} = \sqrt{\frac{l(R^2 + 2b^2)}{b^2 g}} \qquad (22.6.35)$$

The final angular velocity of the Yo-Yo is then (using Eq. (22.6.26) for the z-component of the angular acceleration),

$$\omega_f = \alpha_z \Delta t = \sqrt{\frac{4gl}{(R^2 + 2b^2)}}, \qquad (22.6.36)$$

in agreement with Eq. (22.6.34).

Example 21.4 Cylinder Rolling Down Inclined Plane

A uniform cylinder of outer radius R and mass M with moment of inertia about the center of mass, $I_{\text{cm}} = (1/2)M R^2$, starts from rest and rolls without slipping down an incline tilted at an angle β from the horizontal. The center of mass of the cylinder has dropped a vertical distance h when it reaches the bottom of the incline Figure 21.10. Let g denote the gravitational constant. The coefficient of static friction between the cylinder and the surface is μ_s. What is the magnitude of the velocity of the center of mass of the cylinder when it reaches the bottom of the incline?

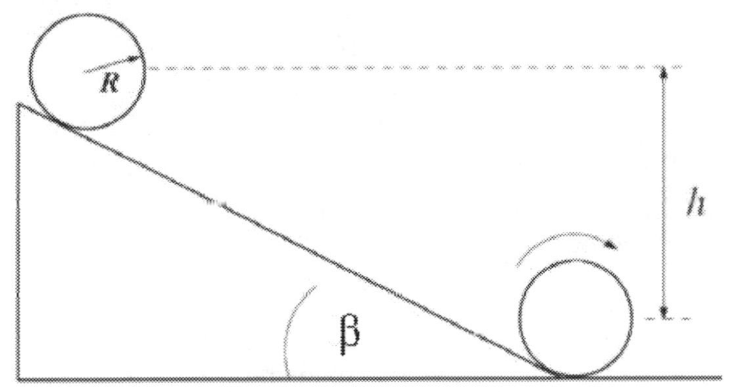

Figure 21.10 Example 21.4

Solution: We shall solve this problem three different ways.

1. Apply the torque condition about the center of mass and the force law for the center of mass motion.
2. Apply the energy methods.
3. Use torque about a fixed point that lies along the line of contact between the cylinder and the surface,

First Approach: Rotation about center of mass and translation of center of mass

We shall apply the torque condition (Eq. (22.4.4)) about the center of mass and the force law (Eq. (22.4.1)) for the center of mass motion. We will first find the acceleration and hence the speed at the bottom of the incline using kinematics. The forces are shown in Figure 21.11.

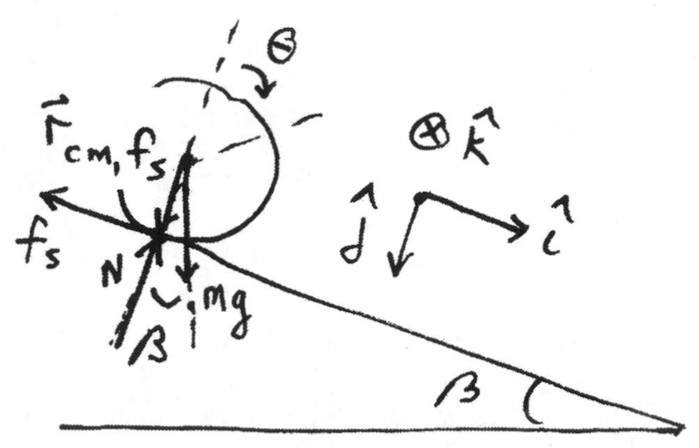

Figure 21.11 Torque diagram about center of mass

Choose $x=0$ at the point where the cylinder just starts to roll. Newton's Second Law, applied in the x- and y-directions in turn, yields

$$Mg\sin\beta - f_s = Ma_x, \tag{22.6.37}$$
$$-N + Mg\cos\beta = 0. \tag{22.6.38}$$

Choose the center of the cylinder to compute the torque about (Figure 21.10). Then, the only force exerting a torque about the center of mass is the friction force, therefore the rotational equation of motion is

$$f_s R = I_{\text{cm}}\alpha_z. \tag{22.6.39}$$

Use $I_{\text{cm}} = (1/2)MR^2$ and the kinematic constraint for the no-slipping condition $\alpha_z = a_x / R$ in Eq. (22.6.39) to solve for the magnitude of the static friction force yielding

$$f_s = (1/2)Ma_x. \tag{22.6.40}$$

Substituting Eq. (22.6.40) into Eq. (22.6.37) yields

$$Mg\sin\theta - (1/2)Ma_x = Ma_x, \tag{22.6.41}$$

which we can solve for the acceleration

$$a_x = \frac{2}{3}g\sin\beta. \tag{22.6.42}$$

In the time t_f it takes to reach the bottom, the displacement of the cylinder is $x_f = h/\sin\beta$. The x-component of the velocity v_x at the bottom is $v_{x,f} = a_x t_f$. Thus $x_f = (1/2)a_x t_f^{\,2}$. After eliminating t_f, we have $x_f = v_{x,f}^{\,2}/2a_x$, so the magnitude of the velocity when the cylinder reaches the bottom of the inclined plane is

$$v_{x,f} = \sqrt{2a_x x_f} = \sqrt{2((2/3)g\sin\beta)(h/\sin\beta)} = \sqrt{(4/3)gh}. \qquad (22.6.43)$$

Note that if we substitute Eq. (22.6.42) into Eq. (22.6.40) the magnitude of the frictional force is

$$f_s = (1/3)Mg\sin\beta. \qquad (22.6.44)$$

In order for the cylinder to roll without slipping

$$f_s \le \mu_s Mg\cos\beta. \qquad (22.6.45)$$

Combining Eq. (22.6.44) and Eq. (22.6.45) we have the condition that

$$(1/3)Mg\sin\beta \le \mu_s Mg\cos\beta \qquad (22.6.46)$$

Thus in order to roll without slipping, the coefficient of static friction must satisfy

$$\mu_s \ge \frac{1}{3}\tan\beta. \qquad (22.6.47)$$

Second Approach: Energy Methods

We shall use the fact that the energy of the cylinder-earth system is constant since the static friction force does no work.

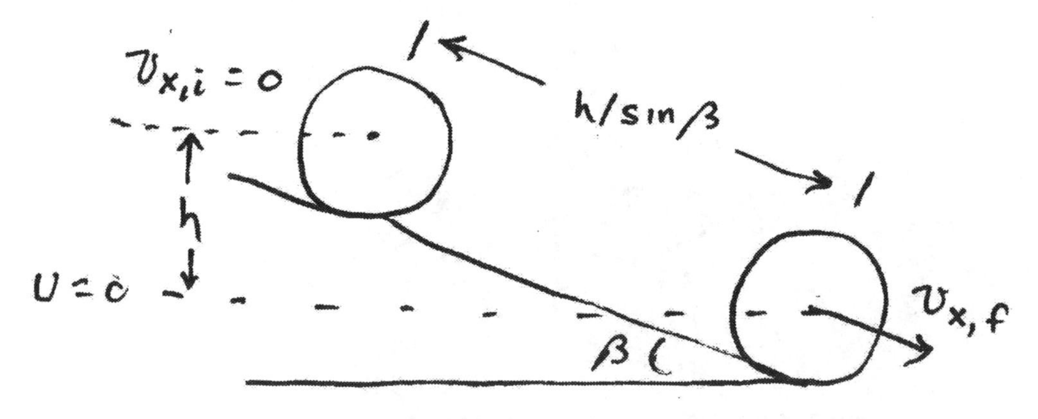

Figure 21.12 Energy diagram for cylinder

Choose a zero reference point for potential energy at the center of mass when the cylinder reaches the bottom of the incline plane (Figure 21.12). Then the initial potential energy is

$$U_i = Mgh. \tag{22.6.48}$$

For the given moment of inertia, the final kinetic energy is

$$\begin{aligned} K_f &= \frac{1}{2} M v_{x,f}{}^2 + \frac{1}{2} I_{cm} \omega_{z,f}{}^2 \\ &= \frac{1}{2} M v_{x,f}{}^2 + \frac{1}{2}(1/2) MR^2 (v_{x,f} / R)^2 \\ &= \frac{3}{4} M v_{x,f}{}^2. \end{aligned} \tag{22.6.49}$$

Setting the final kinetic energy equal to the initial gravitational potential energy leads to

$$Mgh = \frac{3}{4} M v_{x,f}{}^2. \tag{22.6.50}$$

The magnitude of the velocity of the center of mass of the cylinder when it reaches the bottom of the incline is

$$v_{x,f} = \sqrt{(4/3)gh}, \tag{22.6.51}$$

in agreement with Eq. (22.6.43).

Third Approach: Torque about a fixed point that lies along the line of contact between the cylinder and the surface

Choose a fixed point P that lies along the line of contact between the cylinder and the surface. Then the torque diagram is shown in Figure 21.13.

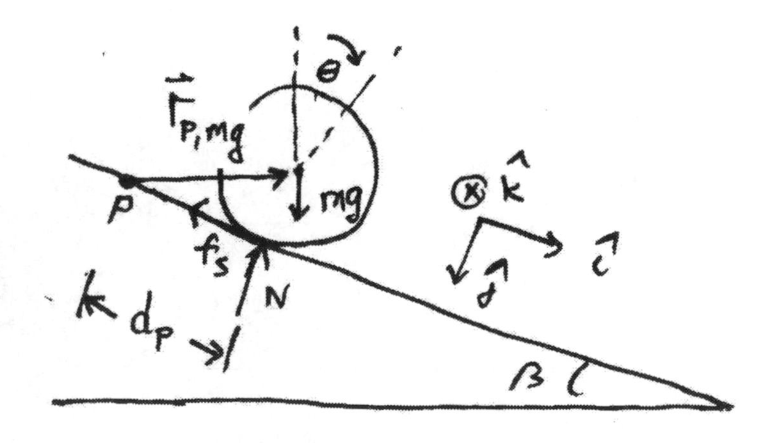

Figure 21.13 Torque about a point along the line of contact

The gravitational force $M\vec{\mathbf{g}} = Mg\sin\beta\,\hat{\mathbf{i}} + Mg\cos\beta\,\hat{\mathbf{j}}$ acts at the center of mass. The vector from the point P to the center of mass is given by $\vec{\mathbf{r}}_{P,mg} = d_p\,\hat{\mathbf{i}} - R\,\hat{\mathbf{j}}$, so the torque due to the gravitational force about the point P is given by

$$\begin{aligned}\vec{\boldsymbol{\tau}}_{P,Mg} &= \vec{\mathbf{r}}_{P,Mg} \times M\vec{\mathbf{g}} = (d_p\,\hat{\mathbf{i}} - R\,\hat{\mathbf{j}}) \times (Mg\sin\beta\,\hat{\mathbf{i}} + Mg\cos\beta\,\hat{\mathbf{j}}) \\ &= (d_p Mg\cos\beta + RMg\sin\beta)\hat{\mathbf{k}}.\end{aligned} \tag{22.6.52}$$

The normal force acts at the point of contact between the cylinder and the surface and is given by $\vec{\mathbf{N}} = -N\,\hat{\mathbf{j}}$. The vector from the point P to the point of contact between the cylinder and the surface is $\vec{\mathbf{r}}_{P,N} = d_p\,\hat{\mathbf{i}}$. Therefore the torque due to the normal force about the point P is given by

$$\vec{\boldsymbol{\tau}}_{P,N} = \vec{\mathbf{r}}_{P,N} \times \vec{\mathbf{N}} = (d_p\,\hat{\mathbf{i}}) \times (-N\,\hat{\mathbf{j}}) = -d_p N\,\hat{\mathbf{k}}. \tag{22.6.53}$$

Substituting Eq. (22.6.38) for the normal force into Eq. (22.6.53) yields

$$\vec{\boldsymbol{\tau}}_{P,N} = -d_p Mg\cos\beta\hat{\mathbf{k}}. \tag{22.6.54}$$

Therefore the sum of the torques about the point P is

$$\vec{\boldsymbol{\tau}}_P = \vec{\boldsymbol{\tau}}_{P,Mg} + \vec{\boldsymbol{\tau}}_{P,N} = (d_p Mg\cos\beta + RMg\sin\beta)\hat{\mathbf{k}} - d_p Mg\cos\beta\hat{\mathbf{k}} = Rmg\sin\beta\hat{\mathbf{k}}. \tag{22.6.55}$$

The angular momentum about the point P is given by

$$\begin{aligned}\vec{\mathbf{L}}_P &= \vec{\mathbf{L}}_{\text{cm}} + \vec{\mathbf{r}}_{P,\text{cm}} \times M\vec{\mathbf{V}}_{cm} = I_{\text{cm}}\omega_z\hat{\mathbf{k}} + (d_P\,\hat{\mathbf{i}} - R\,\hat{\mathbf{j}}) \times (Mv_x)\,\hat{\mathbf{i}} \\ &= (I_{\text{cm}}\omega_z + RMv_x)\,\hat{\mathbf{k}}\end{aligned}. \tag{22.6.56}$$

The time derivative of the angular momentum about the point P is then

$$\frac{d\vec{\mathbf{L}}_P}{dt} = (I_{\text{cm}}\alpha_z + RMa_x)\,\hat{\mathbf{k}}. \tag{22.6.57}$$

Therefore the torque law about the point P, becomes

$$RMg\sin\beta\hat{\mathbf{k}} = (I_{\text{cm}}\alpha_z + RMa_x)\hat{\mathbf{k}}. \tag{22.6.58}$$

Using the fact that $I_{\text{cm}} = (1/2)MR^2$ and $\alpha_x = a_x / R$, the z-component of Eq. (22.6.58) is then

$$RMg\sin\beta = (1/2)MRa_x + Rma_x = (3/2)MRa_x\,. \qquad (22.6.59)$$

We can now solve Eq. (22.6.59) for the x-component of the acceleration

$$a_x = (2/3)g\sin\beta\,, \qquad (22.6.60)$$

in agreement with Eq. (22.6.42).

Example 21.5 Bowling Ball

A bowling ball of mass m and radius R is initially thrown down an alley with an initial speed v_i, and it slides without rolling but due to friction it begins to roll (Figure 21.14). The moment of inertia of the ball about its center of mass is $I_{\text{cm}} = (2/5)mR^2$. Using conservation of angular momentum about a point (you need to find that point), find the speed v_f and the angular speed ω_f of the bowling ball when it just starts to roll without slipping?

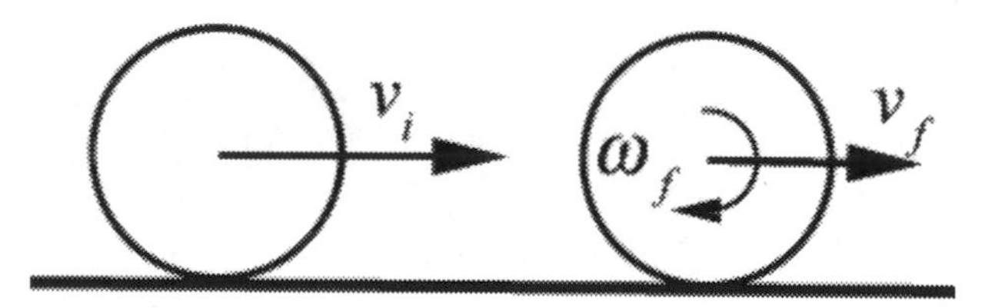

Figure 21.14 Example 21.5

Solution: We begin introducing coordinates for the angular and linear motion. Choose an angular coordinate θ increasing in the clockwise direction. Choose the positive $\hat{\mathbf{k}}$ unit vector pointing into the page in Figure 21.15.

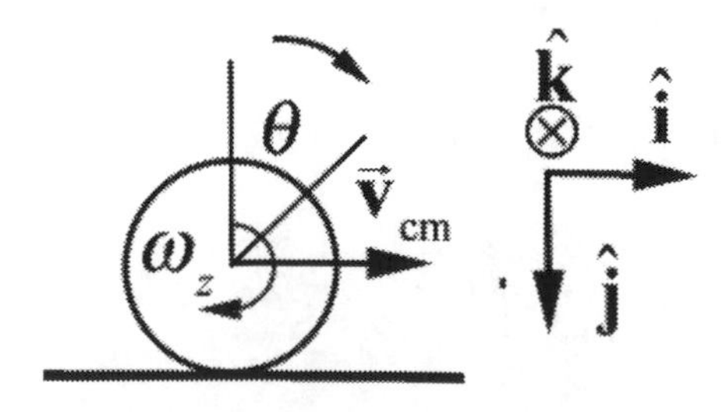

Figure 21.15 Coordinate system for ball

Then the angular velocity vector is $\vec{\boldsymbol{\omega}} = \omega_z\hat{\mathbf{k}} = d\theta/dt\,\hat{\mathbf{k}}$, and the angular acceleration vector is $\vec{\boldsymbol{\alpha}} = \alpha_z\hat{\mathbf{k}} = d^2\theta/dt^2\,\hat{\mathbf{k}}$. Choose the positive $\hat{\mathbf{i}}$ unit vector pointing to the right in Figure 21.15. Then the velocity of the center of mass is given by $\vec{\mathbf{v}}_{\text{cm}} = v_{\text{cm},x}\hat{\mathbf{i}} = dx_{\text{cm}}/dt\,\hat{\mathbf{i}}$, and the acceleration of the center of mass is given by $\vec{\mathbf{a}}_{\mathrm{cm}} = a_{\mathrm{cm},x}\hat{\mathbf{i}} = d^2x_{\mathrm{cm}} / dt^2\,\hat{\mathbf{i}}$. The free-body force diagram is shown in Figure 21.16.

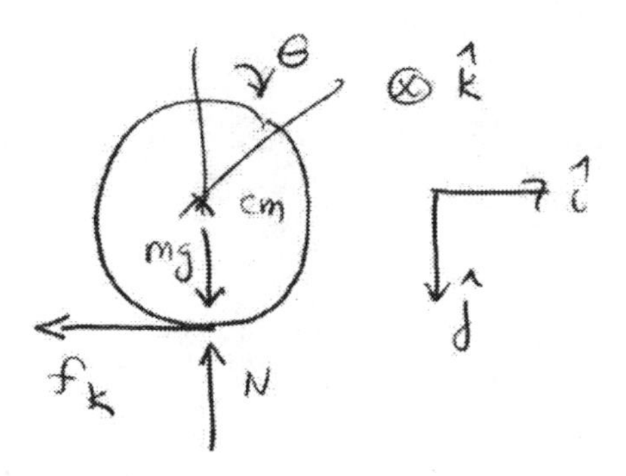

Figure 21.16 Free-body force diagram for ball

At $t = 0$, when the ball is released, $\vec{\mathbf{v}}_{cm,0} = v_0\hat{\mathbf{i}}$ and $\vec{\boldsymbol{\omega}}_0 = \vec{\mathbf{0}}$, so the ball is skidding and hence the frictional force on the ball due to the sliding of the ball on the surface is kinetic friction, hence acts in the negative $\hat{\mathbf{i}}$-direction. Because there is kinetic friction and non-conservative work, mechanical energy is not constant. The rotational equation of motion is $\vec{\boldsymbol{\tau}}_S = d\vec{\mathbf{L}}_S / dt$. In order for angular momentum about some point to remain constant throughout the motion, the torque about that point must also be zero throughout the motion. As the ball moves down the alley, the contact point will move, but the frictional force will always be directed along the line of contact between the bowling bowl and the surface. Choose any fixed point S along the line of contact then

$$\vec{\boldsymbol{\tau}}_{S,f_k} = \vec{\mathbf{r}}_{S,f_k} \times \vec{\mathbf{f}}_k = \vec{\mathbf{0}} \qquad (22.6.61)$$

because $\vec{\mathbf{r}}_{S,f_k}$ and $\vec{\mathbf{f}}_k$ are anti-parallel. The gravitation force acts at the center of mass hence the torque due to gravity about S is

$$\vec{\boldsymbol{\tau}}_{S,mg} = \vec{\mathbf{r}}_{S,mg} \times m\vec{\mathbf{g}} = dmg\hat{\mathbf{k}}, \qquad (22.6.62)$$

where d is the distance from S to the contact point between the ball and the ground. The torque due to the normal force about S is

$$\vec{\boldsymbol{\tau}}_{S,N} = \vec{\mathbf{r}}_{S,N} \times m\vec{\mathbf{g}} = -dN\hat{\mathbf{k}}, \qquad (22.6.63)$$

with the same moment arm d. Because the ball is not accelerating in the $\hat{\mathbf{j}}$-direction, from Newton's Second Law, we note that $mg - N = 0$. Therefore

$$\vec{\boldsymbol{\tau}}_{S,N} + \vec{\boldsymbol{\tau}}_{S,mg} = d(mg - N)\hat{\mathbf{k}} = \vec{\mathbf{0}}. \qquad (22.6.64)$$

There is no torque about any fixed point S along the line of contact between the bowling bowl and the surface; therefore the angular momentum about that point S is constant,

$$\vec{\mathbf{L}}_{S,i} = \vec{\mathbf{L}}_{S,f}. \tag{22.6.65}$$

Choose one fixed point S along the line of contact (Figure 21.17).

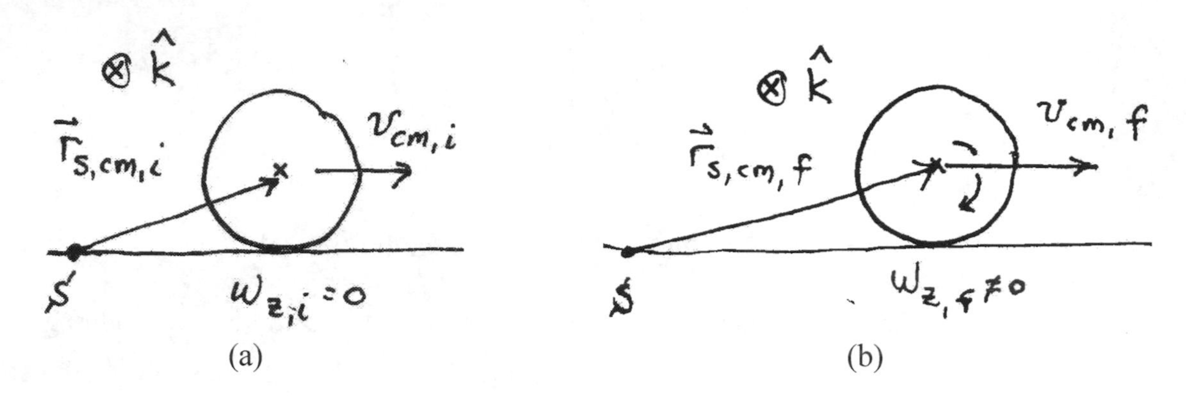

Figure 21.17 Angular momentum about S: (a) initial, (b) final

The initial angular momentum about S is only due to the translation of the center of mass (Figure 21.17a),

$$\vec{\mathbf{L}}_{S,i} = \vec{\mathbf{r}}_{S,\text{cm},i} \times m\vec{\mathbf{v}}_{\text{cm},i} = mRv_{\text{cm},i}\hat{\mathbf{k}}. \tag{22.6.66}$$

In Figure 21.17b, the ball is rolling without slipping. The final angular momentum about S has both a translational and rotational contribution

$$\vec{\mathbf{L}}_{S,f} = \vec{\mathbf{r}}_{S,\text{cm},f} \times m\vec{\mathbf{v}}_{\text{cm},f} + I_{cm}\vec{\boldsymbol{\omega}}_f = mRv_{\text{cm},f}\hat{\mathbf{k}} + I_{\text{cm}}\omega_{z,f}\hat{\mathbf{k}}. \tag{22.6.67}$$

When the ball is rolling without slipping, $v_{\text{cm},f} = R\omega_{z,f}$ and also $I_{\text{cm}} = (2/5)mR^2$. Therefore the final angular momentum about S is

$$\vec{\mathbf{L}}_{S,f} = (mR + (2/5)mR)v_{\text{cm},f}\hat{\mathbf{k}} = (7/5)mRv_{\text{cm},f}\hat{\mathbf{k}}. \tag{22.6.68}$$

Equating the z-components in Eqs. (22.6.66) and (22.6.68) yields

$$mRv_{\text{cm},i} = (7/5)mRv_{\text{cm},f}, \tag{22.6.69}$$

which we can solve for

$$v_{\text{cm},f} = (5/7)v_{cm,i}. \tag{22.6.70}$$

The final angular velocity vector is

$$\vec{\boldsymbol{\omega}} = \omega_{z,f}\hat{\mathbf{k}} = \frac{v_{\text{cm},f}}{R}\hat{\mathbf{k}} = \frac{5v_{\text{cm},i}}{7R}\hat{\mathbf{k}} \ . \qquad (22.6.71)$$

We could also solve this problem by analyzing the translational motion and the rotational motion about the center of mass. Gravity exerts no torque about the center of mass, and the normal component of the contact force has a zero moment arm; the only force that exerts a torque is the frictional force, with a moment arm of R (the force vector and the radius vector are perpendicular). The frictional force should be in the negative x-direction. From the right-hand rule, the direction of the torque is into the page, and hence in the positive z-direction. Equating the z-component of the torque to the rate of change of angular momentum about the center of mass yields

$$\tau_{cm} = R f_k = I_{cm}\alpha_z , \qquad (22.6.72)$$

where f_k is the magnitude of the kinetic frictional force and α_z is the z-component of the angular acceleration of the bowling ball. Note that Eq. (22.6.72) results in a positive z-component of the angular acceleration, which is consistent with the ball tending to rotate as indicated Figure 21.15. The frictional force is also the only force in the horizontal direction, and will cause an acceleration of the center of mass,

$$a_{cm,x} = -f_k / m. \qquad (22.6.73)$$

Note that the x-component of the acceleration will be negative, as expected. Now we need to consider the kinematics. The bowling ball will increase its angular speed as given in Eq. (22.6.72) and decrease its linear speed as given in Eq. (22.6.73),

$$\begin{aligned} \omega_z(t) &= \alpha_z t = \frac{Rf_k}{I_{\text{cm}}}t \\ v_{\text{cm},x}(t) &= v_{\text{cm},i} - \frac{f_k}{m}t. \end{aligned} \qquad (22.6.74)$$

As soon as the ball stops slipping, the kinetic friction no longer acts, static friction is zero, and the ball moves with constant angular and linear velocity. Denote the time when this happens as t_f. At this time the rolling without slipping condition, $\omega_z(t_f) = v_{\text{cm},x}(t_f) / R$, holds and the relations in Eq. (22.6.74) become

$$\begin{aligned} R^2 \frac{f_k}{I_{cm}} t_f &= v_{\text{cm},f} \\ v_{\text{cm},i} - \frac{f_k}{m} t_f &= v_{\text{cm},f}. \end{aligned} \qquad (22.6.75)$$

We can now solve the first equation in Eq. (22.6.75) for t_f and find that

$$t_f = \frac{I_{\text{cm}}}{f_k R^2} v_{\text{cm},f}. \tag{22.6.76}$$

We now substitute Eq. (22.6.76) into the second equation in Eq. (22.6.75) and find that

$$\begin{aligned} v_{\text{cm},f} &= v_{\text{cm},i} - \frac{f_k}{m} \frac{I_{\text{cm}}}{f_k R^2} v_{\text{cm},f} \\ v_{\text{cm},f} &= v_{\text{cm},i} - \frac{I_{\text{cm}}}{m R^2} v_{\text{cm},f}. \end{aligned} \tag{22.6.77}$$

The second equation in (22.6.77) is easily solved for

$$v_{\text{cm},f} = \frac{v_0}{1 + I_{\text{cm}} / mR^2} = \frac{5}{7} v_{\text{cm},i}, \tag{22.6.78}$$

agreeing with Eq. (22.6.70) where we have used $I_{\text{cm}} = (2/5) m R^2$ for a uniform sphere.

Example 21.6 Rotation and Translation Object and Stick Collision

A long narrow uniform stick of length l and mass m lies motionless on ice (assume the ice provides a frictionless surface). The center of mass of the stick is the same as the geometric center (at the midpoint of the stick). The moment of inertia of the stick about its center of mass is I_{cm}. A puck (with putty on one side) has the same mass m as the stick. The puck slides without spinning on the ice with a velocity of $\vec{\mathbf{v}}_i$ toward the stick, hits one end of the stick, and attaches to it (Figure 21.18). You may assume that the radius of the puck is much less than the length of the stick so that the moment of inertia of the puck about its center of mass is negligible compared to I_{cm}. (a) How far from the midpoint of the stick is the center of mass of the stick-puck combination after the collision? (b) What is the linear velocity $\vec{\mathbf{v}}_{\text{cm},f}$ of the stick plus puck after the collision? (c) Is mechanical energy conserved during the collision? Explain your reasoning. (d) What is the angular velocity $\vec{\boldsymbol{\omega}}_{\text{cm},f}$ of the stick plus puck after the collision? (e) How far does the stick's center of mass move during one rotation of the stick?

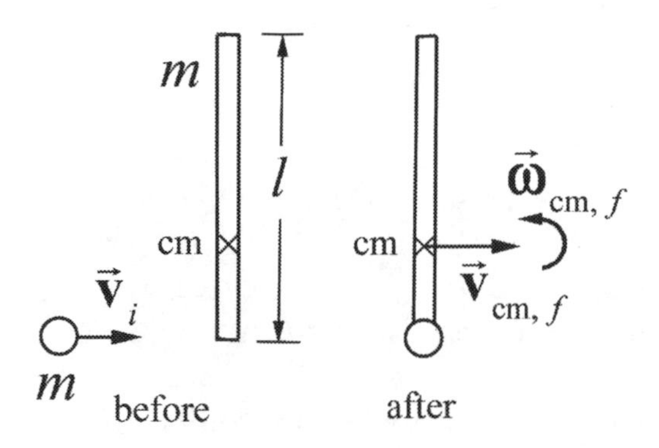

Figure 21.18 Example 21.6

Solution: In this problem we will calculate the center of mass of the puck-stick system after the collision. There are no external forces or torques acting on this system so the momentum of the center of mass is constant before and after the collision and the angular momentum about the center of mass of the puck-stick system is constant before and after the collision. We shall use these relations to compute the final angular velocity of the puck-stick about the center of mass.

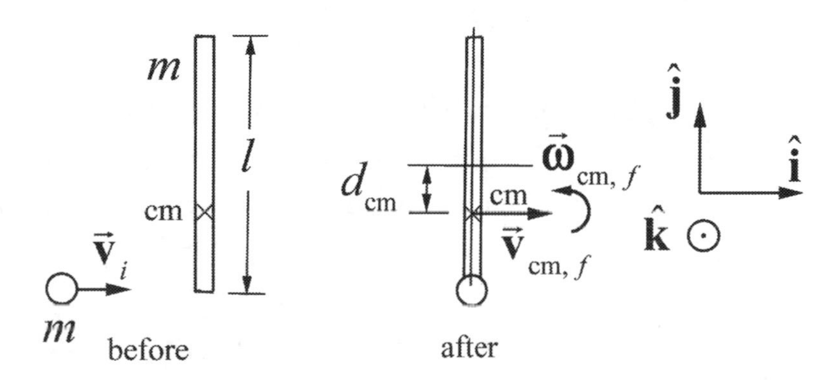

Figure 21.19 Center of mass of the system

a) With respect to the center of the stick, the center of mass of the stick-puck combination is

$$d_{\text{cm}} = \frac{m_{\text{stick}} d_{\text{stick}} + m_{\text{puck}} d_{\text{puck}}}{m_{\text{stick}} + m_{\text{puck}}} = \frac{m(l/2)}{m+m} = \frac{l}{4}. \tag{22.6.79}$$

where we are neglecting the radius of the puck (Figure 21.19).

b) During the collision, the only net forces on the system (the stick-puck combination) are the internal forces between the stick and the puck (transmitted through the putty). Hence, the linear momentum is constant. Initially only the puck had linear momentum $\vec{\mathbf{p}}_i = m\vec{\mathbf{v}}_i = mv_i\hat{\mathbf{i}}$. After the collision, the center of mass of the system is moving with velocity $\vec{\mathbf{v}}_{\text{cm},f} = v_{\text{cm},f}\hat{\mathbf{i}}$. Equating initial and final linear momenta,

$$mv_i = (2m)v_{\text{cm},f} \Rightarrow v_{\text{cm},f} = \frac{v_i}{2}. \qquad (22.6.80)$$

The direction of the velocity is the same as the initial direction of the puck's velocity.

c) The forces that deform the putty do negative work (the putty is compressed somewhat), and so mechanical energy is not conserved; the collision is totally inelastic.

d) Choose the center of mass of the stick-puck combination, as found in part a), as the point S about which to find angular momentum. This choice means that after the collision there is no angular momentum due to the translation of the center of mass. Before the collision, the angular momentum was entirely due to the motion of the puck,

$$\vec{\mathbf{L}}_{S,i} = \vec{\mathbf{r}}_{\text{puck}} \times \vec{\mathbf{p}}_i = (l/4)(mv_i)\hat{\mathbf{k}}, \qquad (22.6.81)$$

where $\hat{\mathbf{k}}$ is directed out of the page in Figure 21.19. After the collision, the angular momentum is

$$\vec{\mathbf{L}}_{S,f} = I_{\text{cm},f}\,\omega_{cm,f}\,\hat{\mathbf{k}}, \qquad (22.6.82)$$

where $I_{\text{cm},f}$ is the moment of inertia about the center of mass of the stick-puck combination. This moment of inertia of the stick about the new center of mass is found from the parallel axis theorem and the moment of inertia of the puck, which is $m(l/4)^2$. Therefore

$$I_{\text{cm},f} = I_{\text{cm, stick}} + I_{\text{cm, puck}} = (I_{\text{cm}} + m(l/4)^2) + m(l/4)^2 = I_{\text{cm}} + \frac{ml^2}{8}. \qquad (22.6.83)$$

Inserting this expression into Eq. (22.6.82), equating the expressions for $\vec{\mathbf{L}}_{S,i}$ and $\vec{\mathbf{L}}_{S,f}$ and solving for $\omega_{\text{cm},f}$ yields

$$\omega_{\text{cm},f} = \frac{m(l/4)}{I_{\text{cm}} + ml^2/8} v_i. \qquad (22.6.84)$$

If the stick is uniform, $I_{\text{cm}} = ml^2/12$ and Eq. (22.6.84) reduces to

$$\omega_{\text{cm},f} = \frac{6}{5}\frac{v_i}{l}. \qquad (22.6.85)$$

It may be tempting to try to calculate angular momentum about the contact point C, where the putty hits the stick. If this is done, there is no initial angular momentum, and after the collision the angular momentum will be the sum of two parts, the angular momentum of the center of mass of the stick and the angular moment about the center of the stick,

$$\vec{\mathbf{L}}_{C,f} = \vec{\mathbf{r}}_{\text{cm}} \times \vec{\mathbf{p}}_{\text{cm}} + I_{\text{cm}}\vec{\boldsymbol{\omega}}_{\text{cm},f}\,. \tag{22.6.86}$$

There are two crucial things to note: First, the speed of the center of mass is not the speed found in part b); the rotation must be included, so that $v_{\text{cm}} = v_i/2 - \omega_{\text{cm},f}(l/4)$. Second, the direction of $\vec{\mathbf{r}}_{\text{cm}} \times \vec{\mathbf{p}}_{\text{cm}}$ with respect to the contact point C is, from the right-hand rule, *into* the page, or the $-\hat{\mathbf{k}}$-direction, opposite the direction of $\vec{\boldsymbol{\omega}}_{\text{cm},f}$. This is to be expected, as the sum in Eq. (22.6.86) must be zero. Adding the $\hat{\mathbf{k}}$-components (the only components) in Eq. (22.6.86),

$$-(l/2)m(v_i/2 - \omega_{\text{cm},f}(l/4)) + I_{\text{cm}}\omega_{\text{cm},f} = 0\,. \tag{22.6.87}$$

Solving Eq. (22.6.87) for $\omega_{\text{cm},f}$ yields Eq. (22.6.84).

This alternative derivation should serve two purposes. One is that it doesn't matter which point we use to find angular momentum. The second is that use of foresight, in this case choosing the center of mass of the system so that the final velocity does not contribute to the angular momentum, can prevent extra calculation. It's often a matter of trial and error ("learning by misadventure") to find the "best" way to solve a problem.

e) The time of one rotation will be the same for all observers, independent of choice of origin. This fact is crucial in solving problems, in that the angular velocity will be the same (this was used in the alternate derivation for part d) above). The time for one rotation is the period $T = 2\pi/\omega_f$ and the distance the center of mass moves is

$$\begin{aligned} x_{\text{cm}} &= v_{\text{cm}}T = 2\pi\frac{v_{\text{cm}}}{\omega_{\text{cm},f}} \\ &= 2\pi\frac{v_i/2}{\left(\dfrac{m(l/4)}{I_{\text{cm}} + ml^2/8}\right)v_i} \\ &= 2\pi\frac{I_{\text{cm}} + ml^2/8}{m(l/2)}. \end{aligned} \tag{22.6.88}$$

Using $I_{\text{cm}} = ml^2/12$ for a uniform stick gives

$$x_{\text{cm}} = \frac{5}{6}\pi l\,. \tag{22.6.89}$$

Chapter 22 Three Dimensional Rotations and Gyroscopes

Chapter 22 Three Dimensional Rotations and Gyroscopes

Hypothesis: The earth, having once received a rotational movement around an axis, which agrees with its axis on the figure or only differs from it slightly, will always conserve this uniform movement, and its axis of rotation will always remain the same and will be directed toward the same points of the sky, unless the earth should be subjected to external forces which might cause some change either in the speed of rotational movement or in the position of the axis of rotation.[1]

Leonhard Euler

22.1 Introduction to Three Dimensional Rotations

Most of the examples and applications we have considered concerned the rotation of rigid bodies about a fixed axis. However, there are many examples of rigid bodies that rotate about an axis that is changing its direction. A turning bicycle wheel, a gyroscope, the earth's precession about its axis, a spinning top, and a coin rolling on a table are all examples of this type of motion. These motions can be very complex and difficult to analyze. However, for each of these motions we know that if there a non-zero torque about a point S, then the angular momentum about S must change in time, according to the rotational equation of motion,

$$\vec{\boldsymbol{\tau}}_S = \frac{d\vec{\mathbf{L}}_S}{dt}. \qquad (22.1.1)$$

We also know that the angular momentum about S of a rotating body is the sum of the orbital angular momentum about S and the spin angular momentum about the center of mass.

$$\vec{\mathbf{L}}_S = \vec{\mathbf{L}}_S^{\text{orbital}} + \vec{\mathbf{L}}_{\text{cm}}^{\text{spin}}. \qquad (22.1.2)$$

For fixed axis rotation the spin angular momentum about the center of mass is just

$$\vec{\mathbf{L}}_{\text{cm}}^{\text{spin}} = I_{\text{cm}}\vec{\boldsymbol{\omega}}_{\text{cm}}. \qquad (22.1.3)$$

where $\vec{\boldsymbol{\omega}}_{\text{cm}}$ is the angular velocity about the center of mass and is directed along the fixed axis of rotation.

[1] L. Euler, Recherches sur la precession des equinoxes et sur la nutation de l'axe de la terre (Research concerning the precession of the equinoxes and of the nutation of the earth's axis), *Memoires de l'academie des sciences de Berlin* 5, 1751, pp. 289-325

22.1.1 Angular Velocity for Three Dimensional Rotations

When the axis of rotation is no longer fixed, the angular velocity will no longer point in a fixed direction.

For an object that is rotating with angular coordinates $(\theta_x, \theta_y, \theta_z)$ about each respective Cartesian axis, the ***angular velocity*** *of an object that is rotating about each axis is defined to be*

$$\begin{aligned}\vec{\boldsymbol{\omega}} &= \frac{d\theta_x}{dt}\hat{\mathbf{i}} + \frac{d\theta_y}{dt}\hat{\mathbf{j}} + \frac{d\theta_z}{dt}\hat{\mathbf{k}} \\ &= \omega_x\hat{\mathbf{i}} + \omega_y\hat{\mathbf{j}} + \omega_z\hat{\mathbf{k}}\end{aligned} \tag{22.1.4}$$

This definition is the result of a property of very small (infinitesimal) angular rotations in which the order of rotations does matter. For example, consider an object that undergoes a rotation about the x-axis, $\vec{\boldsymbol{\omega}}_x = \omega_x\hat{\mathbf{i}}$, and then a second rotation about the y-axis, $\vec{\boldsymbol{\omega}}_y = \omega_y\hat{\mathbf{j}}$. Now consider a different sequence of rotations. The object first undergoes a rotation about the y-axis, $\vec{\boldsymbol{\omega}}_y = \omega_y\hat{\mathbf{j}}$, and then undergoes a second rotation about the x-axis, $\vec{\boldsymbol{\omega}}_x = \omega_x\hat{\mathbf{i}}$. In both cases the object will end up in the same position indicated that $\vec{\boldsymbol{\omega}}_x + \vec{\boldsymbol{\omega}}_y = \vec{\boldsymbol{\omega}}_y + \vec{\boldsymbol{\omega}}_x$, a necessary condition that must be satisfied in order for a physical quantity to be a vector quantity.

Example 22.1 Angular Velocity of a Rolling Bicycle Wheel

A bicycle wheel of mass m and radius R rolls without slipping about the z-axis. An axle of length b passes through its center. The bicycle wheel undergoes two simultaneous rotations. The wheel circles around the z-axis with angular speed Ω and associated angular velocity $\vec{\boldsymbol{\Omega}} = \Omega_z\hat{\mathbf{k}}$ (Figure 22.1). Because the wheel is rotating without slipping, it is spinning about its center of mass with angular speed ω_{spin} and associated angular velocity $\vec{\boldsymbol{\omega}}_{\text{spin}} = -\omega_{\text{spin}}\hat{\mathbf{r}}$.

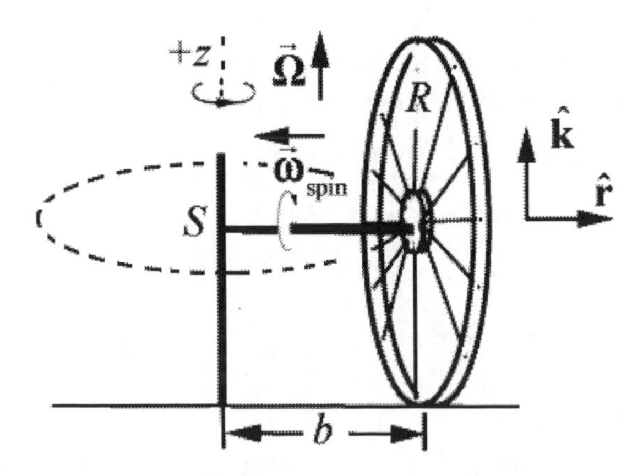

Figure 22.1 Example 22.1

The angular velocity of the wheel is the sum of these two vector contributions

$$\vec{\boldsymbol{\omega}} = \Omega\,\hat{\mathbf{k}} - \omega_{\text{spin}}\hat{\mathbf{r}}\ . \tag{22.1.5}$$

Because the wheel is rolling without slipping, $v_{\text{cm}} = b\Omega = \omega_{\text{spin}} R$ and so $\omega_{\text{spin}} = b\Omega / R$. The angular velocity is then

$$\vec{\boldsymbol{\omega}} = \Omega\,(\hat{\mathbf{k}} - (b/R)\hat{\mathbf{r}})\ . \tag{22.1.6}$$

The orbital angular momentum about the point S where the axle meets the axis of rotation (Figure 22.1), is then

$$\vec{\mathbf{L}}_S^{\text{orbital}} = bmv_{\text{cm}}\hat{\mathbf{k}} = mb^2\Omega\,\hat{\mathbf{k}}\,. \tag{22.1.7}$$

The spin angular momentum about the center of mass is more complicated. The wheel is rotating about both the z-axis and the radial axis. Therefore

$$\vec{\mathbf{L}}_{\text{cm}}^{\text{spin}} = I_z\Omega\,\hat{\mathbf{k}} + I_r\omega_{\text{spin}}(-\hat{\mathbf{r}})\,. \tag{22.1.8}$$

Therefore the angular momentum about S is the sum of these two contributions

$$\begin{aligned}\vec{\mathbf{L}}_S &= mb^2\Omega\,\hat{\mathbf{k}} + I_z\Omega\,\hat{\mathbf{k}} + I_r\omega_{\text{spin}}(-\hat{\mathbf{r}}) \\ &= (mb^2\Omega + I_z\Omega)\,\hat{\mathbf{k}} - I_r(b\,\Omega/R)\hat{\mathbf{r}}.\end{aligned} \tag{22.1.9}$$

Comparing Eqs. (22.1.6) and (22.1.9), we note that the angular momentum about S is not proportional to the angular velocity.

22.2 Gyroscope

A toy gyroscope of mass m consists of a spinning flywheel mounted in a suspension frame that allows the flywheel's axle to point in any direction. One end of the axle is supported on a pylon a distance d from the center of mass of the gyroscope. Choose polar coordinates so that the axle of the gyroscope flywheel is aligned along the r-axis and the vertical axis is the z-axis (Figure 22.2 shows a schematic representation of the gyroscope).

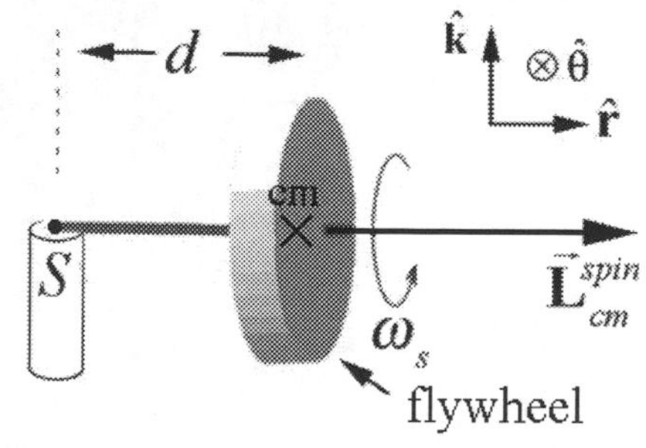

Figure 22.2 A toy gyroscope.

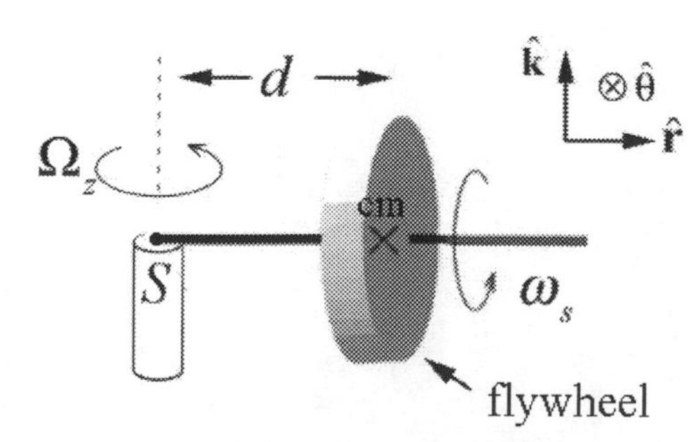

Figure 22.3 Angular rotations

The flywheel is spinning about its axis with a ***spin angular velocity***,

$$\vec{\boldsymbol{\omega}}_s = \omega_s \hat{\mathbf{r}}, \qquad (22.2.1)$$

where ω_s is the radial component and $\omega_s > 0$ for the case illustrated in Figure 22.2.

When we release the gyroscope it undergoes a very surprising motion. Instead of falling downward, the center of mass rotates about a vertical axis that passes through the contact point S of the axle with the pylon with a ***precessional angular velocity***

$$\vec{\boldsymbol{\Omega}} = \Omega_z \hat{\mathbf{k}} = \frac{d\theta}{dt}\hat{\mathbf{k}}, \qquad (22.2.2)$$

where $\Omega_z = d\theta / dt$ is the z-component and $\Omega_z > 0$ for the case illustrated in Figure 22.3. Therefore the angular velocity of the flywheel is the sum of these two contributions

$$\vec{\boldsymbol{\omega}} = \vec{\boldsymbol{\omega}}_s + \vec{\boldsymbol{\Omega}} = \omega_s \hat{\mathbf{r}} + \Omega_z \hat{\mathbf{k}}. \qquad (22.2.3)$$

We shall study the special case where the magnitude of the precession component $|\Omega_z|$ of the angular velocity is much less than the magnitude of the spin component $|\omega_s|$ of the spin angular velocity, $|\Omega_z| << |\omega_s|$, so that the magnitude of the angular velocity $|\vec{\boldsymbol{\omega}}| \simeq |\omega_s|$ and Ω_z and ω_s are nearly constant. These assumptions are collectively called the ***gyroscopic approximation***.

The force diagram for the gyroscope is shown in Figure 22.4a. The gravitational force acts at the center of the mass and is directed downward, $\vec{\mathbf{F}}^g = -mg\,\hat{\mathbf{k}}$. There is also a contact force, $\vec{\mathbf{F}}^c$, between the end of the axle and the pylon. It may seem that the contact force, $\vec{\mathbf{F}}^c$, has only an upward component, $\vec{\mathbf{F}}^v = F_z\,\hat{\mathbf{k}}$, but as we shall soon see there must also be a radial inward component to the contact force, $\vec{\mathbf{F}}^r = F_r\,\hat{\mathbf{r}}$, with $F_r < 0$, because the center of mass undergoes circular motion.

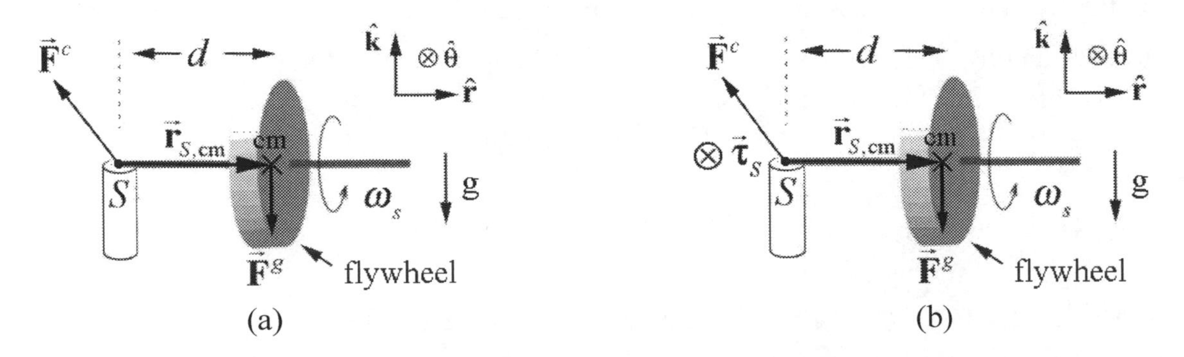

Figure 22.4 (a) Force diagram and (b) torque diagram for the gyroscope

The reason that the gyroscope does not fall down is that the vertical component of the contact force exactly balances the gravitational force

$$F_z - mg = 0 \ . \tag{22.2.4}$$

What about the torque about the contact point S? The contact force acts at S so it does not contribute to the torque about S; only the gravitational force contributes to the torque about S (Figure 22.5b). The direction of the torque about S is given by

$$\vec{\boldsymbol{\tau}}_S = \vec{\mathbf{r}}_{S,\text{cm}} \times \vec{\mathbf{F}}_{\text{gravity}} = d\,\hat{\mathbf{r}} \times mg(-\hat{\mathbf{k}}) = d\,mg\hat{\boldsymbol{\theta}}, \tag{22.2.5}$$

and is in the positive $\hat{\boldsymbol{\theta}}$-direction. However we know that if there a non-zero torque about S, then the angular momentum about S must change in time, according to

$$\vec{\boldsymbol{\tau}}_S = \frac{d\vec{\mathbf{L}}_S}{dt}. \tag{22.2.6}$$

The angular momentum about the point S of the gyroscope is given by

$$\vec{\mathbf{L}}_S = \vec{\mathbf{L}}_S^{\text{orbital}} + \vec{\mathbf{L}}_{\text{cm}}^{\text{spin}}. \tag{22.2.7}$$

The orbital angular momentum about the point S is

$$\vec{\mathbf{L}}_S^{\text{orbital}} = \vec{\mathbf{r}}_{S,cm} \times m\vec{\mathbf{v}}_{cm} = d\,\hat{\mathbf{r}} \times md\Omega_z\,\hat{\boldsymbol{\theta}} = md^2\Omega_z\,\hat{\mathbf{k}}. \tag{22.2.8}$$

The magnitude of the orbital angular momentum about S is nearly constant and the direction does not change. Therefore

$$\frac{d}{dt}\vec{\mathbf{L}}_S^{\text{orbital}} = \vec{\mathbf{0}}. \tag{22.2.9}$$

The spin angular momentum includes two terms. Recall that the flywheel undergoes two separate rotations about different axes. It is spinning about the flywheel axis with spin angular velocity $\vec{\boldsymbol{\omega}}_s$. As the flywheel precesses around the pivot point, the flywheel rotates about the z-axis with precessional angular velocity $\vec{\boldsymbol{\Omega}}$ (Figure 22.5). The spin angular momentum therefore is given by

$$\vec{\mathbf{L}}_{\text{cm}}^{\text{spin}} = I_r\omega_s\hat{\mathbf{r}} + I_z\Omega_z\,\hat{\mathbf{k}}, \tag{22.2.10}$$

where I_r is the moment of inertia with respect to the flywheel axis and I_z is the moment of inertia with respect to the z-axis. If we assume the axle is massless and the flywheel is uniform with radius R, then $I_r = (1/2)mR^2$. By the perpendicular axis theorem $I_r = I_z + I_y = 2I_z$, hence $I_z = (1/4)mR^2$.

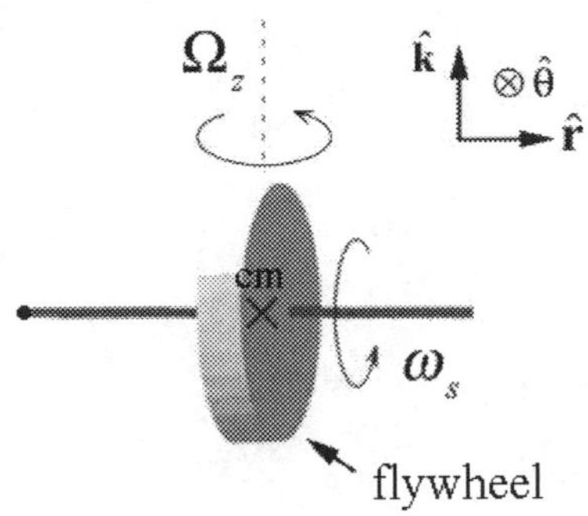

Figure 22.5: Rotations about center of mass of flywheel

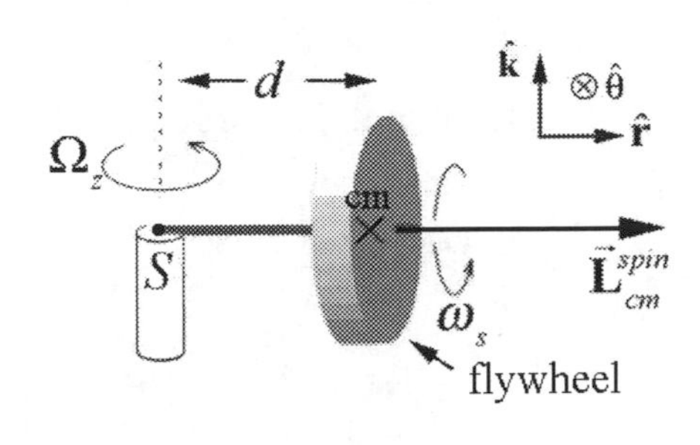

Figure 22.6 Spin angular momentum.

Recall that the gyroscopic approximation holds when $|\Omega_z| << |\omega_s|$, which implies that $I_z\Omega_z << I_r\omega_s$, and therefore we can ignore the contribution to the spin angular momentum from the rotation about the vertical axis, and so

$$\vec{\mathbf{L}}^{\text{spin}}_{\text{cm}} \simeq I_{\text{cm}}\omega_s\hat{\mathbf{r}}. \qquad (22.2.11)$$

(The contribution to the spin angular momentum due to the rotation about the z-axis, $I_z\Omega_z\,\hat{\mathbf{k}}$, is nearly constant in both magnitude and direction so it does not change in time, $d(I_z\Omega_z\,\hat{\mathbf{k}})/dt \simeq \vec{\mathbf{0}}$.) Therefore the angular momentum about S is approximately

$$\vec{\mathbf{L}}_S \simeq \vec{\mathbf{L}}^{\text{spin}}_{\text{cm}} = I_{\text{cm}}\omega_s\hat{\mathbf{r}}. \qquad (22.2.12)$$

Our initial expectation that the gyroscope should fall downward due to the torque that the gravitational force exerts about the contact point S leads to a violation of the torque law. If the center of mass did start to fall then the change in the spin angular momentum, $\Delta\vec{\mathbf{L}}^{\text{spin}}_{\text{cm}}$, would point in the negative z-direction and that would contradict the vector aspect of Eq. (22.2.6). Instead of falling down, the angular momentum about the center of mass, $\vec{\mathbf{L}}^{\text{spin}}_{\text{cm}}$, must change direction such that the direction of $\Delta\vec{\mathbf{L}}^{\text{spin}}_{\text{cm}}$ is in the same direction as torque about S (Eq. (22.2.5)), the positive $\hat{\boldsymbol{\theta}}$-direction.

Recall that in our study of circular motion, we have already encountered several examples in which the direction of a constant magnitude vector changes. We considered a point object of mass m moving in a circle of radius r. When we choose a coordinate system with an origin at the center of the circle, the position vector $\vec{\mathbf{r}}$ is directed radially outward. As the mass moves in a circle, the position vector has a constant magnitude but changes in direction. The velocity vector is given by

$$\vec{\mathbf{v}} = \frac{d\vec{\mathbf{r}}}{dt} = \frac{d}{dt}(r\,\hat{\mathbf{r}}) = r\frac{d\theta}{dt}\hat{\boldsymbol{\theta}} = r\omega_z\hat{\boldsymbol{\theta}} \qquad (22.2.13)$$

and has direction that is perpendicular to the position vector (tangent to the circle), (Figure 22.7a)).

Figure 22.7 (a) Rotating position and velocity vector; (b) velocity and acceleration vector for uniform circular motion

For uniform circular motion, the magnitude of the velocity is constant but the direction constantly changes and we found that the acceleration is given by (Figure 22.7b)

$$\vec{\mathbf{a}} = \frac{d\vec{\mathbf{v}}}{dt} = \frac{d}{dt}(v_\theta \hat{\boldsymbol{\theta}}) = v_\theta \frac{d\theta}{dt}(-\hat{\mathbf{r}}) = r\omega_z \omega_z(-\hat{\mathbf{r}}) = -r\omega_z^2 \hat{\mathbf{r}}\,. \qquad (22.2.14)$$

Note that we used the facts that

$$\begin{aligned} \frac{d\hat{\mathbf{r}}}{dt} &= \frac{d\theta}{dt}\hat{\boldsymbol{\theta}}, \\ \frac{d\hat{\boldsymbol{\theta}}}{dt} &= -\frac{d\theta}{dt}\hat{\mathbf{r}} \end{aligned}, \qquad (22.2.15)$$

in Eqs. (22.2.13) and (22.2.14). We can apply the same reasoning to how the spin angular changes in time (Figure 22.8).

The time derivative of the spin angular momentum is given by

$$\frac{d\vec{\mathbf{L}}_S}{dt} = \frac{d\vec{\mathbf{L}}^{\text{spin}}_{\text{cm},\omega_s}}{dt} = \left|\vec{\mathbf{L}}^{\text{spin}}_{\text{cm},\omega_s}\right| \frac{d\theta}{dt}\hat{\boldsymbol{\theta}} = \left|\vec{\mathbf{L}}^{\text{spin}}_{\text{cm},\omega_s}\right| \Omega_z\,\hat{\boldsymbol{\theta}} = I_r \omega_s \hat{\boldsymbol{\theta}}\,. \qquad (22.2.16)$$

where $\Omega_z = d\theta / dt$ is the z-component and $\Omega_z > 0$. The center of mass of the flywheel rotates about a vertical axis that passes through the contact point S of the axle with the pylon with a precessional angular velocity

$$\vec{\boldsymbol{\Omega}} = \Omega_z \hat{\mathbf{k}} = \frac{d\theta}{dt}\hat{\mathbf{k}}\,, \qquad (22.2.17)$$

Substitute Eqs. (22.2.16) and (22.2.5) into Eq. (22.2.6) yielding

$$d\,mg\hat{\boldsymbol{\theta}} = \left|\vec{\mathbf{L}}^{\text{spin}}_{\text{cm}}\right| \Omega_z \hat{\boldsymbol{\theta}}\,. \tag{22.2.18}$$

Solving Equation (22.2.18) for the z-component of the precessional angular velocity of the gyroscope yields

$$\Omega_z = \frac{d\,mg}{\left|\vec{\mathbf{L}}^{\text{spin}}_{\text{cm}}\right|} = \frac{d\,mg}{I_{\text{cm}}\,\omega_{\text{s}}}\,. \tag{22.2.19}$$

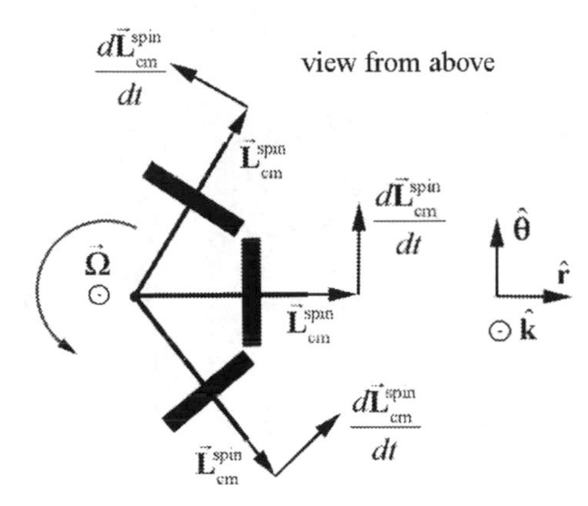

Figure 22.8 Time changing direction of the spin angular momentum

22.3 Why Does a Gyroscope Precess?

Why does a gyroscope precess? We now understand that the torque is causing the spin angular momentum to change but the motion still seems mysterious. We shall try to understand why the angular momentum changes direction by first examining the role of force and impulse on a single rotating particle and then generalize to a rotating disk.

22.3.1 Deflection of a Particle by a Small Impulse

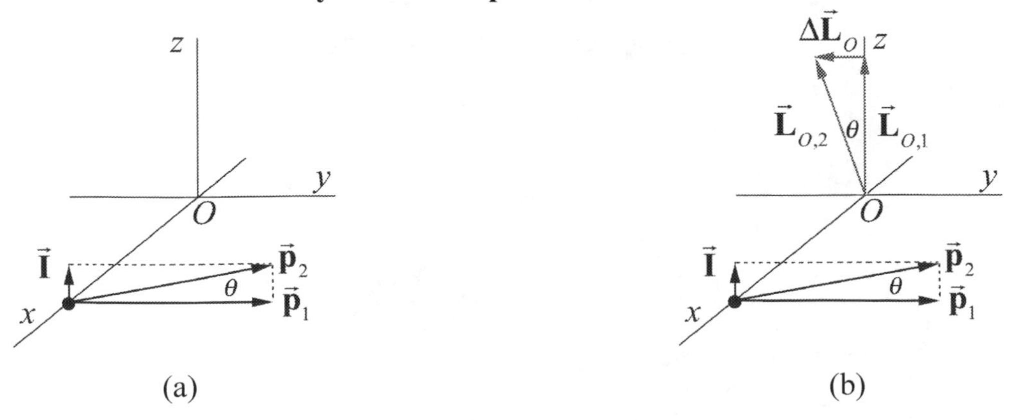

Figure 22.9 (a) Deflection of a particle by a small impulse, (b) change in angular momentum about origin

We begin by first considering how a particle with momentum $\vec{\mathbf{p}}_1$ undergoes a deflection due to a small impulse (Figure 22.9a). If the impulse $\left|\vec{\mathbf{I}}\right| << \left|\vec{\mathbf{p}}_1\right|$, the primary effect is to rotate the momentum $\vec{\mathbf{p}}_1$ about the x-axis by a small angle θ, with $\vec{\mathbf{p}}_2 = \vec{\mathbf{p}}_1 + \Delta\vec{\mathbf{p}}$. The application of $\vec{\mathbf{I}}$ causes a change in the angular momentum $\vec{\mathbf{L}}_{O,1}$ about the origin O, according to the torque equation, $\Delta\vec{\mathbf{L}}_O = \vec{\boldsymbol{\tau}}_{\text{ave},O}\Delta t = (\vec{\mathbf{r}}_O \times \vec{\mathbf{F}}_{\text{ave}})\Delta t$. Because $\vec{\mathbf{I}} = \Delta\vec{\mathbf{p}} = \vec{\mathbf{F}}_{\text{ave}}\Delta t$, we have that $\Delta\vec{\mathbf{L}}_O = \vec{\mathbf{r}}_O \times \vec{\mathbf{I}}$. As a result, $\Delta\vec{\mathbf{L}}_O$ rotates about the x-axis by a small angle θ, to a new angular momentum $\vec{\mathbf{L}}_{O,2} = \vec{\mathbf{L}}_{O,1} + \Delta\vec{\mathbf{L}}_O$. Note that although $\vec{\mathbf{L}}_O$ is in the z-direction, $\Delta\vec{\mathbf{L}}_O$ is in the negative y-direction (Figure 22.9b).

22.3.2 Effect of Small Impulse on Tethered Object

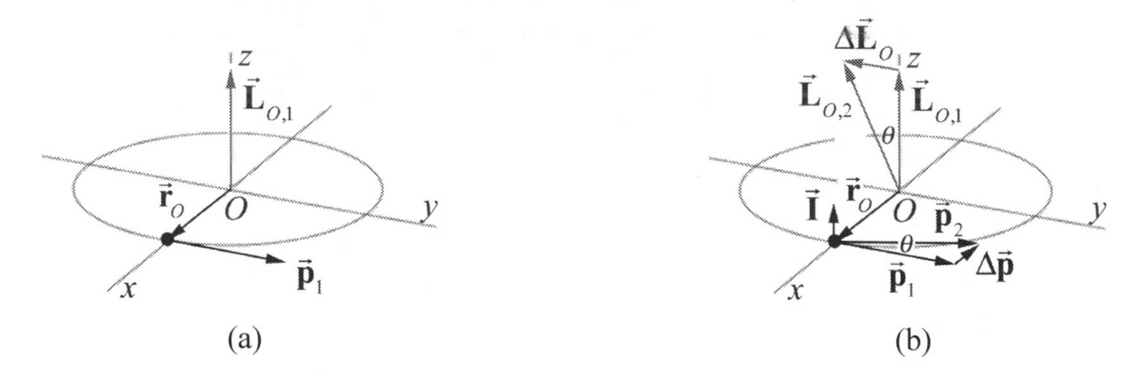

Figure 22.10a Small impulse on object undergoing circular motion, (b) change in angular momentum

Now consider an object that is attached to a string and is rotating about a fixed point O with momentum $\vec{\mathbf{p}}_1$. The object is given an impulse $\vec{\mathbf{I}}$ perpendicular to $\vec{\mathbf{r}}_O$ and to $\vec{\mathbf{p}}_1$. Neglect gravity. As a result $\Delta\vec{\mathbf{L}}_O$ rotates about the x-axis by a small angle θ (Figure 22.10a). Note that although $\vec{\mathbf{I}}$ is in the z-direction, $\Delta\vec{\mathbf{L}}_O$ is in the negative y- direction (Figure 22.10b). Note that although $\vec{\mathbf{I}}$ is in the z-direction, the plane in which the ball moves also rotates about the x-axis by the same angle (Figure 22.11).

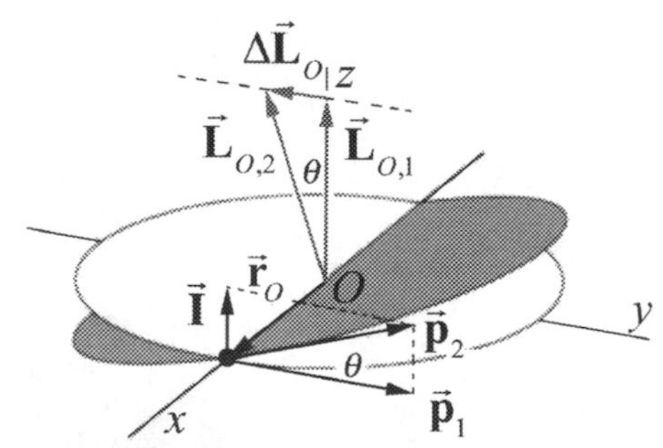

Figure 22.11 Plane of object rotates about x-axis

Example 22.2 Effect of Large Impulse on Tethered Object

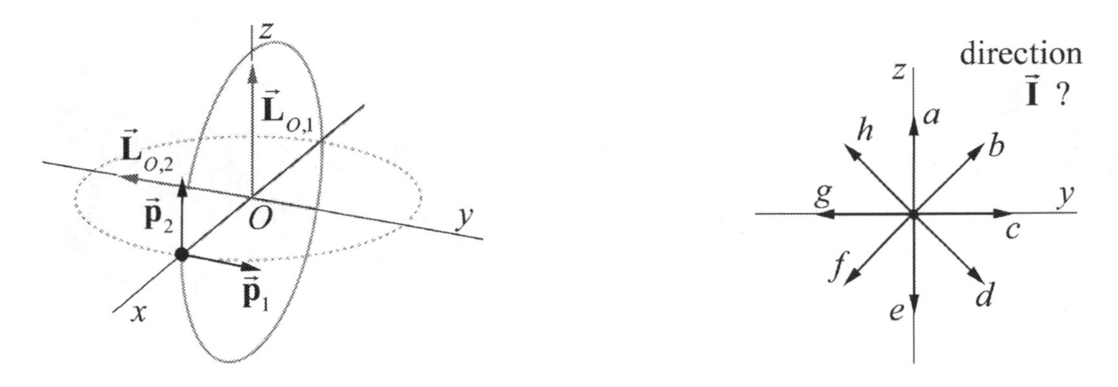

Figure 22.12 Example 22.2

What impulse, $\vec{\mathbf{I}}$, must be given to the ball in order to rotate its orbit by 90 degrees as shown without changing its speed (Figure 21.12)?

Solution: h. The impulse $\vec{\mathbf{I}}$ must halt the momentum $\vec{\mathbf{p}}_1$ and provide a momentum $\vec{\mathbf{p}}_2$ of equal magnitude along the z-direction such that $\vec{\mathbf{I}} = \Delta\vec{\mathbf{p}}$.

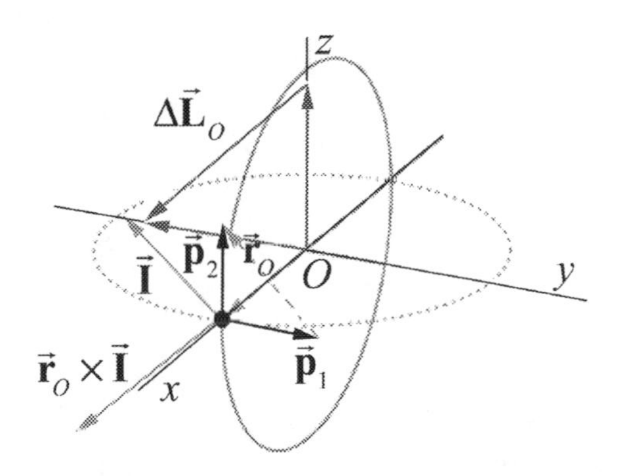

Figure 22.13 Impulse and torque about O

The angular impulse about O must be equal to the change in angular momentum about O

$$\vec{\boldsymbol{\tau}}_O \Delta t = \vec{\mathbf{r}}_O \times \vec{\mathbf{I}} = (\vec{\mathbf{r}}_O \times \Delta\vec{\mathbf{p}}) = \Delta\vec{\mathbf{L}}_O \tag{22.3.1}$$

The change in angular momentum, $\Delta\vec{\mathbf{L}}_O$, due to the torque about O, cancels the z-component of $\vec{\mathbf{L}}_O$ and adds a component of the same magnitude in the negative y-direction (Figure 22.13).

22.3.3 Effect of Small Impulse Couple on Baton

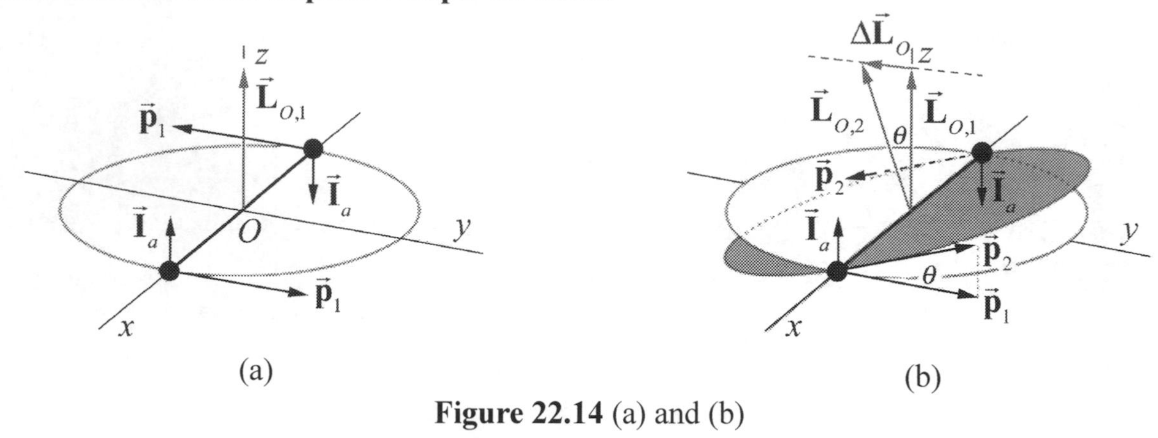

Figure 22.14 (a) and (b)

Now consider two equal masses at the ends of a massless rod, which spins about its center. We apply an impulse couple to insure no motion of the center of mass. Again note that the impulse couple is applied in the z-direction (Figure 22.14a). The resulting torque about O lies along the negative y-direction and the plane of rotation tilts about the x-axis (Figure 22.14b).

22.3.4 Effect of Small Impulse Couple on Massless Shaft of Baton

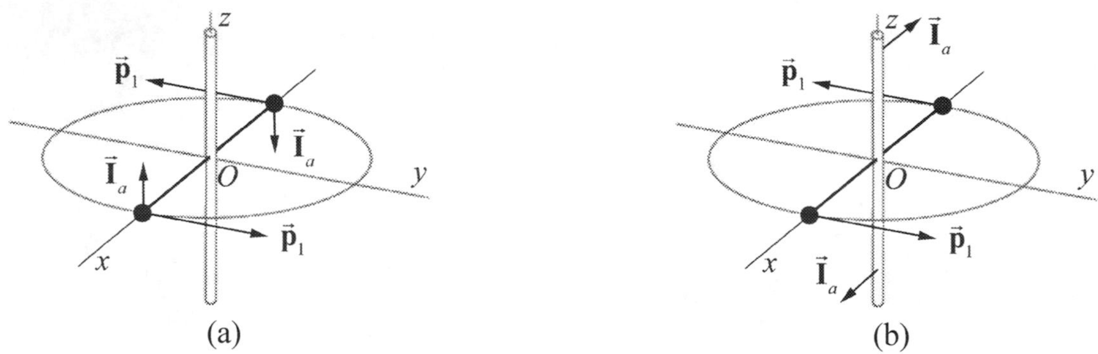

Figure 22.15 Apply impulse couple to (a) objects and (b) shaft

Instead of applying the impulse couple $\vec{I}_a$ to the masses (Figure 21.15a), one could apply the impulse couple $\vec{I}_a$ to the shaft (Figure 22.15b) to achieve the same result.

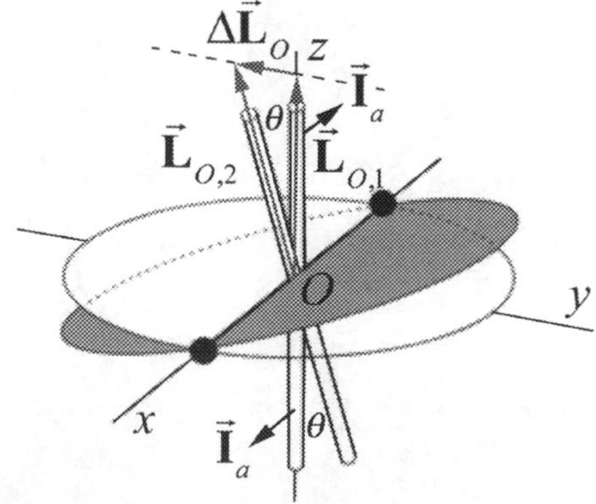

Figure 22.16 Twisting shaft causes shaft and plane to rotate about x-axis

Twisting the shaft around the y-axis causes the shaft and the plane in which the baton moves to rotate about the x-axis.

22.3.5 Effect of a Small Impulse Couple on a Rotating Disk

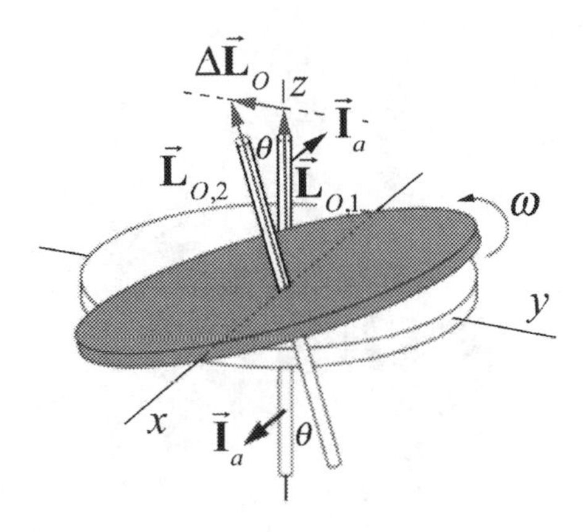

Figure 22.17 Impulse couple causes a disk to rotate about the x-axis.

Now let's consider a rotating disk. The plane of a rotating disk and its shaft behave just like the plane of the rotating baton and its shaft when one attempts to twist the shaft about the y-axis. The plane of the disk rotates about the x-axis (Figure 22.17). This unexpected result is due to the large pre-existing angular momentum about O, $\vec{\mathbf{L}}_{O,1}$, due to the spinning disk. It does not matter where along the shaft the impulse couple is applied, as long as it creates the same torque about O.

22.3.6 Effect of a Force Couple on a Rotating Disk

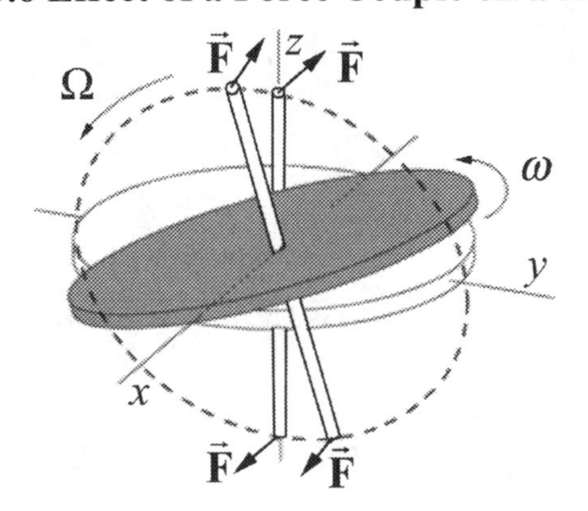

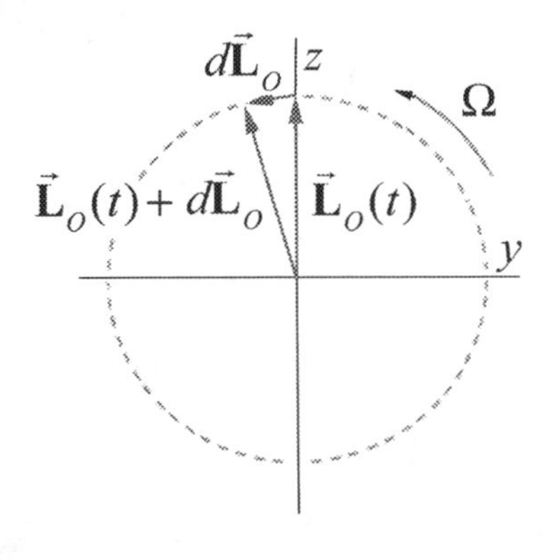

Figure 22.18 A series of small impulse couples causes the tip of the shaft to execute circular motion about the x-axis

A series of small impulse couples, or equivalently a continuous force couple (with force $\vec{\mathbf{F}}$), causes the tip of the shaft to execute circular motion about the x-axis (Figure 22.18). The magnitude of the angular momentum about O changes according to

$\left|d\vec{\mathbf{L}}_O\right| = \left|\vec{\mathbf{L}}_O\right|\Omega\, dt = I\omega\,\Omega\, dt$. Recall that torque and changing angular momentum about O are related by $\vec{\boldsymbol{\tau}}_O = d\vec{\mathbf{L}}_O / dt$. Therefore $\left|\vec{\boldsymbol{\tau}}_O\right| = \left|\vec{\mathbf{L}}_O\right|\Omega = I\omega\,\Omega$. The precession rate of the shaft is the ratio of the magnitude of the torque to the angular momentum $\Omega = \left|\vec{\boldsymbol{\tau}}_O\right| / \left|\vec{\mathbf{L}}_O\right| = \left|\vec{\boldsymbol{\tau}}_O\right| / I\omega$.

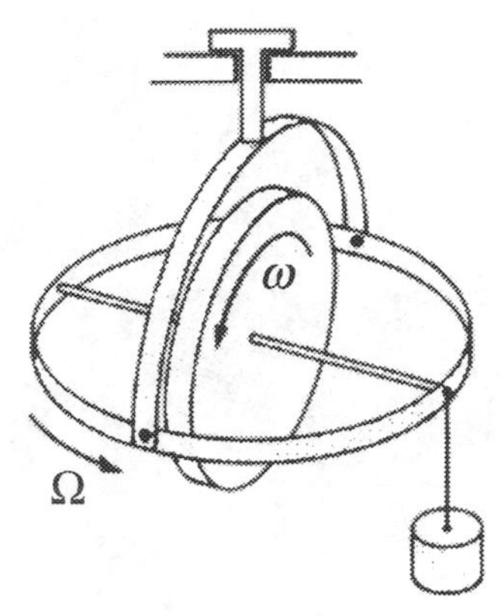

Figure 22.19 Precessing gyroscope with hanging object

Thus we can explain the motion of a precessing gyroscope in which the torque about the center of mass is provided by the force of gravity on the hanging object (Figure 22.19).

22.3.7 Effect of a Small Impulse Couple on a Non-Rotating Disc

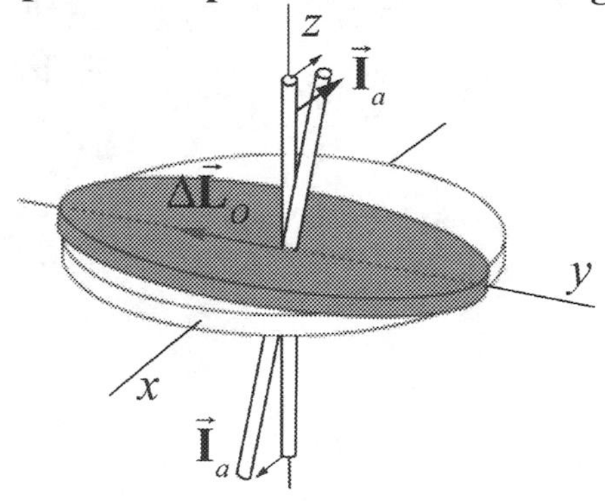

Figure 22.20 Impulse couple on non-rotating disk causes shaft to rotate about negative y -axis.

If the disk is not rotating to begin with, $\Delta\vec{\mathbf{L}}_O$ is also the final $\vec{\mathbf{L}}_O$. The shaft moves in the direction it is pushed (Figure 22.20).

22.4 Worked Examples

Example 22.3 Tilted Toy Gyroscope

A wheel is at one end of an axle of length d. The axle is pivoted at an angle ϕ with respect to the vertical. The wheel is set into motion so that it executes uniform precession; that is, the wheel's center of mass moves with uniform circular motion with z -component of precessional angular velocity Ω_z. The wheel has mass m and moment of inertia I_{cm} about its center of mass. Its spin angular velocity $\vec{\boldsymbol{\omega}}_s$ has magnitude ω_s and is directed as shown in Figure 22.21. Assume that the gyroscope approximation holds, $|\Omega_z| << \omega_s$. Neglect the mass of the axle. What is the z-component of the precessional angular velocity Ω_z? Does the gyroscope rotate clockwise or counterclockwise about the vertical axis (as seen from above)?

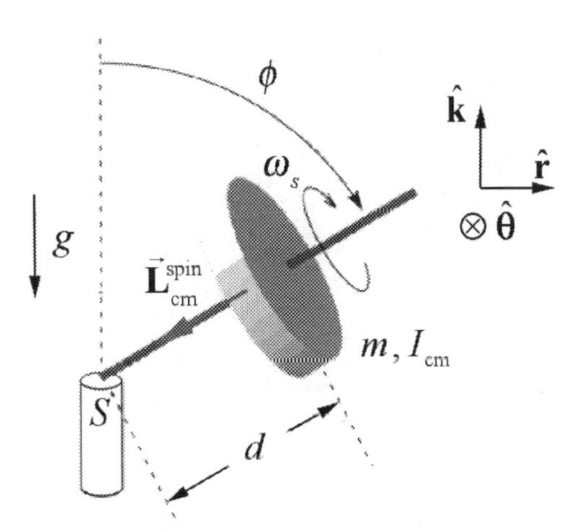

Figure 22.21 Example 22.3

Solution: The gravitational force acts at the center of mass and is directed downward, $\vec{\mathbf{F}}^g = -mg\,\hat{\mathbf{k}}$. Let S denote the contact point between the pylon and the axle. The contact force between the pylon and the axle is acting at S so it does not contribute to the torque about S. Only the gravitational force contributes to the torque. Let's choose cylindrical coordinates. The torque about S is

$$\vec{\boldsymbol{\tau}}_S = \vec{\mathbf{r}}_{S,\text{cm}} \times \vec{\mathbf{F}}^g = (d\sin\phi\,\hat{\mathbf{r}} + d\cos\phi\,\hat{\mathbf{k}}) \times mg(-\hat{\mathbf{k}}) = mgd\sin\phi\,\hat{\boldsymbol{\theta}}, \quad (22.4.1)$$

which is into the page in Figure 22.21. Because we are assuming that $|\Omega_z| << \omega_s$, we only consider contribution from the spinning about the flywheel axle to the spin angular momentum,

$$\vec{\boldsymbol{\omega}}_s = -\omega_s \sin\phi\,\hat{\mathbf{r}} - \omega_s \cos\phi\,\hat{\mathbf{k}} \quad (22.4.2)$$

The spin angular momentum has a vertical and radial component,

$$\vec{\mathbf{L}}^{\text{spin}}_{\text{cm}} = -I_{\text{cm}}\omega_s \sin\phi\,\hat{\mathbf{r}} - I_{\text{cm}}\omega_s \cos\phi\,\hat{\mathbf{k}}\,. \quad (22.4.3)$$

We assume that the spin angular velocity ω_s is constant. As the wheel precesses, the time derivative of the spin angular momentum arises from the change in the direction of the radial component of the spin angular momentum,

$$\frac{d}{dt}\vec{\mathbf{L}}_{\text{cm}}^{\text{spin}} = -I_{\text{cm}}\omega_s \sin\phi \frac{d\hat{\mathbf{r}}}{dt} = -I_{\text{cm}}\omega_s \sin\phi \frac{d\theta}{dt}\hat{\boldsymbol{\theta}}. \quad (22.4.4)$$

where we used the fact that

$$\frac{d\hat{\mathbf{r}}}{dt} = \frac{d\theta}{dt}\hat{\boldsymbol{\theta}}. \quad (22.4.5)$$

The z-component of the angular velocity of the flywheel about the vertical axis is defined to be

$$\Omega_z \equiv \frac{d\theta}{dt}. \quad (22.4.6)$$

Therefore the rate of change of the spin angular momentum is then

$$\frac{d}{dt}\vec{\mathbf{L}}_{\text{cm}}^{\text{spin}} = -I_{\text{cm}}\omega_s \sin\phi\, \Omega_z\, \hat{\boldsymbol{\theta}}. \quad (22.4.7)$$

The torque about S induces the spin angular momentum about S to change,

$$\vec{\boldsymbol{\tau}}_S = \frac{d\vec{\mathbf{L}}_{\text{cm}}^{\text{spin}}}{dt}. \quad (22.4.8)$$

Now substitute Equation (22.4.1) for the torque about S, and Equation (22.4.7) for the rate of change of the spin angular momentum into Equation (22.4.8), yielding

$$mgd\sin\phi\, \hat{\boldsymbol{\theta}} = -I_{\text{cm}}\omega_s \sin\phi\, \Omega_z\, \hat{\boldsymbol{\theta}}. \quad (22.4.9)$$

Solving Equation (22.2.18) for the z-component of the precessional angular velocity of the gyroscope yields

$$\Omega_z = -\frac{d\,mg}{I_{\text{cm}}\,\omega_s}. \quad (22.4.10)$$

The z-component of the precessional angular velocity is independent of the angle ϕ. Because $\Omega_z < 0$, the direction of the precessional angular velocity, $\vec{\Omega} = \Omega_z \hat{\mathbf{k}}$, is in the negative z-direction. That means that the gyroscope precesses in the clockwise direction when seen from above (Figure 21.22).

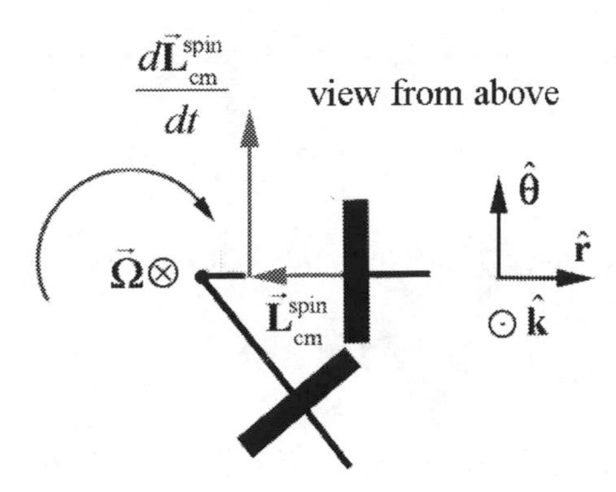

Figure 21.22 Precessional angular velocity of tilted gyroscope as seen from above

Both the torque and the time derivative of the spin angular momentum point in the $\hat{\boldsymbol{\theta}}$-direction indicating that the gyroscope will precess clockwise when seen from above in agreement with the calculation that $\Omega_z < 0$.

Example 22.4 Gyroscope on Rotating Platform

A gyroscope consists of an axle of negligible mass and a disk of mass M and radius R mounted on a platform that rotates with angular speed Ω. The gyroscope is spinning with angular speed ω. Forces F_a and F_b act on the gyroscopic mounts. What are the magnitudes of the forces F_a and F_b (Figure 22.22)? You may assume that the moment of inertia of the gyroscope about an axis passing through the center of mass normal to the plane of the disk is given by I_{cm}.

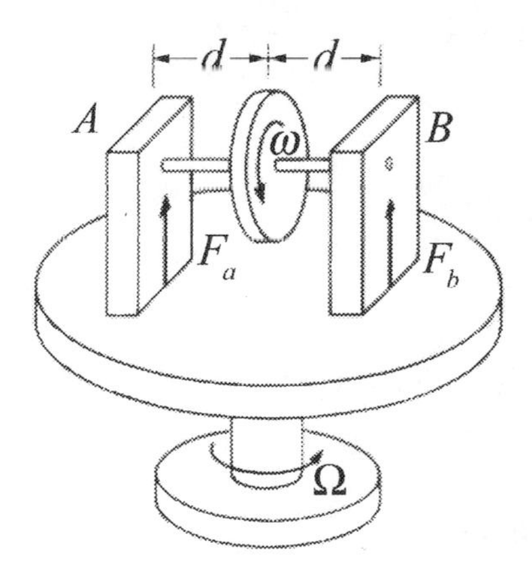

Figure 22.22 Example 22.4

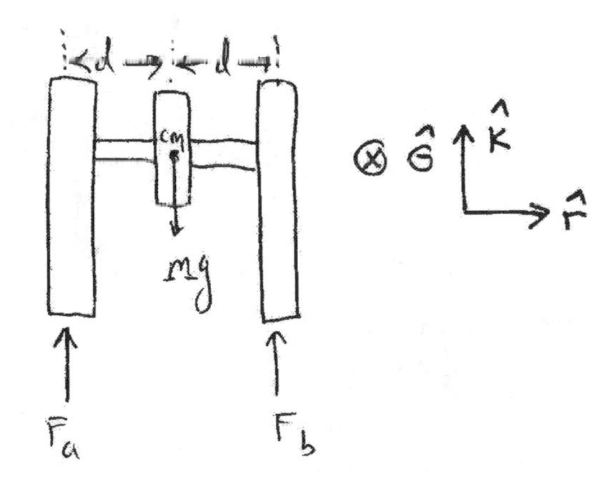

Figure 22.23 Free-body force diagram

Solution: Figure 22.23 shows a choice of coordinate system and force diagram on the gyroscope. The vertical forces sum to zero since there is no vertical motion

$$F_a + F_b - Mg = 0 \qquad (22.4.11)$$

Using the coordinate system depicted in the Figure 22.23, torque about the center of mass is

$$\vec{\boldsymbol{\tau}}_{\text{cm}} = d(F_a - F_b)\hat{\boldsymbol{\theta}} \tag{22.4.12}$$

The spin angular momentum is (gyroscopic approximation)

$$\vec{\mathbf{L}}_{\text{cm}}^{\text{spin}} \simeq I_{\text{cm}}\omega\,\hat{\mathbf{r}} \tag{22.4.13}$$

Looking down on the gyroscope from above (Figure 2.23), the radial component of the angular momentum about the center of mass is rotating counterclockwise.

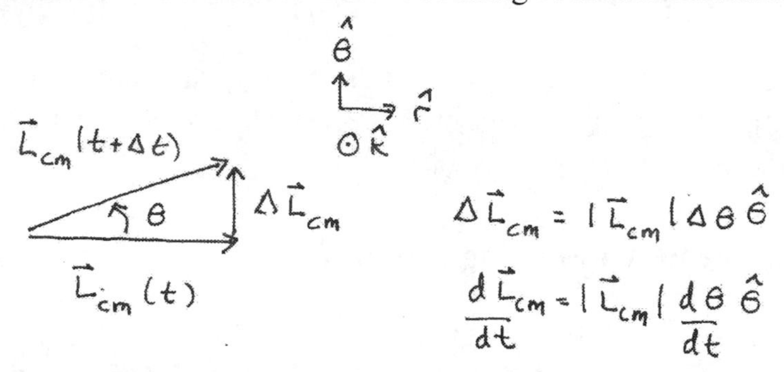

Figure 22.24 Change in angular momentum

During a very short time interval Δt, the change in the spin angular momentum is $\Delta\vec{\mathbf{L}}_{\text{cm}}^{\text{spin}} = I_{\text{cm}}\omega\Delta\theta\,\hat{\boldsymbol{\theta}}$, (Figure 22.24). Taking limits we have that

$$\frac{d\vec{\mathbf{L}}_{\text{cm}}^{\text{spin}}}{dt} = \lim_{\Delta t \to 0}\frac{\Delta\vec{\mathbf{L}}_{\text{cm}}^{\text{spin}}}{\Delta t} = \lim_{\Delta t \to 0} I_{\text{cm}}\omega\frac{\Delta\theta}{\Delta t}\hat{\boldsymbol{\theta}} = I_{\text{cm}}\omega\frac{d\theta}{dt}\hat{\boldsymbol{\theta}} \tag{22.4.14}$$

We can now apply the torque law

$$\vec{\boldsymbol{\tau}}_{\text{cm}} = \frac{d\vec{\mathbf{L}}_{\text{cm}}^{\text{spin}}}{dt}. \tag{22.4.15}$$

Substitute Eqs. (22.4.12) and (22.4.14) into Eq. (22.4.15) and just taking the component of the resulting vector equation yields

$$d(F_a - F_b) = I_{\text{cm}}\omega\,\Omega_z. \tag{22.4.16}$$

We can divide Eq. (22.4.16) by the quantity d yielding

$$F_a - F_b = \frac{I_{\text{cm}}\omega\,\Omega_z}{d}. \tag{22.4.17}$$

We can now use Eqs. (22.4.17) and (22.4.11) to solve for the forces F_a and F_b,

$$F_a = \frac{1}{2}\left(Mg + \frac{I_{\text{cm}}\,\omega\,\Omega_z}{d} \right) \qquad (22.4.18)$$

$$F_b = \frac{1}{2}\left(Mg - \frac{I_{\text{cm}}\,\omega\,\Omega_z}{d} \right). \qquad (22.4.19)$$

Note that if $\Omega_z = Mgd / I_{\text{cm}}\omega$ then $F_b = 0$ and one could remove the right hand support in the Figure 22.22. The simple pivoted gyroscope that we already analyzed Section 22.2 satisfied this condition. The forces we just found are the forces that the mounts must exert on the gyroscope in order to cause it to move in the desired direction. It is important to understand that the gyroscope is exerting equal and opposite forces on the mounts, i.e. the structure that is holding it. This is a manifestation of Newton's Third Law.

Example 22.5 Grain Mill

In a mill, grain is ground by a massive wheel that rolls without slipping in a circle on a flat horizontal millstone driven by a vertical shaft. The rolling wheel has mass M, radius b and is constrained to roll in a horizontal circle of radius R at angular speed Ω (Figure 22.25). The wheel pushes down on the lower millstone with a force equal to twice its weight (normal force). The mass of the axle of the wheel can be neglected. What is the precessional angular frequency Ω?

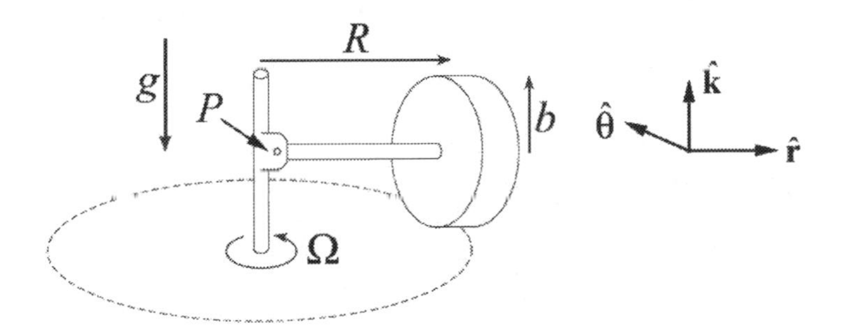

Figure 22.25 Example 22.5

Solution: Figure 22.5 shows the pivot point along with some convenient coordinate axes. For rolling without slipping, the speed of the center of mass of the wheel is related to the angular spin speed by

$$v_{cm} = b\omega\,. \qquad (22.4.20)$$

Also the speed of the center of mass is related to the angular speed about the vertical axis associated with the circular motion of the center of mass by

$$v_{cm} = R\Omega\,. \qquad (22.4.21)$$

Therefore equating Eqs. (22.4.20) and (22.4.21) we have that

$$\omega = \Omega R / b . \qquad (22.4.22)$$

Assuming a uniform millwheel, $I_{\text{cm}} = (1/2)Mb^2$, the magnitude of the horizontal component of the spin angular momentum about the center of mass is

$$L_{\text{cm}}^{\text{spin}} = I_{\text{cm}}\omega = \frac{1}{2}Mb^2\omega = \frac{1}{2}\Omega\, MRb . \qquad (22.4.23)$$

The horizontal component of $\vec{\mathbf{L}}_{\text{cm}}^{\text{spin}}$ is directed inward, and in vector form is given by

$$\vec{\mathbf{L}}_{\text{cm}}^{\text{spin}} = -\frac{\Omega\, MRb}{2}\hat{\mathbf{r}} . \qquad (22.4.24)$$

The axle exerts both a force and torque on the wheel, and this force and torque would be quite complicated. That's why we consider the forces and torques on the axle/wheel combination. The normal force of the wheel on the ground is equal in magnitude to $N_{WG} = 2mg$ so the third-law counterpart; the normal force of the ground on the wheel has the same magnitude $N_{GW} = 2mg$. The joint (or hinge) at point P therefore must exert a force $\vec{\mathbf{F}}_{\text{H,A}}$ on the end of the axle that has two components forces an inward force $\vec{\mathbf{F}}_2$ to maintain the circular motion and a downward force $\vec{\mathbf{F}}_1$ to reflect that the upward normal force is larger in magnitude than the weight (Figure 22.26).

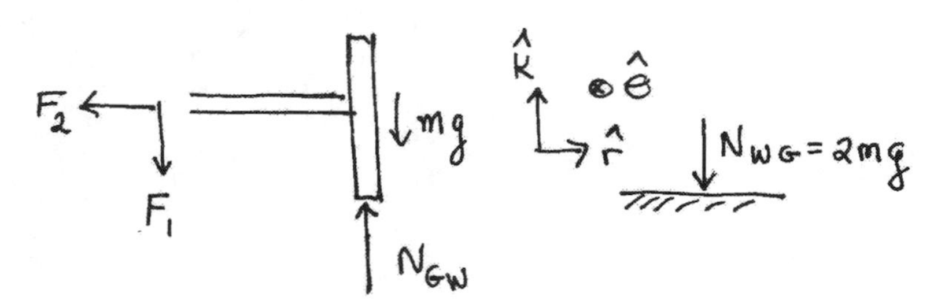

Figure 22.26 Free-body force diagram on wheel

About point P, $\vec{\mathbf{F}}_{\text{H,A}}$ exerts no torque. The normal force exerts a torque of magnitude $N_{GW}R = 2mgR$, directed out of the page, or, in vector form, $\vec{\tau}_{P,N} = -2mgR\hat{\boldsymbol{\theta}}$. The weight exerts a toque of magnitude MgR, directed into the page, or, in vector form, $\vec{\tau}_{P,mg} = MgR\hat{\boldsymbol{\theta}}$. The torque about P is then

$$\vec{\tau}_P = \vec{\tau}_{P,N} + \vec{\tau}_{P,mg} = -2mgR\hat{\boldsymbol{\theta}} + MgR\hat{\boldsymbol{\theta}} = -MgR\hat{\boldsymbol{\theta}} . \qquad (22.4.25)$$

As the wheel rolls, the horizontal component of the angular momentum about the center of mass will rotate, and the inward-directed vector will change in the negative $\hat{\boldsymbol{\theta}}$-direction. The angular momentum about the point P has orbital and spin decomposition

$$\vec{\mathbf{L}}_P = \vec{\mathbf{L}}_P^{\text{orbital}} + \vec{\mathbf{L}}_{\text{cm}}^{\text{spin}} . \qquad (22.4.26)$$

The orbital angular momentum about the point P is

$$\vec{\mathbf{L}}_P^{\text{orbital}} = \vec{\mathbf{r}}_{P,cm} \times m\vec{\mathbf{v}}_{cm} = R\,\hat{\mathbf{r}} \times mb\Omega\,\hat{\boldsymbol{\theta}} = mRb\Omega_z\,\hat{\mathbf{k}} . \qquad (22.4.27)$$

The magnitude of the orbital angular momentum about P is nearly constant and the dircction does not change. Therefore

$$\frac{d\vec{\mathbf{L}}_P^{\text{orbital}}}{dt} = \vec{\mathbf{0}} . \qquad (22.4.28)$$

Therefore the change in angular momentum about the point P is

$$\frac{d\vec{\mathbf{L}}_P}{dt} = \frac{d\vec{\mathbf{L}}_{\text{cm}}^{\text{spin}}}{dt} = \frac{d}{dt}\left(\frac{\Omega\, MRb}{2}(-\hat{\mathbf{r}})\right) = \frac{1}{2}\Omega\, MRb\Omega(-\hat{\boldsymbol{\theta}}) , \qquad (22.4.29)$$

where we used Eq. (22.4.24) for the magnitude of the horizontal component of the angular momentum about the center of mass. This is consistent with the torque about P pointing out of the plane of Figure 22.26. We can now apply the rotational equation of motion,

$$\vec{\boldsymbol{\tau}}_P = \frac{d\vec{\mathbf{L}}_P}{dt} . \qquad (22.4.30)$$

Substitute Eqs.(22.4.25) and (22.4.29) into Eq. (22.4.30) yielding

$$MgR(-\hat{\boldsymbol{\theta}}) = \frac{1}{2}\Omega^2 MRb(-\hat{\boldsymbol{\theta}}). \qquad (22.4.31)$$

We can now solve Eq. (22.4.31) for the angular speed about the vertical axis

$$\Omega = \sqrt{\frac{2g}{b}} . \qquad (22.4.32)$$

Recall that in the gyroscopic approximation, we assume that the magnitude of the spin angular velocity $|\vec{\boldsymbol{\omega}}_s| = |\omega_s|$ is constant. In this approximation, as the flywheel precesses, the spin angular momentum changes its direction.

Chapter 23 Simple Harmonic Motion

Chapter 23 Simple Harmonic Motion

...Indeed it is not in the nature of a simple pendulum to provide equal and reliable measurements of time, since the wide lateral excursions often made may be observed to be slower than more narrow ones; however, we have been led in a different direction by geometry, from which we have found a means of suspending the pendulum, with which we were previously unacquainted, and by giving close attention to a line with a certain curvature, the time of the swing can be chosen equal to some calculated value and is seen clearly in practice to be in wonderful agreement with that ratio. As we have checked the lapses of time measured by these clocks after making repeated land and sea trials, the effects of motion are seen to have been avoided, so sure and reliable are the measurements; now it can be seen that both astronomical studies and the art of navigation will be greatly helped by them. The curved line that a nail fixed to the circumference of a running wheel traces out in air by the continued rotation of the wheel has been given the name cycloid by our geometers of the day, and on account of other things, a great many of its properties have been diligently pondered over. As we have said, we have considered this curve on account of its ability to measure time, and we have learned that it is completely trustworthy and there is so much of the art of insisting on repetition present in this curve. ... In the first place it is not necessary to go beyond the great Galileo and to establish firmly the teachings of the descent of a body under gravity; and just as in his teachings the highest peak is the most desired, so we have found that this is a property of the cycloid.[1]

Christian Huygens

23.1 Introduction: Periodic Motion

There are two basic ways to measure time: by duration or periodic motion. Early clocks measured duration by calibrating the burning of incense or wax, or the flow of water or sand from a container. Our calendar consists of years determined by the motion of the sun; months determined by the motion of the moon; days by the rotation of the earth; hours by the motion of cyclic motion of gear trains; and seconds by the oscillations of springs or pendulums. In modern times a second is defined by a specific number of vibrations of radiation, corresponding to the transition between the two hyperfine levels of the ground state of the cesium 133 atom.

Sundials calibrate the motion of the sun through the sky including seasonal corrections. A clock escapement is a device that can transform continuous movement into discrete movements of a gear train. The early escapements used oscillatory motion to stop

---

[1] Christian Huygens, *The Pendulum Clock or The Motion of Pendulums Adapted to Clocks By Geometrical Demonstrations,* tr Ian Bruce, p. 1.

and start the turning of a weight-driven rotating drum. Soon, complicated escapements were regulated by pendulums, the theory of which was first developed by the physicist Christian Huygens in the mid 17th century. The accuracy of clocks was increased and the size reduced by the discovery of the oscillatory properties of springs by Robert Hooke. By the middle of the 18th century, the technology of timekeeping advanced to the point that William Harrison developed timekeeping devices that were accurate to one second in a century.

23.3.1 Simple Harmonic Motion

One of the most important examples of periodic motion is ***simple harmonic motion*** (SHM), in which some physical quantity varies sinusoidally. Suppose a function of time has the form of a sine wave function,

$$y(t) = A\sin(2\pi t / T) = A\sin(2\pi f t), \tag{23.1.1}$$

where $A > 0$ is the ***amplitude*** (maximum value). The function $y(t)$ varies between A and $-A$, because a sine function varies between $+1$ and -1. A plot of $y(t)$ *vs.* time is shown in Figure 23.1 (with $A = 3$ and $T = \pi$).

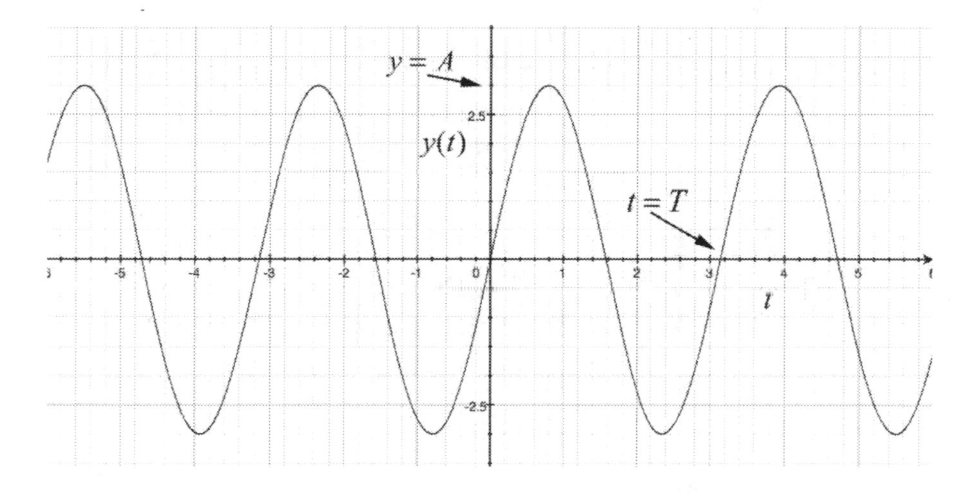

Figure 23.1 Sinusoidal function of time

The sine function is periodic in time. This means that the value of the function at time t will be exactly the same at a later time $t' = t + T$, where T is the ***period***. That the sine function satisfies the periodic condition can be seen from

$$y(t+T) = A\sin\left[\frac{2\pi}{T}(t+T)\right] = A\sin\left[\frac{2\pi}{T}t + 2\pi\right] = A\sin\left[\frac{2\pi}{T}t\right] = y(t). \tag{23.1.2}$$

The ***frequency***, f, is defined to be

$$f \equiv 1/T. \tag{23.1.3}$$

The SI unit of frequency is inverse seconds, $\left[\mathrm{s}^{-1}\right]$, or hertz $[\mathrm{Hz}]$. The ***angular frequency*** of oscillation is defined to be

$$\omega_0 \equiv 2\pi / T = 2\pi f\,, \qquad (23.1.4)$$

and is measured in radians per second. (The angular frequency of oscillation is denoted by ω_0 to distinguish from the angular speed $\omega = |d\theta / dt|$.) One oscillation per second, $1\,\mathrm{Hz}$, corresponds to an angular frequency of 2π $\mathrm{rad \cdot s^{-1}}$. (Unfortunately, the same symbol ω is used for angular speed in circular motion. For uniform circular motion the angular speed is equal to the angular frequency but for non-uniform motion the angular speed is not constant. The angular frequency for simple harmonic motion is a constant by definition.)

23.2 Simple Harmonic Motion

Our first example of a system that demonstrates simple harmonic motion is a spring-object system on a frictionless surface, shown in Figure 23.2

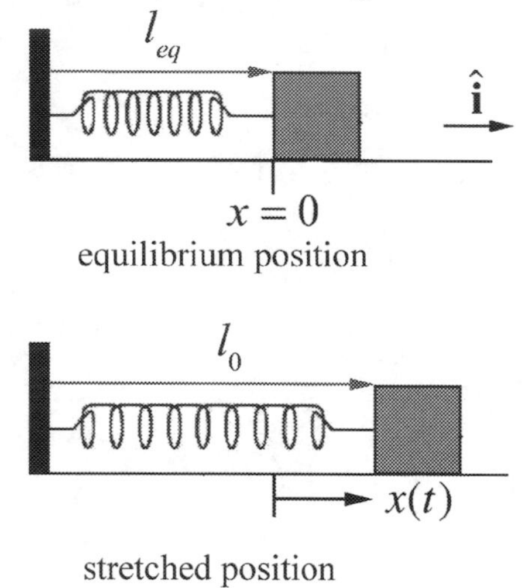

Figure 23.2 Spring-object system

The object is attached to one end of a spring. The other end of the spring is attached to a wall at the left in Figure 23.2. Assume that the object undergoes one-dimensional motion. The spring has a spring constant k and equilibrium length l_{eq}. Choose the origin at the equilibrium position and choose the positive x-direction to the right in the Figure 23.2. In the figure, $x > 0$ corresponds to an extended spring, and $x < 0$ to a compressed spring. Define $x(t)$ to be the position of the object with respect to the equilibrium position. The force acting on the spring is a linear restoring force, $F_x = -k\,x$ (Figure 23.3). The initial conditions are as follows. The spring is initially stretched a distance l_0 and given some initial speed v_0 to the right away from the equilibrium position. The initial position of the stretched spring from the equilibrium position (our choice of origin) is $x_0 = (l_0 - l_{eq}) > 0$ and its initial x-component of the velocity is $v_{x,0} = v_0 > 0$.

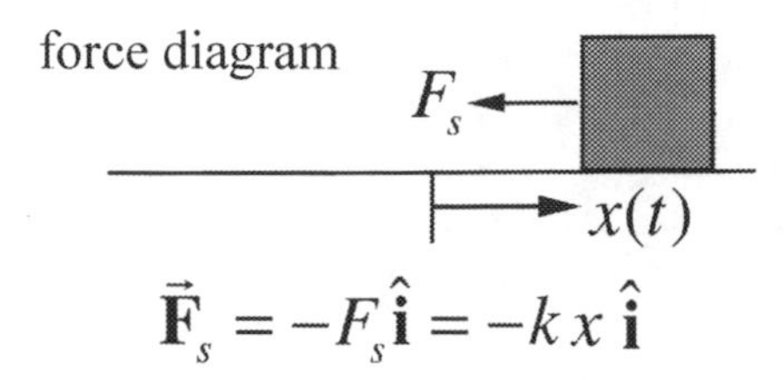

Figure 23.3 Free-body force diagram for spring-object system

Newton's Second law in the x-direction becomes

$$-k\,x = m\frac{d^2x}{dt^2}. \tag{23.2.1}$$

This equation of motion, Equation (23.2.1), is called the ***simple harmonic oscillator equation*** (SHO). Because the spring force depends on the distance x, the acceleration is not constant. Equation (23.2.1) is a second order linear differential equation, in which the second derivative of the dependent variable is proportional to the negative of the dependent variable,

$$\frac{d^2x}{dt^2} = -\frac{k}{m}x. \tag{23.2.2}$$

In this case, the constant of proportionality is k/m,

Equation (23.2.2) can be solved from energy considerations or other advanced techniques but instead we shall first guess the solution and then verify that the guess satisfies the SHO differential equation (see Appendix 22.3.A for a derivation of the solution).

We are looking for a position function $x(t)$ such that the second time derivative position function is proportional to the negative of the position function. Since the sine and cosine functions both satisfy this property, we make a preliminary guess that our position function is given by

$$x(t) = A\cos((2\pi/T)t) = A\cos(\omega_0 t), \tag{23.2.3}$$

where ω_0 is the angular frequency (as of yet, undetermined). (In Equation (23.2.3), the constant A is not necessarily the amplitude of the motion; A is the amplitude for the case $x_0 > 0$, $v_0 = 0$.)

We shall now find the condition that the angular frequency ω_0 must satisfy in order to insure that the function in Eq. (23.2.3) solves the simple harmonic oscillator equation, Eq. (23.2.1). The first and second derivatives of the position function are given by

$$\begin{aligned}\frac{dx}{dt} &= -\omega_0 A\sin(\omega_0 t)\\ \frac{d^2x}{dt^2} &= -\omega_0^2 A\cos(\omega_0 t) = -\omega_0^2 x.\end{aligned} \tag{23.2.4}$$

Substitute the second derivative, the second expression in Equation (23.2.4), and the position function, Equation (23.2.3), into the SHO Equation (23.2.1), yielding

$$-\omega_0^2 A\cos(\omega_0 t) = -\frac{k}{m} A\cos(\omega_0 t). \tag{23.2.5}$$

Equation (23.2.5) is valid for all times provided that

$$\omega_0 = \sqrt{\frac{k}{m}}. \tag{23.2.6}$$

The period of oscillation is then

$$T = \frac{2\pi}{\omega_0} = 2\pi\sqrt{\frac{m}{k}}. \tag{23.2.7}$$

One possible solution for the position of the block is

$$x(t) = A\cos\left(\sqrt{\frac{k}{m}}\,t\right), \tag{23.2.8}$$

and therefore by differentiation, the x-component of the velocity of the block is

$$v_x(t) = -\sqrt{\frac{k}{m}}\,A\sin\left(\sqrt{\frac{k}{m}}\,t\right). \tag{23.2.9}$$

Note that at $t = 0$, the position of the object is $x_0 \equiv x(t=0) = A$ since $\cos(0) = 1$ and the velocity is $v_{x,0} \equiv v_x(t=0) = 0$ since $\sin(0) = 0$. The solution in (23.2.8) describes an object that is released from rest at an initial position $A = x_0$ but does not satisfy the initial velocity condition, $v_x(t=0) = v_{x,0} \neq 0$. We can try a sine function as another possible solution,

$$x(t) = B\sin\left(\sqrt{\frac{k}{m}}t\right). \qquad (23.2.10)$$

This function also satisfies the simple harmonic oscillator equation because

$$\frac{d^2x}{dt^2} = -\frac{k}{m}B\sin\left(\sqrt{\frac{k}{m}}t\right) = -\omega_0^{\ 2}x, \qquad (23.2.11)$$

where $\omega_0 = \sqrt{k/m}$. The x-component of the velocity associated with Eq. (23.2.10) is

$$v_x(t) = \frac{dx}{dt} = \sqrt{\frac{k}{m}}B\cos\left(\sqrt{\frac{k}{m}}t\right). \qquad (23.2.12)$$

The proposed solution in Eq. (23.2.10) has initial conditions $x_0 \equiv x(t=0) = 0$ and $v_{x,0} \equiv v_x(t=0) = (\sqrt{k/m})B$, thus $B = v_{x,0}/\sqrt{k/m}$. This solution describes an object that is initially at the equilibrium position but has an initial non-zero x-component of the velocity, $v_{x,0} \neq 0$.

23.2.1 General Solution of Simple Harmonic Oscillator Equation

Suppose $x_1(t)$ and $x_2(t)$ are both solutions of the simple harmonic oscillator equation,

$$\begin{aligned} \frac{d^2}{dt^2}x_1(t) &= -\frac{k}{m}x_1(t) \\ \frac{d^2}{dt^2}x_2(t) &= -\frac{k}{m}x_2(t). \end{aligned} \qquad (23.2.13)$$

Then the sum $x(t) = x_1(t) + x_2(t)$ of the two solutions is also a solution. To see this, consider

$$\frac{d^2x(t)}{dt^2} = \frac{d^2}{dt^2}(x_1(t) + x_2(t)) = \frac{d^2x_1(t)}{dt^2} + \frac{d^2x_2(t)}{dt^2}. \qquad (23.2.14)$$

Using the fact that $x_1(t)$ and $x_2(t)$ both solve the simple harmonic oscillator equation (23.2.13), we see that

$$\begin{aligned} \frac{d^2}{dt^2}x(t) &= -\frac{k}{m}x_1(t) + -\frac{k}{m}x_2(t) = -\frac{k}{m}(x_1(t) + x_2(t)) \\ &= -\frac{k}{m}x(t). \end{aligned} \qquad (23.2.15)$$

Thus the *linear combination* $x(t) = x_1(t) + x_2(t)$ is also a solution of the SHO equation, Equation (23.2.1). Therefore the sum of the sine and cosine solutions is the *general solution*,

$$x(t) = C\cos(\omega_0 t) + D\sin(\omega_0 t), \qquad (23.2.16)$$

where the constant coefficients C and D depend on a given set of initial conditions $x_0 \equiv x(t=0)$ and $v_{x,0} \equiv v_x(t=0)$ where x_0 and $v_{x,0}$ are constants. For this general solution, the x-component of the velocity of the object at time t is then obtained by differentiating the position function,

$$v_x(t) = \frac{dx}{dt} = -\omega_0 C\sin(\omega_0 t) + \omega_0 D\cos(\omega_0 t). \qquad (23.2.17)$$

To find the constants C and D, substitute $t = 0$ into the Equations (23.2.16) and (23.2.17). Because $\cos(0) = 1$ and $\sin(0) = 0$, the initial position at time $t = 0$ is

$$x_0 \equiv x(t=0) = C. \qquad (23.2.18)$$

The x-component of the velocity at time $t = 0$ is

$$v_{x,0} = v_x(t=0) = -\omega_0 C\sin(0) + \omega_0 D\cos(0) = \omega_0 D. \qquad (23.2.19)$$

Thus

$$C = x_0 \text{ and } D = \frac{v_{x,0}}{\omega_0}. \qquad (23.2.20)$$

The position of the object-spring system is then given by

$$x(t) = x_0\cos\left(\sqrt{\frac{k}{m}}t\right) + \frac{v_{x,0}}{\sqrt{k/m}}\sin\left(\sqrt{\frac{k}{m}}t\right) \qquad (23.2.21)$$

and the x-component of the velocity of the object-spring system is

$$v_x(t) = -\sqrt{\frac{k}{m}}x_0\sin\left(\sqrt{\frac{k}{m}}t\right) + v_{x,0}\cos\left(\sqrt{\frac{k}{m}}t\right). \qquad (23.2.22)$$

Although we had previously specified $x_0 > 0$ and $v_{x,0} > 0$, Equation (23.2.21) is a valid solution of the SHO equation for any values of x_0 and $v_{x,0}$.

Example 23.1 Phase and Amplitude

Show that $x(t) = C\cos\omega_0 t + C\sin\omega_0 t = A\cos(\omega_0 t + \phi)$, where $A = (C^2 + D^2)^{1/2} > 0$, and $\phi = \tan^{-1}(-D / C)$.

Solution: Use the identity $A\cos(\omega_0 t + \phi) = A\cos(\omega_0 t)\cos(\phi) - A\sin(\omega_0 t)\sin(\phi)$. Thus $C\cos(\omega_0 t) + D\sin(\omega_0 t) = A\cos(\omega_0 t)\cos(\phi) - A\sin(\omega_0 t)\sin(\phi)$. Comparing coefficients we see that $C = A\cos\phi$ and $D = -A\sin\phi$. Therefore $(C^2 + D^2)^{1/2} = A^2(\cos^2\phi + \sin^2\phi) = A^2$. We choose the positive square root to insure that $A > 0$, and thus

$$A = (C^2 + D^2)^{1/2} \tag{23.2.23}$$

$$\tan\phi = \frac{\sin\phi}{\cos\phi} = \frac{-D / A}{C / A} = -\frac{D}{C},$$

$$\phi = \tan^{-1}(-D / C). \tag{23.2.24}$$

Thus the position as a function of time can be written as

$$x(t) = A\cos(\omega_0 t + \phi). \tag{23.2.25}$$

In Eq. (23.2.25) the quantity $\omega_0 t + \phi$ is called the ***phase***, and ϕ is called the ***phase constant.*** Because $\cos(\omega_0 t + \phi)$ varies between $+1$ and -1, and $A > 0$, A is the amplitude defined earlier. We now substitute Eq. (23.2.20) into Eq. (23.2.23) and find that the amplitude of the motion described in Equation (23.2.21), that is, the maximum value of $x(t)$, and the phase are given by

$$A = \sqrt{x_0^2 + (v_{x,0} / \omega_0)^2}. \tag{23.2.26}$$

$$\phi = \tan^{-1}(-v_{x,0} / \omega_0 x_0). \tag{23.2.27}$$

A plot of $x(t)$ vs. t is shown in Figure 23.4a with the values $A = 3$, $T = \pi$, and $\phi = \pi / 4$. Note that $x(t) = A\cos(\omega_0 t + \phi)$ takes on its maximum value when $\cos(\omega_0 t + \phi) = 1$. This occurs when $\omega_0 t + \phi = 2\pi n$ where $n = 0, \pm 1, \pm 2, \cdots$. The maximum value associated with $n = 0$ occurs when $\omega_0 t + \phi = 0$ or $t = -\phi / \omega_0$. For the case shown in Figure 23.4a where $\phi = \pi / 4$, this maximum occurs at the instant $t = -T / 8$. Let's plot $x(t) = A\cos(\omega_0 t + \phi)$ vs. t for $\phi = 0$ (Figure 23.4b). For $\phi > 0$, Figure 23.4a shows the plot $x(t) = A\cos(\omega_0 t + \phi)$ vs. t. Notice that when $\phi > 0$, $x(t)$ is shifted to the left compared with the case $\phi = 0$ (compare Figures 23.4a with 23.4b). The function $x(t) = A\cos(\omega_0 t + \phi)$ with $\phi > 0$ reaches it's a maximum value at an earlier time than the function $x(t) = A\cos(\omega_0 t)$. The difference in phases for these two cases is $(\omega_0 t + \phi) - \omega_0 t = \phi$ and ϕ is sometimes referred to as the *phase shift*. When $\phi < 0$, the function $x(t) = A\cos(\omega_0 t + \phi)$ reaches it's a maximum value at a later time $t = T / 8$ than the function $x(t) = A\cos(\omega_0 t)$ as shown in Figure 23.4c.

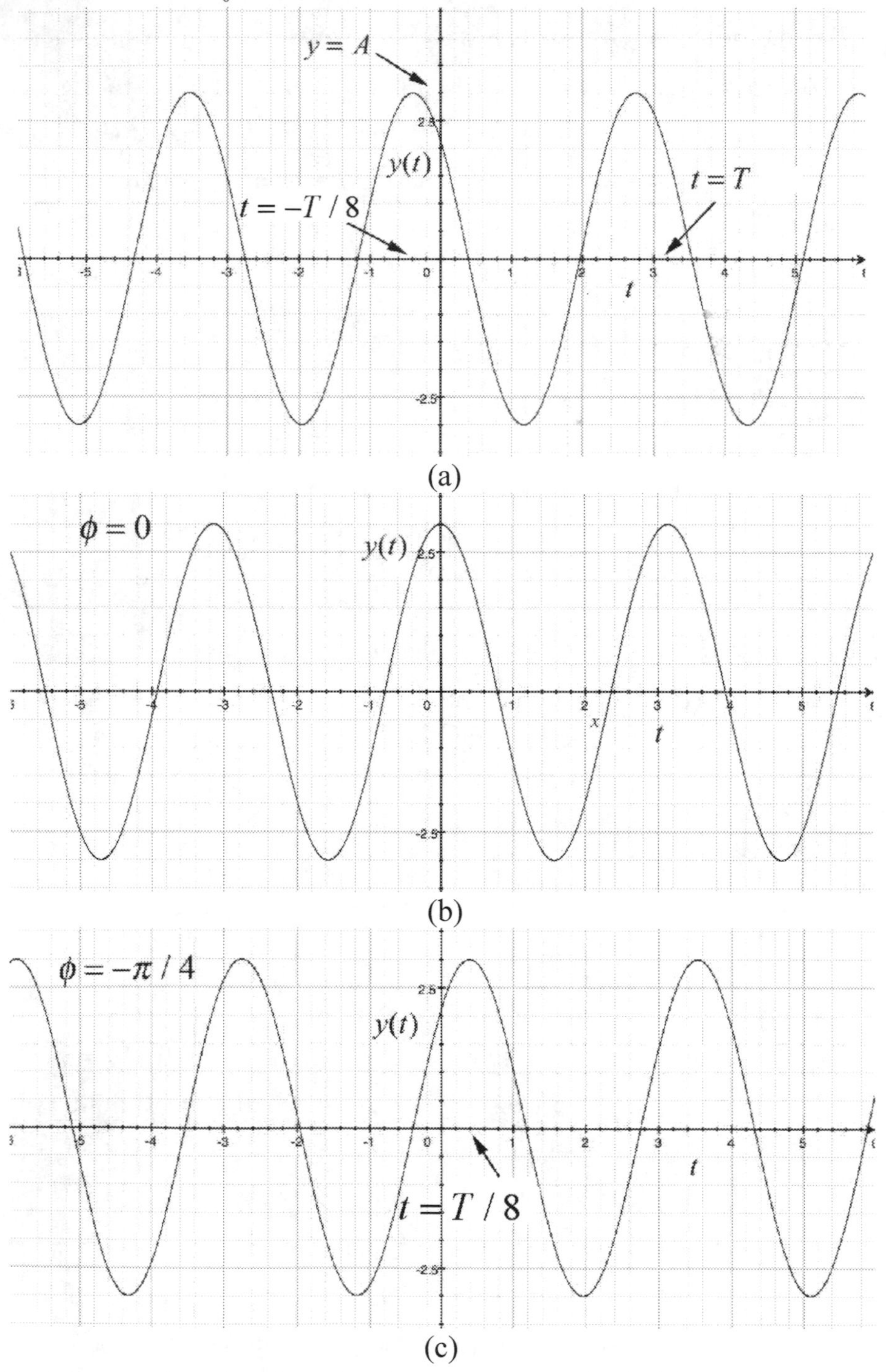

Figure 23.5 Phase shift of $x(t) = A\cos(\omega_0 t + \phi)$ (a) to the left by $\phi = \pi / 4$, (b) no shift $\phi = 0$, (c) to the right $\phi = -\pi / 4$

Example 23.2 Block-Spring System

A block of mass m is attached to a spring and is free to slide along a horizontal frictionless surface. At $t = 0$, the block-spring system is stretched an amount $x_0 > 0$ from the equilibrium position and is released from rest, $v_{x,0} = 0$. What is the period of oscillation of the block? What is the velocity of the block when it first comes back to the equilibrium position?

Solution: The position of the block can be determined from Equation (23.2.21) by substituting the initial conditions $x_0 > 0$, and $v_{x,0} = 0$ yielding

$$x(t) = x_0 \cos\left(\sqrt{\frac{k}{m}}\,t\right), \tag{23.2.28}$$

and its x-component of its velocity is given by Equation (23.2.22),

$$v_x(t) = -\sqrt{\frac{k}{m}}\,x_0 \sin\left(\sqrt{\frac{k}{m}}\,t\right). \tag{23.2.29}$$

The angular frequency of oscillation is $\omega_0 = \sqrt{k/m}$ and the period is given by Equation (23.2.7),

$$T = \frac{2\pi}{\omega_0} = 2\pi\sqrt{\frac{m}{k}}\,. \tag{23.2.30}$$

The block first reaches equilibrium when the position function first reaches zero. This occurs at time t_1 satisfying

$$\sqrt{\frac{k}{m}}\,t_1 = \frac{\pi}{2}, \quad t_1 = \frac{\pi}{2}\sqrt{\frac{m}{k}} = \frac{T}{4}. \tag{23.2.31}$$

The x-component of the velocity at time t_1 is then

$$v_x(t_1) = -\sqrt{\frac{k}{m}}\,x_0 \sin\left(\sqrt{\frac{k}{m}}\,t_1\right) = -\sqrt{\frac{k}{m}}\,x_0 \sin(\pi/2) = -\sqrt{\frac{k}{m}}\,x_0 = -\omega_0\,x_0 \tag{23.2.32}$$

Note that the block is moving in the negative x-direction at time t_1; the block has moved from a positive initial position to the equilibrium position.

23.3 Energy and the Simple Harmonic Oscillator

Let's consider the block-spring system of Example 23.2 in which the block is initially stretched an amount $x_0 > 0$ from the equilibrium position and is released from rest, $v_{x,0} = 0$. We shall consider three states: state 1, the initial state; state 2, at an arbitrary time in which the position and velocity are non-zero; and state 3, when the object first comes back to the equilibrium position. We shall show that the mechanical energy is constant throughout the motion. Choose the equilibrium position for the zero point of the potential energy.

State 1: all the energy is stored in the object-spring potential energy, $U_1 = (1/2)k\,x_0^2$. The object is released from rest so the kinetic energy is zero, $K_1 = 0$. The total mechanical energy is then

$$E_1 = U_1 = \frac{1}{2}k\,x_0^2. \tag{23.3.1}$$

State 2: at some time t, the position and x-component of the velocity of the object are given by

$$\begin{aligned} x(t) &= x_0 \cos\left(\sqrt{\frac{k}{m}}t\right) \\ v_x(t) &= -\sqrt{\frac{k}{m}}x_0 \sin\left(\sqrt{\frac{k}{m}}t\right). \end{aligned} \tag{23.3.2}$$

The kinetic energy is

$$K_2 = \frac{1}{2}mv^2 = \frac{1}{2}k\,x_0^2 \sin^2\left(\sqrt{\frac{k}{m}}\,t\right), \tag{23.3.3}$$

and the potential energy is

$$U_2 = \frac{1}{2}k\,x^2 = \frac{1}{2}k\,x_0^2 \cos^2\left(\sqrt{\frac{k}{m}}\,t\right). \tag{23.3.4}$$

The mechanical energy is the sum of the kinetic and potential energies

$$\begin{aligned} E_2 &= K_2 + U_2 = \frac{1}{2}mv^2 + \frac{1}{2}k\,x^2 \\ &= \frac{1}{2}k\,x_0^2\left(\cos^2\left(\sqrt{\frac{k}{m}}\,t\right) + \sin^2\left(\sqrt{\frac{k}{m}}t\right)\right) \\ &= \frac{1}{2}k\,x_0^2, \end{aligned} \tag{23.3.5}$$

where we used the identity that $\cos^2\omega_0 t + \sin^2\omega_0 t = 1$, and that $\omega_0 = \sqrt{k/m}$.

The mechanical energy in state 2 is equal to the initial potential energy in state 1, so the mechanical energy is constant. This should come as no surprise; we isolated the object-spring system so that there is no external work performed on the system and no internal non-conservative forces doing work.

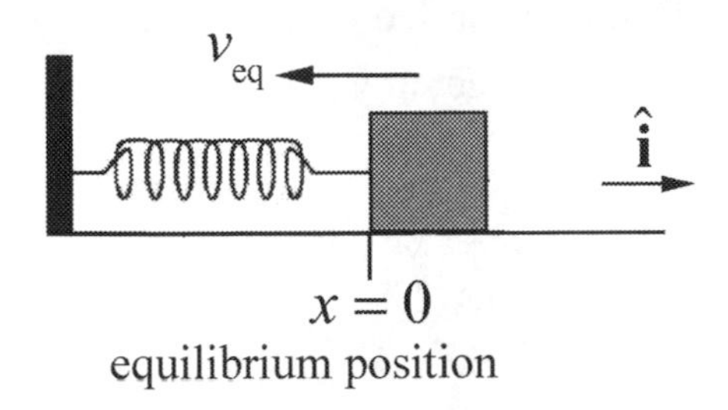

Figure 23.5 State 3 at equilibrium and in motion

State 3: now the object is at the equilibrium position so the potential energy is zero, $U_3 = 0$, and the mechanical energy is in the form of kinetic energy (Figure 23.5).

$$E_3 = K_3 = \frac{1}{2} m v_{eq}^2 . \tag{23.3.6}$$

Because the system is closed, mechanical energy is constant,

$$E_1 = E_3 . \tag{23.3.7}$$

Therefore the initial stored potential energy is released as kinetic energy,

$$\frac{1}{2} k x_0^2 = \frac{1}{2} m v_{eq}^2 , \tag{23.3.8}$$

and the x-component of velocity at the equilibrium position is given by

$$v_{x,eq} = \pm\sqrt{\frac{k}{m}}\, x_0 . \tag{23.3.9}$$

Note that the plus-minus sign indicates that when the block is at equilibrium, there are two possible motions: in the positive x-direction or the negative x-direction. If we take $x_0 > 0$, then the block starts moving towards the origin, and $v_{x,eq}$ will be negative the first time the block moves through the equilibrium position.

23.3.1 Simple Pendulum Force Approach

A pendulum consists of an object hanging from the end of a string or rigid rod pivoted about the point P. The object is pulled to one side and allowed to oscillate. If the object has negligible size and the string or rod is massless, then the pendulum is called a ***simple pendulum***. Consider a simple pendulum consisting of a massless string of length l and a point-like object of mass m attached to one end, called the *bob*. Suppose the string is fixed at the other end and is initially pulled out at a small angle θ_0 from the vertical and released from rest (Figure 23.6). Neglect any dissipation due to air resistance or frictional forces acting at the pivot.

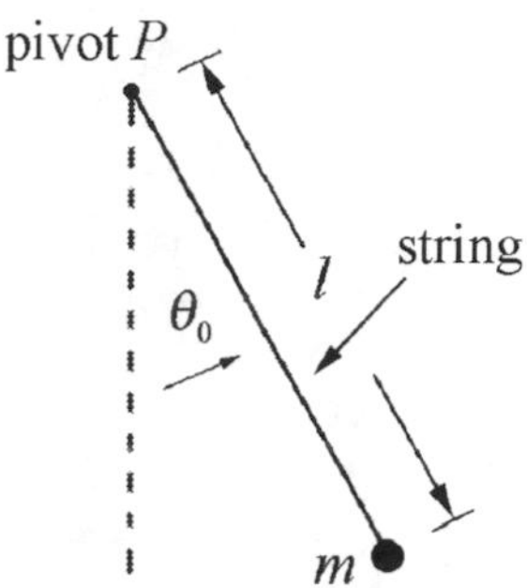

Figure 23.6 Simple pendulum

Let's choose polar coordinates for the pendulum as shown in Figure 23.7a along with the free-body force diagram for the suspended object (Figure 23.7b). The angle θ is defined with respect to the equilibrium position. When $\theta > 0$, the bob is has moved to the right, and when $\theta < 0$, the bob has moved to the left. The object will move in a circular arc centered at the pivot point. The forces on the object are the tension in the string $\vec{\mathbf{T}} = -T\,\hat{\mathbf{r}}$ and gravity $m\vec{\mathbf{g}}$. The gravitation force on the object has $\hat{\mathbf{r}}$- and $\hat{\boldsymbol{\theta}}$-components given by

$$m\vec{\mathbf{g}} = mg(\cos\theta\,\hat{\mathbf{r}} - \sin\theta\,\hat{\boldsymbol{\theta}})\,. \tag{23.3.10}$$

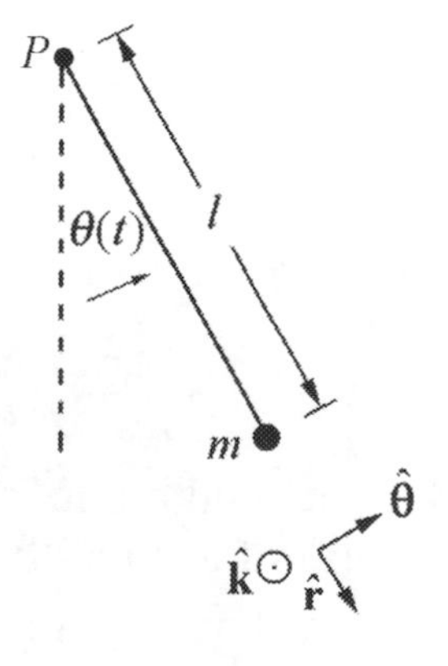

Figure 23.7 (a) Coordinate system

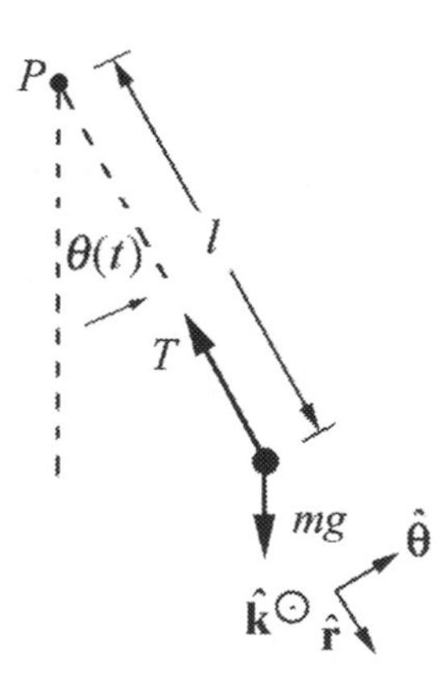

Figure 23.7 (b) free-body force diagram

Our concern is with the tangential component of the gravitational force,

$$F_\theta = -mg\sin\theta\,. \tag{23.3.11}$$

The sign in Equation (23.3.11) is crucial; the tangential force tends to restore the pendulum to the equilibrium value $\theta = 0$. If $\theta > 0$, $F_\theta < 0$ and if $\theta < 0$, $F_\theta > 0$, where we are assuming that the angle θ is restricted to the range $-\pi < \theta < \pi$. In both instances the tangential component of the force is directed towards the equilibrium position. The tangential component of acceleration is

$$a_\theta = l\alpha = l\frac{d^2\theta}{dt^2}\,. \tag{23.3.12}$$

Newton's Second Law, $F_\theta = m a_\theta$, yields

$$-mgl\sin\theta = ml^2\frac{d^2\theta}{dt^2}\,. \tag{23.3.13}$$

We can rewrite this equation is the form

$$\frac{d^2\theta}{dt^2} = -\frac{g}{l}\sin\theta\,. \tag{23.3.14}$$

This is not the simple harmonic oscillator equation although it still describes periodic motion. In the limit of small oscillations, $\sin\theta \cong \theta$, Eq. (23.3.13) becomes

$$\frac{d^2\theta}{dt^2} \cong -\frac{g}{l}\theta. \tag{23.3.15}$$

This equation is similar to the object-spring simple harmonic oscillator differential equation

$$\frac{d^2x}{dt^2} = -\frac{k}{m}x\,. \tag{23.3.16}$$

By comparison with Eq. (23.2.6) the angular frequency of oscillation for the pendulum is approximately

$$\omega_0 \simeq \sqrt{\frac{g}{l}}\,, \tag{23.3.17}$$

with period

$$T = \frac{2\pi}{\omega_0} \simeq 2\pi\sqrt{\frac{l}{g}}\,. \tag{23.3.18}$$

The solutions to Eq. (23.3.15) can be modeled after Eq. (23.2.21). With the initial conditions that the pendulum is released from rest, $\frac{d\theta}{dt}(t=0)=0$, at a small angle $\theta(t=0)=\theta_0$, the angle the string makes with the vertical as a function of time is given by

$$\theta(t)=\theta_0\cos(\omega_0 t)=\theta_0\cos\left(\frac{2\pi}{T}t\right)=\theta_0\cos\left(\sqrt{\frac{g}{l}}t\right). \tag{23.3.19}$$

The z-component of the angular velocity of the bob is

$$\omega_z(t)=\frac{d\theta}{dt}(t)=-\sqrt{\frac{g}{l}}\theta_0\sin\left(\sqrt{\frac{g}{l}}t\right). \tag{23.3.20}$$

Keep in mind that the component of the angular velocity $\omega_z = d\theta / dt$ changes with time in an oscillatory manner (sinusoidally in the limit of small oscillations). The angular frequency ω_0 is a parameter that describes the system. The z-component of the angular velocity $\omega_z(t)$, besides being time-dependent, depends on the amplitude of oscillation θ_0. In the limit of small oscillations, ω_0 does not depend on the amplitude of oscillation.

The fact that the period is independent of the mass of the object follows algebraically from the fact that the mass appears on both sides of Newton's Second Law and hence cancels. Consider also the argument that is attributed to Galileo: if a pendulum, consisting of two identical masses joined together, were set to oscillate, the two halves would not exert forces on each other. So, if the pendulum were split into two pieces, the pieces would oscillate the same as if they were one piece. This argument can be extended to simple pendula of arbitrary masses.

23.3.2 Simple Pendulum Energy Approach

We can use energy methods to find the differential equation describing the time evolution of the angle θ. When the string is at an angle θ with respect to the vertical, the gravitational potential energy (relative to a choice of zero potential energy at the bottom of the swing where $\theta = 0$ as shown in Figure 23.8) is given by

$$U = mgl(1-\cos\theta) \tag{23.3.21}$$

The θ-component of the velocity of the object is given by $v_\theta = l(d\theta / dt)$ so the kinetic energy is

$$K=\frac{1}{2}mv^2=\frac{1}{2}m\left(l\frac{d\theta}{dt}\right)^2. \tag{23.3.22}$$

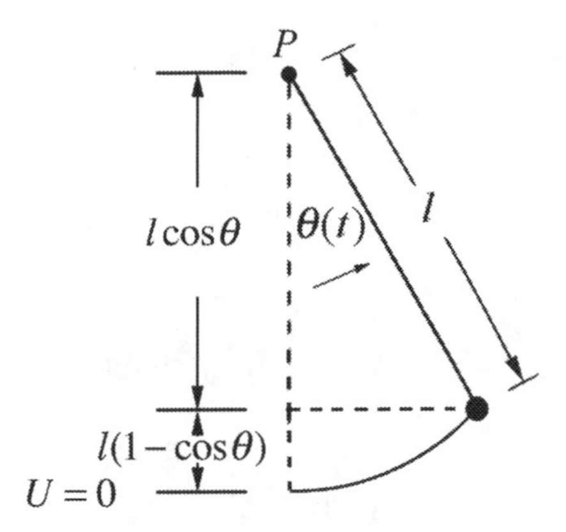

Figure 23.9 Energy diagram for simple pendulum

The mechanical energy of the system is then

$$E = K + U = \frac{1}{2}m\left(l\frac{d\theta}{dt}\right)^2 + mgl\left(1-\cos\theta\right). \qquad (23.3.23)$$

Because we assumed that there is no non-conservative work (i.e. no air resistance or frictional forces acting at the pivot), the energy is constant hence

$$\begin{aligned} 0 &= \frac{dE}{dt} = \frac{1}{2}m2l^2\frac{d\theta}{dt}\frac{d^2\theta}{dt^2} + mgl\sin\theta\frac{d\theta}{dt} \\ &= ml^2\frac{d\theta}{dt}\left(\frac{d^2\theta}{dt^2} + \frac{g}{l}\sin\theta\right). \end{aligned} \qquad (23.3.24)$$

There are two solutions to this equation; the first one $d\theta / dt = 0$ is the equilibrium solution. That the z-component of the angular velocity is zero means the suspended object is not moving. The second solution is the one we are interested in

$$\frac{d^2\theta}{dt^2} + \frac{g}{l}\sin\theta = 0, \qquad (23.3.25)$$

which is the same differential equation (Eq. (23.3.13)), we found using the force method.

We can find the time t_1 that the object first reaches the bottom of the circular arc by setting $\theta(t_1) = 0$ in Eq. (23.3.19)

$$0 = \theta_0\cos\left(\sqrt{\frac{g}{l}}t_1\right). \qquad (23.3.26)$$

This zero occurs when the argument of the cosine satisfies

$$\sqrt{\frac{g}{l}}t_1 = \frac{\pi}{2}. \tag{23.3.27}$$

The z-component of the angular velocity at time t_1 is therefore

$$\frac{d\theta}{dt}(t_1) = -\sqrt{\frac{g}{l}}\theta_0 \sin\left(\sqrt{\frac{g}{l}}t_1\right) = -\sqrt{\frac{g}{l}}\theta_0 \sin\left(\frac{\pi}{2}\right) = -\sqrt{\frac{g}{l}}\theta_0. \tag{23.3.28}$$

Note that the negative sign means that the bob is moving in the negative $\hat{\boldsymbol{\theta}}$-direction when it first reaches the bottom of the arc. The θ-component of the velocity at time t_1 is therefore

$$v_\theta(t_1) \equiv v_1 = l\frac{d\theta}{dt}(t_1) = -l\sqrt{\frac{g}{l}}\theta_0 \sin\left(\sqrt{\frac{g}{l}}t_1\right) = -\sqrt{lg}\theta_0 \sin\left(\frac{\pi}{2}\right) = -\sqrt{lg}\theta_0. \tag{23.3.29}$$

We can also find the components of both the velocity and angular velocity using energy methods. When we release the bob from rest, the energy is only potential energy

$$E = U_0 = mgl(1 - \cos\theta_0) \cong mgl\frac{\theta_0^2}{2}, \tag{23.3.30}$$

where we used the approximation that $\cos\theta_0 \cong 1 - \theta_0^2/2$. When the bob is at the bottom of the arc, the only contribution to the mechanical energy is the kinetic energy given by

$$K_1 = \frac{1}{2}mv_1^{\,2}. \tag{23.3.31}$$

Because the energy is constant, we have that $U_0 = K_1$ or

$$mgl\frac{\theta_0^2}{2} = \frac{1}{2}mv_1^{\,2}. \tag{23.3.32}$$

We can solve for the θ-component of the velocity at the bottom of the arc

$$v_{\theta,1} = \pm\sqrt{gl}\,\theta_0. \tag{23.3.33}$$

The two possible solutions correspond to the different directions that the bob can have when at the bottom. The z-component of the angular velocity is then

$$\frac{d\theta}{dt}(t_1) = \frac{v_1}{l} = \pm\sqrt{\frac{g}{l}}\theta_0, \tag{23.3.34}$$

in agreement with our previous calculation.

If we do not make the small angle approximation, we can still use energy techniques to find the θ-component of the velocity at the bottom of the arc by equating the energies at the two positions

$$mgl(1-\cos\theta_0)=\frac{1}{2}mv_1^{\,2}, \tag{23.3.35}$$

$$v_{\theta,1}=\pm\sqrt{2gl(1-\cos\theta_0)}. \tag{23.3.36}$$

23.4 Worked Examples

Example 23.3 Gravitational Simple Harmonic Oscillator

Two identical point-like objects each of mass m_1 are fixed in place and separated by a distance $2d$. A third object of mass m_2 is free to move and lies on the perpendicular bisector of the line connecting the two objects at a distance $y(t)$ above the midpoint of that line (Figure 23.9). The objects interact gravitationally. Ignore all other interactions. (a) Show that the equation of motion for the object of mass m_2 reduces to a simple harmonic oscillator equation in the limit that $y(t) << d$. (b) Assume that at $t=0$, the object of mass m_2 is at the equilibrium position but has an initial upward speed v_0. Write down an expression for the subsequent motion $y(t)$. Assume that $y(t) << d$.

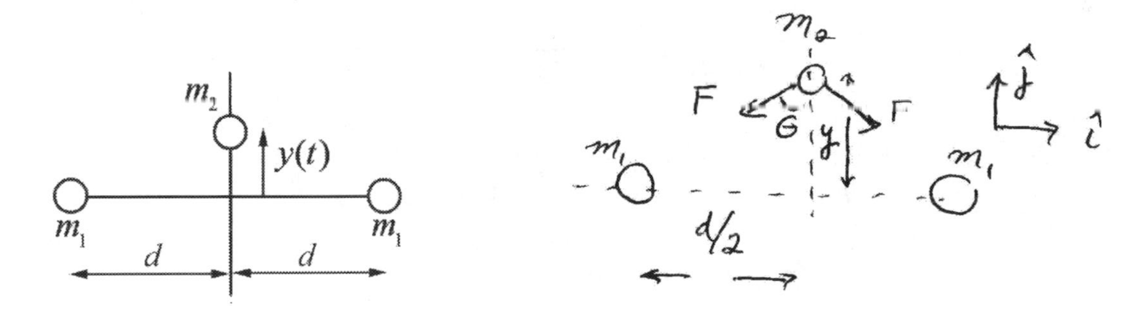

Figure 23.9 Example 23.3

Figure 23.10 Free-body force diagram on object with mass m_2

Solution: The free body force diagram on the object of mass m_2 is shown in Figure 23.10. The y–component of force acting on the object is

$$F_y=-\frac{Gm_1m_2}{((d/2)^2+y^2)}2\cos\theta.$$

Using the Pythagorean theorem (see Figure 23.09), we have that

$$\cos\theta = \frac{y}{((d/2)^2 + y^2)^{1/2}}.$$

The y-component of force is

$$F_y = -\frac{2Gm_1m_2y}{((d/2)^2 + y^2)^{3/2}}.$$

Therefore Newton's Second Law yields the equation of motion

$$m_2\frac{d^2y}{dt^2} = -\frac{2Gm_1m_2y}{((d/2)^2 + y^2)^{3/2}}.$$

When $y(t) << d$ the equation of motion becomes

$$\frac{d^2y}{dt^2} \simeq -\frac{16Gm_1}{d^3}y.$$

This is a simple harmonic oscillation equation with solution

$$y(t) = y_0\cos(\omega_0 t) + \frac{v_{y,0}}{\omega_0}\sin(\omega_0 t)$$

where

$$\omega_0 = \sqrt{\frac{16Gm_1}{d^3}}.$$

The initial conditions are $y_0 = 0$ and $v_{y,0} = v_0 > 0$. Therefore

$$y(t) = \frac{v_0}{\omega_0}\sin(\omega_0 t),$$

$$v_y(t) = v_0\cos(\omega_0 t).$$

Example 23.4 U-Tube

A U-tube open at both ends to atmospheric pressure P_0 is filled with an incompressible fluid of density ρ. The cross-sectional area A of the tube is uniform and the total length of the fluid column is L. A piston is used to depress the height of the liquid column on one side by a distance x_0, (raising the other side by the same distance) and then is quickly removed (Figure 23.11). What is the angular frequency of the ensuing simple harmonic motion? Neglect any resistive forces and at the walls of the U-tube.

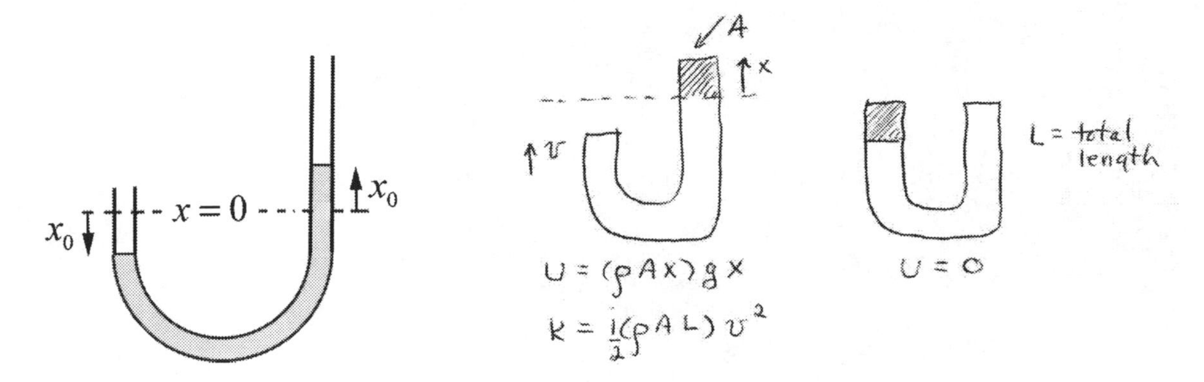

Figure 23.11 Example 23.4

Figure 23.12 Energy diagram for water

Solution: We shall use conservation of energy. First choose as a zero for gravitational potential energy where the water levels are equal on both sides of the tube. When the piston on one side depresses the fluid, it rises on the other. At a given instant in time when a portion of the fluid of mass $\Delta m = \rho A x$ is a height x above the equilibrium height (Figure 23.12), the potential energy of the fluid is given by

$$U = \Delta m g x = (\rho A x) g x = \rho A g x^2 .$$

At that same instant the entire fluid of length L and mass $m = \rho A L$ is moving with speed v, so the kinetic energy is

$$K = \frac{1}{2} m v^2 = \frac{1}{2} \rho A L v^2 .$$

Thus the total energy is

$$E = K + U = \frac{1}{2} \rho A L v^2 + \rho A g x^2 .$$

By neglecting resistive force, the mechanical energy of the fluid is constant therefore

$$0 = \frac{dE}{dt} = \rho A L v \frac{dv}{dt} + 2 \rho A g x \frac{dx}{dt} . \tag{23.4.1}$$

If we just consider the top of the fluid above the equilibrium position on the right arm in Figure 23.13, we rewrite Eq. (23.4.1) as

$$0 = \frac{dE}{dt} = \rho A L v_x \frac{dv_x}{dt} + 2 \rho A g x \frac{dx}{dt} ,$$

where $v_x = dx / dt$. We now rewrite the energy condition using $dv_x / dt = d^2 x / dt^2$ as

$$0 = v_x \rho A \left(L \frac{d^2x}{dt^2} + 2gx \right).$$

This condition is satisfied when $v_x = 0$, i.e. the equilibrium condition or when

$$0 = L \frac{d^2x}{dt^2} + 2gx .$$

This last condition can be written as

$$\frac{d^2x}{dt^2} = -\frac{2g}{L}x .$$

This last equation is the simple harmonic oscillator equation. Using the same mathematical techniques as we used for the spring-block system, the solution for the height of the fluid above the equilibrium position is given by

$$x(t) = B\cos(\omega_0 t) + C\sin(\omega_0 t),$$

where

$$\omega_0 = \sqrt{\frac{2g}{L}}$$

is the angular frequency of oscillation. The x-component of the velocity of the fluid on the right hand side of the U-tube is given by

$$v_x(t) = \frac{dx(t)}{dt} = -\omega_0 B\sin(\omega_0 t) + \omega_0 C\cos(\omega_0 t).$$

The coefficients B and C are determined by the initial conditions. At $t = 0$, the height of the fluid is $x(t = 0) = B = x_0$. At $t = 0$, the speed is zero so $v_x(t = 0) = \omega_0 C = 0$, hence $C = 0$. The height of the fluid above the equilibrium position on the right hand side of the U-tube as a function of time is thus

$$x(t) = x_0 \cos\left(\sqrt{\frac{2g}{L}} t \right).$$

23.5 Small Oscillations

Any potential energy function that is quadratic will undergo simple harmonic motion,

$$U(x) = U_0 + \frac{1}{2}k(x - x_{eq})^2. \tag{23.5.1}$$

where k is a "spring constant", x_{eq} is the equilibrium position, and the constant U_0 just depends on the choice of reference point x_{ref} for zero potential energy, $U(x_{ref})=0$,

$$0=U(x_{ref})=U_0+\frac{1}{2}k(x_{ref}-x_{eq})^2. \qquad (23.5.2)$$

Therefore the constant is

$$U_0=-\frac{1}{2}k(x_{ref}-x_{eq})^2. \qquad (23.5.3)$$

The minimum of the potential x_0 corresponds to the point where the x-component of the force is zero,

$$\left.\frac{dU}{dx}\right|_{x=x_0}=2k(x_0-x_{eq})=0 \Rightarrow x_0=x_{eq}, \qquad (23.5.4)$$

corresponding to the equilibrium position. Therefore the constant is $U(x_0)=U_0$ and we rewrite our potential function as

$$U(x)=U(x_0)+\frac{1}{2}k(x-x_0)^2. \qquad (23.5.5)$$

Now suppose that a potential energy function is not quadratic but still has a minimum at x_0. For example, consider the potential energy function

$$U(x)=-U_1\left(\left(\frac{x}{x_1}\right)^3-\left(\frac{x}{x_1}\right)^2\right), \qquad (23.5.6)$$

(Figure 23.13), which has a stable minimum at x_0.

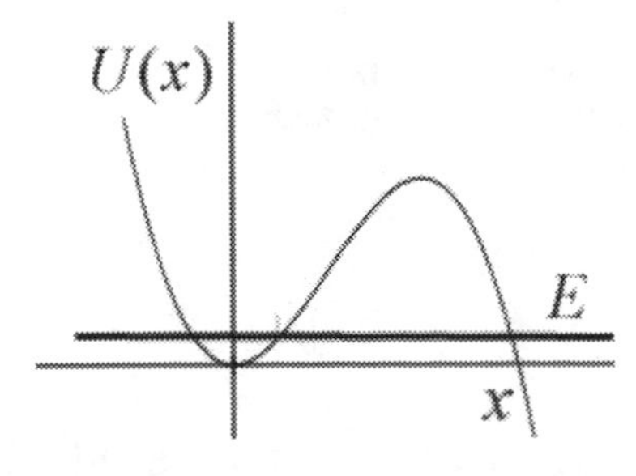

Figure 23.13 Potential energy function with stable minimum and unstable maximum

When the energy of the system is very close to the value of the potential energy at the minimum $U(x_0)$, we shall show that the system will undergo small oscillations about the minimum value x_0. We shall use the Taylor formula to approximate the potential function as a polynomial. We shall show that near the minimum x_0, we can approximate the potential function by a quadratic function similar to Eq. (23.5.5) and show that the system undergoes simple harmonic motion for small oscillations about the minimum x_0.

We begin by expanding the potential energy function about the minimum point using the Taylor formula

$$U(x) = U(x_0) + \frac{dU}{dx}(x_0)(x - x_0) + \frac{1}{2!}\frac{d^2U}{dx^2}(x_0)(x - x_0)^2 + \frac{1}{3!}\frac{d^3U}{dx^3}(x_0)(x - x_0)^3 + \cdots \quad (23.5.7)$$

where $\frac{1}{3!}\frac{d^3U}{dx^3}(x_0)(x - x_0)^3$ is third order term in that it is proportional to $(x - x_0)^3$, and $\frac{d^2U}{dx^2}(x_0)$ and $\frac{dU}{dx}(x_0)$ are constants. If x_0 is the minimum of the potential energy, then the linear term is zero, because

$$\frac{dU}{dx}(x_0) = 0 \quad (23.5.8)$$

and so Eq. ((23.5.7)) becomes

$$U(x) \simeq U(x_0) + \frac{1}{2}\frac{d^2U}{dx^2}(x_0)(x - x_0)^2 + \frac{1}{3!}\frac{d^3U}{dx^3}(x_0)(x - x_0)^3 + \cdots \quad (23.5.9)$$

For small displacements from the equilibrium point such that such $\left|x - x_0\right| << 1$, the third order term and higher order terms are very small and can be ignored. Then the potential energy function is approximately a quadratic function,

$$U(x) \simeq U(x_0) + \frac{1}{2}\frac{d^2U}{dx^2}(x_0)(x - x_0)^2 = U(x_0) + \frac{1}{2}k_{eff}(x - x_0)^2 \quad (23.5.10)$$

where we define k_{eff}, the ***effective spring constant***, by

$$k_{eff} \equiv \frac{d^2U}{dx^2}(x_0). \quad (23.5.11)$$

Because the potential energy function is now a quadratic function, the system will undergo simple harmonic motion for small displacements from the minimum with a force given by

$$F_x = -\frac{dU}{dx} = -k_{eff}(x - x_0). \quad (23.5.12)$$

At $x = x_0$, the force is zero

$$F_x(x_0) = \frac{dU}{dx}(x_0) = 0. \quad (23.5.13)$$

We can determine the period of oscillation by substituting Eq. (23.5.12) into Newton's Second Law

$$-k_{eff}(x - x_0) = m_{eff}\frac{d^2x}{dt^2} \quad (23.5.14)$$

where m_{eff} is the ***effective mass.*** For a two-particle system, the effective mass is the reduced mass of the system.

$$m_{eff} = \frac{m_1 m_2}{m_1 + m_2} \equiv \mu_{red}, \quad (23.5.15)$$

Eq. (23.5.14) has the same form as the spring-object ideal oscillator. Therefore the angular frequency of small oscillations is given by

$$\omega_0 = \sqrt{\frac{k_{eff}}{m_{eff}}} = \sqrt{\frac{\frac{d^2U}{dx^2}(x_0)}{m_{eff}}}.$$

Example 23.5 Small Oscillations

A system with effective mass m has a potential energy given by

$$U(x) = U_0\left(-2\left(\frac{x}{x_0}\right)^2 + \left(\frac{x}{x_0}\right)^4\right),$$

where U_0 and x_0 are positive constants and $U(0) = 0$. (a) Find the points where the force on the particle is zero. Classify them as stable or unstable. Calculate the value of $U(x)/U_0$ at these equilibrium points. (b) If the particle is given a small displacement from an equilibrium point, find the angular frequency of small oscillation.

Solution: (a) A plot of $U(x)/U_0$ as a function of x/x_0 is shown in Figure 22.14. The force on the particle is zero at the minimum of the potential energy,

$$0 = \frac{dU}{dx} = U_0\left(-4\left(\frac{1}{x_0}\right)^2 x + 4\left(\frac{1}{x_0}\right)^4 x^3\right) \Rightarrow x^2 = x_0^{\;2}.$$

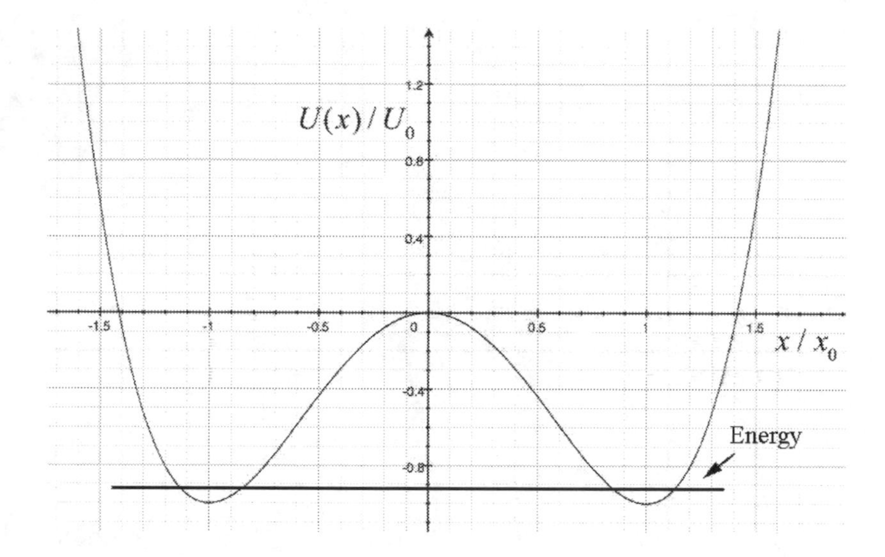

Figure 22.14 Plot of $U(x)/U_0$ as a function of x/x_0

The equilibrium points are at $x = \pm x_0$ which are stable and $x = 0$ which is unstable. The second derivative of the potential energy is given by

$$\frac{d^2U}{dx^2} = U_0\left(-4\left(\frac{1}{x_0}\right)^2 + 12\left(\frac{1}{x_0}\right)^4 x^2\right).$$

If the particle is given a small displacement from $x = x_0$ then

$$\frac{d^2U}{dx^2}(x_0) = U_0\left(-4\left(\frac{1}{x_0}\right)^2 + 12\left(\frac{1}{x_0}\right)^4 x_0^2\right) = U_0\frac{8}{x_0^2}.$$

(b) The angular frequency of small oscillations is given by

$$\omega_0 = \sqrt{\frac{d^2U}{dx^2}(x_0)/m} = \sqrt{\frac{8U_0}{mx_0^2}}.$$

Example 23.6 Lennard-Jones 6,12 Potential and Small Oscillations

A commonly used potential energy function to describe the interaction between two atoms is the Lennard-Jones 6,12 potential

$$U(r) = U_0\left[(r_0/r)^{12} - 2(r_0/r)^6\right]; \ r > 0,$$

where r is the distance between the atoms. Find the angular frequency of small oscillations about the stable equilibrium position for two identical atoms bound to each other by the Lennard-Jones interaction. Let m denote the effective mass of the system of two atoms.

Solution: The equilibrium points are found by setting the first derivative of the potential energy equal to zero,

$$0 = \frac{dU}{dr} = U_0\left[-12r_0^{12}r^{-13} + 12r_0^{6}r^{-7}\right].$$

The equilibrium points occur when $r = r_0$. The second derivative of the potential energy function is

$$\frac{d^2U}{dr^2} = U_0\left[+(12)(13)r_0^{12}r^{-14} - (12)(7)r_0^{6}r^{-8}\right].$$

Evaluating this at $r = r_0$ yields

$$\frac{d^2U}{dr^2}(r_0) = 72U_0 r_0^{-2}.$$

The angular frequency of small oscillation is therefore

$$\omega_0 = \sqrt{\frac{d^2U}{dr^2}(r_0)/m} = \sqrt{72U_0/mr_0^{2}}.$$

Appendix 23A Solution to Simple Harmonic Oscillator Equation

In our analysis of the solution of the simple harmonic oscillator equation of motion, Equation (23.2.1),

$$-k\,x = m\frac{d^2x}{dt^2}, \tag{23.A.1}$$

we assumed that the solution was a linear combination of sinusoidal functions,

$$x(t) = A\cos(\omega_0 t) + B\sin(\omega_0 t), \tag{23.A.2}$$

where $\omega_0 = \sqrt{k/m}$. We shall show that Eq. (23.A.2) is indeed a solution.

Assume that the mechanical energy of the spring-object system is given by the constant E. Choose the reference point for potential energy to be the unstretched position of the spring. Let x denote the amount the spring has been compressed ($x < 0$) or stretched ($x > 0$) from equilibrium at time t and denote the amount the spring has been compressed or stretched from equilibrium at time $t = 0$ by $x(t = 0) \equiv x_0$. Let $v_x = dx/dt$ denote the x-component of the velocity at time t and denote the x-component of the velocity at time $t = 0$ by $v_x(t = 0) \equiv v_{x,0}$. The constancy of the mechanical energy is then expressed as

$$E = K + U = \frac{1}{2}k\,x^2 + \frac{1}{2}m\,v^2. \tag{23.A.3}$$

We can solve Equation (23.A.3) for the square of the x-component of the velocity,

$$v_x^{\,2} = \frac{2E}{m} - \frac{k}{m}x^2 = \frac{2E}{m}\left(1 - \frac{k}{2E}x^2\right). \tag{23.A.4}$$

Taking square roots, we have

$$\frac{dx}{dt} = \sqrt{\frac{2E}{m}}\sqrt{1 - \frac{k}{2E}x^2}\,. \tag{23.A.5}$$

(why we take the positive square root will be explained below).

Let $a \equiv \sqrt{2E/m}$ and $b \equiv k/2E$. It's worth noting that a has dimensions of velocity and b has dimensions of $[\text{length}]^{-2}$. Equation (23.A.5) is separable,

$$\frac{dx}{dt} = a\sqrt{1-bx^2}$$
$$\frac{dx}{\sqrt{1-bx^2}} = a\,dt. \qquad (23.A.6)$$

We now integrate Eq. (23.A.6),

$$\int \frac{dx}{\sqrt{1-bx^2}} = \int a\,dt \; . \qquad (23.A.7)$$

The integral on the left in Equation (23.A.7) is well known, and a derivation is presented here. We make a change of variables $\cos\theta = \sqrt{b}\,x$ with the differentials $d\theta$ and dx related by $-\sin\theta\, d\theta = \sqrt{b}\,dx$. The integration variable is

$$\theta = \cos^{-1}(\sqrt{b}\,x)\,. \qquad (23.A.8)$$

Equation (23.A.7) then becomes

$$\int \frac{-\sin\theta\, d\theta}{\sqrt{1-\cos^2\theta}} = \int \sqrt{ba}\,dt\,. \qquad (23.A.9)$$

This is a good point at which to check the dimensions. The term on the left in Equation (23.A.9) is dimensionless, and the product $\sqrt{ba}$ on the right has dimensions of inverse time, $[\text{length}]^{-1}[\text{length}\cdot\text{time}^{-1}] = [\text{time}^{-1}]$, so $\sqrt{ba}\,dt$ is dimensionless. Using the trigonometric identity $\sqrt{1-\cos^2\theta} = \sin\theta$, Equation (23.A.9) reduces to

$$\int d\theta = -\int \sqrt{ba}\,dt\,. \qquad (23.A.10)$$

Although at this point in the derivation we don't know that $\sqrt{ba}$, which has dimensions of frequency, is the angular frequency of oscillation, we'll use some foresight and make the identification

$$\omega_0 \equiv \sqrt{ba} = \sqrt{\frac{k}{2E}}\sqrt{\frac{2E}{m}} = \sqrt{\frac{k}{m}}\,, \qquad (23.A.11)$$

and Equation (23.A.10) becomes

$$\int_{\theta=\theta_0}^{\theta} d\theta = -\int_{t=0}^{t} \omega_0\,dt\,. \qquad (23.A.12)$$

After integration we have

$$\theta - \theta_0 = -\omega_0\,t\,, \qquad (23.A.13)$$

where $\theta_0 \equiv -\phi$ is the constant of integration. Because $\theta = \cos^{-1}(\sqrt{b}\, x(t))$, Equation (23.A.13) becomes

$$\cos^{-1}(\sqrt{b}\, x(t)) = -(\omega_0 t + \phi)\,. \qquad (23.A.14)$$

Take the cosine of each side of Equation (23.A.14), yielding

$$x(t) = \frac{1}{\sqrt{b}}\cos(-(\omega_0 t + \phi)) = \sqrt{\frac{2E}{k}}\cos(\omega_0 t + \phi)\,. \qquad (23.A.15)$$

At $t = 0$,

$$x_0 \equiv x(t = 0) = \sqrt{\frac{2E}{k}}\cos\phi\,. \qquad (23.A.16)$$

The x-component of the velocity as a function of time is then

$$v_x(t) = \frac{dx(t)}{dt} = -\omega_0\sqrt{\frac{2E}{k}}\sin(\omega_0 t + \phi)\,. \qquad (23.A.17)$$

At $t = 0$,

$$v_{x,0} \equiv v_x(t = 0) = -\omega_0\sqrt{\frac{2E}{k}}\sin\phi\,. \qquad (23.A.18)$$

We can determine the constant ϕ by dividing the expressions in Equations (23.A.18) and (23.A.16),

$$-\frac{v_{x,0}}{\omega_0 x_0} = \tan\phi \qquad (23.A.19)$$

Thus the constant ϕ can be determined by the initial conditions and the angular frequency of oscillation,

$$\phi = \tan^{-1}\left(-\frac{v_{x,0}}{\omega_0 x_0}\right). \qquad (23.A.20)$$

Use the identity

$$\cos(\omega_0 t + \phi) = \cos\omega_0 t\cos\phi - \sin\omega_0 t\sin\phi \qquad (23.A.21)$$

to expand Equation (23.A.15) yielding

$$x(t) = \sqrt{\frac{2E}{k}}\cos\omega_0 t\cos\phi = \sqrt{\frac{2E}{k}}\sin\omega_0 t\sin\phi\,, \qquad (23.A.22)$$

and substituting Equations (23.A.16) and (23.A.18) into Eq. (23.A.22) yields

$$x(t) = x_0 \cos\omega_0 t + \frac{v_{x,0}}{\omega_0} \sin\omega_0 t , \qquad (23.A.23)$$

agreeing with Eq. (23.2.21).

So, what about the missing $\pm$ that should have been in Equation (23.A.5)? Strictly speaking, we would need to redo the derivation for the block moving in different directions. Mathematically, this would mean replacing ϕ by $\pi - \phi$ (or $\phi - \pi$) when the block's velocity changes direction. Changing from the positive square root to the negative *and* changing ϕ to $\pi - \phi$ have the collective action of reproducing Eq. (23.A.23).

Chapter 24 Physical Pendulum

Chapter 24 Physical Pendulum

.... I had along with me....the Descriptions, with some Drawings of the principal Parts of the Pendulum-Clock which I had made, and as also of them of my then intended Timekeeper for the Longitude at Sea.[1]

John Harrison

24.1 Introduction

We have already used Newton's Second Law or Conservation of Energy to analyze systems like the spring-object system that oscillate. We shall now use torque and the rotational equation of motion to study oscillating systems like pendulums and torsional springs.

24.1.1 Simple Pendulum Torque Approach

Recall the simple pendulum from Chapter 23.3.1.The coordinate system and force diagram for the simple pendulum is shown in Figure 24.1.

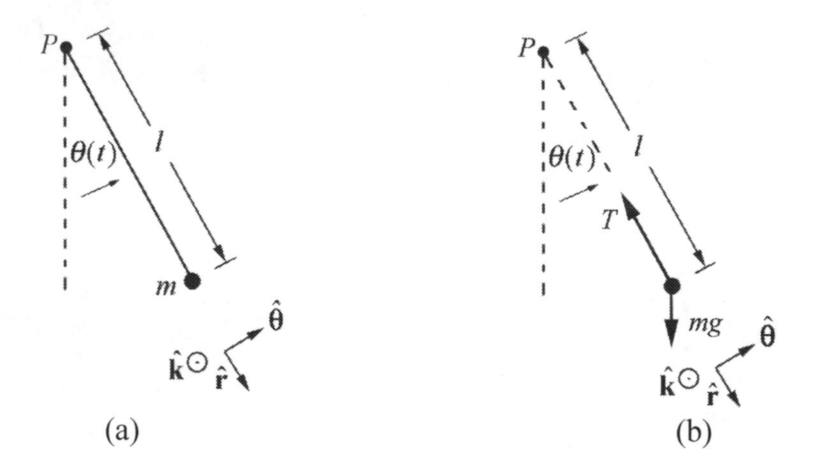

Figure 24.1 (a) Coordinate system and (b) torque diagram, for simple pendulum

The torque about the pivot point P is given by

$$\vec{\boldsymbol{\tau}}_P = \vec{\mathbf{r}}_{P,m} \times m\vec{\mathbf{g}} = l\,\hat{\mathbf{r}} \times mg(\cos\theta\,\hat{\mathbf{r}} - \sin\theta\hat{\boldsymbol{\theta}}) = -l\,mg\sin\theta\,\hat{\mathbf{k}} \qquad (24.1.1)$$

The z-component of the torque about point P

$$(\tau_P)_z = -mgl\sin\theta\,. \qquad (24.1.2)$$

---

[1] J. Harrison, *A Description Concerning Such Mechanisms as will Afford a Nice, or True Mensuration of Time;...*(London, 1775), p.19.

When $\theta > 0$, $(\tau_P)_z < 0$ and the torque about P points in the negative $\hat{\mathbf{k}}$-direction (into the plane of Figure 24.1b), when $\theta < 0$, $(\tau_P)_z > 0$ and the torque about P points in the positive $\hat{\mathbf{k}}$-direction (out of the plane of Figure 24.1b). The moment of inertia of a point mass about the pivot point P is $I_P = ml^2$. The rotational equation of motion is then

$$\begin{aligned}(\tau_P)_z &= I_P\alpha_z \equiv I_P\frac{d^2\theta}{dt^2} \\ -mgl\sin\theta &= ml^2\frac{d^2\theta}{dt^2}.\end{aligned} \tag{24.1.3}$$

Thus we have

$$\frac{d^2\theta}{dt^2} = -\frac{g}{l}\sin\theta\,. \tag{24.1.4}$$

agreeing with Eq. 23. 3.14. When the angle of oscillation is small, we may use the small angle approximation

$$\sin\theta \cong \theta\,; \tag{24.1.5}$$

and Eq. (24.1.4) reduces to the simple harmonic oscillator equation

$$\frac{d^2\theta}{dt^2} \cong -\frac{g}{l}\theta\,. \tag{24.1.6}$$

We have already studied the solutions to this equation in Chapter 23.3. A procedure for determining the period when the small angle approximation does not hold is given in Appendix 24A.

24.2 Physical Pendulum

A ***physical pendulum*** consists of a rigid body that undergoes fixed axis rotation about a fixed point S (Figure 24.2).

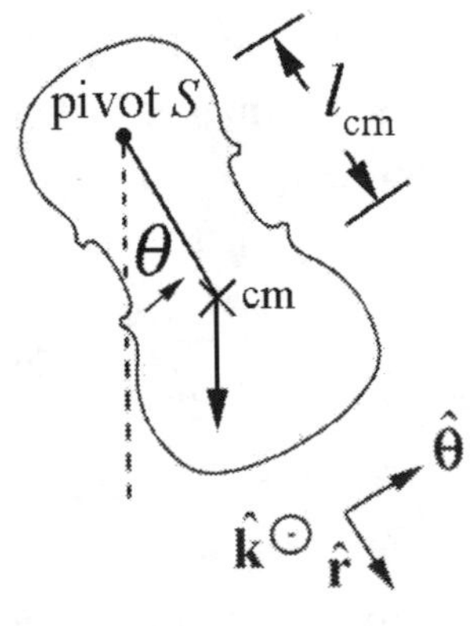

Figure 24.2 Physical pendulum

The gravitational force acts at the center of mass of the physical pendulum. Denote the distance of the center of mass to the pivot point S by is a l_{cm}. The torque analysis is nearly identical to the simple pendulum. The torque about the pivot point S is given by

$$\vec{\boldsymbol{\tau}}_S = \vec{\mathbf{r}}_{S,\text{cm}} \times m\vec{\mathbf{g}} = l_{\text{cm}}\hat{\mathbf{r}} \times mg(\cos\theta\,\hat{\mathbf{r}} - \sin\theta\,\hat{\boldsymbol{\theta}}) = -l_{\text{cm}}mg\sin\theta\,\hat{\mathbf{k}}. \qquad (24.2.1)$$

Following the same steps that led from Equation (24.1.1) to Equation (24.1.4), the rotational equation for the physical pendulum is

$$-mgl_{\text{cm}}\sin\theta = I_S\frac{d^2\theta}{dt^2}. \qquad (24.2.2)$$

As with the simple pendulum, for small angles $\sin\theta \approx \theta$, Equation (24.2.2) reduces to the simple harmonic oscillator equation

$$\frac{d^2\theta}{dt^2} \simeq -\frac{mgl_{\text{cm}}}{I_S}\theta. \qquad (24.2.3)$$

The equation for the angle $\theta(t)$ is given by

$$\theta(t) = A\cos(\omega_0 t) + B\sin(\omega_0 t), \qquad (24.2.4)$$

where the angular frequency is given by

$$\omega_0 \approx \sqrt{\frac{mgl_{\text{cm}}}{I_S}} \quad \text{(physical pendulum)}, \qquad (24.2.5)$$

and the period is

$$T = \frac{2\pi}{\omega_0} \cong 2\pi\sqrt{\frac{I_S}{mgl_{\text{cm}}}} \quad \text{(physical pendulum)}. \qquad (24.2.6)$$

It is sometimes convenient to express the moment of inertia about the pivot point S in terms of l_{cm} using the parallel axis theorem, $I_S = ml_{\text{cm}}^2 + I_{\text{cm}}$, with the result that

$$T \cong 2\pi\sqrt{\frac{l_{\text{cm}}}{g} + \frac{I_{\text{cm}}}{mgl_{\text{cm}}}} \quad \text{(physical pendulum)}. \qquad (24.2.7)$$

Thus, if the object is "small" in the sense that $I_{\text{cm}} << ml_{\text{cm}}^2$, the expressions for the physical pendulum reduce to those for the simple pendulum. The z-component of the angular velocity is given by

$$\omega_z(t) = \frac{d\theta}{dt}(t) = -\omega_0 A\sin(\omega_0 t) + \omega_0 B\cos(\omega_0 t). \tag{24.2.8}$$

The coefficients A and B can be determined form the initial conditions by setting $t = 0$ in Eqs. (24.2.4) and (24.2.8) resulting in the conditions that

$$\begin{aligned} A &= \theta(t=0) \equiv \theta_0 \\ B &= \frac{\omega_z(t=0)}{\omega_0} \equiv \frac{\omega_{z,0}}{\omega_0}. \end{aligned} \tag{24.2.9}$$

Therefore the equations for the angle $\theta(t)$ and $\omega_z(t) = \frac{d\theta}{dt}(t)$ are given by

$$\theta(t) = \theta_0 \cos(\omega_0 t) + \frac{\omega_{z,0}}{\omega_0}\sin(\omega_0 t), \tag{24.2.10}$$

$$\omega_z(t) = \frac{d\theta}{dt}(t) = -\omega_0\theta_0 \sin(\omega_0 t) + \omega_{z,0}\cos(\omega_0 t). \tag{24.2.11}$$

24.3 Worked Examples

Example 24.1 Oscillating rod

A physical pendulum consists of a uniform rod of length d and mass m pivoted at one end. The pendulum is initially displaced to one side by a small angle θ_0 and released from rest with $\theta_0 << 1$. Find the period of the pendulum. Determine the period of the pendulum using (a) the torque method and (b) the energy method.

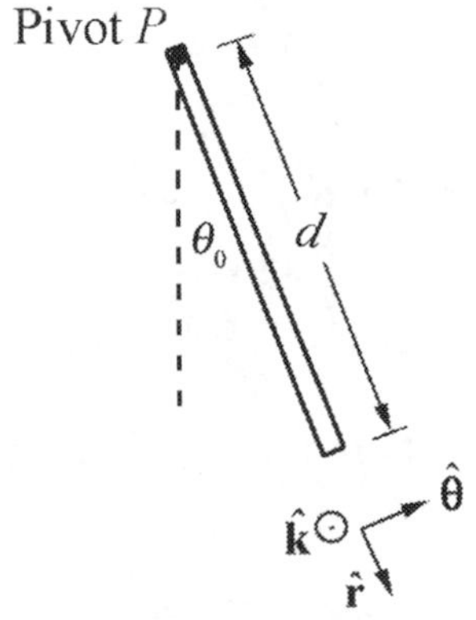

Figure 24.3 Oscillating rod

(a) Torque Method: with our choice of rotational coordinate system the angular acceleration is given by

$$\vec{\alpha} = \frac{d^2\theta}{dt^2}\hat{\mathbf{k}}\,. \qquad (24.3.1)$$

The force diagram on the pendulum is shown in Figure 24.4. In particular, there is an unknown pivot force and the gravitational force acts at the center of mass of the rod.

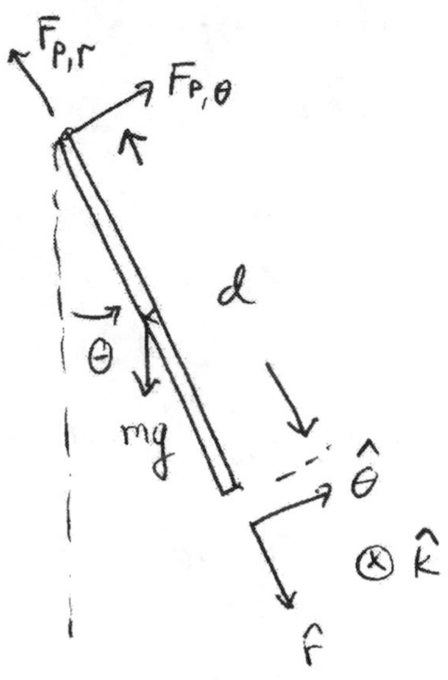

Figure 24.4 Free-body force diagram on rod

The torque about the pivot point P is given by

$$\vec{\boldsymbol{\tau}}_P = \vec{\mathbf{r}}_{P,cm} \times m\vec{\mathbf{g}}\;. \qquad (24.3.2)$$

The rod is uniform, therefore the center of mass is a distance $d/2$ from the pivot point. The gravitational force acts at the center of mass, so the torque about the pivot point P is given by

$$\vec{\boldsymbol{\tau}}_P = (d/2)\hat{\mathbf{r}} \times mg(-\sin\theta\,\hat{\boldsymbol{\theta}} + \cos\hat{\mathbf{r}}) = -(d/2)mg\sin\theta\,\hat{\mathbf{k}}\,. \qquad (24.3.3)$$

The rotational equation of motion about P is then

$$\vec{\boldsymbol{\tau}}_P = I_P\vec{\alpha}\,. \qquad (24.3.4)$$

Substituting Eqs. (24.3.3) and (24.3.1) into Eq. (24.3.4) yields

$$-(d/2)mg\sin\theta\,\hat{\mathbf{k}} = I_P\frac{d^2\theta}{dt^2}\hat{\mathbf{k}}\,. \qquad (24.3.5)$$

When the angle of oscillation is small, we may use the small angle approximation

$$\sin\theta \cong \theta\,. \qquad (24.3.6)$$

Then Eq. (24.3.5) becomes

$$\frac{d^2\theta}{dt^2} + \frac{(d/2)mg}{I_P}\theta = 0, \tag{24.3.7}$$

which is a simple harmonic oscillator equation. The angular frequency of oscillation for the pendulum is approximately

$$\omega_0 \simeq \sqrt{\frac{(d/2)mg}{I_P}}. \tag{24.3.8}$$

The moment of inertia of a rod about the end point P is $I_P = (1/3)md^2$ therefore the angular frequency is

$$\omega_0 \simeq \sqrt{\frac{(d/2)mg}{(1/3)md^2}} = \sqrt{\frac{(3/2)g}{d}} \tag{24.3.9}$$

with period

$$T = \frac{2\pi}{\omega_0} \simeq 2\pi\sqrt{\frac{2}{3}\frac{d}{g}}. \tag{24.3.10}$$

(b) Energy Method: take the zero point of gravitational potential energy to be the point where the center of mass of the pendulum is at its lowest point, that is, $\theta = 0$ (Figure 24.5).

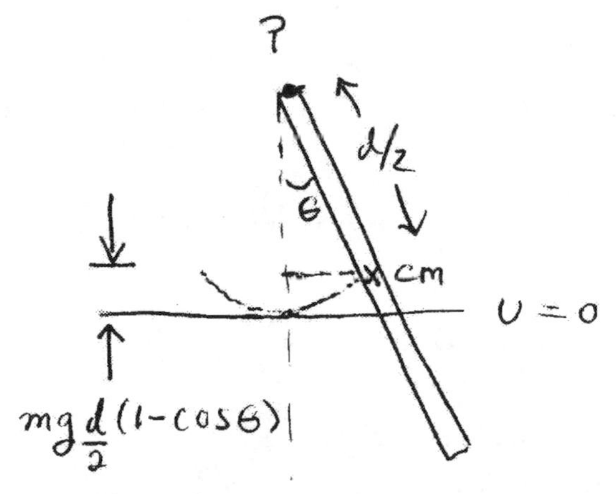

Figure 24.5 Energy diagram for rod

When the pendulum is at an angle θ the potential energy is

$$U = mg\frac{d}{2}(1-\cos\theta). \tag{24.3.11}$$

The kinetic energy of rotation about the pivot point is

$$K_{rot} = \frac{1}{2} I_p \omega^2 . \tag{24.3.12}$$

The mechanical energy is then

$$E = U + K_{rot} = m g \frac{d}{2}(1 - \cos\theta) + \frac{1}{2} I_p \omega^2 , \tag{24.3.13}$$

with $I_p = (1/3)md^2$. Note that $\omega^2 = (\omega_z)^2$. There are no non-conservative forces acting (by assumption), so the mechanical energy is constant, and therefore the time derivative of energy is zero

$$0 = \frac{dE}{dt} = m g \frac{d}{2} \sin\theta \frac{d\theta}{dt} + I_p \omega_z \frac{d\omega_z}{dt} . \tag{24.3.14}$$

Recall that $\omega_z = d\theta / dt$ and $\alpha_z = d\omega_z / dt = d^2\theta / dt^2$ so Eq. (24.3.14) becomes

$$0 = \omega_z \left(m g \frac{d}{2} \sin\theta + I_p \frac{d^2\theta}{dt^2} \right) . \tag{24.3.15}$$

There are two solutions, $\omega_z = 0$, in which the rod remains at the bottom of the swing and

$$0 = m g \frac{d}{2} \sin\theta + I_p \frac{d^2\theta}{dt^2} . \tag{24.3.16}$$

Using the small angle approximation, we obtain the simple harmonic oscillator equation (Eq. (24.3.7))

$$\frac{d^2\theta}{dt^2} + \frac{m g (d/2)}{I_p} \theta \simeq 0 . \tag{24.3.17}$$

Example 24.3 Torsional Oscillator

A disk with moment of inertia about the center of mass I_{cm} rotates in a horizontal plane. It is suspended by a thin, massless rod. If the disk is rotated away from its equilibrium position by an angle θ, the rod exerts a restoring torque about the center of the disk with magnitude given by $\tau_{\text{cm}} = \gamma\, \theta$ (Figure 24.6). At $t = 0$, the disk is released from rest at an angular displacement of θ_0. Find the subsequent time dependence of the angular displacement $\theta(t)$.

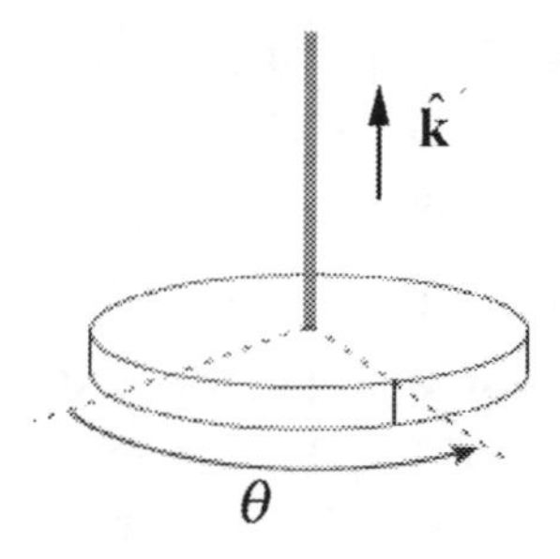

Figure 24.6 Example 24.3

Solution: Choose a coordinate system such that $\hat{\mathbf{k}}$ is pointing upwards, then the angular acceleration is given by

$$\vec{\boldsymbol{\alpha}} = \frac{d^2\theta}{dt^2}\hat{\mathbf{k}}. \tag{24.3.18}$$

The torque about the center of mass is given in the statement of the problem as a restoring torque, therefore

$$\vec{\boldsymbol{\tau}}_{\text{cm}} = -\gamma\,\theta\,\hat{\mathbf{k}}. \tag{24.3.19}$$

The z-component of the rotational equation of motion is

$$-\gamma\,\theta = I_{\text{cm}}\frac{d^2\theta}{dt^2}. \tag{24.3.20}$$

This is a simple harmonic oscillator equation with solution

$$\theta(t) = A\cos(\omega_0 t) + B\sin(\omega_0 t) \tag{24.3.21}$$

where the angular frequency of oscillation is given by

$$\omega_0 = \sqrt{\gamma / I_{\text{cm}}}. \tag{24.3.22}$$

The z-component of the angular velocity is given by

$$\omega_z(t) = \frac{d\theta}{dt}(t) = -\omega_0 A\sin(\omega_0 t) + \omega_0 B\cos(\omega_0 t). \tag{24.3.23}$$

The initial conditions at $t = 0$, are that $\theta(t=0) = A = \theta_0$, and $(d\theta / dt)(t=0) = \omega_0 B = 0$. Therefore

$$\theta(t) = \theta_0\cos(\sqrt{\gamma / I_{cm}}\; t). \tag{24.3.24}$$

Example 24.4 Compound Physical Pendulum

A compound physical pendulum consists of a disc of radius R and mass m_d fixed at the end of a rod of mass m_r and length l (Figure 24.7a). (a) Find the period of the pendulum. (b) How does the period change if the disk is mounted to the rod by a frictionless bearing so that it is perfectly free to spin?

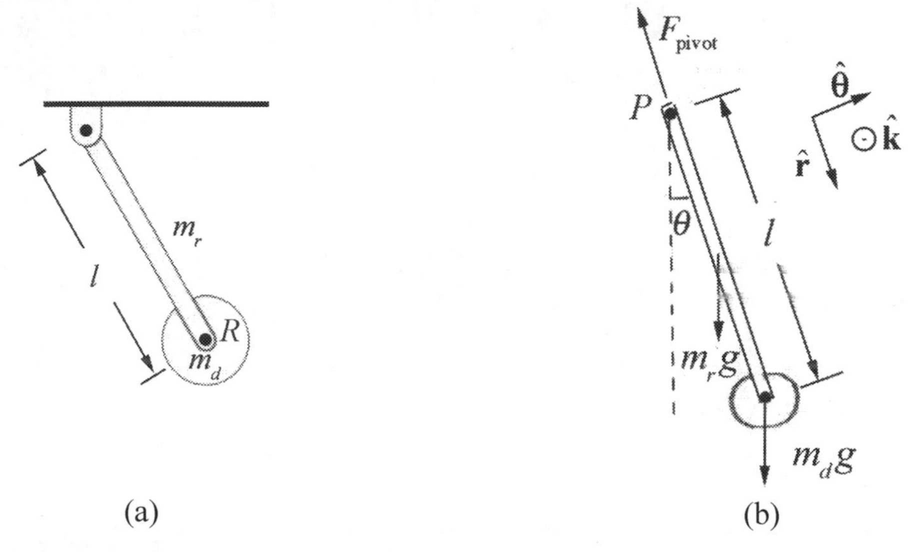

Figure 24.7 (a) Example 24.4 (b) Free-body force diagram

Solution: We begin by choosing coordinates. Let $\hat{\mathbf{k}}$ be normal to the plane of the motion of the pendulum pointing out of the plane of the Figure 24.7b. Choose an angle variable θ such that counter clockwise rotation corresponds to a positive z-component of the angular velocity. Thus a torque that points into the page has a negative z-component and a torque that points out of the page has a positive z-component. The free-body force diagram on the pendulum is also shown in Figure 24.7b. In particular, there is an unknown pivot force, the gravitational force acting at the center of mass of the rod, and the gravitational force acting at the center of mass of the disk. The torque about the pivot point is given by

$$\vec{\boldsymbol{\tau}}_P = \vec{\mathbf{r}}_{P,\mathrm{cm}} \times m_r\vec{\mathbf{g}} + \vec{\mathbf{r}}_{P,disc} \times m_d\vec{\mathbf{g}} . \qquad (24.3.25)$$

Recall that the vector $\vec{\mathbf{r}}_{P,cm}$ points from the pivot point to the center of mass of the rod a distance $l/2$ from the pivot. The vector $\vec{\mathbf{r}}_{P,disk}$ points from the pivot point to the center of mass of the disk a distance l from the pivot. Torque diagrams for the gravitational force on the rod and the disk are shown in Figure 24.8. Both torques about the pivot are in the negative $\hat{\mathbf{k}}$-direction (into the plane of Figure 24.8) and hence have negative z-components,

$$\vec{\boldsymbol{\tau}}_P = -((l/2)m_r + m_d l)g\sin\theta\,\hat{\mathbf{k}} . \qquad (24.3.26)$$

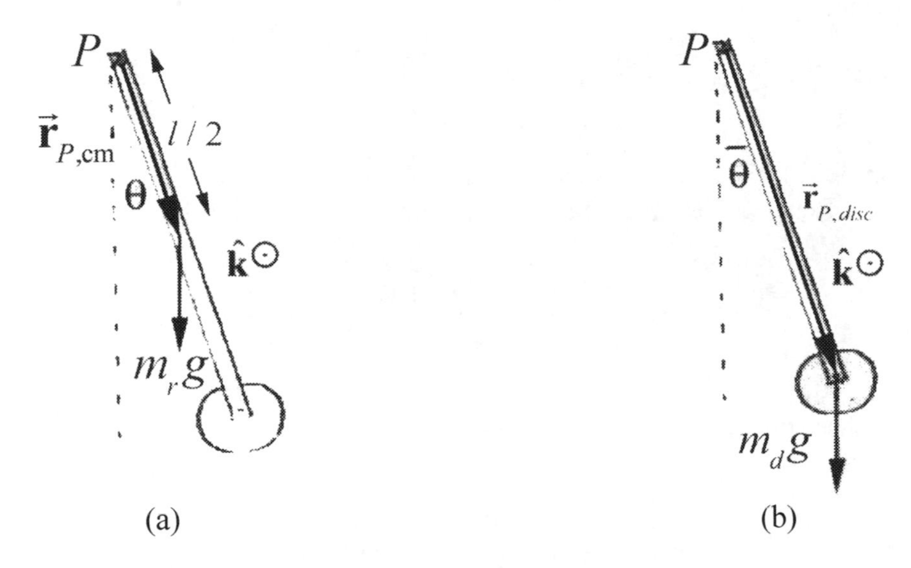

Figure 24.8 Torque diagram for (a) center of mass, (b) disk

In order to determine the moment of inertia of the rigid compound pendulum we will treat each piece separately, the uniform rod of length d and the disk attached at the end of the rod. The moment of inertia about the pivot point P is the sum of the moments of inertia of the two pieces,

$$I_P = I_{P,rod} + I_{P,disk}. \tag{24.3.27}$$

We calculated the moment of inertia of a rod about the end point P (Chapter 16.3.3), with the result that

$$I_{P,rod} = \frac{1}{3}m_r l^2. \tag{24.3.28}$$

We can use the parallel axis theorem to calculate the moment of inertia of the disk about the pivot point P,

$$I_{P,disk} = I_{cm,disk} + m_d l^2. \tag{24.3.29}$$

We calculated the moment of inertia of a disk about the center of mass (Example 16.3) and determined that

$$I_{cm,disk} = \frac{1}{2}m_d R^2. \tag{24.3.30}$$

The moment of inertia of the compound system is then

$$I_P = \frac{1}{3}m_r l^2 + m_d g l^2 + \frac{1}{2}m_d R^2. \tag{24.3.31}$$

Therefore the rotational equation of motion becomes

$$-((1/2)m_r + m_d)gl\sin\theta\,\hat{\mathbf{k}} = (((1/3)m_r + m_d)l^2 + (1/2)m_d R^2)\frac{d^2\theta}{dt^2}\hat{\mathbf{k}}. \quad (24.3.32)$$

When the angle of oscillation is small, we can use the small angle approximation $\sin\theta \simeq \theta$. Then Eq. (24.3.32) becomes a simple harmonic oscillator equation

$$\frac{d^2\theta}{dt^2} \simeq -\frac{((1/2)m_r + m_d)gl}{((1/3)m_r + m_d)l^2 + (1/2)m_d R^2}\theta. \quad (24.3.33)$$

Eq. (24.3.33) describes simple harmonic motion with an angular frequency of oscillation when the disk is fixed in place given by

$$\omega_{fixed} = \sqrt{\frac{((1/2)m_r + m_d)gl}{((1/3)m_r + m_d)l^2 + (1/2)m_d R^2}}. \quad (24.3.34)$$

The period is

$$T_{fixed} = \frac{2\pi}{\omega_{fixed}} \simeq 2\pi\sqrt{\frac{((1/3)m_r + m_d)l^2 + (1/2)m_d R^2}{((1/2)m_r + m_d)gl}}. \quad (24.3.35)$$

(b) If the disk is not fixed to the rod, then it will not rotate as the pendulum oscillates. Therefore it does not contribute to the moment of inertia. Notice that the pendulum is no longer a rigid body. So the total moment of inertia is only due to the rod and the disk treated as a point like object.

$$I_P = \frac{1}{3}m_r l^2 + m_d l^2. \quad (24.3.36)$$

Therefore the period of oscillation is given by

$$T_{free} = \frac{2\pi}{\omega_{free}} \simeq 2\pi\sqrt{\frac{((1/3)m_r + m_d)l^2}{((1/2)m_r + m_d)gl}}. \quad (24.3.37)$$

Comparing Eq. (24.3.37) to Eq. (24.3.35), we see that the period is smaller when the disk is free and not fixed. From an energy perspective we can argue that when the disk is free, it is not rotating about the center of mass. Therefore more of the gravitational potential energy goes into the center of mass translational kinetic energy than when the disk is free. Hence the center of mass is moving faster when the disk is free so it completes one period is a shorter time.

Appendix 24A Higher-Order Corrections to the Period for Larger Amplitudes of a Simple Pendulum

In Section 24.1.1, we found the period for a simple pendulum that undergoes small oscillations is given by

$$T = \frac{2\pi}{\omega_0} \cong 2\pi\sqrt{\frac{l}{g}} \qquad \text{(simple pendulum)}.$$

How good is this approximation? If the pendulum is pulled out to an initial angle θ_0 that is not small (such that our first approximation $\sin\theta \cong \theta$ no longer holds) then our expression for the period is no longer valid. We shall calculate the first-order (or higher-order) correction to the period of the pendulum.

Let's first consider the mechanical energy, a conserved quantity in this system. Choose an initial state when the pendulum is released from rest at an angle θ_i; this need not be at time $t = 0$, and in fact later in this derivation we'll see that it's inconvenient to choose this position to be at $t = 0$. Choose for the final state the position and velocity of the bob at an arbitrary time t. Choose the zero point for the potential energy to be at the bottom of the bob's swing (Figure 24A.1).

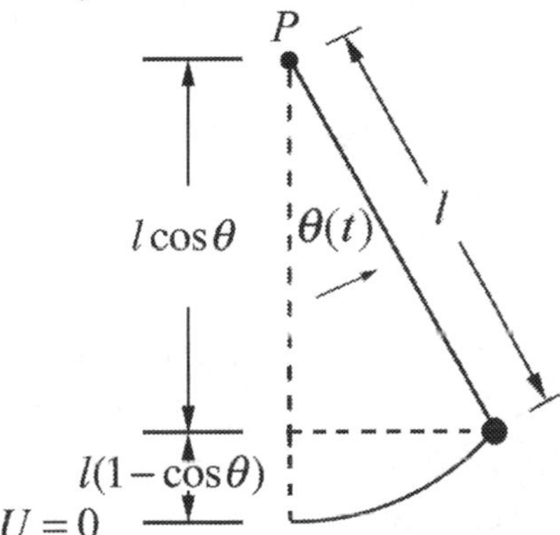

Figure 24A.1 Energy diagram for simple pendulum

The mechanical energy when the bob is released from rest at an angle θ_i is

$$E_i = K_i + U_i = mgl(1 - \cos\theta_i). \qquad (24.A.1)$$

The tangential component of the velocity of the bob at an arbitrary time t is given by

$$v_\theta = l\frac{d\theta}{dt}, \qquad (24.A.2)$$

and the kinetic energy at that time is

$$K_f = \frac{1}{2} m v^2 = \frac{1}{2} m \left(l \frac{d\theta}{dt} \right)^2 . \qquad (24.A.3)$$

The mechanical energy at time t is then

$$E_f = K_f + U_f = \frac{1}{2} m \left(l \frac{d\theta}{dt} \right)^2 + m g l (1 - \cos\theta) . \qquad (24.A.4)$$

Because the tension in the string is always perpendicular to the displacement of the bob, the tension does no work, we neglect any frictional forces, and hence mechanical energy is constant, $E_f = E_i$. Thus

$$\begin{aligned} \frac{1}{2} m \left(l \frac{d\theta}{dt} \right)^2 + m g l (1 - \cos\theta) &= m g l (1 - \cos\theta_i) \\ \left(l \frac{d\theta}{dt} \right)^2 &= 2 \frac{g}{l} (\cos\theta - \cos\theta_i). \end{aligned} \qquad (24.A.5)$$

We can solve Equation (24.A.5) for the angular velocity as a function of θ,

$$\frac{d\theta}{dt} = \sqrt{\frac{2g}{l}} \sqrt{\cos\theta - \cos\theta_i} . \qquad (24.A.6)$$

Note that we have taken the positive square root, implying that $d\theta / dt \geq 0$. This clearly cannot always be the case, and we should change the sign of the square root every time the pendulum's direction of motion changes. For our purposes, this is not an issue. If we wished to find an explicit form for either $\theta(t)$ or $t(\theta)$, we would have to consider the signs in Equation (24.A.6) more carefully.

Before proceeding, it's worth considering the difference between Equation (24.A.6) and the equation for the simple pendulum in the simple harmonic oscillator limit, where $\cos\theta \simeq 1 - (1/2)\theta^2$. Then Eq. (24.A.6) reduces to

$$\frac{d\theta}{dt} = \sqrt{\frac{2g}{l}} \sqrt{\frac{\theta_i^2}{2} - \frac{\theta^2}{2}} . \qquad (24.A.7)$$

In both Equations (24.A.6) and (24.A.7) the last term in the square root is proportional to the difference between the initial potential energy and the final potential energy. The final potential energy for the two cases is plotted in Figures 24A.2 for $-\pi < \theta < \pi$ on the left and $-\pi / 2 < \theta < \pi / 2$ on the right (the vertical scale is in units of mgl).

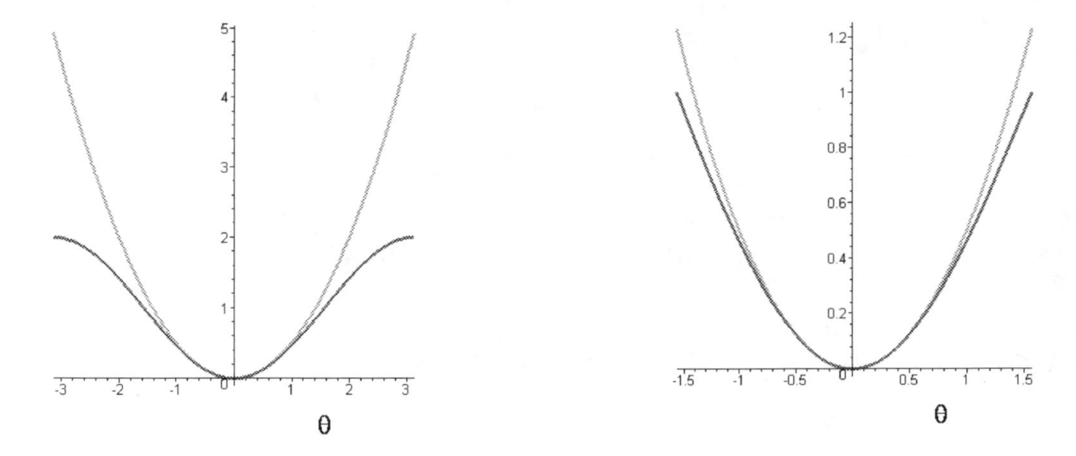

Figures 24A.2 Potential energies as a function of displacement angle

It would seem to be to our advantage to express the potential energy for an arbitrary displacement of the pendulum as the difference between two squares. We do this by first recalling the trigonometric identity

$$1-\cos\theta = 2\sin^2(\theta/2) \tag{24.A.8}$$

with the result that Equation (24.A.6) may be re-expressed as

$$\frac{d\theta}{dt} = \sqrt{\frac{2g}{l}}\sqrt{2(\sin^2(\theta_i/2)-\sin^2(\theta/2))}\,. \tag{24.A.9}$$

Equation (24.A.9) is separable,

$$\frac{d\theta}{\sqrt{\sin^2(\theta_i/2)-\sin^2(\theta/2)}} = 2\sqrt{\frac{g}{l}}\,dt \tag{24.A.10}$$

Rewrite Equation (24.A.10) as

$$\frac{d\theta}{\sin(\theta_i/2)\sqrt{1-\dfrac{\sin^2(\theta/2)}{\sin^2(\theta_i/2)}}} = 2\sqrt{\frac{g}{l}}\,dt\,. \tag{24.A.11}$$

The ratio $\sin(\theta/2)/\sin(\theta_i/2)$ in the square root in the denominator will oscillate (but *not* with simple harmonic motion) between -1 and $+1$, and so we will make the identification

$$\sin\phi = \frac{\sin(\theta/2)}{\sin(\theta_i/2)}\,. \tag{24.A.12}$$

Let $b = \sin(\theta_i / 2)$, so that

$$\begin{aligned} \sin\frac{\theta}{2} &= b\sin\phi \\ \cos\frac{\theta}{2} &= \left(1-\sin^2\frac{\theta}{2}\right)^{1/2} = (1-b^2\sin^2\phi)^{1/2}. \end{aligned} \tag{24.A.13}$$

Eq. (24.A.11) can then be rewritten in integral form as

$$\int \frac{d\theta}{b\sqrt{1-\sin^2\phi}} = 2\int \sqrt{\frac{g}{l}}\, dt\,. \tag{24.A.14}$$

From differentiating the first expression in Equation (24.A.13), we have that

$$\begin{aligned} &\frac{1}{2}\cos\frac{\theta}{2}\; d\theta = b\cos\phi\; d\phi \\ &d\theta = 2b\frac{\cos\phi}{\cos(\theta/2)}d\phi = 2b\frac{\sqrt{1-\sin^2\phi}}{\sqrt{1-\sin^2(\theta/2)}}d\phi \\ &= 2b\frac{\sqrt{1-\sin^2\phi}}{\sqrt{1-b^2\sin^2\phi}}d\phi. \end{aligned} \tag{24.A.15}$$

Substituting the last equation in (24.A.15) into the left-hand side of the integral in (24.A.14) yields

$$\int \frac{2b}{b\sqrt{1-\sin^2\phi}}\frac{\sqrt{1-\sin^2\phi}}{\sqrt{1-b^2\sin^2\phi}}d\phi = 2\int \frac{d\phi}{\sqrt{1-b^2\sin^2\phi}}\,. \tag{24.A.16}$$

Thus the integral in Equation (24.A.14) becomes

$$\int \frac{d\phi}{\sqrt{1-b^2\sin^2\phi}} = \int \sqrt{\frac{g}{l}}\, dt\,. \tag{24.A.17}$$

This integral is one of a class of integrals known as ***elliptic integrals***. We will encounter a similar integral when we solve the *Kepler Problem* in Chapter 25, where the orbits of an object under the influence of an inverse square gravitational force are determined. We find a power series solution to this integral by expanding the function

$$(1-b^2\sin^2\phi)^{-1/2} = 1 + \frac{1}{2}b^2\sin^2\phi + \frac{3}{8}b^4\sin^4\phi + \cdots. \tag{24.A.18}$$

The integral in Equation (24.A.17) then becomes

$$\int\left(1+\frac{1}{2}b^2\sin^2\phi+\frac{3}{8}b^4\sin^4\phi+\cdots\right)d\phi=\int\sqrt{\frac{g}{l}}\,dt\,. \tag{24.A.19}$$

Now let's integrate over one period. Set $t=0$ when $\theta=0$, the lowest point of the swing, so that $\sin\phi=0$ and $\phi=0$. One period T has elapsed the second time the bob returns to the lowest point, or when $\phi=2\pi$. Putting in the limits of the ϕ-integral, we can integrate term by term, noting that

$$\begin{aligned}\int_0^{2\pi}\frac{1}{2}b^2\sin^2\phi\,d\phi&=\int_0^{2\pi}\frac{1}{2}b^2\frac{1}{2}(1-\cos(2\phi))\,d\phi\\&=\frac{1}{2}b^2\frac{1}{2}\left(\phi-\frac{\sin(2\phi)}{2}\right)\Bigg|_0^{2\pi}\\&=\frac{1}{2}\pi b^2=\frac{1}{2}\pi\sin^2\frac{\theta_i}{2}.\end{aligned} \tag{24.A.20}$$

Thus, from Equation (24.A.19) we have that

$$\begin{aligned}\int_0^{2\pi}\left(1+\frac{1}{2}b^2\sin^2\phi+\frac{3}{8}b^4\sin^4\phi+\cdots\right)d\phi&=\int_0^{T}\sqrt{\frac{g}{l}}\,dt\\2\pi+\frac{1}{2}\pi\sin^2\frac{\theta_i}{2}+\cdots&=\sqrt{\frac{g}{l}}\,T\end{aligned}, \tag{24.A.21}$$

We can now solve for the period,

$$T=2\pi\sqrt{\frac{l}{g}}\left(1+\frac{1}{4}\sin^2\frac{\theta_i}{2}+\cdots\right). \tag{24.A.22}$$

If the initial angle $\theta_i<<1$ (measured in radians), then $\sin^2(\theta_i/2)\simeq\theta_i^2/4$ and the period is approximately

$$T\cong2\pi\sqrt{\frac{l}{g}}\left(1+\frac{1}{16}\theta_i^2\right)=T_0\left(1+\frac{1}{16}\theta_i^2\right), \tag{24.A.23}$$

where

$$T_0=2\pi\sqrt{\frac{l}{g}} \tag{24.A.24}$$

is the period of the simple pendulum with the standard small angle approximation. The first order correction to the period of the pendulum is then

$$\Delta T_1 = \frac{1}{16}\theta_i^2 T_0 . \tag{24.A.25}$$

Figure 24A.3 below shows the three functions given in Equation (24.A.24) (the horizontal, or red plot if seen in color), Equation (24.A.23) (the middle, parabolic or green plot) and the numerically-integrated function obtained by integrating the expression in Equation (24.A.17) (the upper, or blue plot) between $\phi = 0$ and $\phi = 2\pi$ as a function of $k = \sin(\theta_0 / 2)$. The plots demonstrate that Equation (24.A.24) is a valid approximation for small values of θ_i, and that Equation (24.A.23) is a very good approximation for all but the largest amplitudes of oscillation. The vertical axis is in units of $\sqrt{l/g}$. Note the displacement of the horizontal axis.

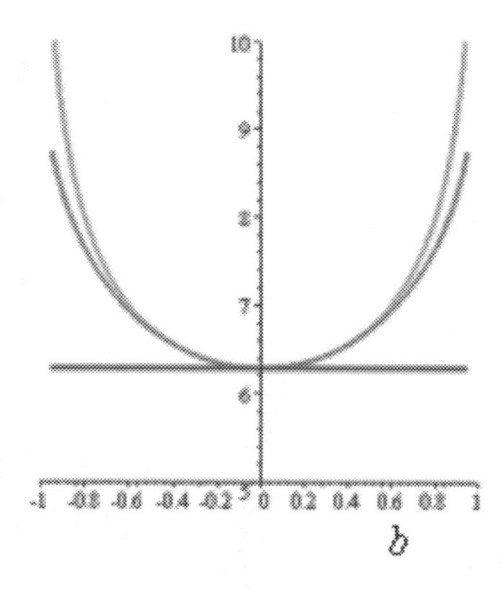

Figure 24A.3 Pendulum Period Approximations as Functions of Amplitude

Chapter 25 Celestial Mechanics

Chapter 25 Celestial Mechanics

...and if you want the exact moment in time, it was conceived mentally on 8th March in this year one thousand six hundred and eighteen, but submitted to calculation in an unlucky way, and therefore rejected as false, and finally returning on the 15th of May and adopting a new line of attack, stormed the darkness of my mind. So strong was the support from the combination of my labour of seventeen years on the observations of Brahe and the present study, which conspired together, that at first I believed I was dreaming, and assuming my conclusion among my basic premises. But it is absolutely certain and exact that "the proportion between the periodic times of any two planets is precisely the sesquialterate proportion of their mean distances ..." [1]

Johannes Kepler

25.1 Introduction: The Kepler Problem

Johannes Kepler first formulated the laws that describe planetary motion,

I. Each planet moves in an ellipse with the sun at one focus.

II. The radius vector from the sun to a planet sweeps out equal areas in equal time.

III. The period of revolution T of a planet about the sun is related to the major axis A of the ellipse by $T^2 = k\,A^3$ where k is the same for all planets.[2]

The third law was published in 1619, and efforts to discover and solve the equation of motion of the planets generated two hundred years of mathematical and scientific discovery. In his honor, this problem has been named ***the Kepler Problem.***

When there are more than two bodies, the problem becomes impossible to solve exactly. The most important "three-body problem" in the 17th and 18th centuries involved finding the motion of the moon, due to gravitational interaction with both the sun and the earth. Newton realized that if the exact position of the moon were known, the longitude of any observer on the earth could be determined by measuring the moon's position with respect to the stars.

In the eighteenth century, Leonhard Euler and other mathematicians spent many years trying to solve the three-body problem, and they raised a deeper question. Do the small contributions from the gravitational interactions of all the planets make the planetary system unstable over long periods of time? At the end of 18th century, Pierre

[1] Kepler, Johannes, *Harmonice mundi* Book 5, Chapter 3, trans. Aiton, Duncan and Field, p. 411

[2] As stated in *An Introduction to Mechanics*, Daniel Kleppner and Robert Kolenkow, McGraw-Hill, 1973, p 401.

Simon Laplace and others found a series solution to this stability question, but it was unknown whether or not the series solution converged after a long period of time. Henri Poincaré proved that the series actually diverged. Poincaré went on to invent new mathematical methods that produced the modern fields of differential geometry and topology in order to answer the stability question using geometric arguments, rather than analytic methods. Poincaré and others did manage to show that the three-body problem was indeed stable, due to the existence of periodic solutions. Just as in the time of Newton and Leibniz and the invention of calculus, unsolved problems in celestial mechanics became the experimental laboratory for the discovery of new mathematics.

25.2 Planetary Orbits

We now commence a study of the Kepler Problem. We shall determine the equation of motion for the motions of two bodies interacting via a gravitational force (two-body problem) using both force methods and conservation laws.

25.2.1 Reducing the Two-Body Problem into a One-Body Problem

We shall begin by showing how the motion of two bodies interacting via a gravitational force (two-body problem) is mathematically equivalent to the motion of a single body acted on by an external central gravitational force, where the mass of the single body is the *reduced mass* μ,

$$\frac{1}{\mu}=\frac{1}{m_1}+\frac{1}{m_2}\Rightarrow\mu=\frac{m_1 m_2}{m_1+m_2}. \qquad (25.2.1)$$

Once we solve for the motion of the reduced body in this ***equivalent one-body problem***, we can then return to the real two-body problem and solve for the actual motion of the two original bodies. The reduced mass was introduced in Chapter 13 Appendix A of these notes. That appendix used similar but slightly different notation from that used in this chapter.

Consider the gravitational interaction between two bodies with masses m_1 and m_2 as shown in Figure 25.1.

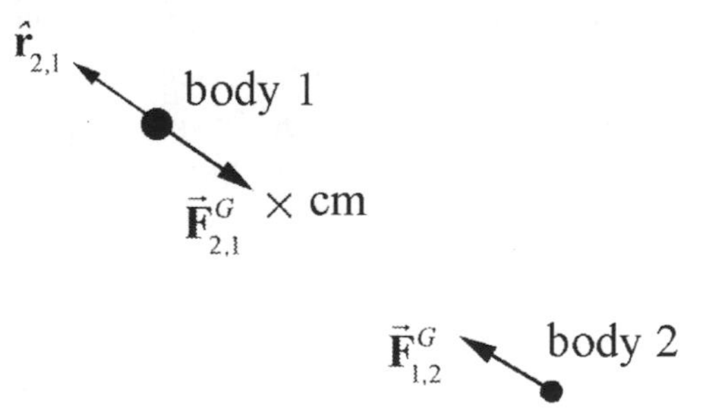

Figure 25.1 Gravitational force between two bodies

Choose a coordinate system with a choice of origin such that body 1 has position $\vec{\mathbf{r}}_1$ and body 2 has position $\vec{\mathbf{r}}_2$ (Figure 25.2). The *relative position vector* $\vec{\mathbf{r}}$ pointing from body 2 to body 1 is $\vec{\mathbf{r}} = \vec{\mathbf{r}}_1 - \vec{\mathbf{r}}_2$. We denote the magnitude of $\vec{\mathbf{r}}$ by $|\vec{\mathbf{r}}| = r$, where r is the distance between the bodies, and $\hat{\mathbf{r}}$ is the unit vector pointing from body 2 to body 1, so that $\vec{\mathbf{r}} = r\,\hat{\mathbf{r}}$.

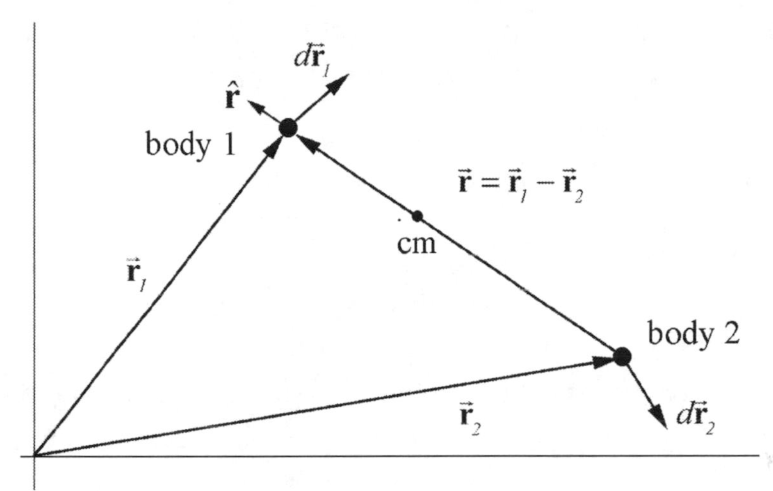

Figure 25.2 Coordinate system for the two-body problem

The force on body 1 (due to the interaction of the two bodies) can be described by Newton's Universal Law of Gravitation

$$\vec{\mathbf{F}}_{2,1} = -F_{2,1}\,\hat{\mathbf{r}} = -G\frac{m_1 m_2}{r^2}\,\hat{\mathbf{r}}. \tag{25.2.2}$$

Recall that Newton's Third Law requires that the force on body 2 is equal in magnitude and opposite in direction to the force on body 1,

$$\vec{\mathbf{F}}_{1,2} = -\vec{\mathbf{F}}_{2,1}. \tag{25.2.3}$$

Newton's Second Law can be applied individually to the two bodies:

$$\vec{\mathbf{F}}_{2,1} = m_1\frac{d^2\vec{\mathbf{r}}_1}{dt^2}, \tag{25.2.4}$$

$$\vec{\mathbf{F}}_{1,2} = m_2\frac{d^2\vec{\mathbf{r}}_2}{dt^2}. \tag{25.2.5}$$

Dividing through by the mass in each of Equations (25.2.4) and (25.2.5) yields

$$\frac{\vec{\mathbf{F}}_{2,1}}{m_1} = \frac{d^2\vec{\mathbf{r}}_1}{dt^2}, \tag{25.2.6}$$

$$\frac{\vec{\mathbf{F}}_{1,2}}{m_2} = \frac{d^2\vec{\mathbf{r}}_2}{dt^2}. \tag{25.2.7}$$

Subtracting the expression in Equation (25.2.7) from that in Equation (25.2.6) yields

$$\frac{\vec{\mathbf{F}}_{2,1}}{m_1} - \frac{\vec{\mathbf{F}}_{1,2}}{m_2} = \frac{d^2\vec{\mathbf{r}}_1}{dt^2} - \frac{d^2\vec{\mathbf{r}}_2}{dt^2} = \frac{d^2\vec{\mathbf{r}}}{dt^2}. \tag{25.2.8}$$

Using Newton's Third Law, Equation (25.2.3), Equation (25.2.8) becomes

$$\vec{\mathbf{F}}_{2,1}\left(\frac{1}{m_1} + \frac{1}{m_2}\right) = \frac{d^2\vec{\mathbf{r}}}{dt^2}. \tag{25.2.9}$$

Using the *reduced mass* μ, as defined in Equation (25.2.1), Equation (25.2.9) becomes

$$\begin{aligned} \frac{\vec{\mathbf{F}}_{2,1}}{\mu} &= \frac{d^2\vec{\mathbf{r}}}{dt^2} \\ \vec{\mathbf{F}}_{2,1} &= \mu\frac{d^2\vec{\mathbf{r}}}{dt^2}, \end{aligned} \tag{25.2.10}$$

where $\vec{\mathbf{F}}_{2,1}$ is given by Equation (25.2.2).

Our result has a special interpretation using Newton's Second Law. Let μ be the mass of a ***single body*** with position vector $\vec{\mathbf{r}} = r\,\hat{\mathbf{r}}$ with respect to an origin O, where $\hat{\mathbf{r}}$ is the unit vector pointing from the origin O to the single body. Then the equation of motion, Equation (25.2.10), implies that the single body of mass μ is under the influence of an attractive gravitational force pointing toward the origin. So, the original two-body gravitational problem has now been reduced to an equivalent one-body problem, involving a single body with mass μ under the influence of a central force $\vec{\mathbf{F}}^G = -\mathrm{F}_{2,1}\,\hat{\mathbf{r}}$. Note that in this reformulation, there is no body located at the central point (the origin O). The parameter r in the two-body problem is the relative distance between the original two bodies, while the same parameter r in the one-body problem is the distance between the single body and the central point. Also, this reduction generalizes to all central forces.

25.3 Energy and Angular Momentum, Constants of the Motion

The equivalent one-body problem has two constants of the motion, energy E and the angular momentum L about the origin O. Energy is a constant because in our original two-body problem, the gravitational interaction was an internal conservative force. Angular momentum is constant about the origin because the only force is directed towards the origin, and hence the torque about the origin due to that force is zero (the vector from the origin to the single body is anti-parallel to the force vector and $\sin\pi = 0$). Because angular momentum is constant, the orbit of the single body lies in a plane with the angular momentum vector pointing perpendicular to this plane.

In the plane of the orbit, choose polar coordinates (r, θ) for the single body (see Figure 25.3), where r is the distance of the single body from the central point that is now taken as the origin O, and θ is the angle that the single body makes with respect to a chosen direction, and which increases positively in the counterclockwise direction.

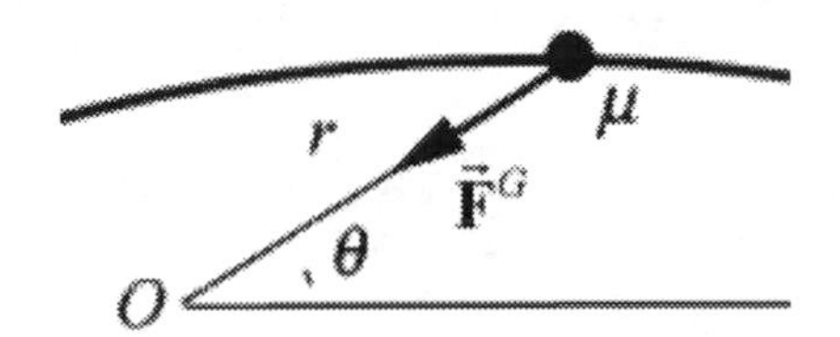

Figure 25.3 Coordinate system for the orbit of the single body

There are two approaches to describing the motion of the single body. We can try to find both the distance from the origin, $r(t)$ and the angle, $\theta(t)$, as functions of the parameter time, but in most cases explicit functions can't be found analytically. We can also find the distance from the origin, $r(\theta)$, as a function of the angle θ. This second approach offers a spatial description of the motion of the single body (see Appendix 25A).

25.3.1 The Orbit Equation for the One-Body Problem

Consider the single body with mass μ given by Equation (25.2.1), orbiting about a central point under the influence of a radially attractive force given by Equation (25.2.2). Since the force is conservative, the potential energy (from the two-body problem) with choice of zero reference point $U(\infty) = 0$ is given by

$$U(r) = -\frac{G m_1 m_2}{r}. \tag{25.3.1}$$

The total energy E is constant, and the sum of the kinetic energy and the potential energy is

$$E = \frac{1}{2}\mu v^2 - \frac{G m_1 m_2}{r}. \tag{25.3.2}$$

The kinetic energy term $\mu v^2 / 2$ is written in terms of the mass μ and the relative speed v of the two bodies. Choose polar coordinates such that

$$\vec{\mathbf{v}} = v_r\,\hat{\mathbf{r}} + v_\theta\hat{\boldsymbol{\theta}},$$
$$v = |\vec{\mathbf{v}}| = \left|\frac{d\vec{\mathbf{r}}}{dt}\right|, \tag{25.3.3}$$

where $v_r = dr/dt$ and $v_\theta = r(d\theta/dt)$. Equation (25.3.2) then becomes

$$E = \frac{1}{2}\mu\left[\left(\frac{dr}{dt}\right)^2 + \left(r\frac{d\theta}{dt}\right)^2\right] - \frac{G\,m_1\,m_2}{r}. \tag{25.3.4}$$

The angular momentum with respect to the origin O is given by

$$\vec{\mathbf{L}}_O = \vec{\mathbf{r}}_O \times \mu\vec{\mathbf{v}} = r\hat{\mathbf{r}} \times \mu(v_r\,\hat{\mathbf{r}} + v_\theta\hat{\boldsymbol{\theta}}) = \mu r v_\theta \hat{\mathbf{k}} = \mu r^2\frac{d\theta}{dt}\hat{\mathbf{k}} \equiv L\hat{\mathbf{k}} \tag{25.3.5}$$

with magnitude

$$L = \mu r\,v_\theta = \mu r^2\frac{d\theta}{dt}. \tag{25.3.6}$$

We shall explicitly eliminate the θ dependence from Equation (25.3.4) by using our expression in Equation (25.3.6),

$$\frac{d\theta}{dt} = \frac{L}{\mu r^2}. \tag{25.3.7}$$

The mechanical energy as expressed in Equation (25.3.4) then becomes

$$E = \frac{1}{2}\mu\left(\frac{dr}{dt}\right)^2 + \frac{1}{2}\frac{L^2}{\mu r^2} - \frac{G\,m_1\,m_2}{r}. \tag{25.3.8}$$

Equation (25.3.8) is a separable differential equation involving the variable r as a function of time t and can be solved for the first derivative dr/dt,

$$\frac{dr}{dt} = \sqrt{\frac{2}{\mu}}\left(E - \frac{1}{2}\frac{L^2}{\mu r^2} + \frac{G\,m_1\,m_2}{r}\right)^{\frac{1}{2}}. \tag{25.3.9}$$

Equation (25.3.9) can in principle be integrated directly for $r(t)$. In fact, doing the integrals is complicated and beyond the scope of this book. The function $r(t)$ can then, in principle, be substituted into Equation (25.3.7) and can then be integrated to find $\theta(t)$.

Instead of solving for the position of the single body as a function of time, we shall find a geometric description of the orbit by finding $r(\theta)$. We first divide Equation (25.3.7) by Equation (25.3.9) to obtain

$$\frac{d\theta}{dr} = \frac{\frac{d\theta}{dt}}{\frac{dr}{dt}} = \frac{L}{\sqrt{2\mu}} \frac{\left(1/r^2\right)}{\left(E - \frac{L^2}{2\mu r^2} + \frac{G m_1 m_2}{r}\right)^{1/2}}. \qquad (25.3.10)$$

The variables r and θ are separable;

$$d\theta = \frac{L}{\sqrt{2\mu}} \frac{\left(1/r^2\right)}{\left(E - \frac{L^2}{2\mu r^2} + \frac{G m_1 m_2}{r}\right)^{1/2}} dr\ . \qquad (25.3.11)$$

Equation (25.3.11) can be integrated to find the radius as a function of the angle θ; see Appendix 25A for the exact integral solution. The result is called the ***orbit equation*** for the reduced body and is given by

$$r = \frac{r_0}{1 - \varepsilon \cos\theta} \qquad (25.3.12)$$

where

$$r_0 = \frac{L^2}{\mu G m_1 m_2} \qquad (25.3.13)$$

is a constant (known as the ***semilatus rectum***) and

$$\varepsilon = \left(1 + \frac{2 E L^2}{\mu (G m_1 m_2)^2}\right)^{\frac{1}{2}} \qquad (25.3.14)$$

is the ***eccentricity*** of the orbit. The two constants of the motion, angular momentum L and mechanical energy E, in terms of r_0 and ε, are

$$L = (\mu G m_1 m_2 r_0)^{1/2} \qquad (25.3.15)$$

$$E = \frac{G m_1 m_2 (\varepsilon^2 - 1)}{2 r_0}. \qquad (25.3.16)$$

The orbit equation as given in Equation (25.3.12) is a general ***conic section*** and is perhaps somewhat more familiar in Cartesian coordinates. Let $x = r\cos\theta$ and $y = r\sin\theta$, with $r^2 = x^2 + y^2$. The orbit equation can be rewritten as

$$r = r_0 + \varepsilon r \cos\theta\ . \qquad (25.3.17)$$

Using the Cartesian substitutions for x and y, rewrite Equation (25.3.17) as

$$(x^2 + y^2)^{1/2} = r_0 + \varepsilon x . \tag{25.3.18}$$

Squaring both sides of Equation (25.3.18),

$$x^2 + y^2 = r_0^{\,2} + 2\varepsilon x r_0 + \varepsilon^2 x^2 . \tag{25.3.19}$$

After rearranging terms, Equation (25.3.19) is the general expression of a conic section with axis on the x-axis,

$$x^2(1-\varepsilon^2) - 2\varepsilon x r_0 + y^2 = r_0^{\,2} . \tag{25.3.20}$$

(We now see that the horizontal axis in Figure 25.3 can be taken to be the x-axis).

For a given $r_0 > 0$, corresponding to a given nonzero angular momentum as in Equation (25.3.12), there are four cases determined by the value of the eccentricity.

Case 1: when $\varepsilon = 0$, $E = E_{\min} < 0$ and $r = r_0$, Equation (25.3.20) is the equation for a *circle*,

$$x^2 + y^2 = r_0^{\,2} . \tag{25.3.21}$$

Case 2: when $0 < \varepsilon < 1$, $E_{\min} < E < 0$, Equation (25.3.20) describes an *ellipse*,

$$y^2 + Ax^2 - Bx = k . \tag{25.3.22}$$

where $A > 0$ and k is a positive constant. (Appendix 25C shows how this expression may be expressed in the more traditional form involving the coordinates of the center of the ellipse and the semi-major and semi-minor axes.)

Case 3: when $\varepsilon = 1$, $E = 0$, Equation (25.3.20) describes a *parabola*,

$$x = \frac{y^2}{2r_0} - \frac{r_0}{2} . \tag{25.3.23}$$

Case 4: when $\varepsilon > 1$, $E > 0$, Equation (25.3.20) describes a *hyperbola*,

$$y^2 - Ax^2 - Bx = k , \tag{25.3.24}$$

where $A > 0$ and k is a positive constant.

25.4 Energy Diagram, Effective Potential Energy, and Orbits

The energy (Equation (25.3.8)) of the single body moving in two dimensions can be reinterpreted as the energy of a single body moving in one dimension, the radial direction r, in an ***effective potential energy*** given by two terms,

$$U_{\text{eff}} = \frac{L^2}{2\mu r^2} - \frac{G m_1 m_2}{r}. \tag{25.4.1}$$

The energy is still the same, but our interpretation has changed,

$$E = K_{\text{eff}} + U_{\text{eff}} = \frac{1}{2}\mu\left(\frac{dr}{dt}\right)^2 + \frac{L^2}{2\mu r^2} - \frac{G m_1 m_2}{r}, \tag{25.4.2}$$

where the ***effective kinetic energy*** K_{eff} associated with the one-dimensional motion is

$$K_{\text{eff}} = \frac{1}{2}\mu\left(\frac{dr}{dt}\right)^2. \tag{25.4.3}$$

The graph of U_{eff} as a function of $u = r / r_0$, where r_0 as given in Equation (25.3.13), is shown in Figure 25.4. The upper red curve is proportional to $L^2 / (2\mu r^2) \sim 1/2r^2$. The lower blue curve is proportional to $-Gm_1 m_2 / r \sim -1/r$. The sum U_{eff} is represented by the middle green curve. The minimum value of U_{eff} is at $r = r_0$, as will be shown analytically below. The vertical scale is in units of $-U_{\text{eff}}(r_0)$. Whenever the one-dimensional kinetic energy is zero, $K_{\text{eff}} = 0$, the energy is equal to the effective potential energy,

$$E = U_{\text{eff}} = \frac{L^2}{2\mu r^2} - \frac{G m_1 m_2}{r}. \tag{25.4.4}$$

Recall that the potential energy is defined to be the negative integral of the work done by the force. For our reduction to a one-body problem, using the effective potential, we will introduce an ***effective force*** such that

$$U_{\text{eff},B} - U_{\text{eff},A} = -\int_A^B \vec{\mathbf{F}}^{\text{eff}} \cdot d\vec{\mathbf{r}} = -\int_A^B F_r^{\text{eff}}\, dr \tag{25.4.5}$$

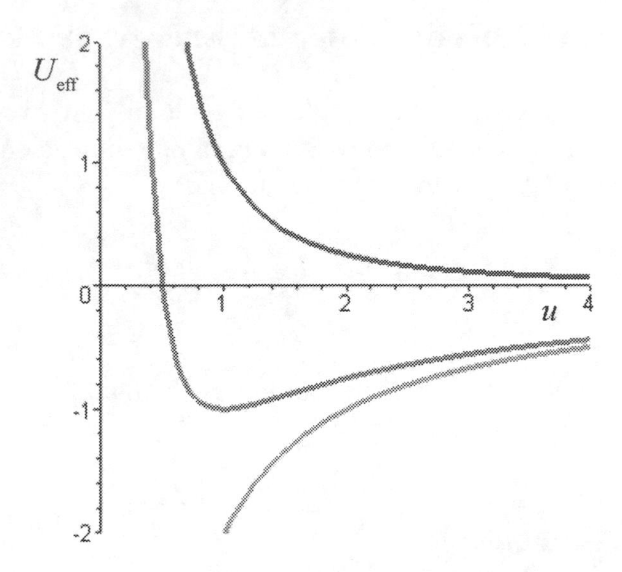

Figure 25.4 Graph of effective potential energy

The fundamental theorem of calculus (for one variable) then states that the integral of the derivative of the effective potential energy function between two points is the effective potential energy difference between those two points,

$$U_{\text{eff},B} - U_{\text{eff},A} = \int_A^B \frac{dU_{\text{eff}}}{dr} dr \tag{25.4.6}$$

Comparing Equation (25.4.6) to Equation (25.4.5) shows that the radial component of the effective force is the negative of the derivative of the effective potential energy,

$$F_r^{\text{eff}} = -\frac{dU_{\text{eff}}}{dr} \tag{25.4.7}$$

The effective potential energy describes the potential energy for a reduced body moving in one dimension. (Note that the effective potential energy is only a function of the variable r and is independent of the variable θ). There are two contributions to the effective potential energy, and the radial component of the force is then

$$F_r^{\text{eff}} = -\frac{d}{dr}U_{\text{eff}} = -\frac{d}{dr}\left(\frac{L^2}{2\mu r^2} - \frac{G m_1 m_2}{r}\right) \tag{25.4.8}$$

Thus there are two "forces" acting on the reduced body,

$$F_r^{\text{eff}} = F_{r,\,\text{centifugal}} + F_{r,\,\text{gravity}}\,, \tag{25.4.9}$$

with an ***effective centrifugal force*** given by

$$F_{r,\text{centrifugal}} = -\frac{d}{dr}\left(\frac{L^2}{2\mu r^2}\right) = \frac{L^2}{\mu r^3} \tag{25.4.10}$$

and the centripetal gravitational force given by

$$F_{r,\text{ gravity}} = -\frac{G m_1 m_2}{r^2}. \tag{25.4.11}$$

With this nomenclature, let's review the four cases presented in Section 25.3.

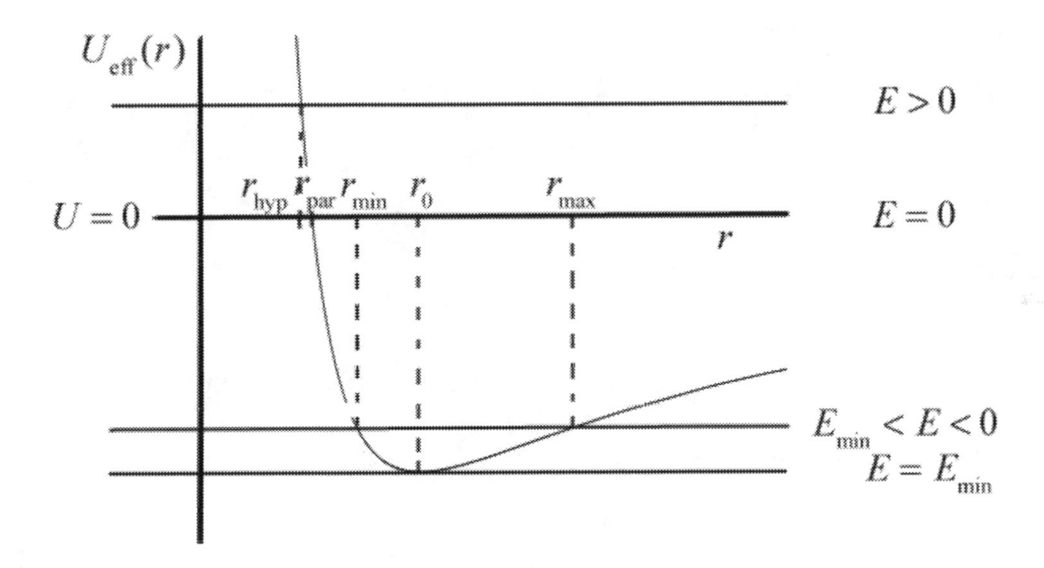

Figure 25.5 Plot of $U_{\text{eff}}(r)$ vs. r with four energies corresponding to circular, elliptic, parabolic, and hyperbolic orbits

25.4.1 Circular Orbit $E = E_{\text{min}}$

The lowest energy state, E_{min}, corresponds to the minimum of the effective potential energy, $E_{\text{min}} = (U_{\text{eff}})_{\text{min}}$. We can minimize the effective potential energy

$$0 = \left.\frac{dU_{\text{eff}}}{dr}\right|_{r=r_0} = -\frac{L^2}{\mu r_0^{3}} + \frac{G m_1 m_2}{r_0^{2}}. \tag{25.4.12}$$

and solve Equation (25.4.12) for r_0,

$$r_0 = \frac{L^2}{G m_1 m_2}, \tag{25.4.13}$$

reproducing Equation (25.3.13). For $E = E_{\text{min}}$, $r = r_0$ which corresponds to a circular orbit.

25.4.2 Elliptic Orbit $E_{min} < E < 0$

For $E_{min} < E < 0$, there are two points r_{min} and r_{max} such that $E = U_{eff}(r_{min}) = U_{eff}(r_{max})$. At these points $K_{eff} = 0$, therefore $dr/dt = 0$ which corresponds to a point of closest or furthest approach (Figure 25.6). This condition corresponds to the minimum and maximum values of r for an elliptic orbit.

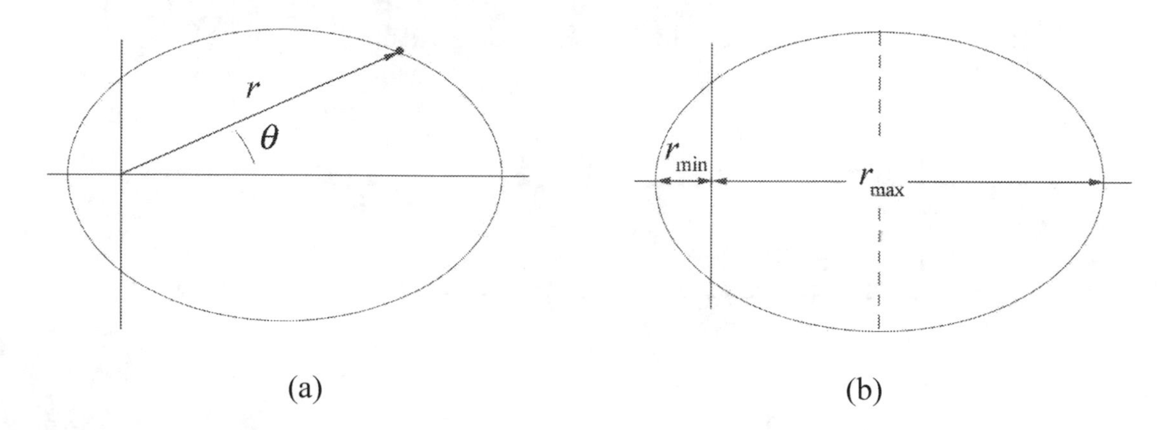

Figure 25.6 (a) elliptic orbit, (b) closest and furthest approach

The energy condition at these two points

$$E = \frac{L^2}{2\mu r^2} - \frac{G m_1 m_2}{r}, \quad r = r_{min} = r_{max}, \tag{25.4.14}$$

is a quadratic equation for the distance r,

$$r^2 + \frac{G m_1 m_2}{E} r - \frac{L^2}{2\mu E} = 0. \tag{25.4.15}$$

There are two roots

$$r = -\frac{G m_1 m_2}{2E} \pm \left(\left(\frac{G m_1 m_2}{2E} \right)^2 + \frac{L^2}{2\mu E} \right)^{1/2}. \tag{25.4.16}$$

Equation (25.4.16) may be simplified somewhat as

$$r = -\frac{G m_1 m_2}{2E} \left(1 \pm \left(1 + \frac{2L^2 E}{\mu (G m_1 m_2)^2} \right)^{1/2} \right) \tag{25.4.17}$$

From Equation (25.3.14), the square root is the eccentricity ε,

$$\varepsilon = \left(1 + \frac{2EL^2}{\mu(G m_1 m_2)^2}\right)^{\frac{1}{2}}, \tag{25.4.18}$$

and Equation (25.4.17) becomes

$$r = -\frac{G m_1 m_2}{2E}(1 \pm \varepsilon). \tag{25.4.19}$$

A little algebra shows that

$$\begin{aligned} \frac{r_0}{1-\varepsilon^2} &= \frac{L^2 / \mu G m_1 m_2}{1 - \left(1 + \dfrac{2L^2 E}{\mu(G m_1 m_2)^2}\right)} \\ &= \frac{L^2 / \mu G m_1 m_2}{-2L^2 E / \mu(G m_1 m_2)^2} \\ &= -\frac{G m_1 m_2}{2E}. \end{aligned} \tag{25.4.20}$$

Substituting the last expression in (25.4.20) into Equation (25.4.19) gives an expression for the points of closest and furthest approach,

$$r = \frac{r_0}{1-\varepsilon^2}(1 \pm \varepsilon). \tag{25.4.21}$$

The minus sign corresponds to the distance of closest approach,

$$r \equiv r_{\text{min}} = \frac{r_0}{1+\varepsilon} \tag{25.4.22}$$

and the plus sign corresponds to the distance of furthest approach,

$$r \equiv r_{\text{max}} = \frac{r_0}{1-\varepsilon}. \tag{25.4.23}$$

25.4.3 Parabolic Orbit $E = 0$

The effective potential energy, as given in Equation (25.4.1), approaches zero ($U_{\text{eff}} \to 0$) when the distance r approaches infinity ($r \to \infty$). When $E = 0$, as $r \to \infty$, the kinetic energy also approaches zero, $K_{\text{eff}} \to 0$. This corresponds to a parabolic orbit (see Equation (25.3.23)). Recall that in order for a body to escape from a planet, the body must have an energy $E = 0$ (we set $U_{\text{eff}} = 0$ at infinity). This *escape velocity* condition corresponds to a parabolic orbit. For a parabolic orbit, the body also has a distance of closest approach. This distance r_{par} can be found from the condition

$$E = U_{\text{eff}}(r_{\text{par}}) = \frac{L^2}{2\mu r_{\text{par}}^{\;2}} - \frac{G m_1 m_2}{r_{\text{par}}} = 0. \tag{25.4.24}$$

Solving Equation (25.4.24) for r_{par} yields

$$r_{\text{par}} = \frac{L^2}{2\mu G m_1 m_2} = \frac{1}{2} r_0; \tag{25.4.25}$$

the fact that the minimum distance to the origin (the *focus* of a parabola) is half the semilatus rectum is a well-known property of a parabola (Figure 25.5).

25.4.4 Hyperbolic Orbit $E > 0$

When $E > 0$, in the limit as $r \to \infty$ the kinetic energy is positive, $K_{\text{eff}} > 0$. This corresponds to a hyperbolic orbit (see Equation (25.3.24)). The condition for closest approach is similar to Equation (25.4.14) except that the energy is now positive. This implies that there is only one positive solution to the quadratic Equation (25.4.15), the distance of closest approach for the hyperbolic orbit

$$r_{\text{hyp}} = \frac{r_0}{1+\varepsilon}. \tag{25.4.26}$$

The constant r_0 is independent of the energy and from Equation (25.3.14) as the energy of the single body increases, the eccentricity increases, and hence from Equation (25.4.26), the distance of closest approach gets smaller (Figure 25.5).

25.5 Orbits of the Two Bodies

The orbit of the single body can be circular, elliptical, parabolic or hyperbolic, depending on the values of the two constants of the motion, the angular momentum and the energy. Once we have the explicit solution (in this discussion, $r(\theta)$) for the single body, we can find the actual orbits of the two bodies.

Choose a coordinate system as we did for the reduction of the two-body problem (Figure 25.7).

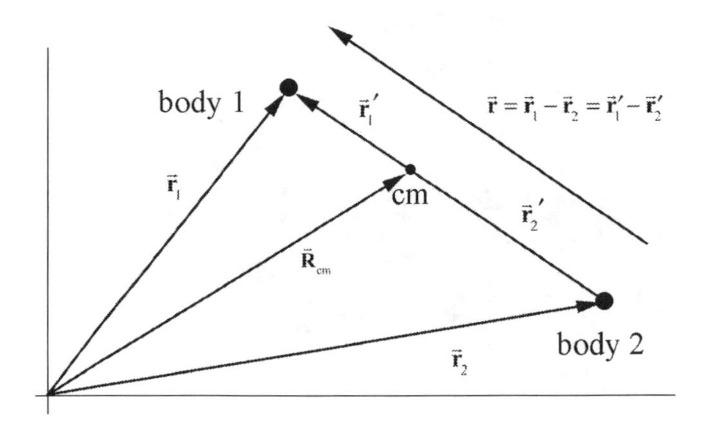

Figure 25.7 Center of mass coordinate system

The center of mass of the system is given by

$$\vec{\mathbf{R}}_{cm} = \frac{m_1 \vec{\mathbf{r}}_1 + m_2 \vec{\mathbf{r}}_2}{m_1 + m_2}. \tag{25.4.27}$$

Let $\vec{\mathbf{r}}_1'$ be the vector from the center of mass to body 1 and $\vec{\mathbf{r}}_2'$ be the vector from the center of mass to body 2. Then, by the geometry in Figure 25.6,

$$\vec{\mathbf{r}} = \vec{\mathbf{r}}_1 - \vec{\mathbf{r}}_2 = \vec{\mathbf{r}}_1' - \vec{\mathbf{r}}_2', \tag{25.4.28}$$

and hence

$$\vec{\mathbf{r}}_1' = \vec{\mathbf{r}}_1 - \vec{\mathbf{R}}_{cm} = \vec{\mathbf{r}}_1 - \frac{m_1 \vec{\mathbf{r}}_1 + m_2 \vec{\mathbf{r}}_2}{m_1 + m_2} = \frac{m_2(\vec{\mathbf{r}}_1 - \vec{\mathbf{r}}_2)}{m_1 + m_2} = \frac{\mu}{m_1}\vec{\mathbf{r}}. \tag{25.4.29}$$

A similar calculation shows that

$$\vec{\mathbf{r}}_2' = -\frac{\mu}{m_2}\vec{\mathbf{r}}. \tag{25.4.30}$$

Thus each body undergoes a motion about the center of mass in the same manner that the single body moves about the central point given by Equation (25.3.12). The only difference is that the distance from either body to the center of mass is shortened by a factor μ / m_i. When the orbit of the single body is an ellipse, then the orbits of the two bodies are also ellipses, as shown in Figure 25.8. When one mass is much smaller than the other, for example $m_1 << m_2$, then the reduced mass is approximately the smaller mass,

$$\mu = \frac{m_1 m_2}{m_1 + m_2} \cong \frac{m_1 m_2}{m_2} = m_1. \tag{25.4.31}$$

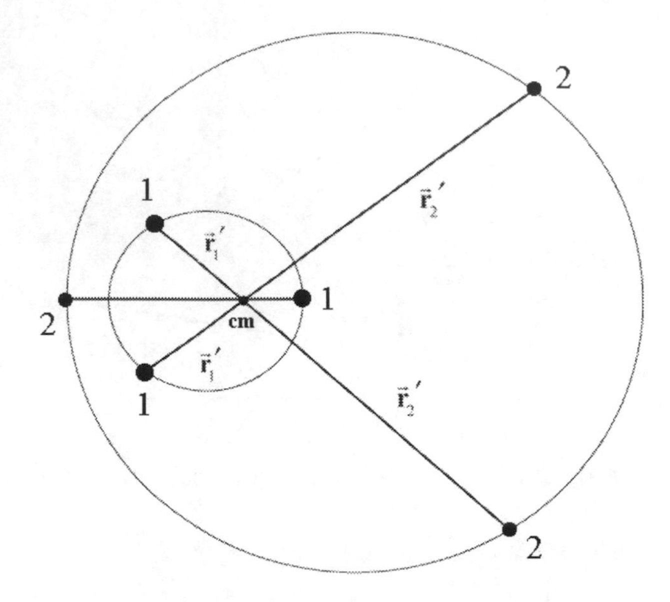

Figure 25.8 The elliptical motion of bodies interacting gravitationally

The center of mass is located approximately at the position of the larger mass, body 2 of mass m_2. Thus body 1 moves according to

$$\vec{\mathbf{r}}_1' = \frac{\mu}{m_1}\vec{\mathbf{r}} \cong \vec{\mathbf{r}}\,, \tag{25.4.32}$$

and body 2 is approximately stationary,

$$\vec{\mathbf{r}}_2' = -\frac{\mu}{m_2}\vec{\mathbf{r}} - \frac{m_1}{m_2}\vec{\mathbf{r}} \cong \vec{\mathbf{0}}\,. \tag{25.4.33}$$

25.6 Kepler's Laws

25.6.1 Elliptic Orbit Law

I. Each planet moves in an ellipse with the sun at one focus.

When the energy is negative, $E < 0$, and according to Equation (25.3.14),

$$\varepsilon = \left(1 + \frac{2\,E\,L^2}{\mu(G\,m_1\,m_2)^2}\right)^{\frac{1}{2}} \tag{25.5.1}$$

and the eccentricity must fall within the range $0 \le \varepsilon < 1$. These orbits are either circles or ellipses. Note the elliptic orbit law is only valid if we assume that there is only one central force acting. We are ignoring the gravitational interactions due to all the other bodies in the universe, a necessary approximation for our analytic solution.

25.6.2 Equal Area Law

II. The radius vector from the sun to a planet sweeps out equal areas in equal time.

Using analytic geometry in the limit of small $\Delta\theta$, the sum of the areas of the triangles in Figure 25.9 is given by

$$\Delta A = \frac{1}{2}(r\,\Delta\theta)r + \frac{(r\,\Delta\theta)}{2}\Delta r \tag{25.5.2}$$

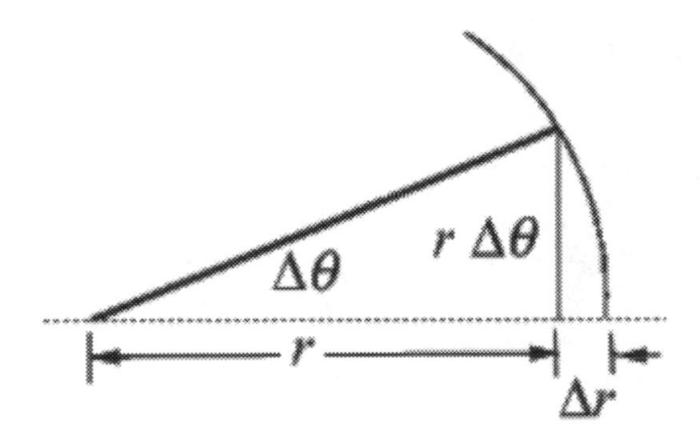

Figure **25.9** Kepler's equal area law.

The average rate of the change of area, ΔA, in time, Δt, is given by

$$\Delta A = \frac{1}{2}\frac{(r\,\Delta\theta)r}{\Delta t} + \frac{(r\,\Delta\theta)}{2}\frac{\Delta r}{\Delta t}. \tag{25.5.3}$$

In the limit as $\Delta t \to 0$, $\Delta\theta \to 0$, this becomes

$$\frac{dA}{dt} = \frac{1}{2}r^2\frac{d\theta}{dt}. \tag{25.5.4}$$

Recall that according to Equation (25.3.7) (reproduced below as Equation (25.6.5)), the angular momentum is related to the angular velocity $d\theta / dt$ by

$$\frac{d\theta}{dt} = \frac{L}{\mu r^2} \tag{25.5.5}$$

and Equation (25.5.4) is then

$$\frac{dA}{dt} = \frac{L}{2\mu}. \tag{25.5.6}$$

Because L and μ are constants, the rate of change of area with respect to time is a constant. This is often familiarly referred to by the expression: *equal areas are swept out in equal times* (see Kepler's Laws at the beginning of this chapter).

25.6.3 Period Law

III. The period of revolution T of a planet about the sun is related to the major axis A of the ellipse by $T^2 = k\,A^3$ where k is the same for all planets.

When Kepler stated his period law for planetary orbits based on observation, he only noted the dependence on the larger mass of the sun. Because the mass of the sun is much greater than the mass of the planets, his observation is an excellent approximation.

In order to demonstrate the third law we begin by rewriting Equation (25.5.6) in the form

$$2\mu \frac{dA}{dt} = L. \quad (25.5.7)$$

Equation (25.5.7) can be integrated as

$$\int_{\text{orbit}} 2\mu\, dA = \int_0^T L\, dt\,, \quad (25.5.8)$$

where T is the period of the orbit. For an ellipse,

$$\text{Area} = \int_{\text{orbit}} dA = \pi\, ab\,, \quad (25.5.9)$$

where a is the semi-major axis and b is the semi-minor axis (Figure 25.10).

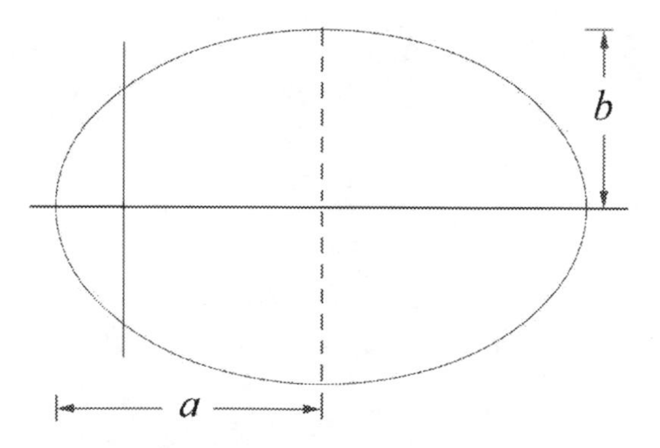

Figure 25.10 Semi-major and semi-minor axis for an ellipse

Thus we have

$$T = \frac{2\mu\pi\, ab}{L}. \quad (25.5.10)$$

Squaring Equation (25.5.10) then yields

$$T^2 = \frac{4\pi^2\mu^2 a^2 b^2}{L^2}. \quad (25.5.11)$$

In Appendix 25B, the angular momentum is given in terms of the semi-major axis and the eccentricity by Equation (25.B.20). Substitution for the angular momentum into Equation (25.5.11) yields

$$T^2 = \frac{4\pi^2\mu^2 a^2 b^2}{\mu G m_1 m_2 a(1-\varepsilon^2)}. \tag{25.5.12}$$

In Appendix 25B, the semi-minor axis is given by Equation (25.B.17) which upon substitution into Equation (25.5.12) yields

$$T^2 = \frac{4\pi^2\mu^2 a^3}{\mu G m_1 m_2}. \tag{25.5.13}$$

Using Equation (25.2.1) for reduced mass, the square of the period of the orbit is proportional to the semi-major axis cubed,

$$T^2 = \frac{4\pi^2 a^3}{G(m_1 + m_2)}. \tag{25.5.14}$$

25.7 Worked Examples

Example 25.1 Elliptic Orbit

A satellite of mass m_s is in an elliptical orbit around a planet of mass $m_p >> m_s$. The planet is located at one focus of the ellipse. The satellite is at the distance r_a when it is furthest from the planet. The distance of closest approach is r_p (Figure 25.11). What is (i) the speed v_p of the satellite when it is closest to the planet and (ii) the speed v_a of the satellite when it is furthest from the planet?

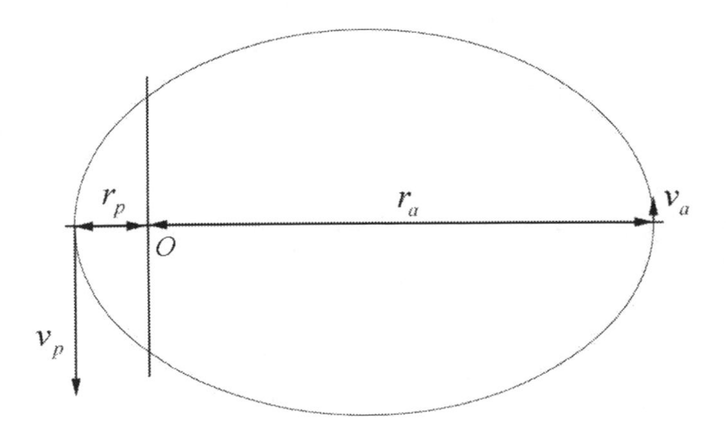

Figure 25.11 Example 25.1

Solution: The angular momentum about the origin is constant and because $\vec{\mathbf{r}}_{O,a} \perp \vec{\mathbf{v}}_a$ and $\vec{\mathbf{r}}_{O,p} \perp \vec{\mathbf{v}}_p$, the magnitude of the angular momentums satisfies

$$L \equiv L_{O,p} = L_{O,a}. \qquad (25.6.1)$$

Because $m_s << m_p$, the reduced mass $\mu \cong m_s$ and so the angular momentum condition becomes

$$L = m_s r_p v_p = m_s r_a v_a \qquad (25.6.2)$$

We can solve for v_p in terms of the constants G, m_p, r_a and r_p as follows. Choose zero for the gravitational potential energy, $U(r=\infty)=0$. When the satellite is at the maximum distance from the planet, the mechanical energy is

$$E_a = K_a + U_a = \frac{1}{2} m_s v_a^{\,2} - \frac{G m_s m_p}{r_a}. \qquad (25.6.3)$$

When the satellite is at closest approach the energy is

$$E_p = \frac{1}{2} m_s v_p^{\,2} - \frac{G m_s m_p}{r_p}. \qquad (25.6.4)$$

Mechanical energy is constant,

$$E \equiv E_a = E_p\,, \qquad (25.6.5)$$

therefore

$$E = \frac{1}{2} m_s v_p^{\,2} - \frac{G m_s m_p}{r_p} = \frac{1}{2} m_s v_a^{\,2} - \frac{G m_s m_p}{r_a}. \qquad (25.6.6)$$

From Eq. (25.6.2) we know that

$$v_a = (r_p / r_a) v_p. \qquad (25.6.7)$$

Substitute Eq. (25.6.7) into Eq. (25.6.6) and divide through by $m_s/2$ yields

$$v_p^{\,2} - \frac{2Gm_p}{r_p} = \frac{r_p^{\,2}}{r_a^{\,2}} v_p^{\,2} - \frac{2Gm_p}{r_a}. \qquad (25.6.8)$$

We can solve this Eq. (25.6.8) for v_p:

$$
\begin{aligned}
&v_p^{\,2}\left(1-\frac{r_p^{\,2}}{r_a^{\,2}}\right)=2Gm_p\left(\frac{1}{r_p}-\frac{1}{r_a}\right)\Rightarrow\\
&v_p^{\,2}\left(\frac{r_a^{\,2}-r_p^{\,2}}{r_a^{\,2}}\right)=2Gm_p\left(\frac{r_a-r_p}{r_pr_a}\right)\Rightarrow\\
&v_p^{\,2}\left(\frac{(r_a-r_p)(r_a+r_p)}{r_a^{\,2}}\right)=2Gm_p\left(\frac{r_a-r_p}{r_pr_a}\right)\Rightarrow\\
&v_p=\sqrt{\frac{2Gm_pr_a}{(r_a+r_p)r_p}}.
\end{aligned}
\tag{25.6.9}
$$

We now use Eq. (25.6.7) to determine that

$$
v_a=(r_p/r_a)v_p=\sqrt{\frac{2Gm_pr_p}{(r_a+r_p)r_a}}. \tag{25.6.10}
$$

Example 25.2 The Motion of the Star SO-2 around the Black Hole at the Galactic Center

The UCLA Galactic Center Group, headed by Dr. Andrea Ghez, measured the orbits of many stars within $0.8''\times 0.8''$ of the galactic center. The orbits of six of those stars are shown in Figure 25.12.

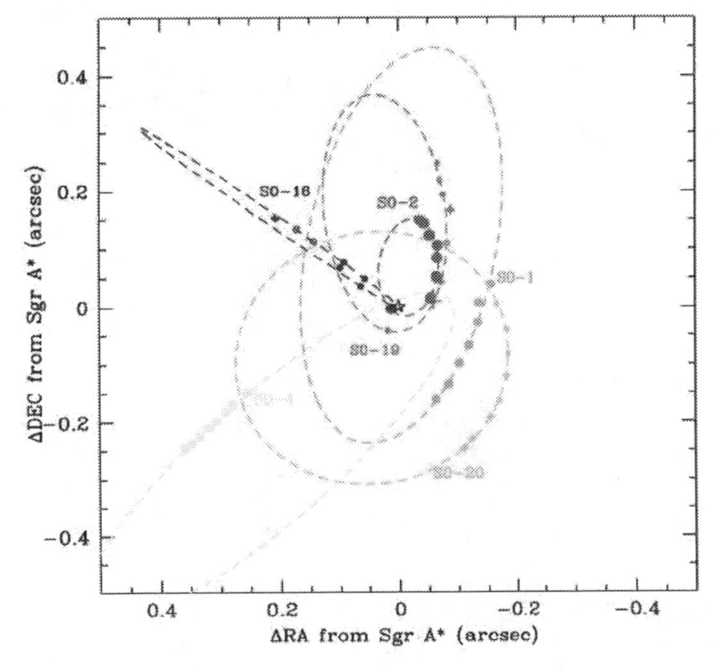

Figure 25.12 Obits of six stars near black hole at center of Milky Way galaxy.

We shall focus on the orbit of the star S0-2 with the following orbit properties given in Table 25.1[3]. Distances are given in astronomical units, $1\,\text{au} = 1.50 \times 10^{11}\,\text{m}$, which is the mean distance between the earth and the sun.

Table 25.1 Orbital Properties of S0-2

Star	Period (yrs)	Eccentricity	Semi-major axis (10^{-3}arc sec)	Periapse (au)	Apoapse (au)
S0-2	15.2 (0.68/0.76)	0.8763 (0.0063)	120.7 (4.5)	119.5 (3.9)	1812 (73)

The period of S0-2 satisfies Kepler's Third Law, given by

$$T^2 = \frac{4\pi^2 a^3}{G(m_1 + m_2)}, \qquad (25.6.11)$$

where m_1 is the mass of S0-2, m_2 is the mass of the black hole, and a is the semi-major axis of the elliptic orbit of S0-2. (a) Determine the mass of the black hole that the star S0-2 is orbiting. What is the ratio of the mass of the black hole to the solar mass? (b) What is the speed of S0-2 at periapse (distance of closest approach to the center of the galaxy) and apoapse (distance of furthest approach to the center of the galaxy)?

Solution: (a) The semi-major axis is given by

$$a = \frac{r_p + r_a}{2} = \frac{119.5\,\text{au} + 1812\,\text{au}}{2} = 965.8\,\text{au}. \qquad (25.6.12)$$

In SI units (meters), this is

$$a = 965.8\,\text{au}\frac{1.50 \times 10^{11}\,\text{m}}{1\,\text{au}} = 1.45 \times 10^{14}\,\text{m}. \qquad (25.6.13)$$

The mass m_1 of the star S0-2 is much less than the mass m_2 of the black hole, and Equation (25.6.11) can be simplified to

$$T^2 = \frac{4\pi^2 a^3}{G\,m_2}. \qquad (25.6.14)$$

Solving for the mass m_2 and inserting the numerical values, yields

[3] A.M.Ghez, et al., Stellar Orbits Around Galactic Center Black Hole, preprint arXiv:astro-ph/0306130v1, 5 June, 2003.

$$m_2 = \frac{4\pi^2 a^3}{GT^2}$$

$$= \frac{(4\pi^2)(1.45 \times 10^{14}\,\text{m})^3}{(6.67 \times 10^{-11}\,\text{N}\cdot\text{m}^2\cdot\text{kg}^{-2})((15.2\,\text{yr})(3.16 \times 10^7\,\text{s}\cdot\text{yr}^{-1}))^2} \qquad (25.6.15)$$

$$= 7.79 \times 10^{34}\,\text{kg}.$$

The ratio of the mass of the black hole to the solar mass is

$$\frac{m_2}{m_{\text{sun}}} = \frac{7.79 \times 10^{34}\,\text{kg}}{1.99 \times 10^{30}\,\text{kg}} = 3.91 \times 10^6. \qquad (25.6.16)$$

The mass of black hole corresponds to nearly four million solar masses.

(b) We can use our results from Example 25.1 that

$$v_p = \sqrt{\frac{2Gm_2 r_a}{(r_a + r_p)r_p}} = \sqrt{\frac{Gm_2 r_a}{ar_p}} \qquad (25.6.17)$$

$$v_a = \frac{r_p}{r_a} v_p = \sqrt{\frac{2Gm_2 r_p}{(r_a + r_p)r_a}} = \sqrt{\frac{Gm_2 r_p}{ar_a}}, \qquad (25.6.18)$$

where $a = (r_a + r_b)/2$ is the semi-major axis. Inserting numerical values,

$$v_p = \sqrt{\frac{Gm_2}{a}\frac{r_a}{r_p}}$$

$$= \sqrt{\frac{(6.67 \times 10^{-11}\,\text{N}\cdot\text{m}^2\cdot\text{kg}^{-2})(7.79 \times 10^{34}\,\text{kg})}{(1.45 \times 10^{14}\,\text{m})}\left(\frac{1812}{119.5}\right)} \qquad (25.6.19)$$

$$= 7.38 \times 10^6\,\text{m}\cdot\text{s}^{-1}.$$

The speed v_a at apoapse is then

$$v_a = \frac{r_p}{r_\text{a}} v_p = \left(\frac{1812}{119.5}\right)(7.38 \times 10^6\,\text{m}\cdot\text{s}^{-1}) = 4.87 \times 10^5\,\text{m}\cdot\text{s}^{-1}. \qquad (25.6.20)$$

Example 25.3 Central Force Proportional to Distance Cubed

A particle of mass m moves in plane about a central point under an attractive central force of magnitude $F = br^3$. The magnitude of the angular momentum bout the central point is equal to L. (a) Find the effective potential energy and make sketch of effective potential energy as a function of r. (b) Indicate on a sketch of the effective potential the total energy for circular motion. (c) The radius of the particle's orbit varies between r_0 and $2r_0$. Find r_0.

Solution: a) The potential energy, taking the zero of potential energy to be at $r = 0$, is

$$U(r) = -\int_0^r (-br'^3)\,dr' = \frac{b}{4}r^4$$

The effective potential energy is

$$U_{\text{eff}}(r) = \frac{L^2}{2mr^2} + U(r) = \frac{L^2}{2mr^2} + \frac{b}{4}r^4.$$

A plot is shown in Figure 25.13a, including the potential (yellow, right-most curve), the term $L^2/2m$ (green, left-most curve) and the effective potential (blue, center curve). The horizontal scale is in units of r_0 (corresponding to radius of the lowest energy circular orbit) and the vertical scale is in units of the minimum effective potential.

b) The minimum effective potential energy is the horizontal line (red) in Figure 25.13a.

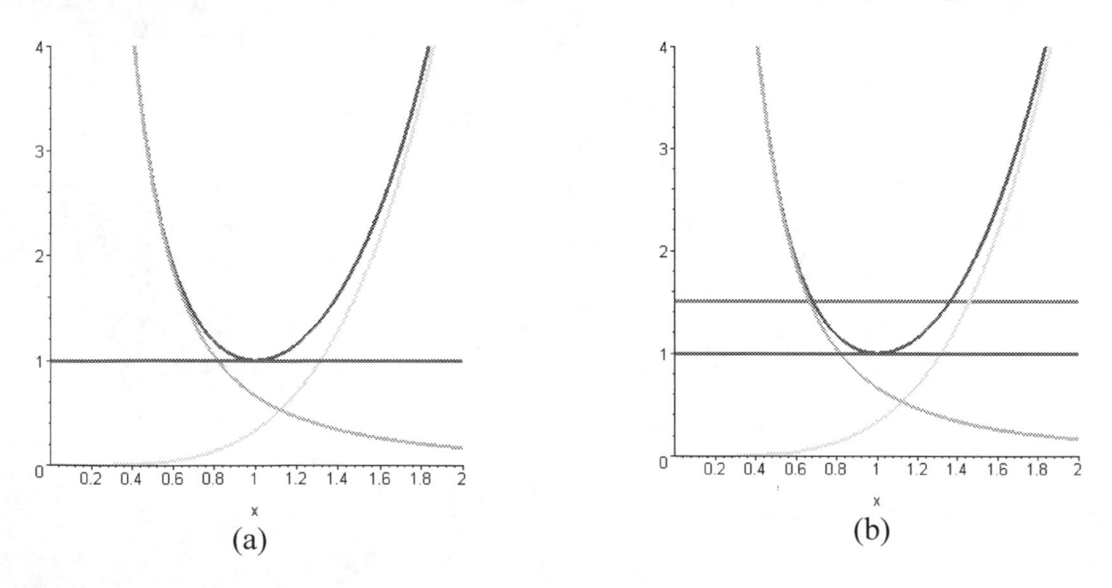

Figure 25.13 (a) Effective potential energy with lowest energy state (red line), (b) higher energy state (magenta line)

c) We are trying to determine the value of r_0 such that $U_{\text{eff}}(r_0) = U_{\text{eff}}(2r_0)$. Thus

$$\frac{L^2}{mr_0^2} + \frac{b}{4}r_0^4 = \frac{L^2}{m(2r_0)^2} + \frac{b}{4}(2r_0)^4.$$

Rearranging and combining terms, we can then solve for r_0,

$$\frac{3}{8}\frac{L^2}{m}\frac{1}{r_0^2} = \frac{15}{4} b r_0^4$$

$$r_0^6 = \frac{1}{10}\frac{L^2}{mb}.$$

In the plot in Figure 25.13b, if we could move the red line up until it intersects the blue curve at two point whose value of the radius differ by a factor of 2, those would be the respective values for r_0 and $2r_0$. A graph, showing the corresponding energy as the horizontal magenta line, is shown in Figure 25.13b.

Example 25.4 Transfer Orbit

A space vehicle is in a circular orbit about the earth. The mass of the vehicle is $m_s = 3.00 \times 10^3$ kg and the radius of the orbit is $2R_e = 1.28 \times 10^4$ km. It is desired to transfer the vehicle to a circular orbit of radius $4R_e$ (Figure 24.14). The mass of the earth is $M_e = 5.97 \times 10^{24}$ kg. (a) What is the minimum energy expenditure required for the transfer? (b) An efficient way to accomplish the transfer is to use an elliptical orbit from point A on the inner circular orbit to a point B on the outer circular orbit (known as a Hohmann transfer orbit). What changes in speed are required at the points of intersection, A and B?

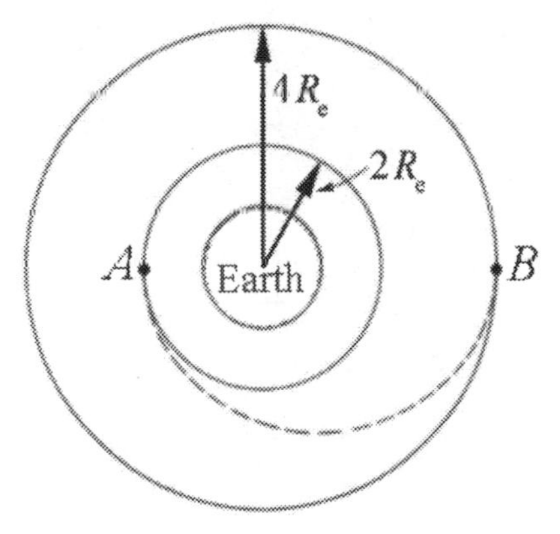

Figure 24.12 Example 25.5

Solution: (a) The mechanical energy is the sum of the kinetic and potential energies,

$$\begin{aligned} E &= K + U \\ &= \frac{1}{2} m_s v^2 - G\frac{m_s M_e}{R_e}. \end{aligned} \quad (25.6.21)$$

For a circular orbit, the orbital speed and orbital radius must be related by Newton's Second Law,

$$
\begin{aligned}
F_r &= m a_r \\
-G\frac{m_s M_e}{R_e^{\,2}} &= -m_s\frac{v^2}{R_e} \Rightarrow \\
\frac{1}{2}m_s v^2 &= \frac{1}{2}G\frac{m_s M_e}{R_e}.
\end{aligned} \tag{25.6.22}
$$

Substituting the last result in (25.6.22) into Equation (25.6.21) yields

$$
E = \frac{1}{2}G\frac{m_s M_e}{R_e} - G\frac{m_s M_e}{R_e} = -\frac{1}{2}G\frac{m_s M_e}{R_e} = \frac{1}{2}U(R_e). \tag{25.6.23}
$$

Equation (25.6.23) is one example of what is known as the ***Virial Theorem***, in which the energy is equal to (1/2) the potential energy for the circular orbit. In moving from a circular orbit of radius $2R_e$ to a circular orbit of radius $4R_e$, the total energy increases, (as the energy becomes less negative). The change in energy is

$$
\begin{aligned}
\Delta E &= E(r = 4R_e) - E(r = 2R_e) \\
&= -\frac{1}{2}G\frac{m_s M_e}{4R_e} - \left(-\frac{1}{2}G\frac{m_s M_e}{2R_e}\right) \\
&= \frac{1}{8}G\frac{m_s M_e}{4R_e}.
\end{aligned} \tag{25.6.24}
$$

Inserting the numerical values,

$$
\begin{aligned}
\Delta E &= \frac{1}{8}G\frac{m_s M_e}{R_e} = \frac{1}{4}G\frac{m_s M_e}{2R_e} \\
&= \frac{1}{4}(6.67\times10^{-11}\,\mathrm{m^3\cdot kg^{-1}\cdot s^{-2}})\frac{(3.00\times10^{3}\,\mathrm{kg})(5.97\times10^{24}\,\mathrm{kg})}{(1.28\times10^{4}\,\mathrm{km})} \\
&= 2.3\times10^{10}\,\mathrm{J}.
\end{aligned} \tag{25.6.25}
$$

b) The satellite must increase its speed at point A in order to move to the larger orbit radius and increase its speed again at point B to stay in the new circular orbit. Denote the satellite speed at point A while in the circular orbit as $v_{A,i}$ and after the speed increase (a "rocket burn") as $v_{A,f}$. Similarly, denote the satellite's speed when it first reaches point B as $v_{B,i}$. Once the satellite reaches point B, it then needs to increase its speed in order to continue in a circular orbit. Denote the speed of the satellite in the circular orbit at point B by $v_{B,f}$. The speeds $v_{A,i}$ and $v_{B,f}$ are given by Equation

(25.6.22). While the satellite is moving from point A to point B in the elliptic orbit (that is, during the transfer, after the first burn and before the second), both mechanical energy and angular momentum are conserved. Conservation of energy relates the speeds and radii by

$$\frac{1}{2}m_s(v_{A,f})^2 - G\frac{m_s m_e}{2R_e} = \frac{1}{2}m_s(v_{B,i})^2 - G\frac{m_s m_e}{4R_e}. \tag{25.6.26}$$

Conservation of angular momentum relates the speeds and radii by

$$m_s v_{A,f}(2R_e) = m_s v_{B,i}(4R_e) \Rightarrow v_{A,f} = 2v_{B,i}. \tag{25.6.27}$$

Substitution of Equation (25.6.27) into Equation (25.6.26) yields, after minor algebra,

$$v_{A,f} = \sqrt{\frac{2}{3}\frac{G M_e}{R_e}}, \quad v_{B,i} = \sqrt{\frac{1}{6}\frac{GM_e}{R_e}}. \tag{25.6.28}$$

We can now use Equation (25.6.22) to determine that

$$v_{A,i} = \sqrt{\frac{1}{2}\frac{G M_e}{R_e}}, \quad v_{B,f} = \sqrt{\frac{1}{4}\frac{G M_e}{R_e}}. \tag{25.6.29}$$

Thus the change in speeds at the respective points is given by

$$\begin{aligned} \Delta v_A &= v_{A,f} - v_{A,i} = \left(\sqrt{\frac{2}{3}} - \sqrt{\frac{1}{2}}\right)\sqrt{\frac{G M_e}{R_e}} \\ \Delta v_B &= v_{B,f} - v_{B,i} = \left(\sqrt{\frac{1}{4}} - \sqrt{\frac{1}{6}}\right)\sqrt{\frac{G M_e}{R_e}}. \end{aligned} \tag{25.6.30}$$

Substitution of numerical values gives

$$\Delta v_A = 8.6 \times 10^2 \,\mathrm{m \cdot s^{-2}}, \quad \Delta v_B = 7.2 \times 10^2 \,\mathrm{m \cdot s^{-2}}. \tag{25.6.31}$$

Appendix 25A Derivation of the Orbit Equation

25A.1 Derivation of the Orbit Equation: Method 1

Start from Equation (25.3.11) in the form

$$d\theta = \frac{L}{\sqrt{2\mu}} \frac{(1/r^2)}{\left(E - \dfrac{L^2}{2\mu r^2} + \dfrac{G m_1 m_2}{r}\right)^{1/2}} dr \,. \qquad (25.A.1)$$

What follows involves a good deal of hindsight, allowing selection of convenient substitutions in the math in order to get a clean result. First, note the many factors of the reciprocal of r. So, we'll try the substitution $u = 1/r$, $du = -(1/r^2)\,dr$, with the result

$$d\theta = -\frac{L}{\sqrt{2\mu}} \frac{du}{\left(E - \dfrac{L^2}{2\mu} u^2 + G m_1 m_2 u\right)^{1/2}} \,. \qquad (25.A.2)$$

Experience in evaluating integrals suggests that we make the absolute value of the factor multiplying u^2 inside the square root equal to unity. That is, multiplying numerator and denominator by $\sqrt{2\mu}/L$,

$$d\theta = -\frac{du}{\left(2\mu E / L^2 - u^2 + 2(\mu G m_1 m_2 / L^2) u\right)^{1/2}} \,. \qquad (25.A.3)$$

As both a check and a motivation for the next steps, note that the left side $d\theta$ of Equation (25.A.3) is dimensionless, and so the right side must be. This means that the factor of $\mu G m_1 m_2 / L^2$ in the square root must have the same dimensions as u, or length^{-1}; so, define $r_0 \equiv L^2 / \mu G m_1 m_2$. This is of course the *semilatus rectum* as defined in Equation (25.3.12), and it's no coincidence; this is part of the "hindsight" mentioned above. The differential equation then becomes

$$d\theta = -\frac{du}{(2\mu E / L^2 - u^2 + 2u / r_0)^{1/2}} \,. \qquad (25.A.4)$$

We now rewrite the denominator in order to express it terms of the eccentricity.

$$d\theta = -\frac{du}{\left(2\mu E / L^2 + 1/r_0^{\,2} - u^2 + 2u/r_0 - 1/r_0^{\,2}\right)^{1/2}}$$

$$= -\frac{du}{\left(2\mu E / L^2 + 1/r_0^{\,2} - (u - 1/r_0)^2\right)^{1/2}} \qquad (25.A.5)$$

$$= -\frac{r_0\, du}{\left(2\mu E r_0^{\,2} / L^2 + 1 - (r_0\, u - 1)^2\right)^{1/2}}.$$

We note that the combination of terms $2\mu E r_0^{\,2} / L^2 + 1$ is dimensionless, and is in fact equal to the square of the eccentricity ε as defined in Equation (25.3.13); more hindsight. The last expression in (25.A.5) is then

$$d\theta = -\frac{r_0\, du}{\left(\varepsilon^2 - (r_0\, u - 1)^2\right)^{1/2}}. \qquad (25.A.6)$$

From here, we'll combine a few calculus steps, going immediately to the substitution $r_0\, u - 1 = \varepsilon \cos\alpha$, $r_0\, du = -\varepsilon \sin\alpha\, d\alpha$, with the final result that

$$d\theta = -\frac{-\varepsilon \sin\alpha\, d\alpha}{\left(\varepsilon^2 - \varepsilon^2 \cos^2\alpha\right)^{1/2}} = d\alpha, \qquad (25.A.7)$$

We now integrate Eq. (25.A.7) with the very simple result that

$$\theta = \alpha + \text{constant}. \qquad (25.A.8)$$

We have a choice in selecting the constant, and if we pick $\theta = \alpha - \pi$, $\alpha = \theta + \pi$, $\cos\alpha = -\cos\theta$, the result is

$$r = \frac{1}{u} = \frac{r_0}{1 - \varepsilon\cos\theta}, \qquad (25.A.9)$$

which is our desired result, Equation (25.3.11). Note that if we chose the constant of integration to be zero, the result would be

$$r = \frac{1}{u} = \frac{r_0}{1 + \varepsilon\cos\theta} \qquad (25.A.10)$$

which is the same trajectory reflected about the "vertical" axis in Figure 25.3, indeed the same as rotating by π .

25A.2 Derivation of the Orbit Equation: Method 2

The derivation of Equation (25.A.9) in the form

$$u = \frac{1}{r_0}(1 - \varepsilon \cos\theta) \tag{25.A.11}$$

suggests that the equation of motion for the one-body problem might be manipulated to obtain a simple differential equation. That is, start from

$$\begin{aligned} \vec{\mathbf{F}} &= \mu \vec{\mathbf{a}} \\ -G\frac{m_1 m_2}{r^2}\hat{\mathbf{r}} &= \mu\left(\frac{d^2r}{dt^2} - r\left(\frac{d\theta}{dt}\right)^2\right)\hat{\mathbf{r}}. \end{aligned} \tag{25.A.12}$$

Setting the components equal, using the constant of motion $L = \mu r^2 (d\theta / dt)$ and rearranging, Eq. (25.A.12) becomes

$$\mu\frac{d^2r}{dt^2} = \frac{L^2}{\mu r^3} - \frac{Gm_1m_2}{r^2}. \tag{25.A.13}$$

We now use the same substitution $u = 1/r$ and change the independent variable from t to r, using the chain rule twice, since Equation (25.A.13) is a second-order equation. That is, the first time derivative is

$$\frac{dr}{dt} = \frac{dr}{du}\frac{du}{dt} = \frac{dr}{du}\frac{du}{d\theta}\frac{d\theta}{dt}. \tag{25.A.14}$$

From $r = 1/u$ we have $dr/du = -1/u^2$. Combining with $d\theta/dt$ in terms of L and u, $d\theta/dt = Lu^2/\mu$, Equation (25.A.14) becomes

$$\frac{dr}{dt} = -\frac{1}{u^2}\frac{du}{d\theta}\frac{Lu^2}{\mu} = -\frac{du}{d\theta}\frac{L}{\mu}, \tag{25.A.15}$$

a very tidy result, with the variable u appearing linearly. Taking the second derivative with respect to t,

$$\frac{d^2r}{dt^2} = \frac{d}{dt}\left(\frac{dr}{dt}\right) = \frac{d}{d\theta}\left(\frac{dr}{dt}\right)\frac{d\theta}{dt}. \tag{25.A.16}$$

Now substitute Eq. (25.A.15) into Eq. (25.A.16) with the result that

$$\frac{d^2r}{dt^2} = -\frac{d^2u}{d\theta^2}\left(u^2\frac{L^2}{\mu^2}\right). \tag{25.A.17}$$

Substituting into Equation (25.A.13), with $r = 1/u$ yields

$$-\frac{d^2u}{d\theta^2}u^2\frac{L^2}{\mu} = \frac{L^2}{\mu}u^3 - Gm_1m_2u^2. \tag{25.A.18}$$

Canceling the common factor of u^2 and rearranging, we arrive at

$$-\frac{d^2u}{d\theta^2} = u - \frac{\mu G m_1 m_2}{L^2}. \tag{25.A.19}$$

Equation (25.A.19) is mathematically equivalent to the simple harmonic oscillator equation with an additional constant term. The solution consists of two parts: the angle-independent solution

$$u_0 = \frac{\mu G m_1 m_2}{L^2} \tag{25.A.20}$$

and a sinusoidally varying term of the form

$$u_{\mathrm{H}} = A\cos(\theta - \theta_0), \tag{25.A.21}$$

where A and θ_0 are constants determined by the form of the orbit. The expression in Equation (25.A.20) is the ***inhomogeneous solution*** and represents a circular orbit. The expression in Equation (25.A.21) is the ***homogeneous solution*** (as hinted by the subscript) and must have two independent constants. We can readily identify $1/u_0$ as the *semilatus rectum* r_0, with the result that

$$\begin{aligned} u = u_0 + u_{\mathrm{H}} &= \frac{1}{r_0}\left(1 + r_0 A(\theta - \theta_0)\right) \Rightarrow \\ r = \frac{1}{u} &= \frac{r_0}{1 + r_0 A(\theta - \theta_0)}. \end{aligned} \tag{25.A.22}$$

Choosing the product $r_0 A$ to be the eccentricity ε and $\theta_0 = \pi$ (much as was done leading to Equation (25.A.9) above), Equation (25.A.9) is reproduced.

Appendix 25B Properties of an Elliptical Orbit

25B.1 Coordinate System for the Elliptic Orbit

We now consider the special case of an elliptical orbit. Choose coordinates with the central point located at one focal point and coordinates (r,θ) for the position of the single body (Figure 25B.1a). In Figure 25B.1b, let a denote the semi-major axis, b denote the semi-minor axis and x_0 denote the distance from the center of the ellipse to the origin of our coordinate system.

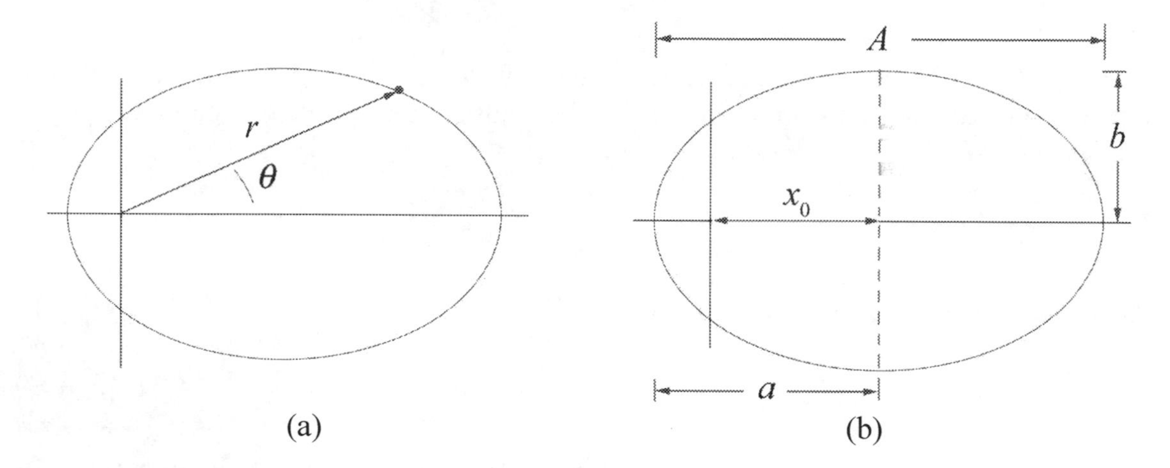

Figure 25B.1 (a) Coordinate system for elliptic orbit, (b) semi-major axis

25B.2 The Semi-major Axis

Recall the orbit equation, Eq, (25.A.9), describes $r(\theta)$,

$$r(\theta) = \frac{r_0}{1 - \varepsilon\cos\theta}. \tag{25.B.1}$$

The major axis $A = 2a$ is given by

$$A = 2a = r_a + r_p. \tag{25.B.2}$$

where the distance of furthest approach r_a occurs when $\theta = 0$, hence

$$r_a = r(\theta = 0) = \frac{r_0}{1-\varepsilon}, \tag{25.B.3}$$

and the distance of nearest approach r_p occurs when $\theta = \pi$, hence

$$r_p = r(\theta = \pi) = \frac{r_0}{1+\varepsilon}. \tag{25.B.4}$$

Figure 25B.2 shows the distances of nearest and furthest approach.

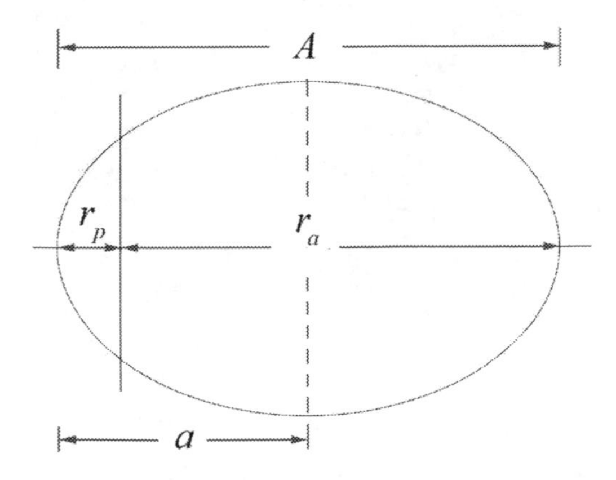

Figure 25B.2 Furthest and nearest approach

We can now determine the semi-major axis

$$a = \frac{1}{2}\left(\frac{r_0}{1-\varepsilon} + \frac{r_0}{1+\varepsilon}\right) = \frac{r_0}{1-\varepsilon^2}. \tag{25.B.5}$$

The *semilatus rectum* r_0 can be re-expressed in terms of the semi-major axis and the eccentricity,

$$r_0 = a(1-\varepsilon^2). \tag{25.B.6}$$

We can now express the distance of nearest approach, Equation (25.B.4), in terms of the semi-major axis and the eccentricity,

$$r_p = \frac{r_0}{1+\varepsilon} = \frac{a(1-\varepsilon^2)}{1+\varepsilon} = a(1-\varepsilon). \tag{25.B.7}$$

In a similar fashion the distance of furthest approach is

$$r_a = \frac{r_0}{1-\varepsilon} = \frac{a(1-\varepsilon^2)}{1-\varepsilon} = a(1+\varepsilon). \tag{25.B.8}$$

25B.2.3 The Location x_0 of the Center of the Ellipse

From Figure 25B.3a, the distance from a focus point to the center of the ellipse is

$$x_0 = a - r_p. \tag{25.B.9}$$

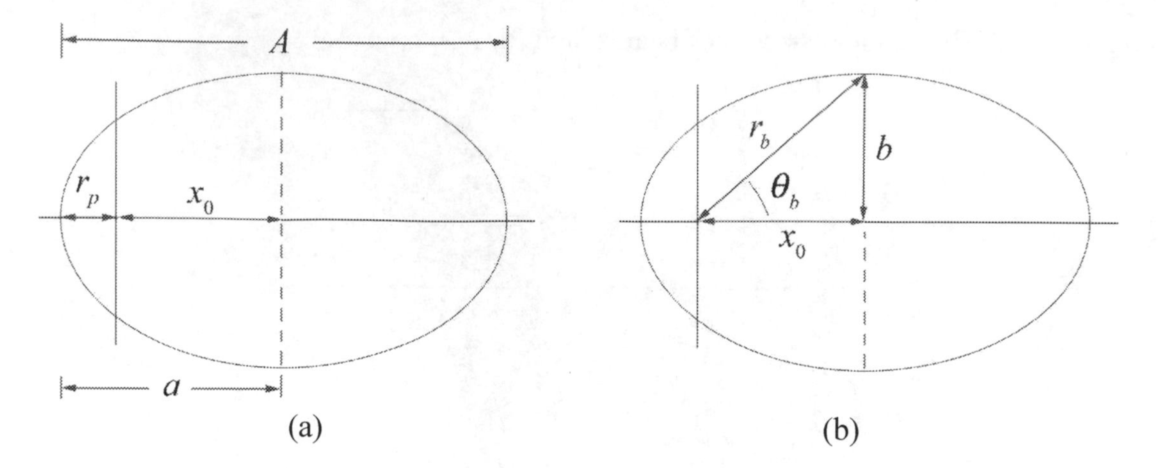

Figure 25B.3 Location of the center of the ellipse and semi-minor axis.

Using Equation (25.B.7) for r_p, we have that

$$x_0 = a - a(1-\varepsilon) = \varepsilon a. \tag{25.B.10}$$

25B.2.4 The Semi-minor Axis

From Figure 25B.3b, the semi-minor axis can be expressed as

$$b = \sqrt{(r_b^{\,2} - x_0^{\,2})}\,, \tag{25.B.11}$$

where

$$r_b = \frac{r_0}{1-\varepsilon\cos\theta_b}. \tag{25.B.12}$$

We can rewrite Eq. (25.B.12) as

$$r_b - r_b\,\varepsilon\cos\theta_b = r_0. \tag{25.B.13}$$

The horizontal projection of r_b is given by (Figure 25B.2b),

$$x_0 = r_b\cos\theta_b, \tag{25.B.14}$$

which upon substitution into Eq. (25.B.13) yields

$$r_b = r_0 + \varepsilon\, x_0. \tag{25.B.15}$$

Substituting Equation (25.B.10) for x_0 and Equation (25.B.6) for r_0 into Equation (25.B.15) yields

$$r_b = a(1-\varepsilon^2) + a\varepsilon^2 = a. \tag{25.B.16}$$

The fact that $r_b = a$ is a well-known property of an ellipse reflected in the geometric construction, that the sum of the distances from the two foci to any point on the ellipse is a constant. We can now determine the semi-minor axis b by substituting Eq. (25.B.16) into Eq. (25.B.11) yielding

$$b = \sqrt{\left(r_b^{\;2} - x_0^{\;2}\right)} = \sqrt{a^2 - \varepsilon^2 a^2} = a\sqrt{1-\varepsilon^2}\,. \qquad (25.B.17)$$

25B.2.5 Constants of the Motion for Elliptic Motion

We shall now express the parameters a, b and x_0 in terms of the constants of the motion L, E, μ, m_1 and m_2. Using our results for r_0 and ε from Equations (25.3.13) and (25.3.14) we have for the semi-major axis

$$\begin{aligned} a &= \frac{L^2}{\mu G m_1 m_2} \frac{1}{(1-(1+2EL^2/\mu(Gm_1 m_2)^2))} \\ &= -\frac{Gm_1 m_2}{2E} \end{aligned}\,. \qquad (25.B.18)$$

The energy is then determined by the semi-major axis,

$$E = -\frac{Gm_1 m_2}{2a}. \qquad (25.B.19)$$

The angular momentum is related to the *semilatus rectum* r_0 by Equation (25.3.13). Using Equation (25.B.6) for r_0, we can express the angular momentum (25.B.4) in terms of the semi-major axis and the eccentricity,

$$L = \sqrt{\mu G m_1 m_2 r_0} = \sqrt{\mu G m_1 m_2 a(1-\varepsilon^2)}\,. \qquad (25.B.20)$$

Note that

$$\sqrt{(1-\varepsilon^2)} = \frac{L}{\sqrt{\mu G m_1 m_2 a}}, \qquad (25.B.21)$$

Thus, from Equations (25.3.14), (25.B.10), and (25.B.18), the distance from the center of the ellipse to the focal point is

$$x_0 = \varepsilon a = -\frac{Gm_1 m_2}{2E}\sqrt{\left(1+2EL^2/\mu\left(Gm_1 m_2\right)^2\right)}, \qquad (25.B.22)$$

a result we will return to later. We can substitute Eq. (25.B.21) for $\sqrt{1-\varepsilon^2}$ into Eq. (25.B.17), and determine that the semi-minor axis is

$$b = \sqrt{aL^2 / \mu G m_1 m_2}\,. \tag{25.B.23}$$

We can now substitute Eq. (25.B.18) for a into Eq. (25.B.23), yielding

$$b = \sqrt{aL^2 / \mu G m_1 m_2} = L\sqrt{-\frac{G m_1 m_2}{2E} / \mu G m_1 m_2} = L\sqrt{-\frac{1}{2\mu E}}\,. \tag{25.B.24}$$

25B.2.6 Speeds at Nearest and Furthest Approaches

At nearest approach, the velocity vector is tangent to the orbit (Figure 25B.4), so the magnitude of the angular momentum is

$$L = \mu r_p v_p\,, \tag{25.B.25}$$

and the speed at nearest approach is

$$v_p = L / \mu r_p\,. \tag{25.B.26}$$

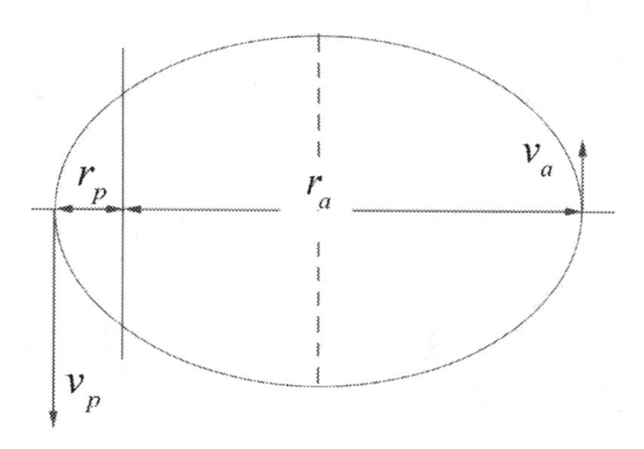

Figure 25B.4 Speeds at nearest and furthest approach

Using Equation (25.B.20) for the angular momentum and Equation (25.B.7) for r_p, Equation (25.B.26) becomes

$$v_p = \frac{L}{\mu r_p} = \frac{\sqrt{\mu G m_1 m_2 (1-\varepsilon^2)}}{\mu a (1-\varepsilon)} = \sqrt{\frac{G m_1 m_2 (1-\varepsilon^2)}{\mu a (1-\varepsilon)^2}} = \sqrt{\frac{G m_1 m_2 (1+\varepsilon)}{\mu a (1-\varepsilon)}}\,. \tag{25.B.27}$$

A similar calculation show that the speed v_a at furthest approach,

$$v_a = \frac{L}{\mu r_a} = \frac{\sqrt{\mu G m_1 m_2 (1-\varepsilon^2)}}{\mu a (1+\varepsilon)} = \sqrt{\frac{G m_1 m_2 1-\varepsilon^2}{\mu a (1+\varepsilon)^2}} = \sqrt{\frac{G m_1 m_2 \left(1-\varepsilon\right)}{\mu a (1+\varepsilon)}}\,. \tag{25.B.28}$$

Appendix 25C Analytic Geometric Properties of Ellipses

Consider Equation (25.3.20), and for now take $\varepsilon < 1$, so that the equation is that of an ellipse. We shall now show that we can write it as

$$\frac{(x - x_0)^2}{a^2} + \frac{y^2}{b^2} = 1, \tag{25.C.1}$$

where the ellipse has axes parallel to the x- and y-coordinate axes, center at $(x_0, 0)$, semi-major axis a and semi-minor axis b. We begin by rewriting Equation (25.3.20) as

$$x^2 - \frac{2\varepsilon r_0}{1-\varepsilon^2}x + \frac{y^2}{1-\varepsilon^2} = \frac{r_0^2}{1-\varepsilon^2}. \tag{25.C.2}$$

We next complete the square,

$$\begin{gathered}
x^2 - \frac{2\varepsilon r_0}{1-\varepsilon^2}x + \frac{\varepsilon^2 r_0^2}{(1-\varepsilon^2)^2} + \frac{y^2}{1-\varepsilon^2} = \frac{r_0^2}{1-\varepsilon^2} + \frac{\varepsilon^2 r_0^2}{(1-\varepsilon^2)^2} \Rightarrow \\
\left(x - \frac{\varepsilon r_0}{1-\varepsilon^2}\right)^2 + \frac{y^2}{1-\varepsilon^2} = \frac{r_0^2}{(1-\varepsilon^2)^2} \Rightarrow \\
\frac{\left(x - \dfrac{\varepsilon r_0}{1-\varepsilon^2}\right)^2}{\left(r_0/(1-\varepsilon^2)\right)^2} + \frac{y^2}{(r_0/\sqrt{1-\varepsilon^2})^2} = 1.
\end{gathered} \tag{25.C.3}$$

The last expression in (25.C.3) is the equation of an ellipse with semi-major axis

$$a = \frac{r_0}{1-\varepsilon^2}, \tag{25.C.4}$$

semi-minor axis

$$b = \frac{r_0}{\sqrt{1-\varepsilon^2}} = a\sqrt{1-\varepsilon^2}, \tag{25.C.5}$$

and center at

$$x_0 = \frac{\varepsilon r_0}{(1-\varepsilon^2)} = \varepsilon a, \tag{25.C.6}$$

as found in Equation (25.B.10).

Fundamental Physical Constants

Quantity	Symbol	Value
Avogadro's number	N_A	$6.02214129(27)\times10^{23}$ / mol
Boltzmann's constant	k_B	$1.3806488(13)\times10^{-23}$ J/K
Coulomb constant	$k_e = 1/4\pi\varepsilon_0$	$8.987551787\cdots\times10^{9}\ \mathrm{N\cdot m^2\cdot C^{-2}}$
Elementary charge	e	$1.602176565(35)\times10^{-19}$ C
Electron mass	m_e	$9.10938215(45)\times10^{-31}$ kg
Gravitational constant	G	$6.67384(80)\cdots\times10^{-11}\ \mathrm{N\cdot m^2\cdot kg^{-2}}$
Neutron mass	m_n	$1.674927351(74)\times10^{-27}$ kg
Permeability of free space	μ_0	$4\pi\times10^{-7}\ \mathrm{T\cdot m/A}$
Permittivity of free space	$\varepsilon_0 = 1/\mu_0 c^2$	$8.854187817\cdots\times10^{-12}\ \mathrm{C^2/N\cdot m^2}$
Planck's constant	h	$6.62606957(29)\times10^{-34}\ \mathrm{J\cdot s}$
Proton mass	m_p	$1.672621777(74)\times10^{-27}$ kg
Speed of light	c	$2.99792458\times10^{8}\ \mathrm{m\cdot s^{-1}}$

Astronomical Data

Earth	**Value**
Solar mass	$(1.98855 \pm 0.00025) \times 10^{30}$ kg
Earth mass	5.97219×10^{24} kg
Earth mean radius	6.371009×10^{6} m
Mean solar day	8.6400×10^{4} s
Earth stellar day	$8.6164098903691 \times 10^{4}$ s
Earth sidereal day	$8.616409053083288 \times 10^{4}$ s
Earth-Sun Orbit	
Aphelion	$1.52098232 \times 10^{11}$ km
Perihelion	$1.470098290 \times 10^{11}$ km
Eccentricity	0.01671123
Orbital Period	3.15581495×10^{7} s
Astronomical unit [au]	$1.49597870700 \times 10^{11}$ m
Moon	
Moon orbital period	6.3258468×10^{5} s
Moon mass	7.3477×10^{22} kg
Moon mean radius	1.73710×10^{6} m
Moon orbital period	2.3605847×10^{6} s
Moon synodic period	2.5514429×10^{6} s